Relationship

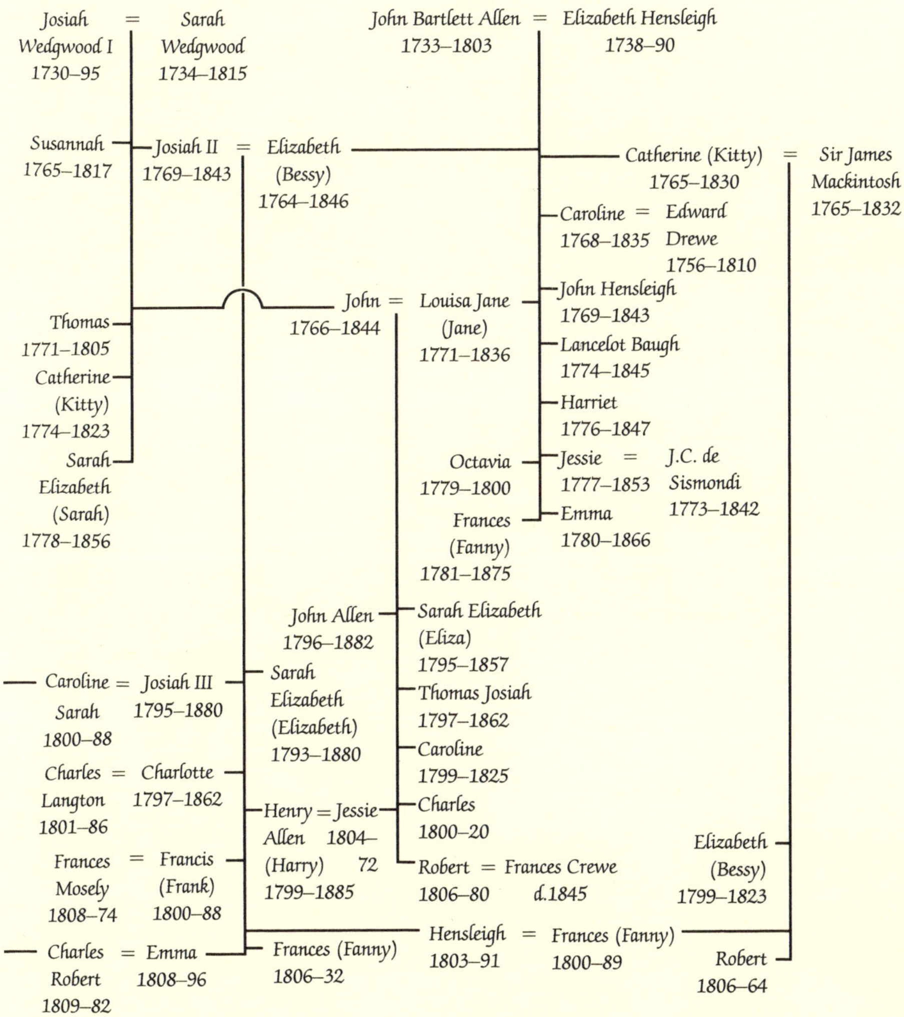

THE CORRESPONDENCE OF

CHARLES DARWIN

Editors

FREDERICK BURKHARDT† JAMES A. SECORD

SHEILA ANN DEAN SAMANTHA EVANS

SHELLEY INNES ALISON M. PEARN

PAUL WHITE

Associate Editors

ANNE SCHLABACH BURKHARDT

ROSEMARY CLARKSON

RUTH GOLDSTONE

SAM KUPER

MURIEL PALMER

ELIZABETH SMITH

ELLIS WEINBERGER

This edition of the Correspondence of Charles Darwin is sponsored by the American Council of Learned Societies. Its preparation is made possible by the co-operation of Cambridge University Library and the American Philosophical Society.

The Advisory Committee for the edition, appointed by the Management Board, has the following members:

Gillian Beer	Tim Birkhead
Stephen Bourne	Daniel V. Grossman
Sandra Herbert	Garrett Herman
Richard Keynes	Gene Kritsky

Support for editing has been received from the Alfred P. Sloan Foundation, the Andrew W. Mellon Foundation, the John Templeton Foundation, the National Endowment for the Humanities, the National Science Foundation, the Pew Charitable Trusts, the British Academy, the British Ecological Society, the Cambridge University Press, the Isaac Newton Trust, the Natural Environment Research Council, the Royal Society of London, the Stifterverband für die Deutsche Wissenschaft, and the Wellcome Trust. The National Endowment for the Humanities funding of the work was under grants nos. RE-23166-75-513, RE27067-77-1359, RE-00082-80-1628, RE-20166-82, RE-20480-85, RE-20764-89, RE-20913-91, RE-21097-93, RE-21282-95, RZ-20018-97 RE-20393-99, and RZ-20849-02; the National Science Foundation funding of the work was under grants Nos. SOC-75-15840, SOC-76-82775, SES-7912492, SES-8517189, SBR-9020874, SBR-9616619, SES-0135528, and SES-0646230. Any opinions, findings, conclusions, or recommendations expressed in this publication are those of the editors and do not necessarily reflect the views of the grantors.

Erasmus Alvey Darwin with Charles Darwin's sons, early 1870s.
Left to right: Leonard, George, Erasmus, Francis (seated) and William.

THE CORRESPONDENCE OF
CHARLES DARWIN

VOLUME 18 1870

CAMBRIDGE
UNIVERSITY PRESS

CAMBRIDGE UNIVERSITY PRESS
Cambridge, New York, Melbourne, Madrid, Cape Town, Singapore,
São Paulo, Delhi, Dubai, Tokyo

Cambridge University Press
The Edinburgh Building, Cambridge CB2 8RU, UK

Published in the United States of America by Cambridge University Press, New York

www.cambridge.org
Information on this title: www.cambridge.org/9780521768894

First published 2010

Citation:
Burkhardt, Frederick, *et al.*, eds. 2010. *The Correspondence of Charles Darwin.*
Vol 18. Cambridge: Cambridge University Press

Printed in the United Kingdom at the University Press, Cambridge

A catalogue record for this publication is available from the British Library

ISBN 978-0-521-76889-4 hardback

Dedicated to

THE BONITA TRUST

for its generosity in funding the Darwin and Gender project

The Darwin Correspondence Project gratefully acknowledges the essential support of the following donors:

Patrons

The Bonita Trust
Golden Family Foundation
Jim and Hilary Potter

Affiliates

Bern Dibner†
William T. Golden†
Kathleen Smith†

Friends

Jane Burkhardt
Pamela Davis
Florence Fearrington and James Needham
Gerald and Sue Friedman
John C. Greene
Daniel V. Grossman and Elizabeth Scott Andrews
Lawrence K. Grossman
Shirley Grossman, M.D.
Mary S. Hopkins
Robert McNeil
Michael Mathews
Victor Niederhoffer
Wendy L. Thompson
Daniel J. Wright

CONTENTS

ILLUSTRATIONS

CALENDAR LIST OF LETTERS

The following list is in the order of the entries in the *Calendar of the correspondence of Charles Darwin.* The first part includes all those letters that are listed in the *Calendar* for the year 1870, and those that have been redated into 1870. The second part includes the letters published in the supplement to this volume. Alongside the *Calendar* numbers are the corrected dates of each letter. A date or comment printed in italic type indicates that the letter has been omitted from this volume.

Letters acquired after the publication of the first edition of the *Calendar,* in 1985, have been given numbers corresponding to the chronological ordering of the original *Calendar* listing with the addition of an alphabetical marker. Many of these letters are summarised in a 'Supplement' to a new edition of the *Calendar* (Cambridge University Press, 1994). The marker 'f' denotes letters acquired after the second edition of the *Calendar* went to press in 1994.

5662. 1 Nov [1870]
6576. 25 Jan [1870]
6579. 26 Jan [1870]
6582. 29 Jan [1870]
6595. [23 Oct 1870?]
6611f. *12 Feb [1870–82]*
6617. 16 Feb [1870]
6646. [7 Mar 1870]
6650. 10 Mar [1870]
6788. 20 June [1870]
6964. [31 May 1870]
6993f. 17 Nov [1870]
7021. 5 Dec [1870]
7040. *[1868–70?]. Appears in Supplement to this volume.*
7041. *[1867–72]. To be published in supplement to vol. 20.*
7042. *[1867–72]. To be published in supplement to vol. 20.*
7043. *[1867–72]. To be published in supplement to vol. 20.*
7044. *[24 Jul 1869]*
7045. *[1865–6]. Appears in Supplement to this volume.*
7046. [20 April 1870]
7047. *[Sept 1868 – Aug 1870]. Appears in Supplement to this volume.*
7048. *[April 1871?]*
7049. [October 1870 or later]
7050. [after 1 July 1870]
7051. *[before 30 June 1869]*
7052. *[11–31 Mar 1868]. Appears in Supplement to this volume.*
7053. [28 Nov 1870]
7054. *Cancelled: draft of 7385 (on 7053).*
7055. [1870]
7056. *[Oct 1869 or later]. To be published in supplement to vol. 20.*
7057. *[1871 or later]*
7058. *[1867–72]. To be published in supplement to vol. 20.*
7059. *[1867–72]. To be published in supplement to vol. 20.*
7060. *[1867–72]. To be published in supplement to vol. 20.*
7061. *[1867–72]. To be published in supplement to vol. 20.*
7062. *[1867–72]. To be published in supplement to vol. 20.*
7063. *[January 1871]*
7064. *Cancel. Copy of part of 7707.*
7065. *[1867–8]. Appears in Supplement to this volume.*
7065f. *Cancel. Not a letter.*
7066. 19 [Dec 1870]
7067. 21 [Dec 1870]
7068. Jan 1870
7069 [*c.* 8 or 9 Apr 1870]
7070. 4 Jan 1870
7071. *Enclosure to 7070.*
7072. 5 Jan [1870]
7073. [before 8 Jan 1870]
7074. 8 Jan 1870

7380. 22 Nov [1870]
7381. 22 Nov 1870
7382. 24 Nov 1870
7383. 27 Nov [1870]
7384. 27 Nov 1870
7385. 29 Nov 1870
7386. 30 Nov 1870
7386a. 30 Nov [1870]
7386b. *Cancel: draft of 10764f.*
7387. 5 Dec 1870
7387f. 5 Dec 1870
7388. 7 Dec 1870
7389. 12 Dec [1870]
7390. 15 Dec 1870
7391. 15 Dec 1870
7392. [after 19 Dec 1870]
7393. [after 15 Dec 1870]
7394. 17 Dec [1870]
7395. 17 Dec 1870
7396. 17 Dec 1870
7397. [after 17 Dec 1870]
7398. 19 Dec 1870
7399. 20 Dec 1870
7400. 20 Dec [1870]
7401. 25 Dec [1870]
7402. 26 Dec [1870]
7403. 29 Dec 1870
7434. 10 Jan [1870 or 1871]
7718a. 21 Apr [1870]
7765f. [22 May 1870]
10495. 3 May 1870
13060. 22 Feb [1870]
13857. 13 Mar [1870?]

SUPPLEMENT, 1835–68

289f. 29 Dec 1835
421f. [before 7 July 1838]
497f. *[before 14 Mar 1839]. Essentially same as 421f.*
562g. 30 Mar 1840
565f. 1[8] May [1840]
565g. 5 May [1840]
605f. 1 Sept 1842
653f. 25 Nov [1840–6]
667f. 7 Apr [1843]
672f. 24 [Apr 1843]
691f. [8 Sept 1843]
701f. 9 Oct [1843]
705f. [14 Oct 1843]
710f. 1 Nov [1843]
711f. [9 Nov 1843]
715f. 22 Nov [1843]
716f. 1 Dec [1843]
720f. 7 Dec [1843]
734f. 5 Feb [1844]
736f. 23 Feb [1844]
767f. 31 July 1844
968f. 31 Mar [1846]
1032f. 25 Nov [1846]
1108f. 14 Aug [1847]
1111f. [22 Aug? 1847]
1188f. [10 July 1848]
1223f. [before 14 Mar 1849]
1232f. 2 Mar [1849]
1267f. 4 Nov [1851]
1385f. 6 Jan 1851
1466h. 30 Dec [1851]
1485f. 5 Aug [1852]
1554f. 22 Feb [1855]
1554g. [after 22 Feb 1855]
1845f. 25 Mar [1856]
1970a. 8 Oct [1856?]
2020f. 21 Dec [1856]
2142f. 23 Sept [1857–9?]
2346f. 20 Oct [1858]
2346g. [21 Oct 1858]
2346h. 22 Oct [1858]
2347f. 23 Oct 1858
2348f. [27 Oct 1858]
2454f. 26 Apr [1859]
2499. 1 Oct [1846?–71]
2564f. 1 Dec [1859]
2753f. 9 Apr [1860]
2755f. 10 Apr [1860]
2832f. 14 June [1860]
2906f. 6 Sept [1860]
2950f. 15 Oct [1860]
3012f. 6 Dec [1860]
3019f. 11 Dec [1860]
3103f. [28 Mar – 14 Apr 1861]
3141. 7 May [1861–8?]
3206. 11 July [1861]
3215a. 20 July [1861]
3285f. 14 Oct [1861]
3414f. 27 Nov [1861]
3633f. 4 Dec [1862]
3693f. 21 Aug [1862]
3751f. 4 Oct [1862]
3993f. 17 Feb 1863
4185f. 25 May [1863]
4185g. [26 or 27 May 1863]
4340. 12 Nov [1860–8]
4580f. 1 Aug [1864 or 1865]
4673. 16 Nov [1861–8]
4726. [1863–6?]
4747a. 18 Jan [1865]
4813f. 14 Apr [1865]
4988. 31 Jan [1863]

5019. 28 Feb [1862]
5030. 10 Mar [1863]
5038f. 23 Mar 1866
5080a. 6 May [1864]
5127. [1 May 1864]
5333. [19 May 1864]
5389. 2 Feb [1864]
5401. 16 Feb [1868]
5418f. 24 Feb [1862–9]
5481g. 2 Apr [1867]
5563. 6 June [1863]
5618f. 30 [Aug 1867]
5633f. [24 Sept – 31 Oct 1867]
5758. [1868?]
5618f. 30 [Aug 1867]
5758. [1868?]
5765. [21 July 1863 – Aug 1870?]
6027. 31 Jan [1863]
6039. 22 Mar [1867]
6319. 16 Aug [1860?]
6403. [after 24 Feb 1867]
6433. 27 Oct [1867]
6513. 26 Dec [1866]
6520. [before 10 Apr 1856]
6538. [1857–61?]
6806. 27 June [1862]
6847f. 31 July 1869
7040. [1868–70?]
7045. [1865–6]
7047. [Sept 1868 – Aug 1870]
7052. [11–31 Mar 1868]
7065. [1867–8?]
7301. 15 Aug [1859 or later]
7312. 30 Aug [1867–70]
7409. [after 29 April 1869?]
13780. *[4 Nov 1867]. Published in vol. 14 under date [4 Nov 1866?].*
13793. *[after 12 Oct 1866]*
13804. 12 [May 1861 – Apr 1863]
13808. [before 1 Mar 1866]
13817. [1860–70?]
13825. *21 Feb [1862 or later]. To be published in a future supplement.*
13828. 10 Dec [1842–5 or 1855–68]
13873. 23 Jan [1843–6 or 1856–69]
13877f. 11 Mar [1862–9]
13878. 29 Mar [1862–9]
13879. 20 June [1861–8]
13880. 24 Aug [1861–8]
13881. 25 Nov [1861–8]
13881a. 31 Dec [1861–8]

INTRODUCTION

The year 1870 is aptly summarised by the brief entry Darwin made in his journal: 'The whole of the year at work on the Descent of Man & Selection in relation to Sex'. Always precise in his accounting, Darwin reckoned that he had started writing on 4 February 1868, only five days after the publication of his previous book, *Variation in animals and plants under domestication.* In fact, *Descent* was the culmination of over three decades of observations and reflections on human ancestry, including the origin of language, mind, morals, and religious temperament. Research on sexual selection and emotional expression had predominated in recent years, and the information gathered on each of these topics was far more extensive than Darwin had anticipated. As a result, *Descent*, like *Variation*, would require two large volumes, and by June Darwin gave up the idea of including the material on emotion; it would eventually appear as a separate book in 1872 (*Expression of the emotions in man and animals*). The year was otherwise coloured by controversies, including vigorous objections to the application of natural selection to humans from Alfred Russel Wallace and St George Jackson Mivart, and heated debates sparked by Darwin's proposed election to the French Academy of Sciences and his nomination for an honorary degree at Oxford. As usual, Darwin tried to avoid involving himself in such controversies, saving his energies for his scientific work and his family. As he was completing corrections to the final proofs of *Descent* in December, he wrote to his friend Charles Lyell, 'thank all the powers above & below, I shall be a man again & not a horrid grinding machine' (letter to Charles Lyell, 25 December [1870]).

Darwin began receiving proofs of some of the illustrations for *Descent* as early as January, including woodcuts of the heads of chameleons and the tail feathers of snipe, much to his delight: 'The sight of the proofs has pleased me more than anything which has happened to me for some weeks' (letter to Albert Günther, 13 January [1870]). Darwin was still working hard on parts of the text, in particular his chapters on the moral sense and the comparative mental powers of humans and animals. He turned, as he had before, to his daughter Henrietta for commentary, sending her parts of the manuscript while she was on holiday in France and Italy: 'After reading once right through, the more time you can give up for deep criticism or corrections of style, the more grateful I shall be' (letter to H. E. Darwin, [8 February 1870]). She had previously read proof-sheets of *Variation* and *Orchids*, the latter when she was just eighteen years of age. CD clearly expected her to make a considerable contribution, instructing her to write any long corrections on separate slips of paper pinned to the relevant page of manuscript. He worried that parts of the book were 'too like a

Sermon: who w[d] ever have thought that I sh[d] turn parson?' (letter to H. E. Darwin, [8 February 1870]). Henrietta disagreed: 'Certainly to have you turned Parson will be a change— I expect I shall want it enlarging not contracting—cos I think *you* think an apology is wanting for writing abt any thing so unimportant as the mind of man!' (letter from H. E. Darwin, [after 8 February 1870]). Darwin was also encouraged to expand his discussion of mind and morals by the religious writer and philanthropist Frances Power Cobbe. At Cobbe's suggestion, Darwin read some of Immanuel Kant's *Metaphysics of ethics*, and he remarked on the contrast between the 'great philosopher looking exclusively into his own mind', and himself, 'a degraded wretch looking from the outside thro' apes & savages at the moral sense of mankind' (letter to F. P. Cobbe, 23 March [1870?]). Cobbe accused Darwin of smiling in his beard with such remarks, and urged, 'are you never going to unite your lines of thought & let us see how metaphysics & physics form one great philosophy?' (letter from F. P. Cobbe, 28 March [1870?]).

Despite Cobbe's plea, most of Darwin's scientific attention in 1870 was devoted to the 'physical' side of human descent. On 7 March 1870, Darwin made a note on the shape of human ears: 'W. has seen the tips in women & men. When he was making his figure of Puck he went to Z garden and made drawings of ears of monkeys & shortly afterwards he saw a man with tip & instantly recognized their signification' (DAR 80: B120). The reference here is to the sculptor Thomas Woolner, and to his statue of Puck, the mischievous fairy in Shakespeare's *A midsummer night's dream*. Darwin obtained a sketch of a human ear from Woolner showing the unusual feature of a pointed tip projecting inward from the folded margin. Darwin, who had posed for the sculptor in 1868, an experience he described as 'purgatory' made lighter by Woolner's 'wonderfully pleasant' manner, named the feature after the famous artist. 'The "Woolnerian tip" is worth anything to me', he wrote in thanks for the drawing (*Correspondence* vol. 16, letter to J. D. Hooker, 26 November [1868]; this volume, letter to Thomas Woolner, 10 March [1870]). Darwin included Woolner's sketch in *Descent*, and discussed the 'tip' as a rudimentary organ, describing its frequency and variability in humans (*Descent* 1: 22–3).

A more troubling anatomical feature for Darwin was the platysma myoides, a band of muscle in the neck extending from the collar bone to the lower part of the cheek. Darwin had read various accounts of the contraction of this muscle in persons experiencing fear or terror. He studied the photographic album by the French physiologist Guillaume Benjamin Amand Duchenne, whose research involved the mechanical production of expressions by applying electrodes directly to the face and neck. Duchenne had dubbed the platysma the 'muscle of fright', and one of his photographs, later used by Darwin in *Expression*, showed a man whose platysma was severely contracted, his face exhibiting extreme horror. Darwin lent Duchenne's album to the asylum director James Crichton-Browne, who had become one of Darwin's most avid observers of facial expression. Browne sent a lengthy account of the movements of the platysma muscle in his asylum patients, but it did not confirm Duchenne's findings (letter from James Crichton-Browne, 15 March 1870). Indeed,

Darwin noted the same longitudinal furrows radiating on the side of the neck of his son Francis when he was playing the flute. Exasperated, Darwin turned to the physician and eye-specialist William Ogle, requesting him to observe the muscle in patients who were having trouble breathing. 'This muscle', he complained, 'is the bane of existence!' (letter to William Ogle, 9 November 1870).

Darwin's research on emotions continued to draw on observations from a variety of domains, from the colonial frontier to the household, from the photographer's studio to the zoological garden. He received more replies to his questionnaire on expression, including four lengthy letters from the explorer William Winwood Reade, who had led an expedition to Africa in search of the source of the Niger river. Reade was sceptical of Darwin's view that standards of beauty were variable across different cultures and claimed that Africans admired the racial characteristics of Europeans. The nose, he remarked, was 'the only doubtful point ... girls have been heard to say "I dont want to marry him he's got no nose"'. Reade noted the strong preference of native peoples for black skin, but attributed this to their belief that all demons and spirits were white (letter from W. W. Reade, 9 November 1870).

Keen for more evidence of the continuity of expressions across species, Darwin asked the zoo-keeper at Regent's Park to study the squirrel monkey's screams: 'does it wrinkle up the skin round the eyes like a Baby always does? ... Could you make it scream without hurting it much?' (letter to A. D. Bartlett, 5 January [1870]). Darwin made a similar request of a London photographer: 'I have been trying to get a [photograph] of a young baby screaming or crying badly; but I fear he will not succeed' (letter to James Crichton-Browne, 8 June [1870]). Darwin's queries were part of an investigation into the physiology of weeping that had begun with observations on his own children as early as 1839. Darwin now recruited family members with infants, including his niece Lucy Wedgwood, who sent a sketch of a baby's brows (letter from L. C. Wedgwood, [5 May 1870]). He also wrote to a leading Dutch ophthalmologist, Frans Cornelis Donders. In order to fulfil Darwin's request, Donders undertook painstaking experiments on the operation of the eye muscles in weeping. He wrote to Darwin when his work was interrupted by the death of his daughter in childbirth: 'We are living with a sorrow which it seems can never end, and ... I have found it hard to return to the laboratory and to attend to our occupations. ... I have examined the circulation of the eye with some care, and all I am lacking is the inclination to finish my note on this subject' (letter from F. C. Donders, 17 May 1870).

While Darwin was still revising his manuscript of *Descent*, the public debate over human evolution grew more heated. Alfred Russel Wallace had expressed reservations about the application of natural selection to the development of the higher intellectual faculties of humans the previous year (see *Correspondence* vol. 17, letter to A. R. Wallace, 14 April 1869). His views were presented more fully in a collection of essays published in April 1870 (Wallace 1870a). Wallace wrote to Darwin in January warning him of the impending publication. Darwin joked about Wallace's own intellectual descent: 'I groan over Man— you write like a metamorphosed (in retrograde direction) naturalist' (letter to A. R. Wallace, 26 January [1870]). Despite their increasing theoretical differences, both men worked as they had in the past to sustain goodwill and mutual respect. Wallace's new book, titled *Contributions to the theory of natural selection*, was dedicated to Darwin. When he received the book, Darwin was full of praise for Wallace's 'modesty and candour'. 'I hope it is a satisfaction to you to reflect,—& very few things in my life have been more satisfactory to me—that we have never felt any jealousy towards each other, though in one sense rivals' (letter to A. R. Wallace, 20 April [1870]). Darwin alluded here to the famous events of 1858, when Wallace had communicated his own version of the theory of descent by natural selection in a letter to Darwin, prompting much anxiety on Darwin's part over his priority of discovery.

While Darwin and Wallace remained on amicable terms, other supporters of Darwinian theory were concerned about the consequences of Wallace's book. Henry Walter Bates urged Darwin to respond to it directly in the form of a review: 'I have been having some conversation with the Editor of the "Academy" about Mr Wallace's last book & the appearance of backsliding from the Darwinian theory which it contains. … The views of friend Wallace are so plausible & suit so well wide-spread prejudices that you no doubt think with me they ought to be controverted. But who is to criticise them? No one but yourself' (letter from H. W. Bates, 20 May 1870). Darwin very rarely used the periodical press to reply to critics, preferring to address what he considered to be important objections in the revised editions of his own publications. To Bates, he pleaded lack of time and weak health: 'my chief reason is that I really have not a grain of spare strength. … You wd. not readily believe how often & urgently I am pressed to write articles … & it is an immense relief to me to be able to say that I *never* write reviews' (letter to H. W. Bates, [22 May 1870]).

Another set of objections to the extension of natural selection to human evolution came from the zoologist St George Jackson Mivart. A protégé of Thomas Henry Huxley, Mivart had established a reputation as a leading comparative anatomist through his work on primates. In July 1869, Mivart published the first of a series of articles, titled 'Difficulties of the theory of natural selection', in the Roman Catholic journal the *Month*. He argued that a number of physical structures and homologies could not have arisen through the operation of natural selection alone, and that these were better accounted for by an underlying design. Darwin commented on Mivart's essay in a letter to William Henry Flower: 'I am glad you noticed the curiously false argument in the Month. … I c^{d} have answered, I think satisfactorily, many of the objections advanced in this article; but my whole time w^{d} be wasted if I once began to answer objectors' (letter to W. H. Flower, 25 March [1870]). In his letters to Mivart, Darwin remained cordial and supportive, praising his earlier anatomical papers and urging him to resume his research: 'I hope that you will continue your wonderfully laborious & valuable labours on the Primates' (letter to St G. J. Mivart, 23 April [1870]). He also tried to recruit Mivart's expertise, and inquired after a point of human and ape anatomy, hoping to cite Mivart as an authority in *Descent*. Mivart's reply, however, underscored their fundamental differences on human ancestry: 'assuming that zoological classification should be Anatomical—Man forms only a family of the higher division of the Primates. But if we introduce into the consideration his intellectual, moral & religious nature I am convinced he differs more from an Anthropoid Ape than such an Ape differs from a lump of granite' (letter from St G. J. Mivart, 22 April 1870). Mivart hinted that his criticism of descent theory was not directed at Darwin himself, but at more aggressive champions of Darwinism: 'For my part I shall never feel anything but gratitude & sincere esteem for the author of "natural selection" but I heartily execrate some who make use of that theory simply as a weapon of offence against higher interests and as a means of impeding Man's advance towards his "end" whatever may have been his "origin"' (letter from St G. J. Mivart, 25 April 1870). In his critical essays (later revised as *Genesis of species* (Mivart 1871)), Mivart tried to carve out a position between the new evolutionary biology that was being promoted by Huxley and others as an instrument of social reform, and another constituency to which he was equally loyal, the Roman Catholic Church. The relationship between Mivart and Darwin would grow more strained in the next few years and eventually reach breaking point.

Darwin's name and theory served a number of polemical and political ends. In France, Darwin was proposed for a corresponding membership of the elite Academy of Sciences, prompting an extended debate. The nomination was made by Henri Milne-Edwards and Armand de Quatrefages, both leading zoologists in Paris. Quatrefages had just completed a book, *Charles Darwin et ses précurseurs français* (Quatrefages 1870), that gave a detailed account, as well as criticism, of Darwin's theory. On receiving the book, Darwin remarked, 'many of your strictures are severe enough, but all are given with perfect courtesy & fairness. I can truly say I would rather be criticised by you in this manner than praised by many others' (letter to Armand de Quatrefages, 28 May [1870]). Quatrefages had corresponded with Darwin regularly since 1859, and viewed his relationship with Darwin as embodying the virtues of scientific community, placing truth above individual or factional differences. 'Yes, I dare to say it,' he wrote, 'we are both pursuing truth and that alone must establish links between us which are stronger than the causes of discord' (letter from Armand de Quatrefages, 30 March 1870). In proposing Darwin for election, Quatrefages hoped to bring the same non-partisan spirit into the French Academy, 'rest assured that I shall be a zealous and convinced advocate. ... Our very dissensions will give more weight to my words'. In fact, the election served as an occasion for attacks not only on Darwin's theory, but on his scientific methods and his status as a naturalist. 'It is being said at the Academy', Quatrefages complained, '... that you had done no more than collect natural history specimens and that you had called on assistants to describe them' (letter from Armand de Quatrefages, 18 July 1870). The assertion had been made by Emile Blanchard, who added that Darwin's pigeon studies were unscientific, and that his various theories, from coral island formation to transformism, were either unoriginal or false. Another detractor, Léonce Elie de Beaumont, referred to Darwin's work as 'la *science mousseuse*' (a science of froth or bubbles). Quatrefages himself seemed uncertain about some of these charges, and he seemed unaware that Darwin had published any of his geological research from the *Beagle* voyage.

Darwin lost the election by a narrow margin. The defeat was seized upon as a matter of national pride by the Belgian zoologist Edouard van Beneden when the Belgian Academy elected Darwin an associate member later that year. 'In this matter the Belgians have taught the members of the Institut de France a lesson,' he wrote. 'May all the learned bodies of the world protest, like the Academy of Belgium, against the debate they dared instigate on the question of your scientific merits. Your name summarises the whole scientific movement of recent years and your immortal work is above all attacks' (letter from Edouard van Beneden, 17 December 1870). A more parochial controversy occurred in Oxford, when Darwin was chosen to receive an honorary degree. Among the other candidates had been Thomas Huxley, who wrote to Darwin about the heated debate that had taken place in the ruling body of the University, and the role played by a leading Anglican conservative, Edward Bouverie Pusey: 'There seems to have been a tremendous shindy in the Hebdomadal board about certain persons who were proposed; and I am told that Pusey came to London to ascertain from a trustworthy friend who were the blackest heretics out of the list proposed—and that he was glad to assent to your being doctored, when he got back—in order to keep out seven devils worse than that first!' (letter from T. H. Huxley, 22 June 1870). In the event, Darwin did not receive the degree because it could only be awarded in person, and he declined to attend the ceremony on grounds of weak health: 'I could have travelled to Oxford,' he confessed, 'but could no more have withstood the excitement of a commemoration than I could a ball at Buckingham Palace' (letter to B. J. Sulivan, 30 June [1870]).

Though Darwin was mostly occupied with the revision of *Descent* and research on expression, his attention was frequently diverted by correspondents to topics of long-standing interest. The Italian botanist Federico Delpino sent a description of a pollination mechanism in *Lotus siliquosus* (now *L. maritimus*), a sample of which had been collected by one of Darwin's sons: 'This . . . apparatus, which I have called "*papilionaceous pump apparatus*" is rather marvellous. It is similar to the instrument with which one makes pasta at the vermicelli-makers.' Bees settling onto the carina of the flowers caused the pistil and filaments to press upon the pollen-mass, which was then ejected through a small hole in the carina and attached to the insects (letter from Federico Delpino, 20 May 1870). Darwin encouraged the German naturalist Hermann Müller to continue his work on the adaptation of insect structures, in particular the mouth-parts of bees, to the forms of flowers: 'As far as I know, no one has carefully observed the structure of insects in relation to flowers, although so many have now attended to the converse relation' (letter to Hermann Müller, 14 March 1870). Darwin received a string of letters from his cousin Francis Galton, reporting on his efforts to alter the colour of rabbits by transfusing the blood of one variety into another and observing the colour of the offspring. The experiments, designed to test Darwin's hypothesis of pangenesis, sometimes involved unusual procedures: 'I hope to succeed in making the ears of two young rabbits grow together & to mix their circulations by breaking adjacent veins into one' (letter from Francis Galton, 25 June 1870). Occasionally Galton reported success: 'Good rabbit news.! One of the latest litters has a white forefoot' (letter from Francis Galton, 12 May 1870). But in general the results were discouraging. Darwin's pangenesis theory also prompted speculation among some of his correspondents about the influence that males could have, both upon later generations of offspring not sired by them, and upon the females with whom they had mated. Some thought that male 'gemmules' (the material of hereditary transmission in Darwin's theory) might be transmitted to the female by the circulation of blood between the mother and foetus during pregnancy. As a case in point, John Jenner Weir described the offspring of a mare that had once mated with a quagga as resembling a donkey; he also added: 'It may be fanciful on my part but I cannot but think that my theory accounts for what has been so often noticed that persons long married grow like each other' (letter from J. J. Weir, 17 March 1870). With Joseph Dalton Hooker, Darwin discussed the recent experiments of Henry Charlton Bastian, which Bastian claimed showed evidence of the spontaneous generation of life from inorganic matter: 'Spontaneous generation seems almost as great a puzzle as preordination; I cannot persuade myself that such a multiplicity of organisms can have been produced, like crystals, in Bastian's solutions of the same kind' (letter to J. D. Hooker, 12 July [1870]). Bastian's results were criticised in a presidential address to the British Association for the Advancement of Science by Thomas Huxley, who had identified bits of the moss *Sphagnum* in Bastian's supposedly empty and sealed test tubes. Huxley's address also discussed recent experiments by Louis Pasteur and John Tyndall that provided evidence for the existence of germs.

Some of the wider implications of evolutionary theory for medicine and society were explored in a campaign that Darwin initiated to gather information about the potentially harmful effects of cousin marriage. Darwin wanted to test the prevalent assumption that marriages to first cousins were injurious to offspring. He wrote initially probably to Henry Hussey Vivian, an MP and fellow of the Geological Society: 'I have every cause to believe that the time will soon come, when this subject will be ... highly important for the welfare of mankind' (letter to [H. H. Vivian?], [April or May 1870?]). Vivian contacted the home secretary, Henry Austin Bruce, about the possibility of inserting a question in the 1871 census about cousin marriage. Darwin believed that if such statistics could be obtained, they could be correlated with figures regarding the frequency of deafness, blindness, and other conditions in the offspring of consanguineous marriages. He enlisted the support of William Farr, a specialist in medical statistics who worked in the registrar-general's office, in drafting a memorandum. He asked his neighbour, the naturalist John Lubbock, who was now MP for Maidstone, to present the proposal to the House of Commons. Lubbock read Darwin's letter before the Commons on 26 July, 'As you are aware, I have made experiments on the subject during several years; & *it is my clear conviction that there is now ample evidence of the existence of a great physiological law, rendering an enquiry with reference to mankind of much importance*' (letter to John Lubbock, 17 July 1870). The motion to amend the Census was defeated by a vote of ninety-two to forty-five.

As was usual in periods of reasonably good health, Darwin seldom took breaks from his scientific work except to receive friends and visit family. He confided to his cousin William Darwin Fox, 'I never pass 6 hours without a fit of extreme discomfort, & so I shall go on to the last of my uncomfortable days' (letter to W. D. Fox, 18 February [1870]). But he had resumed horse-riding and, subject to the usual qualifications, Darwin's health was generally good. He did consult Henry Bence Jones, his physician since 1865, regarding 'pins & needles'. Jones replied, 'some stopped molecular action I suppose which is somehow related to your indigestion. Altho' the manner is very dark to me and to every one else I suspect' (letter from H. B. Jones, 2 August 1870). Darwin had visits from a number of scientific friends and colleagues at Down, including Alexander Agassiz and his family, Anton Dohrn, Albert Günther, Joseph Hooker, Rudolf Albert von Kölliker, Alfred Newton, Robert Swinhoe, and Vladimir and Sofia Kovalevsky. In November, he received a permanent guest in the form of a Scottish deerhound puppy, the pride and joy of George Cupples, who had written to Darwin regularly over the course of the year to apprise him of the dog's progress and fine breeding: 'the father is descended from Sir Walter Scott's celebrated "Maida"' (letter from George Cupples, 17 September 1870). Darwin reassured Cupples after the dog arrived, 'Bran is thriving & growing at a wonderful rate—coat sleek, & not too fat. Plays much with Polly & enjoys English life' (postcard to George Cupples, 27 November [1870]). In addition to receiving many visitors, Darwin was also away from Down more than usual, staying on three occasions in London with his brother Erasmus, a week in Surrey and at Ightam Mote, in Kent, and nearly a fortnight with his son William in Southampton, and making a four-day visit to Cambridge. While in Cambridge, he toured the Woodwardian Museum with his former professor of geology, Adam Sedgwick. 'He utterly prostrated me,' Darwin wrote to Hooker, '... & I have not recovered the exhaustion yet. Is it not humiliating to be thus killed by a man of 86' (letter to J. D. Hooker, 25 May [1870]). On learning of this, Sedgwick replied: 'I was truly grieved to find that my joy at seeing you again was almost too robust for your state of nerves. ... I only speak honest truth when I say that I was overflowing with joy when I saw you; & saw you in the midst of a dear family party & solaced at every turn by the loving care of a dear Wife & Daughters. How different from my position—that of a very old man, living in cheerless solitude!' (letter from Adam Sedgwick, 30 May 1870).

Darwin continued to take great pride in his children, and to worry about any potential waywardness on their part. He reflected on the responsibilities of finding a course in life for one's sons to Hooker: 'God knows it is puzzle enough in every case whatever; & what our Boy Horace is to do, I know no more than the man in the moon' (letter to J. D. Hooker, 25 May [1870]). Horace had entered Trinity College, Cambridge, in the autumn of 1870. Darwin was concerned that his poor health in previous years had left him behind, and arranged for him to have private instruction in mathematics. He wrote to Horace's tutor at Trinity to request that he be excused from attending college lectures for the time being (letter to [E. W. Blore], [October 1870 or later]). Leonard continued to have great success in the army. He gained a commission in the Royal Engineers, obtaining the second highest marks among the candidates. Francis completed his studies at Cambridge, taking third place in first-class honours in the natural sciences tripos in December. He had fallen into debt, however, and had kept the matter secret for some months. Darwin was very stern in his advice: 'I have never known a man who was too idle to attend to his affairs & accounts, who did not get into difficulties; & he who habitually is in money difficulties, very rarely keeps scrupulously honourable. . . . My father, who was the wisest man I ever knew, thought it the duty of every man, young & old, to keep an account of his money; & I very unwillingly obeyed him; for I was not always so bothersome an old fellow as I daresay I appear to you' (letter to Francis Darwin, 18 October [1870]). The affection and concern of the father for his children were reciprocated. George, who was now a fellow at Trinity College, Cambridge, wrote to Henrietta of his concern that his father, despite suffering a bad fall the previous year, continued to ride the same horse that had thrown him (letter from G. H. Darwin to H. E. Darwin, [21–2 February 1870] (DAR 251: 2243)). With some difficulty, he had found a mare so tame that she was ridden by a blind man and used at a school for timid riders. 'She has a canter like an armchair', he informed his father (letter from G. H. Darwin, [3 February 1870 or earlier]). George devoted considerable effort to securing the loan of the animal for his father, but the horse did not suit Darwin, and he evidently continued to ride the dangerous Tommy.

Darwin's European correspondence was disrupted for part of the year owing to the outbreak of war. France had declared war on Prussia on 19 July 1870, and French troops had crossed the border into Prussia at the end of that month. But despite suffering heavy casualties, the Prussian armies won several decisive victories in August. Darwin followed the progress of the war in the press, and to his German correspondents he declared his allegiance to the Prussian cause. 'I cannot express too strongly how I rejoice at the wonderful success of Germany,' he wrote to Julius Victor Carus, '& I have not hitherto met a single person who has not entirely participated in this feeling' (letter to J. V. Carus, 18 August 1870). Carus described how he had to abandon his work to care for the wounded, and how he found some 'philosophic comfort' in viewing the war as a 'necessary consequence of natural conditions', 'a most dreadful 'struggle for existence'' (letter from J. V. Carus, 2 October 1870). Carus, who had already arranged to translate *Descent* into German, assured Darwin that the war would not interfere with the success of his book in Germany: 'People are anxious to find a safe refuge in science'.

As completion of the book drew near, Darwin began to prepare for its reception, and for the disapproval that would greet his conclusions about the 'origin of man'. 'I can most truly say', he wrote to his cousin William Darwin Fox, 'that I have written nothing without deliberate consideration & acquiring all the knowledge which I possibly could.— I will send it, as I know well that you are a charitable man & do not without good evidence believe in bad motives in others' (letter to W. D. Fox, 15 November [1870]). Fox reassured him, however, that descent from a monkey was no longer so hard to contemplate, for he had recently learned from the antics of the orang-utans at the Zoological Gardens in Regent's Park that beasts shared the same vices as humans: 'the Lady Monkey from the Andamans—drinks & smokes like a

Christian; & evidently the Gentleman w[d] thrash, if not kill the Lady, if he had an opportunity' (letter from W. D. Fox, 18 [November 1870]).

This volume contains a supplement of more than one hundred letters that have been discovered since the volumes they should have appeared in were published, or that have been redated to before 1870. It also contains some letters with wide date spans that were probably written before 1870. The supplement includes letters to William Kemp between 1840 and 1843, concerning Kemp's geological discoveries in Scotland; letters to the Arctic explorer John Richardson thanking him for barnacle specimens; a letter to Charles Kingsley written shortly after the publication of *Origin*; and letters to Henry Fawcett thanking him for his favourable review of *Origin*, and reflecting on the methodological and theoretical difficulties of Darwin's early research on species.

ACKNOWLEDGMENTS

The editors are grateful to the late George Pember Darwin and to William Huxley Darwin for permission to publish the Darwin letters and manuscripts. They also thank the Syndics of Cambridge University Library and other owners of manuscript letters who have generously made them available.

The work for this edition has been supported by grants from the National Endowment for the Humanities (NEH), the National Science Foundation (NSF), and the Wellcome Trust. The Alfred P. Sloan Foundation, the Pew Charitable Trusts, and the Andrew W. Mellon Foundation provided grants to match NEH funding, and the Mellon Foundation awarded grants to Cambridge University that made it possible to put the entire Darwin correspondence into machine-readable form. Research and editorial work have also been supported by grants from the Royal Society of London, the British Academy, the British Ecological Society, the Isaac Newton Trust, the Jephcott Charitable Trust, the Natural Environment Research Council, the John Templeton Foundation, and the Wilkinson Charitable Foundation. The Stifterverband für die Deutsche Wissenschaft provided funds to translate and edit Darwin's correspondence with German naturalists.

Cambridge University Library, the American Philosophical Society (APS), and Cornell University have generously made working space and many services available to the editors.

Since the project began in 1975, the editors have been fortunate in benefiting from the interest, experience, and practical help of many people, and hope that they have adequately expressed their thanks to them individually as the work proceeded.

Without the expert help of John L. Dawson of the Literary and Linguistic Computing Centre of Cambridge University, the initial computerisation of the correspondence would not have been possible. Beauty Bapiro, Iain Burke, Peter Dunn, Ray Horne, Patricia Killiard, Merina Tuladhar, and Tomasz Waldoch have helped maintain and extend our access to the Library's electronic resources, and provided essential technical support. This volume has been typeset using LaTeX. We are grateful to Matthew Daws for his expert help in developing the current typesetting procedures, which built on those established by Simon Buck, and maintained and developed by Robin Fairbairns. Gratitude is expressed to the late C. A. Tripp for providing the editors with an invaluable tool for searching texts.

English Heritage has responded most generously to requests for information and for material from the collections at Down House, Downe. We are particularly grateful to Cathy Power, curator of Down House and Tori Reeve, former curator. Richard Darwin Keynes kindly made available Darwin family material in his possession.

Ursula Mommens has also provided letters and other materials that belonged to her grandfather, Francis Darwin.

Institutions and individuals all over the world have given indispensable help by making available photocopies or digital images of Darwin correspondence and other manuscripts in their collections. Those who furnished copies of letters for this volume can be found in the List of provenances. The editors are indebted to them, and to the many people who have provided information about the locations of particular letters.

The editors make daily use of the incomparable facilities of Cambridge University Library and have benefited greatly from its services and from the help and expertise of its staff, particularly the staff of the Manuscripts Department. We are especially grateful to the University Librarian, Anne Jarvis, and to her predecessors Peter K. Fox and Frederick W. Ratcliffe, and to the Keeper of Manuscripts and Archives, Patrick Zutshi, for their generous support. Other members of the library's staff who frequently respond to the editors' requests are: Marjolein Allen, Wendy Aylett, Frank Bowles, Kathleen Cann, Louise Clarke, Colin Clarkson, Jacqueline Cox, Les Goodey, John Hall, Tony Harper, Morag Law, Stephen Lees, David Lowe, Peter Meadows, William Noblett, Adam Perkins, John Reynolds, Jayne Ringrose, Mark Scudder, Clive Simmonds, Nicholas Smith, and Anne Taylor. The fetchers in the Rare Books reading room have also patiently dealt with the editors' often complex requirements, as have the staff of the Map Room.

At the American Philosophical Society Library, a splendid collection of Darwiniana and works in the history of science has been available to the editors since the inception of the project. The editors have benefited from the co-operation of the late Whitfield J. Bell Jr, former secretary of the society, and the late Edward Carter II, Robert S. Cox, Roy C. Goodman, and Martin L. Leavitt, all of the APS Library.

Thanks are due to the faculty and staff of the Department of Science and Technology Studies, Cornell University, for accommodating one of the Project's editors, and to staff at Cornell University Libraries for help with research enquiries.

The editors would like to acknowledge the assistance of Rodney Dennis, Jennie Rathbun, and Susan Halpert of the Houghton Library, Constance Carter of the Science Division of the Library of Congress, and Judith Warnement and Jean Cargill of the Gray Herbarium of Harvard University, who have all been exceptionally helpful in providing material from the collections in their charge.

In Britain, the editors have received assistance from Lynda Brooks (librarian), Gina Douglas (former librarian), and Ben Sherwood of the Linnean Society of London; and from Michele Losse and Kiri Ross Jones of the Royal Botanic Gardens, Kew. We would also like to thank Anne Barrett, college archivist at the Imperial College of Science, Technology and Medicine; and Douglas Ferguson and Candace Guite of Christ's College, Cambridge.

Among the others who advise and assist the editors in their work are Nick Gill, Randal Keynes, Gene Kritsky, Carl F. Miller, Jim Moore, Garry J. Tee, David West,

Béatrice Willis, and Leonard Wilson. The editors are also pleased to acknowledge the invaluable support of the members of the Project's Advisory Committee.

For help with particular enquiries in volume 18, the editors would like to thank, besides those already mentioned, Helen Elliott of the Galashiels Historical Society, Susan Fisher of the Royal Bank of Scotland Group plc., Melanie Gusteyn of the National Library of South Africa, Martin Kjellgren, Antony Larkum, Richard Milner, and Alistair Sponsel.

Thanks are also due to all former staff of the Darwin Correspondence Project, including: Sarah Benton, Charlotte Bowman, Heidi Bradshaw, Janet Browne, P. Thomas Carroll, Mario Di Gregorio, Rhonda Edwards, Kate Fletcher, Hedy Franks, Joy Harvey, Arne Hessenbruch, Thomas Junker, David Kohn, Jyothi Krishnan-Unni, Kathleen Lane, Sarah Lavelle, Anna K. Mayer, William Montgomery, Perry O'Donovan, Stephen V. Pocock, Duncan Porter, Marsha L. Richmond, the late Peter Saunders, Andrew Sclater, Anne Secord, Tracey Slotta, the late Sydney Smith, Emma Spary, Nora Carroll Stevenson, Jonathan R. Topham, Tyler Veak, and Sarah Wilmot.

We are most grateful to Margot Levy for providing the index to the current volume, and look forward to continuing to work with Margot in the future.

Finally, we would like to acknowledge the contribution of colleagues who left the staff during the preparation of this volume, Sheila Ann Dean and Ellis Weinberger. We thank them for their many services, and wish them well in all they undertake in the future.

PROVENANCES

The following list gives the locations of the original versions of the letters printed in this volume. The editors are grateful to all the institutions and individuals listed for allowing access to the letters in their care. Access to material in DAR 261 and DAR 263, formerly at Down House, Downe, Kent, England, is courtesy of English Heritage.

G. R. Agassiz ed. 1913 (publication)
Alexander Autographs (dealers)
J. David Archibald (private collection)
American Philosophical Society, Philadelphia, Pennsylvania, USA
Archives, New York Botanical Garden, Bronx, New York, USA
Armacost Library, University of Redlands, California, USA
Balfour Library, Department of Zoology, University of Cambridge, Cambridge, England
Catherine Barnes, Historical Autographs & Documents (dealer)
Anna Barone (private collection)
Hans A. O. Baumann (private collection)
Bayerische Staatsbibliothek, Munich, Germany
Berg Collection, New York Public Library, Astor, Lennox and Tilden Foundations, New York, USA
Bibliothèque de Genève, Geneva, Switzerland
Bodleian Library, Oxford, England
Bonham's (dealers)
Sir John Paget Bowman (private collecion)
British Library, London, England
Julian Browning Autographs (dealer)
Cambridge University Library, Manuscripts Department, Cambridge, England
Christie's (dealers)
Ralph Colp, Jr (private collection)
M. D. Conybeare (private collection)
Ruth Cramond and David Cramond (private collection)
CUL. *See* Cambridge University Library
DAR. *See* Cambridge University Library
Department of Rare Books, Special Collections and Preservation, River Campus Libraries, University of Rochester, New York, USA
Ernst-Haeckel-Haus, Friedrich-Schiller-Universität, Jena, Germany
Jeremy and Helen Evans (private collection)

Edward Ford (private collection)
John Gittleman (private collection)
Stephen Jay Gould (private collection)
Gray Herbarium of Harvard University, Cambridge, Massachusetts, USA
Gunther 1975 (publication)
Jonathan Haile (private collection)
Dr N. Hammond (private collection)
B. J. Harrison (private collection)
Harvey Cushing/John Hay Whitney Medical Library, Yale University, New Haven, Connecticut, USA
Heritage Auction Galleries (dealers)
Eilo Hildebrand (private collection of facsimiles)
History of Science Collections, University of Oklahoma Libraries, Norman, Oklahoma, USA
House of Commons papers (publication)
Barlow Humphreys (private collection)
The Huntington Library, San Marino, California, USA
Imperial College of Science, Technology, and Medicine, London, England
Invercargill City Libraries, Invercargill, New Zealand
Isle of Wight Record Office, Newport, Isle of Wight, England
Dr Hildegard Jaeger (private collection)
Journal of Botany
Karpeles Manuscript Library Museums, USA
Kettle ed. 1871 (publication)
C. C. Kohler (dealer)
Launceston Library (State Library of Tasmania), Launceston, Tasmania, Australia
Lawrences Auctioneers, Crewkerne, Somerset, England
Ronald Levine, Modern 1st Editions (dealer)
Linnean Society of London, Piccadilly, London, England
Dr W. N. Livingstone (private collection)
Lower Mississippi Valley Collections, Louisiana State University Libraries, Baton Rouge, Louisiana, USA
Lyon & Turnbull (dealers)
Maggs Bros (dealers)
Melvill 1900 (publication)
The Michael Faraday Museum at the Royal Institution of Great Britain, London, England
Mitchell Library, State Library of New South Wales, Sydney, N.S.W., Australia
Murch 1893 (publication)
Muséum National d'Histoire Naturelle: Bibliothèque Cryptogamie, Paris, France
Trustees of the National Library of Scotland, Edinburgh, Scotland
National Library of South Africa, Cape Town, South Africa
Parke Bernet (dealer)

Piasa SA, Paris (dealers)
Profiles in History (dealers)
Bernard Quaritch (dealer)
R & R Enterprises Autograph Auctions (dealers)
Max Rambod (dealer)
The Rare Book & Manuscript Library of the University of Illinois at Urbana-Champaign, Urbana, Illinois, USA
Trustees of the Royal Botanic Gardens, Kew, Richmond, Surrey, England
The Royal Society, London, England
David Schulson (dealer)
Shrewsbury School Library, Shrewsbury, England
Smithsonian Institution Libraries, Washington DC, USA
Jim Somerville (private collection)
Sotheby's (dealers)
Staatsbibliothek zu Berlin – Preußischer Kulturbesitz, Berlin, Germany
J. A. Stargardt (dealer)
Stecher and Klavins (dealers)
Sulivan family (private collection)
Swann (dealer)
Thomas Fisher Rare Book Library, University of Toronto, Toronto, Ontario, Canada
The United Kingdom Hydrographic Office, Taunton, England
University of Birmingham, Birmingham, England
Uppsala University Library, Uppsala, Sweden
Alan Wedgwood (private collection)
Wellcome Library, London, England
John Wilson Manuscripts (dealer)
Zoological Society of London, Regent's Park, London, England

A NOTE ON EDITORIAL POLICY

The first and chief objective of this edition is to provide complete and authoritative texts of Darwin's correspondence. For every letter to or from Darwin, the text that is available to the editors is always given in full. The editors have occasionally included letters that are not to or from Darwin if they are relevant to the published correspondence. Volumes of the *Correspondence* are published in chronological order. Occasional supplements will be published containing letters that have come to light or have been redated since the relevant volumes of the *Correspondence* appeared. Letters that can only be given a wide date range, in some instances spanning several decades, are printed in the supplement following the volume containing letters at the end of their date range. The first such supplement was in volume 7 and included letters from 1828 to 1857. The second supplement was in volume 13, and included letters from 1822 to 1864.

Dating of letters and identification of correspondents

In so far as it is possible, the letters have been dated, arranged in chronological order, and the recipients or senders identified. Darwin seldom wrote the full date on his letters and, unless the addressee was well known to him, usually wrote only 'Dear Sir' or 'Dear Madam'. After the adoption of adhesive postage stamps in the 1840s, the separate covers that came into use with them were usually not preserved, and thus the dates and the names of many recipients of Darwin's letters have had to be derived from other evidence. The notes made by Francis Darwin on letters sent to him for his editions of his father's correspondence have been helpful, as have matching letters in the correspondence, but many dates and recipients have had to be deduced from the subject-matter or references in the letters themselves.

Transcription policy

Whenever possible, transcriptions have been made from manuscripts. If the manuscript was inaccessible but a photocopy or other facsimile version was available, that version has been used as the source. In many cases, the editors have had recourse to Francis Darwin's large collection of copies of letters, compiled in the 1880s. Other copies, published letters, or drafts have been transcribed when they provided texts that were otherwise unavailable.

The method of transcription employed in this edition is adapted from that described by Fredson Bowers in 'Transcription of manuscripts: the record of variants',

Studies in Bibliography 29 (1976): 212–64. This system is based on accepted principles of modern textual editing and has been widely adopted in literary editions.

The case for using the principles and techniques of this form of textual editing for historical and non-literary documents, both in manuscript and print, has been forcefully argued by G. Thomas Tanselle in 'The editing of historical documents', *Studies in Bibliography* 31 (1978): 1–56. The editors of the *Correspondence* followed Dr Tanselle in his conclusion that a 'scholarly edition of letters or journals should not contain a text which has editorially been corrected, made consistent, or otherwise smoothed out' (p. 48), but they have not wholly subscribed to the statement made earlier in the article that: 'In the case of notebooks, diaries, letters and the like, whatever state they are in constitutes their finished form, and the question of whether the writer "intended" something else is irrelevant' (p. 47). The editors have preserved the spelling, punctuation, and grammar of the original, but they have found it impossible to set aside entirely the question of authorial intent. One obvious reason is that in reading Darwin's writing, there must necessarily be reliance upon both context and intent. Even when Darwin's general intent is clear, there are cases in which alternative readings are, or may be, possible, and therefore the transcription decided upon must to some extent be conjectural. Where the editors are uncertain of their transcription, the doubtful text has been enclosed in italic square brackets.

A major editorial decision was to adopt the so-called 'clear-text' method of transcription, which so far as possible keeps the text free of brackets recording deletions, insertions, and other alterations in the places at which they occur. Darwin's changes are, however, recorded in the back matter of the volume, under 'Manuscript alterations and comments', in notes keyed to the printed text by paragraph and line number. All lines above the first paragraph of the letter (that is, date, address, or salutation) are referred to as paragraph '0'. Separate paragraph numbers are used for subscriptions and postscripts. This practice enables the reader who wishes to do so to reconstruct the manuscript versions of Darwin's autograph letters, while furnishing printed versions that are uninterrupted by editorial interpolations. The Manuscript alterations and comments record all alterations made by Darwin in his letters and any editorial amendments made in transcription, and also where part of a letter has been written by an amanuensis; they do not record alterations made by amanuenses. No attempt has been made to record systematically all alterations to the text of copies of Darwin letters included in the correspondence, but ambiguous passages in copies are noted. The editors believe it would be impracticable to attempt to go further without reliable information about the texts of the original versions of the letters concerned. Letters to Darwin have been transcribed without recording any of the writers' alterations unless they reflect significant changes in substance or impede the sense; in such cases footnotes bring them to the reader's attention.

Misspellings have been preserved, even when it is clear that they were unintentional: for instance, 'lawer' for 'lawyer'. Such errors often indicate excitement or haste and may exhibit, over a series of letters, a habit of carelessness in writing to a particular correspondent or about a particular subject.

Capital letters have also been transcribed as they occur except in certain cases, such as 'm', 'k', and 'c', which are frequently written somewhat larger than others as initial letters of words. In these cases an attempt has been made to follow the normal practice of the writers.

In some instances that are not misspellings in a strict sense, editorial corrections have been made. In his early manuscripts and letters Darwin consistently wrote 'bl' so that it looks like 'lb' as in 'albe' for 'able', 'talbe' for 'table'. Because the form of the letters is so consistent in different words, the editors consider that this is most unlikely to be a misspelling but must be explained simply as a peculiarity of Darwin's handwriting. Consequently, the affected words have been transcribed as normally spelled and no record of any alteration is given in the textual apparatus. Elsewhere, though, there are misformed letters that the editors have recorded because they do, or could, affect the meaning of the word in which they appear. The main example is the occasional inadvertent crossing of 'l'. When the editors are satisfied that the intended letter was 'l' and not 't', as, for example, in 'stippers' or 'istand', then 'l' has been transcribed, but the actual form of the word in the manuscript has been given in the Manuscript alterations and comments.

If the only source for a letter is a copy, the editors have frequently retained corrections made to the text when it is clear that they were based upon comparison with the original. Francis Darwin's corrections of misreadings by copyists have usually been followed; corrections to the text that appear to be editorial alterations have not been retained.

Editorial interpolations in the text are in square brackets. Italic square brackets enclose conjectured readings and descriptions of illegible passages. To avoid confusion, in the few instances in which Darwin himself used square brackets, they have been altered by the editors to parentheses with the change recorded in the Manuscript alterations and comments. In letters to Darwin, square brackets have been changed to parentheses silently.

Material that is irrecoverable because the manuscript has been torn or damaged is indicated by angle brackets; any text supplied within them is obviously the responsibility of the editors. Occasionally, the editors are able to supply missing sections of text by using ultraviolet light (where text has been lost owing to damp) or by reference to transcripts or photocopies of manuscript material made before the damage occurred.

Words and passages that have been underlined for emphasis are printed in italics in accordance with conventional practice. Where the author of a letter has indicated greater emphasis by underlining a word or passage two or more times, the text is printed in bold type.

Paragraphs are often not clearly indicated in the letters. Darwin and others sometimes marked a change of subject by leaving a somewhat larger space than usual between sentences; sometimes Darwin employed a longer dash. In these cases, and when the subject is clearly changed in very long stretches of text, a new paragraph has been started by the editors without comment. The beginnings of letters, valedictions, and postscripts are also treated as new paragraphs regardless of whether they appear as new paragraphs in the manuscript. Special manuscript devices delimiting sections or paragraphs, for example, blank spaces left between sections of text and lines drawn across the page, are treated as normal paragraph indicators and are not specially marked or recorded unless their omission leaves the text unclear.

Occasionally punctuation marking the end of a clause or sentence is not present in the manuscript, but the author has made his or her intention clear by allowing, for example, extra space or a line break to function as punctuation. In such cases, the editors have inserted an extra space following the sentence or clause to set it off from the following text.

Additions to a letter that run over into the margins, or are continued at its head or foot, are transcribed at the point in the text at which the editors believe they were intended to be read. The placement of such an addition is only recorded in a footnote if it seems to the editors to have some significance or if the position at which it should be transcribed is unclear. Enclosures are transcribed following the letter.

The hand-drawn illustrations and diagrams that occur in some letters are reproduced as faithfully as possible and are usually positioned as they were in the original text. In some cases, however, it has been necessary to reduce the size of a diagram or enhance an outline for clarity; any such alterations are recorded in footnotes. The location of diagrams within a letter is sometimes changed for typesetting reasons. Tables have been reproduced as close to the original format as possible, given typesetting constraints.

Some Darwin letters and a few letters to Darwin are known only from entries in the catalogues of book and manuscript dealers or mentions in other published sources. Whatever information these sources provide about the content of such letters has been reproduced without substantial change. Any errors detected are included in footnotes.

Format of published letters

The format in which the transcriptions are printed in the *Correspondence* is as follows:

1. *Order of letters.* The letters are arranged in chronological sequence. A letter that can be dated only approximately is placed at the earliest date on which the editors believe it could have been written. The basis of a date supplied by the editors is given in a footnote unless it is derived from a postmark, watermark, or endorsement that is recorded in the physical description of the letter (see section 4, below). Letters with the same date, or with a range of dates commencing with that date, are printed in the alphabetical order of their senders or recipients unless their contents dictate a clear alternative order. Letters dated only to a year or a range of years precede letters that are dated to a particular month or range of months, and these, in turn, precede those that are dated to a particular day or range of dates commencing with a particular day.

2. *Headline.* This gives the name of the sender or recipient of the letter and its date. The date is given in a standard form, but those elements not taken directly from the letter text are supplied in square brackets. The name of the sender or

recipient is enclosed in square brackets only where the editors regard the attribution as doubtful.

3. *The letter text.* The transcribed text follows as closely as possible the layout of the source, although no attempt is made to produce a type-facsimile of the manuscript: word-spacing and line-division in the running text are not adhered to. Similarly, the typography of printed sources is not replicated. Dates and addresses given by authors are transcribed as they appear, except that if both the date and the address are at the head of the letter they are always printed on separate lines with the address first, regardless of the manuscript order. If no address is given on a letter by Darwin, the editors have supplied one, when able to do so, in square brackets at the head of the letter. Similarly, if Darwin was writing from an address different from the one given on the letter, his actual location is given in square brackets. Addresses on printed stationery are transcribed in italics. Addresses, dates, and valedictions have been run into single lines to save space, but the positions of line-breaks in the original are marked by vertical bars.

4. *Physical description.* All letters are complete and in the hand of the sender unless otherwise indicated. If a letter was written by an amanuensis, or exists only as a draft or a copy, or is incomplete, or is in some other way unusual, then the editors provide the information needed to complete the description. Postmarks, endorsements, and watermarks are recorded only when they are evidence for the date or address of the letter.

5. *Source.* The final line provides the provenance of the text. Some sources are given in abbreviated form (for example, DAR 140: 18) but are listed in full in the List of provenances unless the source is a published work. Letters in private collections are also indicated. References to published works are given in author–date or short-title form, with full titles and publication details supplied in the Bibliography at the end of the volume.

6. *Darwin's annotations.* Darwin frequently made notes in the margins of the letters he received, scored significant passages, and crossed through details that were of no further interest to him. These annotations are transcribed or described following the letter text. They are keyed to the letter text by paragraph and line numbers. Most notes are short, but occasionally they run from a paragraph to several pages, and sometimes they are written on separate sheets appended to the letter. Extended notes relating to a letter are transcribed whenever practicable following the annotations as 'CD notes'.

Quotations from Darwin manuscripts in footnotes and elsewhere, and the text of his annotations and notes on letters, are transcribed in 'descriptive' style. In this method the alterations in the text are recorded in brackets at the places where they occur. For example:

'See Daubeny ['vol. 1' *del*] for *descriptions of volcanoes in [*interl*] S.A.' *ink*

means that Darwin originally wrote in ink 'See Daubeny vol. 1 for S.A.' and then deleted 'vol. 1' and inserted 'descriptions of volcanoes in' after 'for'. The asterisk before 'descriptions' marks the beginning of the interlined phrase, which ends at the bracket. The asterisk is used when the alteration applies to more than the immediately preceding word. The final text can be read simply by skipping the material in brackets. Descriptive style is also used in the Manuscript alterations and comments.

Editorial matter

Each volume is self-contained, having its own index, bibliography, and biographical register. A chronology of Darwin's activities covering the period of each volume and translations of foreign-language letters are supplied, and additional appendixes give supplementary material where appropriate to assist the understanding of the correspondence. A cumulative index is planned once the edition is complete. References are supplied for all persons, publications, and subjects mentioned, even though some repetition of material in earlier volumes is involved.

If the name of a person mentioned in a letter is incomplete or incorrectly spelled, the full, correct form is given in a footnote. Brief biographies of persons mentioned in the letters, and dates of each correspondent's letters to and from Darwin in the current volume, are given in the Biographical register and index to correspondents. Where a personal name serves as a company name, it is listed according to the family name but retains its original order: for example, 'E. Schweizerbart'sche Verlagsbuchhandlung' is listed under 'S', not 'E'.

Short titles are used for references to Darwin's books and articles and to collections of his letters (e.g., *Descent*, 'Parallel roads of Glen Roy', *LL*). They are also used for some standard reference works and for works with no identifiable author (e.g., *Alum. Cantab.*, *Wellesley index*, *DNB*). For all other works, author–date references are used. References to the Bible are to the authorised King James version unless otherwise stated. Words not in *Chambers dictionary* are usually defined in the footnotes with a source supplied. The full titles and publication details of all books and papers referred to are given in the Bibliography. References to archival material, for instance that in the Darwin Archive at Cambridge University Library, are not necessarily exhaustive.

Darwin and his correspondents writing in English consistently used the term 'fertilisation' for the processes that are now distinguished as fertilisation (the fusion of female and male gametes) and pollination (the transfer of pollen from anther to stigma); the first usage known to the editors of a distinct term for pollination in English was in 1873 (letter from A. W. Bennett, 12 July 1873 (*Calendar* no. 8976)). 'Fertilisation' in Darwin's letters and publications often, but not always, can be regarded as referring to what is now termed pollination. In the footnotes, the editors, where possible, have used the modern terms where these can assist in explaining the details of experimental work. When Darwin or his correspondents are quoted directly, their original usage is never altered.

The editors use the abbreviation 'CD' for Charles Darwin throughout the footnotes. A list of all abbreviations used by the editors in this volume is given on p. xlii.

The Wedgwood and Darwin

Robert Waring Darwin = Susannah Wedgwood
1766–1848 1765–1817

- Henry Parker = Marianne
  1788–1856 1798–1858
  - Robert b.1825
  - Henry 1827–92
  - Francis 1829–71
  - Charles b.1831
  - Mary Susan 1836–93
- Caroline Sarah = Josiah III (Jos)
  1800–88 1795–1880
  - Sophy Marianne 1838–9
  - Katherine Elizabeth Sophy (Sophy) 1842–1911
  - Margaret Susan 1843–1937
  - Lucy Caroline 1846–1919
- Susan Elizabeth 1803–66
- Erasmus Alvey 1804–81
- Charles Robert = Emma 1808–96
  1809–82
  - William Erasmus 1839–1914
  - Anne Elizabeth 1841–51
  - Mary Eleanor Sept.–Oct. 1842
  - Henrietta Emma (Etty) 1843–1927
  - George Howard 1845–1912
  - Elizabeth 1847–1926
  - Francis (Frank) 1848–1925
  - Leonard 1850–1943
  - Horace 1851–1928
  - Charles Waring 1856–8
- Charles Langton = Emily Catherine (Catherine) 1810–66
  1801–86

Families up to 1870

Josiah Wedgwood II 1769–1843 = Elizabeth (Bessy) Allen 1764–1846

- Sarah Elizabeth (Elizabeth) 1793–1880
- Mary Anne 1796–8
- Charlotte 1797–1862 = Charles Langton 1801–86
  - Edmund 1841–75
- Henry Allen (Harry) 1799–1885 = Jessie Wedgwood 1804–72
  - Louise Frances 1834–1903
  - Caroline b.1836
  - John Darwin 1840–70
  - Anne Jane 1841–77
  - Arthur 1843–1900
  - Rowland 1847–1921
- Frances Mosley 1808–74 = Francis (Frank) 1800–88
  - Godfrey 1833–1905
  - Amy 1835–1910
  - Cicely Mary 1837–1917
  - Clement 1840–89
  - Lawrence 1844–1913
  - Constance Rose 1846–1903
  - Mabel Frances 1852–1930
- Hensleigh 1803–91 = Frances (Fanny) Mackintosh 1800–89
  - Frances Julia (Snow) 1833–1913
  - James Mackintosh (Bro) 1834–64
  - Ernest Hensleigh 1838–98
  - Katherine Euphemia (Effie) 1839–1931
  - Alfred Allen 1842–92
  - Hope Elizabeth 1844–1934
- Frances (Fanny) 1806–32

ABBREVIATIONS

AL	autograph letter
ALS	autograph letter signed
DS	document signed
LS	letter in hand of amanuensis, signed by sender
LS(A)	letter in hand of amanuensis with additions by sender
Mem	memorandum
(S)	signed with sender's name by amanuensis
CD	Charles Darwin
CUL	Cambridge University Library
DAR	Darwin Archive, Cambridge University Library
del	deleted
illeg	illegible
interl	interlined
underl	underlined

TRANSCRIPTION CONVENTIONS

[some text]	'some text' is an editorial insertion
*[*some text*]*	'some text' is the conjectured reading of an ambiguous word or passage
[some text]	'some text' is a description of a word or passage that cannot be transcribed, e.g., '*3 words illeg*'
⟨ ⟩	word(s) destroyed
⟨some text⟩	'some text' is a suggested reading for a destroyed word or passage
⟨*some text*⟩	'some text' is a description of a destroyed word or passage, e.g., '*3 lines excised*'

THE CORRESPONDENCE OF CHARLES DARWIN
1870

From Federico Delpino[1] [1870][2]

Onorevole Signore!

Permetta che le offra in omaggio anche un altro piccolo lavoro "l'applicazione della Teoria Darwiniana ai fiori e agl'insetti visitatori dei fiori"[3]

Tanto l'autore che il traduttore[4] contano tra i di Lei discepoli, più convinti e persuasi delle grandi verità esposte nella Dottrina che da Lei prende il nome. Ella voglia nella sua bontà gradire il tenue dono.

Forse occuperà men ben⟨ ⟩ il suo tempo, se si risolve a leggere questo libretto; non ostante mi sarebbe estremamente caro che leggesse la mia annotazione che si trova a pagg. 31. 32.

In quest' annotazione io ho sommariamente esposto le ragioni che mi hanno reso profondamente persuaso della variabilità indefinita degli esseri, comecchè in perfetto accordo colle teorie vitalistiche e teleologiche da noi propugnate.

Riceva di nuovo l'attestato della mia altissima stima e riconoscenza. | Ossequentissimo discepolo | Federico Delpino | Professor di botanica a Vallombrosa.

(Firenze per Vallombrosa)

DAR 162: 141

[1] For a translation of this letter, see Appendix I.
[2] The date is established by the reference to Delpino trans. 1870a (see n. 3, below).
[3] Delpino refers to Delpino trans. 1870a (The application of Darwinian theory to flowers and to the insects that visit flowers), an annotated translation of H. Müller 1869. There is an annotated copy in the Darwin Pamphlet Collection–CUL.
[4] Delpino refers to himself and to Hermann Müller.

From Charles Ottley Groom Napier January 1870

20 Maryland Road | Paddington | W.
Janry 1870

To Ch. Darwin Esqre F.R.S.

My dear Sir

By desire of the late Lord Brougham[1] I send you a copy of my new work *The Book of Nature and the Book of Man*, will you accept it in remembrance of his Lordship who had a very high opinion of the author of "Natural Selection" to which he was latterly a convert.[2]

I regret much the delay in forwarding the book which is only just published I should have liked to have sent it in the lifetime of Lord Brougham. Let it now be an olive leaf snatched from his tomb

I remain yours most respectfully | C O Groom Napier

DAR 165: 228

1 Henry Peter Brougham died in 1868. Groom Napier was later exposed as an imposter; it is doubtful whether he knew Brougham, or whether Brougham wrote the preface to Groom Napier 1870, as Groom Napier claimed.

2 There is no copy of Groom Napier 1870 in the Darwin Libraries at CUL or Down.

From William Turner 3 January 1870

6 Eton Terrace | Edinburgh
Jan^{r}. 3^{d}. 1870

My Dear Sir

Kölliker of Würzburg and Langer of Vienna are the best authorities on the development of the mammae & the glands of the skin.[1]

They both agree in describing the mammary gland as developed before the nipple—[2] Kölliker figures in a 5 months human embryo the first indications of a mamma, and the horny layer of the epidermis is prolonged over its surface without any appreciable elevation corresponding to a nipple—[3]

Langer states that a distinct lens shaped body may be recognised in the mammary region of a human embryo, $7\frac{1}{2}$ centimetres long, which is the rudimentary gland, but at this early stage no trace whatever of mammary ducts can be recognized in it—[4]

In a female human embryo 10 centimetres long, he could not trace mammary ducts in the lens shaped body—

When hair first begins to protrude through the skin of the embryo then the mammary ducts may be seen in different stages of development.

Langer concludes that the first development of the mammary gland is connected with the existence of a special body, in which the ducts arise independently, & are not involutions of the external skin—

At first the ducts possess one common orifice— on the surface of the skin, but with the development of the nipple each of the main ducts opens independently on the surface—

It is clear therefore that the nipple is a structure developed in connection with the system of excretory ducts, and it is equally clear that the ducts are not developed until the position & general form of the mammæ have been mapped out in the skin—

I must apologize for the delay in replying to your queries but about the time that I received your letter a great whale came ashore here, and the dissection which I undertook so absorbed my time and thoughts that I fear many matters which should have been attended to were for a time at least forgotten—[5]

I remain | Yours sincerely | W Turner

DAR 80: B158–9

CD ANNOTATIONS
1.1 Kölliker] *underl red crayon*
1.1 Langer] *underl red crayon*
3.3 ducts] *underl red crayon*

1 Turner refers to Rudolf Albert von Kölliker and Karl von Langer.
2 CD cited Turner, on the authority of Kölliker and Langer, on this point in *Descent* 1: 209.
3 See Kölliker 1855, pp. 558–61.
4 See Langer 1852.
5 CD's letter to Turner has not been found. The whale was beached near Longniddry on the Firth of Forth, about twelve miles from Edinburgh, on 4 November (see *The Times*, 6 November 1869, p. 9, 17 November 1869, p. 5).

To Anton Dohrn 4 January 1870

Down. | Beckenham | *Kent. S.E.*
Jan 4 1870

My dear Sir

On the receipt of your letter I wrote at once to the Director of our Zoolog. Gardens, about the Limulus.[1] He informs me that they now have only one old adult male; but when they had both sexes eggs were *never* observed. Your best chance w^{d} be to write to some Zoologist on the shores of the United States; I cannot help you in this, as my sole correspondent, Dana,[2] is now much out of health.

I enclose a separate letter about your scheme, which has my good wishes; but I am sure that you estimate my influence & judgment much too highly. The opinion of naturalists who have visited the coast for some special investigation w^{d} be worth far more than mine. I fear your plan will cost you much loss of time in writing letters & making arrangements. I w^{d} suggest to you to delay attempting so great an addition as the formation of a scientific library.

Lastly let me thank you for your very kind & strong expressions towards myself, & for your information with respect to your present views on embryology.

Forgive me for suggesting one caution; as Demosthenes said, "action, action, action" was the soul of eloquence, so is caution almost the soul of science.[3]

Pray bear in mind that if a naturalist is once considered, though unjustly, as not quite trust worthy, it takes long years before he can recover his reputation for accuracy.

Pray forgive me for this caution. You ask about my health; I cannot say much in its praise, but as long as I live the life of a hermit I am able to work some hours daily.

With the most sincere good wishes | I remain my dear Sir | yours very faithfully | Ch. Darwin

[Enclosure]

Down. | Beckenham | *Kent. S.E.*
Jan 4 1870

My dear Sir

In your letter of Dec. 30 you ask me for my opinion with respect to founding an acquarium with the necessary apparatus, at some favorable station such as Messina, for scientific researches.

As far as my judgement goes, I can feel no doubt that at present embryological investigations on the lower marine animals are of the utmost importance; & for this purpose your scheme offers obvious facilities. If Johann Müller c^{d} have left the larvæ of the Echinoderms in an acquarium well provided with a flow of sea-water, he c^{d} have examined them after considerable intervals of time, or he might have arranged with some other naturalist to follow out & mature his great discovery.[4] Therefore if sufficient funds can be obtained to construct & keep up an acquarium, & if it is found practicable to regulate its use amongst old & new subscribers or strangers, you will, no doubt have the good wishes of every naturalist in Europe. For the sake of shewing my own good will, I shall at any time be happy to subscribe the small sum of five pounds sterling towards the necessary expenses

Pray believe me my dear Sir | yours very faithfully | Charles Darwin

To Dr Anton Dohrn

Bayerische Staatsbibliothek München (Ana 525. Ba 697)

1 See *Correspondence* vol. 17, letter from Anton Dohrn, 30 December 1869. The superintendent of the Zoological Society's gardens in Regent's Park, London, was Abraham Dee Bartlett. CD's correspondence on this subject with Bartlett has not been found; however, see the letter to A. D. Bartlett, 5 January [1870]. *Limulus* is a horseshoe crab.

2 James Dwight Dana.

3 The anecdote about Demosthenes is related in Pseudo-Plutarch, *Lives of the ten orators* (Demosthenes). 'Action' is sometimes rendered 'delivery'.

4 Johannes Peter Müller discovered the distinctive larvae of various different classes of echinoderms (see Winsor 1976, pp. 106–13).

To Abraham Dee Bartlett 5 January [1870][1]

Down. | Beckenham | *Kent. S.E.*
Jan 5th

My dear Sir

Many thanks about Limulus.[2] I am going to ask another favour, but I do *not want to trouble you to answer it by letter.*

When the Callithrix sciureus screams violently does it wrinkle up the skin round the eyes like a Baby always does? When thus screaming do the eyes become suffused with moisture? Will you ask Sutton to observe carefully.— Could you make it scream without hurting it much?[3]

I sh^{d}. be *truly* obliged some time for this information, when in Spring I come to Gardens,—or Sutton c^{d}. write to me.—[4]

Your's very faithfully | Ch. Darwin

Archives of the New York Botanical Garden (Charles Finney Collection) (ser. 14)

1 The year is established by the relationship between this letter and the letter to Anton Dohrn, 4 January 1870.
2 See letter to Anton Dohrn, 4 January 1870. Bartlett's letter to CD has not been found.
3 *Callithrix sciureus*, the squirrel monkey, is now *Saimiri sciureus*. CD discussed the secretion of tears by screaming monkeys in *Expression*, pp. 166–7, but did not mention *Callithrix*. CD also refers to Seth Sutton.
4 CD stayed in London from 5 to 10 March 1870 (CD's 'Journal' (Appendix II)). No letter from Bartlett or Sutton on the subject has been found.

To Charles Boner [before 8 January 1870][1]

I have just received your two works, and have made some use of them for my present book; and I should have made more use of them had I received them earlier.[2] You describe the grand scenery of the Tyrol most graphically, and it makes me long to be strong and young again to ramble over the mountains. My next book will not be ready for a considerable time, though part is finished.[3]

I am very much obliged to you for your very interesting letter, and especially glad to hear about the co-existence of varieties in the wild state.[4] The fact relating to the variation of the wild boar is quite new to me.[5] I hope I shall get your book soon,[6] and with my best thanks, I remain, dear sir, &c., | Charles Darwin.

Kettle ed. 1871, p. 77

1 The date is established by the relationship between this letter and the letter from Charles Boner, 8 January 1870.
2 CD refers to Boner 1860 (*Chamois hunting in the mountains of Bavaria and in the Tyrol*) and Boner 1861 (*Forest creatures*). Neither has been found in the Darwin Libraries at CUL or Down, but CD cited them both in *Descent*.
3 *Descent* was published in 1871.
4 See *Correspondence* vol. 17, letters from Charles Boner, 25 November [1869] and [December 1869 – early January 1870].
5 See *Correspondence* vol. 17, letter from Charles Boner, 25 November [1869].
6 CD refers to Boner 1865 (*Transylvania: its products and its people*; see letter to Charles Boner, 20 January [1870]). There is a copy in the Darwin Library–Down. This sentence of the letter, which survives only as a published transcription, may have come from a letter written after Boner's letter of 8 January 1870, in which he told CD he was sending the book. See *Correspondence* vol. 19, letter from R. M. Kettle, 10 March [1871], for Rosa Mackenzie Kettle's request to use some sentences of CD's in Kettle ed. 1871.

From Charles Boner 8 January 1870

Munich.
January 8. 1870

My dear Sir,

I need hardly say that it gave me very particular satisfaction to hear that a book of mine had afforded you some information worthy your attention.[1] It pleased me, too, to know that the Chamois Hunting interested you and awakened in you a wish for mountain air & mountain scenery. There is nothing like it.

It is very grievous that you should be in such a state of health as you describe.[2] I am sorry to hear of it, and can well sympathise with you as I have myself been since many month suffering much pain from a severe attack of Sciatic, which has left behind all sorts of ills, among others a state of the muscles & nerves of the thighs, which makes it almost impossible for me to move about. For one who hitherto was as healthy as an oak tree, & for whom no fatigue was too great, to be chained thus like a prisoner is a sad trial. What must be so painful for you is the inability to work. One grieves at evey moment lost, especially when, as in your case, there is so much to be accomplished; thoughts, views, opinions, discoveries that one wishes still to be able to record. And unfortunately the excitement which this inability to work must cause, can do you no good, but harm rather, as it irritates continually. I sincerely hope that your illness may leave you hope of ultimate, of speedy, recovery. Independent of the sympathy one may feel for you individually, we are all interested in your getting well & strong again, for much as you have already given to Science, we are greedy for additional contributions

As you were so obliging to say that the Chamois book had interested you, I have presumed to send you a volume about a country which possibly may not be well known to you, & which is in many respects very interesting.[3] Perhaps sometimes when you can do nothing else, when you are unable to work and still would care to have some light occupation, you might in one or other of the pages of "Transylvania" find a subject-matter that amused you, and thus helped to while away a weary hour. Were it to be so I should be more glad than I can say.

There are some few notices which I collected about the gypsies that to me seemed curious. Perhaps you may find them so too.

I should not have ventured to offer you anything of such small worth as a book of mine, had it not seemed to me by your letter that the 2 volumes you had read had really interested you.

With very sincere wishes for improvement in your health | Believe me, dear Sir, | Your faithful servant | Charles Boner.

To Charles Darwin, Esq | &c &c &c

DAR 160: 239

[1] See letter to Charles Boner, [before 8 January 1870].

[2] This information was probably in a part of CD's letter to Boner of [before 8 January 1870] that has not been preserved.

[3] Boner refers to Boner 1865 (see letter to Charles Boner, [before 8 January 1870] and n. 6).

To Richard Kippist 10 January [1870 or 1871][1]

Down. | Beckenham | *Kent. S.E.*
Jan. 10th

My dear Sir

I have duplicate copies of some Books & others useless to me, some of which may be worth accepting by the Linn. Soc;[2] so I will send them by Carrier on Thursday *morning.*—

Will you kindly have ready for Carrier any parts of Linn. Transact. due to me.— Also kindly oblige me by sending order on next page for Phil: Transact.[3] & have them packed in same parcel with Linn. Transact.—

My dear sir | Yours very faithfully | Ch. Darwin

Linnean Society of London

1 The years are established by the form of the address, with 'Bromley' struck through and 'Beckenham' added by hand; this form was used by CD between April 1869 and May 1871.

2 CD refers to the Linnean Society of London. The books have not been identified.

3 *Philosophical Transactions of the Royal Society of London.*

From Adolf Reuter 11 January 1870

Pfauen-Insel near Potsdam
the 11.1.70.

Dear Sir!

I suppose, that you have been very much astonished to receive some months ago a small box from Germany with different fruits of oranges, a specimen of wine[1] and a sort of beans without any letter.—

Dearest Sir I took me the liberty to send it to you and I did have also the intention to send you afterwards a letter with some explications but I did have the misfortune to fall down in our Oranges-house, so that I was forced to stop in the bed for some weeks. In this time I did receive also the order to change my place and to govern another garden, what was a very great trouble for me.—

I am now on the Pfauen Insel (Isle of Pavons), one and a half hour of Potsdam, a very interesting place, a fine Island populated with more than sixty pavons[2] and other birds in a quit wild state, decorated by different trees, shrubs and also fine collections of Palms.

Because the Isle is very isolated, so it will be the greatest pleasure for me to collect so much as possible for you.—

The reason for what I did send you the fruits, was to show you a small collection of our cultivated Oranges-varietys and mostly for the one monstrosity, where perhaps a citron and an orange is joined in one fruit; all the fruits of this tree are in the same manner but some more or less monstrous than the others.—[3]

The Ink-wine is interesting because his leaves are red coloured and also no other wine variety has quit red colouring sap as it is the case with the last.[4]

The bean with the blue fruits has rose flowers and is perhaps a hybrid between Phaseolus vulgaris and any species of Dolichos?—[5]

Their origin is the garden of Mr. Lucas, Pomologist at Reutlingen in Würtenberg.[6]

I was very much enjoyed indeed, as you told me in your last letter, that you and also our adorated Dr. Hooker from Kew was astonished about the specimens which

I did send you—[7] You are indeed to indulgent with me and I am always anxious that I molest you to often!—

With my sincerest thank for all the signs of your favour and kindness I have the honour to be my Dearest Sir | your | most obedient servant | Ad. Reuter.

My address is now:

Ad. Reuter
Pfauen-Insel near Potsdam.)

DAR 176: 127

[1] Wine: i.e. vine. See n. 4, below.
[2] Pavons: i.e. peacocks (*Pavo cristatus*).
[3] CD had discussed crosses between oranges and lemons in *Variation* 1: 399, 2: 365.
[4] Reuter probably refers to *Passiflora suberosa*, then known as the ink vine, now the corky passionflower, devil's pumpkin, or indigo berry (Halliday 1837, p. 398; *ITIS*). The sap of this plant was generally black.
[5] *Phaseolus vulgaris* is the French bean. *Dolichos* is a genus in the same family (Fabaceae); the number of species now included in *Dolichos* has been greatly reduced from that which Reuter would have known. For more on the genus as it was recognised at this time, see Bentham and Hooker 1862–83, 1: 540–1.
[6] Reuter refers to Eduard Lucas.
[7] See *Correspondence* vol. 17, letter to Adolf Reuter, 27 September 1869. Reuter refers to Joseph Dalton Hooker.

From Anton Dohrn 13 January 1870

Jena.
13th. January 1870.

My dear Sir!

I thank You again and again for all the kindness, You have shown me by Your letter![1]

Of course, I never intended to ask You the smallest sum of money in favour of my plan,—and I hope I shall carry it on without making use of your Kindness and liberality. I have got that idea of the Zoological Stations into my head, and I think I'll not get rid of it, but by building at least one.

I am also very thankful for Your kind caution. But I hope, I shall not have committed such grave things as would justify anybody to withdraw confidence from me. It was rather, because I thought it right to name an error an error, that I wrote to You as I did,—[2] but certainly there are in our Science a quantity of much heavier faults withdrawn and hidden by those who committed them, by preparing soft envelops for them, so that they might go down easier and without making any noise. This could never be my way,— therefore perhaps I shall risk more;— but there is a good deal of confidence within me, that makes me open and free.

The Editors of the Journal of Microscopical Science have applied to me, for an abstract of my present investigations. I have declined to give one, before I have finished them. But then I'll not fail to do so, and I'll send You a copy, that You might see, what a state the Embryology of Arthropods is in.[3] I trust, You will be satisfied. There is an immense quantity of labour spent in that chapter since some years. My endeavouring has been, to show as clear as possible principles and method of genealogical investigation, by applying them to Crustacea and Insects. And I hope the conclusiveness of the results, I have reached at, will then justify as well principles as method.

With regard to Limulus, I thank You too for the trouble You have kindly taken in my behalf. I had already written to a young Naturalist in Salem (Massachusetts)[4] and he has promised, to send me Embryonic Stages in spirits. Besides that I saw lastly two Limulus in the Berlin-Aquarium in, as one must believe—copula; and the owner of the Aquarium[5] has promised me to tell me in the very moment, that eggs will be found. So I hope, I shall get to my purpose.

I cannot finish this letter without thanking You once more Your last letter, especially the caution You placed on its end. Your words are heavier within my ears than anybody's else; so I'll always remember them, as soon as I shall risk some dangerous things in our Science.

With the most sincere good wishes for your health | Yours very faithfully | Anton Dohrn

Charles Darwin Esq.

DAR 162: 205

1 See letter to Anton Dohrn, 4 January 1870.
2 Dohrn refers to his letter to CD of 30 December 1869 (*Correspondence* vol. 17), in which he said that some of his former views on the developmental history of arthropods had been completely mistaken.
3 Dohrn did not publish in the *Quarterly Journal of Microscopical Science.*
4 Alpheus Spring Packard (Groeben 1982, p. 93 n. 31).
5 Alfred Edmund Brehm (Groeben 1982, p. 93 n. 32).

To Albert Günther 13 January [1870]

Down. | Beckenham | *Kent. S.E.*
Jan 13th

My dear Dr. Günther

I am very much obliged to you. The sight of the proofs has pleased me more than anything which has happened to me for some weeks.[1] Pray tell Mr Ford[2] (or show him this note) that I am delighted with their appearance. I hope you are pleased with them, for they all illustrate facts which I give wholly on your authority. I declare I think the sexual differences in the Chæmelions look more wonderful in the figures than in the real specimens.[3] They will make my chapter capital. Pray tell Mr Ford that I am excessively pleased, also, with the feathers, which look far better than I thought possible.[4] I long to see those of the Argus Pheasant & Peacock—[5] There ought to be a drawing of the head of *female* salmon.—[6]

Accept my cordial thanks for all your extraordinary kindness

Yours very sincerely | Ch. Darwin

The Callionymus also is beautiful.— I lately saw some one who said he did not believe fishes ever differed much sexually![7]

Endorsement: '1870'
Shrewsbury School

[1] CD refers to the proofs of the woodcuts for *Descent.*
[2] George Henry Ford.
[3] See *Descent* 2: 35, fig. 35, and 36, fig. 36.
[4] See figures in *Descent* 2: 63–6.
[5] See *Descent* 2: 137, fig. 53; 143, fig. 56; 144, fig. 57; 146, fig. 58; 149, fig. 59.
[6] See *Descent* 2: 5, fig. 27.
[7] The woodcuts of *Callionymus lyra* (the dragonet) show a male and a female fish.

From Albert Günther [after 13 January 1870][1]

Private

My dear Sir

I am very glad to hear that you are pleased with the proofs; I suppose you will have the remainder in a week or two, & I am looking forward for the appearance of your work with great impatience.[2]

You are very kind in asking after my boy; indeed, I should have added a line about him in my last note, only I thought you had to think about so many things that I would not intrude upon you with my domestic affairs.[3]

Shortly after I had seen you my solicitor informed me that the Counsel of the Respondents finding their case unjustifiable, had refused to go on with it. Consequently the child had to be given up, & the Court dispensing with my personal appearance, simply found the respondents liable into costs

It so happened that after the case had been settled, it came out that the same man who refused me my child, had persuaded my poor wife a few hours before her death behind my back, to ask me to send the child to Scotland! However, enough of it; the mere thought of such conduct makes me still very bitter. I have the comfort to know that there was no guile in my wife, & I am justified in forgetting her relations.[4]

The boy is a fine strong child, doing very well under the faithful care of my mother,[5] & being a source of great comfort to me. Fathers, they say, are talkative about their youngsters; therefore I will not expose myself; there is nothing extraordinary about the boy except that he is born on the anniversary of the *death* of Cuvier[6]

Yours faithfully | AG.

DAR 165: 244

[1] The date is established by the relationship between this letter and the letter to Albert Günther, 13 January [1870].
[2] Günther refers to proofs of some of the woodcuts for *Descent* (see letter to Albert Günther, 13 January [1870]). *Descent* was published in January 1871.
[3] Robert William Theodore Gunther, Günther's son, was born 23 August 1869; his mother, Roberta, died a few days later (*ODNB*). Günther's 'last note' has not been identified. No letter survives in which CD asked after Robert.
[4] Evidently the family of Günther's wife, who was from Scotland, had taken charge of her child, and Günther had taken them to court.
[5] Eleanora Louise Günther.
[6] Georges Cuvier.

To Adolf Reuter 15 January [1870][1]

Down. | Beckenham | *Kent. S.E.*
Jan 15th.

My dear Sir

I am very sorry to hear of your accident, which seems to have been of a serious nature. I received the fruit all safely, and guessed that the specimens were sent by you. I was surprised at the vine with red leaves, and admired much the beans of the Phaseolus[2]

Your new residence sounds very delightful, and I hope that it will suit you in every respect.

Pray believe me My dear Sir | yours very faithfully | Ch. Darwin

LS(A)
Staatsbibliothek zu Berlin – Preußischer Kulturbesitz, Slg. Darmstaedter Lc 1859 Darwin, Charles, Bl. 220/221

1 The year is established by the relationship between this letter and the letter from Adolf Reuter, 11 January 1870.
2 See letter from Adolf Reuter, 11 January 1870.

To Albert Günther 16 January [1870][1]

Down. | Beckenham | *Kent. S.E.*
Jan 16th

My dear Dr. Gunther

Do you feel inclined as yet to leave home? A little change wd be good for you. Newton of Cambridge (& I hope Hooker & Swinhoe) is coming here next Saturday Jan. 22d to dine, & to stay till Monday morning. Will you join us: it wd give me great pleasure if you can prevail on yourself to come.—[2] I enclose memorandum about best trains.—[3]

Yours very sincerely, | C. Darwin

Many thanks for your last note.—[4]

Shrewsbury School

1 The year is established by the reference to a visit to Down by Alfred Newton, Robert Swinhoe, and Albert Günther on 22 January (see n. 2, below).
2 According to Emma Darwin's diary (DAR 242), Günther, Newton, and Swinhoe came to Down on 22 January and Joseph Dalton Hooker on 23 January.
3 The memorandum has not been found.
4 CD may refer to the letter from Albert Günther, [after 13 January 1870].

To Richard Kippist 16 January [1870 or 1871][1]

Down. | Beckenham | *Kent. S.E.*
Jan 16th

My dear Sir

I am sorry to have caused you trouble.— The text & the Plates was sent by mistake, & I shd be glad of its return, as one of my sons wishes to keep the Plates, which,, however, w[d] be of very little use to the Soc, as they are chiefly a curiosity in the art of engraving.— Kindly give volume specified to Royal Soc.—[2]

My dear sir | Yours sincerely | Ch. Darwin

Linnean Society of London

[1] The years are established by the form of the address, with 'Bromley' struck through and 'Beckenham' added by hand; this form was used by CD between April 1869 and May 1871.

[2] The text, plates, son, and volume have not been identified; the text and plates may have been among the books referred to in the letter to Richard Kippist, 10 January [1870 or 1871]. CD refers to the Linnean Society and the Royal Society of London.

From Henry Walter Bates 17 January 1870

15 Whitehall Place
Jan 17 1870

My dear Mr Darwin

I have sent to the post stamped & registered, your packet of Manuscript, which I hope will reach you safely. You will see my suggested alterations all marked lightly in pencil in their places & will notice that they nearly all relate to the orthography of scientific names of Insects—a very small Matter![1]

As to the *facts* there is only one which I do not clearly understand: it is the rudiments of horns in ♀ of Onitis furcifer. I conclude, however, that you consider the indented or flattened (retuse) front part of thorax as the rudiment of horn formation. Well, it is so in effeminate males of Phanæus onthophagus & so forth & I presume you are correct in so expressing the fact with regard to O. furcifer; but still some qualification in terms seems necessary.[2]

Other facts I have noted are additions to your stores. Of course I would not think of giving you new facts of those classes in which you have a superabundance already. But in one case—sexual dissimilarity of colours in Coleoptera,—you have clearly not enough. I have recently been working at *Longicorns*, a group in which an enormous amount of modification for mere ornament has taken place & I find that there are here many cases of sexual disparity of colours. The genera Mallaspis Pyrodes & Esmeralda, very large & beautiful Prionidæ Longicorns offer the most striking examples. In Esmeralda the two sexes have been placed in different genera partly on account of difference of colour. Many other genera in the tribe offer less striking cases.[3]

The reasoning I had not ought to touch & have not touched in the M.S.— I almost always agree with you & have less scruple on this account to suggest a modification of your view of females not being made dull-coloured by selection, which will bring your opinion & Wallace's nearly into harmony.[4] It is this—the necessity of females being dull-coloured for protection is true, but they have not been *made* dull from former brighter hues, but have simply been *kept* dull by natural selection steadily eliminating all tendency to brightness. This will not disagree with your clenching

& true argument against Wallace that females of a genus are truer in colour to the generic type than males.[5]

Yours sincerely | H W Bates

DAR 82: A44–5

CD ANNOTATIONS
1.1 I have ... Matter! 1.4] *crossed pencil*
2.2 I conclude, ... necessary. 2.6] *crossed pencil*
3.6 The genera ... beautiful 3.7] *scored pencil*
3.7 Prionidæ ... of colour. 3.9] *double scored pencil*
4.1 The reasoning ... males. 4.9] *crossed pencil*
Top of letter: '*40 Bartholomew R*[d]*, Kentish Town—*' *pencil*

1 CD had evidently sent his chapters on the secondary sexual characters of insects for *Descent* (*Descent* 1: 341–423) to Bates for comment.

2 In *Descent* 1: 372–3, CD argued that the thoracic crest of the female *Onitis furcifer* was indeed the rudiment of the projection proper to the male. *Onitis furcifer* is now *Chironitis furcifer*.

3 CD added Bates's information about sexual difference in colours in Prionidae to *Descent* 1: 367–8. Many species formerly classified as *Mallaspis* now are placed in other genera within the tribe Mallaspini (family Cerambycidae, the longhorn or longicorn beetles), to which all the genera mentioned belong. The family Prionidae is now the subfamily Prioninae (family Cerambycidae).

4 CD and Alfred Russel Wallace had discussed since 1867 the relative influence of protection and sexual selection in determining colour differences between the sexes of various organisms (see *Correspondence* vol. 15, letter from A. R. Wallace, 24 February [1867]).

5 See *Descent* 1: 404.

To Charles Boner 20 January [1870][1]

Down. | Beckenham | *Kent. S.E.*
Jan. 20th

My dear Sir

I am very much obliged for your extremely kind note & for your valuable present of your work on Transylvania, received this morning.[2] I do not think I ever read a word about this country, & am indeed ashamed to confess that I had to look at a map to be sure where it lay. Therefore, as soon as I have finished some books in hand, I will begin your volume. You must indeed feel most severely your present state of health, which I am truly sorry to hear of.— My own health has been slowly & gradually failing for so many years, that I can hardly imagine the sense of former vigour & feats of endurance. As long as I can use my mind for a few hours daily I must be content, & must enjoy the glorious scenes of nature, as described by you & others. With every hope that you may perfectly recover & observe the habits of wild animals again, I beg leave to remain with my best thanks for your kind present. Yours faithfully | Ch. Darwin

Staatsbibliothek zu Berlin – Preußischer Kulturbesitz, Slg. Darmstaedter Lc 1859 Darwin, Charles, Bl. 203/204

[1] The year is established by the relationship between this letter and the letter from Charles Boner, 8 January 1870.

[2] See letter from Charles Boner, 8 January 1870. CD refers to Boner 1865; there is a copy in the Darwin Library–Down.

From Thomas Henry Huxley 21 January 1870

26 Abbey Place
Jan 21. 1870

My dear Darwin

It is hard to resist an invitation of yours—but I dine out on Saturday;[1] & next week three Evenings are abolished by Societies of one sort or another— And there is that horrid Geological address looming in the future—![2]

I am afraid I must deny myself at present

I am glad you liked the sermon— Did you see the 'Devonshire Mans', attack in the Pall Mall?.

I have been wasting my time in polishing that worthy off— I would not have troubled myself about him, if it were not for the political bearing of the Celt question just now.—[3]

My wife[4] sends her love to all your circle | Ever Yours | T. H. Huxley

DAR 166: 323

[1] CD had evidently invited Huxley to Down for the same weekend as he hoped Albert Günther, Joseph Dalton Hooker, and Robert Swinhoe would come (see letter to Albert Günther, 16 January [1870]).

[2] Huxley gave his presidential address to the Geological Society of London on 18 February 1870 (T. H. Huxley 1870b).

[3] Huxley had given a lecture on 9 January 1870 on 'The forefathers and forerunners of the English people' under the auspices of the National Sunday League; the text of the lecture was published in the *Pall Mall Gazette*, 10 January 1870, pp. 8–9. In the lecture, Huxley argued that populations of mixed Celtic and Teutonic (or Saxon) ancestry occurred throughout large parts of Britain and Ireland, so that differences in the behaviour of 'a Tipperary man' and 'a Devonshire man' had nothing to do with blood. He cautioned, 'the sooner we leave off drawing political distinctions between Celts and Saxons the better'; and added that there was in any case no sufficient proof of the existence of any difference other than language between Celt and Teuton. His reasoning was criticised in a letter from 'A Devonshire man', published in the *Pall Mall Gazette*, 18 January 1870, p. 6; Huxley published a rejoinder in the *Pall Mall Gazette*, 21 January 1870, p. 6. The Act for the Disestablishment of the Irish Church had been passed in 1869, and a Land Bill, to improve the conditions of Irish tenants, was under discussion.

[4] Henrietta Anne Huxley.

From Alfred Russel Wallace 22 January 1870

9, St Mark's Crescent. N.W.
Jan. 22nd. 1870

Dear Darwin

My paper on Geolog. Time having been in type nearly *two months*, & not knowing when it will appear, I have asked for a proof to send you, Huxley & Lyell.[1] The latter part only, contains what I think is *new*, & I shall be anxious to hear if it at all helps to get over your difficulties.[2] I have been lately revising & adding to my various papers bearing on the "Origin of Species" &c. and am going to print them in a volume immediately, under the title of— *"Contributions to the Theory of Natural Selection: A series of Essays."*

In the last, I put forth my heterodox opinions as to Man, & even venture to attack the Huxleyian philosophy![3]

Hoping you are quite well & are getting on with your "Man" book,[4]

Believe me Dr. Darwin | Yours very faithfully | Alfred R. Wallace

Charles Darwin Esq.

P.S. When you have read the *proof* & done with it, may I beg you to return it to me. | ARW.

DAR 106: B90–1

[1] Wallace refers to Wallace 1870b, Thomas Henry Huxley, and Charles Lyell.

[2] Wallace 1870b, 'The measurement of geological time', was published in two parts in *Nature*, on 17 February and 3 March 1870. In the second part, Wallace argued that species change had proceeded much more rapidly in the period before 60,000 years ago than it had since, owing to the stabilising of the climate in the recent period, and set the beginning of the Cambrian period at 24,000,000 years before the present; this was well within the 100,000,000 years before the present that William Thomson estimated that the earth had been habitable.

[3] In the last chapter of his *Contributions to the theory of natural selection*, Wallace criticised Thomas Henry Huxley's views on the emergence of consciousness from matter (Wallace 1870a, p. 362).

[4] *Descent.*

From Alice Bonham-Carter to Emma Darwin 25 January [1870][1]

Ravensbourne, | Beckenham, Kent.
Jany 25

Dear Mrs Darwin—

I enclose two sentences with my translation of them by Henrietta's desire, she having sent them to me this morning,—& perhaps you may find something of news in her letter[2]

We are much obliged for the loan of yours & Bessies, & Elinor will send them tomorrow—if not tonight[3] She is just back from her campaign in search of her Committee, & has many letters to write tonight I have been to see the Smedley party—& on Thursday I go to Cumberland Pl. to see Julia before her departure to Hitchin—[4] I have got the "Climbing Plants" back from Roberts & will *bring* it over.[5]

I

It is only with (on) these premisses that we could account satisfactorily for the true morphological significance of the floral organs of the Merantaceæ in spite of the extreme perturbation to which they were exposed[6]

II

This edge (lip) embraces the anther of the upper stamen (filament) which remains thus held subtended (stretched) towards the lower part, in a forced position—[7]

Hoping that Bessie got home quite comfortably

Believe me | y[rs] aff[tly] | Alice B C.

DAR 160: 240

[1] The year is established by the source of the translated sentences (see n. 6, below).

[2] The reference is to Henrietta Emma Darwin.

[3] Bonham-Carter refers to Elizabeth Darwin and Elinor Mary Bonham-Carter.

[4] One Cumberland Place, London, was the home of Hensleigh Wedgwood. Since his daughter Frances Julia Wedgwood was usually called Snow, the reference may be to Julia Smith, Bonham-Carter's aunt.

[5] The reference is to *Climbing plants*. Roberts has not been identified.

[6] The translated sentence is from a paper by Federico Delpino (Delpino 1869b, p. 296). A translation of the paper appeared in two parts in *Scientific Opinion* in the issues dated 2 and 9 February 1870 (Delpino 1870b). See also letter from Federico Delpino, 28 February 1870. CD's copy of the original paper is in the Darwin Pamphlet Collection–CUL; the first few pages have been translated in pencil in the margin.

[7] The description relates to the floral mechanism in Marantaceae (Delpino 1869b, p. 298). During the bud stage, pollen is shed onto the back of the style behind the stigma. When the flower opens (anthesis), a cucullate staminode encloses the style, which is held under tension. The staminode possesses a lateral appendage that acts as a trigger when depressed, releasing the style, which then springs forward bringing the stigma into contact with pollen on the pollinator and then depositing fresh pollen on the same spot. CD later described the feature in *Thalia dealbata* (letter to G. H. Darwin, 11 [July 1878] (*Calendar* no. 11601)).

From Gilbert William Child 26 January [1870][1]

Elmhurst, | *G[t] Missenden.*
Jany 26

Dear Sir

I am sending you by this post a copy of the 2nd edition of my Essays wh: has now been out about 2 months[2] I have rather hesitated to do so thinking it might seem rather absurd to send a second so comparatively soon after the first, but really the book is now so much more what I first meant it to be & the essence of it is so much contained in the 4th Essay wh: did not appear at all before that I am induced to indulge myself in the pleasure of sending it to you[3] This I can do with the more satisfaction in that there is now no temporary or special reason for its publication

I remain dear Sir | faithfully yours | Gilbert. W. Child

DAR 161: 141

[1] The year is established by the reference to Child 1869 (see n. 2, below).

[2] Child refers to the second edition of *Essays on physiological subjects* (Child 1869). The book was published in November 1869 (*Publishers' Circular*, 15 November 1869, p. 747). The first edition had been published in 1868 (Child 1868). There are two copies of Child 1869 in the Darwin Library–Down and one of Child 1868.

[3] The fourth essay in Child 1869 had the title 'Some aspects of the theory of evolution'; it included a discussion of the relation of evolution to religious belief.

To A. R. Wallace 26 January [1870][1]

Down. | Beckenham | *Kent. S.E.*
Jan 26th

My dear Wallace

I have been very much struck, by your whole article (returned by this post), especially on the rate of denudation, for the still glaciated surfaces have of late much perplexed me.[2] Also *especially* on the lesser mutations of climate during the last 60,000 years; for I quite think with you no cause so powerful in inducing specific changes through the consequent migrations. Your argument wd be somewhat strengthened about organic changes having been formerly more rapid, if Sir W. Thompson is correct that physical changes were formerly more violent & abrupt.[3]

The whole subject is so new & vast, that I suppose you hardly expect anyone to be at once convinced, but that he shd. keep your view before his mind & let it ferment. This, I think, everyone will be forced to do.— I have not as yet been able to digest the fundamental notion of the shortened age of the sun & Earth. Your whole paper seems to me admirably clear & well put.— I may remark that Rütimeyer has shown that several wild mammals in Switzerland since neolithic period have had their dentition & I *think* general size **slightly** modified.[4] I cannot believe that Isthmus of Panama has been open since commencement of Glacial period; for notwithstanding the Fishes so few shell, crustacean & according to Agassiz not one echinoderm is common to the sides.—[5] I am very glad you are going to publish all your papers on Nat Selection: I am sure you are right, & that they will do our cause much good.

But I groan over Man— you write like a metamorphosed (in retrograde direction) naturalist, & you the author of the best paper that ever appeared in Anth. Review![6]

Eheu Eheu Eheu[7] | Your miserable friend | C. Darwin

British Library Board (Add 46434 ff. 198–9)

1 The year is established by the relationship between this letter and the letter from A. R. Wallace, 22 January 1870.

2 CD refers to Wallace 1870b (see letter from A. R. Wallace, 22 January 1870).

3 See letter from A. R. Wallace, 22 January 1870 and n. 2; CD refers to William Thomson and W. Thomson 1862.

4 Ludwig Rütimeyer mentioned the diminishing size of wild animals alongside the increasing size of domestic ones at prehistoric lake-dwellings in Switzerland in Rütimeyer 1861, p. 30 (see also *Variation* 2: 427).

5 In Wallace 1870b, p. 401, Wallace argued that before or during the glacial period, part of central America had been submerged, since a third of the known species of marine fish were identical on both sides of the isthmus of Panama. CD refers to Louis or Alexander Agassiz, and to A. Agassiz *et al.* 1869. There is an annotated offprint of A. Agassiz *et al* 1869 in the Darwin Pamphlet Collection–CUL; it is a set of reports from the *Bulletin of the Museum of Comparative Zoology.* The concluding report is by Louis Agassiz.

6 In an earlier paper Wallace had suggested that natural selection could not explain the origin of the moral and higher intellectual faculties of humans or some of their physical characteristics, such as the absence of body hair ([Wallace] 1869b, pp. 391–3; see also *Correspondence* vol. 17, letter to A. R. Wallace, 14 April 1869). Wallace had written a paper on the origin of human races for the *Anthropological Review* in 1864, in which he argued that human intelligence, moral qualities, and social instincts would have

been to the benefit of the community from their earliest appearance and would therefore have become subject to the force of natural selection (Wallace 1864, p. clxii).

7 *Eheu*: alas (Latin).

From John Brodie Innes 28 January 1870

HYBRID COW.— *There is at present a hybrid cow among a herd at the home farm of Beaufort—a cross apparently between a deer and a cow. Except in the colour, which is black, the animal is identical in appearance with the one lately sold by Mr Maclean, V.S., to the proprietor of Wombwell's Menagerie.*[1] *Strange to say, both animals were bred in or near the same parish—Ardclach*[2]*—and in many respects bear a similar history. Mr Lawson, manager at Beaufort,*[3] *bought the cow last autumn among a herd of cattle, and it has since been seen and admired by many. The hybrid cow in Wombwell's menagerie was calved at Lochindorb,*[4] *and was exhibited by Mr M'Donald, Blervie,*[5] *at the Forres Christmas Cattle Show in 1868.*

Dear Darwin

The above is from the Elgin Courant of this day 28th. Janry. 1870.[6] I have not seen the animal. It is satisfactory to know that the former one, which I told you of before, and which seemed if not a hybrid, a wonderfully good imitation of one has gone to a menagerie where perhaps in life or death the truth of its origin may be tested. Some old red deer seems to have taken to erratic courses about the ancient haunts of the Wolf of Badenoch[7]

Faithfully Yours | J Brodie Innes

DAR 167: 27

1 Mr S. McLean, a veterinary surgeon at Inverness, had bought the supposed deer–cow hybrid from Mr McDonald of Blervie at the Forres Cattle Show on 7 December 1868 (see *Correspondence* vol. 16, letters from J. B. Innes, 7 December 1868 and 14 December 1868, and letter from S. McLean to J. B. Innes, 15 December 1868). On the history of Wombwell's Menagerie, which had three branches, see Bostock 1927.

2 Ardclach is a hamlet eleven miles south-south-west of Forres.

3 Mr Lawson has not been further identified.

4 Lochindorb Castle is south of Forres.

5 McDonald of Blervie has not been further identified. Blervie castle is five miles south-east of Forres.

6 Innes glued the clipping to the top of his letter.

7 The 'wolf of Badenoch' was Alexander Stewart, earl of Buchan, who occupied Lochindorb from about 1371 (*ODNB*).

From William Boyd Dawkins 29 January [1870][1]

R. S. Evans Esq,[2] | Belvedere Park, Lessness Heath | Kent.

29. Jan 1869.

My dear Sir,

I just write you a line about the skeletons in the Cave at Perth y chwari near Corwen, because I have lighted on a fact of considerable ethnological value. The human skulls are indisputably of the River-bed type (of Huxley),[3] the tibiæ are strongly platycnemic,[4] and thus the bandy-legged-ness of the River-bed race is established. The bandy-legged race occurs in Algeria, S. France, & Spain, as well as in the laterite of India (.— Last year Dr Oldham[5] sent me a scrap big enough to swear by.) and now must be extended to Britain. I believe that once they occupied a large portion of Europe. Busk and I are laying our heads together, and I think that we shall make something out of the investigation.[6] At Easter I am to have another dig in Denbigh, and after that the Craven caves are in hand as you will see by the enclosed.[7]

Wishing you a happy new year, I am Dear Sir | Yours truly | W. Boyd Dawkins

Charles Darwin Esq.

DAR 162: 122

1 The year is established by the relationship between this letter and the letter from W. B. Dawkins, 19 October 1869 (*Correspondence* vol. 17). Dawkins must have written '1869' in error.

2 Robert Speke Evans was Dawkins's father-in-law.

3 Thomas Henry Huxley had noticed common features in prehistoric skulls found mainly in river-bed sites in Britain and Ireland (T. H. Huxley 1862).

4 Platycnemia: flattening of the tibia (*OED*).

5 Thomas Oldham.

6 George Busk and Dawkins described the human remains at Perthichwareu in the *Journal of the Ethnological Society* (Dawkins and Busk 1870).

7 The Craven caves are near Settle, North Yorkshire. The enclosure has not been found, but probably related to the formation of a committee to undertake the exploration of the caves, beginning with the Victoria Cave; Dawkins made a preliminary report in *Nature*, 21 April 1870, pp. 628–9. On Dawkins's research on the platycnemic bones, see also Lucas 2007.

To James Crichton-Browne 31 January [1870][1]

Down
Jan. 31,

My dear Sir

I am very sorry to trouble you again, but I should be very much obliged if you could soon return to me Duchenne, with, I hope, some notes, as your former notes were of such extreme interest to me.[2] Please address parcel "Ch. Darwin, Esq., Orpington Station, Kent, Per S.E. Railway".

Yours very faithfully | Ch. Darwin

P.S. If not asking too much trouble, will you let me have a line that I may know when parcel is sent off.

DAR 143: 329

1 The year is established by the relationship between this letter and the letter to James Crichton-Browne, 8 June 1869 (*Correspondence* vol. 17).

2 CD had lent Crichton-Browne his copy of Duchenne 1862 (see *Correspondence* vol. 17, letter to James Crichton-Browne, 8 June 1869 and n. 9). For Crichton-Browne's earlier notes, see *ibid.*, letter from Henry Maudsley, 20 May 1869, enclosure 2, and letter from James Crichton-Browne, 1 June 1869. Crichton-Browne had not written to CD since June 1869 owing to illness (see letter from James Crichton-Browne, 15 March 1870).

From D. Appleton & Co. to Asa Gray 1 February 1870

Statement of Sales of Darwins Origin of Species
to 1st Feby *1870*
by D. APPLETON & CO. for account of Asa Gray[1]

On hand last account,	70	*On hand this date,*	433
Printed since,	1000	*Given away,*	
		Sold to date,	637
	1070		1070

Sold 637 *Copies.* $\$2^{50}$ *Rate* 5% — *$* 79.63
Febry 1/69 Balance — 219.36
$ 298.99

Document
DAR 159: 86

1 This statement gives sales figures for the US edition of *Origin.* Gray had negotiated the original terms of publication of the book with Appleton & Co. (see *Correspondence* vol. 8, Appendix IV). Gray may have enclosed the statement in his letter of 14 February 1870.

From George Howard Darwin [1 February 1870 or earlier][1]

14 Arl. St. | Picc.[2]

My Dear Father,

I went to a horse dealers on Wednesday to see if they had anything that wd suit you, but the man assured me that no London dealer wd allow his horses to go on trial, but he told me the best way wd be to hire a horse from a riding school & go on changing until you were suited.[3]

I went this m^{g} to the Grosvenor Riding School—where they keep about fifty horses & I saw there a horse about 14.2 hands up to about 12 stone & wh. was so quiet that it had been ridden by a blind man.

I think the best way wd be for me to take a turn in the park on this or any other of their beasts wh. they think wd be suitable & then you to hire the one or ones I think best for a week or fortnight. I expect you wd have to pay rather higher than the value of the horse as the don't profess to sell & they say that this beast is of great value to the riding school for timid riders. We can talk over it tomorrow

Your affectionate son | G. H. Darwin

DAR 210.2: 13

1 The date is established by the relationship between this letter and the letter from G. H. Darwin, [3 February 1870 or earlier], and on the assumption that this letter could not have been written on a Wednesday (2 February 1870 was a Wednesday).

2 Fourteen Arlington Street, Piccadilly.

[3] George was trying to find a new horse for CD to replace Tommy, who had thrown CD in April 1869 (letter from G. H. Darwin to H. E. Darwin, [21–2 February 1870] (DAR 251: 2243); Emma Darwin's diary (DAR 242)).

From Mary Susan Owen 2 February [1870][1]

Astley— Shrewsbury
Feb 2d

My dear Uncle Charles—

You will be surprised at receiving a letter from me, but I am writing to ask a favour of you.—

Would you have any objection to being God-father to our small child no 3? it would be very kind of you if you have none, but I know some people do object, for I have received a refusal about it— I hope you won't mind my asking you, for it really is but a form after all— I hope you & Aunt Emma are pretty well, I hear of you occassionally from the Leith Hill girls—[2] they tell me that Henrietta is abroad—[3]

With best love to yourself & Aunt Emma

Believe me | Yr. affcate. niece | M. M. Owen.

DAR 173: 45

[1] The year is established by the reference to Maud Owen's christening. She was born 3 December 1869 (*Darwin pedigree*).

[2] Sophy and Lucy Caroline Wedgwood lived at Leith Hill Place in Surrey.

[3] Henrietta Emma Darwin was travelling in the south of France and Italy (letter from H. E. Darwin to Emma Darwin, 1 February [1870] (DAR 245: 31)).

From Robert Swinhoe 2 February 1870

33 Oakley Square. SW.
2 Feby. 1870.

My dear Mr. Darwin,

You told me the other day that Mr. Galton had received no replies to his circular in China on Hereditary Genius.[1] I have turned up one in "Notes & Queries".[2] I send you separately the copy of "N. & Q." in which it occurs. Perhaps Mr. Galton would like to see it. Kindly return me the Pamphlet when finished with.

I sent you copies of my Paper on the Formosan Rat.[3]

I was greatly refreshed by my visit to Down and have really felt much better ever since.[4]

With kind regards to all, | I remain, | Your's very sincerely. | Robert Swinhoe

DAR 177: 333

[1] Francis Galton's *Hereditary genius* was published in 1869 (Galton 1869). Swinhoe had visited Down from 22 to 24 January 1870 (Emma Darwin's diary (DAR 242)).

[2] Galton enquired in *Notes and Queries on China and Japan* 2 (1868): 120 whether the history of public competition in China tended to show that certain families were 'eminently endowed with scholastic ability'. A reply appeared in the issue for October 1868, and Galton cited it in Galton 1869, p. 335. No other reply has been found.

[3] Swinhoe refers to his paper, 'On a new rat from Formosa', first published in the *Proceedings of the Zoological Society of London* for 1864 (Swinhoe 1864). There are no offprints of the paper in the Darwin Pamphlet Collection–CUL.

[4] According to Emma Darwin's diary (DAR 242), Albert Günther, Alfred Newton, and Swinhoe came to Down on 22 January 1870.

From G. H. Darwin [3 February 1870 or earlier][1]

Thursday evg.

My dear Father,

When I first went to the riding school I asked one of the men whether he thought they wd sell the horse & he said that he thought very likely they wd.[2]

But I believe they do not profess to sell & so perhaps if I was to go & ask first of all they might say plump out no—but if you hired the horse & liked it I think they wd most likely sell. If however you wd rather not hire without knowing certainly if they'd sell I will go & inquire.

The horse stands I shd guess about $\frac{1}{2}$ a hand higher than Tommy & is up to a greater weight, but that you wd'nt mind. I rode her (its a mare) for about 20 m. in the park; she has a canter like an armchair and a very fair walk—but is a little wooden in the trot—tho' not at all a high action.— Except Tommy I never rode a quieter horse— certainly she does not go as entirely without the rein as Tommy—but unless she changes when she gets to Down you wo'nt often come across a quieter beast. Of course I ca'nt really judge for you.—

DAR 210.2: 14

[1] The date is established by the relationship between this letter and the letter from G. H. Darwin, [4 February 1870].

[2] See letter from G. H. Darwin, [1 February 1870 or earlier].

From G. H. Darwin [4 February 1870][1]

14 Arlington St.
Friday

My dear Father,

I have just been having a ride on the cob that I told you of.[2] I shd say that he is very little smaller than Tommy; he has got a very nice walk, trot & canter & is brisk tho' extremely quiet—& unless he turns out differently I shd think he was the very animal for you. And now for the bad side— the price is 50 guineas & they wo'nt let him out for less than a month for wh. they charge 10 guineas (to be prepaid) & to be subtracted from the purchase money. But lest this shd scare you too much from trying him—if you find he is not suitable I wd take him back to London & keep him on my own hook including share of 10-er for the rest of the month & try how riding suited me.—[3] I have written a letter for you to sign (if you like) from wh. you will see the arrangements I have made provisionally for his coming down.— If the waggonette is mouldering in the yard or if the horses happen to be at Orpington Station at 5.50 tomorrow you may send to meet the 4.12 train.

Your affectionate Son | George Darwin

The address is
Mr. Allen[4]
70 Seymour Place
Bryanstone Square
I told Allen I wd let him know tomorrow so you must write by return.—
G. D

[Enclosure][5]

Down | Beckenham | Kent
Sat. Feb. 5. 70

Sir,

I write to inform you that I wish to hire the cob, which my son rode yesterday, for 4 weeks— with the option of purchasing him during that time.— I enclose a cheque for £10.10s for wh. I beg you will be so good as to return me a receipt.

I should wish the cob to be sent by the train leaving Charing Cross at 10.30 AM to Orpington Station, South Eastern R.y on Monday Feb 7th; & my groom shall meet that train at Orpington.—

I am Sir | Yours obediently

Mr. Allen

DAR 210.2: 15–16

CD ANNOTATIONS
Enclosure:
1.2 time] 'for 50 gn, if I agree to purchase this c or *[*another*] [10 or 11 words illeg, del]*', *interl after, pencil*
1.3 £10.10s] *comma added after, pencil*
1.3 for wh.] *crossed pencil*
1.3 return] *crossed pencil*; 'send' *above, pencil*
1.3 receipt] 'by my servant who shall call at your stable for the horse on Monday next'[6] *added after, pencil*

1 The date is established by the date of the enclosure.

2 The horse that George describes here is apparently a different one from that described in his letter of [3 February 1870 or earlier].

3 The horse did not suit CD, and George kept him for the remainder of the month. George wrote to Henrietta Emma Darwin: 'I have been moving heaven & earth to get a horse for Father— my plan was to go to Riding Schools & hire their quietest hack— as dealers won't give a trial— I got one down that I thought wd do, but unfortly he turns out as John say 'too much of a orse for master'. He is a most charming little cob. As they wd only let him for a month & not less, I am going to have the riding

of him in town for the remaining fortnight. I have been doing all I can to stop Father riding Tommy, as I'm sure he's not safe— but I'm afraid he's going on at present.' (Letter from G. H. Darwin to H. E. Darwin, [21–2 February 1870] (DAR 251: 2243).)

4 Frederick Allen.

5 The enclosure is in George's hand with amendments by CD.

6 A payment to Allen of £10 10*s.* is recorded in CD's Account books–banking account (Down House MS) on 5 February 1870.

To the American Philosophical Society 5 February 1870

Down. | Beckenham | *Kent. S.E.*
Feb. 5th. 1870

Sir

I beg leave to acknowledge the receipt of your letter of Oct. 15th, in which you inform me that the American Philosophical Society has conferred on me the distinguished honour of electing me a Member; for this honour I beg permission to return my most sincere thanks.[1]

I have the honour to remain | Sir | Your most obedient servant | Ch. Darwin

To the | Secretary

American Philosophical Society (378)

1 See *Correspondence* vol. 17, Appendix III.

From M. S. Owen 7 February [1870][1]

Astley Shrewsbury
Feb 7th.

My dear Uncle Charles—

Many thanks for your kind letter & we are very much obliged to you for consenting to be god father— I never thought at all of asking you to stand in person, as I knew that was impossible any how—[2]

The little lady has been christened now & we have called her Maud as being a short & not very common name— When one meets with a refusal about the sponsorship, it makes one hesitate about asking others, though I suppose Sponsors are necessary..

With love to yourself & Aunt Emma | Believe me | Yr. affcate. niece | M. M Owen.

DAR 173: 46

1 The year is established by the relationship between this letter and the letter from M. S. Owen, 2 February [1870].

2 See letter from M. S. Owen, 2 February [1870]. CD's letter to Owen has not been found.

To Henrietta Emma Darwin [8 February 1870][1]

My dear H.

Please read the Ch. first **right through** without a pencil in your hand, that you may judge of general scheme; as, also, I particularly wish to know whether parts are extra tedious; but remember that M.S is always *much* more tedious than print.— The object of Ch. is simply comparison of mind in men & animals: in the next chapt. I discuss progress of morals &c.— Some sentences are at back of Page marked thus @.—[2]

I do not send foot-notes, as I have no copy & they are almost wholly mere authorities.— After reading once right through, the more time you can give up for deep criticism or corrections of style, the more grateful I shall be.— Please make any long corrections on separate slips of paper, leaving narrow blank edge, & pin them to margin of each sheet, so that I can turn each back, & read whilst still attached to its proper page.— This will save me a world of troubles Heaven only knows what you will think of the whole, for I cannot conjecture.— You are a very good girl indeed to undertake the job.—

Your affect Father | C. Darwin

(I suspect that here & there style will want a good deal of improvement, though I hope greater part fair.—)

(I fear parts are too like a Sermon: who w[d] ever have thought that I sh[d]. turn parson?)

British Library Board (Add 58373 ff. 1–2)

1 The date is established by the relationship between this letter and a letter from Emma Darwin to H. E. Darwin, [8 February 1870] (DAR 219.9: 72), in which Emma says she is enclosing a letter from CD.

2 Henrietta had been reading parts of the manuscript of *Descent* at least since she left England in January 1870 (letter from Emma Darwin to H. E. Darwin, [18 January 1870] (DAR 219.9: 69)). With this letter CD had presumably sent the second chapter on the comparison of the mental powers of humans and animals (*Descent* 1: 70–106). Henrietta was on holiday in France and Italy (see letter to W. D. Fox, 18 February [1870] and n. 10). She had also read the proof-sheets of *Variation* (see *Correspondence* vol. 15, letter to H. E. Darwin, 26 July [1867]).

From H. E. Darwin [after 8 February 1870][1]

Dear F.

Thanks for your note abt the M.S.S. I will obey all your instructions—& I shall be extremely interested in reading it—so you needn't thank me for it—& here what time I spend in my own room is so very undisturbed it goes much further.[2] When I know no human being will come after me & lay out my plans for the day & stick to them. Certainly to have you turned Parson will be a change— I expect I shall want it enlarging not contracting —cos I think *you* think an apology is wanting for writing abt any thing so unimportant as the mind of man!

AL
DAR 245: 33b

[1] The date is established by the relationship between this letter and the letter to H. E. Darwin, [8 February 1870].

[2] See letter to H. E. Darwin, [8 February 1870]. Henrietta refers to the manuscript of *Descent*. From the letter from H. E. Darwin to Emma Darwin, 17 February 1870 (DAR 245: 33a), it seems that Henrietta was in Genoa.

From Berry Benson 9 February 1870

Augusta, Ga.
Feb. 9th. 1870.

Mr. Charles Darwin.
Dear Sir:

I have read your work on "The Origin of Species" with a great deal of interest, and beg to express my gratitude in a simple way for opening to me the gate to such broad fields of knowledge and thought, hitherto so basely explored.

In your chapter on "Geographical Distribution" (XI.) on the subject, "Means of Dispersal", (Page 318 of the edition published by D. Appleton & Co. N.Y.) is the following:[1]

"It would be a great error to argue that because a well-stocked island like Gt. Britain, has not, as far is known (and it would be very difficult to prove this,) received within the last few centuries, through occasional means of transport, immigrants from Europe, or any other Continent, that a poorly stocked island, though standing more remote from the mainland, would not receive colonists by similar means."

To put in your hands proof, (if true,) that your island *has* received *one* colonist in quite recent times, I have taken the trouble to copy the accompanying account of a waterplant, from Wells' Annual of Scientific Discovery for 1854.[2] which I remembered to have read, when I came across the passage in your book above quoted.

If you have ere this had your attention called to the same, or if it may not prove useful to you, as I had been led to hope, pray excuse my having so far intruded on your valuable time. But if the account does have a direct bearing on the point in question, and proves serviceable to you, please accept it as the partial expression of the obligation due you, for the pleasure & satisfaction derived from studying your inestimable work.

Very Truly Yours. | Berry Benson.

[Enclosure]

Extract from Wells' Annual of Scientific Discovery. 1854.
Page 334. Trübner & Co, London.
"Curious Fresh Water Plant.

A recent number of the Edinburgh Journal contains a description of a new and singular fresh-water plant, which has recently appeared in the inland waters of Gt.

Britain, filling her canals, lakes and streams, and threatening, unless a remedy be speedily discovered, to interfere seriously with, perhaps to destroy, all or most of her facilities for inland navigation and drainage.[3] This terrible foe is a species of aquatic weed, the first specimen of which was discovered in the year 1842, in the lake of Dunse Castle, Berwickshire, Scotland. The appearance of the little vegetable stranger excited considerable interest among botanists, and the discovery was duly noted in various scientific publications; speciments were sent to a few individuals, and then—it was forgotten. But the plant was of too aspiring a character to live and blush unseen in British waters. In 1847, it was found growing in great abundance, and closely matted together, in the reservoir adjoining the Foxton Locks, on the Canal in Leicestershire. This discovery induced a re-examination of the original plant, when it was found that the long-neglected water-weed had travelled out of the lake, and was making its way down the Whiteadder to join the Tweed. In the same season it was found in great profusion in a tributary of the Trent, in Nottinghamshire, and it was also discovered to exist in "dense masses, and great abundance" in the Watford Locks, Northamptonshire. In 1849, it appeared in Derbyshire and Staffordshire, "forming very large submerged masses, of a striking appearance," in the Trent and adjoining canals. In 1850, it was found near Rugby, in Warwickshire; and in 1851, it was discovered in the rivers Ouse and Cam, near Cambridge.

The botanical interest which was at first excited by the discovery of large quantities of this new plant in various districts, at nearly the same time, soon gave place to a feeling of serious alarm prompted by the injurious effects of its unparalleled increase. A year after it was first noticed in the Cam, the stream near the Colleges was so blocked that extra horses were required to draw barges through the vegetable mass. A year after its appearance at Ely, the railway dock became so choked with the weed that boats could not enter until several tons of it had been removed. In many places fishermen were obliged to discontinue setting long lines, or using nets, because the weed either carried them away or stripped them of the fish. It is evident that this plant, styled by Mr. Babington, of Cambridge, the *anacharis alsinastrum*, cannot a great while have been a native of Gt. Britain.[4]

If so, its remarkable prolific powers would long since have brought it into notice. And it is a noteworthy fact that the anacharis is dioecious,—that is, the male and female flowers grow on separate individuals,—and all the plants known to exist in England are females. This fact affords an almost positive proof that only one stem or seed of a female plant was the progenitor of all the anacharis in Gt. Britain; and this seed was possibly introduced from Canada in the crevice of some one of the many logs which are annually conveyed across the ocean to England, and of which so many have been used in the great rail-road works at or about Rugby. It is thought that in the clear, swift-flowing rivers of America, the weed would not form the immense masses which characterise its growth in the sluggish and—as they contain a greater amount of inorganic animal and vegetable matter,—more nourishing waters of the English rivers and canals. But our readers will inquire how—as only the female anacharis has been discovered in England, and it is therefore unable to propagate seed in that country,—how it contrives to extend itself so rapidly and widely, and wherein are its prolific powers vested. These questions are easily answered; its leaves, which grow in threes, around a slender stem, are studded with minute teeth which cause them to cling to every object with which they come in contact, and the stem is so very brittle that whenever the plant is disturbed pieces are broken off; and as every fragment of the stem is capable of becoming an independent plant, producing roots and leaves, and extending itself indefinitely in every direction, it is evident that the anacharis must be in an almost continual state of reproduction. All the localities in which this singular plant seemed to appear almost simultaneously, are reducible to two,—Dunse Lake, in Berwickshire; and the Foxton Locks, in Leicestershire. It probably originated in the Foxton Locks, and was afterward introduced into the former place. The Foxton Locks are in direct communication with nearly all the English localities of the plant, and a single sprig of the anacharis would in a very short time inoculate any connected water-system, from one end to the other. The plant was introduced by scientific men into the Cam, and other places, originally for experimental purposes, and is of such a nature that, clinging to the bottoms of vessels or boats, it might easily inoculate other streams and waters.

In the case of the Cam, in the short space of four years, it multiplied so, from a single stem, as to impede both navigation and drainage. It has become a great and growing evil; and the attention of scientific men— of practical men— is loudly demanded to the subject of its extinction."

DAR 160: 148

1 Benson refers to *Origin* US ed.: in the first English edition, the pages referred to are 364–5.

2 The *Annual of Scientific Discovery; or Year-Book of Facts in Science and Art* was published in Boston, Mass., and was edited by David Ames Wells and others.

3 The reference is to *Chambers's Edinburgh Journal* n.s. 19 (1853): 372–3.

4 *Anacharis alsinastrum* is now *Elodea canadensis*. For CD's interest in the plant, see *Correspondence* vol. 5, *Correspondence* vol. 7, letter to J. D. Hooker, 31 December [1858], and *Correspondence* vol. 8, letter to J. S. Henslow, 2 April [1860] and nn. 2 and 3, and letter to William Marshall, 9 April [1860]. Benson also refers to Charles Cardale Babington.

To Alfred Newton 9 February [1870][1]

Down. | Beckenham | *Kent. S.E.*
Feb. 9th

Dear Newton

I suppose it w^d be universally held extremely wrong for a defendant to write to a Judge to express his satisfaction at a judgment in his favour; & yet I am going thus to act.— I have just read what you have said in the Record about my Pigeon chapters, & it has gratified me beyond measure.[2] I have sometimes felt a little disappointed that the labour of so many years seemed to be almost thrown away, for you are the first man, capable of forming a judgment (excepting partly Quatrefages), who seems to have thought anything of this part of my work.[3]

The amount of labour, correspondence, & care, which the subject cost me, is more than you could well suppose.— I thought the article in the Athenæum, written I have no doubt by Owen, was very unjust;[4] but now I feel amply repaid, & I cordially thank you for your sympathy & too warm praise.— What labour you have bestowed on your part of the Record! I ought to be ashamed to speak of my amount of work.—

I thoroughily enjoyed the Sunday which you & the others spent here,[5] &

I remain | Dear Newton | Yours very sincerely | Ch. Darwin

DAR 185: 90

1 The year is established by the reference to Newton's review of *Variation* in *Zoological Record* (see n. 2, below).

2 Newton wrote the section on birds in the *Zoological Record*. In the volume for 1868 he included a brief description of *Variation*, and summarised CD's chapters on pigeons. He concluded that these were 'two of the most remarkable chapters ever written on any zoological subject', and added, 'few reflective men will deny the utility of such an accumulation of facts relating to one species, and none will presume to question the ability with which they are presented' (*Zoological Record* 5 (1868): 33, 94–6). The *Zoological Record* for 1868 was published at the end of 1869.

3 For Armand de Quatrefages's response to *Variation*, see *Correspondence* vol. 16, letter from Armand de Quatrefages, 4 March 1868, and *Correspondence* vol. 17, letter from Armand de Quatrefages, 29 March 1869. Quatrefages discussed *Variation* in Quatrefages 1868–9, pp. 209–25, and CD's pigeon work in *ibid.*, pp. 211–14; he concluded that in this work CD had rendered science a signal service (*ibid.*, p. 114).

4 The review of *Variation* in the *Athenæum* was in fact by John Robertson ([Robertson] 1868), not Richard Owen. The attribution is based on the publisher's marked copy of the *Athenæum* (City University Library, London), and on *LL* 3: 110.

5 Newton, Robert Swinhoe, Joseph Dalton Hooker, and Albert Günther were all at Down on Sunday 23 January 1870 (Emma Darwin's diary (DAR 242)).

From George P. Bissell 10 February 1870

Salem, Oregon, United States of America,
Feb, 10th 1870.

Charles Darwin; M.A.
Dear Sir.

I have just finished reading your first and smaller work, on The Origin of Species,[1] &c. and would herein express my pleasure at the natural and scientific view of the subject therein advocated.

In the interest of Science, I suppose that I am justified in addressing you, for the purpose of calling your attention to some few facts, which possibly bear affirmatively on the points discussed.

First, I would call attention to the Humming birds of Puget Sound. Your map will show you that Puget sound lies near the Pacific Ocean, adjacent to Vancouvers Island, British Columbia.[2]

A description of these Humming-birds will be found among the papers of the Smithsonian Institute.[3] I can not indicate more nearly, for I am but a cursory reader, not a special student. Now the particular fact is this, that there are about (40) forty varieties, or species of these birds on the western coast of the continent of America, and only three or four species or varieties on the eastern Coast, notwithstanding that they are birds of passage.[4]

2d. The Bobolink, or Carolina rice bird, is never seen on the Pacific coast, though very common on the Atlantic coast. It is my impression, but I am not certain, that it is never seen west of the Alleghany Mountains.[5] I do not know of any works wherein this is mentioned.

3d I can not find that the American Bison ever ranged east of the Aleghany Mountains, and it is certain that they suddenly passed west of the Mississipi River during the present century, abandoning forever their former range on the East of that river.

At about the same period, they suddenly contracted their range from the west of the Rocky Mountains, over a space of probably about 700 or 800 miles of latitude. This fact again I have never seen mentioned, but from my own observation, and from information derived from old hunters, I know it to be true. And I judge that the contraction on the eastern limit, could not have been caused by the pressure of civilized man on their flank, because we know that the almost simultaneous contraction of the western limit was not induced by such a cause. Records of the contraction of the eastern limit may be found in books.[6]

4th. A short distance west of the Missouri River the Oak and the Hickory suddenly cease, although they are very plentiful to the eastward.

The same holds true of other plants. The termination is abrupt, but the line of termination is waved.[7]

5th The Hudson River (in the eastern part of the state of New York) is, and always has been, the eastern limit of the Turkey buzzard, and of the opossum. The same river is also the western limit of some small animals, and of some plants.[8] But it is quite singular and striking in the case of the Turkey buzzard. These are facts which you can verify in the reading of an extensive library, especially of American Authors.

I conceive that they have some bearing on the theory which you advance, and offer them solely in the interest of science.

George P. Bissell. M.D.

Charles Darwin M.A. | Down, Broomley, Kent, England

DAR 205.3: 261 (Letters)

CD ANNOTATIONS

4.5 three or four ... mentioned. 5.4] *crossed blue crayon*
8.1 4th. . . . is waved. 9.2] *crossed blue crayon*
Verso of last page: 'Range of Bison— (Rivers)' *blue crayon*

1 Bissell refers to either the first American edition of *Origin* (*Origin* US ed.), or to one of the first five British editions (*Origin*).

2 Puget Sound, on the north-western coast of Washington State, is connected to the Pacific Ocean by the Strait of Juan de Fuca; Vancouver Island of British Columbia extends to the north-west (*Columbia gazetteer of the world*).

3 The publication has not been identified.

[4] In 1870, a single species of humming-bird (the ruby-throated) was known on the east coast of the United States, and only six (the rufous, callipe, Costa's, Anna's, Allen's, and black-chinned) were known on the west coast north of Mexico (Dennis Paulson, personal communication, Slater Museum of Natural History, University of Puget Sound).

[5] The present distribution of the bobolink (*Dolichonyx oryzivorous*) extends as far west as eastern and south-central Washington, and eastern Oregon (S. G. Martin and Gavin 1995). The Allegheny Mountains, part of the Appalachian Mountain Range of the eastern United States, extend south-west from central Pennsylvania to south-western Virginia.

[6] Bison or buffalo (*Bison bison*) once inhabited the mid-Atlantic states, from New York State to Florida, and extended westward to eastern Oregon. They were most plentiful in the Great Plains; however, their numbers diminished notably in the nineteenth century due to hunting and shrinking habitat. On the historic range of bison, and on their nineteenth-century demise, see Lott 2002, pp. 67–76, 167–9, 170–7. For a list of early written accounts on bison, see Garretson 1938, pp. 233–42.

[7] Oak-hickory forests are found only in the eastern half of the United States (Kricher 1998, pp. 60, 82). The Missouri River flows south-east from Montana to its confluence with the Mississippi River north of St Louis. It forms the western boundary of Iowa, beyond which oak-hickory forests are uncommon.

[8] By 1840, the turkey-buzzard or turkey-vulture (*Cathartes aura*) was said to not have been seen further north or east of New Jersey (Audubon 1840–4, vol. 1, plate 2); it is now widespread throughout the United States. Early European settlers found no American opossums (*Didelphis*) north of present-day Virginia and Ohio; they have since continued spreading northward from South America (D. W. MacDonald ed. 2006, 1: 16).

From Robert Swinhoe 10 February 1870

33 Oakley Square SW.
10 February, 1870.

My dear Mr. Darwin,

I must apologise for troubling you again so soon.[1] You told me some time ago of a gentleman who made a study of the skulls of Cattle, and was doing good work with them. Could you kindly inform me where I could find this gentleman?[2] I have a series of Skulls of the small Chinese Yellow Cow[3] together with a photograph of a fine Bull of the same breed and I should like much to get him to examine them before I put them into the Museum.

Your's very truly, | Robert Swinhoe.

DAR 177: 334

[1] See letter from Robert Swinhoe, 2 February 1870.

[2] The reference has not been identified.

[3] Swinhoe probably refers to a specimen of Southern yellow cattle, a variety of zebu cattle, *Bos taurus indicus*.

From Alfred Newton 11 February 1870

Magd: Coll:
11 Febr. 1870.

My dear Mr. Darwin,

The gratification of which you are pleased to write is I can assure you mutual—[1] I am *exceedingly glad* to find that my notice of your investigations with regard to Pigeons satisfied you—but I cannot refrain from saying that the admirably lucid manner in which you placed the results before your readers rendered the task an easy one—indeed you will see that nearly every phrase used is your own—

I sometimes think that the labour expended by the Recorders is very inadequately appreciated by the scientific public— therefore your praise is doubly welcome— Almost every author considers that he is unfairly dealt by by his work being subjected to compression, and you are nearly the first who has told me that he was content now with the result as it appears in the 'Record'.

I shall not soon forget the pleasure I enjoyed during my late visit to you and I am rejoiced to learn that you were none the worse for the excitement which simultaneous occurrence of so many *raræ aves* was likely to produce—[2]

Dr. McCann's pamphlet of which I promised to let you know the particulars is entitled 'Anti-Darwinism' & is published by Bryce & Co., Glasgow— It is no real criticism of your theory but a virulent attack on Hooker & Huxley—from a metaphysical point of view which I in my ignorance cannot reach.[3]

I have been reading the article you told me of in the 'North British Review'—the geological argument & that based on Sir W. Thomson's researches are out of my line, but the assertions & reasoning with respect to the Zoological aspect are I think to be fully & entirely answered—[4] Pray make my compliments to Mrs. Darwin & your daughters.[5]

& believe me | yours very truly | Alfred Newton

DAR 172: 48

[1] See letter to Alfred Newton, 9 February [1870]. CD refers to Newton's notice of *Variation*, and in particular his work on pigeons, in the *Zoological Record* for 1868.

[2] Newton, Robert Swinhoe, Albert Günther, and Joseph Dalton Hooker visited Down between 22 and 24 January 1870 (Emma Darwin's diary (DAR 242)). *Rarae aves*: rare birds (Latin).

[3] James M'Cann's pamphlet (M'Cann 1869) was the published version of a paper he read at the meeting of the British Association for the Advancement of Science in Exeter in 1869. It included as its final section 'Professor Huxley's reply', which was actually M'Cann's criticism of remarks made by Thomas Henry Huxley after the reading of the paper.

[4] Newton refers to [Tait] 1869, and to William Thomson; see also *Correspondence* vol. 17, letter to J. D. Hooker, 24 July [1869].

[5] Emma, Henrietta Emma, and Elizabeth Darwin.

From Asa Gray 14 February 1870

Cambridge, Mass.
Feb. 14. 70

My Dear Darwin

Being eve of post-day we respond at once to yours of the 27th January—which arrived this very morning—lest you should send us down to posterity with a fabulous dog-story.[1]

I well remember telling you of our "Max" and his habit of washing cat-fashion—which you suggested might have come from being brought up with a cat, and I think I told you that I had not been able to learn definitely whether that was the case or no. Here, you see, by some shuffling of memory, a suggestion of what might explain a fact has taken the place of the fact itself. I am curious to know if it be true,—for it is the only explanation I can think of.

As to your question which I brought home from you to Agassiz, I spoke to him of it the first time I saw him, and soon sent him your written mem. I fear it is too late now to get the answer from him,—unless he mends more decidedly than he is now doing.[2] About a month ago A. had a serious and peculiar attack of trouble in the brain, which we fear betokens serious mischief, and has laid him up wholly. He is getting better slowly, but his state is critical & doubtful. I advise you to forward the query to Alexr. Agassiz. (letters will reach thro' Baring Brothers, & Co. 8, Bishopgate St. within.); he can either inform you or find out.[3] J. Wyman[4] is now in Italy, or on the way, or I could learn from him.

So you have effected a treaty with Appleton of New York.[5] The bait you offered—of a book sure to make a great stir, and sell hugely—was more than he could resist. He ought to have done it from a sense of propriety.

I trust you have some of the slender leaved *Drosera*, I sent thro' Hooker.[6]

Well, our homeward voyage was not a nice one, especially for Mrs. Gray,[7] and it now seems a long time ago. I dropped at once into a world of work—but am not killing myself. The main struggle for existence will come in spring, when my duties crowd on me dreadfully.

It gave us both very great pleasure to see again Mrs. Darwins well-known handwriting, and your signature.

I knew you would be pleased with young Agassiz and his Yankee wife.[8] I wish his health were better.—and I do hope your own will be such that you can next summer see and know my *trump* of a colleague J. Wyman.

Ever Yours truly, | A. Gray

DAR 165: 173

1 CD's letter to Gray of 27 January 1870 has not been found. See also letter from J. L. Gray, 14 February 1870.

2 Louis Agassiz suffered a severe breakdown after giving an address on 14 September 1869 and was ill for ten months (Marcou 1896, pp. 175, 177–8). The written memorandum has not been found. See also letter from Asa Gray, 27 February and 1 March 1870.

3 Louis Agassiz's son Alexander was travelling in Europe (G. R. Agassiz ed. 1913, pp. 97–114). Bishopsgate Street within: i.e. within the walls of the City of London.

4 Jeffries Wyman.

5 D. Appleton & Co. had agreed to reprint the fifth edition of *Origin* and *Descent* (*Correspondence* vol. 17, letter from Charles Layton, 23 December 1869). The head of the firm was William Henry Appleton.

6 *Drosera linearis*, the slenderleaf sundew, is found in North America. Gray also refers to Joseph Dalton Hooker.

7 Jane Loring Gray.

8 Alexander and Anna Russell Agassiz visited Down on or shortly before 1 December 1869 (see *Correspondence* vol. 17, letter to Fritz Müller, 1 December [1869]).

From Jane Loring Gray 14 February 1870

Botanic Garden, Cambridge
Feb. 14— '70—

My dear Mr. Darwin,

Dr. Gray says, "You write & tell about the dog!"—[1] And indeed it was only a supposition of mine that he was suckled by a cat, from his queer tricks when he came to us, a young dog of about 7 months old— He then would chase his own tail for sport; but I have heard of other dogs doing that— But he still keeps up the trick of washing his face with his paws, & will sit as demurely as any old tabby, licking one paw & rubbing his face, & then changing to the other— I am glad if he has any tricks worth noticing, for he is a stupid little doggie at learning anything new, & has nothing but an affectionate heart & some beauty to recommend him— He came to us through one or two transfers, so I cannot know much about his puppyhood; but the lady who gave him to us is coming here this week, & I will ask her if she can find out anything about it—

I wish you had told us something how you were, but I hope you are as well at least as when we last saw you— And finding winter not too severe upon you—

As for us I can't say that having no winter one year, makes coming back to it any more delightful— Snow & ice & slush & mud, for we are enough on the shore to have our cold turns soon followed by a thaw, are delightful to young people, who can skate & coast & walk & ride, but I have got to an age when I prefer dry roads & green trees & fresh turf!—

Dr. Gray is busy as ever; & astonishing his old friends & puzzling them very much as an "Arab Sheykh," as our Greek Professor, Mr. Sophocles calls him—[2] He seems very well—

I wish I could tell you & Mrs. Darwin what delightful recollections we keep of you all & your charming home—[3] I assure you I am a woman of consideration that I have seen Mr. Darwin!—

Pray remember us most cordially to all your circle— | Very faithfully | Yours, Jane L. Gray

DAR 80: 162–3

CD ANNOTATIONS
1.3 about 7 months] *underl blue crayon*
1.3 He … doing that— 1.4] *del blue crayon*
1.6 I am … puppyhood; 1.9] *crossed blue crayon*
2.1 I wish … circle— 6.1] *crossed blue crayon*
Top of letter: 'To stay all *[month]* | p. 14' *blue crayon*

[1] See letter from Asa Gray, 14 February 1870.

[2] Jane Loring Gray refers to Evangelinus Apostolides Sophocles, Harvard professor of ancient, Byzantine, and modern Greek. The Grays had travelled in Egypt in 1869, and become 'specially fond of Arabs' (*Correspondence* vol. 17, letter from Asa and J. L. Gray, 8 and 9 May 1869).

[3] The Grays had visited Down on 28 August 1869 (Emma Darwin's diary (DAR 242)).

From Edward Caldwell Rye 14 February 1870

10, Lower Park Fields, | Putney, | London, SW.
14th. Feb. 1870.

Dear Sir,

In your most interesting "Notes on the ⟨fertiliz⟩ation of Orchids", in Ann. & Mag. of ⟨Nat.⟩ Hist., Sep. 1869[1] (which I have just ⟨care⟩fully re-perused with the view of extracting ⟨ma⟩tter for the *Coleoptera* of Zoological Record, 1869, now in my hands), I find no specification of any species of beetles except *Malthodes brevicollis*; and, therefore, venture to draw your attention (if you have not, indeed, already observed the same yourself) to the publication, about the same time as your Notes, by Girard, in the Annales d. l. Soc. Entom. ⟨d⟩e France, 1869, Bulletin, p. xxxi, of ⟨c⟩ertain observations, from which it appears ⟨th⟩at two species of *Longicornes*, viz., ⟨*Strangal*⟩*ia nigra*, Linn., and *S. atra*, Fab., ⟨assist⟩ in the fecundation of *Orchidaceæ*, ⟨by d⟩etaching glutinous pollen.[2] Possibly, M. ⟨Girard⟩ is already a correspondent of yours: ⟨but, u⟩nder any circumstances, you will ⟨doubtless⟩ excuse my writing thus, although ⟨*pers*⟩*onally* unknown to you. My connection ⟨with Mr.⟩ G. R. Waterhouse emboldens me to this.[3]

⟨Yours tr⟩uly— | ⟨E. C. Rye.⟩

DAR 176: 228

1 'Fertilization of orchids'.

2 CD cited Maurice Girard's remarks (from *Annales de la Société Entomologique de France* 4th ser. 9 (1869): xxxi–xxxii) in *Orchids* 2d ed., p. 16 n. *Strangalia atra* is now *S. aethiops*.

3 George Robert Waterhouse was Rye's father-in-law.

From William Darwin Fox 15 February [1870][1]

Broadlands | Sandown | I. Wight
Feb 15

My dear Darwin

It is so long since I have heard anything of you and yours, that I trouble you with a few lines, which perhaps Mrs Darwin will answer for you—to ask how you are this very trying Winter. We have all been here since October and are become quite settled here, tho' at present ⟨ ⟩ Delamere up,[2] and some ⟨ ⟩ of us, possibly all—may go there in Summer.

We have all been laid up with this extraordinary cold weather. Sandown is a complete ⟨tem⟩ple full of the Winds, and this East one has been very biting.

What was the accident from riding, which Mrs Darwin alluded to, with you. I hope you have quite got over it, and that it has not prevented your riding as usual, as that seemed to do you so much good.[3]

I hope George[4] got quite strong and well again.

I must try to make the acquaintance of your Banker Son at Southampton[5] when days get a little longer & brighter ⟨ ⟩ I never thought Sou⟨thampt⟩on a very healthy ⟨p⟩lac⟨e⟩ ⟨ ⟩ I used to live in ⟨those⟩ parts.

What old fellows we get— you however are the younger of the two.

I should very much enjoy seeing you again, are you likely to be in Town this Spring? I suppose I shall be there for a short time next month, if able to go. How is Mrs Wedgwood— are her daughters[6] ⟨ ⟩ But I will not go on bothering your poor Head with questions. Give this to Mrs Darwin & I shall feel immensely grateful if she will a⟨nswer⟩ it, when she has time, ⟨and⟩ will tell me all about ⟨ ⟩ and your belongings.

In the meanwhile, with our kindest regards | Believe me | Ever your attached Friend | W. D Fox

DAR 164: 190

[1] The year is established by the reference to CD's riding accident (see n. 3, below).
[2] Delamere Rectory, Northwich, Cheshire, was Fox's previous residence.
[3] CD was thrown by his horse, Tommy, in April 1869 (Emma Darwin's diary (DAR 242)).
[4] George Howard Darwin.
[5] William Erasmus Darwin.
[6] Caroline Sarah Wedgwood was CD's sister; her daughters were Sophy Wedgwood, Margaret Susan Vaughan Williams, and Lucy Caroline Wedgwood.

From G. W. Child 16 February [1870][1]

Elmhurst, | *G^t. Missenden.*
Feb 16

Dear Sir

I have again read your chapter on the good effects of crossing since I received your note of Jany 27—[2] It appears to me that the questions with wh: that chaptr & also my Essay deal are 3.[3]

(1) Is it *proved* that any amount of close breeding *alone* will produce degeneracy I think you admit that it is not at the beginning of your chapter, & this is all that I contend for—[4] The evidence is conflicting, but it appears to me that in sound logic one positive instance in wh: it can be shewn that an extreme degree of close-breeding may exist without any discoverable degeneracy out weighs any number of cases in wh: degeneracy is found to exist in asmuch as the antecedents in all such cases are so complicated that it is impossible to shew that the close breeding per se is in cause of the deterioration

I think that in considering those cases in wh: breeds of domestic animals have been found to degenerate greatly when close-bred, sufficient attention has not always been paid to the absolute uniformity of external conditions (climate, locality, food &c) under which they have often been kept for many generations— The cases wh: you mention (p 125 &c) in wh: breeders have been able to go on with the same blood when kept at different & distant establishments seem to lend some force to this suggestion— We know that a mere change of diet or change of air will often produce a good effect & that if I mistake not on animals as well as on man.

(2) The positive evidence of improvement by crossing produces much more effect on my mind than the negative— But even with regard to this I am inclined to think that in the case of domestic animals breeders sometimes speak of what is merely an improvement with reference to their special purposes (fattening qualities mere size, the like) as if it were a real advance in the organization of the creature— And again I think that some of the above suggestions might go some way to account for the improvement when it really exists by suggesting that it is due rather to the two animals brought together to breed having lived under more or less widely differing circumstances than to their coming of different stocks—

Finally (3) with regard to the extension of the doctrine of the deleteriousness of close-breeding up to the case of the marriage of blood relations among mankind wh: is indeed the proper subject of my Essay published first in 1863.[5] It seems to me that to maintain this view in the face of the Pedigrees in the herd book & such facts as you mention at p 115 is almost paradoxical.[6] It is like saying that when a drug taken in a large dose is productive of no discoverable effects whatever it is highly probable that if taken in a small dose it might act as a dangerous poison— May I not fairly answer with the old logicians that de non apparentibus et de non existentibus eadem est ratio?[7] This last point therefore is the only one on wh: my slight investigations of this subject leads me to a conclusion greatly different from that to which your incomparably deeper one has led you—

Pray excuse my troubling you with so long a letter & believe me dear Sir | faithfully yours | Gilbert. W. Child

C. Darwin Es &c &c

DAR 161: 142

[1] The year is established by the relationship between this letter and the letter from G. W. Child, 26 January [1870].

[2] CD's letter to Child has not been found. Child refers to chapter 17 of *Variation* (*Variation* 2: 114–44).

[3] Child refers to Essay 1, Marriages of consanguinity, in his *Essays on physiological subjects* (Child 1869).

[4] CD wrote, 'The evil results from close interbreeding are difficult to detect, for they accumulate slowly, and differ much in degree with different species … several causes interfere with our detecting the evil—such as the deterioration being very gradual, and the difficulty of distinguishing between such direct evil and the inevitable augmentation of any morbid tendencies which may be latent or apparent in the related parents' (*Variation* 2: 114, 115).

[5] Child's essay on marriages of consanguinity was first published in the *Westminster Review* 80 (1863): 88–109.

[6] In *Variation* 2: 117–18, CD discussed the apparently advantageous effects of close breeding in cattle and sheep.

[7] *De non apparentibus et de non existentibus eadem est ratio*: concerning the invisible and the non-existent, the argument is the same.

From Fritz Müller 16 February 1870

As soon as the *Passiflora*, which was visited by humming birds in my garden, flowers again next spring, I shall try fully to answer Mr Farrer's questions.—[1]

I enclose a dried flower of this *Passiflora*; the long flower-stalk is dependent, but curved upwards at the end, so that the flower itself is upright.— You see, there is space enough for a hovering humming-birds head between the anthers and the entrance to the (nectarless) nectary.

There is now flowering with me a white Passiflora with dependent flowers, which is visited by humble-bees.—

The enclosed Passiflora-seeds may perhaps be acceptable to Mr Farrer; at least they are from wild, not hybridised plants.[2]

Fritz Müller

Itajahy, Febr. 16. 1870.

Incomplete
Linnean Society of London

[1] CD had sent Thomas Henry Farrer an extract from one of Müller's letters in which Müller said that *Passiflora* flowers growing in his garden were visited by humming-birds; Farrer had sent CD a couple of questions for Müller (*Correspondence* vol. 17, letter to T. H. Farrer, [27 November 1869], letter from T. H. Farrer, 28 November 1869, and letter to Fritz Müller, 1 December [1869]).

[2] CD sent this portion of Müller's letter and the seeds to Farrer with his letter to Farrer of 13 [May 1870].

From William Swale 16 February [1870?][1]

Ornithology in Canterbury, New Zealand
American Blight Bird.

Ornithologists state here that this Bird belongs to the family Luscindæ and probably to the genus Zosteropes.[2] They also state that it is a small Migratory Bird. I think I am able to contradict this or nearly so for I am certain it has become a settler here like myself.[3] It is called by the Maories "Kanohi-mowhiti". It is with us winter and summer and very fond of ripe fruit when in season. In winter they fly in flocks from Apple tree to tree devouring the Aphis and cleaning the trees and all the carbuncles to all appearance but unfortunately the Blight finds its way again up the stem and branches from the bole of the trees during summer. Its little cranium is a very destructive one at any rate and I will give you the benifit of my experience of the onslaught, he and she, makes on my stone fruit. Soon as ripe cherries are fit to gather they make a perfect skeleton of the stones and leave them hanging; while these last in season they are always walking into them with their sharp pointed beaks or bills. Morello Cherries they finish up with. Their next favourite fruits are Green Gages and Apricots—these they make dreadful slaughter amongst to the tune of destroying bushels. They only chirp a little when in mischief. I can hear them where I now sit in my garden. They build their little tiny nests in very secluded nooks. I find a nest occasionally in a Peach tree which makes me think they must feed their young families—on the flesh of stone fruit.

Dear Sir, I thought I could not do better than send you this brief sketch of this little mischevious bird; feeling in my mind that it would be welcomly received. I have, of late, very often been reading of your Voyage in the Beagle.[4]

Dear Sir, receive this from your old Correspondent[5] | William Swale | Avonside Botanic Garden | Ch. Church, Canterbury, N.Z. Feb 16

To | Ch. Darwin Esq.

Our Native plants, in this the Middle Island,[6] stand their ground—and very firmly too. Take this hint, Dear Sir, and be cautious what you write forthwith about N.Z. vegetation in your "works"[7] A stubborn fact goes a long way with you I know.

DAR 177: 324

1 The date is conjectured from an early archivist's note (possibly by Francis or Henrietta Emma Darwin) on the letter.

2 The blight-bird, or silvereye, *Zosterops lateralis*, is now in the family Zosteropidae. In the mid nineteenth century, the family was disputed, but some authorities placed it in Luscinidae (see Taylor 1870, p. 608).

3 Two articles were published on *Zosterops lateralis* in the *Transactions and Proceedings of the New Zealand Institute* 2 (1869): 61–2, and 3 (1870): 79. According to the second article, the bird had first been recorded in New Zealand in 1832, and appeared in greater numbers in 1856. The first article argued that the bird was a recent colonist (because of its habit of building suspended nests); the second argued that it was indigenous to the South Island and had recently spread north.

4 *Journal of researches.*

5 Swale had corresponded with CD in 1858 (see *Correspondence* vol. 7, letter from William Swale, 13 July 1858).

6 Middle Island is now the South Island of New Zealand.

7 CD had written in *Origin*, pp. 201–2, that the endemic plants of New Zealand were retreating before the advance of plants introduced from Europe.

From William Thierry Preyer [before 17 February 1870][1]

interest for your theory, viz:

1) The chemical and other differences of the blood-crystals (hæmoglobine) according to the species from which the blood is taken— this astonishing fact—proves that natural selection acts not only on the morphological development, but even on the purely chemical constitution of organs. The red colouring matter of the blood, although it always shews the same spectrum, is a different substance in every species.[2]

2) The intimate relation between the sensations of sound & colour corroborate your statements (Variation etc II, p. 328, 1868).[3]

3) The occurrence of sulphuric acid in a free state in the glands of Dolium galea confirmed by my analysis is a case of wonderful adaptation (not yet quite understood). It may be compared to the exceptional development of electric organs (in torpedo etc).[4]

4) The noxious effects of prussic acid on animals are very instructive shewing: 1º. how great are the individual differences within the same species; 2º. surprising

resemblances between the symptoms of warm-blooded vertebrates & true insects and again between cold-blooded vertebrates & mollusca.[5]

You would very much oblige me by occasionally informing me what relation of yours is Robert Waring Darwin mentioned by Göthe (Farbenlehre, Works Vol XXXIX, p. 404–408). He wrote "On the ocular spectra of light and colours" (phil. trans. Vol. 76. Nov. 1785) reprinted in your celebrated grandfathers "Zoonomia".[6]

I should very much like to know whose lectures you heard at Edinburgh & Cambridge besides Dr. Grants & Rev. Henslows.[7] & which book or books made deep impressions on you in youth.

But this letter is already too long.

I remain | your's truly | Wm. Preyer

Incomplete
DAR 261.11: 21

CD ANNOTATIONS
3.1 2) The intimate ... etc). 4.4] *crossed pencil*
6.1 You ... long. 8.1] *crossed pencil*

1 The date is established by the relationship between this letter and the letter to W. T. Preyer, 17 February [1870].

2 Preyer sent CD a copy of his paper on haemoglobin (Preyer 1868); there is a copy in the Darwin Pamphlet Collection–CUL.

3 Preyer sent CD a copy of his paper on sound and colour (Preyer 1870a); there is a copy in the Darwin Pamphlet Collection–CUL. In *Variation* 2: 328, CD wrote, 'The organs of sight and hearing are generally admitted to be homologous, both with each other and with the various dermal appendages; hence these parts are liable to be abnormally affected in conjunction.'

4 The snail *Dolium galea* secretes sulphuric acid. For a brief account of Preyer's work, see *Sitzungberichte der Niederrheinischen Gesellschaft für Natur- und Heilkunde* (1866): 6–9. CD discussed electric organs in fish in *Origin*, pp. 192–3.

5 Preyer published on this subject in Preyer 1869.

6 Preyer refers to R. W. Darwin 1786, and to Erasmus Darwin's *Zoonomia* (E. Darwin 1794–6). For Johann Wolfgang von Goethe's *Farbenlehre*, see Goethe 1989; the discussion of Robert Waring Darwin is on pp. 875–8 of this edition.

7 CD had mentioned Robert Edmond Grant and John Stevens Henslow in his reply to Preyer's biographical questions (see *Correspondence* vol. 17, letter to W. T. Preyer, [before 21 March 1869]).

To Gustav Jäger 17 February 1870

Feb. 17— 1870

Dear Sir

You will see, by the enclosed letter that I wrote to you in September to express my feelings about your work; and since which time I have received another copy of your book from yourself.[1] I addressed the enclosed letter to the Zoological Gardens at Vienna; & it seems to have travelled all over Germany, & was returned to me this morning by the Post-Office.—[2] I will address this to your publisher at Stuttgart.— I fear that you must have thought me very ungrateful to have taken no notice of your kind present, & this causes me much pain.—

Believe me dear Sir with much respect. | Yours very faithfully | Ch. Darwin

Dr Hildegard Jaeger (private collection)

1 See *Correspondence* vol. 17, letter to Gustav Jäger, 9 September 1869. There is an annotated copy of Jäger's book, *Darwin'sche Theorie und ihre Stellung zur Moral und Religion* (Jäger [1869]) in the Darwin Library–CUL (see *Marginalia* 1: 429); CD's other copy is now in the library of the Linnean Society of London.

2 Jäger had left the Zoological Gardens of Vienna in 1866 and in 1867 became professor of zoology and anthropology at the Hohenheim Academy near Stuttgart (*DBE*).

To W. T. Preyer 17 February [1870][1]

Down. | Beckenham | *Kent. S.E.*
Feb. 17

My dear Sir

I am very much obliged for your extremely kind letter & for your several presents.[2] Although your appreciation of my work is certainly too high, yet it is very encouraging to me, especially as yesterday I read two pamphets, just published in England, in which every form of abuse is heaped on me.[3] I am called, for instance, a "filthy dreamer".— You seem to be doing splendid work in physiology, the noblest of sciences, as I have long thought it. What you say about the differences of the blood-crystals is truly astonishing.[4] I am also much interested by what you say of the different effect of Prussic acid on different individuals of the same species: I remember some years ago wishing in vain for information on this head. I think it arose from observing how differently in quickness (whether due to rate of respiration or to direct action of the poison I could not tell) the vapour acted on different insects. I remember Bees died instantly, but (I think,) it was a Longicorn beetle which resisted all effects for an astonishing time.—[5]

R. W. Darwin is my Father, but I believe he was greatly aided in his paper on optics by his Father Erasmus D.[6]

I have really nothing of interest about myself, but as you desire it, I will scribble whatever occurs to me.— I derived *no* advantage from the Lectures at Edinburgh, for they were infinitely dull & cured me of any taste for Geology for 3 years. Dr Grant[7] was not a Professor, but worked at zoology out of pure love, & his Society was a great encouragement. I used to amuse myself with examining marine animals, but I did so solely for amusement. I believe I was the first person who ever saw the earliest locomotive egg-like state of a Bryozoon: I showed it to Grant, who stated so at the meeting of the Wernerian Nat. Hist. Soc. & this little discovery was an immense encouragement.—[8] I was disgusted at anatomy & attended only 2 or 3 Lectures & this has been ever since an irreparable loss to me.— When I went to Cambridge, I became a most enthusiastic collector of Coleoptera; but again only for amusement. If any one told me the name of a Beetle, I thought I knew all that anyone could desire; & I believe I never ever looked even at the oral organs of any insect! Yet I worked like a slave at collecting. Henslow's Society was a great charm & benefit to me, & I liked much his Lecture on Botany.[9] All my early life I was mad for collecting, minerals, shells, plants, Bird-skins have all had their turn. Near the end of my Cambridge life Henslow persuaded me to begin Geology. I was always very fond of observing the habits of Birds, & White's Nat. Hist. of Selborne,[10] thus had much influence on my mind. But of all books, Humboldt's Travels had by far the greatest influence—[11] I read large parts over & over again.— I had nearly managed to get a party to go to the Canary Isl^{ds}, when the offer of joining the *Beagle* was made to me & joyfully accepted.[12] I suppose, however, no ever started worse prepared than I was except as a mere collector. I knew nothing of anatomy, & had never read any systematic work on Zoology— I had never touched a compound microscope & had begun Geology for only about 6 months. But I took out plenty of Books[13] & worked as hard as I could & dissected roughly all sorts of the lower marine animals. Here I felt fearfully the want of practice & knowledge. My education in fact began on board the Beagle. I remember nothing previously which deserved to be called education except some experimental work at chemistry when a school-boy with my Brother.—[14] No doubt collecting largely in so many branches had improved my powers of observation.—

I never wrote so much about myself in my life, & I hope it may be worth your reading, but I doubt.—

Believe me, my dear Sir | Yours sincerely | Ch. Darwin

I do not know whether you will care to see extracts from my letters printed by Prof. Henslow, but I send a copy by this Post.—[15]

Ralph Colp, Jr (private collection)

1 The year is established by the relationship between material in this letter and Preyer's article on CD in *Ausland*, published on 2 April 1870 (Preyer 1870b). See also *Correspondence* vol. 17, letter to W. T. Preyer, [before 21 March 1869], and letter from W. T. Preyer, 21 March 1869.

2 See letter from W. T. Preyer, [before 17 February 1870]. Preyer enclosed a copy of Preyer 1868, of which there is a copy in the Darwin Pamphlet Collection–CUL, and possibly also Preyer 1869.

3 The pamphlets have not been identified.

4 See Preyer 1868.

5 See Preyer 1869. CD had used prussic acid gas to kill pigeons and insects (see *Correspondence* vol. 5, letter to W. D. Fox, 22 [July 1855]). He had also been interested in the effects of prussic acid (also known as hydrocyanic acid) on the insectivorous plant *Drosera* (see *Correspondence* vol. 10, letter to J. D. Hooker, 26 September [1862], and *Insectivorous plants*, pp. 195–6).

6 See letter from W. T. Preyer, [before 17 February 1870]. CD refers to Robert Waring Darwin, Erasmus Darwin, and R. W. Darwin 1786.

7 Robert Edmond Grant.

8 CD identified the organs of motion of the ova of *Flustra*; his observations were read by Grant at meetings of the Plinian and Wernerian Societies in Edinburgh in 1827 (see DAR 118 (*Collected papers* 2: 285–91); see also *Correspondence* vol. 1, Appendix I n. 14).

9 John Stevens Henslow had been professor of botany at Cambridge. On their relationship, see Walters and Stow 2001.

10 CD refers to Gilbert White's *Natural history of Selborne* (White 1825).

11 CD refers to Alexander von Humboldt and to Humboldt 1814–29.

12 See *Correspondence* vol. 1.

13 For the books CD took on the *Beagle*, see *Correspondence* vol. 1, Appendix IV.

14 CD refers to Erasmus Alvey Darwin (see *Correspondence* vol. 1).

15 CD refers to [J. S. Henslow ed.] 1835.

To W. D. Fox 18 February [1870][1]

Down. | Beckenham | *Kent. S.E.*
Feb. 18th

My dear Fox

I was very glad to receive your letter with some news of yourself.[2] We hope (but it is not quite certain) to be at 6. Q. Anne St on March 4th & stay a week there, & I shall be very glad to see you there once again.—[3] But if not, or indeed in any case, will you not come down here & pay us a visit?— We shall all be heartily glad to see you.— From what I remember of Sandown, I can quite believe it the "temple of the winds". This horrid cold has made me miserable: nothing will keep me decently warm. I have resumed my rides, & am able to do a good deal of writing every day, but I never pass 6 hours without a fit of extreme discomfort, & so I shall go on to the last of my uncomfortable days. Yet with all my discomfort I am very happy, thanks to my dear wife & good children.

They are all pretty well, but George[4] is not strong, & I greatly fear he will find the Bar too severe work. I regret this much for he is such an indomitable worker & has so clear a head that I think he wd be very successful. He dined lately at Mr Bristowe's,[5] who has been very kind to him & advised him.— I hope you will get to know William at Southampton, who flourishes in his business & has got a little House at Basset near the common.—[6] Frank, (nor 3) is going to be educated as a Doctor, & is going next year to try for Honours in the Science tripos.[7] Leonard is head man, ie, Divisional officer, at Woolwich, & will be sure of his commission as Engineer this time next year.—[8] Horace is rather an invalid, at a tutor in Suffolk & has great mathematical aptitude, but is I fear too weak to do much[9] Henrietta is staying in S. of France, partly for health & partly for pleasure, & Bessy alone is at home—[10]

By Heavens I have told you enough about us all! One of Caroline's daughters is married to a clergyman, son of Judge Williams.[11]

I hope we shall meet. Farewell my dear old friend. | Yours affectionately | Ch. Darwin

Isle of Wight Record Office (Ac 2008/79)

1 The year is established by the relationship between this letter and the letter from W. D. Fox, 15 February [1870].

2 See letter from W. D. Fox, 15 February [1870].

3 CD stayed with his brother, Erasmus Alvey Darwin, at 6 Queen Anne Street, London, from 5 to 10 March 1870 (CD's 'Journal' (Appendix II)).

4 George Howard Darwin.

5 Probably Henry Fox Bristowe, Fox's nephew.

6 William Erasmus Darwin was a banker in Southampton.

7 Francis Darwin was at Trinity College, Cambridge; he graduated with first class honours in the natural sciences tripos in 1870, and then studied medicine at St George's Hospital, London (*ODNB*).

8 Leonard Darwin was at the Royal Military Academy at Woolwich, and was commissioned in the Royal Engineers in January 1871 (RMA Gentleman Cadet Register (RMAS archives, Sandhurst collection)).

9 Horace Darwin's tutor was Rolla Charles Meadows Rouse.

10 Henrietta Emma Darwin was by this time in Italy (letter from H. E. Darwin to Emma Darwin, 1 February [1870] (DAR 245: 31)). CD also refers to Elizabeth Darwin.

[11] Caroline Sarah Wedgwood's daughter Margaret Susan married Arthur Charles Vaughan Williams, the son of Edward Vaughan Williams, first judge of common pleas.

From Vladimir Onufrievich Kovalevsky 20 February [1870][1]

Münich— | Gluckstrasse N 6.
Febr. 20.—

Dear Sir

It is an eternity that I had no news of You, and if from my part I did not write it was for not forcing You into an answer and robbing in this way Your time. The whole winter I worked in the great palaeontological Museum at Münich but chiefly composed of the mesosoic fauna from the Trias to the newest tertiary, I think it is one of the largest collections existing for the Jura and for the much discussed and yet not settled Titonic fauna, holding an intermediate position between the Portland and the Neocomien, perhaps a continental marine equivalent for your Purbeck.[2] The Belemnites, Ammonitidae[3] and the Brachiopoda of the Titonic group have a decidedly cretaceous character and the whole fauna is in many respects intermediate between Jura and the Cretaceous formation,—as for instance the prodigious quantity of large bivalves holding a place between the Diceras arietina of the Upper Jura and the Caprinas of the Untere Kreide.[4] The whole fauna is even undetermined and not named yet, only the Brachiopoda were published by Suess.[5] I am quite certain that we will find there many facts supporting Your theory, although he is now so firmly settled as not to require support. As one of the wonderful illustration of it could serve the splendid specimens of the Triassic Turrilites with sutural lobes of the Ceratites existing in a very rich private collection of a certain D.r Fischer at Munich, they were published by Hauer under the name of *Cochloceras*.[6] They are very rare, and I doubt if the British Museum have specimens of them. By and by the fauna of the Trias, so poor and realy abominable everywhere, is growing every year richer in the Alps, and this striking difference at a distance of not more than 150 miles (f.i. between Würtemberg and the Bavarian Alps) is really astonishing if we consider that the Alps were not raised at the time. Till yet I could not find even attempts to explain this difference, most of the palaeontologist content themselves with the phrase "of the special Alpine facies" but such opinion is little more than á dowright nonsense, seeing that the Alps did not exist at the time, and the whole fauna bears not only what is called a distinct facies, but is quite strikingly distinct.— I have heard that somebody, I think Semper, but am not sure of it, is going to attack your theory of the coral reefs, so is Wagner in his new Travels in America attacking the Selection theory, but as it seems to me on false grounds.[7]

As a laughable instance of the Fruitfulness of Your theory, that gives life to everything it touches, I shall relate to You the following anecdote. Some time ago I wrote to my brother[8] (he is now Professor of Zoology at Kiev) something about the failure of the Suez canal[9] and made a joke, that at all events the canal will be in 100 years one of the chief means to test Your theory; some fisches, the larvae of echinoderms, mollusca, the eggs of all sort of animals will certainly be transported by millions from the Red Sea to the Mediterranean adhering to the bottom of the ships, they will come into new conditions, intercross with Meditteranean species an so on,—in 100, nay in 50 years, both seas (having now not more than ten species common) will have hundreds of species in common, new species, perhaps new genera will rise,— this for my joke;— my brother took the matter so earnestly that he insisted on the Counsil of the University for sending immediately a man to explore with the utmost accuracy the fauna of both seas, and especially to determine what species are now common to both,— the man, a certain Mr. Boretzky is already gone.—[10] But the exploration of deep sea fauna is such a difficult and expensive work that certainly one man cannot achieve it, it would be therefore very desirably that England should send a good dredging expedition with the same purpose. There was a Frenchman sent from Paris with the same objects but did nothing good; if they dont make haste one of precious occasion for settling important questions on distribution of faunas will be lost for ever.—[11]

How is progressing the work on Man,[12] I hope to receive some scheets in April by Your Kindness, my wife[13] send her salute to Mrs. Darwin and the ladies,—we were separated the whole winter, she working at her mathematics at Heidelberg, and I sticking in the Museum of Munich, in three weeks the vacation begins and we go to Nizza, in the autumn We hope once more to see You in England[14]

Yours truly | W. Kowalevsky

DAR 169: 60

1 The year is established by the reference to Sofia Vasilyevna Kovalevskaya's being in Heidelburg; she was there between May 1869 and mid-1870 (Koblitz 1983, pp. 88, 97).

2 The Titonic or Tithonian age is the latest stage of the Jurassic period. The Portlandian is a stage of the Upper Jurassic in Britain. The Neocomian epoch is early Cretaceous. The Purbeckian is a stage spanning the Upper Jurassic and Lower Cretaceous in Britain. (*Collins dictionary of geology*.)

3 The former family Ammonitidae is almost coincident with the modern subclass Ammonoidea (Arkell 1950, p. 361 n. 8).

4 *Untere Kreide*: Lower Cretacaeous. *Caprina* is a genus of fossil bivalves.

5 Kovalevsky refers to Eduard Suess and to Suess 1854 and 1855.

6 Ammonoids (extinct cephalopods) are classified in three groups according to the form of their sutures; the ceratites have sutures with frilled lobes. Turillites are Cretaceous ammonoids with sharp-spired shells on which the whorls barely touch. (Challinor 1978; *Collins dictionary of geology*.) Kovalevsky refers to Franz von Hauer and Hauer 1860; Dr Fischer has not been identified. *Cochloceras* is a genus in the order Ceratitida, subclass Ammonoidea.

7 Kovalevsky refers to Karl Gottfried Semper, Moritz Wagner, and M. Wagner 1870. In M. Wagner 1870, p. 434, Wagner claimed, in opposition to CD, that isolation was the only cause of speciation.

8 Alexander Onufrievich Kovalevsky.

9 The Suez canal opened successfully in November 1869 (*The Times*, 19 November 1869, p. 10).

10 Mr Boretzky has not been identified.

11 The Frenchman has not been identified. The movement of species between the Red Sea and the Mediterranean via the Suez Canal is now known as Lessepsian migration, after the developer of the canal, Ferdinand de Lesseps.

12 *Descent*.

13 Sofia Kovalevskaya.

[14] Nizza: Nice. Kovalevsky had visited CD from 30 September to 1 October 1869 (Emma Darwin's diary (DAR 242)).

To Joseph Dalton Hooker 21 February [1870][1]

Down. | Beckenham | *Kent. S.E.*
Feb. 21

My dear Hooker

I read yesterday the Notes on Round Isl^{d}. which I owe to you.—[2] Was there ever such an enigma? If in the course of a week or two you can find time to let me hear what you think, I sh^{d}. very much like to hear. Or we hope to be at Erasmus on March 4th for a week;[3] would there be any chance of your coming to Luncheon then? What a case it is.— Palms, Screw-Pines, 4 snakes, not one living on main island, Lizards, insects & not one Land Bird. But above everything such a proportion of individual monocotyledons! The conditions do not seem very different from the tuff Galapagos isl^{d};[4] but as far as I remember very few monocotyledons there. Then again the isl^{d} seems to have been elevated. I wonder much whether it stands out in the line of any oceanic current, which does not so forcibly strike the main island?

But why, oh why, sh^{d} so many cotyledons have come there; or why sh^{d}. they have survived there more than on main island, if once connected?[5] So again I cannot conceive that 4 snakes sh^{d} have become extinct in Mauritius & survived on Round Island. For a moment I thought that the Mauritius might be the newer island; but the *enormous* degradation which the outer ring of rocks has undergone flatly contradicts this; & the marine remains on summit of Round Isl^{d}. indicates the isl^{d} to be comparatively new; unless indeed they are fossil & extinct marine remains.—[6] Do tell me what you think. There never was such an enigma.— I rather lean to separate immigration, with of course subsequent modification; some forms, of course, also coming from Mauritius. Speaking of Mauritius, reminds me that I was so much pleased day before yesterday by reading a review of book on Geology of St. Helena, by officer who knew nothing of my hurried observations, but confirms nearly all that I have said on general structure of the Islands & on its marvellous denudation.[7] The Geology of that island was like a novel.—

Yours affectionately | C. Darwin

DAR 94: 164–6

[1] The year is established by the reference to Henry Barkly's 'Notes on the flora and fauna of Round Island' (see n. 2, below).

[2] Hooker had evidently sent CD proof-sheets of Barkly's 'Notes on the flora and fauna of Round Island'. The essay was read at a meeting of the Royal Society of Arts and Sciences of Mauritius on 30 December 1869 and published in the *Mauritius almanac and colonial register for 1870* (Barkly 1870). The proof-sheets, corrected by Barkly and annotated by CD, are in the Darwin Pamphlet Collection–CUL. See also *Correspondence* vol. 17, enclosure to letter from J. D. Hooker to Emma Darwin, 29 March 1869.

[3] The Darwins stayed with Erasmus Alvey Darwin from 5 March to 12 March 1870 (Emma Darwin's diary (DAR 242)).

[4] Round Island, like the Galápagos Islands, is volcanic.

[5] Barkly noted that the ratio of monocotyledons to dicotyledons on Round Island was 12 to 14, the established norms in the tropics being 1 to 5 (Barkly 1870, p. 95).
[6] Barkly had found madrepores (a type of coral) on the summit of Round Island (Barkly 1870, p. 101).
[7] CD refers to *The geology of St Helena*, by John Ryder Oliver (Oliver 1869); the review has not been identified. CD discussed St Helena in chapter 4 of *Volcanic islands.*

To Robert Garner 22 February [1870–1][1]

[thanking him for sending him 'the hybrid gathered from a place to which I am so much attached as Maer,'[2] discussing its likely sterility or otherwise, and suggesting what Garner might investigate further.]

Incomplete[3]
Bernard Quaritch Ltd, December 1996

[1] The year range is established by the form of the address described in the catalogue entry (Down House, with Bromley altered by hand to Beckenham). CD used this form of address from April 1869 until May 1871. In the catalogue, the year is given as [1864], on the basis of a note made on the letter.
[2] Maer Hall in Staffordshire had been the home of Josiah Wedgwood II and Elizabeth Wedgwood, Emma Darwin's parents.
[3] The original letter is complete and is described in the sale catalogue as being two pages long.

To V. O. Kovalevsky 22 February [1870][1]

Down. | Beckenham. | *Kent. S.E.*
Feb. 22[d]

My dear Sir

I received a very large box full of beautiful tea from Russia yesterday, & no doubt from you.— It is very kind of you, but you are really too magnificent. I suppose you are at Heidelberg, but I have always regretted, since your visit here,[2] that I forgot to ask you to [. . .]

[. . .] As for myself—my life is as regular & monotonous as a clock. I make sure, but wofully slow progress, with my new book, & do not yet in the least know when I shall go to press.[3] I doubt whether I shall be ready before the end of the year [. . .]

[. . .] I saw lately an allusion to some paper confirming his remarkable discovery about the chorda dorsalis in the lance of Ascidia [. . .][4]

Incomplete[5]
Swann (dealer) (1 October 1953); J. A. Stargardt (17 March 1995)

[1] The year is established by the reference to Heidelberg; see letter from V. O. Kovalevsky, 20 February [1870] and n. 1.
[2] Kovalevsky visited Down from 30 September to 1 October 1869 (Emma Darwin's diary (DAR 242)).
[3] *Descent* was published at the beginning of 1871.

[4] CD refers to Kovalevsky's brother, Alexander Onufrievich Kovalevsky, and to A. Kovalevsky 1866. There is a heavily annotated copy of A. O. Kovalevsky's article in the Darwin Pamphlet Collection–CUL. The paper confirming Kovalevsky's discovery was Kupffer 1870. CD discussed the findings in *Descent* 1: 205. *Chorda dorsalis*: the notochord, a cartilaginous rod forming the primitive basis of the spinal column (*OED*).

[5] The first page of the letter, down to 'ask you to', is transcribed from a facsimile; the rest is from the dealers' transcriptions. The original letter is complete and is described in the sale catalogue as being four pages long.

To James Orton 24 February [1870][1]

Down. | Beckenham. | *Kent. S.E.*
Feb. 24th

My dear Sir

I received the day before yesterday your "Andes & the Amazon", & most sincerely thank you for your present.— I feel deeply sensible of the honour which you have conferred on me by your kindly expressed Dedication, & for the frequent notices in your text of my Journal.[2] You will make this old book of mine quite well known in your great country.

Your work is most elegantly illustrated & got up: I have as yet had time to read only as far as page 105, but I like extremely all that I have read, & I look forward with much pleasure to each evening for continuing it

Your account of Quito is quite new to me & your descriptions bring the scenes before my eyes with much power.[3] It has caused me much regret that I failed to procure information about the Horse's tooth: I wrote a second time & called on Prof. Huxley.[4] Each time I am sure he intended to compare the tooth & give his judgment; but in fact he undertakes much more than mortal man can do, & is half-worked to death.— I hope & fully expect that your book will be in every way highly successful.

With cordial thanks, believe me | Dear Sir | Yours truly obliged | Ch. Darwin

History of Science Collections, University of Oklahoma Libraries

[1] The year is established by the reference to Orton 1870 (see n. 2, below).

[2] CD refers to Orton 1870, and to his own *Journal of researches*. There is an annotated copy of Orton 1870 in the Darwin Library–CUL (see *Marginalia* 1: 644). The dedication reads: 'To Charles Darwin, M.A., F.R.S., F.L.S., F.G.S., whose profound researches have thrown so much light upon every department of science, and whose charming "Voyage of the Beagle" has so pleasantly associated his name with our southern continent, these sketches of the Andes and the Amazon are, by permission, most respectfully dedicated.'

[3] See Orton 1870, pp. 55–84.

[4] Orton had sent CD a tooth from the fossil skull of a horse that he had found in the valley of Quito in Ecuador, hoping to have the species identified (see *Correspondence* vol. 17, letter from James Orton, 8 September 1869). CD first wrote to Thomas Henry Huxley about the find in his letter of 21 September [1869] (*ibid.*). CD's second letter has not been found. CD was last in London from 1 to 9 November 1869 (*ibid.*, Appendix II).

To John Lubbock 26 February [1870][1]

Down. | Beckenham. | *Kent. S.E.*
Febr 26

My dear Lubbock

Although you will be overwhelmed with congratulations I must write to say how heartily I rejoice over your success. Your speech at Maidstone struck me as quite excellent, & I fully expect to see you a great man in Parliament, as you are in Science.[2] But even in the moment of triumph I must let one little groan escape me for poor deserted Science. Anyhow I know that you will always love your first-born child, & not despise her for the sake of gaudy Politicks—

I wrote to ask you a question about savages & suicide, before I had heard of Maidstone; otherwise of course I w^d not have troubled you.[3] If in the course of a few weeks, you can inform me, I sh^d. be glad, but the point is not very important for me.

Once again I do most sincerely congratulate you & especially Lady Lubbock[4]

Ever most truly yours | Ch. Darwin

DAR 261.7:5 (English Heritage 88205930)

1 The year is established by the references to Lubbock's election as MP for Maidstone (see n. 2, below).

2 *The Times*, 26 February 1870, p. 8, announced Lubbock's election as MP for Maidstone, and gave an account of his speech.

3 CD's query has not been found.

4 Ellen Frances Lubbock.

From Asa Gray 27 February and 1 March 1870

Feb. 27, 1870

Dear Darwin

You see by this that I have followed up your request, and that there is no answer at present. I only hope that A. may be well enough to take a shot at his fish next summer.[1] But it is far from certain.

Now news; I am working away at what I am fittest for—study of groups of N. Amer. plants one by one,— slow work, but pleasant.[2]

My wife[3] sends best regards—

Ever Yours | A. Gray

P.S. March. 1.

Here comes another note from Mrs. A.—evidently intended to be sent on to you. Agassiz evidently regrets having abused you in former times.[4]

[Enclosure]

stay in England—

Agassiz says "ask D^r. Gray to tell Darwin that I cannot answer his question for I have never succeeded in securing one of the fishes sitting upon the eggs separately from others. They never bite at the hook at that time and as their nests are generally crowded it is impossible catch a single fish with a net. I'll try next summer by shooting at them".

Mr Agassiz is fitter in these last few days and though his progress

DAR 82: B80

CD ANNOTATIONS
1.1 You see ... times. 6.2] *crossed blue crayon*
Enclosure:
2.1 Agassiz says ... generally crowded 2.4] *scored blue crayon*
3.1 Mr Agassiz ... progress] *crossed blue crayon*

1 See letter from Asa Gray, 14 February 1870 and n. 2. Gray's letter is written on the back of the enclosure; he refers to Louis Agassiz.

2 Gray had first begun work on a planned flora of North America, in collaboration with John Torrey, in 1833. A portion of the work was published between 1838 and 1843 (Torrey and Gray 1838–43), but it remained unfinished. Although Gray for many years planned to complete the work, he ultimately abandoned it, beginning to publish instead his *Synoptical flora of North America* (A. Gray 1878–84), which also remained unfinished (Dupree 1959, pp. 385–93).

3 Jane Loring Gray.

4 The second note from Elizabeth Agassiz has not been found. Louis Agassiz had published a negative review of *Origin* and in 1866 had mounted an expedition to Brazil to try to disprove CD's theory; see *Correspondence* vols. 10 and 14, and Lurie 1960, pp. 297–8, 345–53.

From John Lubbock 27 February [1870][1]

High Elms
27 Feb.

Dear Mr. Darwin

I must apologise for leaving your note so long unanswered, but I am sure you will excuse my having done so.[2]

Tanner in his "Captivity" mentions that the Redskins sometimes commit suicide,[3] but as far as I can remember it is very unusual among savages.[4] The Fiji & Japanese customs are of course familiar to you. Other races are very indifferent to life, but that is not the same thing.

I do not think Savages ever regard suicide as a crime.

I send you the Abstract of the Phil. Trans for 1773, & hope it may perhaps answer your purpose.[5] Pray keep it as long as you like.

I hope Mrs. Darwin is better.[6]

Believe me | Yours most sincerely | John Lubbock

C Darwin Esq

DAR 80: 166–7

CD ANNOTATIONS
1.1 I must ... done so. 1.2] *crossed pencil*
4.1 I send ... better. 5.1] *crossed pencil*
Top of letter: ' p. 89 of my M. S.'[7] *pencil*

1 The year is established by the relationship between this letter and the letter to John Lubbock, 26 February [1870].
2 See letter to John Lubbock, 26 February [1870]. CD's note has not been found.
3 Lubbock refers to John Tanner and Tanner 1830.
4 In *Descent* 1: 94, CD cited Lubbock for the observation that suicide was rarely practised by the 'lower barbarians'.
5 This volume of the *Philosophical Transactions* contained an article on singing birds that CD referred to in *Descent* 1: 55 n. 33 (Barrington 1773).
6 The entry for Monday 14 February 1870 in Emma Darwin's diary (DAR 242) reads, 'In bed till Thursday'.
7 CD refers to the manuscript of *Descent*.

From Federico Delpino[1] 28 February 1870

Firenze
addì 28. Febbrajo 1870.

Celeberrimo uomo!

Ho ricevuto per insigne cortesia della S.V. i due numeri della "scientific opinion", ove Ella si è piaciuta di far pubblicare, tradotto in inglese, il saggio sulle Marantacee.[2] Gradisca i miei più sentiti ringraziamenti.

Ella scorgerà con piacere, credo, che in tutti i miei lavori, per quanto poco ne possa essere il merito intrinseco, non ostante è mio intendimento sempre di rendere omaggio e di sviluppare la dottrina che la S.V. ha saputo con tanta potenza di ingegno fondare e costituire. Altri chiameranno ipotesi la teoria che dalla S.V. prende il nome; ma io con piena convinzione non esito a propugnarla, non già come teoria ma come *dottrina di fatto*. Tutte le parti di cui si compone la scienza fitologica, se si ammette la dottrina della trasformazione, sono piane e intelligibili; ammettendo invece la teoria avversaria, nulla più si comprende—

Vi sono poi alcune parti, come sarebbe la teratologia, la quale a dirittura non ha senso se non che colla dottrina della trasformazione— La esistenza degli organi rudimentarii poi è una prova assoluta della verità di detta dottrina— La S.V. avrà pertanto in me un discepolo, quanto poco abile per se stesso a seguire con successo la strada da Lei tracciata, tanto però risoluto, convinto, e nelle sue convinzioni tenace ed immutabile.

Ora io divido coll'intiero mondo scientifico il vivo desiderio che presto siano per vedere la luce i trattati "sull'uomo" e "sulle variazioni degli esseri allo stato di Natura".[3] È facile il pronosticare il grande successo che saranno per avere i trattati medesimi, e come tanto bene staranno a fianco degl'immortali scritti "sulla origine delle specie" e "sulle variazioni allo stato di domesticità"

In questa aspettativa, io ho l'onore di dichiararmi della S.V. con profonda stima e vivissima ammirazione

ossequentissimo discepolo | Federico Delpino

DAR 162: 146

1 For a translation of this letter, see Appendix I.

2 Delpino refers to the translation of his article on the biology and genealogy of the Marantaceae that appeared in *Scientific Opinion* on 2 and 9 February 1870 (Delpino 1870b).

3 Delpino refers to *Descent*, and to the 'second work' that CD discussed in *Variation* 1: 4–8, but which was never published.

From V. O. Kovalevsky 28 February [1870][1]

Dear Sir!

Your letter adressed to Heidelberg was delivered there to my wife and so I received it some days ago at Munich.[2] As You are kind enough to take an interest in our doings I schall give You some details of our common plans. My brother[3] book now a leave from Russia for 18 months and is now in the position of a boy before the long vacation not knowing how to make the best of it; he intended to go to Naples or Cherso[4] to finish the developement of the Bonelia commenced two years ago, but not finished as it was not the season of developement. The thing is very interesting (if his conclusion about the supposed males is correct), as it present us with such a complete dimorphism, the male and female at first glance seeming to belong to diferrent groups or even classes of worms. As it may interest You, and the little notice of it is printed in Russian I send You the Plate and some explanation.[5] You certainly know the very interesting paper of Lacase Dudier in the Annales des Sciences Natur. on the Bonelia and the splendid figures given there,[6] Lacase had in hands about a hundred Bonelias, but always females; Smarda who also wrote an article about these Sipunculidae[7] could never detect a male, so was it the first time with my brother; he found at a lucky place a whole colony of Bonelias, but all were females and after mutilating hundreds he gave it away in despair; but returning to the same place in September 68 and beginning a new massacre he saw in the outer bag or infundibulum of the sexual organ of the animal (female) in every specimen some strange planarias, very small, from 2 to 7 in every female. They were covered with cilia and seemed quite like Small Turbellarias with a white patch or line in the centre, this white line contained spermatazoa. As at this special period literarly every female had these planarias living in her sexual organs there is great probability that they are the males. If the case will stand future investigation it will certainly be one of the best instance of dimorphism, the female being a Sipunculus and the male a Planarian like animal; so as I said his intention was to go to Naples, but as it seems he changed his mind and intends to go to a new place where no naturalist worked before, the little bay is called Thor and lays on the Sinai peninsula not very far from Suez, facing rich coral reefs. I, for my part am against his undetaking the journey before October, as in the summer the pilgrims going from Africa to Mekka generally bring with them pest, cholera and what not.—[8]

My wife wished to go to Berlin, and the Rector the know physiologist Dubois Reymond[9] was one of our advocates, and a better one is difficult to have, but the matematicians were against it, holding to the strict sense of the law. Generally the matematicians are the worst thing living, not only in Germany but also in our country and I hope in England. One of the principal opponents to the admission of ladies was Kunt,[10] a true pure, who at one of the Natural Science Congress proposed a toast "for the theory of numbers (Zahlenteorie) as a part of mathematics that till the present was not besmeared with practical application". So are they all, more or less. But happily Kirchhow took a great interest in her and promised to give her a special investigation about light in the summer, so she remains a sum̄er at Heidelberg.[11] For my part I am quite incertain what to do, going to Vienna and working with the assistance of Tehermak and Suess,[12] or going to England and working alone whithout direction. I think I'll take the latter course; firstly because with Your kindness to me I will have as good assistance as can be; scondly because I will have the use of the best collections existing.—

The one point which most interest me is, can I have trough a member the use of the Library of the Geological Society, not only reading there, but taking the books out to my lodging as my principal workink time is in the night. I think I shall take the liberty to write about this point to Mr. Huxley, as the free use of the Library is one of the Chief point which determine my going to London.—[13] The vacation commences this Sunday, and Monday I hope to break off and go to Nizza with my wife and one of her friends, a young lady about her age but "learned in the law", as she is pursuing now law studies at Heidelberg.[14] What will Your ladies say of such an ultra blue stocking.

The principal attraction of the coast of Nizza for me is the rich fauna of mollusks, the splendid series of deposits from the trias to the upper Chalk and especially the wonderful new Wealden, only not of the Jurrasic but of the Upper Cretaceous period with a developement of 600 feet of freschwater sediments and lignites, at least so is it affirmed by Coquand and I am very restless to see the wonder with my own eyes and especially to collect and study its fauna.[15] In case You write me, please to adress *Nizza, poste restante*. With Your permission I write You from Nizza.— My compliments to Mrs. Darwin and the ladies

Yours very truly | W. Kowalevsky

Munich | 28 February

[Enclosure]

1. Grown up female of *Bonelia* always sticking with her pyriform and in an aperture of a rock (Lacase— Ann. des Sc. Nat. vol X.)

2. Bonelia cut from the dorsal side *g.* ova in the Leibeshöhle, *q.* anal aperture; *m.* uterus; p. inpundibulum trough which the ova pass from the F cavity of the thorax into the uterus *v* bag where the planarias live

3. Supposed male magnified 70 diam; *rs* receptaculum seminis *t.* infundibulum; h. granular mass clothing the cavity, from which undeveloped spermatozoa are cast of; s. glomercili of spermatozoans swimming freely in the cavity

4. Infundibulum of the receptae. sem. (magnified 500 diam). with flimmerzellen

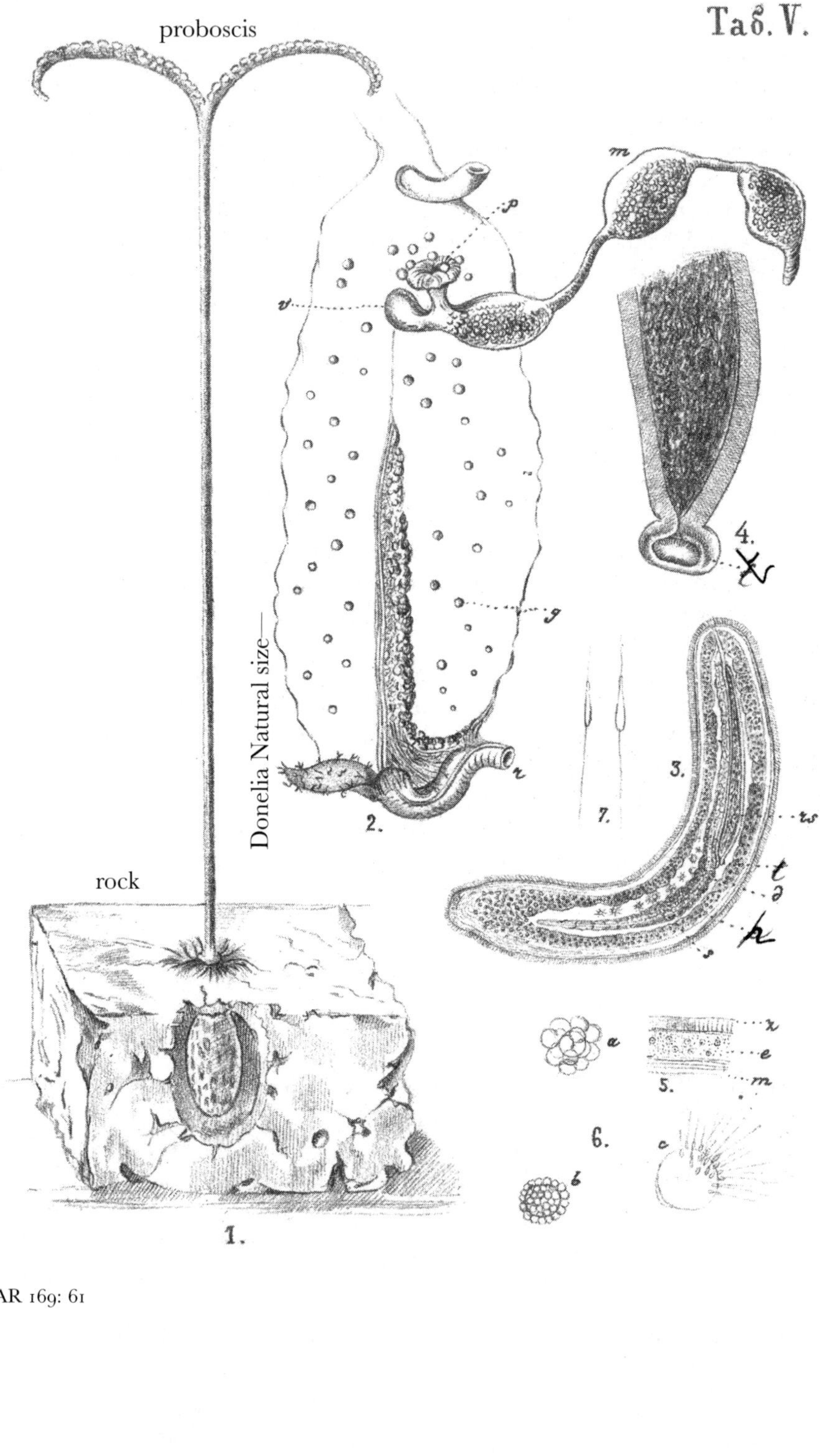
Tab. V.
proboscis
rock
Donelia Natural size—
1.
2.
3.
4.
5.
6.
7.
m
p
v
g
r
ws
t
d
k
s
a
b
c
x
e
m

DAR 169: 61

1 The year is established by the reference to Sofia Vasilyevna Kovalevskaya's being in Heidelburg; she was there between May 1869 and mid-1870 (Koblitz 1983, pp. 88, 97).
2 Letter to V. O. Kovalevsky, 22 February [1870]. Kovalevsky's wife was Sofia Kovalevskaya.
3 Alexander Onufrievich Kovalevsky.
4 Cherso: now Cres, in Croatia.
5 See enclosure. *Bonelia* is an incorrect subsequent spelling of *Bonellia*, a genus of spoon worms.
6 Kovalevsky refers to Félix Joseph Henri de Lacaze-Duthiers and Lacaze-Duthiers 1858. The article had four plates, showing the general morphology, and the digestive, reproductive, and nervous systems of the animal.
7 *Bonellia* is now in the family Bonelliidae in the phylum Echiura. The family Sipunculidae is now placed in the phylum Sipuncula (peanut worms). Kovalevsky refers to Ludwig Karl Schmarda's description of *Bonellia viridis* (Schmarda 1852).
8 Tor, on the south-west coast of the Suez Peninsula, served as a quarantine station for pilgrims returning from Mecca (*Columbia gazetteer of the world*).
9 Emil Heinrich Du Bois-Reymond.
10 Kovalevsky refers to August Kundt.
11 Gustav Robert Kirchhoff was a professor at Heidelberg (*DSB*).
12 Gustav Tschermak and Edward Suess.
13 Thomas Henry Huxley was president of the Geological Society of London (*ODNB*).
14 Nizza: Nice. Kovalevsky also refers to Anna Mikhailovna Evrainova.
15 Kovalevsky refers to Henri Coquand.

To H. E. Darwin [March?] 1870[1]

Spring 1870

My dear Hen.

I have worked through, (& it is hard work) half of the 2[d] Chapter on mind, & your corrections & suggestions are *excellent*.[2] I have adopted the greater number, & I am sure that they are very great improvements.— Some of the transpositions are most just. You have done me real good service; but by Jove how hard you must have worked & how thoroughily you have mastered my M.S. I am pleased with this Chapter now that it comes fresh to me.—

Your affectionate, admiring & obedient | Father, C. D.

All is as clear as daylight— your plan of putting corrections saves me a world of trouble, by just as much as it must have caused you.— N.B. you **can** write, I see, a perfectly clear hand, as in *all* the corrections.—

DAR 185: 58

1 The month is conjectured from the reference to spring, and from the relationship between this letter, the letter to H. E. Darwin, [8 February 1870], and the letter from Emma Darwin to H. E. Darwin, [25 March 1870] (DAR 219.9: 82), in which Emma asks Henrietta to write when she receives a manuscript. This manuscript was presumably that of the chapter or chapters of *Descent* following those on the mental powers of humans and animals.
2 CD had asked Henrietta, who was travelling in France and Italy, to read the manuscript of *Descent*; see letter to H. E. Darwin, [8 February 1870], and letter from H. E. Darwin, [after 8 February 1870].

From Hensleigh Wedgwood [March 1870][1]

1 Cumberland Place, | Regent's Park. N.W.

"Here the grey man expressed his astonishment and disapprobation by a prolonged whistle"

—Wenderholme II. 91.[2]

This is the interjection whew! But why does disagreeable surprise so express itself? I think it is rather when something turns out the contrary of what one expects.

The fox scheming to catch a goose would picture to himself the goose in some of its usual haunts, then he would imagine some way of pouncing upon it,

Incomplete
DAR 181: 56

CD ANNOTATIONS
3.1 This … itself?] *scored blue crayon*
4.1 The fox … it, 4.2] *crossed pencil*
Top of letter: 'H. Wedgwood | March 1870' *pencil*

1 The date is established by CD's annotation at the top of the letter.
2 Hamerton 1869.

From Charles Edward Cutts Birchall Appleton 2 March [1870][1]

St. John's Coll. Oxford
Mar 2

"*The Academy*."

Dear Sir,

You were kind enough to send us a valuable notice of your forthcoming work for the first number of the *Academy*,[2] & I now write to ask you to be so kind as to send me a note or two of anything important which may have recently come before you & not yet been published, or about any researches you may be engaged in &c.

I send the enclosed circular round every where in Europe—& hope to get this side of our enterprize very complete in time.[3] Any suggestions as to the conduct of the 'Academy' wh. wld make it more valuable e.g. to yourself, I shd. be very glad to receive.

An occasional review of a book, say once in a year or so from your pen—I scarcely like to ask for—yet it wd. delight Mr. Murray greatly, & be of the very greatest service to us. Do you care to say anything about Murphy's book which touches upon your subjects?[4]

Should you be able to let me have a few lines for March, I ought to get it by Sunday morning here. | Believe me dear Sir | Yours very faithfully | C. Appleton | (Ed.)

C. Darwin Esq FRS

DAR 159: 78

1 The year is established by the reference to the notice of *Descent* in the *Academy*, 9 October 1869, and by the relationship between this letter and the letter from C. E. C. B. Appleton, 18 May [1870].

2 See *Academy* 1 (1869–70): 15–16. The notice appeared under the heading 'Scientific notes' and described the book as consisting of three parts: I. The descent of man; II. On sexual selection; and III. On expression of the emotions.

3 The circular has not been found in the Darwin Archive–CUL.

4 CD's reply has not been found, but he probably refused; see the letter from C. E. C. B. Appleton, 18 May [1870]. John Murray was one of the publishers of the *Academy.* Appleton also refers to Joseph John Murphy and his *Habit and intelligence in their connexion with the laws of matter and force* (Murphy 1869). This book was not reviewed in the *Academy.* CD discussed Alfred Russel Wallace's review of Murphy 1869 (Wallace 1869d) in his letter to A. R. Wallace, 5 December [1869] (*Correspondence* vol. 17).

From George Henslow 3 March 1870

8. Robertson Terrace | Hastings
March 3/70.

My dear Mr Darwin,

I have been going to write to you some time ago—but several things intervened—to thank you for yr letter containing directions about an experiment on Lapageria.[1] I found it was rather late in the season—& could not get more than about 20 blossoms fit for trial. I, however, dusted about half with their own pollen & the rest with pollen from diff^t. fls. & cut out their stamens. The perianths of *all* faded much the same time. & the pistils have *all* remained on, but at present look all alike. M^r H[2] tells me that they do not commence swelling under 6 months, so that I cannot ascertain the result for some time to come. I hope to repeat the exp^t. more fully next season when the blossoms are abund^t. The *white Var* is a doubtful affair—its history is unknown Mr H. says he believes it has sent forth *red blossoms* in some body else's possession!

All his plants are from cuttings: but seedlings are more profitable if reard with a little care.

Y^rs faithfully | Geo Henslow

P.S. I have been sent down here to '*rusticate*' by the Metropol^n Ry Co,[3] for giving me concussion of brain in an accident— getting all right again

DAR 166: 171

1 CD's letter to Henslow has not been found; however, see *Correspondence* vol. 17, letter from George Henslow, 22 November 1869.

2 Mr H. may be Andrew Henderson, a nurseryman in St John's Wood; a Mr Henderson of St John's Wood had told Henslow that *Lapageria rosea* never set seed unless artificially impregnated; see *Correspondence* vol. 17, letter from George Henslow, 22 November 1869.

3 Metropolitan Railway Company.

From J. D. Hooker [7 March 1870][1]

Kew.
Monday.

Dear old Darwin

I went a pilgrimage to town yesterday, at 1 PM on purpose to see you— on my way in, I remembered that I forgot your brother's number, & so called at the Lyell's to ask it—, & finding that they knew nothing of you, I *took for granted* that you had not come up & I went & spent the evening with Mrs Boott instead.[2] So your letter this morning disgusted me.[3] I am so dreadfully busy, that I doubt if I can get in this week, do tell me how long you stay in town— Can you come here?— I do want a talk about Barkly's letter & other things; so let me hear your movements.[4]

I do not give much for the Botanical results of Ronde Island as depending on Monocots $\pm$ Dicots, but the segregation of new forms & the Zoolog are most wonderful.[5]

Do back me up in my attempt to get Lyell to reduce his forthcoming little book to *Elementary Principles* & a compress of a volume like Huxleys Elementary Physiology—or not much bigger—[6] A book like that to teach masters & children the *principles* of the Science in a concrete form is what is wanted, & it must be small & concrete. I have been coaching Charlie[7] in Huxley which is admirable and the boy took to it (during his holidays) with actual avidity.— Just such a book on Geology is quite as much as should be *compulsory* on *all* youths with a liberal Education.

My Brit. Flora is a frightful drag—[8] how difficult it is to do simple books well, & how little show they make for it.— I think Lyell must go to work in a different *spirit* or he will not succeed. He should go to work as the Veteran Phil. expanding the fundamental Principles of his own grand system—& not as a Scientific Expositor of data & facts. He tells me that "this", & "that" requires modification explanation expansion & so forth— I tell him & such this'[s] & that'[s] should have no place in such a book: they are for proficients, advanced students & so forth— —but I bore you—we must meet some how—

Just write *one line* to say how long you will be in town—

Ever yrs affec | J D Hooker

Grove is disgusted at your being disquieted about Sir W Thomson— tell George from me not to sit upon you—with his mathematics—[9] When I threatened your tropical cooling views with the facts of the physicists, you snubbed me & the *facts* sweetly, over & over again—[10] & now because a scare crow of x + y has been raised on the *self same facts* you boo hoo— Take another dose of Huxley's penultimate G. S. address[11] & send George back to College.

A queer man—Crompton of Norwich, is too modest to send you enclosed, & asks me to do so—[12] please tear it up— I do not want it again.

DAR 103: 42–5

1 The date is established by the relationship between this letter and the letter to J. D. Hooker, 21 February [1870], and by the reference to CD's being in London. The Darwins came to London on 5 March 1870 (CD's 'Journal' (Appendix II)). The first Monday after 5 March was 7 March.

2 Hooker refers to Erasmus Alvey Darwin, who lived at 6 Queen Anne Street, Charles and Mary Elizabeth Lyell, and Mary Boott.

3 CD's letter to Hooker has not been found.

4 Hooker had evidently sent CD copies of Henry Barkly's letters to him concerning his visit to Round Island (near Mauritius) dated 17 November 1869 and 13 January 1870 (Archives of the Royal Botanic Gardens, Kew, DC 188: 148–68). See also *Correspondence* vol. 17, enclosure to letter from J. D. Hooker

to Emma Darwin, 29 March 1870, and this volume, letter to J. D. Hooker, 21 February [1870]. The Darwins returned to Kent on 12 March (Emma Darwin's diary (DAR 242)). Hooker and CD did not meet during this visit (see letter to Asa Gray, 15 March 1870).

5 See letter to J. D. Hooker, 21 February [1870]. Hooker refers to Round Island near Mauritius.

6 Hooker refers to Lyell's *Student's elements of geology* (Lyell 1871); also to Thomas Henry Huxley and T. H. Huxley 1866.

7 Charles Paget Hooker.

8 Hooker refers to his *Student's flora of the British Islands* (Hooker 1870).

9 Hooker refers to William Robert Grove, William Thomson, and George Howard Darwin. CD had evidently written to Hooker about Thomson's calculations of the age of the earth; see also *Correspondence* vol. 17, letter from Charles Lyell, 2 November 1869.

10 For Hooker's objection to CD's suggestion that temperate plants had crossed through the tropics during a global cool period, see *Correspondence* vol. 6, letter from J. D. Hooker, 9 November 1856.

11 In his 1869 address to the Geological Society of London (T. H. Huxley 1869b), Huxley challenged the accuracy of Thomson's calculations of the cooling of the earth.

12 The enclosure has not been found: it may have been Joseph Crompton's first presidential address to the Norfolk and Norwich Naturalists' Society, delivered on 27 April 1869 (*Transactions of the Norfolk and Norwich Naturalists' Society* 1 (1869–70): 13–18). John Gunn, Hooker's uncle by marriage, was a member of the society.

To A. R. Wallace 7 March [1870?][1]

6 Queen Anne St. | W.
March 7th.

My dear Wallace

Though I have nothing particular to say, I sh^{d}. very much like quarter of an hours talk with you. *So if I do* **not** *hear to contrary* I will venture to call about 10 oclock on Wednesday morning; & I sh^{d} think you will have finished breakfast at that late hour.

Ever yours | C. Darwin

David Schulson (Cat. 46, June 1988)

1 The year is conjectured from the fact that CD was in London from 5 to 12 March 1870 (Emma Darwin's diary (DAR 242)). In 1870, 7 March was a Monday.

To J. D. Hooker 8 March [1870][1]

6. Queen Anne St | W.
March 8th

My dear Hooker

Thanks for your very nice letter.[2] You must not think *for a moment* of coming here on purpose, though it will be a great disappointment not seeing you.— You will own it was rather stupid of you not just to look in here, & see if we were arrived.— I sh^{d}. very much enjoy, as I always have, coming to Kew, but I fear the exertion: it almost always knocks me up.— Erasmus desires me to say that dinner is at 7 & luncheon at 1 every day—sh^{d} you be able to come.[3] But mind I won't have your blood on my head & I do **not** ask you to come.

I have no time to write & cannot analyse my feelings, but I sh$^{d.}$ like to see you Sir J. H., & I do not care so much what sort of K$^{d.}$—[4] I must also differ from you, (though this will make you angry), & I do not think that you owe your position in any degree as a man of science to your Father.—[5]

I have forgotten the main object of my note which is to say that we return home early on Saturday morning.

If you write to Barkly[6] urge on him to ascertain

(1) whether R^{d} Is^{d}[7] is a volcano, & what the fossil remains are.—whether surface or embedded.—

(2) Depth of Channel

(3) Currents—, whether drift wood large seeds—wrecks &c on any sloping part of coast.

We must meet some other time.—

Ever yours affect | C. Darwin

Did you have time to read Delpino on Marantaceæ?—[8]

DAR 94: 167–8

[1] The year is established by the relationship between this letter and the letter from J. D. Hooker, [7 March 1870].

[2] See letter from J. D. Hooker, [7 March 1870].

[3] CD was staying with his brother, Erasmus Alvey Darwin.

[4] Hooker had declined a knighthood (see *Correspondence* vol. 17, letter from J. D. Hooker, 14 November 1869).

[5] See *Correspondence* vol. 17, letter from J. D. Hooker, 21 November 1869. Hooker's father was William Jackson Hooker.

[6] Henry Barkly. See letter from J. D. Hooker, [7 March 1870] and n. 4.

[7] Round Island, Mauritius.

[8] CD refers to Federico Delpino and Delpino 1870b. See also letter from Federico Delpino, 28 February 1870.

From St George Jackson Mivart 8 March [1870][1]

6 Pelham Crescent | Hastings
March 8th (Tuesday evening)

My Dear Sir,

I hope this letter may be in time to save you from making a journey to North Bank for nothing— I am stopping here for a short time to give my wife a little repose after the anxiety attending the recent death of my dear Mother—[2] We purpose returning in about a week & then I need hardly assure you how *much* pleasure I shall feel in receiving a visit from you—

With best regards | believe me, | My dear Sir | Your's most truly | S$^{t.}$ G. Mivart.

DAR 171: 183

[1] The year is established by the reference to the death of Mivart's mother (see n. 2, below).

[2] Mivart's mother, Caroline Georgina Mivart, died in the first quarter of 1870 (*BMD* (*Death index*)). Mivart's wife was called Mary Anne. North Bank, Regent's Park, London, was Mivart's usual address.

From Hermann Müller 8 March 1870

Lippstadt
8/3 70.

My dear Sir!

I am very much obliged to you for your assistance in obtaining Mr Fr. Smith's willingness to revise my Westfalian bees.[1] My wishes on this head are perfectly accomplished by your kind mediation, and now I can begin more confidently to elaborate the matter that occupies me since two years.

From a paper, I lately have sent to you, you will have seen, that I intended at first to try an employment of your doctrine on flowers and insects visiting flowers in general.[2] But this whole matter being extremely extensive, I have resolved now to elaborate previously only a single section of it, namely the employment of your doctrine on the bees, being adapted in their parts of mouth and in their pollen-collecting apparatus to their nourishment gained from flowers.—[3]

So turning my attention chiefly to the bees of my neighbourhood, I have found, that this family of insects also in many other regards is highly abundant of facts decisive for your theorie. It is, for instant, very rich of doubtful forms ranked by one Zoologist as a species and by another as a variety, it is very rich of species which differ but very slightly from each other, and the differences, in the one case judged as sufficient for the distinction of species, in the other case looked at as mere varieties, blend into each other in an insensible series.— In the family of bees further there are found many surprising instances of variation of instincts—

I have encountered also many pecularities in the one sex, that can only be acquired by inheritance from the other sex etc. These and other observations, occasionally made, have determinated myself, to exceed, considering the family of bees, the exact limits of my original thema and to treat in its various regards the employment of your theorie—on bees

Till the end of this year I hope to have finished the first memoir on this subject.[4]

Nevertheless I will have, also during next summer, an eye on all facts relating to the whole thema, and I would be very much obliged to you, if you would communicate to me all your scruples and objections against my statements in the paper lately sent to you.

As you have corrected repeatedly, in former cases, my opinions, I hope, you will also in this case do so.

With the greatest respect | I remain, | my dear Sir, | yours very faithfully | H Müller.

DAR 171: 296

CD ANNOTATIONS

Top of letter: 'Bot Zeitung' *pencil*

Verso of last page: 'Send to Delpino'[5] | 'Axel'[6] *pencil*; 'Delpino' *pencil, encircled pencil*; 'Calculate number of flowers each species | Müller | insertion | [original] keep | Angræcum'[7] *pencil*

1 Müller had asked CD to forward to Frederick Smith a letter asking Smith to identify some of Müller's Westphalian bees (see *Correspondence* vol. 17, letter from Hermann Müller, 12 November 1869). No letter from CD to Müller on the subject has been found.

[2] There is an annotated copy of Müller's paper, 'Die Anwendung der Darwin'schen Lehre auf Blumen und blumen-besuchende Insekten' (The application of Darwinian theory to flowers and flower-visiting insects; H. Müller 1869), in the Darwin Pamphlet Collection–CUL.
[3] Müller had also studied the adaptation of Syrphidae (hoverfly) mouthparts to various sizes of pollen (see *Correspondence* vol. 15, letter from Hermann Müller, 23 October 1867).
[4] Müller's paper, 'Anwendung der Darwin'schen Lehre auf Bienen' (Application of Darwinian theory to bees), was published in 1872 (H. Müller 1872).
[5] Federico Delpino translated and annotated H. Müller 1869 (Delpino 1870a). See letter from Federico Delpino, [1870].
[6] Possibly a reference to the Swedish botanist Johan Severin Axell.
[7] On the pollination of *Angraecum sesquipedale*, see *Correspondence* vol. 15, letter from A. R. Wallace, 1 October [1867] and n. 5.

To Thomas Woolner 10 March [1870][1]

Down. | *Bromley.* | *Kent. S.E.* [6 Queen Anne Street]
March 10th

My dear Mr Woolner

Very many thanks for the drawing: it does excellently.—[2] The "Woolnerian tip" is worth anything to me.—[3]

Yours very sincerely | Ch Darwin

Bodleian Library

[1] The year is established by the relationship between this letter and a note in DAR 80: B113 dated 'March 7 (70)': 'W. has seen the tips in women & men. When he was making his figure of Puck he went to Z garden and made drawings of ears of monkeys & shortly afterwards he saw a man with tip & instantly recognized their signification'.
[2] Woolner's drawing of a human ear showing the unusual feature of a pointed tip ('Woolner's tip') on the outer ear projecting from the inwardly folded margin is in DAR 80: B120.
[3] CD reproduced the drawing and discussed the feature in *Descent* 1: 22–3.

From Charles Wicksted to Georgina Tollet[1] 13 March [1870?][2]

Shakenhurst, Bewdley.
March 13th.

My Dearest Georgina,

I have submitted Charles Darwin's questions about Foxes to a jury of fox-hunters assembled here, on the occasion of the Ludlow Fox hounds having met here yesterday, & their experiences quite agree with mine— Neither I nor any of them can remember ever having seen a hound catch hold of a fox by the brush, or ever seen the common movement (of a dog (that puts his tail between his legs when frightened) performed by a fox either in a tame[3]

AL incomplete
DAR 181: 97

CD ANNOTATIONS
1.4 or ever ... tame 1.6] *scored blue crayon*

1.6 either in a tame] 'or wild state' *added pencil*
Top of letter: 'M[r] Wickstead, kept Fox-Hounds for 20 years.—' *pencil*

[1] The correspondents are identified by CD's annotation and by the address. Charles Wicksted, a noted fox-hunter and breeder of hounds, lived at Shakenhurst in Worcestershire, and was the brother of Georgina Tollet.

[2] The year is conjectured on the assumption that the letter is a reply to a question CD submitted after reading an article on foxes in *Land and Water*, 6 November 1869 (see n. 3, below). Charles Wicksted died on 4 June 1870 (*Burke's landed gentry* 1879).

[3] CD's questions have not been found. In *Expression*, p. 125, CD refers to an article in the 6 November 1869 issue of *Land and Water* that claimed that foxes, however tame, never exhibited expressive movements, such as tucking their tails, when frightened. CD stated of foxes: 'I have been assured that when frightened they never tuck in their tails' (*Expression*, p. 126).

To Hermann Müller 14 March 1870

Down, Beckenham, Kent
March 14. 1870.

My dear Sir

I think you have set yourself a new, very interesting and difficult line of research.[1] As far as I know, no one has carefully observed the structure of insects in relation to flowers, although so many have now attended to the converse relation. As I imagine few or no insects are adapted to suck the nectar or gather the pollen of any single family of plants, such striking adaptations can hardly, I presume, be expected in insects as in flowers.—

Yours very sincerely | Ch. Darwin

Copy
DAR 146: 432

[1] See letter from Hermann Müller, 8 March 1870.

From James Crichton-Browne 15 March 1870

West Riding Asylum, | Wakefield.
15th. March. 1870

My dear Sir

I am ashamed to write to you, having delayed so long to thank you for your kind present— a copy of your incomparable book on the origin of species, & to send you the information which I promised upon several subjects. I trust however that you will hold my offence extenuated if not excused when I explain to you that since I last wrote I have been prostrated by serious & protracted illness from which I am only now recovering & have suffered two family afflictions of a most distressing character.[1] My time & attention have been absorbed by my infirmities & sorrows together with my very onerous routine duties, so that I have kept postponing— all unimperative occupations from week to week & month to month. I hope you will pardon my apparent negligence after this apology & receive me once more into the number of your correspondents. I promise to be more prompt & punctual in future & to do my best to aid you in your important researches.

Along with this I send you some notes on 'blushing, on the Platysma myoides, & on the 'grief muscle.' Tell me if they meet your wishes & if not direct my observations into the proper channel.

On friday next I shall despatch by rail carefully packed Duchennes Plates respecting which I have made a few notes that shall be sent along with them.[2]

Is there any other way in which I can be of service to you? I am most anxious to make amends for the past & to show my gratitude for your present which it would be an impertenence even to eulogise.

With profound respect | I am | Yours most faithfully & obediently | J. Crichton Browne

Charles Darwin Esq | &c. &c.

[Enclosure]

⟨*third of a page excised*⟩ The second case was that of E. ⟨B⟩. P. aet. 18. Single ... from Leeds a girl of hysterical tendencies, who had suffered from Epilepsy from the period of puberty, when she was terribly frightened by a fight which took place in her presence, & whose mind had become gradually enfeebled & unhinged, so that she had lost much of her acquired knowledge, & had attempted suicide & violence against her relatives. The change to the Asylum seemed to interrupt her mental lethargy. When examined she was tolerably intelligent & answered questions correctly, the whole of her face being of a dull purplish red hue. On uncovering her chest it was thought at first that she was suffering from Scarlet fever, as over the breasts & sternum there was a purplish red appearance similar to that on the face, the true character of which was only recognised when it suddently vanished in the course of a few minutes & as suddenly again presented itself in a few minutes more. The blushing seemed to depend upon internal conditions or emotions, as there was nothing in external circumstances to account for its evanesence & reappearance. This girl improved greatly under treatment & was ultimately discharged.

In several other cases, all epileptics, I have noticed, scattered blotches of blushing & red mottling over the chest & breasts, & in all of these cases I have also noticed the cerebral maculæ of *Trousseau*. In all of them when gentle markings were made with a pencil or the finger nail, on the thorax or abdomen, the portion of skin touched, in less than half a minute, became suffused with a bright red colour, which spread to some distance on each side & persisted for several minutes perhaps ten or fifteen. This condition which undoubtedly indicates a deep modification of the vascularity of the skin, is produced in severe cases by the slightest touch so that the pressure of a finger may produce a diffused blush.[3] **Trousseau** has recorded cases in which blushing occurred under such circumstances on the thighs as well as the abdomen & chest.[4]

Sir Charles Bell says that blushing extends on to the breasts.[5]

A lady friend tells me that when ashamed or agitated she blushes on her face neck, wrists & hands.— the exposed portions of the skin.[6]

I have sometimes thought that I could trace, in the progressive development of blushing over the cheeks neck &c, the propagation of an influence, inhibitory or otherwise to the *three* cervical ganglia of the sympathetic system the relative diffusion of the blush depending upon the number of these involved. The number of my observations do not however justify me in calling this anything but a speculation.

I possess a large mass of details as to blushing in morbid conditions, after injury & galvanization of the sympathetic &c. which can be forwarded, if desired.

II.
The Platysma Myoides

I have had an opportunity lately of watching the action of the Platysma Myoides in the case of a patient— T. B. a labourer from Sheffield, labouring under General Paralysis. Whenever this man attempted to speak the whole of the muscles of his face were thrown into a state of violent agitation. The *eyebrows* were twitched upwards with amazing rapidity, the eyelids were quickly approximated & separated the nostrils & cheeks trembled, the lips shook, & the whole of the skin of the neck from the lower jaw to the clavicle, (as far back as a line drawn from the ear to the outter fourth of the clavicle,) was twitched, & thrown into folds mostly transverse but a few near the mesial line, & the external margin of the Platysma, longitudinal. It is impossible to describe in words the singular effects produced by these contortions of the countenance, which invariably followed any effort at articulation, mastication or the protrusion of the tongue. The skin of the neck, trembled, & quivered strongly & this quivering could be felt going on between the fingers, when the skin was lifted up from the subjacent muscles & structures of the neck. The transverse folds were short, few exceeding an inch & a half in length but numerous while the longitudinal furrows were longer. So violent at times was the action in the skin of the neck, that I am satisfied that there must have been hypertrophy of the of the fibres of the platysma myoides. At times the outline of the Platysma could be traced by the agitation of the skin This man has since died of organic disease of the brain. I find that one of my assistants, knowing nothing of my researches, has entered in the case book.— "*When he opens his mouth the platysma myoides* is thrown *into a state of violent spasm*".

In several old emaciated men, I have remarked slight twitching or trembling of the skin of the neck, when the chest is exposed, & a perceptible drawing up of the skin over the clavicle, (from the chest to the neck) when the mouth is opened & the tongue put out. This seems to me to be effected by the outter fibres of the platysma, attached to the skin over the pectoralis major which are directly continuous with those attached to the angle of the mouth & lower lip, & which must be stretched by the movement referred to.

I have been quite unable to connect the action of the platysma with any emotional condition.[7] In the old ematiated men alluded to above the general mental state when the muscle twitched was impatience & annoyance at my interference, & observations.

The grief muscle.

I have recorded persistent contraction of this muscle, in three cases during the past winter—all cases of hypochondriacal melancholia, the most *selfish*, and to the uncultivated mind therefore the most intense form of grief.

R. S. aet. 51. widow. Tadcaster. fancied that she had passed all her viscera by stool, & that her chest & belly were entirely empty. She was in constant mental agony.—worst in the morning—beat her semi-closed hands rhythmically together for hours. & wore an expression of great distress The corrugators of the eyebrows were permanently contracted, the eyes widely opened, the upper eyelid, having an angular bend about its middle from which its margin was straight to the inner angle of the eye, & having above it a rounded fullness. This condition continued for months but R. S. has now recovered; & her face has resumed its natural expression the margin of the upper lid is now rounded & the fullness has disappeared[8]

J. S. aet 61 from Huddersfield believes that her head is burnt & that there is not a drop of blood in her body. She has an agonized expression. There is marked elevation of the outter third of the left eyebrow, the skin of the forehead over it being gathered into a multitude of minute wrinkles; the corresponding part of the right eyebrow depressed & overhanging. The corrugatores superciliorum, powerfully contracted The eyeballs much exposed, the upper eyelids having an aprupt angular arch rather to the inside of their central points; a rounded kite-shaped fullness over the upper lid: angles of the mouth pulled down; constant restless movements, rubbing & wringing of the hands, with twisting of the body from side to side, the neck being held rigid as if apprehensions were entertained as to any movement of the head.

I have not had time to read over or correct what I have written | J. C. B. | West Riding Asylum | Wakefield. | 15th. March 70

DAR 161: 310, DAR 161: 323/2–5

CD ANNOTATIONS
Enclosure:
1.1 The second case] *after opening square bracket, blue crayon*
1.1 The second case ... head 14.11] *crossed pencil*
1.1 The second case ... hue. 1.8] 'Not new' *added in margin, pencil, circled pencil*
2.1 In several ... fifteen. 2.6] '☞ Cerebral diseases seem to give tendency to skin to redden—slight scratch' *added in margin, pencil*
2.1 I have noticed ... over. 2.2] *scored blue crayon*
2.2 red mottling] *underl blue crayon*
2.3 Maculæ] 'Maculæ' *added above pencil*
2.10 on the thighs ... chest 2.11] *scored blue crayon*
2.10 thighs as well] *underl blue crayon*
4.2 the exposed ... skin.] *scored blue crayon*
4.2 exposed portions] *underl blue crayon*
6.1 injury] 'injury' *added above pencil*
8.2 under ... Paralysis. 8.3] *scored & underl blue crayon*
8.4 The *eyebrows* ... tongue. 8.12] 'All tend shook' *added in margin, blue crayon*
8.6 the whole ... clavicle, 8.7] *underl blue crayon*

8.8 mostly transverse] *underl blue crayon*
8.9 longitudinal.] *underl blue crayon*
8.11 invariably followed] *underl blue crayon*
8.11 articulation,] *underl blue crayon*
8.17 hypertrophy of the] 'of the' *del blue crayon*
8.12 "*When* ... *spasm*". 8.21] *double scored blue crayon*
9.3 (from ... out. 9.4] *double scored blue crayon*
10.2 In the old ... observations. 10.3] *double scored blue crayon*
12.1 persistent contraction] *underl pencil*
13.1 passed ... viscera] *underl pencil*
13.3 beat ... semi-closed] *underl pencil*
13.3 rhythmically ... hours. 13.4] *underl pencil*
13.5 eyelid] *underl pencil*
13.7 This condition ... rounded 13.9] *scored pencil*
13.7 continued for months] *underl pencil*
13.9 lid] *underl pencil*
14.1 J.S.] *after opening square bracket, pencil*
14.1 her head] 'her' *underl pencil*
14.1 J.S. ... expression. 14.2] '♀' *in margin, pencil*
14.6 aprupt angular] 'abrupt angular' *added above pencil*
14.8 angles] 'Angle' *added above pencil*
14.8 angles ... down;] *scored blue crayon*; 'angles of mouth depressed' *added in margin, pencil*
14.8 the mouth ... down;] *underl blue crayon*
Top of third page of encl: 'B' *added pencil*
Top of seventh page of encl: 'B' *added pencil*

1 CD had arranged for Crichton-Browne to be sent a copy of the fifth edition of *Origin* (see *Correspondence* vol. 17, letter to James Crichton-Browne, 22 May 1869). For CD and Crichton-Browne's earlier correspondence, see *Correspondence* vol. 17. Crichton-Browne's youngest brother, Vincent de Paul Browne, died on 1 February 1870 (*The Times*, 5 February 1870, p. 1). Also in 1870, his father, William Alexander Francis Browne, became blind as a result of injuries received when he was thrown from his carriage (*ODNB*).

2 CD had lent Crichton-Browne his copy of the 'Atlas' to Duchenne 1862; he asked for it to be sent back in his letter of 31 January [1870].

3 CD recorded Crichton-Browne's remarks on blotches and cerebral maculae (also known as meningitic streaks: see Landau ed. 1986) in epileptics in *Expression*, p. 326. Crichton-Browne refers to Armand Trousseau.

4 Trousseau 1868–72, 1: 459–61.

5 Bell 1844, p. 95.

6 CD cited Crichton-Browne for this information in *Expression*, p. 315.

7 CD cited Crichton-Browne's observation in *Expression*, p. 300. For CD's discussion of the contraction of the platysma myoides muscle, see *Expression*, pp. 298–303.

8 CD described this case in *Expression*, p. 185.

From Francis Galton 15 March 1870

42 Rutland Gate SW
March 15 70

My dear Darwin

Very many thanks for the information & books when I have got up the subject I will write again & will in the mean time, take all care of the books.[1]

I shall hope in a week from now to give you some news & by Saturday week definite facts about the rabbits. One litters or has littered today & all looks well with her. two others towards the end of the week, viz: Wed^y. & Saturday.[2]

I grieve to say that my most hopeful one was confined prematurely by 3 days having made no nest & all we knew of the matter was finding blood about the cage & the *head* of one of the litter.

She was transfused from yellow & the buck also from yellow.— Well the head was certainly much lighter than the head of another abortion I had seen. & was certainly *irregularly* colored. being especially darker about the muzzle, but I did not & do not care to build anything about such vague facts. & have not even kept the head.

As soon as ever I know **anything** I will write instantly & first to you. For my part, I am quite sick with expectant hope & doubt.

Ever very sincerely | F. Galton

DAR 105: 5–6

[1] No letter from CD to Galton from this period has been found, but CD had been in London from 5 to 12 March (Emma Darwin's diary (DAR 242)).

[2] Galton was attempting to test CD's hypothesis of pangenesis by transfusing the blood of one variety of rabbit into another and observing the colour of the offspring. See *Correspondence* vol. 17, letter from Francis Galton, 11 December 1869, and Galton 1871. According to CD's hypothesis, each cell in an organism produced particles (gemmules) that were capable of generating new cells; the gemmules circulated throughout the organism until required, and then congregated in the right place in order to reproduce or in some cases reconstruct parts. Galton evidently assumed that all gemmules circulated in the blood, and that transfusion of blood would therefore result in the transferring of hereditary information from the donor to the recipient. For accounts of the experiments, see Pearson 1914–30, 2: 156–66, and Bulmer 2003, pp. 116–18.

To Asa Gray 15 March [1870][1]

Down. | Beckenham. | *Kent. S.E.*
March 15th

My dear Gray

You & M^rs. Gray have been very good, & I thank you for your letters & enclosures.—[2] If Mrs Gray ever learns anything authentic about the nursing of her dog, I am sure she will inform me.— Two days ago my wife read a passage to me from Miss Mitfords life, minutely describing a dog, which had been nursed by a cat & which licked its paws! But as this is a second-hand account, it will not do to quote; & the description of the cat-like habits of this dog was too much even for my capacious gullet.[3]

Whenever you have any communication with Agassiz, pray give him my cordial thanks for his kind message & for his information, which I much feared would prove negative.[4] I wish with all my heart that I could feel that I deserved what Alex. A. says of me.[5]

We have just returned from a week in London, where we went for a rest to me, as I was pretty well worn out.[6] I find the man-essay very interesting but very difficult; & the difficulties of the Moral sense has caused me much labour.[7] But as Hooker disrespectfully says of me, "oh you will wriggle out of anything".[8] But as I tell him, he has become quite as good a wriggler as I am.— He is awfully busy & we did not see him, which was a great loss.

I suppose you will have seen some notice of Round Isl[d] off Mauritius: the diavolo himself would not be able to explain how the plants & animals got distributed on these 2 islands.—[9] I have just moved your Droseras into the Greenhouse; but we do not know whether they are dead or alive.—[10]

Your's ever affectionately | Ch. Darwin

Gray Herbarium of Harvard University (91)

[1] The year is established by the relationship between this letter and the letters from Asa and J. L. Gray, 14 February 1870.

[2] See letters from Asa and J. L. Gray, 14 February 1870.

[3] CD refers to Mary Russell Mitford and Mitford 1870. Mitford had been told of a King Charles spaniel that had been nursed by a cat (Mitford 1870, 2: 276–7):

> He is as afraid of rain as his foster-mother; will never, if possible to avoid it, set his paw in a wet place; licks his feet two or three times a day, for the purpose of washing his face, which operation he performs in the true cattish position, sitting up on his tail; will watch a mousehole for hours together; and has, in short, all the ways, manners, habits, and disposition of his wet-nurse, the cat.

[4] CD refers to Louis Agassiz. See letter from Asa Gray, 27 February and 1 March 1870.

[5] CD refers to Alexander Agassiz, whom he had met in 1869 (see letter to Fritz Müller, 1 December [1869]). It is not known what Agassiz had been saying about CD.

[6] The Darwins had been in London from 5 to 12 March 1870 (Emma Darwin's diary (DAR 242)).

[7] CD had been working on the second and third chapter of *Descent*, 'Comparison of the mental powers of man and the lower animals'.

[8] Joseph Dalton Hooker and CD had long joked about CD's ability to 'wriggle' out of difficulties; see for example *Correspondence* vol. 9, letter from J. D. Hooker, 7 November [1861].

[9] See letter to J. D. Hooker, 21 February [1870] and n. 2. Diavolo: devil.

[10] Gray had sent CD specimens of *Drosera filiformis*, the thread-leaved sundew, in 1869; see letter from Asa Gray, 20 October 1869 and n. 1.

To Edwin Ray Lankester 15 March [1870][1]

Down Beckenham Kent
March 15.

My dear Sir

I do not know whether you will consider me a very troublesome man, but I have just finished your book & cannot resist telling you how the whole has much interested me.—[2] No doubt, as you say, there must be much speculation on such a subject, & certain results cannot be reached; but all your views are highly suggestive, & to my mind that is high praise. I have been all the more interested as I am now writing on closely allied though not quite identical points. I was pleased to see you refer to my much despised child, Pangenesis who, I think, will some day under some better nurse turn out a fine stripling.—[3]

It has also much pleased me to see how thoroughly you appreciate (& I do not think that this is general with the men of science) H. Spencer:[4] I suspect that hereafter he will be looked at as by far the greatest living philosopher in England, perhaps equal to any that have lived.— But I have no business to trouble you with my notions.— With sincere thanks for the interest which your work has given me, I remain, | yours very faithfully, | Ch. Darwin

Photocopy
DAR 249: 92

[1] The year is established by the reference to Lankester's *On comparative longevity in man and the lower animals*, which was published in 1870 (Lankester 1870).

[2] There is an annotated copy of Lankester 1870 in the Darwin Library–CUL (see *Marginalia* 1: 482).

[3] Lankester discussed CD's hypothesis of pangenesis (see *Variation* 2: 357–404) in Lankester 1870, pp. 30–1, 35.

[4] Lankester discussed Herbert Spencer's 'physiological units' as an alternative to CD's 'gemmules' in Lankester 1870, pp. 32, 35–8.

From R. F. Albrecht[1] 16 March 1870

Hochgeehrter Herr!

Augenblicklich bin ich damit beschäftigt auf den Principien Ihrer Theorie der Artenbildung einen Beweis aufzubauen, auf welche Art und Weise die Flugorgane der Vögel entstanden sind. Dazu ist es aber unbedingt nothwendig, daß ich meiner Broschüre einen kurzen Abriss über die Gesetze der Abänderungen beifüge. Dadurch wird es mir aber unmöglich Ihre eigenen Worte zu vermeiden. Ich mochte ew. Wohlgeboren daher ersuchen, mir gefälligst dazu autorisiren zu wollen, weil ich sonst jedenfalls einen Proceß wegen Preßvergehen entgegensehen könnte.— Zugleich theile ich Ew. Wohlgeboren hierdurch mit, daß ich noch eine 2. Broschüre zu veröffentlichen gedenke, jetzt aber noch als Hirngespinst im Kopfe umhertrage, nämlich: "Über die Mittel—oder Uebergangsformen. Nach Beendigung meiner Arbeiten werde ich mir die Ehre nehmen, Ihnen dieselben zuzusenden.[2] Daß Sie den Anhängern Ihrer Theorie in Deutschland wohl zurufen könnten, wie Dr. A. B. Meyer[3] sagt: "Gott beschütze mich vor meinen Freunden, mit meinen Feinden will ich schon fertig werden", kann ich nur bestätigen.

Ich werde mich aber als solcher Freund und Anhänger Ihrer Theorie stellen, denen Sie dies zuzurufen nicht nöthig haben.

Zum Schluß möchte ich Sie noch auffordern, doch alle Ihre Schriften, die Sie überhaupt bis jetzt veröffentlicht haben, zusammen zu stellen und so zu veröffentlichen, und mit den Nachfolgenden ein Gleiches zu thun, damit dieselben einem deutschen, und überhaupt allen Völkern, mehr zugänglich werden, wie dies jetzt leider nicht der Fall ist. Ihrer Antwort entgegen sehend, | zeichne ich hochachtungsvoll | R. F. Albrecht.

Berlin, Bahnhofstraße 3. | Berlin, d. 16. März 1870.

DAR 159: 33

1 For a translation of this letter, see Appendix I. R. F. Albrecht has not been identified.
2 The publications have not been identified.
3 Adolf Bernhard Meyer.

From Francis Galton 17 March 1870

42 Rutland Gate SW
March 17 1870

My dear Darwin.

No good news.—

Bartlet[1] apprised me this morn^{g} that it was a popular prejudice that young rabbits might not be looked at, reasonable care being taken, so we opened 2 boxes & examined the litters. The first contained 4 dead young ones all true silver greys. The second contained 5 lively young ones all true silver greys. One however has a largish light colored patch on its nose, but Bartlet tells me this is not unusual with silver greys as the very tips of their noses are often white. However this patch is somewhat large & there are faint hopes, I think, that it may prove more considerable than Bartlet believes.[2]

I have one more litter yet to come & hope to send you the result by Monday ev^{g} post.[3]

I have coupled a new pair and re-coupled the 2 does whose litters have failed; one of them with a more suitable mate, & expect the following results

Date of expected litter	Buck transfused from rabbit colored as below	Doe transfused from
April 14	hare colored	hare colored
April 16	yellow	yellow
April 16	black & white	black & white

The quantity of blood transfused was only 1.25 per cent of the weight of the rabbits, which is the same thing as 30 oz of blood to an average man. I know this is a very small proportion of the whole amount of blood, but hope by a second operation on the old bucks & by improved operations on all the young ones to get a great deal more of alien qualities into their veins.

Very sincerely yours | Francis Galton.

DAR 105: 7–8

1 Galton refers to Abraham Dee Bartlett, the superintendent of the Zoological Gardens of London. Galton kept his rabbits in the Zoological Gardens partly so that he could be advised by Bartlett (Galton 1871, p. 395).
2 See letter from Francis Galton, 15 March 1870 and n. 2. Galton was trying to 'mongrelise' silver-grey rabbits by introducing blood from other breeds into them and then breeding from them (Galton 1871, p. 396).
3 Galton next wrote on Tuesday 22 March 1870.

From Georg Recht[1] 17 March 1870

Mist. Darvin!

Ich erlaube mir Ihnen die neuesten Resultate meiner Forschungen mitzutheilen, welche ich in einen einzigen Satz zusammenfassen kann—indem ich ausspreche, dass das Gesetz der Traegheit—vis inertiae—das Grundgesetz unserer Mechanik, und damit unserer Maschinenkunde, unserer Physik, unserer Astronomie und all unseres Wissens falsch oder ein Irrthum ist.

Nach diesem Gesetze sind alle Erscheinungen in der Natur Folgen eines primitiven Anstosses, den die Astronomen mit Tangentialkraft, die Chemiker mit Affinitaet & bezeichnen, und da dieser Stoss in der Astronomie nicht nachweisbar ist, denn die Hypothese von Laplace[2] ist unhaltbar, nehmen die Philosophen den Bewegungs—Zustand des Universums als den primitiv gegeben an, i.e. sie nehmen an, dass der gegenwartige Zustand des Universums ein ewiger ist.

Nach diesem erwaehnten Grundgesetze betrachten die Physiker alle Natur erscheinungen als bloss mechanische ohne weitere Thaetigkeit von Kraeften, waehrend sie meinem Grundgesetze nach alle dynamisch sind.

Die Mathematiker naemlich haben aus Arithmetik, Geometrie und einer besondern Mechanik ein eignes Universum construirt, in welchem sie rechnen koennen, und man hat dieses theoretische oder mathematische Universum als congruent oder identisch mit dem physischen Universum angenommen. Das ist der grosse Irrthum. Das Grundgesetz der analytischen oder mathematischen Mechanik ist:

Wenn ein Koerper einmal durch eine aeussere Ursache in den Zustand der Bewegung versetzt ist geradlinig oder rotirend oder beides zugleich—so bleibt er von selbst in diesem Zustande ohne weitere Thaetigkeit einer Kraft.

Diesem Grundgesetze liegen folgende Voraussetzungen unter:

a) Jedes Bewegungsobject nimmt die ihm zugedachte Geschwindigkeit momentan an ohne Zwischenstufen von Geschwindigkeiten;

b) Jedes Bewegungsobject nimmt jede Geschwindigkeit in einem und demselben in allen seinen Puncten seiner ganzen räumlichen Ausdehnung nach an.

Die naehsten Folgerungen aus obigem Grundgesetze aber sind:

1) Jede momentan thaetige Kraft erzeugt eine gleichmaessige geradlinige Bewegung;

2) Jede constante Kraft erzeugt eine beschleunigte, und zwar eine gleichmaessig beschleunigte Bewegung.

Diesen Gesetzen setze ich für physische Körper folgende entgegen:

1) Jeder physische Koerper nimmt eine jede Bewegung nur successiv von Theil zu Theil an, und ist folglich im Zustande der Bewegung *ausgedehnt*;— Beweis für die molekülar artige Zusammensetzung der Koerper, insbesondere aber für die *Verbindung* dieser Moleküle unter einander, welche in eine physikalische und in eine chemische zerfaehlt—

2) Jeder Koerper nimmt eine gewisse Geschwindigkeit nur successiv von Grad zu Grad und zwar von Null aus an.

In Folge des ersten Gesetzes ist jeder Koerper waehrend einer Bewegung elastisch erregt, und, da nicht nur Waerme die Koerper ausdehnt, sondern umgekehrt auch Ausdehnung Waerme bindet, in seinem Waerme zustande veraendert— man kann im luftleeren Raume eine Kugel durch Rotation an ihrem Aequator zum Glühen bringen, waehrend sie an ihren Polen kalt bleibt oder sich sogar abkühlt— Diese elastische Ausdehnung oder Zusammendrüchung, je nachdem die Kraft zieht oder schiebt tritt mit dem Erlöschen der Kraft in Thaetigkeit, verwandelt jede Bewegung eines physischen Koerpers in eine retardirende und bringt ihn in seinen primitiven Raumzustand in Bezug auf Groesse & Form und damit in Bezug auf seinen Waermezustand zurück, womit aber zugleich der Zustand der Ruhe eintritt.

Ich hatte Gelegenheit, Versuche auf der Eisenbahn anzustellen, und habe gefunden, dass die Retardirungszeit eines Zuges einzig und allein von dem elastischen Zustande der Zugstange abhaengt, wobei man die Bewegung derselben oder die Elastizitaetsgrenze wesentlich unterscheiden muss von der Groesse ihrer elastischen Kraft. Durch *Aufeinanderlegen* von Gummi-Ringen vermehrt man nur die Elastizitaetsgrenze, durch *Nebeneinanderlegen* die elastische Kraft.

Nach diesen Erklaerungen tritt nie & nirgens in der Natur eine Kraft verinzelt auf, sondern stetz zwei Kraefte in Verbindung mit einander und zwar eine äussere und eine innere und diese Verbindung ist nur der Gleichgewichtszustand derselben, und zwar nicht labiles, sondern stabiles Gleichgewicht, weil mit der Veraenderung der einen der beiden Kraefte sich zugleich die andere veraendert.

Ist der Anfang der Wirkungen zweier entgegengesetzter Kraefte, die sich das Gleichgewicht halten, ganz glichzeitig, so haben wir Gleichgewicht in Verbindung mit Ruhe, beginnt die eine erst, wenn die andere schon begonnen hat, so ergibt sich ein Zeitatom & damit ein Geschwindigkeitsatom oder Bewegung.

Jeder Körper, der durch eine Kraft in Bewegung versetzt wird, hat in seinen innern Eigenschaften demnach ein Streben zur Ruhe, das sich erst aeussert, wenn die Kraft ihre Wirkung schon begonnen hat, und kommt durch dieses Streben successiv zur Ruhe nach dem Erlöschen der Kraft i.e. wenn die Kraft nicht fortwährend erneuert wird. Darin liegt mein Grundgesetz der Bewegung. Aus diesem folgt:

1) Die gleichmässige Bewegung ist die Folge einer constanten Kraft oder einer continuirlichen *Krafterneuerung.*

2) Die beschleunigte Bewegung Folge einer continuirlichen *Kraftvermehrung.*

Geht man mit diesen Bewegungsgesetzen an die Betrachtung der Natur, so erscheinen alle Raethsel derselben gelöst, alle Geheimnisse derselben liegen aufgedeckt vor uns. Der Individualisirungs-Prozess von der Bildung der Weltkoerper an bis zur Entstehung der einfachsten Zelle herab wird klar, wenn man die Natur des Aetherkoerpers, und die wahre Groesse, Form & Bewegung der Atome desselben, so wie die der Moleküle damit in Verbindung bringt dadurch, dass man die physikalische & die chemische Vereinigung derselben zu besondern Koerpern & Individuen erfasst. Ueber alle diese Dinge bin ich vollstaendig ins Klare gekommen und alle diese Geheimnisse liegen so offen vor mir, p.e. das Wesen oder die Ursache der Schwerkraft, die man bisher bloss in ihren Wirkungen aber nicht in ihren Ursachen erkannte, wie mathematische Lehrsaetze; allein der Raum eines Briefes reicht nicht hin zu deren Auseinandersetzung. Nur für eine Sache erlaube ich mir noch Ihre Geduld in Anspruch zu nehmen, und dieses ist der Begriff der Elastizitaet.

Gewöhnlich versteht man unter Elastizicitaet eines Koerpers das Bestreben desselben, die primitive Groesse & Form, wenn diese seine Raumelemente durch eine aussere Ursache veraendert wurden, wieder herzustellen.

Die Groesse eines Körpers haengt ab von der Groesse seiner Drei Dimensionen, die von einander abhaengig sein können wie p.e. in einer Kugel, oder die auch ganz unabhaengig von einder sind p.e. in einem Ellipsoid mit dri Axen oder in einem Parallelepipedon.

Die Form eines Koerpers besteht in den Winkeln, welche diese seine Dimensionen unter einander bilden.

Demnach ist vor allem zu unterscheiden: 1) Groessenelastizitaet und 2) Formelastizicitaet; Erstere ist entweder bloss einfach linear oder zweifach linear—Flächenelastizitaet—oder dreifach linear—Körperelastizitaet—

Da nun jede Dimension eines Körpers wieder entweder vergrössert oder verkleinert werden kann, so müssen wir jede Elastizitaet wieder unterscheiden in Zug—oder Druck—Elastizitaet oder in positive & negative.

Da aber jeder Körper nicht nur als solcher elastisch ist, sondern auch seine Moleküle, so ist auch die Körper-Elastizitaet zu unterscheiden von seiner Molekülar-Elastizitaet.

Was die Formelastizitaet anbelangt, so laesst sich jeder Winkel vergrössern oder verkleinern, und diese Elastizitaet bedingt eine Biegung des Koerpers, so lange die Winkelebenen unverändert bleiben. Jede solche Ebene kann sich aber auch entweder nach der einen oder nach der andern Seite drehen, und daraus entsteht die doppelte Rotations-Elastizitaet. Die Biegungs- und die Rotations-Elastizitaet bilden zusammen die Formelastizitaet.

Wenn wir nun p.e. einen Körper streken, so aendert sich mit seiner Laengenelastizitaet entweder auch seine Breitenelastizitaet oder nicht; jedenfalls aber ist mit der Längenbewegung eine Veraenderung seiner Molekülarverbindung bedingt, diese müssen sich daher einander nähern und von einander entfernen koennen ohne ihre Verbindung zu verlieren. Diese Molekülarbewegung kann nur Rotation sein, so dass in den Koerpern bei ihrer Veraenderung stets Rotation mit Translation verbunden ist.

Elastizitaet ist latente Geschwindigkeit oder latente Waerme oder Folge von latenten Aether, der damit in das übergeht, was man gewöhnlich unter Stoff oder Materie versteht, während der Aether selbst nur aus unveraenderlichen & daher unelastischen Atomen besteht, welche nur durch ihre Geschwindigkeit und durch ihre Verbindung die Grundlage des Universums bilden, und zwar die sogenannte geistige und die materielle Grundlage.

Was das Streben einer Materie anbelangt, und was als solches mit dem Begriffe der Materie nicht vereinbar ist, so ist darunter ein Bestreben nach Geschwindigkeit oder nach Bewegung zu verstehen und ist daher bedingt durch das Vorhandensein eines Geschwindigkeitsatomes ohne dass Bewegung eintritt. Besteht p.e. in den Molekülen respective in deren Elastizitaeten ein Geschwindigkeitsatom ohne dass dasselbe im Stande ist den Körper zu zertrümern oder in eine Entwicklung überzuführen, so besteht im Körper ein Streben.

Ist ein Koerper p.e. von der Anziehung durch einen andern Koerper ergriffen, so ist er in seiner Molekülarelastizitaet erregt, welche eine Bewegung des Gesammtkörpers herbeiführt, wird diese durch eine Unterlage aufgehoben, so bleibt die erregte Molekularelastizitaet als Streben nach Bewegung.

Was die Elastizitaet in den Naturkörpern anbelangt, so sind die unaeren Körper einfach, die binaeren zweifach, ü. die ternaeren dreifach elastisch oder dreifach d.i. räumlich linear oder nach allen Richtungen linear elastisch. Diese Elastizitaet bezieht sich aber auf die Moleküle, wahrend diese elastische Uterscheidung der Körper dieselben in feste, flüssige und luftförmige zertheilt.

In der organischen Natur sind die festen Körper dargestellt, durch die linear elastischen Pflanzen, die zweifach elastischen durch das Thierreich, die nach allen Richtungen linear elastischen oder luftförmigen durch den Menschen.

Im Thiere bilden die Muskeln des Ruchgrates und der Extremitaeten rechte Winkel daher die Elemente einer Fläche, während Ruchgrat, das kleine Gehirn und das grosse Gehirn in einer einzigen Geraden liegen, daher das Thier geistig nur linear elastisch ist. Der Mensch ist aber körperlich linear und geistig zweifach elastisch, die Ruchgratlinie und die Gehirnlinie stehen auf einander senkrecht.

Zum Schlusse erlaube ich mir nur noch den Begriff der *allgemeinen* Bewegung festzusetzen.

Die erste und die Grundbewegung ist die gleichmaessig geradlinige, die zweite die gleichmaessige Rotation. die allgemeine geradlinige Bewegung geht von einem Puncte aus nach zwei entgegengesetzten Richtungen, die allgemeine Rotation gleichfalls nach zwei entgegengesetzten Richtungen. In der Schraube oder in der Spirallinie sind die beiden Bewegungen als einseitige mit einander verbunden, während in der DoppelSchraube, in welcher sich zwei Schraubenlinien durch entgegengesetzte Rotationen nach den zwei entgegengesetzten Richtungen auch translatorisch fortbewegen, die allgemeine lineare Bewegung vorgestellt ist.

Betrachtet man einen Körper, in welchem jedes Atom mit jedem andern Atome verbunden ist, und traegt die allgemeine lineare Bewegung auf jede Verbindungslinie dieses Körpers über, so hat man einen Körper, in welchem von jedem Puncte, resp. von jedem Atome aus nach allen Richtungen allgemeine lineare Bewegungen stattfinden, und dieser Koerper ist der Aetherkoerper, der damit alle Bedingungen des sogenannten geistigen & materiellen Lebens des Universums enthaelt nebst allen Bedingungen des ganzen Individualisirungsprocesses.

Was die allgemeine lineare Bewegung mit entgegengesetzten Rotationen und entgegengesetzten Translationen anbelangt, so sehen wir dieselbe an jedem Lichstrahl, bei jeder electrischen und jeder Waermebewegung, so wie in jedem chemischen Processe. In jedem einzelnen Lichtstrahl sehen wir Stoff sich fortbewegen nach zwei entgegengesetzten Richtung in Verbindung mit zwei entgegengesetzten Rotationen, die man für Wellenbewegung haelt.

Wir sehen diese Bewegungen aber auch im Thierreiche in den Ubergaengen der Landthiere in Wasserthiere und umgekehrt der Wasserthiere in Landthiere; der Landthiere in Luftthiere und von den Luftthieren in Landthiere; der Wasserthiere in Luftthiere und von den Luftthieren in Wasserthiere abgesehen von der Wanderung der Organe von Gattung zu Gattung. So p.e. beginnt das Geschlechtsleben bei den Menschen mit dem Gesichtssinne, bei den Landthieren dagegen mit dem Gerüchssinne, waehrend es bei den Vögeln mit dem Gehörssinne beginnt. der Gesichtssinn endet bei den Vögeln in den Nierenpuncten. etc. etc.

Euer Hochgeboren würden mich sehr glücklich machen, wenn Sie diesen Zeilen einige Zeit widmen, und dieselben zur Kenntniss der Royale Societe bringen würden. Unendlich glücklich würde ich aber sein, wenn ich eine Antwort über deren Aufnahme erhalten würde. Ich getraue mir kaum zu ahnen, dass mir je das Glück beschieden sein wird, persönlich vor der Royale Societe zu erscheinen und irgend welche an mich gestellte Fragen beantworten zu koennen.[3]

Mit ausgezeichnetster Hochachtung und Verehrung | Euer Hochgeboren | ergebenster Diener | Dr. Recht | in München.

den $\frac{17}{3}$ 70.

DAR 176: 77

1 For a translation of this letter, see Appendix I.

2 In his letter of 1 March 1869 (*Correspondence* vol. 17), Recht had discussed the application of physical principles concerning motion to living systems. Recht refers to the nebular hypothesis formulated by by Pierre Simon, marquis de Laplace, in *Exposition du système du monde* (Laplace 1796).

3 No papers by Recht appear in the *Proceedings of the Royal Society of London*. Recht had asked CD to communicate an earlier monograph (Recht 1869) to a scientific society (see *Correspondence* vol. 17, letter from Georg Recht, 9 February 1869).

From John Jenner Weir 17 March 1870

6 Haddo Villas | Blackheath, SE
17th March 1870

My Dear Sir

Some doubt having been thrown on the subject of the foals of Lord Martins mare mentioned in your work on Domesticated animals, has caused me to be ever on the look out for any similar case.—[1]

I heard the point disputed by Mr. Busk;[2] Sir John Lubbock taking the opposite and, in my opinion, the true view.—

I mentioned to the latter the observations I am about to detail to you & he said they should be published, but I doubt whether they are worth it.—

My friend Mr. Lethbridge of Blackheath, has a horse bred by Lord Mostyn at Ruthin, from a mare which had previously had a foal by a Quagga, and he has been kind enough to afford me every facility for examining it.—[3]

The color of the horse is dun with dark mane, tail & lower part of the legs, it has "a dark stripe down the centre of the back, faint stripes on the forehead between the eyes, plainly striped on the inner side of the fore legs, and rather more faintly on the hind.— I could see no shoulder stripe.—

The mane grows much lower on the forehead than in Equus caballus, tæniopus, onager, hemionus or hemippus, but not so low as in Zebra, Burchellii or Quagga.—[4]

The hoofs are proportionately longer than in the horse, so much so that the Farrier who first shoed it and who knew nothing of its origin, said "Had I not seen I was shoeing a horse I should have thought I was shoeing a donkey"

M$^{r.}$ Lethbridge tells me it is a very intelligent animal, rather mulish in some of its habits, making for instance more use of its ears by moving them, than a horse would.—

My Theory is this I assume the truth of your doctrine of Pangenesis, & the existence of Gemmules.—[5]

Now the fœtus partakes of the specific character of both parents; in the case of a hybrid therefore of two species, and further the gemmules of two species must be existing & reproducing themselves, throughout the whole of the foetus & in every drop of its blood.—

The circulation of the blood of the dam & foetus being during the whole period of gestation identical, does it not necessarily follow that after parturition, the dam should still retain the gemmules which have been produced from the male element in the foetus.—

It appears to me that the fact could not be otherwise & that the female herself by being impregnated by a distinct species has become hybridized.—

My theory indeed leads to the queer paradox, that it is possible for offspring to have more than one male parent

It may be fanciful on my part but I cannot but think that my theory accounts for what has been so often noticed that persons long married grow like each other.—

Trusting you will not consider I claim anything more for myself than the application of your grand idea of Pangenesis.—

Believe me | My Dear Sir | Yours very faithfully | J Jenner Weir

C Darwin Esq$^{r.}$

DAR 181: 81

CD ANNOTATIONS

2.1 Sir … it.— 3.2] *crossed pencil*
4.1 My … the legs, 5.1] *crossed ink*; 'Colour of **parents**' *added blue crayon*
5.4 hind.—] 'legs' *interl ink*
5.4 I could see] *del ink*; 'but no' *added ink*
6.1 lower] 'down' *interl ink*
6.1 than in Equus] 'the Horse,' *added above* 'Equus'; *crossed indicating* 'but not' *to follow* 'the Horse,'
6.2 Burchelii] *del ink*
8.1 M^{r}. … animal] *crossed pencil*
9.1 My Theory … Pangenesis.— 15.2] *crossed pencil*
Top of letter: 'Effect of 1st impregnation by Quaggas like Ld Martins' *pencil*

1 In *Variation* 1: 403–4, CD cited a description by Lord Morton (George Sholto Douglas) of a chestnut Arabian mare who bore a hybrid to a quagga, and subsequently bore two partly dun-coloured, partly striped colts to a black Arabian horse.

[2] George Busk.

[3] Weir refers either to John Christopher Lethbridge or to his son Christopher Lethbridge, and to Edward Mostyn Lloyd-Mostyn, second Baron Mostyn. Ruthin is a town in Denbighshire, Clwyd, Wales. CD cited Weir for this information in *Variation* 2d ed., 1: 435–6.

[4] *Equus caballus*: the horse; *E. onager*: the onager; *E. hemionus*: the Asian wild ass or kulan; *E. hemippus*: now *E. onager.* On *E. taeniopus*, see Harper 1940, who concludes that the name is best regarded as indeterminable. The zebra is *E. zebra*, and the quagga *E. quagga*. *Equus burchellii* is Burchell's zebra.

[5] For CD's hypothesis of pangenesis, see *Variation* 2: 357–404.

To J. J. Weir 17 March [1870]

Down. | Beckenham | *Kent. S.E.*
Mar 17th

My dear Sir

It is my decided opinion that you ought to send an account to some Sci Socy. & I think to the Royal Socy.— I wd. communicate it, if you so decide.—[1] You might give as a preliminary reason the publication in the Transats. of the celebrated Morton case & the Pig case by Mr Giles..—[2] You might also allude to the evident physiological importance of such facts, as bearing on the theory of Generation: whether it wd be prudent to allude to despised Pangenesis, I cannot say; but I fully believe Pan: will have its successful day.— Pray ascertain carefully the colour of the dam & sire.— see about duns in my Book. vol I p. 55[3] The extension of the mane & the form of hoofs are grand new facts.— Is the hair of your horse at all curly: for observe case given by me Vol. 2. p. 325, from Azara of correlation of form of hoof with curly hair?—[4] See, also, in my Book; Vol I p. 55. Vol. 2. p 41 how excessively rare stripes are on faces of horses in England.— Give age of your Horse.—

You are aware that Dr Carpenter & others have tried to account for the effect of a first impregnation, from the influence of the blood of the crossed embryo;[5] but with Physiologists who believe that the reproductive elements are actually formed by the reproductive glands this view is inconsistent.— Pray look at what I have said in Dom: Animals. Vol. I p. 402–405. Against this doctrine: it seems to me more probable that the gemmules affect the *ovaria alone*.— I remember formerly speculating, like you, on the assertion that wives grow like their husbands; but how impossible to eliminate effect of imitation & same habits of life &c.— Your letter has interested me *profoundly*.

In Haste. | Very sincerely your's | Ch Darwin

PS. Since publishing I have heard of additional cases—very good one in regard to Westphalian pigs crossed by English boar, & all subsequent offspring affected given in Illust: landwirth: Zeitung 1868 p. 143.[6]

I have shown that mules are often striped though neither parent may be striped, due to ancient reversion— Now Fritz Müller writes to me from S. Brazil "I have been assured by persons who certainly never had heard of Ld. Morton's mare, that mares which have borne hybrids to an ass, are particularly liable to produce afterwards striped ass-colts."[7] So a previous fertilisation apparently gives to the subsequent offspring a tendency to certain characters, as well as characters actually possessed by the first male.—

In the Reprint, (not *called* a 2^{d} Edition) of my Domestic Animals, I give good additional case of *subsequent* progeny of Hairless dog, being hairy, from effects of first impregnation.—[8]

P.S. 2^{d}.

The suggestion no doubt is superfluous, but you ought, I think, to *measure* extension of mane beyond a line joining front or back of ears & compare with horse.

Also to measure (& give comparison with horse) length, breadth & depth of Hoofs.—

Postmark: MR18 70
American Philosophical Society (Getz 7138)

1 It seems that Weir did not submit his account to the Royal Society of London (see letter from J. J. Weir, 17 March 1870, and letter to J. J. Weir, 14 June [1870]).

2 CD refers to George Sholto Douglas (Lord Morton) and Douglas 1820 and to Daniel Giles and Giles 1820. See also letter from J. J. Weir, 17 March 1870 and n. 1. CD discussed Giles's pig in *Variation* 1: 404; the pig was a black and white Essex sow, and was bred first with a chestnut-coloured boar. Later, she was bred with a black and white boar, and some of the piglets were chestnut coloured.

3 CD discussed the dun colour in horses in *Variation* 1: 55–9.

4 CD refers to Félix d'Azara and Azara 1801.

5 CD refers to William Benjamin Carpenter.

6 CD refers to *Illustrirte landwirthschaftliche Zeitung*, 17 November 1868, p. 143 (see *Variation* 2d ed., 1: 436 n. 152). This journal was published in Leipzig: a copy may have been sent by Hermann Müller, who lived in Westphalia. No copy of the newspaper or article has been found in the Darwin Archive–CUL, and CD did not describe the case in the second edition of *Variation*, although he gave the reference above. The letter in which Müller gave this information has not been found.

7 On stripes appearing on mules and other animals owing to reversion to an ancestral type, see *Variation* 2: 42–3. CD cited the information from Fritz Müller in *Variation* 2d ed., 1: 436 n. 152.

8 See *Variation* (2d printing) 1: 404 n. 138, and *Variation* 2d ed., 1: 436 n. 152. The first edition of *Variation* was reprinted with some additions and corrections in early 1868 after the first printing had sold out (see *Correspondence* vol. 16, letter to John Murray, 9 February [1868]).

To James Crichton-Browne 18 March 1870

Down,
March 18, 1870.

My dear Sir

It had occurred to me that you might be ill, and I was thinking of writing to Dr. Maudsley to enquire.[1] I fully sympathise with you in the heavy losses to which you refer, and quite understand how little you could feel inclined to employ yourself on miscellaneous subjects. I thank you cordially for your extremely kind note and offer of further assistance; but as I know how hard worked you must be, I will try and be reasonable. The essays which you have sent are exactly what I wanted and the fullest which I have received. Your results about blushing closely agree with what Mr. Paget has observed; but all that you tell me about cerebral affections influencing the action of the skin is quite new to me. Mr. Paget after numerous observations has noticed *one* woman who blushed on the femora and other parts of the body, and I must enquire whether she was subject to epilepsy or anything of the kind.[2] I will ask two questions about your blushing cases: does the young lady who is so great a blusher blush over her arms; and does she frequently wear short sleeves? Was the epileptic woman who blushed over the upper part of her bosom a lady, so as to have been in the habit of wearing a low gown? With respect to the platysma I infer from your observations that Duchenne's statement that its contraction expresses great fear or horror is extremely doubtful.[3] Our imagination easily leads us astray on such a subject. But I should be very much obliged if you would attend to this muscle in any patients suffering from extreme terror and horror. With respect to the grief muscles you speak twice or thrice in your MS. of the upper eyelid being angularly drawn upwards; this is quite new to me; and the suspicion has occurred to me that you may have written by mistake eyelid for eyebrow. I daresay I shall receive Duchenne early next week;[4] and I shall then anxiously look for any remarks, however few, which you may have made on any of the plates. Pray believe that I feel grateful for your kindness and all the trouble which you have taken in giving me most valuable information.

Yours sincerely | Ch. Darwin.

Pray never hurry yourself in answering my questions: your apologies made me feel guilty.

DAR 143: 330

1 See letter from James Crichton-Browne, 15 March 1870. CD refers to Henry Maudsley, who had introduced Crichton-Browne to him (see *Correspondence* vol. 17, letter from Henry Maudsley, 20 May 1869).

2 In *Expression*, p. 313, CD wrote that James Paget had never himself seen a blush that extended below the upper part of the chest; however, Paget had heard of a case in which a little girl blushed on her abdomen and the upper part of her legs (*Expression*, p. 314).

3 Guillaume Benjamin Amand Duchenne asserted that the contraction of the platysma, in conjunction with certain other muscles, expressed fear or horror (Duchenne 1862, pp. 95–108).

4 CD had lent Crichton-Browne his copy of the 'Atlas' to Duchenne 1862; see letter from James Crichton-Browne, 15 March 1870 and n. 2.

From Francis Galton 22 March 1870

42 Rutland Gate. S.W.
March 22. 1870

My dear Darwin

Another litter;— this time of 4,— and all of them are true silver greys.— Also, one of the does (mentioned in my last letter as transfused from a black & white) is dead.[1]

My stud now stands as overleaf. I call each silver grey by the name of the colour of the rabbit from which it has been infused. I also give the particulars of my first batch. You will see that there was much less variety in my pairs then, than there is now.

I hope to try a new mode of transfusion upon a wholly new stock, taking younger rabbits and putting much more alien blood into them.

Ever very sincerely y^s. | F. Galton

1st. batch of silver greys littered between March.13 and March.19.

Name of buck	Name of doe	result
yellow No.2	yellow	abortion, Eat up litter all but one head.
ditto	black and white	4 dead young all silver grey
ditto	white	5 healthy young all silver grey but one with a white spot on nose
ditto	Himalya	4 healthy young all silver grey

2nd. batch of silver greys to litter between April 14th. and, say, April 30th.

		litter due
yellow No. 2	yellow	April 16
black & white	grey & white	April 21
Hare coloured	Hare colored	April 14
Yellow No.1	White	} not yet coupled
Angola	Himalya	} not yet coupled

DAR 105: 9–10

[1] See letters from Francis Galton, 15 March 1870 and n. 2, and 17 March 1870.

To Frances Power Cobbe 23 March [1870?][1]

Down Beckenham | Kent.
Mar 23.

My dear Miss Cobbe

It was very good of you to send me nolens volens Kant, together with the other book.[2] I have been extremely glad to look through the former. It has interested me much to see how differently two men may look at the same points, though I fully feel how presumptuous it sounds to put myself even for a moment in the same bracket with Kant;—the one man a great philosopher looking exclusively into his own mind, the other a degraded wretch looking from the outside thro' apes & savages at the moral sense of mankind.

I have glanced thro' the cut pages of Despine, but he appears to me far from being an original & vigorous thinker, but many of his facts about criminals seem very curious.

The book shall be sent to London by our carrier tonight, & thence to you by Parcels Del. Co.[3]

With sincere thanks | pray believe me | yours very sincerely | Ch. Darwin

Huntington Library, Art Gallery and Botanical Gardens, San Marino (CB 385)

[1] The year is conjectured from the fact that CD met Cobbe in Wales in July 1869 (see *Correspondence* vol. 17, letter to J. D. Hooker, 8 July [1869], and Cobbe 1904, pp. 485–7). According to Cobbe, she had met CD before that visit, but her description of the sequence of events suggests that this letter was written in 1870.

[2] Cobbe may have sent CD Immanuel Kant's *Metaphysics of ethics* (Kant 1836). Cobbe had recommended the book to him and offered to lend him a translation in Wales in July 1869; he had declined it, but she sent it anyway (Cobbe 1904, p. 487). CD cited this work in *Descent* 1: 70–1 and n. 3. The 'other book' was probably Prosper Despine's *Psychologie naturelle* (Despine 1868). CD mentioned this book in *Descent* 1: 92 n. 22.

[3] In 1870, 23 March was a Wednesday; the carrier service to London left Down at 2 A.M. on Thursday (*Post Office directory of the six home counties* 1870). CD also refers to the London Parcels Delivery Company.

To Albert Günther 23 March [1870][1]

Down, Beckenham | Kent
March 23

Dear Günther

As I do not know Mr. Ford's address, will you hand him this note, which is written solely to express my unbounded admiration of the woodcuts. I fairly gloat over them. The only evil is that they will make all the other woodcuts look very poor!

They are all excellent, & for the feathers I declare I think it the most wonderful woodcut I ever saw: I cannot help touching it to make sure that it is smooth. How I wish to see the two other & even more important ones of the feathers & the four other Reptiles &c.— Once again accept my very sincere thanks for all your kindness. I am greatly indebted to Mr Ford. Engravings have always hitherto been my greatest misery & now they are a real pleasure to me.—

Yours very sincerely | Ch. Darwin.

I thought I shd have been in press by this time, but my subject has branched off into sub-branches, which have cost me infinite time, & Heaven knows when I shall have all my M.S ready, but I am never idle.—[2]

P.S. Please ask Mr Ford to keep my rough sketches of feathers & my notes, as I am much puzzled what to do about the letters of reference to the different marks & dots.—[3]

Gunther 1975, p. 32

[1] The year is established by the reference to George Henry Ford's woodcuts for *Descent*. Ford seems to have started work on the woodcuts of feathers in September 1869 (see *Correspondence* vol. 17, letter to Albert Günther, 27 September [1869]). *Descent* was published early in 1871 (R. B. Freeman 1977).

[2] CD had originally planned a 'chapter on man' as part of *Variation*: it eventually grew into two works, *Descent* and *Expression* (see *Correspondence* vol. 15, letter to J. D. Hooker, 8 February [1867]).

[3] The illustration that CD refers to is presumably *Descent* 2: 146, fig. 58.

To ? 23 March [1870–1][1]

Down. | Beckenham | *Kent. S.E.*
March 23d

Dear Sir

I feel greatly obliged by your offer, but I really have not the least claim on your kindness, & I would suggest that you w[d] do better to present an additional copy to some one who attends more to Physics than I am able to do.—

I beg you, however, not to think that I am in the least less obliged by your offer, for declining it.

I remain, Dear Sir | Yours faithfully & obliged | Ch Darwin

American Philosophical Society (B/D25.271a)

[1] The year span is established by the form of the address, with Bromley crossed out and Beckenham added by hand; CD used this form between April 1869 and May 1871. The correspondent and the work he sent have not been identified.

To William Henry Flower 25 March [1870][1]

Down Beckenham | Kent
Mar 25

My dear M[r] Flower

I am much obliged to you for having sent me yr Lecture, of which I had heard & was anxious to see. It strikes me as in every way very good & very well written. I am sure I feel much obliged to you for the highly honourable manner in which you mention my name.[2]

The feet of the Koala & Kangoroo is one of the most beautiful cases I ever heard of.[3] Such facts will do more to convince unbelievers than pages of general reasoning. I am glad you noticed the curiously false argument in the Month about the teeth of the dog & Thylacinus: I had selected this in my M.S. as an excellent instance of analogical resemblance.[4] I c[d] have answered, I think satisfactorily, many of the objections advanced in this article; but my whole time w[d] be wasted if I once began to answer objectors.

With very sincere thanks for the great interest & instruction which you have afforded me Pray believe me | yours very sincerely | Ch. Darwin

B. J. Harrison (private collection)

[1] The year is established by the reference to Flower 1870; see n. 2, below.

[2] Flower had given the introductory lecture to the course of comparative anatomy at the Royal College of Surgeons on 14 February 1870; it was published in the *Medical Times and Gazette* for 19 February (Flower 1870). Flower mentioned CD along with Alfred Russel Wallace as the originators of the theory of the origin of species by means of natural selection (Flower 1870, p. 196).

[3] In *Origin* 6th ed., p. 382, CD cited Flower for the information that the hind feet of the kangaroo, the koala, and the bandicoot were constructed on the same plan, despite being used for different purposes. See also Flower 1870, pp. 197–8.

[4] The July 1869 issue of the *Month* included the first part of an anonymous article by St George Jackson Mivart ([Mivart] 1869) titled 'Difficulties of the theory of natural selection'. The author argued (p. 49):

> Nothing can be more remarkable than the identity of structure between certain of the teeth of the large predatory marsupial called the thylacine, or Tasmanian wolf, and those of the common dog. The resemblance is so complete that community of descent is at once most forcibly suggested, and yet on the assumption adopted, the thylacine may be

> closely related genetically to the kangaroo and its allies, but must be separated from the dog by an abyss.

In the sixth edition of *Origin*, published in 1872, CD commented (p. 374):

> Numerous cases could be given of striking resemblances in quite distinct beings between single parts or organs, which have been adapted for the same functions. A good instance is afforded by the close resemblance of the jaws of the dog and Tasmanian wolf or Thylacinus,—animals which are widely sundered in the natural system. But this resemblance is confined to general appearance, as in the prominence of the canines, and in the cutting shape of the molar teeth.

CD added that Flower had come to the same conclusion (*ibid.*, p. 375). See Flower 1870, p. 199.

From F. P. Cobbe 28 March [1870?][1]

March 28th

Dear Mr. Darwin,

I feel very proud of having inveigled you into "looking through" Kant—[2] Though I cannot quite say, like one of his disciples, "God said: Let there be Light, & there was—the Kantian Philosophy" yet I have retained for these twenty years a most lively sense of gratitude to him for helping me to find (or think I found) a stepping stone or two in the Slough of Despond[3]

I more than suspect you of a smile in your beard when you write of him as "a great philosopher looking exclusive into his own mind"— But surely may I not argue that, after all, his mind, & that of another philosopher I could name are things not wholly undeserving of attention,— phenomena quite as much needing to be studied & accounted for, say, as even our beloved dogs? We poor humble learners who would fain be the most docile of your scholars, see one of you driving complacently down the "high priori road", & the other with infinite skill progressing on the solid causeway of material facts— But are you never going to unite your lines of thought & let us see how metaphysics & physics form one great philosophy?—

Pray forgive dear Mr Darwin, my infinite impudence! Though I attended on Saturday a most successful Woman's Rights Meeting[4] I am of opinion that our Ancient privilege of talking nonsense even to those we most deeply honour, is one not to be parted with on any terms!—

I enclose a little notice embodying what I thought the point of those "Cut pages" of Despine—[5] Do not think of acknowledging this or returning *that*—

Most truly your's | Frances Power Cobbe

DAR 161: 186

[1] The year is conjectured from the relationship between this letter and the letter to F. P. Cobbe, 23 March [1870?].

[2] Cobbe refers to Immanuel Kant. See letter to F. P. Cobbe, 23 March [1870?] and n. 2.

[3] The disciple of Kant was Carl Ludwig Fernow (Schaub 1925, p. 13). The quotation as given by Cobbe also appeared in an article on German theology in the *Christian Observer* (1855): 449. Cobbe had been inspired by Kant's *Metaphysics of ethics* (Kant 1836) to write her first book, *The theory of intuitive morals*

([Cobbe] 1855–7). Cobbe wrote the book late at night and published it anonymously to avoid offending her father, who held traditional religious views (*ODNB*).

4 The meeting was held in Hanover Square on 26 March 1870 and reported in the *Times*, 29 March 1870, p. 9. John Stuart Mill and John Morley attended; none of the women present were named in the report.

5 Cobbe probably refers to Despine 1868; see letter to F. P. Cobbe, 23 March [1870?] and n. 2.

From Fritz Müller 29 March 1870

Itajahay, S[a] Catharina, Brazil
March 29. 1870.

My dear Sir

I have been detained from answering sooner your letter of Decbr. 1st, by an excursion into our primeval forests and afterwards by a severe illness of one of my daughters.[1] And now, I must first express my cordial thanks for a copy of N° 1 of "the Academy" and for the German translation of Wallace's exceedingly interesting work on the Malay Archipelago.[2] One of the points, which have interested me most in this work, is the dimorphism and mimicry of several butterflies.[3] My attention having thus been called to our endemic butterflies, I have met also with several cases of close ressemblance of species belonging to distinct genera, which must probably be attributed to mimicry. There is, for instance a very common Pieris (1), and three or four other rather rare species, when seen from above, are more or less closely imitating the colours of this Pieris.—

I am much obliged for your account on Eschscholtzia;[4] the influence of external conditions on the self-fertility of this ⟨*foot of page, probably containing diagram, excised*⟩ species is very curious; it is shown in a striking manner by a plant, which I had raised 1868 from your seeds and which only flowered 1869.—[5] The other plants from the same seeds, which flowered the first year (1868) had retained self-fertility in a small degree; but this plant, which flowered only the second year, and had thus been exposed for a far longer time to our climate, proved perfectly self-sterile; it has been left uncovered, but growing at some distance from the other plants, which flowered at the same time, it produced by itself only a few very poor pods containing not even a single seed. Two flowers fertilised by pollen of a distinct plant yielded large pods, one of which contained 115 seeds.—

I have been much pleased by hearing from Alex. Agassiz, that he is now inclining to your views.[6] I think very high of him as one of the most skilful and conscientious observers. His "Embryology of the Starfish" is a master-piece.—[7]

I gave you, in my last letter, some cases of plants, the first flowers of which had a larger number of petals or stamens than the later ones.[8] I have since found, in a tree of C⟨assia⟩ multijuja, that the first flowers of ⟨*foot of page excised*⟩

Incomplete
DAR 76: B36

CD ANNOTATIONS
1.1 I have . . . Pieris.— 1.11] *crossed blue crayon*

2.7 it has ... seeds.— 2.11] *double scored red crayon*
3.1 I have ... master-piece.— 3.3] *crossed blue crayon*
4.1 I gave ... flowers of 4.3] *crossed ink*

1 Müller refers to the letter to Fritz Müller, 1 December [1869] (*Correspondence* vol. 17). Müller had six surviving daughters (see West 2003).
2 The first issue of the *Academy*, dated 9 October 1869, contained a review of the English translation of Müller's *Für Darwin* (Dallas trans. 1869; *Academy* 1 (1869–70): 14–15). Müller also refers to Wallace 1869c (see *Correspondence* vol. 17, letter to Fritz Müller, 1 December [1869]).
3 See Wallace 1869a, 1: 199–207.
4 See *Correspondence* vol. 17, second enclosure to letter to Fritz Müller, 1 December [1869].
5 See *Correspondence* vol. 16, letters from Fritz Müller, 22 April 1868, 9 September 1868, and 31 October 1868, and *Correspondence* vol. 17, letters from Fritz Müller, 12 January 1869 and 18 October 1869.
6 Müller refers to Alexander Agassiz (see also *Correspondence* vol. 17, letter to Fritz Müller, 1 December [1869]).
7 A. Agassiz 1864.
8 See *Correspondence* vol. 17, letter from Fritz Müller, 18 December 1869.

From H. E. Darwin [30 March 1870][1]

I can swear to Mildred's[2] having the knots in the elevated corners of her eyebrows. The wrinkles were hardly to be seen. The corners of her mouth down. This was before crying, and when she cried the look ceased.[3]

DAR 162: 69

CD ANNOTATIONS
1.1 Mildred's] 'Langton' *interl pencil*
Top of letter: 'March 30— 1870 Henrietta' *added pencil*
End of letter: 'Analogous remarks were made on Margarets child'[4] *added pencil*

1 The date is established by CD's annotation at the top of the letter.
2 Henrietta refers to Mildred Langton, CD's nephew's daughter.
3 CD discussed the expression of grief in children in *Expression*, pp. 190–5.
4 CD refers to his niece Margaret Susan Vaughan Williams. See *Correspondence* vol. 17, letter from M. S. V. Williams to H. E. Darwin, [after 14 October 1869]. Her child was Hervey Wedgwood Williams.

From Armand de Quatrefages[1] 30 March 1870

Paris
30 Mars 1870

Monsieur et cher confrère

Voila bien longtemps que je veux vous écrire et vous dire la joie que m'a fait éprouver votre dernier lettre. Mais—nous disons en francais que l'enfer est pavée de bonnes intentions—et je me suis laissé aller a ajouter une pierre au dallage du sejour infernal. Ma longue absence de l'année dernière m'avait laissé un tel arrière que j'en suis à peine sorti et entrainé d'une chose à l'autre, j'ai négligé bien des devoirs.

Votre lettre, je vous le répète m'a rendu bien heureux en me prouvant que nos dissentiments Scientifiques ne diminuerait pas la cordialité de nos rapports.[2] Je l'esperais bien du reste; car la noblesse de votre caractere qui se montre a chaque page de vos livres me rassurait d'avance. Oui j'ose le dire, nous cherchons egalement la vérité et cela seul doit établir entre nous des liens plus forts que les causes de division résultant de la différence dans nos manières de voir. Aussi je n'hésiterai pas à vous envoyer la nouvelle edition de mes etudes sur le transformisme, qui s'impriment en ce moment en un Volume.[3] Vous les recevrez j'en suis certain dans le même esprit que me les fera envoyer

Je suis du reste très heureux que l'occasion se présente de montrer autrement que par des paroles toute l'estime que je porte a vos travaux. Nous avons à nommer un correspondant dans notre section de l'Académie des Sciences. J'en ai causé il y a quelques jours avec M. Edwards et j'ai vu avec joie que la même pensée nous était venue a tous les deux, celle de Vous présenter comme notre candidat.[4] J'espère bien que nous serons soutenus par nos autres collegues; mais en tout cas, soyez sur que je serai auprès de l'Académie un avocat zélé et convaincu. Or, nos dissentiments mêmes, donneraient au besoin plus de poids a mes paroles et je ne doute pas de votre nomination. Le plaisir de montrer tout ce qu'il y a de valeur dans vos travaux, compensera un peu pour moi le regret que j'éprouve d'avoir été *forcé* de vous combattre sur certains points. Mais les choses en étaient venues a un dégré tel que je ne pouvais plus garder le silence sans paraitre abandonner mes convictions. Cette consideration seule, soyez en bien sur, m'a mis la plume à la main car je n'aime pas les controverses.

Recevez Monsieur et cher confrere l'expression de l'affectueux devouement de votre bien dévoué | De Quatrefages

DAR 175: 6

1 For a translation of this letter, see Appendix I.

2 CD's letter to Quatrefages has not been found.

3 Quatrefages refers to *Charles Darwin et ses précurseurs français* (Quatrefages 1870). There is a copy in the Darwin Library–CUL.

4 Quatrefages refers to Henri Milne-Edwards, who was the head of the zoology and anatomy section of the French Académie des Sciences. CD was nominated six times between 1870 and 1878 for membership of the zoological section of the Académie, but was not elected until 1878, and then it was to the botanical section (Corsi and Weindling 1985, p. 699). Among the other nominees on this occasion were Thomas Henry Huxley, Rudolf Leuckart, and Carl Vogt; the successful candidate was Johann Friedrich von Brandt (*Comptes Rendus Hebdomadaires des Seances de l'Académie des Sciences* 70 (1870): 1418; 71 (1870): 41). See also letter from Armand de Quatrefages, 18 July 1870.

From Francis Galton 31 March 1870

42 Rutland Gate SW
March 31. 1870

My dear Darwin

Better news— decidedly better.

I opened the hutches where the young rabbits are, this morning, and found not only that the white patch on the nose, of which I spoke, had become markedly conspicuous and larger but also that a white vertical bar had begun to appear in the forehead.[1]

On going to the other litter, which I had never before got a proper view of, I found another young one with precisely similar marks. (The male parent was the same in both cases)

I have spent a most unsuccesful morning with new apparatus trying to inject more completely. but I have yet hopes of success by making some alterations

I will return to you Naudin & the 2 pamphlets by *to-morrow's* book post.[2] Very many thanks for them & for all the references. With great reluctance, I feel it would be too much for me to undertake the experiments[3] I am too ignorant of gardening and, living in London with a summer tour in prospect, I dont see my way to a succesful issue: but I hope to practise my eye & get some experience this year which may be of service next year or hereafter.

I congratulate you about the Quagga. taint.[4]

Once more about the rabbits, very many thanks for your hints. I will try more gray blood. Bartlett takes great interest & gives much care.[5] Muries assistant remains & looks after the rabbits. Murie himself looks in now & then.[6]

Very sincerely. F Galton

DAR 105: 11–12

CD ANNOTATIONS
2.2 on the nose, of which] 'Has Barett ever known a case'[7] *added pencil*
2.3 the forehead 2.4] '(White star)' *added pencil*
Top of first page: '*Theological Review* | get one of your Book'[8] *added ink*

1 See letters from Francis Galton, 15 March 1870 and n. 2, and 22 March 1870.
2 CD may have sent Galton his copy of Naudin 1862; there is an annotated copy in the Darwin Library–CUL (see *Marginalia* 1: 638–9). The pamphlets have not been identified.
3 The experiments have not been identified.
4 See letter from J. J. Weir, 17 March 1870. CD must have communicated this information to Galton in a letter that has not been found.
5 Abraham Dee Bartlett. See letter from Francis Galton, 17 March 1870 and n. 1.
6 Galton refers to James Murie, prosector of the Zoological Society. His assistant was Oscar Louis Fraser (see Galton 1871, p. 395).
7 CD may have meant to write 'Bartlett'; see letter from Francis Galton, 8 April 1870.
8 A review by Frances Power Cobbe of Galton's *Hereditary genius* (Galton 1869) appeared in the April issue of the *Theological Review* (Cobbe 1870).

From James Orton 31 March 1870

Vassar College. | *Poughkeepsie, N.Y.,*
Mch. 31st *1870*

My dear Sir

I wish to thank you for your very great favor of Feb. 24th and for speaking so kindly of my little work.[1] I am not blind to its shortcomings, and I trust you will remember it was designed only as a popular description of Equatorial America: details of Science were left to "Bulletins" and "Proceedings".

If "The Andes and the Amazon" possesses any merit, it should be known that it was written under the inspiration of your own delightful "Voyage".[2]

It is not necessary to say that your works are extensively read and thoroughly appreciated in America—

As to the Horse's tooth from the Valley of Quito, I am of course very anxious to secure Professor Huxley's opinion;[3] yet I should be very sorry to add even a straw to the burden of a man so valuable to Science. I expect you to *keep* the specimen; and whenever you obtain a judgment upon it, should be glad to learn the decision. I am curious to know if it is distinct from the species found in the lowlands of eastern South America.

I wrote Mr. Hauxwell—the noted English Collector of thirty years residence on the Amazon and now hunting for me, to examine for *fossil shells* on the Marañon, and Ambiyacu.[4] In his reply received a few days ago, he says: I have collected quite a lot of fossil shells for you about thirty miles below Pebas on the opposite side of the river—some large ones; and have been told they are found at Omaguas up the Maranon near Iquitos, and also up the Ambiyacu above Pebas."[5]

So that my discovery is confirmed— I have not found a single Naturalist in America who holds Agassiz's Theory of the Amazon Valley—[6]

With Sentiments of profound respect, I have the honor to be | Your obedient Servant | James Orton.

Charles Darwin Esq.

DAR 173: 39

CD ANNOTATION
Verso of last page: 'Huxley' *pencil*

1 Orton refers to CD's letter to him of 24 February [1870], and to his *The Andes and the Amazon* (Orton 1870).
2 Orton refers to CD's *Journal of researches*.
3 Orton refers to Thomas Henry Huxley; see letter to James Orton, 24 February [1870] and n. 4.
4 Orton refers to John Hauxwell. He discussed the finding of fossil shells in the Amazon region in Orton 1870, pp. 282–3. See also *Correspondence* vol. 17, letter from James Orton, 4 January 1869. Marañnón is a river in Peru, tributary to the Amazon. The Ambiyacu is a river in the Amazon region.
5 Pebas and Omaguas are towns, and Iquitos a city, in the Loreto province of north-eastern Peru.
6 Louis Agassiz believed that the sandstone deposits in the Amazon had resulted from glacial drift and showed no sign of marine origin (see J. L. R. Agassiz and Agassiz 1868, pp. 250, 411, and 424, and Orton 1870, p. 282).

To A. R. Wallace 31 March [1870][1]

Down Beckenham | Kent
Mar 31

My dear Wallace

Many thanks for the wood-cut, which, judging from the rate at which I crawl on, will hardly be wanted till this time next year. Whether I shall have it reduced, or beg Mr Mcmillan for a stereotype, as you said I might, I have not yet decided—[2]

I heartily congratulate you on your removal being over, & I much more heartily condole with myself at your having left London, for I shall thus miss my talks with you which I always *greatly* enjoy.[3]

I was excessively pleased at your review of Galton, & I agree to every word of it.[4] I must add that I have just re-read yr article in the Anthropol. Rev. & I **defy** you to upset yr own doctrine.[5]

ever yours very sincerely | Ch. Darwin

LS
British Library Board (Add 46434 ff. 200–1)

1 The year is established by the reference to Wallace's review of Francis Galton's *Hereditary genius* (Galton 1869), which was published on 17 March 1870 (Wallace 1870c).

2 CD refers to a woodcut of a skull of a babirusa pig from Wallace's *Malay Archipelago* (Wallace 1869a, 1: 434), which he reproduced in *Descent* 2: 264. Alexander Macmillan was the publisher of the *Malay Archipelago*. CD did not have the image reduced.

3 On 25 March 1870, Wallace had moved with his family to a cottage in Barking, east of London (Raby 2001, p. 206). Wallace and CD may have met in London earlier in March (see letter to A. R. Wallace, 7 March [1870?]).

4 See n. 1, above.

5 CD refers to Wallace's article on the origin of human races (Wallace 1864). There is an annotated copy in the Darwin Pamphlet Collection–CUL. See also letter to A. R. Wallace, 26 January [1870] and n. 6.

To [Henry Hussey Vivian?] [April or May 1870?][1]

Sir

I hope that you will excuse the liberty which I take in addressing you on a subject which I believe to be of scientific importance, & which you might greatly aid, if you shd think fit, to give the *[illeg]* after consulting those competent to form an opinion.— I have been led to study experimentally & theoretically during many years the good effects of crossing & the evil effect of close interbreeding; & the difference in vigour, & size & power *[illeg]* & health of the offspr[ing] in the 2 cases is in most cases extraordinary.— In Chap 17 of my work on the variation of Domestic Anim I have given a resume of my results,—[2] & these might be consulted by any one who wishes to see what is at present *[known]* on the subject. In various European countries & the U. States the most contradictory opinions owing to the *[continued]* want of any reliable data have been found. whether the marriage of near relatives is in any way injurious to man..[3]

Some time ago I went to the of the Registrar general, suggesting that during the next Census, a return should be demanded, whether in each family the parents were related in blood;[4] for it wd then it wd be possible by comparing the average numbers of the *[surviving]* children, of related & not related parents, to judge of the vitality of the children in the 2 cases, or of the fertility of the parents.

I have every cause to believe that the time will soon come, when this subject will be *[emerging]* as highly important for the welfare of mankind, though at present the subject has *[8 words illeg]*.

Draft
DAR 96: 86

1 The addressee and the date are conjectured from the relationship between this draft and the letter from H. H. Vivian, 10 May 1870.
2 Chapter 17 of *Variation* (*Variation* 2: 114–44) is titled 'On the good effects of crossing, and on the evil effects of close interbreeding'. CD's also published his work on crossing in *Cross and self fertilisation* (1876).
3 CD discussed consanguineous marriages in humans in *Variation* 2: 122–4.
4 CD may refer to his correspondence with William Farr, superintendent at the General Register Office (see *Correspondence* vol. 16, letter from William Farr, 21 May 1868). CD's side of the correspondence has not been found. The registrar-general was George Graham.

From Erasmus Alvey Darwin 1 April [1870?][1]

1 Ap.

Dear Charles

In selling you only pay Brokerage, & I have looked at two sales in my acc[ts] when the Brok[e] was $\frac{1}{2}$% Bk on £2533.6.8. on the selling price.

Your financial operations excite my envy beyond words. The L & N W[2] are issuing 4% Deb Stock & you get it free of all charges, but pay the interest that has accrued & so receive the half years interest entire.

Yours ED

DAR 105: B74

1 The year is conjectured from a pencil note on the letter.
2 London and North Western Railway Company. CD sold his stock in this company in 1852 (CD's Investment book (Down House MS)).

To James Crichton-Browne 2 April [1870][1]

Down,
April 2,

My dear Sir

As Duchenne's book is of value to me for the marks and your notes of great value, will you have the great kindness to send me literally one line by Post to tell me when it is despatched.[2] You thought that you should send it off yesterday fortnight, and in case of loss per Rail, it would of course be advisable to make early enquiries. I beg you to forgive me for being so troublesome and I remain

My dear Sir | Yours very faithfully | Ch. Darwin

Copy
DAR 143: 331

1 The year is established by the relationship between this letter and the letter from James Crichton-Browne, 15 March 1870.

2 CD refers to Duchenne 1862; see letter from James Crichton-Browne, 15 March 1870.

From Francis Galton 8 April 1870

5 Bertie Terrace | Leamington
April 8 1870

My dear Darwin

The white nose and vertical bar is, I find, of no import.[1] Bartlett[2] was not accessible the day I found them out but he has since told me they are common varieties & I hear the same from M^r. Royds, the rabbit fancier & judge of poultry shows,[3] from whom I bought them.

Before leaving London last week, I succeeded in infusing 2 per cent of the rabbits weight in alien blood before I had only achieved 1.25 or $\frac{1}{80}$th part which (on the supposition of Huxley that blood constitutes $\frac{1}{10}$th of the whole weight of the body) is only $\frac{1}{8}$th of the blood.[4]

In other words my transfusion hitherto, has given only 1 great-grandparent of mongrel blood to the otherwise pure silver greys. and this is a very small matter.

I do not like to risk another operation on the other jugular of my rabbits till after the forthcoming 3 litters nor till after I have had some more success in the system of more abundant transfusion

I can do nothing with the blood in its natural state it coagulates so quickly, so I defribrinise it.— If I cannot ever succeed in transfusing as much into the rabbits as is necessary to make a fair experiment, I must go to larger animals & try cross circulation with big dogs.

You are very kind in giving me so much valuable advice & so much encouragement.

Miss Cobbe's review is very characteristic. She has not however quite caught what I am driving at in religious matters & which—if the book sh^d be enough read to make it reasonable for me to do so—I shall express more clearly.[5]

Very sincerely yours | Francis Galton

DAR 105: A13–14

1 See letter from Francis Galton, 31 March 1870.

2 Abraham Dee Bartlett was assisting Galton in his experiments transfusing rabbits' blood.

3 Ernest Edmund Molyneux Royds.

4 Thomas Henry Huxley made this statement in T. H. Huxley 1866, p. 74. Fibrin: 'an insoluble protein precipitated as a network of fibres when blood coagulates' (*Chambers*). Galton defibrinised blood by stirring it with a split stick (Galton 1871, p. 396).

5 Frances Power Cobbe had reviewed Galton's *Hereditary genius* (Galton 1869) in the *Theological Review* (Cobbe 1870). No letter from CD to Galton commenting on the review has been found, but see the annotation to the letter from Francis Galton, 31 March 1870 and n. 8. A second edition of *Hereditary genius* was published in 1892, but it was essentially a reprint of the first, with corrections.

From William Winwood Reade [*c.* 8 or 9 April 1870][1]

Swanzy's Factory | Accra | West Coast of Africa

My dear Sir

I read your last letter with very great pleasure. I should consider a letter from *Darwin* a treat anywhere—how much more so out here![2]

Alex Agassiz has the reputation in the U.S. of being a patient & earnest investigator.[3] I knew he differed from his father who has a perfect right to believe in the immutability of species but who has no right (scientifically speaking) to say "we are the children of God not the children of monkeys", & other such catch-penny or catch-parson statements.[4] I am pleased & surprised he has written something contrary to his previously expressed opinions.[5] To tell the truth I feared he was not honest enough, & brave enough to change his assertions even if he changed his opinions. What is a naturalist if he is not sincere? Sincerity is to him what faith hope & charity are to the religieux; if he shows a deficiency in that the most splendid talents can scarcely save him from contempt.

I thought you w^d like the Jollof instance—[6] I need scarcely say that anything I write to you is fully at your disposal. My only fear is that I cannot send you anything worth having—

The hand to mouth expression of astonishment I am told is common on the coast. I made a statement about a girl in her presence w^h. I knew w^d. excite her astonishment & watched her face. She protruded her lips (much in that curious manner in w^h. the chimpanzee does) as if in the act of *blowing*.[7]

I discovered a small fly in the Niger country (ie near Kankan)[8] which makes honey in holes of trees. It does not sting. I found a nest & identified the species, specimens of which I sent to M^r. Swanzy of 122 Cannon S^t. E.C.[9] I also sent an insect which carries a load of rubbish on its back perhaps to imitate dead leaves. Perhaps Bates can tell you something about these insects if they are worth inquiring about, for he knows Swanzy & I have asked him to look at them.[10] I am ignorant of things entomological.

I am going tomorrow to a German mission station 1500 ft above sea-level. There I shall stay for some weeks & recruit.[11] I shall moreover be among Europeans who really live among the natives, speak their language & know something of their inner life. In this country the traveller gets little profit from the experience of residents. Most men come out for a few years; & those who do spend their lives here are usually traders who take no interest in anything but trade. As regards natural history there is not to my knowledge an observer on the coast, except perhaps the present Governor of the Gold Coast who has little time for science—if indeed he can boast of anything beyond a taste for it.[12] Since the days of Adanson naturalists (especially in his branch) have come out from time to time—[13] I see you gleaned one fact of importance from Mann's energetic researches.[14] Do you know if Rohlfs[15] made any discovery of a scientific nature apart from geography?

I hope to have the pleasure of being *personally* questioned by you when I return w^h. will be probably next autumn.[16] In the mean-time any hint as to ethnological inquiry will be welcome to me. I forget whether I told you that I had seen a blue-eyed negress (not an albino nor c^d. she have had European blood) in the interior.[17] This instance I believe stands alone. I hope you will not delay your work too long.[18] The best way & the surest to get information upon the points which are detaining you is I imagine to publish; facts will then stream in you can then add to subsequent editions— I conjecture it is painful to you to put forward anything in an incomplete form; there is a finish about your work in a literary sense which c^d hardly I think have been achieved without much pains. But consider your disciples. Your new work will doubtless prove a revelation to many; and will certainly suggest to ethnologists fresh methods of investigation. I know for my own part that it will be a great loss to me not to have seen your book till after I have left my present field to which I shall never return— I shall have spent probably 3½ or 4 years in Africa; to spend more would be to enslave myself to one idea. I am beginning to understand this race; that is up to a certain point. The complex man of civilization is quite undecipherable: women & savages are a little easier; & there is less variety among them. But still the laws of human nature are difficult to get at, even in their simplest form

Hoping to hear from you soon | I remain | Yours truly | Winwood Reade

The above address will find me.

DAR 176: 36

CD ANNOTATIONS
1.1 I read ... having— 3.3] *crossed blue crayon*
4.1 The hand ... *blowing*. 4.4] *enclosed in square brackets pencil; scored pencil*
5.1 I ... form 7.18] *crossed blue crayon*
Top of letter: 'Africa' *blue crayon*

1 The date range is established by the relationship between this letter and the letter from Reade of 24 April 1870, in which he says that he has been at the Akropong mission station for a fortnight. Akropong is about thirty miles from Accra.

2 CD's letter to Reade has not been found. Reade's career as an explorer and his relationship with CD are discussed in Driver 2001.

3 Alexander Agassiz had visited CD in late November or early December 1869 (see *Correspondence* vol. 17, letter to Fritz Müller, 1 December [1869]).

4 Louis Agassiz had been a vigorous opponent of CD's theory of natural selection (see Lurie 1960, pp. 252–350).

5 CD may have reported to Reade a message received from Louis Agassiz via Elizabeth Agassiz and Asa Gray (see letter from Asa Gray, 27 February and 1 March 1870 and n. 4, and letter to Asa Gray, 15 March [1870]).

6 See *Correspondence* vol. 17, letter from W. W. Reade, 26 December 1869. CD quoted Reade's account of the 'Jollofs' in *Descent* 2: 357. The Wolof people (also spelled Ouolof) now live primarily in Senegal, Gambia, and Mauritania (Appiah and Gates eds. 2005, 5: 430).

7 See *Correspondence* vol. 17, letter from W. W. Reade, 26 December 1869 and n. 4.

8 The city of Kankan is now in Guinea.

9 Andrew Swanzy, who traded with the Gold Coast, had sponsored Reade's explorations (Reade 1873, 2: 352–3). The fly has not been identified.

10 Reade refers to Henry Walter Bates. The insect has not been identified.

11 The mission station was at Akropong (see letter from W. W. Reade, 24 April 1870). Akropong is now in Ghana. The mission station took in invalids; Reade stayed there for two months (see Reade 1873, 2: 125). Recruit: reinvigorate (*Chambers*).

12 The Gold Coast colony had not had a governor since 1866, since when it had been governed from Sierra Leone. The administrator of the government was Herbert Taylor Ussher. (*Colonial Office list* 1870.)

[13] Michel Adanson studied the natural history of Senegal from 1748 to 1754 (*DBF*).
[14] Reade refers to Gustav Mann, and to *Origin* 4th ed., p. 450: 'So again, on the island of Fernando Po, in the Gulf of Guinea, Mr. Mann found temperate European forms first beginning to appear at the height of about five thousand feet.'
[15] Gerhard Friedrich Rohlfs.
[16] Reade arrived back in London in August 1870 (letter from W. W. Reade, 3 September 1870).
[17] See *Correspondence* vol. 17, letter from W. W. Reade, 28 June [1869].
[18] Reade refers to CD's work on *Descent* and *Expression*.

From Albert Günther [*c.* 9 April 1870][1]

has finished now all your woodcuts, so that you may expect them from the woodcutter in a very short time.[2]

The object in my writing today is a request from Prof. Kölliker[3] of Würzburg who at present is in London for a fortnight, & anxious to pay you his respects. As his time is very much taken up by an examination of corals in the Museum, I advised him to ask you whether you would see him on Good-Friday, a day on which the Museum is closed.[4] Thus, if you should be disengaged on that day, & if you & Mrs Darwin are inclined to receive visitors, Kölliker & myself would come by one of the morning-trains, walk from the station over to Down, & after a few hours stay return to London in the afternoon.

Kölliker stays with Sharpey[5] at present, but has promised to shift his quarters to my house during Easter.

I hope you, Mrs Darwin & all your family are quite well, & remain | Yours very sincerely | A Günther

Incomplete
DAR 165: 242b

[1] The date is established by the relationship between this letter and the letter from Günther of 12 April 1870.
[2] Günther refers to the woodcuts for *Descent*, which were being made under his supervision by George Henry Ford (see letter to Albert Günther, 23 March [1870]).
[3] Rudolf Albert von Kölliker.
[4] In 1870, Good Friday fell on 15 April. Günther refers to the British Museum.
[5] William Sharpey.

To John Murray 9 April [1870][1]

Down Beckenham | Kent
Apr 9th

My dear Mr Murray

I had intended last month to write to thank you very much for sending me the Academy; but pray do not trouble yourself, as I take it in myself, & generally find valuable matter for my work.—[2] I am steadily at work on my new book, but I do not suppose I shall be ready for press for several months;[3] as my progress is always crawling half as far backwards as I advance forwards.—

Permit me to add how sincerely grieved I was some time ago to hear of your dreadful loss: I have lost children & know & can feel what the blow is.[4]

Believe me | Yours very sincerely | Ch. Darwin

National Library of Scotland (John Murray Archive) (Ms. 42152 ff. 207–208)

[1] The year is established by the reference to the *Academy* (see n. 2, below).

[2] Murray was the publisher of the *Academy*; the first issue came out in October 1869 and included a notice of *Descent* (see letter from C. E. C. B. Appleton, 2 March [1870]).

[3] CD refers to *Descent*.

[4] Murray's youngest son, Arthur, died of congestion of the lungs while at Eton in 1870 (Paston 1932, p. 149).

From Charles Warren Stoddard 11 April 1870

St Francisco Col
1870 11— April

Charles Darwin. M.A.

My dear Sir

Your profound and vigorous volumes have given me a new interest in life and I feel like thanking you personall for them. When visiting the Sandwich Islands (in 1869) I re-read them with a new interest awakened by the singular freaks of Nature over there.[1] A California peach stone will sprout, develop into a flourishing tree, blossom and give every promise of fruit in those Islands, but the blossoms fall and there has so far been no fruit produced from one. Geraniums and some few other hardy plants thrive in portions of the islands but refuse to blossom at all in the richest soil of other parts of the same Island.

There are a few blonds among the straight-haired Islanders. And I have seen there a leprous woman who had communicated the leprosy to three husbands in succession while she still showed no signs of the plague herself.[2] Will you not come over and see us some day we shall welcome you heartily.

I shall be very proud of your favor if you will send me a line or two in reply to this.

Very truly yours | Chas. Warren Stoddard. | Box 1005: P.O.

The verses are all I have to offer you in the literary way[3]

DAR 177: 258

[1] Stoddard presumably refers primarily to *Variation*. The Sandwich Islands are now known as the Hawaiian Islands.

[2] Stoddard later published on the leper colony on Molokai island, Hawaii (Stoddard [1885]).

[3] The verses have not been found.

From Albert Günther 12 April 1870

Simla Villa | Surbiton
12.4.70

My dear Sir

Kölliker was delighted when I told him you could see him on Good Friday, & I only hope that you will not be put to any inconvenience by our visit.[1]

Trains will run as on Sunday; & we intend to start from Charing Cross at 2.30 P.M., arriving at Orpington at 3.7. We shall delight in a walk over to Down, provided the weather is not too bad, in which case we should be obliged to forgo the pleasure of seeing you.

We shall be glad if you can give us a lift back to the station; but Good Friday is frequently a holiday for servants, & neither I nor Kölliker (who is a mountaineer) would think it a hardship to return *per pedes apostolorum*[2]

Yours sincerely | A Günther

DAR 165: 245

1 Günther refers to Rudolf Albert von Kölliker. See letter from Albert Günther, [*c.* 9 April 1870].

2 *Per pedes apostolorum*: 'on the feet of the apostles'. Down is about $3\frac{1}{2}$ miles from Orpington.

From Julius Victor Carus 16 April [1870][1]

61, Grosvenor Str. | W.
Saturday, Apr 16.

My dear Sir,

I have been in London for some weeks, yet so busy that I could not call on any body. In about four or five days I shall leave England, but shall be back I hope in August. I knew Kölliker went to see you yesterday, but I declined to join him that you might not have too much visitors at once.

If your health allows it I should be very happy if I could see you. Perhaps you are so kind to let me know, what day would suit you best. Would you be so good as to tell me, what station I must take in order to come to your house?

Of the third German edition of your Origin I had the last sheet some few days ago. But it will not be out but after some weeks, as Mr Koch has arranged to give an Index to it.[2]

Believe me | My dear Sir | Yours very sincerely. | J. Victor Carus

DAR 161: 74

1 The year is established by the reference to Rudolf Albert von Kölliker's visit to Down (see letter from Albert Günther, 12 April 1870).

2 Carus refers to Bronn and Carus 1870 and to Eduard Koch (in fact it was the fourth German edition of *Origin*).

From J. V. Carus 18 April 1870

61, Grosvenor Street | London, W.
Apr 18. 1870.

My dear Sir,

I am so sorry I cannot come on Thursday, as I must be back in Leipzig Saturday.[1] It is my fault not to have written to you before. But always trying to get on with the work I undertook I looked upon a visit in your house as a sort of premium after toilsome weeks.

I must then ask you to allow me to come and see you in the autumn.[2]

With the best wishes for your health believe me | Ever yours sincerely, | J. Victor Carus

DAR 161: 75

[1] See letter from J. V. Carus, 16 April [1870].

[2] Carus did not visit CD in the autumn, being prevented by the Franco-Prussian war (see letter from J. V. Carus, 2 October 1870).

To Rudolf Albert von Kölliker 18 April [1870][1]

Down. | Beckenham | *Kent. S.E.*
Apr 18th

My dear Prof. Kölliker

I sent off by Rail some specimen of Virgularia, on Saturday, but I fear they are in a very bad state.[2] I cannot find my dried specimens. I now enclose my notes, which I fear will be of little use, but I think may be relied on.—[3]

I also enclose a memorandum for Prof. Semper; but pray tell him they are only such notes as a school-boy might make.— The specimen may not be worth his examination.—[4]

I think it possible that Alcide d'Orbigny may have given name of Virgularia *[pata*gonica*]*.—[5]

It was very good of you to take the trouble to pay me a visit, & I enjoyed much making your acquaintance.—[6]

Pray believe me | Yours very sincerely | Ch. Darwin

In my old notes I have inter-polated only one new remark between (   )

John Gittleman (private collection)

[1] The year is established by the reference to Kölliker's visit to Down (see n. 6, below).

[2] *Virgularia* is a member of the order Pennatulacea, sea pens.

[3] CD may refer to his zoological notes from the *Beagle* voyage; see DAR 30: 106–11 (R. D. Keynes ed. 2000, pp. 96–100).

[4] CD refers to Karl Gottfried Semper. The notes and specimen have not been identified.

[5] CD discussed *Virgularia patagonica* (now considered a 'nomen nudem', that is, invalid because the name is based on lost or inadequate specimens) in *Journal of researches* 2d ed., pp. 99–100. He refers to Alcide Charles Victor Dessalines d'Orbigny.

[6] Kölliker visited Down on 15 April 1870 (see letters from Albert Günther, [*c.* 9 April 1870] and 12 April 1870).

From G. H. Darwin [20 April 1870][1]

New University Club | *St. James's St*
Wednesday

My dear Father,

I send by this post the French paper that I thought I had lost—[2] I suppose I picked it up without perceiv[g] it;—but it is very odd I did'nt see it when I got in.— I am going across tonight & have just done my packing up & am so hot I do'nt know what to do with myself.— I have bought a new hat in order to avoid waiting for the old one; I couldn't bear wait[g] after scurrying about the whole afternoon to get off.— I sha'nt write till I get to Genoa.—

I am glad I'm going by night the heat wd be so awful in the day.—

Your affec Son | G H Darwin

DAR 210.2: 12

[1] The date is established by the relationship between this letter and letters from Emma Darwin to H. E. Darwin in DAR 219.9: 69, 87, and 88. In a letter dated [18 January 1870] (DAR 219.9: 69), Emma writes that the leaves won't be out when George goes to Genoa; in a letter dated [19–20 April 1870] (DAR 219.9: 87), she writes that Down has had four to five days of 'blazing summery weather'; in a letter dated [20 April 1870] (DAR 219.9: 88), she writes that George leaves on Thursday (which would be 21 April 1870). In 1870, 20 April was a Wednesday.

[2] The paper has not been identified.

From George Rolleston 20 April 1870

April 20. 1870.

My dear Sir,

I write to thankyou for your kind note. Especially am I gratified by your recognition of the heaviness of the labour which the *writing* of the book entailed.[1] Much the same, I am sure, is true of the *reading*, and I therefore beg that you will not waste your valuable time upon it— If however you will look at two "special Phylogenies" of mine which may be found at pages 138 and 152 of the Descriptions I should be glad. They are in the notes—[2]

What I say of the "problematical" character of such "special Phylogenies" at p. xxv. was prompted by the reading of Haeckel's and of Semper's speculations as to the Pedigrees of so many animals.[3] Much of the entire mass of some of Haeckel's writing, is made up of hypotheses of this kind; and as I am sure he is quite wrong in much of his speculation as to ancestry, I think other men may go equally astray wasting time and thought which might be spent otherwise more profitably.

May I ask you to put your pen on that same p. xxv through the words "*the value of the*" which come before the words "Special Phylogenies". The sentence is now ungrammatical, having been made so in the correction of the proofs by the very (under other circumstances) proper substitution of the words "are likely to" for the words "must always"— The obliteration of the words "the value of the" will make the sentence decent which, as it stands, it is not.[4]

I am | Yours very Truly | George Rolleston

DAR 176: 209

[1] CD's note to Rolleston has not been found, but evidently concerned Rolleston 1870b (*Forms of animal life*). There is an annotated copy of Rolleston 1870b in the Darwin Library–CUL (see *Marginalia* 1: 713–14).

2 The 'special phylogenies' were long notes on pp. 38–9 and 152–7 concerning the derivation of the Hirudineae (now the subclass Hirudinea, leeches) and the subkingdom (now subphylum) Echinodermata.

3 In Rolleston 1870b, p. xxv, Rolleston wrote:

> In the face, on the one hand, of our knowledge of the greatness of the unlikeness, which may be compatible with specific identity; and, on the other, of our ignorance of the entirety of the geological record, the value of the special 'Phylogenies,' or hypothetical genealogical pedigrees, reaching far out of modern periods, are likely to remain in the very highest degree arbitrary and problematical.

Rolleston refers to Ernst Haeckel and the genealogies in his *Generelle Morphologie* (Haeckel 1866), and to Karl Gottfried Semper and probably to the first volume, on holothurians, of his *Reisen im Archipel des Philippinen* (Semper *et al.* 1868–1916). See also Nyhart 1995, p. 177.

4 See quotation in n. 3, above.

To Tord Tamerlan Teodor Thorell 20 April 1870

Down. | Beckenham | *Kent. S.E.*
20 April 1870

[Darwin expresses heartfelt thanks to Tord Tamerlan Teodore Thorell, a junior professor of zoology at the University of Upsala, Sweden, for a copy of his *On European Spiders.*][1]

I have looked through it & have admired the wonderful research & labour which you have bestowed on this highly useful work. It has been a great satisfaction to me to observe in one or two passages, especially in an admirable discussion on the principles of classification, that you believe in the evolution of species[2]

Incomplete[3]
Sotheby's, New York, 1997

1 There is a lightly annotated copy of Thorell's *On European spiders* (Thorell 1869–70) in the Darwin Library–CUL (see *Marginalia* 1: 806).

2 CD cited Thorell 1869–70 in *Descent* 1: 315 n. 71. In Thorell 1869–70, p. 42 n. 1, Thorell wrote, 'I believe with DARWIN, "that *propinquity of descent*—the only known cause of similarity of organic beings—is the bond, hidden as it is by various degrees of modification, which is partially revealed to us by our classifications".' The quotation is from *Origin* 4th ed., p. 489.

3 The original letter is complete and is described in the sale catalogue as being two pages long, and signed.

To A. R. Wallace 20 April [1870][1]

Down Beckenham | Kent
Ap 20.

My dear Wallace

I have just received yr book & read the preface. There never has been passed on me, or indeed on any one, a higher eulogium than yours. I wish that I fully deserved it.[2] Your modesty and candour are very far from new to me. I hope it is a satisfaction to you to reflect,—& very few things in my life have been more satisfactory to me—that we have never felt any jealousy towards each other, though in one sense rivals. I believe that I can say this of myself with truth, & I am absolutely sure that it is true of you.

You have been a good Christian to give a list of yr additions, for I want much to read them, & I should hardly have had time just at present to have gone thro' all yr articles. Of course I shall immediately read those that are new or greatly altered, & I will endeavour to be as honest as can reasonably be expected.[3] Your book looks remarkably well got up.

Believe me my dear Wallace | to remain | yours very cordially | Ch. Darwin

LS
British Library Board (Add 46434 ff. 202–3)

1 The year is established by the reference to Wallace's *Contributions to the theory of natural selection* (Wallace 1870a), which was published in 1870.
2 There is an annotated copy of Wallace 1870a in the Darwin Library–CUL (see *Marginalia* 1: 837). On page iv of the preface, Wallace wrote, 'I have felt all my life, and I still feel, the most sincere satisfaction that Mr. Darwin had been at work long before me, and that it was not left for me to attempt to write "The Origin of Species".'
3 In pages v to xi of his preface, Wallace listed the original publication details of each essay in the book, together with a note of the amount of alteration they had received.

From Giovanni Canestrini[1] 21 April 1870

Padova
21/4 1870

Hochgeehrtester Herr,

Ich habe bis jetzt die Antwort auf Ihr werthes Schreiben verzögert, weil ich Ihnen Nachrichten von der italienischen Übersetzung Ihres letzten werthvollen Werkes geben wollte.[2]

Mit Herrn Zanichelli[3] in Bologna sind hinsichtlich der Auflage der Übersetzung Unannehmlichkeiten entstanden und die Übersetzung blieb bis jetzt unthätig. Die Sache ist nun mit einem andern Aufläger abgemacht worden, nämlich mit Herrn Vincenzi[4] in Modena und der Druck ist bereits begonnen. Als Autor der Übersetzung erscheint ein Schüler von mir, nämlich Dr Bonizzi,[5] der auch thatsächlich die letzen Capitel übersetzt hat. Herr Vincenzi in Modena wird nun recht bald den Werth der Stereotipen an Ihren Buchhändler bezahlen. Est ist mir leid, dass dies so spät geschieht.[6]

Sobald die Übersetzung wird erschienen sein, werde ich Sorge tragen, dass eine Copie allsogleich an Sie geschickt werde.

In ein Paar Wochen werde ich so frei sein, Ihnen eine meiner Arbeiten zu schicken, wo ich die *natürliche* Abstammung des Menschen vertheidige, gestützt besonders 1. auf die abnormen Charaktäre des Menschen, 2. auf die rudimentären Organe, 3. auf die sehr alten (vorhistorischen) Schädelformen und 4. auf die embryologischen Charactäre.[7]

Ihre Theorie von der Abstammung der Species hat hier sehr grosse Fortschritte gemacht und gewinnt täglich mehr Anhänger. Wie die Übersetzung Ihres ersten Werkes vieles geleistet hat, so hoffe ich wird auch die des zweiten Werkes vieles thun.[8]

Mit aller Hochachtung zeichne ich mich | ergebenster | D[r]. Joh. Canestrini | Prof. der Universität | Padova

DAR 161: 35

[1] For a translation of this letter, see Appendix I.
[2] CD's letter to Canestrini has not been found. Canestrini refers to *Variation.*
[3] Nicola Zanichelli.
[4] Mr Vincenzi has not been further identified.
[5] Paolo Bonizzi.
[6] In the event, Canestrini's translation of *Variation* (Canestrini trans. 1876, based on the second English edition) was published by Unione Typografico-Editrice in Turin.
[7] There is a copy of Canestrini's *Origine dell'uomo* (Canestrini 1870) in the Darwin Library–Down.
[8] Canestrini and Leonardo Salimbeni had translated *Origin* into Italian in 1864. On the reception of Darwinism in Italy, see Corsi and Weindling 1985 and Pancaldi 1991.

To St G. J. Mivart 21 April [1870][1]

Down. | Beckenham | *Kent. S.E.*
Ap. 21st

My dear Sir

Will you forgive me for troubling you with 2 or 3 questions.—

In some M.S. after discussing the bearing of the amount of modification & the lines of descent on the terms used in classification, & after giving your class. of the Primates in the Phil. Tr., I go on to add, that on genealogical principles alone, & considering whole organisation man probably diverged from the Catarhine stem a little below the branch of the anthropo: apes (i.e. higher up than in your diagram in Phil. Tr.)[2] I have then added in my M.S. that this is your opinion, but I cannot remember whether I derived this from you from conversation or *inferred* it after reading your 2 great papers.—[3] Is this your opinion? & may I say (if so) that you tell me so.— I conceive that the *line of descent* may be as just indicated, & yet from the amount of modification suffered by man, he may perhaps deserve to be called a distinct sub-order or Family.—[4]

Secondly you describe in Zoo. Tr. in your paper on Lemurs (in which, by the way, I found much very interesting to me on rudiments—variability &c) you describe great differences in the shape of muzzles of the genera; but I want to know whether that structure of the nose, which led Owen to use term "Strepsirhine" (not that I understand how the nose is twisted) holds good in all the genera.—[5]

Lastly, Büchner in one of his compilations says Rütimeyer has found a fossil ape uniting Catarhine & platyrhine characters; do you know anything about this? I have seen no such account, & I thought that his eocene monkey was apocryphal.[6]

I left London before your return so could not profit by your kind invitation to call on you.—[7] Pray forgive me for being so troublesome & believe me | Yours very faithfully | Ch. Darwin

Karpeles Manuscript Library Museums

[1] The year is established by the relationship between this letter and the letter from St G. J. Mivart, 22 April 1870.

[2] CD refers to his manuscript for *Descent* and to Mivart's paper 'On the appendicular skeleton of the primates' in the *Philosophical Transactions of the Royal Society of London* (Mivart 1867); Mivart's diagram is on page 425. See *Descent* 1: 185–99.

[3] CD refers to Mivart 1867 (see n. 2, above), and probably to Mivart 1866a, a paper in the *Transactions of the Zoological Society of London* on the appendicular skeleton of *Simia*. *Simia* is a Linnean category roughly equivalent to the modern suborder Haplorrhini except that it excludes the genus *Homo*.

[4] See *Descent* 1: 194–5.

[5] CD refers to Mivart's paper 'On the Lemuroidea' in the *Transactions of the Zoological Society of London* (Mivart 1866c). Some classifications of primates include, as well as catarrhine and platyrrhine species, strepsirrhine (literally 'twisted-nose') species, which have moist tips to their noses and a cleft upper lip bound to the gum (e.g., lorises and lemurs; Allaby ed. 1999). Richard Owen used the term strepsirrhine (or strepsirhine) in Owen 1866–8, 2: 290.

[6] Ludwig Büchner described Ludwig Rütimeyer's discovery of a fossil ape in Büchner 1868, pp. 202 or 204 (different printings of the edition are paginated differently). This passage is scored in CD's annotated copy of Büchner 1868 in the Darwin Library–CUL (see *Marginalia* 1: 97–8). Rütimeyer described the fossil ape *Caenopithecus lemuroides* in Rütimeyer 1862, pp. 88–92. On page 92, Rütimeyer noted that the fossil ape had characteristics in common with modern-day lemurs and New-World (platyrrhine) apes. There is a copy of Rütimeyer 1862 in the Darwin Pamphlet Collection–CUL.

[7] See letter from St G. J. Mivart, 8 March [1870]. CD was in London from 5 to 12 March 1870 (Emma Darwin's diary (DAR 242)).

From St G. J. Mivart 22 April 1870

7 North Bank | N.W.
April 22[d]— 1870.

My dear Sir,

I have really expressed no opinion as to Man's origin nor am I prepared to do so at this moment. The diagram in the Pro. Z. Soc. expresses what I believe to be the degree of resemblance as regards the spinal column *only*. The diagram in the Phil. Trans. expresses what I believe to be the degree of resemblance as regards the appendicular skeleton *only*.[1]

As regards the skull the divergence would be greater than that of the appendicular skeleton in some respects—less in others & not having gone into the question thoroughly I am not able to sketch a diagram on that basis. But if I combined all these we should still have only considered the skeleton & were I to follow this up by a careful comparison of all the organs in all forms we should still only have considered the dead body.

Now I believe fully that this alone being taken into account—& assuming that zoological classification should be Anatomical—Man forms only a family of the higher division of the Primates. But if we introduce into the consideration his intellectual, moral & religious nature I am convinced he differs more from an Anthropoid Ape than such an Ape differs from a lump of granite.

As to the nostrils of the Lemuroidea I do not recollect noticing any exceptional form but it was not a point we went into.[2]

Buchner's undifferentiated Ape is unknown to me but I have been months absent & since my return have not looked up any matters connected with Primates other than human.[3]

Surely such a beast will be found someday—if he has not been found already.

Regretting much that I cannot reply more satisfactorily to your questions & trusting that a visit from you is a pleasure postponed only for a little, I remain, with best regards & wishes | Your's very sincerely | S[t] Geo. Mivart.

DAR 171: 186

[1] Mivart refers to his paper on the axial skeleton of primates in the *Proceedings of the Zoological Society of London* (Mivart 1865) and to his paper 'On the appendicular skeleton of the primates' in the *Philosophical Transactions of the Royal Society of London* (Mivart 1867). There are tree diagrams in Mivart 1865, p. 592, and Mivart 1867a, p. 425, showing the affinities between the axial and appendicular skeletons of the various groups. The axial skeleton is the spinal column; the appendicular skeleton is the skeleton of the limbs, not including the spinal column.

[2] See letter from St G. J. Mivart, 21 April [1870] and n. 5.

[3] See letter from St G. J. Mivart, 21 April [1870] and n. 6. Mivart refers to Ludwig Büchner.

To St G. J. Mivart 23 April [1870][1]

Down. | Beckenham | *Kent. S.E.*
April 23[d]

My dear Sir

I write merely one word to thank you very much for so promptly answering my note.—[2] I cannot conceive what made me write down in my notes what I did about your opinion, but I saw that it was not to be trusted.— I hope that you will continue your wonderfully laborious & valuable labours on the Primates.—[3] Whenever I publish my book I can see that I shall meet with universal disapprobation, if not execration.—[4] The truth is hard to gain, however much one may try.—

With many thanks Pray believe me | Yours sincerely | Ch. Darwin

American Philosophical Society (375)

[1] The year is established by the relationship between this letter and the letter from St G. J. Mivart, 22 April 1870.

[2] See letter to St G. J. Mivart, 21 April [1870], and letter from St G. J. Mivart, 22 April 1870.

[3] CD refers to Mivart 1865, 1866a, 1866b, and 1867.

[4] CD refers to *Descent*.

From W. W. Reade 24 April 1870

Swanzy's Factory | Accra
April 24. '70

My dear Sir

I enclose a specimen of the insect I mentioned to you in my last letter.[1] It carries as you see a number of dead ants on its back. The specimen was taken from the charred end of a trunk of a tree which had fallen across the path & had been burnt in two— There were various chinks & crevices in the charred part of the tree (ie its inner part or heart) & there they resided. Their load being composed completely of black ants assimilated them to their habitat. The "imitation" was almost perfect—

On the upper part of the trunk I found one; but this one had something (a seed or a ball of clay) of brown colour on its back, perhaps to assimilate it to the colour of the tree.

The first I ever saw were near Falaba (200 miles NE of Sierra Leone). They lived in the crevices of a clay hillock. They had sometimes an ant but usually brown leaves on them; I also found one in a path near here, which had no ants, but a morsel of dead leaf—

I kept the enclosed alive with others for some days. When the load is taken off very gently one can see that it is attached to the posterior segment of the body the surface of which I think is slightly hollowed, by a substance resembling a spiders web. In one large specimen (which escaped) I had the load dangling by this thread the true nature of which I c[d]. not discern, not having a lens here. The enclosed died of its own accord.

The insect left to itself soon puts it on again; examining it first, it passes in front of it takes it up with the tip of one hind leg & puts it on its back, & then uses both of the hindermost legs to adjust it properly reminding me forcibly of a native woman "fixing" her child in the usual seat behind her back. It is not particular about taking up the same load; but the first that comes. A big one will often take the load off a small one's back.

On one occasion removing a load I left the insect by itself in the box with nothing but a pellet of paper as enclosed. It put it on. I then put in some of the dead ants &c. It put them on, but did not take off the paper. I then removed all together; separated the paper, & the ants, and the insect resumed not only the ants, but also the paper showing that animals will not only use new material when the old one cannot be procured but having once used that new material *may* continue to use it (however incongruous & under ordinary circumstances unfit for the purpose) after the necessity for so using it is passed. I say "may" not "must". A smaller one in the box at the same time preferred remaining naked till the ants were restored; & when the enclosed insect was stripped again (by another, person) it did not again put on the paper. That is why you do not see it there. It is possible that the load will come off in the letter but you will at all events see the materials.

When I put small ants in the box they went up to the insect without suspicion which seized it with its fore feet & killed it slowly, & clumsily. In one case I saw it pass the ants to its hind foot as if to put it on its back, but did not do so. The movement of the insect is by no means speedy; it has a nodding motion when it walks like "a coach on springs" as a lady cleverly remarked in reference to a Manti out here, whose gait is precisely similar—(Letters from Sierra Leone *Murray*).[2]

I have not been able to get any information about these insects from the natives. I shall make such arrangements here that they can be procured, if ordered.

I have been here at Akropong German mission 1500 ft above sea-level for a fortnight, and go off tomorrow to spend a few days with a German missionary who has lived in his station near the banks of the Volta, for 20 years[3] & is therefore pretty well up in the natives of those parts which are sharply divided from these here; who belong to the same family as the Ashantis & are uncircumcised; the others are circumcised & belong to the Dahomy family—[4] It is therefore a good place for observation, and I hope to get something for you. I shall write to you again before I leave here which will be in June to go up to the Confluence of the Niger if I can get a passage.

I hope you are in good health & that the great work is progressing. I have been smoking too much, eating too much and reading too much—faults of the Germans. Perhaps I am, like the enclosed, "assimilating" myself to my present habitat

I remain | Yours very truly | Winwood Reade

DAR 176: 37

[1] See letter from W. W. Reade, [*c.* 8 or 9 April 1870].

[2] Reade refers to a passage on a praying mantis in Elizabeth Helen Melville's *A residence at Sierra Leone* ([Melville] 1849, p. 217). [Melville] 1849 was published by John Murray.

[3] The missionary has not been identified.

[4] The Ashanti are an ethnic group who are the principal inhabitants of the Ashanti region of present-day Ghana; the Dahomey, or Fon, are the largest ethnic group in Benin, formerly known as Dahomey (*Columbia gazetteer of the world*).

From St G. J. Mivart 25 April 1870

7 North Bank | N.W.
April 25th 1870

My dear Sir,

It is exceedingly kind of you to write again & I feel I owe you some apology for saying more, in my last, than was strictly necessary in order to reply to your questions.[1]

Still it seemed to me that honesty & frankness required me to say what I did & that it was more truly respectful to express the statement than to withhold it.

I very much wish that I could chat with you at leisure over some of these matters and trust, dear Sir, that when you come to town you will kindly afford me the opportunity.

For my part I shall never feel anything but gratitude & sincere esteem for the author of "natural selection" but I heartily execrate some who make use of that theory simply as a weapon of offence against higher interests and as a means of impeding Man's advance towards his "end" whatever may have been his "origin".

In my wanderings about Italy I have been amused and saddened to see our friend Huxley's "Mans place in Nature"[2] for sale at most of the railway stations amongst a crowd of *obscenities*. It was evident that the Vendors counted on what we may term a "tendency to reversion" and I fear not a few of the purchasers would *prefer* to find that man "diverged" *above* the *Cynocephali*.[3]

At present I am mainly occupied with Batrachians but I shall no doubt return to my "first loves" some day.[4]

With kind regards & many thanks believe me | Dear Sir | Your's most truly | St George Mivart.

DAR 171: 187

[1] See letter from St G. J. Mivart, 22 April 1870, and letter to St G. J. Mivart, 23 April [1870].

[2] Mivart refers to Thomas Henry Huxley and T. H. Huxley 1863.

[3] *Cynocephalus* is the flying lemur. Mivart's 'trees' were meant to show affinities between primates in relation to various parts of their skeletons, not genealogical relationships (see letter to St G. J. Mivart, 21 April [1870] and n. 2, and letter from St G. J. Mivart, 22 April and n. 1). Mivart imagines Huxley's followers as wishing the divergence of humans from the rest of the primates to come later rather than earlier.

[4] Mivart's paper on the amphibian order Urodela (now Caudata: salamanders) was read at the Zoological Society of London on 28 April 1870 (Mivart 1870).

From Lucy Caroline Wedgwood [25 April 1870][1]

Monday

Dear Uncle Charles

I have at last made out for certain that my dog *does* raise his lip on the side on which he is attacked, without raising it on the other.

I tried making a horse kick by whipping him behind when riding him several times and always with the same result: he just turned his ears round before kicking but *never once* laid them flat as before biting.[2]

The horse when I scratch him stretches out his nose, opens his mouth a little, and with lips parted &

Incomplete
DAR 181: 59

CD ANNOTATIONS
1.1 I have … other. 1.2] 'Sneering' *in margin, blue crayon*
2.1 I tried … behind] *crossed pencil*
3.1 The horse … & 3.2] *crossed pencil*
Top of letter: 'L. Wedgw' *pencil* 'April 25 1870 | Leith Hill Place'[3] *ink*

[1] The date is established by the annotation. In 1870, 25 April was a Monday.

[2] CD discussed the expressions of dogs and horses in *Expression*, pp. 117–18 and 129–30, but did not mention Wedgwood.

[3] Wedgwood lived at Leith Hill Place in Surrey.

From Francis Galton 26 April 1870

5 Bertie Terrace | Leamington
April 26/70

My dear Darwin

Two more litters & no happy results, the young being all true silver greys.[1]

There ought to have been a third litter but the doe had not kindled. I shall next give a fresh infusion to every one of my old stock and hope to raise the proportion of alien blood in their bodies to at least 3 per cent of their entire weight or, say, 30 per cent of their entire blood.

I am obliged to defer all this for a week or two longer for my mother[2] has been lying at the verge of death for a fortnight & I am wanted by her. She is now a trifle better & her illness—the results of bronchitis—may be less acute for a while & I may be able to get back to London— We have no reasonable hope that she will ever recover even a most moderate degree of health.

Very sincerely yours | Francis Galton

DAR 105: A15–16

[1] Galton was transfusing rabbits' blood into other rabbits to test CD's hypothesis of pangenesis; his last letter was on 8 April 1870.

[2] Violetta Galton.

To Francis Galton [27 April 1870?][1]

M^r D is much obliged & very sorry for the information about the R... He is glad to hear that you keep up your spirits.—[2]

Draft
DAR 105: A16v

[1] The date is conjectured from the relationship between this letter and the letter from Francis Galton, 26 April 1870. This letter is a draft written on the back of the letter of 26 April 1870.

[2] See letter from Francis Galton, 26 April 1870.

From George Cupples 29 April 1870

The Cottage, | Guard Bridge, Fifeshire. | N.B.
April 29/70

My Dear Mr Darwin,

I had your kind letter yesterday afternoon, and was exceedingly glad to see your handwriting again—indeed several times lately had thought of *making* a pretext to write, but for the idea of troubling you to reply (which your great courtesy makes me afraid of.)[1]

It will be a very great pleasure to me to send a Deerhound puppy—which shall certainly be of the very best.[2] I have at present two *brood*-deerhounds, one of them a beautiful young animal, from which I expect to have a litter this summer, excelling everything before produced from the kennel. It will be some months, however, before any are ready—and perhaps, considering your inquiry as to their qualities for the house, I may see reason to recommend rather one with a "dash" of the *Irish Wolfhound* blood in it (supposed to be extinct, but nevertheless still extant and in my power to get.) In a week or two hence I expect to be able to say definitely when one of either kind could be counted upon—and shall write accordingly. Meanwhile as to your question.

No dog can be better for docility, attachment, and good behaviour about a house than the pure Scotch deerhound—but greater intelligence and a more decided turn for guardianship, with more of the mastiff temper, can be obtained in a dog partaking of the old wolf-hound blood, while at the same time the *appearance* of the deerhound is preserved. In case of my finding that a dog-puppy of this strain is available from a friend who breeds them, and has promised me one or two—there would be a saving of time, for in all likelihood these will come into readiness considerably earlier in the season. I shall make a point of seeing about it without delay, and giving full particulars before you make choice.

I look forward eagerly to the appearance of your new work.[3] The bearings of evolution upon sex, and vice versâ, must be of intense interest as treated of by *you*.

I have of late by accident got into epistolary acquaintance with a very remarkable lady, Miss Hennell—whom you have seen, I believe, and whose writings show a devout appreciation of your works.[4] Through her I found occasion to read for the first time the Psychology of Herbert Spencer,[5] and to go on to other books of his just now. Being full of them, I am at this moment tempted as an outsider to make one or two comments en passant—but space, time, and propriety forbid. One remark I may make— Mr Spencer on the one hand, and orthodoxists on the other, seem to me to give themselves a great deal of superfluous discomfort and trouble on account of/about the results of Science. Their disposition to outcries of distress is unintelligible to Transcendentalists, who can quietly take in (I hope) all that is discovered or can be discovered—without the slightest consequent loss of ability to translate it into equivalent terms of the "good old" practical scheme of things.[6]

I am glad to think you are in working health.

Mrs Cupples[7] desires to be remembered to Mrs Darwin—and I remain

My dear Mr Darwin | faithfully yours | George Cupples

Charles Darwin, Esqr. | Down

P.S. I do not feel well, but hope to be better as the Spring advances.

DAR 161: 292

[1] CD's letter to Cupples has not been found.

[2] Cupples had offered CD a deer-hound puppy in his letter of 11–13 May 1868 (*Correspondence* vol. 16).

[3] Cupples refers to *Descent*.

[4] Sara Sophia Hennell mentioned CD in the second volume of her *Present religion as a faith owning fellowship with thought* (Hennell 1865–[87]), but this was not published until 1873. In the first volume she wrote, '*God* . . . means . . . the universal Power residing in creation, human as well as other. . . . *Making* is, no longer instantaneous creation, effected by the fiat of the sovereign will outside nature, but the sustenation of the *growth* that pervades the whole internal constitution of nature' (*ibid*, 1: 13). Hennell believed that conventional religious belief rested on a superstitious regard for symbols, but that ultimately no religious feeling need be lost as a result of using abstract thought to analyse such symbols.

[5] Spencer 1855.

[6] Cupples refers to the transcendentalist philosophy associated with Ralph Waldo Emerson.

[7] Anne Jane Cupples.

To Ferdinand Vandeveer Hayden 2 May [1870][1]

Down. | Beckenham | *Kent. S.E.*
May 2nd.

Dear Sir

I am much obliged to you for your kindness & for the honour which you have done me in sending your Geological Report of the Yellowstone River & your Preliminary Field Report on the Colorado & New Mexico.[2] I had heard of your Geological researches on the Colorado & was anxious to see the conclusions at which you had arrived, & I am therefore especially obliged to you for having sent me your works.

With much respect & my best thanks, I remain, | Dear Sir, | yours faithfully | Charles Darwin

LS
Smithsonian Institution Libraries: Dibner Library of History of Science and Technology

[1] The year is established by the form of the address and the date of the publications mentioned (see n. 2, below). From April 1869 until May 1871, CD used stationery with 'Bromley' crossed out in the printed address and 'Beckenham' written in.

[2] CD refers to Hayden 1869a and 1869b. CD's copy of Hayden 1869a is in the Darwin Pamphlet Collection–CUL; his copy of Hayden 1869b has not been found. Hayden 1869b was submitted on 15 October 1869 (Hayden 1869b, p. 3).

From John Michels 3 May 1870

10 Castlenau Villas | Barnes. S.W
May 3rd. 1870

Dear Sir,

I have taken the liberty of sending for your inspection some drawings, I have lately made of the Pollen of the Geranium and Honeysuckle—[1]

I noticed that all the pollen bore a marked likeness to oneanother, but that with both specimens, there were one or two exceptions of so remarkable a character, that they might be taken for the Pollen of another plant—

The number of pollen grains in each specimen, were about 30 to 40. The exceptions shown were the only ones of that type.—

Is it possible that the varieties in the character of seedlings, owe their origen to the particular seeds being impregnated with these exceptional pollen.—

It appears probable that as the pollen is the sexual part of the part,[2] and any marked peculiarity would be transmitted to the seed impregnated—

For my drawing, I used one of Smith & Becks instrmts and $\frac{1}{5}$ Object glass.[3] They are however enlarged about $\frac{1}{3}$.

I remain dear Sir | Your ob^t St | John Michels

C. Darwin Esq | F.RS.

P.S.— I can show the pollen at any time. JM.

DAR 171: 175

1 The drawings referred to have not been found.
2 Michels meant to write, 'part of the plant'.
3 Michels refers to Smith, Beck & Beck, scientific instrument makers. '$\frac{1}{5}$ object glass' refers to the nominal focal length of the lens used as object glass in the microscope.

To C. W. Stoddard 5 May [1870][1]

Beckenham, Kent

['It is of course a great satisfaction to me to hear that my work has in any way interested an instructed & observing person.' Speaks of flora of the Sandwich (Hawaiian) Islands and concludes: 'There is nothing I shd enjoy so much as to visit California, but I am growing old & my health is weak. . . .[2]

Ch. Darwin']

Incomplete[3]
Parke Bernet, 8 October 1963 (Item: 71)

1 The year is established by the relationship between this letter and the letter from C. W. Stoddard, 11 April 1870. The date is given in the sale catalogue.
2 See letter from C. W. Stoddard, 11 April 1870 and n. 2.
3 The original letter is complete and is described in the sale catalogue as being 2$\frac{1}{4}$ pages long.

From L. C. Wedgwood [5 May 1870][1]

I can perceive no difference between the motion of the horses ears when he turns them before kicking, and when he turns them for any other reason, (except fear which is different)[2]

I think you wanted to know about the babys brows.

When vext at anything & crying (without tears) he does straighten his eyebrows, raising the inner corner a very little

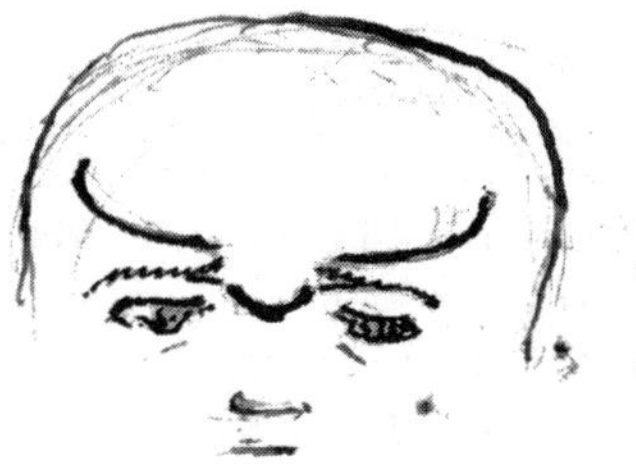

What is most conspicuous is a pucker on the bridge of the nose and 2 lines sloping across the forehead something like that. I have seen the same thing when he has been dazzled in the sunshine too.[3]

My worms have not turned up any earth since I enclosed them[4]

I am with many thanks for y^r. letter | y^r. aff^t. niece | Lucy Wedgwood

Incomplete
DAR 181: 60

CD ANNOTATIONS
1.1 I can ... different) 1.3] *crossed pencil*
2.1 I think ... brows.] '(10 months)' *pencil*
3.3 What ... forehead 3.4] 'Pyramidalis must here act if possible so as to smooth forehead' *pencil*
4.1 My ... letter 5.1] *crossed pencil*
Top of letter: 'May 5— 1870' *pencil*

1 The date is established by CD's annotation.

2 See letter from L. C. Wedgwood, [25 April 1870] and n. 2.

3 In *Expression*, pp. 147–54, CD discussed facial expressions associated with crying in infants, but did not mention Wedgwood. The baby Wedgwood observed may have been her nephew, Hervey Wedgwood Williams.

4 A short notice on worms pulling stones to cover their burrows, written by Wedgwood, had appeared in *Gardeners' Chronicle* in 1868 (see *Correspondence* vol. 16, letter to M. T. Masters, 21 March [1868] and n. 2).

From W. T. Preyer 9 May 1870

Jena, University,
May 9, 1870.

Dear Sir,

I beg you to accept my very best thanks for the most interesting letter you had the kindness to address to me.[1] I would long since have sent you the biographical sketch: "Charles Darwin", eine biographische Skizze with your portrait published in the "Ausland", had I not been absent from here for two months.[2] I only arrived here, coming from Naples, a few days ago. I hope the german article may not displease you. At the same time I take the liberty of sending you a book published by D^r. O. Peschel, in Augsburg, editor of the "Ausland".[3] It is in his name that I subjoin it to my paper.

I have been told that you intend to write a book on the application of your theory to the development of mankind.[4] I should feel very much obliged if you would kindly inform me when this work is to be published, but only in case you can find without difficulty five minutes of leisure for, dear Sir | your's most truly | W^m. Preyer

DAR 174: 68

1 See letter to W. T. Preyer, 17 February [1870].

2 The reference is to Preyer 1870b; there are two copies of the article in the Darwin Pamphlet Collection–CUL.

3 Oscar Ferdinand Peschel had been the director of the weekly *Das Ausland* since 1854. No book by Peschel has been found in the Darwin Library–CUL or the Darwin Library–Down. Preyer may have sent Peschel's monograph *Neue Probleme der vergleichenden Erdkunde als Versuch einer Morphologie der Erdoberfläche* (New problems in comparative geography as an attempt at a morphology of the earth's surface; Peschel 1870).

4 Preyer refers to *Descent.*

From H. H. Vivian 10 May 1870

House of Commons Library
10 May. 1870

Dear Mr Darwin

I have spoken to Mr Bruce[1] about the insertion of a Cousin's Marriage column in the Census Paper:[2] He was quite alive to the importance of the point involved & begged me to assure you that it was receiving their very careful consideration: beyond this I could not get him but it will not be overlooked & if it can be done without great extra labour or unless other returns should appear to have a higher claim I have hopes that your wishes may be granted.

I will not lose sight of it & will be at him again about it when I have a chance; I believe there will be a discussion on the question before long & if you can get any members to back it up, so much the better

Yours faithfully | H Hussey Vivian

DAR 180: 8

1 Henry Austin Bruce was home secretary from 1868 until 1873 (*ODNB*).

2 See letter to [H. H. Vivian?], [April or May 1870?] and nn. 3 and 4.

To H. H. Vivian 11 May [1870][1]

Down Beckenham
May 11—

Dear Mr Vivian

I feel greatly obliged to you for the kind interest which you have shewn on the subject of the census, which interests me greatly. I have no doubt that your personal communication with Mr Bruce will have far more influence than any memorial or any thing that I cd do.[2]

With my sincere thanks I beg leave to remain | yours faithfully | Charles Darwin

P.S. I will endeavour to persuade Sir J. Lubbock[3] to speak on the subject.

LS
Invercargill City Libraries (Alex Robertson Collection, vol. 12)

1 The year is established by the relationship between this letter and the letter from H. H. Vivian, 10 May 1870.

2 See letter from H. H. Vivian, 10 May 1870 and n. 1. CD refers to Henry Austin Bruce.

3 John Lubbock was elected MP for Maidstone in February 1870 (see letter to John Lubbock, 26 February [1870].

From Francis Galton 12 May 1870

42 Rutland Gate SW
May 12 1870

My dear Darwin

Good rabbit news.! One of the latest litters has a white forefoot. It was born April 23rd. but as we do not disturb the young, the forefoot was not observed till to-day. the little things had huddled together shewing only their backs & heads and the foot was never suspected. The mother was injected from a grey and white and the father from a black and white.

This, recollect, is from a transfusion of only $\frac{1}{8}$th part of alien blood in each parent; now, after many unsuccesful experiments, I have greatly improved the method of operation and am beginning on the other jugulars of my stock.[1] Yesterday I operated on 2 who are doing well to-day & who now have $\frac{1}{3}$rd. alien blood in their veins. On Saturday I hope for still greater success. and shall go on at any waste of rabbit life, until I get at least $\frac{1}{2}$ alien blood.

The experiment is not fair to Pangenesis until I do.[2]

We are for the time relieved from anxiety about my poor dear mother,[3] who suffered the agonies of death over & over again, but has strangely pulled through. & is now comfortable though very weak and seriously shaken

Very sincerely yours | Francis Galton

DAR 105: A17–18

1 Galton discussed the difficulties of transfusing blood in rabbits in his letter of 8 April 1870.

2 Galton's experiments were designed to test CD's hypothesis of pangenesis (see letter from Francis Galton, 15 March 1870 and n. 2). For accounts of the experiments, see Galton 1871, Pearson 1914–30, 2: 156–66, and Bulmer 2003, pp. 116–18.

3 Violetta Galton. See letter from Francis Galton, 26 April 1870.

To Fritz Müller 12 May 1870

Down Beckenham | Kent
May 12. 1870

My dear Sir

I thank you for your two letters of Dec 15 & Mar 29, both abounding with curious facts.[1] I have been particularly glad to hear in yr last about the Escholtzia; for I am now rearing crossed & self-fertilized plants, in antagonism to each other, from your semi-sterile plants, so that I may compare their comparative growth with that of the offspring of English fertile plants.[2] I have forwarded your post script about Passiflora with the seeds, to Mr Farrer, who I am sure will be greatly obliged to you; the turning up of the pendant flower plainly indicates some adaptation.[3] When I next go to London I will take up the specimens of butterflies & shew them to Mr Butler of the British Museum, who is a learned lepidopterist & interested on the subject.[4] This reminds me to ask you whether you received my letter about the ticking butterfly described at p. 33 of my Journal of researches; viz whether the sound is in any way sexual? Perhaps the species does not inhabit yr island.[5]

The case described in yr last letter of the trimorphic monocot. Pontederia is grand. I wonder whether I shall ever have time to recur to this subject; I hope I may, for I have a good deal of unpublished material.[6]

I thank you for telling me about the first-formed flowers having additional petal, stamens, *carpels* &c; for it is a possible means of transition of form: it seems also connected with the fact on which I have insisted of *peloric* flowers being so often terminal.[7] As *pelorism* is strongly inherited (I have just got a curious case of this in a legum. plant from India); would it not be worth while to fertilise some of your early flowers having additional organs with pollen from a similar flower, & see whether you c[d] not make a race thus characterised? Some of yr abutilons have germinated, but I have been very unfortunate with most of your seed.[8] You will remember having given me in a former letter an account of a very curious popular belief in regard to the subsequent progeny of asses, which have borne mules; & now I have another case almost exactly like that of Lord Morton's mare, in which it is said the *shape of the hoofs* in the *subsequent* progeny are affected.[9]

(Pangenesis will turn out true some day!)[10]

A few months ago I rec[d] an interesting letter & paper from yr brother, who has taken up a new & good line of investigation, viz the adaptation in insects for the fertilisation of flowers.[11]

The only scientific man I have seen for several months is Kölliker who came here with Günther & whom I liked extremely.[12]

I am working away very hard at my book on man & on sexual selection, but I do not suppose I shall go to press till late in the autumn[13]

Believe me my dear Sir | with many thanks for all yr kindness | yours sincerely | Ch. Darwin

LS(A)
British Library Board (Loan 10:32)

1 See letters from Fritz Müller, 18 December 1869 (*Correspondence* vol. 17) and 29 March 1870. CD evidently wrote '15' in error.

2 In his letter of 29 March 1870, Müller had reported his most recent results with plants of *Eschscholzia californica* raised from seeds sent by CD. CD reported the results of his experiments in *Cross and self fertilisation*, pp. 109–17; 262–3.

3 For the postscript referred to, see the letter from Fritz Müller, 16 February 1870. Müller had enclosed seeds of *Passiflora*. Thomas Henry Farrer had been studying the fertilisation mechanisms of *Passiflora* (see *Correspondence* vol. 17).

4 No letter from Müller mentioning specimens of butterflies has been found. CD refers to Arthur Gardiner Butler.

5 See *Correspondence* vol. 17, letter to Fritz Müller, 8 September [1869] and n. 2. The ticking butterfly was *Papilio feronia* (now *Hamadryas feronia*, the blue cracker), which is found where Müller lived in Santa Catarina province (now state), Brazil. See *Journal of researches* 2d ed., p. 33. Müller moved to the mainland in 1867, but had previously lived at Destêrro (now Florianópolis) on Santa Catarina Island.

[6] See *Correspondence* vol. 17, letter from Fritz Müller, 18 December 1869 and n. 3. Müller published on the three forms of *Pontederia* (now *Eichhornia*, the water hyacinth) in F. Müller 1871. CD discussed it in *Forms of flowers*, pp. 183–7.

[7] In his letter of 18 December 1869 (*Correspondence* vol. 17), Müller had described peloric flowers of *Agapanthus* and *Jussiaea*, a synonym of *Ludwigia*. In *Variation* 2: 345–7, CD discussed the tendency of terminal or central flowers to be peloric.

[8] Müller had sent seeds of *Abutilon* with his letter of 18 October 1869 (*Correspondence* vol. 17; see *ibid.*, letter to Fritz Müller, 1 December [1869]).

[9] The letter in which Müller discussed the offspring of asses has not been found, but see the letter to J. J. Weir, 17 March [1870], in which CD quoted a passage from Müller on the topic. See also the letter from J. J. Weir, 17 March 1870 and n. 1 for more on the mare of Lord Morton (George Sholto Douglas).

[10] CD refers to his hypothesis of pangenesis, a theory of heredity described in *Variation* 2: 357–404.

[11] See letter from Hermann Müller, 8 March 1870 and n. 2.

[12] Rudolf Albert von Kölliker and Albert Günther visited Down on 15 April 1870 (see letter from Albert Günther, 12 April 1870).

[13] CD refers to *Descent*, which went to press on 30 August 1870 (see CD's 'Journal' (Appendix II)).

To Frans Cornelis Donders 13 May [1870][1]

Down | Beckenham | Kent S.E.
May 13th.

My dear Sir,

In about a fortnight's time I shall begin to arrange my notes on Expression: & it would be a great assistance to me if you could spare time to give me some notion of any conclusion at which you may have arrived, on the cause of the contraction of the muscles round the eyes during screaming &c[2]

Secondly, can you give me any explanations why tears are poured out when these same muscles are spasmodically contracted, as during violent choking vomiting &c without any mental emotion such as grief being felt. I find that the *voluntary*, prolonged & strong contraction of these muscles, even with children, does not cause tears to flow, or hardly at all. I hope my questions will not cause you much trouble, as I know your time is taken up by valuable researches.

I have lately read in an English Journal with extreme interest an abstract of your experiments on the rate of the travelling of the nervous powers.[3]

With very sincere respect | pray believe me, my dear Sir, | yours faithfully | Charles Darwin.

Copy
American Philosophical Society (B/D25.316)

[1] The year is established by the relationship between this letter and the letter from F. C. Donders, 17 May 1870.

[2] CD refers to his research for *Expression*; Donders had been conducting experiments to answer questions posed by CD, and reiterated here, concerning Charles Bell's account of the physiology of weeping (see *Correspondence* vol. 17, letter to F. C. Donders, 6 September 1869 and n. 2, and letter from F. C. Donders, 12 November 1869).

[3] CD refers to Donders's development of a technique (now commonly called the 'subtraction method') for measuring the duration of a psychological process by measuring reaction time. Donders's article 'Over de snelheid van psychische processen' (The speed of mental processes; Donders 1868) appeared

in Dutch, German, and French. The English abstract has not been identified, but the article is now available in English (Donders 1969).

To Thomas Henry Farrer 13 [May 1870][1]

Down, Beckenham | Kent
13th

My dear Mr Farrer

Enclosed a partial answer with hopes of further answer.—[2] You had better sow seeds as soon as possible; but I fear they look squashed.— It is curious the turning up of flower; but I am afraid opposed to you.—[3]

I am endeavouring to persuade Mr. Bruce to have inserted in Census query whether in each household the parents are cousins: I am deeply convinced that this is an important subject: if you can influence any member of government, pray do so.[4] Some few M.P.s will take up the question.—

I have given my reasons in a Chapt in 2d. Vol. of my Domestic animals.—[5]

Pray believe me | yours very sincerely | C. Darwin

We were both truly grieved to hear of Mrs. Farrer's[6] most serious illness, though I hope she is now quite well again.

Linnean Society of London

1 The month and year are established by the relationship between this letter and the letter to H. H. Vivian, 11 May [1870].

2 CD enclosed part of the letter from Fritz Müller, 16 February 1870. It is the only section of that letter that has been found.

3 Müller had enclosed seeds of wild plants of *Passiflora* and a dried flower of a species from his garden for Farrer with his letter of 16 February 1870. Earlier, Farrer had concluded that humming-birds would not be able to act as pollinators for *Passiflora*. He argued, 'The inner coronas and the processes belonging to them admit the flexible proboscis of the bee or other insect to the nectary below, but would not admit the stiff beak of the humming bird.' Farrer further argued that the pendant flowers, elastic processes and long tube in *Tacsonia* (a genus now subsumed within *Passiflora*) were better adapted for access by hummingbirds (see *Correspondence* vol. 17, Appendix IV, pp. 593–4).

4 CD refers to Henry Austin Bruce. See letter from H. H. Vivian, 10 May 1870 and n. 1, and letter to H. H. Vivian, 11 May [1870].

5 See *Variation* 2: 123–4.

6 Frances Farrer.

From Patrick Nicol 13 May 1870

Sussex Lunatic Asylum | Haywards Heath,
13th May, 1870

Case of vigorous contractions of the *platysma myoides* muscle under great fear[1]

John Hearsay[2] (admitted March 10th, 1870; aet., 20) an undersized but tolerably well made lad, slightly prognathous, of quiet and very obliging disposition and fair intellectual powers, has been taken on several occasions since admission with an affection of the following kind (lasting for a day or two each time):— He is found, when one is called to see him, sitting in a chair or lying down, not speaking unless addressed, with the lower part of his face, and the front of his neck engaged in a series of spasmodic retractions— The corners of his mouth are drawn at each retraction very much backward, and the skin of the parts mentioned is drawn away from the middle line in front. By palpation the *platysma myoides* in its whole extent, and also the *depressors of the jaw and larynx*, and the *sternomastoid muscle*, may be felt to be contracting at each period of spasm— The mouth is not opened nor is the larynx much depressed. The retractions occur about forty times a minute and the series of them lasts from five minutes to several hours.

When questioned as to his state he answers with some difficulty, a contest seeming to occur between the spasmodic movements and the movements necessary to speech. After a little, questioning him and exacting answers seem to allay the spasm. His replies express an all-absorbing fear for the fate of his immortal soul; he says he fears he has been a dreadful sinner, that there is no more hope for him &c., and occasionally when being questioned he ejaculates "Oh,! Lord ("this word very much prolonged") have mercy on me" "Oh Lord! take me to thyself this night", and similar sentences. About the times of these attacks he reads the Bible very much—

The drinking habits of his parents are supposed to be connected with a certain amount of mental weakness which has been from birth; masturbation and unsteady habits are probably causally connected with the present affection which has been going on for two years.

Note— In the case of *Joseph Ellis* (lately a patient in the West Riding Asylum),[3] who was affected with melancholia & most dread-inspiring delusions, there were strongly marked lines on the lower parts of the cheeks and on the lower jaw, concave forwards and upwards, transverse to the direction of the platysma, and seemingly capable of being caused only by its steady ("tonic") contraction. See also Wakeford's case.[4]

Case of Melancholia, coincident with advancing blindness— Effect on facial expression.—

Richard Emsley[5] (Admitted 8th April 1870; aet 52) a man of moderately strong build but now very shaken, troubled with advancing blindness (to which his insanity is attributed), of very distressed expression of face, is in a state of continual terror and anxiety—except when under the influence of narcotics—believing that he is being, or is about to be torn in pieces by imaginary foes,—among whom the Evil One is prominent.

Facial expression. The notable points are that the wrinkles in the middle part of his forehead (i.e. over the root of the nose) are very well marked and likewise lines running outward from the outer angles of his eyes— The eyelids droop.—[6] The corners of his mouth are drawn back and slightly elevated.

His expression is rather blank when he is not agitated by grief; the latter is expressed by a sort of half-sob.

Case of melancholia with delusion— Facial expression—

Elizabeth Wakeford[7] (admitted 14th December, 1870; aet 46) a tall spare woman, of very quiet, retiring disposition, is affected with deep melancholia, which arose on the death of her husband, and is now much aggravated by the accession of a delusion to the effect that she has been bewitched by a fellow-patient and will never recover.—

Facial expression. The long transverse lines of the forehead are slightly marked, and are visible principally in the middle part— Two perpendicular furrows from the inner ends of the eyebrows, running upwards, are notable, especially (at present) that on the right side. The lines below and on the lower eyelid are well marked. The upper eyelids cover about one fourth of the corneae— The eyebrows run, for half their length, outwards and a very little downwards; for the other half outwards downwards and backwards. There are well marked furrows running downwards and outwards from the outer parts of nostrils and mouth, and there are curving lines on the lower jaw & lower part of the cheek, concave upwards and forwards, transverse to the *platysma* fibres. The lips are considerably appressed. Corners of mouth slightly depressed. The expression as a whole is one of seriousness and trouble.

Case of simple melancholia;—with the effects on the facial expression.—

Sarah Ann Greenfield[8] (Admitted March 10th, 1870; aet. 23) of gentle intelligent disposition, and soft womanly countenance (though with not very regular features) is in a state of melancholia owing to the desertion of her husband— While she is most tractable in every way—takes medicines, employs herself when desired with laundry or needle work, and listens quietly and intelligently to verbal admonitions and encouragements—none of all these seems to move her in the slightest degree from a calm & fixed despair of her being of any more good in this world. She is sinking under an attack of pulmonary consumption.

Facial expression.

In the middle of the forehead there are distinct though narrow transverse lines, the constant (very slight) play of which indicates the activity of the muscle beneath.[9]

Just above the inner end of the eyebrow a slight furrow commences which runs upwards and inwards for a little way ($\frac{1}{4}$inch); this is most marked, as she is being examined, on the left side. The eyebrows run, from their inner ends, outward and a little downward—almost in a straight line— There is well marked drooping of both eyelids so as to conceal nearly half the cornea. There is a shallow but still notable depression running from the inner angle of either eye, downwards and outwards. The downward and outward lines, from the outer part of the nostrils and mouth, are marked sufficiently to aid in giving the sad expression of the face. There is an appearance of tightening and slight drawing down, about the corners of the mouth; and there is considerable appression of the lips.— The cheeks are rather sunk.

The expression is one of disconsolate sadness—imprinted upon the bloom of a full, soft, youthful, almost girlish face— This effect was still more noticeable before the ravages of pulmonary consumption[10] gave a more wasted, and older appearance.—

Answers to "Queries About Expression"; from these and other cases.—[11]

5.—[12] Depression not in every case, often slight. The eyebrows are apparently always raised as described.

7.—[13] In one well marked case the phenomenon was as mentioned— In another the skin above & outside the angle of the mouth was drawn back in a direction between the lines of the buccinators & the elevators of the upper lip.

AMemorandum
DAR 172: 55

CD ANNOTATIONS
2.1 *John Hearsay* ... upper lip. 21.3] *crossed pencil*
2.9 By palpation ... spasm— 2.11] *double scored, blue crayon*
7.1 *Richard Emsley* ... prominent. 7.6] 'A man of 52' *in margin pencil*
7.4 believing ... prominent. 7.6] *scored pencil*
8.1 wrinkles ... forehead 8.2] *underl pencil*
8.2 over ... nose] *underl pencil*
11.4 that she has been bewitched] *underl blue crayon*
12.1 The long ... right side. 12.4] *scored pencil*; 'Dep' *in margin, pencil del pencil*; 'Grief Muscle' *in margin, pencil*
12.8 there are ... appressed. 12.10] *double scored blue crayon*; 'Corner of Mouth' *in margin, pencil*
12.10 Corners of mouth slightly depressed 12.11] *double scored pencil*
13.1 Case of simple] 'E' *top of page pencil*
17.1 Just above ... left side. 17.3] *scored pencil*
17.4 There is well ... eyelids 17.5] 'drooping of eyelids' *in margin, pencil*
17.8 There is an ... corners of mouth; 17.9] *scored pencil*; 'Corners of the mouth' *in margin, pencil*
20.1 5.— ... described. 20.2] 'Grief Muscle' *in margin, pencil; line ruled between para. 20 and para. 21, pencil*
Top of first page: '55 Copies' *ink*; '(Mr Patrick Nicol) at the request of Dr Lockhart Robertson.—'[14]
Top of last page: 'E' *pencil*

1 See *Expression*, pp. 298–303, for CD's discussion of the platysma myoides, a sheet of muscle in the neck extending from the collar bone to the lower part of the cheek. CD cited Patrick Nicol for information on the platysma myoides contained in this letter but did not refer by name to individual patients (*Expression*, p. 300). See also *Correspondence* vol. 17, letter to James Crichton-Browne, 22 May 1869 and n. 4.

2 John Hearsay has not been further identified. Aet: *aetatis*, of the age of (Latin).

3 Nicol had been a medical assistant at the West Riding Asylum, Wakefield, Yorkshire.

4 See n. 7, below.

5 Richard Emsley has not been further identified.

6 CD discussed furrowing of the forehead and obliquity of the eyebrows caused by contraction of the 'grief muscle' in *Expression*, pp. 179–86, and cited Nicol for information contained in this letter (*Expression*, pp. 185–6).

7 Elizabeth Wakeford has not been further identified.

8 Sarah Ann Greenfield has not been further identified.

9 CD described this case in *Expression*, p. 186.

10 Pulmonary consumption: tuberculosis.

11 For CD's printed Queries about expression, see Appendix IV.

12 Question 5 in CD's printed Queries about expression is, 'When in low spirits, are the corners of the mouth depressed, and the inner corner of the eyebrows raised by that muscle which the French call the 'Grief muscle'? ...'. CD refers to Nicol's response in *Expression*, pp. 185–6.

[13] Question 7 in CD's printed Queries about expression is, 'When a man sneers or snarls at another, is the corner of the upper lip over the canine or eye tooth raised on the side facing the man whom he addresses?'. CD refers to Nicol's response in *Expression*, p. 245.

[14] Charles Alexander Lockhart Robertson had recently retired as medical superintendent of the Sussex Lunatic Asylum. Nicol was briefly assistant to his successor (Gardner 1999).

To W. T. Preyer 15 May 1870

May 15. 1870.

My Dear Sir.—

I received two days ago your article, which has gratified me in a very high degree.[1] Every word which you say is dictated in the kindest possible spirit; but your estimation of the value of my work would be considered, I fear, by most persons as very much too high. Nothing so full or, as far as facts are concerned, so true has hitherto been published about me. There are one or two very trifling inaccuracies, i.e, about my shooting and hunting in the Highlands. You give also too bad a character to our roads or rather lanes; they are very narrow and rather bad; but when you come to England, I hope that you will prove by visiting me here that they are not impassable. When you have any communication with D^r. Peschel, pray give him my sincere thanks for his present.[2] I can see that the subject is extremely interesting; but the great difficulty with which I read German is a never-ending grief to me.

You ask about my next book; it is partly on the Descent of Man, but chiefly on Sexual Selection, which latter I have found a very large subject. I suppose I shall go to press late in the autumn; but I make very slow progress.[3]

With cordial thanks for all the honours which you have piled on my head, believe me, my Dear Sir | Yours sincerely | Charles Darwin.

P.S.— Please to give my very kind remembrances to my good friend E. Häckel[4]

Copy
DAR 147: 256–8

[1] See letter from W. T. Preyer, 9 May 1870 and n. 2. Preyer sent two copies of Preyer 1870b.

[2] Preyer had also enclosed a work by Oscar Ferdinand Peschel (see letter from W. T. Preyer, 9 May 1870 and n. 3).

[3] *Descent* went to press on 30 August 1870 (see 'Journal' (Appendix II)).

[4] Ernst Haeckel.

From F. C. Donders[1] 17 May 1870

Utrecht,
17 Mai | 1870

Dear Sir,

Je sens maintenant que j'aurais dû vous donner la triste nouvelle de la perte de notre fille, notre unique enfant, l'objet et le centre de notre vie, notre bonheur et

Frans Cornelis Donders
Courtesy of the Smithsonian Institution Libraries, Washington, DC

notre espoir. Elle est décedée le 5 du mois de Mars, 5 Jours seulement après avoir donné la vie à deux enfans, qui, quoique bien faibles, lui ont survécu jusqu'ici.[2] Nous vivons dans une tristesse qui semble ne pouvoir plus finir, et autant que mon gendre bien-aimé, le Docteur Engelmann,[3] J'ai eu de la peine à rentrer dans le laboratoire et à vaquer à nos occupations. Déjà dans les deux derniers mois de la grossesse, notre fille était téllement souffrante, que nous vivions dans une inquiétude continuelle. Sans cela le petit travail que vous m'aviez fait l'honneur de me confièr, aurait été terminé depuis long-temps. Pourtant je venais justement d'y remettre la main, et vous le recevrez dans peu de jours. J'ai examiné avec quelque soin la circulation de l'oeil, et il ne me faut qu'un peu de disposition pour finir ma note à ce sujet que je desire publier dans mon Journal: en général je crois pouvoir confirmer les vues de Sir Charles Bell.[4] Le Docteur Moore[5] vous enverra la traduction anglaise de cet article. Quant à l'autre question, je ne pense pas que le rapport supposé existe.[6] Dans ma lettre suivante, je formulerai ma reponse. Soyez sûr de l'avoir en mains dans 8 Jours après la reception de celle-ci, et avant ce temps Mr. Moore aura reçu un exemplaire imprimé de mon article sur la première question.

Veuillez agréer l'assurance de mon respect et de mon affection, | Dear Sir, | Votre bien devoué | Donders

DAR 162: 224

1 For a translation of this letter, see Appendix I.
2 Donders's daughter was Marie Engelmann; his wife was Ernestine J. A. Donders. The two babies have not been identified.
3 Theodor Wilhelm Engelmann.
4 CD had asked Donders questions about the contraction of the muscles around the eye, and about Charles Bell's account of the physiology of weeping (see letter to F. C. Donders, 13 May [1870] and n. 2). Donders refers to the *Nederlandsch Lancet*, of which he was editor (*DSB*).
5 William Daniel Moore.
6 Donders refers to CD's query regarding the possibility of contraction of the muscles around the eyes producing tears (see letter to F. C. Donders, 13 May [1870], and *Correspondence* vol. 17, letter to F. C. Donders, 6 September 1869 and n. 3).

From T. H. Farrer 17 May 1870

3 Gloucester Terrace
17 May 1870

My dear Mr Darwin

Your letter arrived just at the time that our great sorrow became certain to us: and I write without scruple because no one, I am sure, will feel more deeply for us than yourself.[1] You knew what she was, and what we have lost.— As yet it all seems a dream. She seemed to me so full of life & so likely to keep loving charge over my little ones long after I was gone—

Dont think either the Passifloras or the statistics intrusive or in the way— Anything external that I can interest myself in is good: and I shall plant the seeds as soon as I get to Abinger—and see what I do about the statistics when I have read your chapter.[2] My kindest remembrances to your family who will I am sure feel for us.

Dont trouble yourself to write again

Sincerely yours | T H Farrer

C Darwin Esq[re] FRS

DAR 164: 62

[1] Farrer refers to the death of his wife, Frances Farrer.

[2] See letter to T. H. Farrer, 13 [May 1870] and nn. 3–5.

From C. E. C. B. Appleton 18 May [1870][1]

New Club, | 9, Spring Gardens. S.W.
18th. May

Dear Sir,

I have found considerable difficulty in getting a thoroughly competent reviewer of M[r]. Wallace's recent book for 'the Academy', because I was unwilling to trouble you with a request to do it.[2] Mr. Bates however, with whom I have been sitting this afternoon has led me to hope that as the book contains a new chapter in which the application of your great theory to man is directly called in question, you might be induced to criticize this portion of it with some detail.[3]

The rest of the book being mainly composed of old essays requires little beyond a description of its contents, & a statement how far the views are up to the mark of present knowledge. The smaller additions & emendations, the references to which are collected together in the preface, sh[d]. be cursorily noticed, but the main stress of the criticism sh[d]. be made to bear upon the heretical chapter. Indeed I w[d]. not mind having this noticed exclusively.

As to time I have no desire to hurry you: take your own time to write down the views which you no doubt have already formed respecting this chapter. But a review from you, & on such a subject will be such a treat for my readers that I really do trust that you will not refuse my request. The little note you sent me for No I was most acceptable & did much good to the Journal.[4]

Believe me D[r] Sir | Very truly yours | C. Appleton | Editor.

C. Darwin Esq.

DAR 159: 79

[1] The year is established by the reference to Wallace 1870a.

[2] Appleton refers to Alfred Russel Wallace and Wallace's book, *Contributions to the theory of natural selection. A series of essays* (Wallace 1870a), published in April 1870 (*Publishers' Circular*, 2 May 1870, p. 268). The *Academy* was a new periodical; the first issue had appeared in October 1869. For CD's praise of an early issue, see *Correspondence* vol. 17, letter to J. D. Hooker, 19 November [1869].

3 The new essay referred to by Henry Walter Bates, 'The limits of natural selection as applied to man', expanded on remarks Wallace had made in an earlier essay review (Wallace 1869b). Wallace concluded that 'superior intelligence has guided the development of man in a definite direction' (Wallace 1870a, p. 359). CD underlined the remark in his copy of the book, which is in the Darwin Library–CUL (see *Marginalia* 1: 837). CD and Wallace had discussed the topic when Wallace 1869b first appeared (see, for example, *Correspondence* vol. 17, letter to A. R. Wallace, 14 April 1869).

4 A notice of CD's forthcoming book, *Descent*, had appeared in the *Academy* 1: 15.

To F. C. Donders 19 May 1870

Down. Beckenham. Kent. | S.E.
May 19th. 1870.

My dear Sir

You have my heartfelt sympathy. If I had lived in London, I should no doubt have heard of your great calamity.[1] I once lost a dear & good girl, & know what a dreadful grief it is; but then I had other children left to love.—[2] Your loss is irreparable, & I feel deeply for you.

It is most kind in you to take at present trouble for me in scientific matters. I shall be very grateful for the translation of your paper & your letter whenever they may arrive: for I feel an exaggerated degree of interest in the subject of expression.—

With the kindest sympathy & respect, believe me | my dear Sir | yours sincerely | Charles Darwin.

Copy
American Philosophical Society (B/D25.317)

1 See letter from F. C. Donders, 17 May 1870. Donders's daughter, Marie Engelmann, had died in March 1870.

2 CD refers to the death of his daughter Annie (Anne Elizabeth Darwin) in 1851 (see *Correspondence* vol. 5).

From H. W. Bates 20 May 1870

Royal Geographic Society | 15, Whitehall Place, S.W.
May 20 1870

My dear Mr Darwin

I have been having some conversation with the Editor of the "Academy" about Mr Wallace's last book & the appearance of backsliding from the Darwinian theory which it contains.[1] Other sincere friends of the pure truth have expressed a little surprise & bewilderment at the same phenomenon. The views of friend Wallace are so plausible & suit so well wide-spread prejudices that you no doubt think with me they ought to be controverted. But who is to criticise them? No one but yourself. I do not think anyone else would have the present insight into the fallacy but yourself: to others it would require much study & labour to marshal the arguments. I said so to Mr Appleton & he begged of me to write to you in support of a request he is going to make to you to write him a short article as review of the book.[2]

When you were last in town I spoke to you about some sentences I had written on Man, interwoven in last chapter of Mrs Somerville's book.[3] It weighs on my conscience to think that you took too much notice of what I said: for I do not really think there is much in the matter worthy of your attention

Yours sincerely | H W Bates

DAR 160: 88

[1] Bates refers to Charles Edward Cutts Birchall Appleton, Alfred Russel Wallace, and Wallace 1870a. See letter from C. E. C. B. Appleton, 18 May [1870] and nn. 2 and 3.

[2] See letter from C. E. C. B. Appleton, 18 May [1870].

[3] CD was in London from 5 to 12 March 1870 (Emma Darwin's diary (DAR 242)). Bates had revised the sixth edition of Mary Somerville's book *Physical geography* (Somerville 1870).

From Federico Delpino[1] 20 May 1870

Reale Museo | di | Fisica e Storia Naturale | di Firenze
20. Maggio 1870—

Onorevole Signore!

Soltanto oggi ho ricevuto la sua pregiata lettera del quindici andante, e mi affretto a rispondere.[2]

Mi rincresce che presentemente nel nostro orto botanico non si trovi alcun seme di *Canna Warszewiczii*;[3] ma oggi stesso scriverò una lettera ai miei amici Direttori dei giardini botanici di Siena, Modena e Napoli, i quali potranno somministrarmene; giacchè vedo che i semi di detta pianta figurano nell'Elenco o Catalogo dei semi pubblicato da quegli stabilimenti per l'anno 1870.

Adunque spero fra qualche giorno poter inviare alla S. V. Ill'mo un pacco dei desiderati semi, di provenienza diversa.

Io ringrazio la S. V. della comunicazione fattami sulla leguminosa raccolta a Ravenna dal suo signor figlio. Essa è il *Lotus siliquosus* di Linneo.[4]

Io non ebbi fin qui occasione di studiare lo apparecchio dicogamico di detta pianta. Studiai bensi e descrissi gli apparecchi del *Lotus corniculatus*, del *Lotus hirsutus* e del *Lotus ornithopodioides*. (*Sugli apparecchi della fecondaz. ecc* p. 25, e *Ulteriori osservazioni ecc* p. 45 e 46). Ora esaminando l'esemplare inviatomi dalla S.V. Ill'ma, ho potuto rilevare che lo apparecchio del *Lotus siliquosus* è affatto simile a quello del *Lotus corniculatus*, e dei restanti.[5]

Tale apparecchio dicogamico, che ho denominato "*apparecchio papilionaceo a pompa*" è assai meraviglioso. Esso è paragonabile a quello strumento con cui si fanno le paste da vermicellajo. Gl'insetti fecondatori, che sono esclusivamente apiarie (*apis*, *eucera*, *anthophora* e *Bombus*), per suggere il miele deprimono la carena, e questa depressione fa sì che la sommità dei filamenti che sono clavati e rigonfii alla cima, agisce come uno stantuffo e spinge fuori il polline in forma di vermicelli che si attaccano allo sterno e allo addome degl'insetti.[6]

Questo apparecchio è stato anche descritto da Severino Axell, a pag. 29—del suo scritto "Om anordningarna ecc" "Sätter sig nu en insekt på kölen, så sänkes denna något, den styfva pistillen och filamenterna verka som en tryckpump på pollenmassen, hvilken utjagas i form af ett litet moln genom hålet i kölen spets".[7] La operazione degl'insetti è qui descritta assai bene, ma vi è una inesattezza, in quanto che dice che il polline esce fuori dalla punta della carena in forma di nuvoletta (moln), mentre più precisamente esce in forma di vermicelli.

Gradisca, Uomo preclaro e sommo, il sincero e rispettoso attestato della mia più alta considerazione—

Suo ossequentissimo discepolo | Federico Delpino

DAR 162: 147

[1] For a translation of this letter, see Appendix I.

[2] CD's letter has not been found.

[3] *Canna warszewiczii* is now *C. indica.* Delpino had described the structure of the flowers of *Canna* in 1867 (Delpino 1867b, p. 23).

[4] It is not known which of CD's sons made the collection, but William Erasmus Darwin was in Italy in April 1870, and according to Henrietta Emma Darwin, visited Florence (letter from H. E. Darwin to Emma Darwin, 14 April [1870]; DAR 219.9: 87). *Lotus siliquosus* is now *L. maritimus.* Delpino refers to Carl von Linné (Carolus Linnaeus).

[5] Delpino refers to two works on dichogamy which he had sent to CD, Delpino 1867b, p. 25 and Delpino 1868–75, part 1, pp. 45–6. *Lotus corniculatus* is bird's-foot trefoil; *L. hirsutus* is hairy canary clover.

[6] See Delpino 1868–75, part 1, pp. 45, 47, 52. *Apis*, *Eucera*, *Anthophora*, and *Bombus* are all genera in the family Apidae, which includes bumble bees, euglossine bees, honey bees, and stingless bees.

[7] Delpino quotes from Johan Severin Axell's monograph *Om anordningarna för de fanerogama växternas befruktning* (On the mechanisms for the fertilisation of phanerogamic plants; Axell 1869, p. 29). The passage reads, 'If now an insect settles on the carina, this is somewhat lowered, the stiff pistil and the filaments work as a pressure pump on the pollen-mass, which is ejected in shape of a small cloud through the hole in the cusp of the carina.'

From Charles Renard[1] 20 May 1870

Société Impériale des Naturalistes de Moscou. | *Moscou, le* $\frac{8}{20}$ Mai *1870*[2]

Monsieur

Vos travaux multipliés et utiles pour l'avancement de la science dont témoignent largement et pleinement Vos recherches nombreuses et importantes faites pendant le voyage de circumnavigation du Beagle, Vos observations géologiques sur les îles volcaniques, Votre monographie des Cirripèdes, Vos observations géologiques sur l'Amérique du Sud et. et. et enfin l'ouvrage, si justement celèbre sur l'origine des espéces, ont fait depuis long temps l'admiration de la Société Impériale des Naturalistes de Moscou.—[3] Elle s'est donc procurée la satisfaction de Vous élire unanimement dans sa séance du 19 Février son membre honoraire et elle m'a chargé de Vous en envoyer le diplome qui va suivre demain ces quelques lignes.—[4]

La Société Impériale des Naturalistes qui possède une bibliothèque très riche, uniquement le resultat des dons de ses membres et de ses relations litéraires (aucune somme est fixée et employée pour son augmentation) serait tres charmée si

Vous étiez dans le cas de lui faire parvenir quelquesunes de Vos publications dont Vous pouvez encore disposer.—

Agréez, je Vous prie, Monsieur, l'expression de la plus haute et de la plus parfaite considération avec laquelle j'ai l'honneur d'être | Monsieur | Votre tout dévoué serviteur | Dr. Renard | sécretaire de la Société I. des Naturalistes

Mon adresse: Son Excellence, M. le Conseiller d'état actuel | Dr. Ch[s]. Renard. | Moscou. Miloutinskoe Peréoulok, maison Askarkhanoff.

à Monsieur le Professeur Charles Darwin à Londres.

DAR 176: 120

1 For a translation of this letter, see Appendix I.
2 Renard gives both the Julian (8 May) and Gregorian (20 May) calendar dates.
3 Renard refers to *Living Cirripedia* (1851), *Living Cirripedia* (1854), *Fossil Cirripedia* (1851), *Fossil Cirripedia* (1854); he also alludes to *Journal of researches*, *Volcanic islands*, *South America*, and *Origin*.
4 The diploma is in DAR 229: 23. See Appendix III.

To H. W. Bates [22 May 1870][1]

Cambridge
Sunday

My dear Bates

I have heard from M[r] Appleton, but I have written to him to say that I am very sorry that I must decline undertaking the Review, which however, I fully grant is very desirable.—[2] I have had no practice & it w[d] take me much time; nor do I know that I shd. succeed. But my chief reason is that I really have not a grain of spare strength, & have come down here now for 2 or 3 days to try & rest.[3] You wd. not readily believe how often & urgently I am pressed to write articles— a week ago a most urgent request from one who had much claim; shortly before another;[4] & it is an immense relief to me to be able to say that I *never* write reviews.— So I am very sorry, but must decline, even though you recommend it—

My dear Bates | Yours very sincerely | Ch Darwin

PS If you can let me see an old Proof of your sentences on Man, I shd **really very much** like to read them—[5]

Julian Browning Autographs (List, September 2000); Maggs Bros 2001, Cat. 1303[6]

1 The date is established by the relationship between this letter, the letter from C. E. C. B. Appleton, 18 May [1870], and the letter from H. W. Bates, 20 May 1870, and by the dates of CD's visit to Cambridge. The only Sunday during CD's trip to Cambridge was 22 May 1870 (see n. 3, below).
2 See letter from C. E. C. B. Appleton, 18 May [1870]. CD's reply to Charles Edward Cutts Birchall Appleton has not been found.
3 CD was in Cambridge from 20 to 24 May 1870 (see 'Journal' (Appendix II)).
4 These letters have not been found.
5 See letter from H. W. Bates, 20 May 1870 and n. 3.
6 A facsimile of the first and last pages appeared on the website of Julian Browning Autographs (giving the text down to and including 'nor do I' and from 'reviews.—' A transcription of the letter text and a facsimile of the valediction and postscript appeared the Maggs Bros catalogue.

From J. D. Hooker [22 May 1870][1]

Royal Gardens Kew
Sunday late

Dear *very* old Darwin

because it is so long since I have written to you that, if alive, you must now be really *very* old.— I have such lots to write to you about myself, that I put off & off the boring you over & over again. Then too I have been worked to death with that British Flora, & all manner of things. & now Willy is back from New Zealand & that costs time & thought—[2] as this last is the best news I shall begin with it— the lad is "fat & well-liking" & apparently not in the least hurt by his rough life & voyagering— he is as nice & simple as ever; & beyond that a very few Colonial phrases have replaced many school ones, & his voice is breaking, he is quite unchanged in mind & manners. I quite expected that some roughness at least would have resulted from contact with rough people, but I do not find a trace of it— Hector certainly managed splendidly for him, I am under a very great obligation to him.[3] Mentally I find little development, but he has ability & if I could only get him to take the smallest interest in any one subject, I should not despair. I have not settled what to do with him: he enjoyed his life, & talks most nicely of all the people he was thrown into company with,—but has no wish to go abroad again.

Have I told you of the jolly tour I took with Huxley to the Eifel, with my boy Charlie, to whom H. has taken a great fancy— we dabbled a little in the Geology, which is most curious, took long walks, ate very heartily & came back quite as well as we went.[4]

I am now in a frightful state of mind. The R.S. have referred to me Dawsons Bakerian lecture & I find it so full of *perfect* trash that I am compelled to recommend it's *non* publication, it will be a knock-down blow to the poor man— The systematic part is very meagre indeed, the vegetable anatomy miserable & often utterly wrong; the affinities more often mere guess work, than not; & as to the theories & speculations, they would make your hair stand on end.[5]

What do you say to the Appalachian chain being "the centre of Plant-distribution from the Carboniferous era onwards" and the **N.E.** corner of that chain being "the point of origin of the Palæozoic Flora".— Dawson's discovery of many fossil plants of the Carboniferous type, low down in the Devonian, is a very excellent one, & if he had but contented himself with the proper summation of that, & accompanied it with sufficient evidence his paper would have been a valuable one but he has overloaded the systematic portion with generic & specific names from the merest scraps of tissue, & utterly misunderstood the structure of the most complete specimens. Then too he has page after page of the wildest speculations upon *Archaic* & *synthetic* types, on *origination of species* by "embryonic retardation & advance", on "conditions of compression & expansion", upon "advanced & postponed maturity"— & on "tendency of synthetic types to become specialized in the direction of their constituents elements" upon "types comparable to a chemists allotropic forms" & so forth. The blessed lecture cost me $2\frac{1}{2}$ days hard work, & I am excessively sorry for the poor man, whose reception was enthusiastic by Lyell & Murchison,[6] & whose fall will be proportionally heavy & headlong. He has brought the paper all the way from Canada & was *invited* to make the Bakerian Lecture of it! Altogether the affair has cut me up terribly, & I would rather have burnt my fingers than performed so hateful a duty.— It is a very small consolation, that Dr. D. has a wonderfully good opinion of—*himself*. He & his wife spent half a day here;— she must have been very pretty, & that, to me, makes matters worse— & worse still, their son is deformed & I know that he too will feel hurt—[7] The curse of Cain will cleave to me.— By the way he pooh-poohed my Greenland paper,—[8] this has only just come into my head, & does not mend matters— for he will if he hears of it, put my sinister report down to spite, whereas I would fain have heaped coals of fire on the poor devils head by a gushing (not crushing) report—

We have had Cinthea Symonds[9] here for a fortnight & I still find her very charming— My Uncle John Gunn has just returned from the Mediterranean, he had collected largely fossils at Marathon & Electra in Greece, but they were all *stolen* off his caleche[10] at Naples!— what a sell for the theives—

Charlie has come out top of his form at Marlborough,[11] & is promoted— I cannot tell you how happy this has made me.

I am at a monograph of *Nepenthes* for DC. Prod. & then shall go to long suspended Gen. Plant.—[12]

Do write me a line & tell me what you are doing & how you all fare.

Ever Yr aff | J D Hooker

DAR 103: 47–50

1 The date is established by the relationship between this letter and the letter to J. D. Hooker, 25 May [1870]. The Sunday before 25 May 1870 was 22 May.

2 The last known letter written by Hooker was that of 21 November 1869 (*Correspondence* vol. 17). William Henslow Hooker had been sent to New Zealand for his health in 1869 (see *Correspondence* vol. 17).

3 James Hector had supervised Hooker's son while he was in New Zealand (see *Correspondence* vol. 17).

4 Hooker refers to Thomas Henry Huxley and Charles Paget Hooker. The Eifel is an undulating plateau in western Germany characterised by deep valleys, extinct volcanoes, and crater lakes (*Columbia gazetteer of the world*).

5 Hooker was one of two referees who reported on John William Dawson's Bakerian lecture of 1870. He sent a highly critical report to George Gabriel Stokes, the secretary of the Royal Society. Dawson's was the first Bakerian lecture not to be published in the *Transactions of the Royal Society of London* (see Sheets-Pyenson 1991, pp. 179–81, 183–4).

6 Charles Lyell and Roderick Impey Murchison.

7 Dawson was professor of geology at McGill University, Montreal. Dawson's wife was Margaret Ann Young Dawson; their son, George Mercer Dawson, was disfigured by the effects of tuberculosis of the spine, which he suffered as a child (*DCB*).

8 Hooker refers to his paper 'Outlines of the distribution of Arctic plants' (Hooker 1860); see *Correspondence* vol. 10, letter from J. D. Hooker, 2 November 1862 and n. 8.

9 Hooker refers to Hyacinth Symonds.

10 Caleche or calash: a light carriage with low wheels and removable hood or top (*OED*).

11 Marlborough College, Wiltshire.

[12] Hooker refers to his monograph 'Nepenthaceae, Cytinaceae', which appeared in the last part of Candolle and Candolle 1824–73 (17: 90–116). Hooker also refers to *Genera plantarum* (Bentham and Hooker 1862–83). The first part of the second volume appeared in 1873.

To Alfred Newton [22 May 1870][1]

Bull Hotel
Sunday

My dear Newton

Many thanks for your kind note.[2] I intend trying to see Sedgwick tomorrow morning as I did not find him today; so if $4\frac{1}{4}$ will suit you to call here & pick me up we will go on together to the Museum.[3]

yours very sincerely | Charles Darwin

LS
Balfour Library, University of Cambridge (held at CUL)

[1] The date is established by the dates of CD's visit to Cambridge (see 'Journal' (Appendix II)). The only Sunday during CD's trip to Cambridge from 20 to 24 May 1870 was 22 May 1870.

[2] Newton's letter has not been found.

[3] CD refers to Adam Sedgwick and the Woodwardian Museum, Cambridge. Sedgwick was Woodwardian Professor of geology and head of the museum (*ODNB*).

To J. D. Hooker 25 May [1870]

Down. | Beckenham. | *Kent. S.E*
May 25

My dear Hooker

I have indeed had time to grow very old since we have written to each other: never so long before.[1] I had intended more than once writing, but I heard indirectly how hard you had been pressed by work & so refrained. I suppose thank God that your British Flora is as good as finished.[2] Now you have rewarded my virtue by a very long & pleasant letter, telling me much that I wished to hear. I am heartily glad that Willy[3] has come back so well & not in the least deteriorated in manners; not that I sh^d. have expected this from all that I have heard of him. It will be a fearful puzzle for you what to do with him now: God knows it is puzzle enough in every case whatever; & what our Boy Horace is to do, I know no more than the man in the moon: he is clever enough; but so often ailing in health, that I doubt whether he can stick to anything. About ourselves I have little news: Henrietta[4] has just returned much strengthened by her 4 months residence at Cannes. Last Friday we all went to the Bull Hotel at Cambridge to see the Boys & for a little rest & enjoyment.—[5] The Backs of the Colleges were simply paradisical. On Monday I saw Sedgwick who was most cordial & kind: in the morning I thought that his mind was enfeebled: in the evening he was brilliant & quite himself. His affection & kindness charmed us all. My visit to him was in one way unfortunate: for after a long sit he proposed to take me to the Museum;[6] & I could not refuse, & in consequence he utterly prostrated me; so that we left Cambridge next morning, & I have not recovered the exhaustion yet. Is it not humiliating to be thus killed by a man of 86, who evidently never dreamed that he was killing me.— As he said to me "Oh I consider you as a mere baby to me".— I saw Newton[7] several times, & several nice friends of Frank.— But Cambridge without dear Henslow was not itself: I tried to get to the two old Houses, but it was too far for me.—[8]

As for myself I have been working away on Man, & as usual running to *much* greater length than I expected. I hope to go to press this Autumn.[9] In the plant line I have been continuing my comparison of crossed & self-fertilised plants, & am coming I think to some interesting results, & some curious anomalies.— Do you happen by any strange chance to have seeds of *Canna Warszewiczi* **matured in some hot country**; they w^d. be of great value to me— I shall, however, receive some from Italy.—[10]

I truly sympathise with you about Dawson's paper: under all the circumstances it is quite horrid for you; but I fancy there are always two referees, which will partly relieve the burthen.[11] It amuses me to hear of his hashing up Cope's unintelligible views about accleration of development & Agassiz on Synthesis, in explanation of the origin of species, as he spoke so contemptuously on any attempt to explain such problems, when my book first appeared. He seemed to me to be the staunchest of creationists.—[12]

Will you not come here some Sunday—or any time, any day w^d suit us, & the sooner the better: you w^d rejoice all our hearts.

I am delighted to hear about Charlies success[13]—our son's success in life is more delightful than our own—

Yours affect. | C. Darwin

Endorsement: '/70'
DAR 94: 169–72

1 See letter from J. D. Hooker, [22 May 1870].

2 CD refers to *The student's flora of the British Islands* (Hooker 1870).

3 William Henslow Hooker. See letter from J. D. Hooker, [22 May 1870] and n. 2.

4 Henrietta Emma Darwin.

5 CD refers to Francis Darwin, who was a student at Trinity College, Cambridge, and to George Howard Darwin, a fellow of that college.

6 Adam Sedgwick and CD had visited the Woodwardian Museum, Cambridge, on 23 May 1870 (see letter to Alfred Newton [22 May 1870]).

7 Alfred Newton.

8 John Stevens Henslow had died in 1861. CD probably refers to the two houses where Henslow lived in Cambridge, the 'Gothic Cottage' in Regent Street and the house overlooking Parker's Piece (see Walters and Stow 2001, p. 108).

9 *Descent* went to press on 30 August 1870 (see CD's 'Journal' (Appendix II)).

10 CD had asked Federico Delpino for seeds of *Canna warszewiczii* (now *C. indica*; see letter from Federico Delpino, 20 May 1870 and n. 3). CD has double scored both margins, as well as double underlining the words 'matured in some hot country'.

11 CD refers to Hooker's critical report on John William Dawson's Bakerian lecture of 1870 (see letter from J. D. Hooker, [22 May 1870] and n. 5). The other referee was Peter Martin Duncan, who gave a generally favourable report on the paper (see Sheets-Pyenson 1991, pp. 181, 184).

[12] Dawson had discussed the evolutionary ideas of Edward Drinker Cope in his paper, 'Modern ideas of derivation' (Dawson 1869, pp. 134–7). CD had scored part of this section in his copy of an offprint of the article (p. 135; p. 15 in CD's copy, which is in the Darwin Pamphlet Collection–CUL). Cope proposed that changes in species were the result of acceleration or retardation during embryogenesis of the individual, but had not explained why or how the alterations would occur, other than crediting the direct intervention of God. Dawson considered Richard Owen's theory of evolutionary change (CD evidently referred to Louis Agassiz in error) to be a synthesis of the theories of Jean Baptiste Lamarck and CD (Dawson 1869, pp. 133–4).

[13] CD refers to Charles Paget Hooker. See letter from J. D. Hooker, [22 May 1870].

From James Philip Mansel Weale [25 May 1870][1]

description.

The Periglossum I was also uncertain about but think it may be "*macrum*' as that species is spoken of as a Kaffrarian plant but I have no description.[2]

Our present house is so small that I have been unable to open any of my collections or I had intended sending you some larvæ of Dactylethra & the bee's nest spoken of before.[3]

I am trying the introduction of English & other grass seeds on my farm & hope they will prove successful. It is most singular to see how some weeds take hold of the ground where manure is in plenty about old sheep kraals especially Xanthium spinosum, Datura stramonium, a solanum & numerous other plants. The French marygold I have never remarked before as a weed but here it grows abundantly & bears magnificent double flowers finer than I have seen in any garden.[4]

Rev[d]. M[r]. Henslow's remarks on Medicago, Indigofera &c. & my own observations on Schizanthus, Muraltia &c I have carefully watched various species of Polygala but have never noticed any movements.[5]

I have been much interested by observing a common species of Termes,[6] the soldier of which has a pointed head. I have often remarked that a species of ant was always very active about their nests seizing the workers, but although much larger, retreated at once from the soldiers. As I had several times put my hand into a nest with the view of ascertaining whether the soldiers could hurt one I was the more surprised as I had never suffered any inconvenience although other species bite severely for their size, & the workers of this species have what look like formidable jaws although they never use them in defence: At last I found out the secret. I saw one

Incomplete[7]
DAR 181: 43

CD ANNOTATIONS
Margin: 'J. P. Mansel Weale May 25— 1870 | Natal.— Instinct' *ink*; 'Soldier Termes' *pencil*

[1] The date is established by CD's annotation.

[2] *Periglossum macrum* is now *Sisyranthus macer*; Kaffraria was a region on the eastern coast of South Africa, between Natal and Cape Colony; it became part of Cape Colony in 1865 (Stewart 1996).

[3] The genus *Dactylethra* is now *Xenopus* (clawed frogs). The reference to Weale's previous mention of a bee's nest has not been identified. However, Weale enclosed with this letter a manuscript paper about a bee's nest that has not been found, and four other manuscript papers, with figures and specimens of the plants described, that are now in the archives of the Linnean Society (MSS SP1250, SP1251, SP1252, SP1253, Weale Misc, drawer 39). See letter to Richard Kippist, 28 July 1870, and letter to J. P. M. Weale, 30 July 1870. The four extant papers were published in the *Journal of the Linnean Society* (Weale 1870a, 1870b, 1870c, and 1870d). None of the figures were published, and Weale 1870c was substantially cut by CD.

[4] *Xanthium spinosum* is the spiny cocklebur; *Datura stramonium* is the thorn apple or jimsonweed. Solanaceae is the potato family. The French marigold is *Tagetes patula*.

[5] George Henslow's observations on *Medicago sativa* (alfalfa) and *Indigofera* were published in the *Journal of the Linnean Society* (*Botany*) 9 (1867): 327–9 and 355–8 (G. Henslow 1865 and 1866). Weale discussed movement in the stamens of *Muraltia* in a letter of 7 July 1867 (*Correspondence* vol. 15); he described irritability of the stamens in *Schizanthus* in a letter of 23 October 1868 (*Correspondence* vol. 16).

[6] *Termes* is a genus of the family Termitidae (higher termites).

[7] The text down to 'in any garden' is on one side of a sheet of paper; the text from 'Rev[d] M[r] Henslow's' and CD's annotations on the other.

From T. H. Farrer 26 May 1870

Abinger Hall, | Reigate.(Post Town) | Gomshall(Station) S.E.R.
26. May/70

My dear Mr Darwin

I am not discouraged by M. Müller's Passiflora.[1] It is very difficult to judge of a dried specimen But so far as I can see the corona belongs to a different type, of which, according to engravings we have some kinds in this country, where the corona is not so regular, so stiff, & so grating like, as in P. Cærulea &c.[2] What it means is a different question.

I have been interested in watching the common Berberis (barberry),[3] much visited by bees, wasps, hornets & flies. Old Sprengel is accurate & quaint as usual, but, I think has not gone quite far enough.[4] He assumes that the stamens move, when touched by insects, so as to bring the open anthers in contact with the viscid edge of the stigma. This however does not explain the undoubtable fact that the stamens do not continue appressed to the stigma, but, under this sun, quickly recover their old places, ready, when there is more nectar and another insect, again to spring forward. If the end is to place the pollen on successive insects, ready to be carried by each to another flower, this double motion becomes intelligible

Sincerely yours | T H Farrer

I have sent some of Mullers seeds to Kew & to the Regents Park Gardens, besides sowing some here[5]

DAR 164: 63

[1] Farrer refers to Fritz Müller and to a dried flower of a *Passiflora* that Müller had sent from Brazil (see letter to T. H. Farrer, 13 [May 1870]).

[2] See letter to T. H. Farrer, 13 [May 1870] and n. 3. *Passiflora caerulea* is the blue or common passion flower.

[3] Farrer refers to *Berberis vulgaris*.

[4] Christian Konrad Sprengel described his observations on *Berberis vulgaris* in Sprengel 1793, pp. 203–6.

[5] Farrer refers to the Royal Botanic Gardens, Kew, and to the Botanic Gardens, Regents Park, London.

From F. C. Donders[1] 27 May 1870

Utrecht,
27 Mai | 1870

My dear Sir,

L'expression de votre affectueuse sympathie m'a été bien chère. Vous qui avez fait la perte douloureuse d'une fille chérie, vous avez, en quelque sorte, la mésure de ce que c'est que de perdre son enfant unique.[2] A côté du vide, aucun espoir de jamais le voir remplir. J'y pense toujours, mais mieux vaut ne pas en parler. Je n'ai donc qu'à vous remercier de vos bonnes paroles de sympathie là où toute consolation doit nécessairement faire défaut. Le courage de bien employer, autant que possible, les jours qui me resteront à vivre, me revient peu-à-peu. C'est vers là que tendent tous mes efforts.

J'ai fait l'article sur l'action des paupières dans les cas de congestion oculaire par la pression exspiratoire. Je l'envoie ce jour même au Docteur Moore, à Dublin, qui a bien voulu se charger de la traduction.[3] En même temps, je vous envoie l'article, en épreuve, en Hollandais. Vous ne serez pas satisfait: vous y trouverez bien des choses, dont vous n'avez guère besoin, et la reponse à la question fundamentale ne vous paraîtra pas assez préremptoire. Mais vous verrez au moins que je n'ai pas traité la question à la légère. Permettez-moi de vous donner un court aperçu de mon travail.

d'Abord j'ai constaté le fait de l'occlusion de paupières dans les excès de pression exspiratoire. Puis j'ai examiné la manière dont cette pression produit les congestions et la dilatation des vaisseaux en général. J'ai étudié ensuite les conditions spéciales, sous lesquelles se trouvent les vaisseaux antérieurs, les vaisseaux intra-oculaires et les vaisseaux retro-oculaires, et j'ai decrit les symptomes produits par la pression exspiratoire dans ces trois ordres de vaisseaux. En attendant, l'occasion se présente d'examiner à quel point la pression palpébrale soit en état de les réprimer.

Je me résume a-peu-près dans les termes suivants:

"Nous nous sommes convaincus que les vaisseaux externes de l'oeil, autant que les vaisseaux intra-oculaires et retro-oculaires, se dilatent sous l'influence de la pression respiratoire immoderée. Nous avons vu que l'occlusion des paupières tend à diminuer ou à réprimer cette dilatation, d'abord par la pression, qu'elles exercent, et, en outre, à ce qu'il semble, en vertu d'une certaine action associée à cette occlusion. Nous avons constaté qu'à chaque pression exspiratoire immoderée les paupières se ferment ou montrent au moins une tendence à se fermer. Nous sommes donc en droit de conclure que l' occlusion des paupières protege les vaisseaux et l'oeil lui- même contre les dangers qui *puissent* profluer d'une pression exspiratoire excessive.

Le point faible c'est—qu'il n'y a pas des indices assez certaines du danger que cette pression fait encourir dans l'état normal des yeux."

J'avais intention d'ajouter à cette lettre la reponse à l'autre question;[4] mais je dois vous prier de vouloir attendre jusqu'à demain. Il est fâcheux que je ne suis pas maître de mon temps.— Veuillez, en attendant, agréer les assurances de mon respect & de ma plus parfaite considération.

Votre tout devoué | Donders

DAR 162: 225

[1] For a translation of this letter, see Appendix I.

[2] Donders refers to Anne Elizabeth Darwin and Marie Engelmann. See letter to F. C. Donders, 19 May 1870 and nn. 1 and 2.

[3] See letter from F. C. Donders, 17 May 1870. Donders refers to his article, 'De beweging van het oog toegelicht met het phaenophthalmotroop' (Donders 1870a), and the English translation by William Daniel Moore, 'On the action of the eye-lids in determination of blood from expiratory effort' (Donders 1870b). CD discussed Donders's conclusions in *Expression*, pp. 160–2.

[4] For CD's second question, see the letter to F. C. Donders, 13 May [1870].

From F. C. Donders[1] 28 May 1870

Utrecht,
28 Mai 1870

My dear Sir,

Je continue la lettre que j'avais l'honneur de vous adresser hier.[2] Vous me permetrez, j'espère, d'entrer dans quelque développement préliminaire qui me facilitera l'exposition de mes idées.

La sécrétion des larmes se fait sous l'influence directe de fibres nerveuses sécretoires, spécifiques, comprises dans la portio major nervi trigémini.[3] Elles s'en séparent comme nerf lacrymal, qui prend son origine de la 1ère et peutêtre aussi de la 2ième branche du nerf V. Les fibres sécrétoires ont une action centrifugale; toutes les autres fibres de ces deux grandes branches semblent avoir une action centripetale seulement: elles sont sensitives. En quelque sorte, c'est donc une anomalie que les fibres, qui président à la sécrétion des larmes, sont contenues dans le même neurilemma.[4] Mais, en réalité, cette disposition n'a aucune conséquence, la fonction du nerf ne dépendant que de son centre d'origine et de sa distribution périphérique. De la même manière, nous voyons quelques fibres se séparer du nerf facial (la chorda tympani),[5] qui président à la secretion de la glande submaxillaire.

Il s'agit donc de trouver dans la moëlle allongée le noyau spécial, c'est-à-dire le centre, à cellules nerveuses, qui, par son action, détermîne l'innervation des fibres nerveuses qui produisent la sécrétion des larmes, et d'étudier ses rapports. D'après ce que nous savons sur les noyaux d'autres nerfs, correspondant à des fonctions déterminées (selon Stilling, Schroeder Van der Kolk, Deiters et autres anatomistes),[6] comme à la fonction de l'organe électrique, au mouvement consensual des yeux, à la contraction de la pupille, etc. etc., nous sommes en droit d'admettre un noyau spécial pour la secretion des larmes, pour la sécrétion de la saline, pour l'accommodation, pour chaque groupe de muscles, dont l'action correspond à une fonction spéciale soit de locomotion soit d'expression, etc.

Or, sur le noyau du nerf lacrymal agissent deux sortes de fibres nerveuses: 1º. des fibres centrales qui sont en rapport direct avec les grands hémisphères, 2º. des fibres, agissant par action reflexes, qui partent de la péripherie et vont abouter au noyau.

Un noyau pareil doit correspondre au muscle orbiculaire, et il est mis en action également 1º. par des fibres centrales, qui déterminent les mouvements volontaires et automatiques, 2º. par des fibres, qui, parties de la péripherie, agissent par reflexe.

Si maintenant l'action du muscle orbiculaire et celle de la glande lacrymale quelquefois sont combinées, il faut que l'action, soit de fibres centrales, soit de fibres reflexes, agisse en même temps sur les deux noyaux. Le rapport pourrait être un tel que l'action, provoquée d'une manière ou d'une autre, dans le muscle détermine la sécretion de larmes, ou vice versa, par action reflexe. C'est là encore une question à resoudre.

Examinons d'abord, s'il y a des rapports entre les *fibres centrales* des *deux* noyaux:

1º. l'innervation volontaire du muscle orbiculaire peut avoir lieu, même à un haut dégré, sans action associée de la glande lacrymale. S'il y a quelque trace de larmes, ce phénomène peut s'expliquer soit par la pression, exercée directement sur la glande et l'évacuant en quelque sorte de son contenu, soit par irritation de l'œil, qui, par reflexe, comme nous verrons, stimule la sécrétion des larmes.

2º. l'innervation automatique de la glande lacrymale peut, à son tour, avoir lieu, indépendamment de celle du muscle orbiculaire. Tout le monde sait qu'un souvenir douleureux fait couler les larmes, sans contraction du muscle orbiculaire, que les enfants, frustrés dans leurs désirs, souvent dirigent sur nous leurs yeux grandement ouverts, pleins de larmes. Si plus tard, les yeux se ferment, c'est probablement par action reflexe, produite par l'irritation de l'œil.

3º. Il arrive que les fibres centrales portent simultanément sur les deux noyaux, p.e. dans un grand chagrin, une douleur violente. On voit surtout les enfants et les personnes nerveuses verser alors des larmes, en fermant les yeux. A la lorsque, l'action réflexe s'y ajoute, mais elle n'est pas primaire.

Quant aux rapports entre les fibres, qui *par action réflexes* portent sur *les deux* noyaux, ils semblent être beaucoup plus fréquents. Ces fibres appartiennent à deux classes de nerfs différentes: a. le nerf optique et b. le nerf de la cinquième paire.

a. Quant au nerf optique, son irritation par une lumière vive produit plus tôt la contraction du muscle orbiculaire que la sécrétion de larmes. Pourtant, il n'est pas rare de voir en même temps la sécrétion augmentée.

Mr. Herkenstein trouva, dans ses expériences sur des animaux, que l'irritation du nerf optique d'un seul oeil provoque la sécrétion de larmes dans les deux yeux; tandisque, au contraire, ce n'est que l'oeil du même côté qui réagit sur une irritation de la muqueuse, dans laquelle se distribue la cinquième paire.[7]

b. L'irritation de certaines branches sensitives de la $5^{\text{ième}}$ paire agit sur les deux noyaux, ou, pour mieux m'exprimer, il y a des points de la periphérie dont l'irritation provoque en même temps, par action réflexe de ses nerfs, la contraction de l'orbiculaire et la sécrétion de larmes. C'est ce que nous trouvons dans l'irritation de la conjonctive oculaire, surtout près de la cornée et de la cornée elle-même,—soit par un simple attouchement, soit par un corps étranger permanent, soit par une inflammation superficielle dans l'ophthalmie à phlyctènes, dite scrophuleux, etc.— J'ai encore observé quelques cas d'une affection bien curieuse: après un attouchement, qui ne produisait ni plaie, ni contusion, par exemple, la friction d'un habit, il se produisait un spasme du muscle orbiculaire et un larmoiement très profus, qui ne duraient qu'une heure peut-être.[8] Mais plus tard, il y avait de temps en temps, quelque fois avec de grandes intervalles, de quelques semaines p.e., des accès violents de spasme du muscle orbiculaire et de sécrétion de larmes, avec rougeur soit primaire soit secondaire de l'oeil.[9] Il y a donc des points à la surface de l'oeil agissant sur les deux noyaux.— Il y en aussi dans la muqueuse du nez. Une irritation de cette muqueuse fait couler les larmes & produit en même temps une contraction d'abord de la partie orbitale seulement du muscle orbiculaire, assez souvent comme préliminaire d'éternûment, qui, faisant explosion, fait fermer les paupières avec force et favorise encore la secrétion de larmes: c'est que, une sécrétion, du reste, periodique, une fois établie par l'irritation nerveuse, est augmentée régulierement par une congestion sanguine.—

Cette action reflexe double d'une irritation de la muqueuse nasale parait correspondre à ce qu'il faut pour éliminer l'irritation—produite par un corps étranger— autant par la courant d'air violent de l'éternûment que par les larmes qui descendent dans le nez.

Si pendant la vomitarition, le muscle orbiculaire se contracte en même temps qu'il y a larmoiement, n'y aurait-il pas le plus souvent irritation de la muqueuse nasale par les substances vomies, ou au moins dans l'arrière-gorge, qui semble aussi déjà agir sur les deux phénomènes.

Dans ces cas d'irritation de la muqueuse nasale, il n'y a pas de la douleur vraie, il n'y a qu'un leger picotement ou une sensation de constriction, dont le premier surtout excite l'éternûment, et dont la dernière nous échappe facilement.

Une irritation du sac lacrymal ne semble produire que le larmoiement. Au moins dans les cas d'irritation inflammatoire chronique les larmes coulent sans que l'oeil soit fermé, et, chose interessante, si l'on détruit dans ce cas le sac lacrymal, aussi qu'une cicatrisation se fait, sans état irrité, la sécrétion de larmes souvent est presque nulle, autant que l'œil est exempt de toute irritation: on ne voit pas des larmes, malgré l'occlusion entière des conduits. Ceci prouve bien que la sécretion de larmes est périodique comme celle de la saline etc.—

Le clignotement ordinaire ne semble donc pas accompagné d'une secrétion des larmes: C'est une action reflexe tant à l'irritation du nerf optique que du nerf de la cinquième paire (surface de l'oeil). Pourtant je n'oserai pas affirmer absolument que la sécrétion de larmes fait absolument défaut dans le clignotement: d'ailleurs nous savons que le méchanisme de l'absorption des larmes par les points lacrymaux est mis en jeu par ce clignotement. Pour le moment, My dear Sir, je ne saurais vous en dire davantage. Vous même, vous avez remarqué que la contraction volontaire du muscle orbiculaire n'a pour ainsi dire, aucun effet sur la sécrétion de larmes. Or, ça ne peut nous étonner, car il est bien sûr que les deux actions ne sont pas dans les rapports de cause à effet: ce n'est que l'irritation produit par la contraction, qui pourrait donner lieu indirectement à quelque larmoiement, et l'irritation des yeux et des paupières mêmes par les larmes, qui pourrait donner lieu à une contraction,—dans les deux cas par action réflexe.

Il faut donc recourir aux autres nerveux, aux noyaux comme Stelling les a appelés.[8] J'ai la conviction qu'un examen anatomique très assidu peut nous faire monter des fibres du nerf lacrymal, jusqu'aux cellules nerveuses, où elles aboutissent: c'est difficile et laborieux, mais ce n'est pas impossible. On attendrait la même chose pour la chorda tympani, pourvu que qu'on veuille: "where there is a will, there is a way". Peut-être je saurai engager un de mes élèves d'entreprendre l'examen.

Toujours il est vrai que la premiere question qui se presente dans l'ordre des idées qui nous occupent, consiste dans la détermination précise des noyaux, en prenant pour point de depart les fibres periphériques. Plus tard se presentera la question plus difficile des rapports de ces noyaux avec les fibres reflexes et les fibres centrales, et leurs rapports mutuels, c'est-à-dire de chaque noyau aux autres. S'il y a manifestation de la vie cérébrale dans l'expression de la figure, c'est qu'il y a correspondence organique parfaite dans les details, chaque muscle ou groupe des muscles ayant leurs cellules nerveuses, sur lesquelles des actions locales, specifiques peut-être, liées à la manifestation psychique, agissent d'une manière harmonieuse.

Pardonnez-moi, je vous en prie, si, en écrivant, je me suis laissé un peu entraîner. Peut-etre çà ou là, il y a allusion à quelque fait physiologique, qui puisse vous servir à fixer vos idées sur la question que vous m'avez fait l'honneur de me poser. Soyez bien sûr, que vous me rendrez heureux en vous addressant à moi sur des points qui sont du domaine de la physiologie, dont votre esprit n'embrasserait pas tous les détails. J'ai pour vos ouvrages une admiration et un respect sans bornes.

Veuillez, my dear Sir, agreer les assurances de cette considération illimitée | Donders

DAR 162: 226

CD ANNOTATIONS
1.1 Je ... idées. 1.3] *crossed pencil*
11.1 Quant] *after opening square bracket, pencil*
14.8 J'ai encore] *after opening square braket, pencil;* 'encore' *del pencil*
14.14 Il y a donc] *after closing square bracket, pencil*
14.19 c'est que, ... sanguine.— 14.21] *scored pencil*
18.4 sans état ... larmes, 18.5] *scored pencil*
19.1 Le clignotement] *after opening square bracket, pencil*

1 For a translation of this letter, see Appendix I.

2 See letter from F. C. Donders, 27 May 1870. Donders there discussed the first of two questions that CD had put to him in the letter to F. C. Donders, 13 May [1870].

3 For CD's question on tears, see the letter to F. C. Donders, 13 May [1870]. The 'portio major nervi trigemini' is the large sensory root of the trigeminal nerve.

4 Donders evidently uses the term 'neurilemma' to refer to the sheath of connective tissue around a nerve or nerve fascicle (*OED*).

[5] 'Chorda tympani': a branch of the facial nerve which traverses the mucous membrane of the tympanum (*OED* s.v. 'tympanichord').
[6] Donders refers to Benedict Stilling, Jacob Ludwig Conrad Schroeder van der Kolk, and Otto Friedrich Karl Deiters.
[7] Ulrich Herzenstein reported these experiments in his article, 'Zur Physiologie der Thränensecretion' (On the physiology of tear secretion; Herzenstein 1867).
[8] CD quoted, in translation, the two preceding sentences in *Expression*, p. 166.
[9] 'Stilling nucleus': a column of large neurons that gives rise to the dorsal spinocerebellar tract (*Stedman's medical eponyms*).

To T. H. Farrer 28 May [1870][1]

Down. | Beckenham | *Kent. S.E.*
May 28th

My dear Farrer

I suppose I must have known that the stamens reversed their former position in Berberis, for I formerly tried experiments with anæsthetics, but I had forgotten the fact, & I quite agree with you that it is a sound argument that the movement is not for self-fertilisation.[2] The N. American barberries (mahonia) offer a good proof to what an extent natural crossing goes on in this genus; for it now almost impossible in the country to procure a true specimen of the 2 or 3 forms originally introduced—[3]

I hope the seeds of Passiflora will germinate, for the turning up of the pendent flower must be full of meaning.[4] I am so glad that you are able to occupy yourself a little with flowers: I am sure it is most wise in you for your own sake & children's sake.—[5]

Some little time ago Delpino wrote to me praising the Swedish book on the fertilisation of plants; as my son George can read a little Swedish, I shd like to have it back for a time just to hear a little what it is about, if you wd. be so kind as to return it by book Post.—[6]

I am going steadily on with my experiments on the comparative growth of crossed & self fertilised plants, & am now coming to some curious anomalies & some interesting results.[7] I forget whether I showed you any of them when you were here for a few hours.—[8] You ought to see them; as they explain at a glance why nature has taken such extraordinary pains to ensure frequent crosses between distinct individuals.

If in the course of the summer, you shd. feel any inclination to come here for a day or two, I hope that you will propose to do so, for we shd. be delighted to see you; though, as you know, I can talk with no one except for a very short time.—

Pray believe me | yours very sincerely | Ch. Darwin

P.S | I remember years ago being surprised at observing that the pollen in the genus Hedaroma (or Darwinia)[9] is almost liquid viscid, & this seemed to me deserving examination. The plant is also rather ornamental & curious.

Linnean Society of London

1 The year is established by the relationship between this letter and the letter from T. H. Farrer, 26 May 1870.
2 See letter from T. H. Farrer, 26 May 1870 and n. 4. No record of the experiments CD mentions has been found.
3 Species of *Mahonia* native to western North America were introduced to Europe in the 1820s (Ross and Auge 2008, p. 22).
4 See letter from T. H. Farrer, 26 May 1870 and n. 1.
5 CD alludes to the recent death of Farrer's wife, Frances Farrer (see letter from T. H. Farrer, 17 May 1870).
6 CD refers to Federico Delpino's remarks on Johan Severin Axell's monograph, Axell 1869 (see letter from Federico Delpino, 20 May 1870 and n. 7). CD had lent his copy to Farrer (see *Correspondence* vol. 17, letter to Federico Delpino, 14 October 1869 and n. 8). CD also refers to George Howard Darwin.
7 CD refers to his research for *Cross and self fertilisation.*
8 The last known visit by Farrer to Down House was sometime shortly before 9 October 1869 (see *Correspondence* vol. 17, letter from T. H. Farrer, 9 October 1869 and n. 6).
9 *Hedaroma* is now subsumed within *Darwinia*, a genus of Australian plants in the family Myrtaceae.

To Armand de Quatrefages 28 May [1870][1]

Down Beckenham | Kent
May 28th.

Dear Sir

I have received and read your volume & am much obliged for your present.[2] The whole strikes me as a wonderfully clear & able discussion, & I was much interested by it to the last page. It is impossible that any account of my views could be fairer, or as far as space permitted fuller, than that which you have given. The way in which you repeatedly mention my name is most gratifying to me. When I finished the second part, I thought that you had stated the case so favourably that you would make more converts on my side than on your own side. On reading the subsequent parts I had to change my sanguine view. In these latter parts many of your strictures are severe enough, but all are given with perfect courtesy & fairness.[3] I can truly say I would rather be criticised by you in this manner than praised by many others. I agree with some of your criticisms, but differ entirely from the remainder; but I will not trouble you with any remarks. I may, however, say that you must have been deceived by the french translation as you infer that I believe that the Parus & the Nuthatch (or **Sitta**) are related by direct filiation: I wished only to show by an imaginary illustration how either instincts or structure might first change.[4] If you had seen *Canis magellanicus* alive, you would have perceived how foxlike its appearance is, or if you had heard its voice, I think, you would never have hazarded the idea that it was a domestic dog run wild;[5] but this does not much concern me. It is curious how nationality influences opinion: a week hardly passes without my hearing of some naturalist in Germany who supports my views, & often puts an exaggerated value on my works; whilst in France I have not heard of a single zoologist except M. Gaudry[6] (and he only partially) who supports my views. But I must have a good many readers, as my books are translated, & I must hope, notwithstanding your strictures, that I may influence some embryo naturalists in France.

You frequently speak of my good faith, & no compliment *can* be more *delightful* to me, but I may return you the compliment with interest for every word which you write bears the stamp of your cordial love for the truth.

Believe me, dear Sir | with sincere respect | yours very faithfully | Charles Darwin

P.S. I am much obliged to you for sending me your lectures in the "Revue des Cours etc" for they have been very useful to me.—[7]

I have also to thank you for your kindness of my son George in Paris.[8]

LS
American Philosophical Society (379)

1 The year is established by CD's reference to receiving Quatrefages's book (see n. 2, below).

2 CD refers to Quatrefages's book *Charles Darwin et ses précurseurs français* (Quatrefages 1870). CD's copy, with the pages uncut, is in the Darwin Library–CUL. The book was a reproduction of a collection of articles that Quatrefages had sent in 1869; CD's annotated copy of an offprint of the articles (Quatrefages 1868–9) is in the Darwin Pamphlet Collection–CUL.

3 The first part of Quatrefages 1870 begins with an exposition of the views of some French naturalists on the transformation of species; the second chapter details CD's theory, while the third focuses on facts that are compatible with the theory. In the second part of the book, Quatrefages presents a detailed critique of Darwinian theory.

4 In *Origin* 4th ed., p. 212, CD noted that he had seen titmice hammering yew seeds to break them 'like a nuthatch'. On page 281, he suggested that natural selection might preserve each slight variation of the titmouse's beak, until it had one 'as well constructed for this purpose as that of the nuthatch'. In Quatrefages 1868–9, p. 63, Quatrefages had questioned whether CD had demonstrated that the nuthatch was the 'grandson' of the titmouse. In the book version, a note was added giving a reference to the section of *Origin* where CD had commented on the titmouse and nuthatch (see *Origin* 4th ed., p. 281), but there was no mention in the note that CD had not, in fact, suggested a genealogical connection between these species (Quatrefages 1870, p. 156).

5 *Canis magellanicus* is now *Lycopex culpaeus magellanicus*, a subspecies of the culpeo, a type of South American fox. Quatrefages had suggested that many supposedly unique species were merely dogs that had become feral and reverted to ancestral characteristics (Quatrefages 1870, pp. 281–4).

6 Albert Gaudry.

7 CD refers to the published version of Quatrefages's lectures on anthropology, which appeared in instalments in the *Revue des Cours Scientifiques de la France et de l'Etranger* (Quatrefages 1867–9). The lectures discussed various aspects of human development; CD referred to them several times in *Descent.*

8 George Howard Darwin had visited Paris in March 1869 (see *Correspondence* vol. 17, letter from Armand de Quatrefages, 29 March 1869 and n. 2).

To Charles Renard 28 May 1870

Down | Beckenham | Kent
May 28 70

Sir,

I beg leave to thank you for your very courteous letter of May 20th. in which you announced to me that the Imperial Society of Naturalists of Moscow has conferred on me the distinguished honour of electing me an Honorary Member.[1] This morning I received your Diploma[2]

I hope that you will express on my part to the Society how deeply I feel gratified by this honour. In accordance with your suggestion I have directed my publisher to forward to your Society the last edition of my Origin of Species and my Journal of researches during the Voyage of the Beagle[3]

With my sincere thanks | I have the honour to remain, | Sir, | Your most obedient Servant | Charles Darwin

LS
Stecher and Klavins 1965

1 See letter from Charles Renard, 20 May 1870.
2 For the diploma, see Appendix III.
3 CD refers to *Origin* 5th ed. and *Journal of researches* (1860). On the books, and the election of CD as an honorary member, see Stecher and Klavins 1965.

From Adam Sedgwick 30 May 1870

Cambridge
May 30 1870

My dear Darwin

Your very kind letter surprised me— Not that I was surprised at the pleasant & very welcome feeling with which it was written.[1] But I could not make out what I had done to deserve the praise of "extraordinary kindness of yourself & family". I would most willingly have done my best to promote the objects of your visit but you gave me no opportunity of doing so. I was truly grieved to find that my joy at seeing you again was almost too robust for your state of nerves, & that my society, after a little while, became oppressive to you. But I do trust that your Cambridge visit has done you no constitutional harm—nay rather that it has done you some good— I only speak honest truth when I say that I was overflowing with joy when I saw you; & saw you in the midst of a dear family party & solaced at every turn by the loving care of a dear Wife & Daughters.[2] How different from my position—that of a very old man, living in cheerless solitude!

May God bless & cheer you all with the comfort of hopeful hearts!—you & your Wife, & your Sons & Daughters!—

You were talking about my style of writing— I send you my last specimen; & it will probably continue to be my last— It is the continuation of a former Pamphlet of which I have not one spare copy— I do not ask you to read it.[3] It is addressed to the old people in my native Dale of Dent, on the outskirts of Westmorland— while standing at the door of the old vicarage I can see down the valley the Lake mountains—Hill Bell at the head of Windermere about 20 miles off—[4] On Thursday next D.V.[5] I am to start for Dent which I have not visited for full two years.— Two years ago I could walk three or four miles with comfort— Now, alas!, I can only hobble about on my stick

I remain your true hearted old friend | A Sedgwick

DAR 177: 128

[1] CD's letter has not been found. CD had visited Cambridge earlier in May and had met with Sedgwick on 23 May 1870 (see letter to J. D. Hooker, 25 May [1870] and n. 6).

[2] Sedgwick refers to Emma Darwin, Henrietta Emma Darwin, and Elizabeth Darwin.

[3] Sedgwick refers to his *Supplement to the memorial of the trustees of Cowgill Chapel* (Sedgwick 1870). CD's copy is in the Darwin Library–Down.

[4] The Yorkshire dale of Dent (now Dentdale) is south-east of the county of Westmorland (both regions are now part of Cumbria).

[5] D.V.: *Deo volente* (God willing; Latin).

From J. D. Hooker [31 May 1870][1]

Kew.
Tuesday

Dear Darwin

I am choking with vanity—& behold the reason of it—enclosed—which I cannot resist the pleasure of sending to you—[2]

I will come & see you as soon as I can, meanwhile your Willy[3] is coming to me on Saturday week I hope.

I am going to send my Willy to Mr La Touche in Salop—a friend of Symonds' & Bentham's, who has brought up young Colenso, & who will have Frank Lyell.[4] I fancy that some of my friends will think I am putting my son into a nest of young Adders!—

Ever yr affec | J D Hooker

[Enclosure]

Thames Ditton
30 May

Dear Dr Hooker

I thank you very much for kindly sending 'The Student's Flora', which I am right glad to see.[5] It will surely prove a success, and be held a standard for many years to come. If asked the prominent characteristics of the S. F. I might sum up in two pair of words:

'Condensed completeness— Judicious Selection.' These characteristics appear alike in the plan of the work as a whole, & in the workmanship in detail.

It is a marvel to me how you have got time for it, considering the many other matters you have to look to, official & individual.

Your Appendix list, as a receptacle for plants which uselessly increase and interfere with the general text of our Floras, makes practical a remark already in print for Part iii of my Compendium Cyb. Brit., which thus, some months hence, will appear as a needless anachronism,— a post-factum suggestion.[6]

You have hit the medium between Bentham's Handbook & Bab[n's] Manual, in the uexata questio of species;—avoiding the opposite fault or defect of each.[7] Babington places too many very questionable species on equality with the unquestioned. In avoiding that extreme, one way, Bentham too much neglects them, as real though subordinate forms. I doubt whether either of them has fairly realized to his own mind the obvious fact that "species" are widely non-equivalents, & require treatment accordingly,—that it is not a drop at once from "true species" to "mere variety", as the expressions usually run.

Of course, opinions will diverge very much in special instances,—between true species, sub-species, variety, variation, etc.— But a great step is gained for the rising botanists by training them through a descriptive Flora to recognize and trace *gradations*. Among others, C. Darwin ought to thank you for that; the stereotype representations of 'species' and 'variety' in Floras, even more than the like in Faunas, being barriers to the acceptance of Darwinic views.

As a sample how opinions will differ in the special instances, I should about as soon call the Pear a sub-species of Apple, as some of the plants you place in that relationship; ee: gr. Chenopodium album & ficifolium, or Hypericum quadrangulum & undulatum.[8] And yet, it must be confessed that some facts side with you, rather than me, in *these* instances. The Chenopodia & Hyperica have been much confused, or one taken for the other; not so, Pears & Apples, either trees or fruit.

Sincerely Yours | Hewett C[l]. Watson

DAR 103: 46; Royal Botanic Gardens, Kew, DC 105: 236

1 The date is established by the date of the enclosure and the relationship between this letter and the letter to J. D. Hooker, 2 [June 1870]. The first Tuesday after 30 May 1870 was 31 May.

2 The enclosure, from Hewett Cottrell Watson, was evidently returned to Hooker, probably with CD's letter of 2 [June 1870].

3 William Erasmus Darwin.

4 William Henslow Hooker had recently returned from New Zealand (see letter from J. D. Hooker, [22 May 1870]). James Digues La Touche was the vicar of Stokesay, Shropshire. Hooker also refers to William Samuel Symonds, George Bentham, Robert John or Francis Ernest Colenso, and Francis Horner Lyell.

5 Watson refers to *The student's flora of the British Islands* (Hooker 1870), published in May 1870 (*Publishers' Circular* (1870): 328).

6 For Hooker's appendix of excluded species, see Hooker 1870, pp. 475–80. Watson refers to his remarks in *A compendium of the Cybele Britannica: or British plants in their geographical relations* (Watson 1870, pp. 425–9). The book appeared in November 1870 (*Publishers' Circular* (1870): 762), but had already been privately published in three parts between 1868 and 1870. CD's annotated copy of the privately published version (Watson 1868–70) is in the Darwin Library–CUL.

7 Watson refers to George Bentham and Bentham 1858 and to Charles Cardale Babington and Babington 1843. Uexata (or vexata) questio: vexed question (Latin).

8 *Chenopodium album* (fat-hen, or white goosefoot) and *C. ficifolium* (figleaf goosefoot) are now considered separate species but are known to hybridise in the wild (*Chenopodium x zahnii*; see Stace 1997, p. 140).

Hooker classified *C. ficifolium* as a subspecies of *C. album* (Hooker 1870, p. 316). The name *Hypericum quadrangulum*, which was at the time sometimes applied to plants that are now recognised as different species, is used by Hooker for *H. maculatum* (imperforate St John's wort) and not for *H. tetrapetalum* (four-petal St John's wort), the valid name for *H. quadrangulum* (see Hooker 1870, p. 69, and Robson 1990, p. 136). *Hypericum undulatum* (wavy St John's wort), classified by Hooker as a subspecies of *H. quadrangulum*, is now generally considered to be a separate species.

To Federico Delpino 1 June [1870][1]

Down. | Beckenham | *Kent. S.E.*
June 1st

Dear Sir

I am greatly obliged for the extraordinary trouble which you have taken in procuring for me the seeds of the Canna.[2] I have received them safely & they will be extremely useful to me. I have so often shown my son the manner of fertilisation of various leguminous plants that I do not think it is likely that he has exaggerated the novelty of the contrivance in the Lotus; & therefore if you have any opportunity I still think it might be worth your looking at[3]

With very sincere thanks for your kindness, I remain with much respect,

Dear Sir, | yours faithfully | Charles Darwin

Anna Barone (private collection)

1 The year is established by the relationship between this letter and the letter from Federico Delpino, 20 May 1870.

2 See letter from Federico Delpino, 20 May 1870.

3 See letter from Federico Delpino, 20 May 1870 and n. 4. The son concerned may have been William Erasmus Darwin.

From Francis Galton 1 June 1870

42 Rutland Gate SW.
June 1. 1870

My dear Darwin

Though I have no new litter to report, & shall only have one before the end of this month, I do not like to let more time go by, without heartily thanking you for your helpful & encouraging letter.[1] I will not trouble you with details now, but simply say that I feel sure unless some unexpected disaster to my stock should arise, I shall have a very complete set of experiments finished before August.

My bucks have been heavily re-transfused & I have a doe in the same state. Also I shall have all the combinations, extreme & intermediate of pure and tranfused bucks with pure & transfused does.

I find I cannot manage pigeons for want of a dove cot & dare not try dogs lest the Zoolog: Gardens shd. be alarmed by the noise and I should be extruded.[2]

But notwithstanding this I can assure you that I have the matter firmly enough in hand. and will be guided by the results, as to the extent of future work.

De-fibrinised blood[3] is my salvation. I literally put into my silver grays during *one operation* as much blood as I can get from the divided throats of two rabbits each the same size as the patient. & I have 3 bucks who have undergone two operations (but unluckily the earlier ones were far less succesful)

Very sincerely yours | Francis Galton

DAR 105: 19–20

[1] CD's letter to Galton has not been found. Galton's last known letter to CD is dated 12 May 1870. Galton was testing CD's hypothesis of pangenesis by transfusing blood from rabbits of one colour into rabbits of another and observing the colour of later offspring. See letter from Francis Galton, 15 March 1870 and n. 2, and Galton 1871.

[2] Galton was carrying out his experiments in the Zoological Gardens, Regent's Park, London.

[3] On the process of 'defibrinisation', see the letter from Francis Galton, 8 April 1870 and n. 4.

To Adam Sedgwick 1 June [1870][1]

Down. | Beckenham | *Kent. S.E.*
June 1

My dear Professor Sedgwick

I must write a line to thank you very much for the present of your last writing.[2]

What I said in my last note was exactly true: I felt your manner towards myself & family extraordinarily kind, & I have not often in my life been more gratified.—[3] Walking round the museum, after conversing with you, tired me a good deal, for I am knocked up with incredible ease; but I was quite right again next morning.[4] I shall always look back with great satisfaction to my last visit (as it will probably prove) to Cambridge.— I hope that you will feel strong, so as to enjoy to a certain extent your visit to your old home.—[5]

With the most sincere respect, believe me | Yours truly obliged | Charles Darwin

Stephen Jay Gould (private collection)

[1] The year is established by the relationship between this letter and the letter from Adam Sedgwick, 30 May 1870.

[2] See letter from Adam Sedgwick, 30 May 1870 and n. 3.

[3] CD's letter to Sedgwick has not been found; however, see the letter to J. D. Hooker, 25 May [1870].

[4] CD refers to the Woodwardian Museum (see letter to Alfred Newton, [22 May 1870] and n. 3).

[5] Sedgwick was planning to visit Dent, in Yorkshire (see letter from Adam Sedgwick, 30 May 1870). Dent is now in Cumbria.

To J. D. Hooker 2 [June 1870]

Down
2[d]

My dear Hooker

I have been glad to see Watson's letter. I must study your manner of arrangement of vars. & subspecies &c.—[1]

No wonder you are proud; I suppose the man is a *very* good judge & he is surely very clever, & considering his temper & delight in finding fault, I sh^{d}. think his note was the greatest complement ever paid to any human being!![2]

Yours affect | C. Darwin

Endorsement: 'June/70'
DAR 94: 174

1 Hewett Cottrell Watson had written to Hooker to praise his *Student's flora of the British Islands* (Hooker 1870); see letter from J. D. Hooker, [31 May 1870] and enclosure.

2 For an earlier disagreement between Hooker and Watson, see *Correspondence* vol. 16, letter from J. D. Hooker, 24 November 1868.

To F. C. Donders 3 June [1870][1]

Down Beckenham | Kent
June 3rd

My dear Sir,

I do not know how to thank you enough for the very great trouble which you have taken in writing at such length, & for your kind expressions towards me.[2] I am particularly obliged for the abstract with respect to Sir C. Bell's views, as I shall now proceed with some confidence; but I am intensely anxious to read your essay in full when translated & published, as I hope, in the Dublin Journal.[3] As you speak of the weak-point in the case, viz; that injuries are not known to follow from the gorging of the eye with blood, I may mention that my son & his friend at a Military Academy, tell me that when they perform certain feats with heads downwards, their faces become purple & veins distended, and that they then feel an uncomfortable sensation in their eyes, but that, as it is necessary for them to see, they cannot protect their eyes by closing the eyelids.[4] The companions of one young man who naturally has very prominent eyes, used to laugh at him when performing his feats, & declare that some day both eyes w^{d} start out of his head.

Your essay on the physiological & anatomical relations between the contraction of the orb-muscles & the secretion of tears is wonderfully clear, & has interested me greatly. I had not thought about irritating substances getting into nose while vomiting; but my clear impression is that mere retching causes tears; I will however try to get this point ascertained.[5] When I reflect that in *vomiting*, (subject to the above doubt) in violent coughing from choking, in yawning, violent laughter, in the violent downward action of the abdominal muscle as during the evacuation of fæces when constipated, & in your very curious case of the spasms,—that in all theses cases, the orb-muscles are strongly and unconsciously contracted; & that at the same time tears often certainly flow, I must think that there is a connection of some kind between these Phenomena; but you have clearly shewn me that the nature of the relation is at present quite obscure.[6]

With the *most cordial* thanks for your great kindness | believe me | my dear Sir | yours sincerely obliged | Charles Darwin

Copy
American Philosophical Society (B/D25.318)

1 The year is established by the relationship between this letter and the letters from F. C. Donders, 27 May 1870 and 28 May 1870.
2 See letters from F. C. Donders, 27 May 1870 and 28 May 1870.
3 In his letter of 27 May 1870, Donders summarised his article in Dutch on the action of the eyelid in modulating blood pressure raised by expiratory effort (Donders 1870a). The article was translated in *Archives of Medicine* (Donders 1870b), not the *Dublin Journal.* See letter from F. C. Donders, 27 May 1870 and n. 3. CD refers to Charles Bell; see letter from F. C. Donders, 17 May 1870 and n. 4.
4 Leonard Darwin was at the Royal Military Academy at Woolwich; the friend mentioned may have been Richard Matthews Ruck.
5 CD refers to Donders's letter of 28 May 1870. He discussed whether mere retching produced tears in *Expression*, p. 165.
6 CD discussed the secretion of tears while laughing, coughing, retching, straining, or yawning in *Expression*, pp. 164–6.

To James Paget 4 June [1870][1]

Down. | Beckenham | *Kent. S.E.*
June 4th

My dear Paget

You have been so kind in helping me that I think you will again.—[2] Can you persuade the resident Doctor in some hospital to observe a person retching *violently*, but throwing nothing from the stomach.— When a person vomits violently, he always contracts the orbicular muscles round the eyes, & tears often flow. I have been discussing with, or rather enquiring from, Prof. Donders of Utrecht about the relation between the spasmodic contraction of the orbicular muscles & the secretion of tears. In the case of vomiting he suspects that irritating matter is injected into the nose & thus causes tears.[3] From my own personal experience I do think that this is the case. Therefore what I want to know is whether in *violent* retching, with nothing ejected from stomach, when the orbiculars are strongly contracted, do the eyes fill with tears or even overflow?—[4]

Pray do not acknowledge this note & believe me | Yours very sincerely | Ch. Darwin

Wellcome Library (Wellcome MS 5703/38)

1 The year is established by the relationship between this letter and the letter to F. C. Donders, 3 June 1870.
2 For Paget's earlier assistance to CD, see *Correspondence* vol. 11, letter to James Paget, 11 March [1863], *Correspondence* vol. 15, letter from James Paget, 9 July 1867, and *Correspondence* vol. 17, letter to James Paget, 29 April [1869].
3 CD refers to Frans Cornelis Donders. See letter from F. C. Donders, 28 May 1870, and letter to F. C. Donders, 3 June [1870] and n. 5.
4 Paget's reply has not been found, but see *Expression*, p. 165.

From W. W. Reade 4 June 1870

cre His Excellency | The Administrator | Lagos
June 4th. '70

My dear Sir

I put my future address above but do not start from here (Accra) till tomorrow. I came down from the hills yesterday—[1]

The other day I took a trip with a German missionary to have a look at the Volta, wh. is the largest river on the Gold Coast; on taking our lunch out of a newspaper the word *Darwinismus* caught my eye. I found it was an article on a work justifying your theory as regards scripture much in the style of Asa Gray's article. The work was I think by a Dr. Jäger.[2] But you have doubtless many correspondents in Germany who keep you au courant with the Darwinian controversy in that country.

After a series of very cautious inquiries I am able to assert that the negro's idea of beauty is the same as ours, & not exactly opposite as so many have supposed— I hope to give you details shd you require them when I return.—[3] They admire a black skin more than the lighter varieties (I do not mean European mixtures but pure native) which are so common here & almost everywhere.[4] But the Jollofs are I think without exception of a *bottle* blackness.— I formerly ascribed this blackness to the fact that they lived in the hottest part of Western Africa; on the Senegal which borders on the Sahara; of course the Sahara itself is hotter: but I was supposing that the greatest amount of heat *with moisture* produced the black skin. Since the black skin is a beauty, however, it is possible that the Jollofs in their process of selection, *rogued* out the reds & yellows.[5] But I have since met with other jet-black tribes. They inhabit a country which is swampy & moist in the rainy season, and *febrile*: intensely hot in the day. Dr. Wells' idea is I believe possible; although the immunity of the negroes from coast fever is by no means so complete as is usually supposed.[6]

Respecting the thick scull. Children a few days after birth are put behind their mothers' backs & exposed bare-headed to the sun. The mortality is great. Would not the children with thick sculls be selected and so on according to your theory? Then the udder-like breast was I think originally produced by long periods of suckling (usually 3 years) & this once established, the peculiar feature would I imagine appear in young women at the suckling period even if they had not suckled— The capacity of cows & goats to give more & more milk as the udder is systematically teased is curious: by milking the goats twice a day instead of only once the quantity is much increased. Among instances of habit you may have noticed the goats' love for any stone log of wood or hillock on wh. he can stand however little it may be above the level.

I have found stone implements here. My attention was called to their existence by a German missionary who said he had sent some to the mission house at Basle but had heard nothing concerning them. I send them by this mail to the cre. of Mr. Swanzy 122 Cannon St City. This is his expedition. I write to apprise Huxley D. Forbes Bates &c. of the fact.[7] I hope it will be an addition to existing facts on the Stone Age but am not sufficiently up in the science to know much about that.

The negroes are also fond of pigeons. I met with them in the house of a chief who lived interior of S^a. Leone[8] beyond the Niger, almost at the boundaries of the S. Leone trade. The travellers who come down to S. Leone often buy English fowls to take home to breed from. At Falaba I noticed a horse w^h. was decidedly above the average. said so to the King who replied that the master of the horse was noted among them for his skill in breeding horses.[9] This I think illustrates y^r statement chapter 1. p. 36 (edit of '66)[10] No people c^d be more pure of European influences than they are. The poverty of the Africans generally prevents them from doing much in the way of rejecting & selecting.

The more I see & hear of these people; the more I am convinced that they are acquainted with the principles of breeding. A West Indian emigrated to Akropong where I have be⟨en⟩ staying: the natives after examining her toe⟨s⟩ said that she came from their country & this was partly corroborated by her recollections of the circumstances of her sale & export. This story may be laughed at; but no people c^d even make such an assertion or dream of making it unless they were in the habit of closely studying race or rather tribe peculiarities The girls of Croboe near Akropong have a reputation for strength & beauty; when for a certain offence against the laws of their country they are banished, the Akropong people are delighted to get them to breed from.

I lately read a learned article by Ehrenberg on earth eating in this country.[11] As far as my investigations go, the earths in question are argillaceous with a pleasant odour & are used by young people as a kind of sweetmeat, & by others as a medicine. But the idea of using earth as a food has not entered their heads *here* at all events. He also writes about the famous red sand of the Cape de Verdes.[12] Certainly there is abundance of red soil from S^a. Leone downwards more than he supposes but would it not require a desert of red sand ⟨to⟩ send off such quantities? There is one desert not yet visited; probably a small one: behind Ashanti, & just where geographers usually place the mythical Kong Mountains.[13]

The astonishment expression of the hand to the mouth is also expressed in words in Croboe. The man who uses this gesture will say "my mouth cleaves to me" (ie to my hand).[14] I hope to be home now in a few months, so any other point of inquiry you must please notify in your next, which ⟨wi⟩ll probably come to hand on my return from the Niger[15]

I remain | My dear Sir | yours very truly | Winwood Reade

DAR 176: 38

CD ANNOTATIONS
1.1 I put … country. 2.6] *crossed blue crayon*
3.1 After … blackness.— 3.6] *crossed pencil*
3.3 I hope … everywhere. 3.5] *scored blue crayon*
3.6 I formerly … level. 4.10] *crossed blue crayon*
6.4 At Falaba] *opening square bracket red crayon*
6.5 King … country 7.4] *scored blue crayon*
6.5 the master] 'owner' *added red crayon*
6.5 of the horse] *del red crayon*
6.5 was noted … horses. 6.6] *enclosed in quotation marks red crayon*

6.6 among them] *del red crayon*
6.7 pure of European influences] *underl red crayon*
7.4 & this ... Mountains. 8.9] *crossed blue crayon*
9.1 The ... Niger 9.5] *scored blue crayon*; *crossed pencil*

CD note:[16]

'p. 1. idea of Beauty' *ink; del pencil*; '2' *added*; 'The immunity of negros p. 2. from fever not so great as supposed.' *ink, enclosed in braces ink, crossed pencil*; 'Celts S. Africa & Egypt' *ink del ink*; '(Dom. Animals.—p. 3 Principles of Breeding by savages'[17] *ink*; 'p. 4. Expression of Astonishment.' *ink del ink*; 'about some negroes at Fal [*interl*] who had never [*above* 'not'] *[3 words illeg]*' *red crayon*; 'W. Winwood Reade' *ink*; 'Selection by Savages' *pencil*; 'who had never seen [*below del* 'seen men of'] Europeans' *pencil*

1 Reade had been staying at a missionary station at Akropong (see letter from W. W. Reade, [*c.* 8 or 9 April 1870] and n. 11).

2 The German missionary has not been identified. The article in the newspaper was presumably a review of Gustav Jäger's *Die Darwin'sche Theorie und ihre Stellung zur Moral und Religion* (Jäger [1869]). See also the letter to Gustav Jäger, 17 February 1870. Reade also refers to Gray's pamphlet, *Natural selection not inconsistent with natural theology* (A. Gray 1861). For more on Gray's pamphlet, see *Correspondence* vol. 9, Appendix III.

3 CD had asked Reade to observe what style of beauty was admired by natives in his letter of 21 May [1868] (*Correspondence* vol. 16). CD reported Reade's views in *Descent* 2: 350.

4 CD reported this observation in *Descent* 2: 346.

5 The Wolof people (also spelled Ouolof) now live primarily in Senegal, Gambia, and Mauritania (Appiah and Gates eds. 2005, 5: 430). On the Wolof's 'process of selection', see *Correspondence* vol. 17, letter from W. W. Reade, 26 December 1869 and n. 7.

6 Reade refers to William Charles Wells and Wells 1818. CD discussed the idea of Wells, and others, that darker skinned peoples were immune to various fevers, poisons, and parasites in *Descent* 1: 242–6. 'Coast fever' is not the cattle disease (East Coast fever), but the illness or illnesses described by CD in *Descent* 1: 243 as 'the fatal intermittent fevers that prevail along, at least, 2600 miles of the shores of Africa, and which annually cause one-fifth of the white settlers to die, and another fifth to return home invalided'.

7 Reade described and provided a woodcut of some of the stone tools in Reade 1873, 2: 167–9. Andrew Swanzy, the London-based merchant who sponsored Reade's 1868 and 1869 expeditions into the Niger region, gave them to the Christy Collection (Reade 1873, 2: 168 n. 1, 352–3). Reade also refers to Thomas Henry Huxley, David Forbes, and Henry Walter Bates. John Lubbock described these and other finds in Lubbock 1870b; according to his account, the implements, mostly axes, were found at Akropong and Aburi on the Gold Coast (now in Ghana) and at Odumassie on the Volta (Lubbock 1870b, p. xciv). Odumassie was in the region of modern-day Akuse, Ghana.

8 Sierra Leone.

9 The King of Falaba was Manga Sewa. CD cited Reade for this story in *Variation* 2d ed., 2: 191.

10 In *Origin* 4th ed., p. 36, CD discussed 'unconscious selection', which he said resulted from 'every one trying to possess and breed from the best individual animals'.

11 Reade refers to Christian Gottfried Ehrenberg and Ehrenberg 1868.

12 Cape Verde is an archipelago off the coast of West Africa. See Ehrenberg 1868, p. 20.

13 On the mythical Kong mountains, see Bassett and Porter 1991.

14 CD cited Reade for this information in *Expression*, p. 289. The expression of astonishment was the first asked about in CD's list of queries about expression, which he had sent to Reade with his letter of 21 May [1868] (see *Correspondence* vol. 16).

15 Reade did not make another expedition to the Niger; in Reade 1873, 2: 509, he explains, 'commercial jealousy prevented me from obtaining a passage, and my work was therefore left in some degree incomplete'.

16 The notes are on a slip of paper that was filed with the letter in the archive. The page numbers are references to the pages of Reade's letter.

17 In *Descent* 2: 358, CD commented, 'negroes fully appreciate the importance of selection in the breeding

of their domestic animals, and I could give from Mr. Reade additional evidence on this head.' See also n. 9, above.

From T. H. Farrer 5 June 1870

3, Gloucester Terrace, | Regent's Park.
5 June/70

My dear Mr Darwin

I could not find any thing to send to F. Müller so I wrote him a long screed of a letter, giving the gist of my notions about the corona &c of Passiflora. I hope he wont be bored.[1]

Sincerely yours | T H Farrer

DAR 164: 64

[1] Fritz Müller had sent Farrer, via CD, seeds and a dried flower of *Passiflora* (see letter from Fritz Müller, 16 February 1870, and letter to Fritz Müller, 12 May 1870). On Farrer's notions about the corona of *Passiflora*, see the letter to T. H. Farrer, 13 [May 1870], n. 3.

From James Crichton-Browne 6 June 1870

West Riding Asylum, | Wakefield.
6. June— 1870.

My dear Sir,

I am ashamed to write to you & infinitely distressed to contemplate all the annoyance & trouble that my negligence may have occasioned you. Enclosure was given to my man-servant to pack up. He did so & placed it in a cupboard where it was again forgotten.[1]

Today I have *myself* seen it despatched by rail carriage—pre-paid. Will you kindly let me have one line to say whether it arrives in safety.

Enclosed in Duchenne (at the beginning) you will find a few crude notes on expression.[2] I promise more, in a little time, although I fear you will scarcely trust to me after all my carelessness. Bear in mind, in extenuation of my faults that I am one of the hardest worked men in her Majestys Dominions. As a rule I *toil* daily from 8. a.m. to 11. pm. contending all the while with bad health & great anxiety.

I send you a photograph of a female patient in the Southern Counties Asylum, Dumfries N.B. under the care of Dr. Gilchrist in whom the bristling of the hair was well seen.[3] The woman was in a tranquil mood when the portrait was taken. When she was agitated—the ascendant emotion being horror—the hair stood out like *wire*.

We are beginning to take large photographs here, the size of Duchennes[4] & will I think secure some interesting observations. I shall send you some. Is there any point connected with expression that you would particularly wish to have illustrated?

With sincere apologies & profound esteem, | I am, | Yours most faithfully | J. Crichton-Browne

Charles Darwin Esq &c &c

[Enclosure]

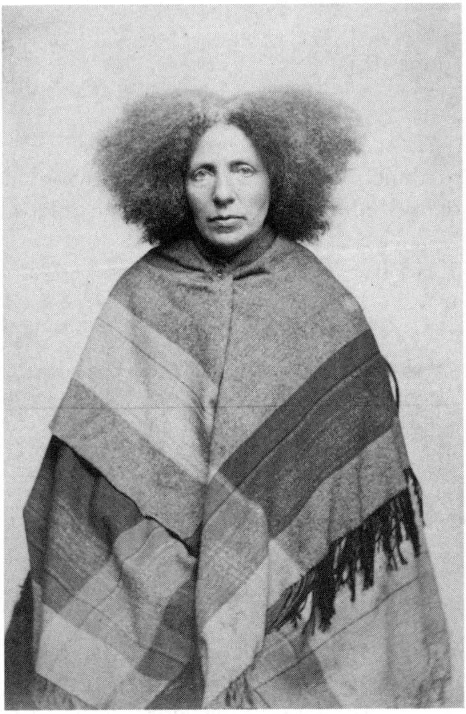

DAR 53.1: 68; DAR 161: 311

[1] Crichton-Browne refers to the 'Atlas' to Duchenne 1862 (see letter to James Crichton-Browne, 2 April [1870]). The man-servant has not been identified.

[2] See memorandum from James Crichton-Browne, [6 June 1870].

[3] The photograph is probably that in DAR 53.1: 68, reproduced in *Expression*, p. 296. See enclosure, above. For an earlier photograph of a woman with bristling hair sent by Crichton-Browne, see *Correspondence* vol. 17, enclosure to letter from Henry Maudsley, 20 May 1869. James Gilchrist had succeeded Crichton-Browne's father, William Alexander Francis Browne, as medical superintendent of the Crichton Royal Institution in 1857; the Southern Counties Asylum was part of the institution. N.B.: North Britain.

[4] Crichton-Browne refers to the large photographs in Guillaume Benjamin Amand Duchenne's 'Atlas' to his work on expression (Duchenne 1862).

From James Crichton-Browne [6 June 1870][1]

With reference to figures 7, 8, 9, 10 & 11.[2] I would venture with great deference to submit that my observations lead me to believe, that awakened attention, especially if accompanied by surprise or wonder is expressed is expressed by the action of the occipito-frontalis, elevation of the eyebrows &c, but that strong, sustained concentrated attention is accompanied by contraction of the corrugators of the eyebrows.

Figures 34 & 23.[3] An exceedingly tragic expression, apparently produced by the powerful contraction of the muscles of the eyebrows, with some elevation of the skin of the forehead & transverse folds. This expression if perfect would produce the so-called 'horse-shoe' on the forehead about which Sir Walter Scott speaks in 'Redgauntlet'.[4] Mrs. Scott Siddons the actress[5] has the power of producing these lines on the forehead with singular precision. She tells me that all her family have been remarkable for this power. The lines referred are if I remember rightly well seen in Sir Joshua Reynolds portrait of the great Mrs. Siddons as the Tragic Muse.[6] My brother Mr. Balfour Browne informs me that the last descendent of the Griersons of Largg (the Redgauntlets of Sir Walter Scott) prides herself on possessing the family peculiarity, the power of producing in a striking manner, the horse-shoe on the forehead.[7]

Figures 16. 17. 18.[8] The action of the pyramidalis nasi does not convey to my mind any idea of an expressive expression. It suggests rather *painful attention.* In cases of profound melancholia I have frequently seen it combined in persistent action with the corrugatores superciliorum, notwithstanding Duchennes statement that they are antagonistic.[9]

Figures 19. & 20. The muscles of the eyebrows are constantly seen in energetic action in cases of melancholia. The lines due to the habitual contraction of this muscle are most characteristic of the physiognomy of melancholia, especially hypochondriacal-melancholia, in which grief & anxiety are felt respecting bodily health & conditions.[10] Along with the contraction of the eyebrows in such cases there is a peculiar acute arching of the upper eylid which I am at a loss to understand.

Figures 30. 31. 32. & 33.[11] The action of the muscles included in this group is singularly well illustrated in a form of mental disease, well known as the General Paralysis of the insane. "In this malady there is almost invariably optimism, delusions, as to wealth, rank, grandeur &c,— insane joyousness, benevolence & profusion, while its very earliest physical symptom is trembling at the corners of the mouth & outter corners of the eyes. This is a well recognised fact. Constant tremulous agitation of the inferior palpebral & great zygomatic muscles is pathognomic of the earlier stages of general paralysis. The countenance has a pleased, self-complacent & benevolent expression: As the disease advances other muscles become involved but until complete fatuity is reached, the prevailing expression is that of *feeble* benevolence."[12]

Figure 34. To my thinking this is not a grimace but a genuine expression—mirth suppressed by voluntary effort—as when chiding a child for a ludicrous offence.[13]

Figure 38. Presents not the faintest trace of any lascivious feeling but rather contempt disgust, meanness.[14]

Figure 43.[15] The action of the triangularis ori is well seen in young children in whom the angles of the the mouth are constantly depressed, as the preliminary of tears. Along with the pulling down of the angles of the mouth, there is some pouting of the lower lip.

Why does not Duchenne deal with other muscles, very influential in the expression of the emotions—such as those regulating the movements of the eyeballs, the buccinator[16] (precisely analogous to the other facial muscles, in that it is attached to the lips &c.) used in laughter, & the masseter;[17] which occasions the grinding & gnashing of the teeth, in extreme rage & despair.

AMemorandum
DAR 161: 323, 323/1

CD ANNOTATIONS
1.1 With ... forehead. 2.12] *crossed pencil*
1.4 but that ... eyebrows. 1.5] 'I suspect 'attention" does not express what we sh^{d} call reflexion' *pencil*
2.1 Figures 34 & 23.] 'Grief Muscles' *added above, pencil*
2.1 tragic] 'tragic' *added above, pencil*
2.3 This expression ... 'Redgauntlet'. 2.5] 'Grief-Muscle Horse-shoe shape' *pencil*
2.5 Scott Siddons] *underl pencil*
2.5 M^{rs}. Scott Siddons ... power. 2.7] *scored pencil*; 'Hereditary in' *pencil*
2.8 My brother ... forehead. 2.12] *scored pencil*; 'Heredity' *pencil*
3.1 Figures] 'Pyramidalis nasi' *added above, pencil*
3.1 Figures ... expression. 3.2] *scored pencil*; 'I fully agree' *pencil*
3.3 melancholia] *underl pencil*
3.3 I have ... antagonistic. 3.5] *scored pencil*; 'in insane melancholic person' *pencil*; 'in persistent action with corrugator' *ink*
4.1 Figures ... understand. 4.6] *crossed pencil*
4.1 Figures 19. & 20.] 'Grief muscles' *pencil*
4.2 The lines ... especially 4.3] *underl pencil*
4.2 due] 'due' *added above, pencil*
4.5 Along] 'Along' *added above, pencil*
4.6 arching] 'arching' *added above, pencil*
5.1 Figures 30. . . . 33.] 'Zygomatic' *pencil*
5.2 well known ... insane. 5.3] 'common in' *pencil*
5.2 General ... insane. 5.3] *underl pencil*
5.3 optimism] 'optimism' *added above, pencil*
5.4 joyousness] 'joyousness' *added above, pencil*
5.4 profusion] 'profusion' *added above, pencil*
5.5 outter corners 5.6] 'outer corners' *added above, pencil*
5.6 Constant ... paralysis. 5.8] *scored & underl pencil*
5.10 fatuity] 'fatuity' *added above, pencil*
5.10 benevolence."] *before closing square bracket, pencil*
5.10 *End of para.*: 'I can send photograths illustrating this.' *ink, in Crichton-Browne's hand, del ink, probably by Crichton-Browne, scored pencil*
6.1 Figure 34. . . . offence. 6.2] *crossed pencil*
7.1 Figure 38. . . . meanness. 7.2] 'I quite agree' *pencil*
7.1 lascivious] *underl pencil*
8.1 The action ... tears. 8.3] 'depression triangularis ori' *pencil*
8.1 in young] *underl pencil*
8.3 pouting] *underl pencil*; 'pouting' *added below, pencil*
9.2 such as ... eyeballs,] 'Eye-balls in veneration' *pencil*
9.3 buccinator] *underl pencil*
9.4 used in laughter] *underl pencil*
9.4 masseter] *underl pencil*; 'masseter' *pencil*
9.4 grinding & gnashing 9.5] *underl pencil*
9.5 teeth] 'teeth' *added above, pencil*

Top of 1st and 3d page of memorandum: 'A' *blue crayon*
Top of 4th page: 'A' *ink*
End of memorandum: 'June 6 1870— D[r] J. Crichton Browne' *pencil*

1 The date is established by the annotation and by the relationship between this letter and the letter from James Crichton-Browne, 6 June 1870.

2 Crichton-Browne refers to photographs in the 'Atlas' accompanying Duchenne 1862. Figures 7 to 11 are a set illustrating the 'muscle de l'attention', the occipito-frontalis or frontal muscle (see *Expression*, pp. 24–5, for diagrams of facial muscles).

3 Crichton-Browne probably meant to write 24 and 23. Figures 23 and 24 are part of a set showing the action of the 'muscle du douleur', grief muscle (corrugator supercilii).

4 See W. Scott 1832, 1: 334:

> The furrows of the brow above the eyes became livid and almost black, and were bent into a semicircular, or rather elliptical form, above the junction of the eyebrows. I had heard such a look described in an old tale of *diablerie*, which it was my chance to be entertained with not long since; when this deep and gloomy contortion of the frontal muscles was not unaptly described, as forming the representation of a small horseshoe.

5 Mary Frances Scott-Siddons.

6 Reynolds's portrait of Sarah Siddons as the tragic muse, painted in 1783 and 1784, is now at the Huntington Library, San Marino, California.

7 Crichton-Browne refers to John Hutton Balfour Browne and probably to a relative of his late godmother, Elizabeth Crichton, a descendant of Robert Grierson, first baronet of Lag and Rockhall. Grierson ('the persecutor') is mentioned in Walter Scott's *Redgauntlet* (W. Scott 1832, 1: 194 n. 6), and was the prototype for Scott's Sir Robert Redgauntlet (*ODNB*).

8 Figures 16 to 18 are a set demonstrating the 'muscle de l'aggression' (pyramidalis nasi).

9 Guillaume Benjamin Amand Duchenne made this statement in the text to the 'Atlas' to Duchenne 1862, p. 20.

10 Figures 19 and 20 are part of the same set as 23 and 24 (see n. 3, above). CD cited Crichton-Browne on this point in *Expression*, p. 185.

11 Figures 30 to 33 are part of a set showing the 'muscles de la joie et de la bienveillance' or muscles of joy and benevolence (the zygomatic and the lower orbicularis palpebrarum).

12 CD quoted the section in quotation marks in *Expression*, p. 205, but did not cite any other original source.

13 Figure 34 is part of the set showing the 'muscles de la joie et de la bienveillance' (see n. 11, above); according to the text, it shows a 'grimace'.

14 Figure 38 is part of a set showing the 'muscle de la lascivité' or muscle of lasciviousness (muscle transverse du nez or nasalis muscle; this muscle is not shown in CD's diagrams in *Expression*, pp. 24–5).

15 Figure 43 is part of a set showing the 'muscle de la tristesse' or muscle of sadness (triangularis oris or depressor anguli oris), and according to the text shows disgust.

16 The buccinator is a cheek muscle used in chewing and blowing (*Chambers*).

17 The masseter is a muscle that raises the lower jaw (*Chambers*).

From Robert Cecil 7 June 1870

Hatfield House, | Hatfield, | Herts.
June. 7. 1870.

Sir

It is my pleasing duty to inform you that the University of Oxford proposes to express its deep admiration for your great achievements in the field of Natural Science, by conferring upon you an Honorary Degree at the approaching Commemoration.

I trust that it will not be disagreeable to you to accept this token of their esteem: & that you will be able to be present in the theatre for that purpose.[1]

The degree will be conferred on the morning of Tuesday June the 21st. The ceremony will be over before one.

I am | Your obedient Servant | Salisbury.[2]

DAR 177: 8

1 Cecil refers to the Sheldonian Theatre in Oxford; it was built to provide a secular venue for meetings and ceremonies of the University of Oxford. CD declined on the grounds of ill health; see *LL* 3: 126 n. Henry Parry Liddon later proposed that the degree be awarded *in absentia*, but the vote to do so was tied and not passed. See Browne 2002, pp. 337–9.

2 Cecil was the marquess of Salisbury, and chancellor of Oxford University.

To James Crichton-Browne 8 June [1870][1]

Down,
June 8

My dear Sir

Duchenne arrived this morning all safe.[2] The loss of the book was beginning to cause me trouble, but I assure you I felt more annoyment at troubling you so much and so often than at the want of the book. Considering how hard you are worked and that you have causes of anxiety, I have more reason to apologise to you, than you to me for the accidental hiding of the book and forgetfulness of your servant. I have just been reading your remarks with very great interest: you always tell me exactly the things which I am anxious to hear. I agree with all that you say, and am particularly pleased at your remarks on the pyramidal of the nose and the so-called muscle of lasciviousness. I believe it to be all fancy.[3] In order to test Duchenne's plates I have shown the most characteristic (hiding any indication of what they were meant to express) to between 20 and 30 persons of all kinds, and have recorded their answers: when all or nearly all agree in their answer, I trust him.[4] Now, I believe, not *one* person understood the supposed meaning of the contracted pyramidal! As for the lascivious muscle, I did not think it worth exhibiting. I have been very glad to see the photograph of the woman with bristling hair: I suppose I might, if I wished, have a wood-cut from it: she looks like a Papuan.[5] You propose to send me a photograph of a case of "*general paralysis* of *the insane*", and I should be very glad to see it: I have been trying to get a London Photographer to make me one of a young baby screaming or crying badly; but I fear he will not succeed.[6] I much want a woodcut of a baby in this state. I presume it will be hopeless, from constant movement, to get an insane person photographed, whilst crying bitterly. Should you ever have time to send me any more notes, I can assure you that they are *fully* appreciated by me. My present book has grown so large, that I am going to take the MS. to London to see how big a book it will make; and perhaps I shall print this first, and retain what I am *now* writing on expression for a separate essay, which I will print as soon as I have got the rest of my MS. printed off.[7] To return to the Photographs; if ever you get one of a person in a paroxysm of fear or horror, I should much like to see it. Have you ever noticed whether the alæ of the nostrils are then raised or distended? With the most sincere thanks for all your assistance, I remain

My dear Sir | yours very faithfully | Ch. Darwin

Heaven only knows whether my essay will be worth the trouble which I have caused to many of my kind friends.

DAR 143: 332

[1] The year is established by the relationship between this letter and the letter from James Crichton-Browne, 6 June 1870.

[2] CD refers to the 'Atlas' to Duchenne 1862; see letter from James Crichton-Browne, 6 June 1870.

[3] See memorandum from James Crichton-Browne, [6 June 1870] and nn. 8 and 14.

[4] In *Expression*, p. 14, CD explained that he showed some of Guillaume Benjamin Amand Duchenne's photographs of an old man whose expressions had been produced by galvanic apparatus to 'above twenty educated persons of various ages and both sexes'.

[5] See plate on p. 154. CD reproduced the photograph in *Expression*, p. 296.

[6] See memorandum from James Crichton-Browne, [6 June 1870], annotations to paragraph 5. CD eventually had photographs of babies and young children crying taken by Oscar Gustaf Rejlander of London and Adolph Diedrich Kindermann of Hamburg (see *Expression*, plate 1 (facing p. 148)). However, CD seems not to have made contact with Rejlander until 1871 (see *Correspondence* vol. 19, letter to James Crichton-Browne, 7 April [1871]).

[7] CD had initially intended his work on expression to form part of *Descent*; it was eventually published as a separate book (*Expression*). CD visited London from 24 June to 1 July 1870 (CD's 'Journal' (Appendix II)).

From Benjamin Collins Brodie 9 June 1870

Cowley house, | *Oxford.*
June 9. 1870.

Dear Sir.

As I hear that you may possibly come to Oxford at the approaching Commemoration for the purpose of receiving the degree, which we hope to have the honour of conferring upon you.[1] I venture to ask whether you would come and stay with me and make my house your home by bringing with you Mrs Darwin and your daughter.[2] I have, I know, no particular right to make such a request. But I hope the common bond of scientific pursuits will justify the freedom which I take, and also so many of your friends are my friends also that I cannot but say it will be a great pleasure to me, if you will allow me to become known to you.[3]

I am, | very truly yours | BC Brodie

Charles Darwin Esq[r].

DAR 160: 315

[1] See letter from Robert Cecil, 7 June 1870 and n. 1.

[2] Benjamin Collins Brodie (1817–80) refers to Emma and presumably Henrietta Emma Darwin. CD had consulted Brodie about Henrietta's health in August 1856 (Emma Darwin's diary (DAR 242)).

[3] Brodie was the Wayneflete Professor of chemistry at the University of Oxford (*ODNB*). Joseph Dalton Hooker had visited him in 1869 (see *Correspondence* vol. 17, letter from J. D. Hooker, 14 November 1869).

From John Murray 10 June [1870][1]

50, Albemarle St. | *W.*
June 10.

My Dear Sir

Judging from what you have told me from time to time about your new book, I sh^d^ not be surprised if you were in that state of forwardness—that you might allow me to announce it as a forthcoming publication in my next Quarterly List. If this sh^d^ turn out to be the case, I sh^d^ be obliged by your favoring me with a form of Title—if I am mistaken, & you are not nearly ready to go to press—I pray you to excuse me for thus troubling you, at least I shall understand silence to mean that the time has not yet come. I hope your health is good & think our Artists have executed well—the few woodcuts you have put in hand—[2]

I am my Dear Sir | Yours very faithfully | John Murray

Charles Darwin Esq

DAR 171: 375

[1] The date is established by a pencil annotation on the letter and by the relationship between this letter and the letter to John Murray, 9 April [1870].

[2] CD had planned to speak to Murray about woodcuts for *Descent* in November 1869 (see *Correspondence* vol. 17, letter to John Murray, [after 18 September 1869] and n. 5). He also had woodcuts made by George Henry Ford and Edward A. Smith (see *Correspondence* vol. 17).

From St G. J. Mivart 11 June 1870

7 North Bank | N.W.
June 11th. 1870

My dear Sir.

I should be very much obliged if you would kindly tell me by what action you think the curious resemblance of the Bee, Spider & Fly Ophrys to the several insects has been produced—[1] It is I think far too marked & striking to be accidental!

I have also great difficulty in seeing by what possible action of natural or sexual selection the beauty of bivalves can have been produced and I should be grateful for any hint from you on the subject which (unless you tell me I may quote) I shall consider confidential[2]

I feel sure you will kindly excuse my thus troubling you & remain

My dear Sir | Always Your's very truly | St G Mivart

C. Darwin Esqr.

DAR 171: 188

[1] The bee, spider, and fly ophrys are, in CD's *Orchids*, *Ophrys apifera*, *O. aranifera*, and *O. muscifera*. *Ophrys aranifera* is a synonym of *O. sphegodes*; *O. muscifera* is a synonym of *O. insectifera*.

[2] Mivart discussed the appearance of the bee, fly, and spider orchids, and of bivalve shellfish, in his book on the theory of natural selection, Mivart 1871, pp. 54–5. In *Orchids*, pp. 68–9, CD had written, 'Robert Brown imagined that the flowers resembled bees in order to deter insects from visiting them; I cannot think this probable. The equal or greater resemblance of the Fly Ophrys to an insect does not deter the visits of some unknown insect, which, in that species, are indispensable for the act of fertilisation.'

From David Forbes 13 June 1870

11, York Place, | Portman Square, | London. W.
13 June. 1870

My dear Mr Darwin

I have just returned from a two months residence in Poland and on Tuesday 21st. am to give my lecture on the Aymara Indians—[1] It would give me profound pleasure to see you there if your health would permit you, for I believe I shall bring forward some very important points in support of your views— I took all my notes with me to Poland and accomplished there what I am sure I should not have done here, ie a somewhat complete memoir on the subject— I think it will be about 90 pages or so of 8vo print—and if I am not overrating its importance, I believe it contains some rather important data in fact I was surprised myself at the number of observations when I came to string them together. My intention was if time had allowed it to have placed the MS in your hands before the lecture but I have only just got back and find I have to copy out fair, some of it and fill in some references &c which I had not with me—so I shall not be able to do so— the measurements are all put in proper form and I have made out a sketch so as to compare them with other Nations. black & white—proving I think the case about the altitude &c completely & several other differences for which I at present can assign no reason[2]

Mr Squeer the Nicaraguan–Peruvian traveller is now in Paris and comes over the morning of my lecture expressly for it & has sent already a number of illustrations which will assist me—and I think that there will be others who can speak on the subject.—[3] I shall of course only give an abstract of the more interesting and important conclusions vivâ voce—or otherwise it would take up far too much time—

I do not know whether your book is as yet so far advanced that it is too late to supply you with any data but now that I have got this matter off my hands at last—I think you will find in my MS. better or rather more correct data than what you noted down in the course of our conversation when I had to trust mainly to recollections[4]

Trusting that your health is better now that we have such fine weather | I remain my dear Mr Darwin | Yours most sincerely | David Forbes

DAR 164: 144

[1] Forbes's lecture took place at the Ethnological Society of London and was published as Forbes 1870. CD visited London only from 24 June (CD's 'Journal' (Appendix II)).

[2] In Forbes 1870a, pp. 213–15 (see also *ibid.*, pp. 277–81), Forbes printed a table comparing relative average proportions of 'normal' Aymara men from high cold mountains; two 'abnormal' Aymara men from hot valleys; two Englishmen, one 'Saxon' and one 'Celtic', and three 'Menas negroes' from West Africa (presumably male). Forbes speculated that the larger chest measurement of the Aymara was due to the fact that they normally lived in the high tablelands of Bolivia and Peru, where the air was more rarefied, so that they needed greater lung capacity (*ibid.*, p. 218). He also considered that various other differences between the highland and lowland Aymara were due to the change in altitude (*ibid.*, p. 222). There is an offprint of Forbes 1870a, inscribed by the author, in the Darwin Pamphlet Collection–CUL. It has one annotation on the back cover ('p. 16 blushing').

[3] The discussion following Forbes's paper, including remarks by Ephraim George Squier, is summarised in the *Journal of the Ethnological Society of London* following the paper.

[4] It is not known when Forbes and CD had a conversation about the Aymara Indians. CD cited Forbes on the Aymara in *Descent* 1: 119, 246, 2: 320, 322, 348, and cited Forbes 1870a in *Descent* 1: 119 n. 32. Most of the information that CD cites is in the letter from David Forbes, 26 March 1868 (*Correspondence* vol. 16). CD also cited Forbes for information in Forbes 1870a and in his letter in *Expression*, pp. 232, 318, 336.

To J. D. Hooker [13 June 1870?][1]

Seeds ripened in Algiers: imported seed would be of no use.—

Iberis amara
— umbellata, especially any red var. as Kermiseana[2]
Nolana prostrata[3]
Hibiscus Africanus[4]
Canna Warzewizi[5]
Mimulus luteus, garden var. yellowish or whitish flowers, blotched with purplish-red or orange-brown.[6]

C. Darwin

Memorandum
American Philosophical Society

[1] The date is conjectured on the supposition that the memorandum was handed to Hooker when he visited Down on 13 June 1870 (letter to St G. J. Mivart, 13 June [1870]). See also the letters to J. D. Hooker, 25 May [1870] and [29 June 1870], and the letter from J. D. Hooker, 10 July 1870.

[2] The variety Kermesiana (as CD spells it in *Cross and self fertilisation*, p. 103), has not been identified. CD raised plants of a red variety of *Iberis umbellata* from seed bought in England, and crossed them with pollen from plants raised from seed sent from Algiers (*Cross and self fertilisation*, p. 105).

[3] *Nolana prostrata* is now *N. humifusa*. CD had carried out crossing experiments with *N. prostrata* in 1862 and 1867 (*Cross and self fertilisation*), but mentions no further experiments.

[4] CD carried out crossing experiments with *Hibiscus africanus* (now *H. trionum*; *Cross and self fertilisation*, pp. 140–1), but did not mention using seed from Algiers.

[5] *Canna warszewiczii* is now *C. indica*. CD carried out crossing experiments with *C. warszewiczii* (*Cross and self fertilisation*, pp. 230–3), but did not mention using seed from Algiers.

[6] After the list and before the signature, Hooker wrote, 'A little of each is enough.'

To St G. J. Mivart 13 June [1870][1]

Down. | Beckenham | *Kent. S.E.*
June 13th

My dear Sir

I have often reflected over the two cases to which you refer. When working on Orchids, I have often & long watched the species of Ophrys, because R. Brown thought that their supposed likeness to insects served to keep them away.[2] This might possibly be advantageous to the Bee Ophrys; but *certainly* not to the others, as they absolutely require the aid of insects for fertilisation, though they get this aid but rarely or rather rarely.—

I believe (& Dr Hooker[3] who happens to be here agrees with me) that the resemblance is fanciful;[4] the flowers are odd looking & insects are the most natural standard of comparison. Who ever saw a Bee with violet wings like the petals of the Bee Ophrys? The Fly Ophrys is more like. Hooker believes that the Spider ophrys is so called simply from the curved marks on the Labellum like the marks on the backs of some Epeiræ. The Butterfly orchis has hardly any resemblance to a butterfly, & so with some foreign orchids which have received all kinds of fanciful names.—[5]

Your second case seems to me much more difficult: for my own part I believe the beauty of shells is altogether incidental on other causes & of no service whatever to the animals. The form depending on manner & symmetry of growth, & the colour being incidental on the chemical nature of the constituent element of the shell.[6] To explain what I mean,—nothing is more beautiful than arterial blood, yet no one supposes that the colour has been gained for the sake of colour or beauty, though it gives beauty to the cheeks of Caucasian maidens. Or again, & this is a better illustration; the extreme beauty of the Eolidæ is due to their biliary secretions & organs being seen through their transparent tissues; but no one will suppose that their bile has been coloured for beauty, but is incidental on its chemical nature, as much as the recently discovered aniline colours.[7] Nothing in this note can, I think, be of the least service to you, but you are welcome to use it as *[illeg]* from me, if you think fit. I have introduced these illustrations in my M.S. on sexual selection.—[8]

Pray believe me | My dear Sir | Yours sincerely | Ch. Darwin

Pray forgive this untidy & ill-expressed letter from want of time.—

Bearing in mind the magnificent colours of many artificial organic substances made by elements, it wd be wonderful if such had not been produced under nature, without being of any special use: if there had been no such beautiful colours, this fact wd. have had been to have been accounted for—

I think of the beauty & splendour of the withering leaves in an American forest.—

Sotheby's (20–1 July 1988)

1 The year is established by the relationship between this letter and the letter from St G. J. Mivart, 11 June 1870.

2 See Brown 1831, pp. 740–1. CD cited Robert Brown on this point in *Orchids*, pp. 68–9.

3 Joseph Dalton Hooker.

4 Mivart quoted CD's suggestion, without mentioning his name, in Mivart 1871, p. 55 n. 2. Mivart objected, 'The denial, however, of the fact of a resemblance which has struck so many observers, reminds one of the French philosopher's estimate of facts hostile to his theory— "Tant pis pour les faits!"' (That is, 'So much the worse for the facts!')

5 On the bee, fly, and spider ophrys, see the letter from St G. J. Mivart, 11 June 1870 and n. 1. The large butterfly orchis is, in CD's *Orchids*, *Habenaria chlorantha*, now *Platanthera chlorantha*. The lesser butterfly orchis, CD's *H. bifolia*, is now *P. bifolia*.

6 Mivart quoted CD's suggestion about the shells of bivalves, without mentioning his name, in Mivart 1871, p. 55 n. 2. Mivart objected, 'But surely beauty depends on some such matters in *all* cases!'

7 The Eolidae (naked sea-slugs), are now the Aeolidiidae, a family in the order Nudibranchia. The first aniline dye, aniline purple or mauveine, was discovered accidentally by William Henry Perkin when he was attempting to synthesise quinine. It was patented in 1856; in 1869, an economical process for synthesising alazarin (red dye) was developed (see *ODNB* s.v. Perkin, William Henry; see also Garfield 2000 and Christie 2001, pp. 3–7).

8 See *Descent* 1: 323.

From F. V. Hayden 14 June 1870

Washington DC.
June 14th 1870.

Dear Sir

Your very kind letter came duly to hand.[1] I send you a roll of maps today through the Smithsonian Institution—[2] These maps will serve to aid you in following explorers over our far west. I will send you from time to time succh books or maps as you wish to make more clear, the results of examinations in that region. I start soon for Idaho & Montana with $25,000, for another summers labors.[3] Will go along the Union Pacific Railroad to Salt Lake, thence northward to the sources of the Columbia & Missouri rivers.

Hoping that you are well and will long live to work out the grand results of your studies. I remain | your sincere admirer & reader | F. V. Hayden

DAR 166: 123

1 See letter to F. V. Hayden, 2 May [1870].

2 The maps have not been found in the Darwin Archive–CUL.

3 Hayden was carrying out geological and natural history surveys for the US Department of the Interior (*ANB*). In the summer of 1870, he explored Wyoming and Utah territories (Foster 1994, p. 175); for his report, see Hayden 1871.

To R. A. von Kölliker 14 June 1870

Down. | Beckenham Kent. S.E.
June 14. 1870.

My dear Sir.

I received two days ago from the Royal Soc: your extremely kind present of the Memoir on the Pennatuliden, with its numerous beautiful illustrations.[1] Now that I see how you have worked on the subject, I am rather ashamed of having sent you my crude notes, at which you must have smiled.[2]

If you can spare a little time I should be very grateful for some information with respect to one of your former discoveries—[3] Do you know whether the muscles to the quills of the porcupine, and those which raise and cause to vibrate the tail coverts of the Peacock, are striped or unstriped? Whatever they may be, I suppose that the same kind of muscles raise the plumes on the head, and the other plumes which many birds erect under excitement and as an ornament. If these muscles are striped, like those to the vibrissæ of the Carnivora (as is stated on your authority) do they occupy the same position and are they homologous with the unstriped muscles which move ordinary hairs? If this is the case all who believe in the evolution of species must I suppose admit that unstriped muscles can by some unknown means be developed into striped muscles.[4] Can you believe in the possibility of such a process of development? I should be greatly obliged for a few words in answer; especially with respect to the muscles to the quills of the porcupine & to the plumes of such birds as the Peacock.

Permit me in conclusion to say how very much I was gratified & pleased at your little visit to Down[5]

With sincere respect. | Yours very faithfully | Charles Darwin.

Copy
DAR 146: 22

[1] CD probably refers to the first part of Kölliker's paper, 'Anatomisch-systematische Beschreibung der Alcyonarien: I. Die Pennatuliden' (Anatomical-systematic description of the Alcyonaria: I. the Pennatulidae; Kölliker 1869–72). There is an offprint of the first part (Kölliker 1870) in the Darwin Library–Down.

[2] See letter to R. A. von Kölliker, 18 April 1870 and n. 3.

[3] CD refers to Kölliker's isolation of 'unstriped' (smooth) involuntary muscles (*Expression*, pp. 100–1, Kölliker 1847).

[4] See *Expression*, pp. 102–3.

[5] Kölliker visited Down on 15 April 1870 (see letters from Albert Günther, [*c.* 9 April 1870] and 12 April 1870).

To J. J. Weir 14 June [1870][1]

Down. | Beckenham | *Kent. S.E.*
June 14th

My dear Sir

As usual I am going to beg for information. Can you tell me whether any Fringillidæ, or Sylviadæ erect their feathers when frightened or enraged?[2] I want to show that this expression is common to all or most of the families of Birds— I know of this only in the Fowl, Swan, Tropic-bird, Owl, Ruff & Reeve, & Cuckoo—[3] I fancy that I remember having seen nestling birds erect their feathers greatly when looking into nests, as is said to be the case with young cuckoos. I shd much like to know whether nestlings do really thus erect their feathers? I am now at work on expression in animals of all kinds & birds; & if you have any hints I shd be very grateful for them, & you have a rich wealth of facts of all kinds.—

Any cases like the following: the sheldrake pats or dances on the tidal sands to make the sea-worms come out; & when Mr. St John's tame sheldrake came to ask for their dinners, they used to pat the ground; & this I sh[d] call an expression of Hunger & impatience.—[4]

How about the Quagga case?[5]

I am working away as hard as I can on my book,[6] but good Heavens how slow my progress is.— I hope that ⟨you⟩ are well.—

Believe me | Yours very sincerely | Ch. Darwin.

Armacost Library, University of Redlands

[1] The year is established by the relationship between this letter and the letter from J. J. Weir, 27 June 1870.

[2] Fringillidae is the family of finches; Sylviidae is the family of Old World warblers.

[3] See *Expression*, pp. 97–100. The red-tailed tropic bird mentioned by CD in *Expression* as *Phaeton rubricauda* is now *Phaethon rubricauda*; the ruff (male) and the reeve (female) mentioned by CD as *Machetes pugnax* are now *Philomachus pugnax*.

[4] CD recounted this story in *Expression*, pp. 47–8. Charles St John reported the behaviour of the sheldrakes (*Tadorna tadorna*) in St John 1849.

[5] See letter from J. J. Weir, 17 March 1870.

[6] *Descent.*

From Thomas W. Wood 14 June 1870

221 Stanhope Street, | Hampstead Road N.W.
June 14th 1870.

Sir,

I hope you will excuse the liberty I take in writing to you, but, having consulted with my friend M[r] Bates,[1] he has told me that you would take an interest in the facts I wish to communicate.

The letter to the Field enclosed will speak for itself & although I feel convinced of the truth of your theory of the origin of Species I think that such a fact as that mentioned about the Argus cannot be explained by it, but seems to point to (& almost to prove) the existence of a great artistic power.[2] The diagram was badly printed, so I have touched it up a little & can get you the block if you would like to make use of it in your forthcoming volume, in which as I understand from M[r] Bates, you have alluded to my discovery of the disguise assumed by the Orangetip butterfly when at rest.[3] Have you seen it in nature? If not allow me to advise you to search along the hedgerows at dusk on the little white flowers, as I am sure you would be delighted to see the absolute perfection of the imitation of the flower by the butterfly. The season, however, for both flower & butterfly will *soon* be over, May being the best time.

May I also be allowed to call your attention to my article on "the courtship of birds" in *the Student* for April in which you will find some peculiarities of the males illustrated by a coloured picture & woodcuts.[4]

I have the honour to remain | Sir, | Your obedient servant, | T. W. Wood.

Charles Darwin Esq. F.R.S. &c. &c.

DAR 181: 147

[1] Henry Walter Bates.

[2] The enclosure is at DAR 84.1: 174. In the *Field* newspaper, 28 May 1870, p. 457, Wood argued that since when the Argus pheasant displayed his wing-feathers like a fan, those nearest the body stood more upright than the outer ones, the shading of the ball-and-socket ocelli ought to be slightly different on the different feathers in order to preserve the three-dimensional illusion. Wood thought that this was indeed the case. See *Descent* 2: 143–4 n. 48, where CD discussed this point, and said that he could not see any difference in the shading. There is a draft of the note in *Descent* in the hand of an amanuensis in DAR 84.2: 224; it continues with two sentences not included in *Descent*: 'I may however add that on a few of the basal ball & socket ornaments on the outermost feathers the shading did appear slightly difft. N.B. on the outermost secondaries the external ring is broken at the summit.' The Argus pheasant is now the great Argus, *Argusianus argus*.

[3] See *Descent* 1: 394 and n. 6; CD observed that the undersides of the wings of the orange-tip butterfly (*Anthocharis cardamines*) resembled the flower-head of wild parsley, on which it rested at night.

[4] Wood 1870.

To ? 14 June [1870][1]

Down. | Beckenham | *Kent. S.E.*
June 14—

My dear Sir

I shall be in London in about ten days, & will immediately call on you & explain why I cannot at once answer your question.

Yours very faithfully | Charles Darwin

LS
DAR 221.5: 43 (photocopy)

[1] The year is established by the fact that CD was in London from 24 June in 1870 (CD's 'Journal' (Appendix II)). The correspondent has not been identified.

From Louis Rérolle 15 June 1870

Lyon,
15 June 1870

Sir,

I am happy to inform you that the French translation of your book on the fertilisation of Orchids is published, and I have requested Mr Reinwald, the publisher, to send you immediately a copy, which I gratefully beg you to accept.[1]

I have not had the honour to write to you since a long time about this publication, which required a little time more than had been formerly supposed;[2] it is now completed, and I feel highly gratified to have been the means of presenting your valuable work to the French public; I feel convinced that, if my translation is not too inferior, I have done a useful thing in making my countrymen acquainted with another of your works;[3] your approbation alone is wanting to complete my satisfaction.

Should you desire to present a few copies to any friends of yours, I shall feel most happy to send you some.

I have the honour to be, Sir, your most obliged and respectful servant | Louis Rérolle

DAR 176: 132

[1] The French translation of *Orchids* was Rérolle trans. 1870. For earlier letters from Rérolle, see *Correspondence* vol. 17, letters from Louis Rérolle, 17 March 1869, 10 May 1869, and 30 July 1869. Rérolle also refers to Charles-Ferdinand Reinwald.

[2] Rérolle had expected the book to be ready in February 1870 (*Correspondence* vol. 17, letter from Louis Rérolle, 30 July 1869).

[3] *Origin* had been translated by Clémence Auguste Royer (Royer trans. 1862, 1866, 1870) and *Variation* by Jean Jacques Moulinié (Moulinié trans. 1868).

From T. W. Wood 16 June 1870

221 Stanhope Street, | Hampstead Road N.W.
June 16th 1870.

Dear Sir,

I beg to thank you for your very kind and appreciative letter, & with regard to your query about the difference in the shading of the spots of the Argus being an optical illusion, I can assure you that is not the case, the markings being quite fixed & the only illusion is that produced by nature in making flat surfaces appear convex.[1] I have a feather taken from close to the bird's body & also one of those farthest from it; in the former the spots are thus

& in the latter=

Mr Tegetmeier of "the Field" will get the block stereotyped for you without charge & has kindly promised that it shall be ready soon. I told him of your wish that the Gold pheasant's feathers should be cut out & am sorry you cannot find room for them, if in reconsidering the matter you should alter your mind please let me know at once.[2] In "the Student" you will find a sketch of this beautiful bird with the collar expanded.[3] The only condition Mr Tegetmeier mentioned was that the cut should be alluded to as having been taken from "the Field"[4]

I remain, | Dear Sir, | Yours faithfully, | T. W. Wood.

Charles Darwin Esq. F.R.S. &c— &c—

DAR 181: 148

[1] Wood refers to the ocellated spots on the plumage of the argus pheasant; see letter from T. W. Wood, 14 June 1870. The letter to which this is a reply has not been found.

[2] Wood's letter to the *Field*, 28 May 1870, p. 457, included diagrams of feathers; two of them were of collar feathers of the gold pheasant (*Chrysolophus pictus*). Wood was interested in these feathers because the double stripe on each feather had the effect of maintaining the appearance of continuous stripes on the plumage even when the feathers were fluffed up and separated.

[3] Wood 1870, second plate, fig. h.
[4] CD did not use Wood's diagrams from the *Field* in *Descent*.

From Joachim Barrande[1] 19 June 1870

Prague | 419-Kleinseite | Choteksgasse
19 Juin 1870—

Très honoré mâitre,

Au moment où votre aimable lettre du 28 Mai dernier m'est parvenue, je terminais les pages ci-jointes, en réponse à un géologue français, qui m'avait fait part de ses impressions après la lecture de mon travail sur la *distrib^n des Cèphalop. silur.*[2]

Bien que ces pages ne soient pas destinées à être publieés, je me ferais un scrupule de courtoisie de les envoyer à leur destination, sans vous les communiquer. J'espère que vous excuserez mon scrupule, si ces pages renferment quelques circonstances, qui pourraient ne pas avoir attiré jusqu'ici votre attention.

Dans tous les cas, je suis hereux de saisir cette occasion, pour vous exprimer ma sincère et profonde admiration pour tant de travaux variés, par lesquels vous avez enrichi la science et que personne ne saurait apprécier plus haut que moi.

Agréez, très honoré Mâitre, l'expression de mes sentimens les plus distingués.

J Barrande

[Enclosure]

Prague
14 juin 1870—

Copie.

Monsieur et très honoré confrére,

Je loue hautement la loyale franchise avec laquelle vous reconnaissez que les résultats de mes études sur les Céphalop. silur. gênent vos convictions, au sujet de la doctrine de la filiation et de la transformation, que vous avez adoptée.

Puis que vous vous proposez de chercher une solution entièrement satisfaisante du problême de l'apparition successive des êtres organisés, en partant du principe théorique en question, j'appèle votre attention sur $\overline{\text{qq}}$ faits qui semblent rebelles à cette théorie.

D'après les vues actuelles des transformistes, Eozoon serait le premier type connu, ou la première souche de tous les animaux.[3]

Or, vous savez comme moi, combien ce type primitif se rapproche de celui des Polypiers,[4] par tous les élémens de sa structure. On peut dire, qu'entre Eozoon et les Polypiers, il n'y a qu'un pas à faire, dans la voie de la filiation et de la transformation. D'après les idées théoriques, ce pas aurait dû aussi être le premier fait dans cette voie. Les Polypiers auraient donc dû couvrir le fond des mers, après l'époque Eozoique et avant la prédominance des autres types d'une organisation plus élevée.

En dépit de cette vraisemblance théorique, on n'a découvert aucune trace de Polypiers calcaires, ni dans les roches qui renferment Eozoon, ni dans les terrains nommés antéprimordiaux, ni même sur l'horizon de la faune primordiale silurienne, faune qu'on nous dit relativement récente par rapport aux âges Eozoiques. On n'aperçoit les premiers vestiges de ces Polypiers que tardivement, c. à d. dans la faune seconde silurienne et leur grand développemt n'a lieu que dans la faune troisième et dans les âges postérieurs.

Il serait difficile de croire, que leurs traces ont partout disparu dans les formations antérieures à la faune seconde silurienne, car, en raison de leur nature calcaire, ils auraient pu se conserver au moins aussi bien que les restes délicats et microscopiques d'Eozoon.

Par contraste, les dépouilles des Trilobites se montrent avec une fréquence incomparable dans toutes les régions des deux continens, sur l'horizon le plus profond qui ait fourni une faune, c. à d. sur l'horizon de la faune prémordiale silurienne.

Pourquoi l'enveloppe fragile de ces Crustacés a-t-elle joui de ce privilège de conservation, si l'enveloppe massive des Polypiers contemporains ou antérieurs a réellement disparu, dans toutes les contrées explorées.

Si l'on considère, que la faune primordiale se trouve bien développée dans un grand nombre de pay⟨s⟩ géographiquement très espacés, on doit se demander, pourquoi presque tous les descendans d'Eozoon, après mille transformations supposées, durant l'immense série des âges antéprimordiaux, et sous l'influence des circonstances les plus diverses, se sont pour ainsi dire donné le mot, pour aboutir en même temps au même type trilobitique, si bien caractérisé et si concordant sur tout le globe, malgré ses variations locales. Ce fait singulier mérite bien d'être expliqué par les théories.

Remarquons aussi que, dans la faune primordiale on trouve avec les Trilobites quelques autres types, fournissant des individus relativement rares. Ce sont principalement des Brachiopodes et des Ptéropodes, qui représentent les Mollusques à coté des Articulés.

Si tous ces types sont réellement dérivés d'un prototype commun, leur coexistence se manifestant uniformément partout où l'on rencontre la faune primordiale, nous induirait à croire, qu'ils sont les plus rappr⟨o⟩chés par leur organisation, parmi les articulés et les Mollusques, puisqu'ils sont les plus rapprochés par la filiation de leur source commune. Mais les zoologues nous démontreront-ils cette affinité relative entre les Trilobites, Brachiopodes et Ptéropodes?

En admettant la commune descendance de tous ces types, à partir d'un même prototype, il resterait encore la difficulté que la théorie rencontre à chaque pas, c. à d. la disparition inexplicable de toutes les formes intermédiaires entre Eozoon et chacun des types primordiaux. Ces formes auraient dû être représentées par d'innombrables individus, durant l'ère antéprimordiale. La théorie peut-elle rendre compte de cette invariable disparition dans tous les pays, lorsqu'elle invoque partout la conservation d'Eozoon?

Le doute fort que la transformation brusque par accident, qui vous semble promettre une plus heureuse solution que la transformation lente, puisse mieux satisfaire à l'explication de tous ces faits.

Introduire des accidens quelconques comme cause de l'apparition des types successifs, c'est attribuer ces apparitions au hazard. Mais, qui oserait invoquer le hazard comme ordonnateur du développement graduel de la série animale, lorsque, aux yeux de tous les hommes de science, cette série présente les résultats d'un plan unique, merveilleusement exécuté, malgré tous les troubles subis par l'écorce terrestre et par tous les élémens matériels, au milieu desquels la vie a dû se propager.

Je termine en vous souhaitant les plus ingénieuses et les plus heureuses conceptions, dans votre future théorie. Mais vous ne trouverez pas mauvais que je vous répète ce que j'ai déjà exprimé dans mon dernier travail sur les Céphalopodes: Je n'ai pas l'espoir que l'esprit humain parvienne à pénétrer le mystère qui enveloppe la première apparition et la succession des formes de la vie sur notre globe.

Agreez &c | J B

P. S— Vous recevrez prochainement ma *défense des Colonies*—*IV*[5] et vous trouverez que j'ai constaté (p. 145–) ce fait remarquable: que certains Trilobites de Bohême, offrant diverses oscillations de forme, sont revenus à leur forme primitive, après un long intervalle de temps. Ces faits méritent attention, parcequ'ils nous font concevoir, que les modifications de forme produites sous l'influence de l'homme, pourraient n'être réellement que des oscillations temporaires.

DAR 160: 44–5

1 For a translation of this letter, see Appendix I.

2 CD's letter to Barrande of 28 May 1870 has not been found. Barrande refers to Barrande 1870. The geologist has not been identified.

3 Barrande presumably refers to *Eozoon canadense*. In 1864, John William Dawson identified samples taken from pre-Silurian strata in eastern Canada as fossilised Foraminifera, single-celled protists with shells; he named the species *Eozoon canadense*, the 'Dawn animal from Canada' (Dawson 1864). Further samples were sent to William Benjamin Carpenter, an expert on Foraminifera, who confirmed Dawson's interpretation (Carpenter 1864). CD added information on the discovery of *Eozoon canadense* to *Origin* 4th ed., p. 371, as substantiating his claim, made in *Origin*, p. 307, that life existed before the Silurian period. The interpretation of the samples as pre-Silurian fossils remained controversial, however (see, for example, Carpenter 1866, and King and Rowney 1866); and by the end of the century, comparisons with similar, more recent, formations indicated that the samples were mineral in origin (see Schopf 2000).

4 Polypiers (French) or polypidoms (English) are now (in English) polypariums. A polyparium is 'the common stem or supporting structure of a colonial cnidarian (or other colonial invertebrate), to which the individual polyps are attached', or (an obsolete meaning) a colonial cnidarian (*OED*).

5 Barrande refers to the fourth volume of Barrande 1861–81.

From George Cupples 20 June 1870

The Cottage, | Guard Bridge, Fifeshire, | N.B.
June 20/70.

My Dear Mr Darwin,

I write merely to show that I am keeping fully in mind about the Deerhound puppy, with regard to which I can now engage pretty definitely.[1] After an unusual delay in the arrangements of the dogs among themselves, I am now able to count on a litter of puppies by the 15th. August—in which case they will be, I think, the very best I have ever bred, grey in colour, large, and rough. I shall select a dog-puppy, keep it for about 8 weeks, and then send it to you. This will bring it to about the middle of October, at the best age for training to the habits desired. The mild climate of Kent will help it to get strength for the winter. In due time, along with the puppy, I will send full particulars for management &c.

I hope you are now again at the seaside, or somewhere else, recruiting.

I may add that I have from time to time full confirmation of my previous statement that dogs in breeding form *decided preferences* towards each other—often being influenced by *size*, bright colour, and individual character, as well as the foregoing degree of their familiarity with each other.[2] So undoubtedly, to some extent, with poultry—as stated in the article I sent you from All-the-Year-Round, whose assertions in that respect I could have corroborated from actual observation.[3]

With best regards I am | Dear Mr Darwin | ever truly yours | George Cupples

Chas. Darwin, Esqr. | Down.

DAR 83: 142–3

CD ANNOTATION
1.1 I write … &c. 1.9] *crossed pencil*

1 See letter from George Cupples, 29 April 1870. No intervening letter from CD has been found.

2 See *Correspondence* vol. 16, letters from George Cupples, 11–13 May 1868, third enclosure, and 26 May 1868.

3 CD discussed the courtship of fowls in *Descent* 2: 117, but did not mention Cupples or the article from *All the Year Round.* The article was presumably 'Cocks and hens', *All the Year Round*, 14 November 1863, pp. 276–83, in which the anonymous author discussed, amongst other things, the preferences of cocks and hens for one another.

To T. H. Huxley 20 June [1870][1]

Down | Beckenham | *Kent. S.E.*
June 20

My dear Huxley

You were so kind as to say that you would lend me Ecker Icones Phy. (Part with Pl 30) & Bischoff Hund-eies (Part with Pl. 41) to get the embryos copied for engraving—[2] I shd. keep the books only for a week.— Will you have kindness to send them by parcel Delivery Coy, to 6 Queen Anne St Cavendish Sqe. to be there on *Thursday* 23d, when we come to London for a week—[3]

If I do not hear to the contrary, I hope you will let me pay you my early Sunday call on the 26th—

I will caution Mr Ford to be very careful with your Books.—[4]

Ever yours | Ch. Darwin

Imperial College of Science, Technology and Medicine Archives (Huxley 5: 269)

[1] The year is established by the form of the address, with '*Bromley*' crossed out and 'Beckenham' inserted by hand, a form that CD used between April 1869 and May 1871, and by the reference to a week's visit to London in June: the Darwins were in London from Friday 24 June to Friday 1 July 1870 (CD's 'Journal' (Appendix II)).

[2] CD refers to Rudolph Wagner's *Icones physiologicae* (R. Wagner 1851–9), edited by Alexander Ecker, and Theodor Ludwig Wilhelm Bischoff's *Entwicklungsgeschichte des Hunde-Eies* (Developmental history of the dog-egg; Bischoff 1845). CD used the illustrations of human and dog embryos in *Descent* 1: 15; see also *ibid*, p. 16, n. 11, for CD's note on these illustrations.

[3] Six Queen Anne Street, London, was the address of CD's brother, Erasmus Alvey Darwin. In fact, the Darwins seem to arrived in London on the Friday (see n. 1, above).

[4] George Henry Ford made many of the illustrations for *Descent*.

To F. C. Donders 21 June 1870

Down— Beckenham | Kent
June 21. 1870

My dear Sir

I have received from Dr Moore the translation of your paper, & have read it with the greatest interest. After making some extracts I am asked by Dr M. to send it to Dr Beale for publication in the October number of his Journal.[1] Your information has proved incomparably fuller & more decisive than I had ventured to hope. You cannot imagine what an advantage it is to my little essay to have on this head a firm foundation. Now that I see how laborious yr. investigation has been,—I shd have regretted to have incidentally led you to employ so much of your time, did I not feel convinced that physiologists will think the results, independendly of the particular point in question, worth much labour. Several of yr. incidental remarks have also interested me much. I had intended publishing my essay on expression as part of a larger work nearly completed;[2] but I believe that it will be too large, & that I must keep back for a short time, my essay[3]

I do not know whether it is worth mentioning, but I think that the closing of the eyelids serves in part to prevent the eyes suffering from vibrations, I observe that dogs & cats when crushing a bone, close their eyelids. Various kinds of monkeys which do not close their eyelids when screaming close them when sneezing.[4]

With the most cordial thanks for your extreme kindness, believe me my dear Sir | yours very sincerely | Ch. Darwin.

Copy
American Philosophical Society (B/D25.319)

[1] William Daniel Moore had translated Donders's article on the circulation of the eye (Donders 1870a; see letters from F. C. Donders, 17 May 1870 and 27 May 1870). Lionel Smith Beale edited *Archives of Medicine* from 1857 to 1870 (*ODNB*). Moore's translation of Donders's article did appear in *Archives of Medicine* (Donders 1870b); there is an annotated copy of the translation in the Darwin Pamphlet Collection–CUL. CD cited the article in *Expression*, p. 160.

[2] The 'larger work' that CD refers to is *Descent*.

[3] *Expression* was eventually published more than a year after *Descent*, in 1872.

[4] CD mentioned these examples in *Expression*, p. 163.

From T. H. Huxley 22 June 1870

Geological Survey of England and Wales

My dear Darwin

I sent the books to Queen Anne St this morning—[1] Pray keep them as long as you like, as I am not using them—

I am greatly disgusted that you are coming up to London this week—as we shall be out of Town next Sunday—[2] It is the rarest thing in the world for us to be away and you have pitched upon the one day. Cannot we arrange some other day?

I wish you could have gone to Oxford, not for your sake, but for theirs.—[3]

There seems to have been a tremendous shindy in the Hebdomadal board about certain persons who were proposed; and I am told that Pusey came to London to ascertain from a trustworthy friend who were the blackest heretics out of the list proposed—and that he was glad to assent to your being doctored, when he got back—in order to keep out seven devils worst than that first![4]

Ever, oh Coryphæus diabolicus[5] | your faithful follower | T. H. Huxley

June 22. 1870

DAR 166: 322

[1] Huxley refers to R. Wagner 1851–9 and Bischoff 1845; see letter to T. H. Huxley, 20 June [1870] and n. 2.

[2] The Darwins were in London from 24 June to 1 July 1870 (CD's 'Journal' (Appendix II)). See letter to T. H. Huxley, 20 June [1870].

[3] CD had been offered an honorary degree by the University of Oxford but declined to attend the ceremony due to ill health (letter from Robert Cecil, 7 June 1870 and n. 1).

[4] The Hebdomadal Council was the University of Oxford's ruling executive. Huxley himself and John Tyndall had also been nominated for honorary degrees (Browne 2002, p. 338). Edward Bouverie Pusey accepted CD's nomination only in return for Huxley's being dropped (*ibid.*).

[5] The *coryphaeus* is the leader of the chorus in a Greek play.

From Louis Rérolle 22 June 1870

Lyon,
June 22 1870

Sir

I feel very pleased and grateful that my translation should have met with your approbation.[1] I am glad to say that Mr Reinwald will immediately forward at his expense the copies to MM. C. Naudin, de Quatrefages and Planchon, and to the Botanical Society of France, from the author. A copy for Mr Charles Martins has been already sent to him at the Faculté de Médecine of Montpellier.[2]

I intend myself to send a few to the following persons, whose names are probably not unfamiliar to you: MM. Duchartre, Brongniart and Baillon, at Paris; M. Lecoq, prof. at Clermont–Ferrand; M. Godron, prof. at Nancy; and M. Alph. de Candolle, at Geneva.[3]

I remain, Sir, with much respect, your obliged and faithful servant | Louis Rérolle

DAR 176: 133

[1] Rérolle refers to his translation into French of CD's *Orchids* (Rérolle trans. 1870). See letter from Louis Rérolle, 15 June 1870. CD's letter to Rérolle has not been found.
[2] Charles-Ferdinand Reinwald was the publisher of Rérolle trans. 1870. Rérolle also refers to Charles Victor Naudin, Armand de Quatrefages, Jules Emile Planchon, and Charles Frédéric Martins.
[3] Pierre Etienne Simon Duchartre, Adolphe Théodore Brongniart, Ernest Henri Baillon, Henri Lecoq, Dominique Alexandre Godron, and Alphonse de Candolle.

From John and Emma Pender [before 23 June 1870][1]

His Royal Highnesss
The Prince of Wales.
*M*r. *& M*rs. *Pender*
request the honor of
Mr. Charles Darwin's
Company on the Evening of
Thursday *the* 23rd. June *at* 10 *o'c⟨lock⟩*[2]

Invitation card
DAR 53.2: 168v

CD ANNOTATIONS
Recto: 'Keep' *blue crayon circled blue crayon*
Verso: 'Keep' *blue crayon circled blue crayon*; 'Size of *page, Heading & type [*above del* 'Wood-Blocks'] | Descent of Man[3] | for Photographs & folding Plates' *pencil*

[1] The date is established by the day and month mentioned in the invitation and by the year of the Penders' party to celebrate the opening of the telegraph link with India, 1870 (see *ODNB* s.v. Pender, John).
[2] CD had met the prince of Wales (Albert Edward, later Edward VII) at a Royal Society of London soirée on 28 April 1866 (*Emma Darwin* (1915), 2: 184–5; Emma Darwin's diary (DAR 242)). The party at the Penders' London house was to celebrate the opening of a telegraphic link with India, which had been achieved under John Pender's supervision. About 700 people attended, including the prince of Wales. The Darwins did not arrive in London until 24 June.
[3] CD's notes are evidently to do with the publication of *Descent*.

To Ernst Haeckel 23 June [1870][1]

Down. | Beckenham. | *Kent. S.E.*
June 23d

My dear Häckel

I received this morning your present of the new Edit. & of two Essays.[2] I see in one of these you, as usual, pile honours high on my head.— As I leave home tomorrow morning for more than a week, I have thought that I would not delay in thanking you.—[3] I am delighted for the sake of science & the public, & likewise for your own sake, to see the new Edition, which seems very greatly enlarged & I do not doubt in all ways improved. For my own especial self, it makes me groan a little, for my M.S is completed, so that I cannot profit by your improvements & additions:[4] it would, alas, from my wretched knowledge of German take me a very long time to read the new Edition carefully.

In what I am going to send immediately to the Printers, I have tried to *a certain extent* to keep on distinct ground from you, & have much shortened some of the subjects which you have fully treated, & refer my readers to your works.[5] And I hope before very long your Morphologie Gen: will be accessible to English Readers.[6]

I do not know how my Book & your Book will be received by the English: I observe at present that there is some tendency to decry geneological classification.— I have been rather better of late, & have worked harder than usual; but I suppose that my best day's work is not one-third of what you can do. You seem a most indomitable worker.— I have seen no scientific man for a long time, except, about a month ago, Kölliker, whom I liked extremely.[7]

Do not work too hard, & believe me, my dear Hackel with the sincerest esteem— yours very truly | C: Darwin

Ernst-Haeckel-Haus (Bestand A-Abt. 1: 1–52/24)

[1] The year is established by the reference to the second edition of Haeckel's *Natürliche Schopfungsgeschichte* (Haeckel 1870a).

[2] CD refers to the second edition of Haeckel's *Natürliche Schopfungsgeschichte* (Haeckel 1870a); there is a copy in the Darwin Library–Down. One of the articles that Haeckel sent to CD may have been 'Beiträge zur Plastidentheorie', which was published in the issue of *Jenaische Zeitschrift für Medizin und Naturwissenschaft* dated 30 May 1870 (Haeckel 1870b). CD was mentioned on the first page. The other may have been 'Das Leben in den grossten Meerestiefen' (Haeckel 1870c), of which there is a copy in the Darwin Pamphlet Collection–CUL.

[3] CD was in London from 24 June to 1 July 1870 (CD's 'Journal' (Appendix II)).

[4] CD refers to his manuscript for *Descent*. He referred to the second edition of Haeckel's *Natürliche Schopfungsgeschichte* in *Descent* 1: 4.

[5] In *Descent* 1: 4, CD wrote that if Haeckel's *Natürliche Schopfungsgeschichte* (Haeckel 1868) had appeared before his essay had been written, he would probably not have completed it, and that all his conclusions had been confirmed by Haeckel.

[6] Haeckel's *Generelle Morphologie der Organismen* (Haeckel 1866) has not been translated into English to date. See also *Correspondence* vol. 16, letter to Ernst Haeckel, 19 November 1868 and n. 4.

[7] Rudolf Albert von Kölliker visited CD with Albert Günther on 15 April 1870 (letter from Albert Günther 12 April 1870; letter from J. V. Carus, 16 April [1870]).

From Hermann Hoffmann 23 June 1870

Giessen.
23.6.70

Dear Sir!

Seeing from your work on variat. in domest (germ. transl. I 1868 p. 321) that you have probablÿ not seen yourself a cock or hen with swimming-membranes between the toes, I am beg your permission to send you the copÿ of such a case, which I had an opportunitÿ to see living these latter daÿs in the neighbourhood of our town.[1] The tribe, to which this young chicken belongs, is the ordinarÿ one of our peasants, of no particular or qualified charakter; the cocks verÿ often almost of the size and colour of Gallus Bankiva. The little cock was about 5–7 days old, his other limbs and everÿ visible part, the beak, eÿes, front, walk, voice were altogether normal.— The case excited some attention, the hen (mother) having been frequentlÿ covered bÿ a male-duck (anas boschas domesticus).— I see no reason to beleave his efforts in anÿ waÿ successful.[2]

With the highest respects | Yours | H Hoffmann | Prof.

[Enclosure]

Phasianus
Gallus domest. pull. mas.

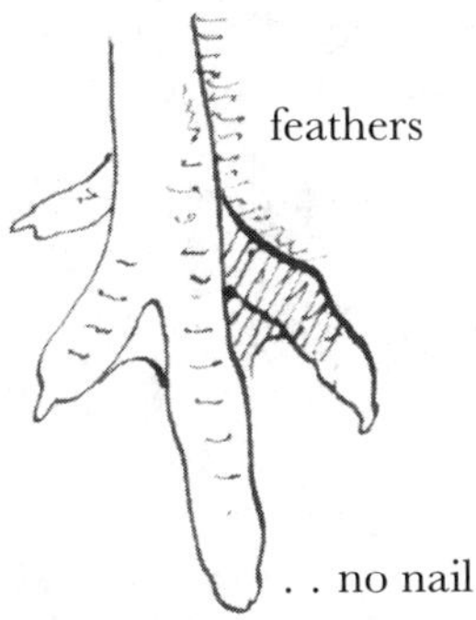

von oben

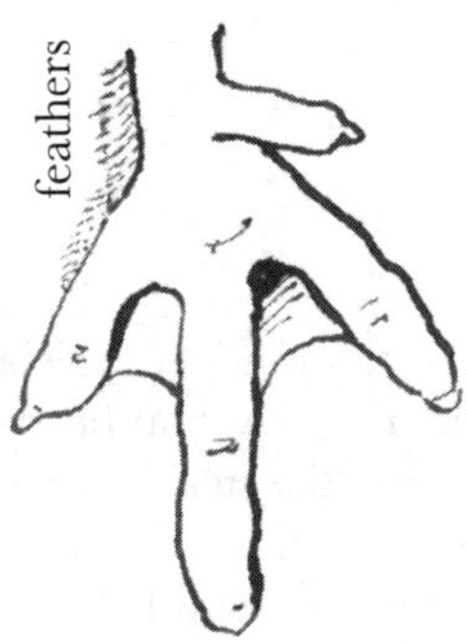

from below

nat. size H. H.

Rödchen near Giessen. 13.6.70

DAR 193: 113–14

[1] Hoffmann refers to the German translation of *Variation*, Carus trans. 1868. In the English edition, the passage referred to is *Variation* 1: 259: 'Golden-spangled Polish fowls are said to have the skin between their toes much developed; Mr. Tegetmeier observed this in one bird, but it was not so in one which I examined.'

[2] CD referred to Hoffmann's sketch in *Variation* 2d ed., 1: 272. *Anas boschas* (now *A. platyrhynchos*, the mallard) is the wild ancestor of most domestic ducks.

[3] *Phasianus gallus* var. *domesticus* is now *Gallus gallus domesticus*, the domestic fowl. *Mas*: male.

[4] *Von oben*: from above (German).

From Osbert Salvin 24 June 1870

Zoological Society of London, | 11, Hanover Square, | London, W.
24 June *1870*

Sir,

I beg leave to inform you that your letter asking permission to have the use of a certain woodblock from this Society's Proceedings was read at the last meeting of the Publication Committee and that it was agreed that your request should be complied with

I enclose herewith an order on Messrs. Taylor and Francis for the use of the wood block in question[1]

I am Sir | Your obt servant | Osbert Salvin.

C. Darwin Esquire

DAR 177: 19

[1] CD's letter to the Zoological Society of London has not been found. In *Descent*, CD used three figures from the *Proceedings of the Zoological Society of London*; fig. 41 (*Descent* 2: 63), fig. 45 (*ibid.*, p. 66), and fig. 65 (*ibid.*, p. 265). Taylor and Francis were the printers of the *Proceedings*.

From Francis Galton 25 June 1870

42 Rutland Gate SW
June 25— 1870

My dear Darwin

A curious and, it may be, very interesting result delays my tranfusion experiments.[1] It is that 2 &, I think all 3, of the does that had been coupled with the largely transfused bucks, proved *sterile*! Of course the sterility may be due to constitutional shock or other minor matters but, it *suggests* the idea that the reproductive elements are in that position of the blood which I did *not* transfuse;—to wit, the **fibrine**.[2] In my earlier experiments, the blood was only partially de-fibrinised,—hence I was able to get a white leg;—but in these later ones it was wholly de-fibrinised. It seems reasonable that the part of the blood which does most in the reparation of injuries, should also be the most rich in the reproductive elements.

Of course I go on with the experiments & modifications of proceedure. Thus, I hope to succeed in making the ears of two young rabbits grow together & to mix their circulations by breaking adjacent veins into one.[3]

I wish I had more to tell you. I have transfused into 32 rabbits, in 6 case twice over—Experimented in 6, and killed 43 others to get at their blood;—making a total of 81 killed & wounded, animals—poor things.

Very sincerely yours | Francis Galton

DAR 105: A21–2

1 See letter from Francis Galton, 15 March 1870 and n. 2.

2 See Galton 1871, pp. 402–3.

3 On Galton's procedure for establishing cross-circulation between rabbits (eliminating the need to defibrinise the blood), see Galton 1871, p. 397.

To Edward Burnett Tylor 25 June 1870

Down, | Beckenham | *Kent. S.E.* [6 Queen Anne Street, London]

June 25 1870

My dear Sir

I want to beg the favour of a little information, which no one can give me if you cannot do so. In your Early Hist. 1865 p. 41. you shew that the Cistercian monks used certain signs for friend & a wise man, & that they use an almost exactly opposite sign for enemy & fool, that is for objects of an opposite nature.[1] You also shew from Burton that certain savages wave their arms in an opposite manner in affirmation & negation; but this case appears a little doubtful for my object, as I suspect that *both* the nod of affirmation & the lateral shake of negation have had a natural origin.[2]

Now what I want to know is whether with the deaf & dumb, when from any cause some sign is used for a certain object, quality &c, whether opposi⟨te⟩ signs are used (apparently merely from being opposite) for objects, qualities &c of an opposite nature[3]

Would you have the kindness to consider this point & permit me to quote the result of your consideration; whether you know of any or many such cases, or after deliberation do not know of any—

Pray excuse me for troubling you & believe me | yours very faithfully— | Ch. Darwin

LS(A)
British Library Board (Add 50254: 33–4)

1 CD refers to Tylor 1865. There is an annotated copy in the Darwin Library–CUL (see *Marginalia* 1: 810–11). Two fingers placed at the right side of the nose signified friend; two fingers at the left side of the nose, enemy; the little finger touched to the tip of the nose signified fool; the forefinger in the same position, wise man. CD mentioned the Cistercians' sign language and cited Tylor in *Expression*, p. 61 and n. 2.

2 Tylor discussed Native American signs of affirmation and negation in Tylor 1865, p. 38, citing Richard Francis Burton's *The city of the saints* (see Burton 1861, p. 127) for his information. CD discussed signs of affirmation and negation in *Expression*, pp. 273–7.

3 No reply from Tylor has been found, but see the letter from W. R. Scott to E. B. Tylor, 28 June 1870. CD discussed deaf-and-dumb sign language and the use of opposition in *Expression*, pp. 61–2, citing W. R. Scott 1870: he commented on how few unequivocal instances could be found.

To Frederick Bates 26 June [1870?][1]

6. Queen Anne St | London W
June 26th

Dear Sir

It is extremely kind of you to remember my request about Trox.[2] Unfortunately I have left home for a week, & the beetles have been forwarded here. I have opened them & they look very lively. But I have no microscopes here or other means of investigation. I have put a few in a separate Box & will feed & take home, & will soon take or send the remainder to your brother. [3]

I am rather in despair how to observe them, for they sham dead when touched, & do not like most other beetles seem at all inclined to stridulate.

With very sincere thanks | Pray believe me | Yours truly obliged | Ch. Darwin

DAR 185: 121

[1] The year is conjectured from the relationship between this letter and the letter to Frederick Bates, 19 June [1868?] (*Correspondence* vol. 16), and from CD's mention of being in London for a week. In 1870, CD was in London from 24 June to 1 July (CD's 'Journal' (Appendix II)). In 1869, he had been in Wales for most of June (*Correspondence* vol. 17, Appendix II).

[2] See *Correspondence* vol. 16, letter to Frederick Bates, 19 June [1868?]. CD discussed the stridulation of *Trox sabulosus* in *Descent* 1: 380, commenting that he himself had failed to discover the organs of stridulation in this beetle (see also *Correspondence* vol. 16, letter to B. D. Walsh, 9 June 1868).

[3] Henry Walter Bates.

From T. W. Wood [after 26 June 1870][1]

I was greatly surprised & disappointed to find by your last letter that you cannot see the Argus as I and some others do.[2] Mr Blyth[3] has just called on me & when I asked him whether he had thought at all on the subject he said he perfectly agreed with me about the markings alluded to. Mr Blyth also said he has had many through his hands while in India alive & dead although he never saw one display his wings. There are he informs me three specimens now living in Europe—1 male & 2 females but all separated. I looked at the Museum specimen on Saturday in company with a young friend who is an amateur artist & *when we stood a few yards away from the bird* we both saw very distinctly the art wonder in nature Still I greatly admire your extrem⟨ ⟩ ⟨ ⟩ ⟨ ⟩ness in not admitting what you do ⟨ ⟩ You do not say how you like my ⟨ ⟩ Student[4] nor anything about the c⟨ ⟩ ⟨ ⟩ pheasant, the peculiarity in which, to ⟨ ⟩ your attention, may be seen in any ⟨ ⟩ period of courtship, & I have made ⟨ ⟩ life shewing the unbroken stripes.

I was at the Gardens on a Sunday morning & saw your name in the book but was disappointed to hear, that you had just left.[5] May I ask the favour of an interview with you when you are in Town again?

& believe me to remain, | Dear Sir, | Very faithfully yours, | T. W. Wood.

C. Darwin Esq. F.R.S.

P.S. Is not this matter one of those things which have been hidden from the wise & prudent & revealed unto babes?

Incomplete
DAR 181: 146

[1] The date is established by the relationship between this letter and the letters from T. W. Wood, 14 June 1870 and 16 June 1870, and by the mention of Wood's missing CD at the Zoological Gardens on Sunday. CD was in London from 24 June to 1 July 1870 (CD's 'Journal' (Appendix II)); the only Sunday in that period was 26 June.

[2] See letters from T. W. Wood, 14 June 1870 and 16 June 1870.

[3] Edward Blyth.

[4] Wood refers to his article on the courtship of birds in the *Student and Intellectual Observer* (Wood 1870; see letter from T. W. Wood, 14 June 1870).

[5] Wood refers to the gardens of the Zoological Society of London at Regent's Park.

From Bartholomew James Sulivan 27 June 1870

Bournemouth
June 27/70

My dear Darwin

I was sorry to see that your health did not allow you to go to Oxford to recieve the honor intended for you, and the almost unanimous expression of feeling with which it would have been accompanied—[1] I hope it was merely fear of the excitement, that prevented your going, and not that your health is too bad to take the journey— I foolishly did myself much harm lately by want of care. I had felt better though my head was not well; and I went to town for two or three days returning by Portsmouth to stay two or three days, and see my old friend the admiral,[2] there as well as some of the new ships. I returned pretty well and therefore fancied my head was strong enough a few days after to go up to a ball at Buckingham Palace.[3] I felt pretty well for the two hours I was talking to old friends & seeing the sights, but when I returned to the hotel could not sleep & by the morning was so ill with pain in the head, fainting feelings &c, that I could hardly stand & with difficulty got to the train the next forenoon to come home, having given my-self a severe shake— I have never been so ill before as I was for some hours. and I have felt the effects ever since. I hope if well enough to go with my wife to see some of our relatives in July and August—

I hope M^rs. Darwin and all your party are well, I think I saw that another of your boys had passed well at Woolwich—[4] What a comfort to you to see them all doing so well. My youngest has lately gone into a good City House as clerk, and seems doing well. He lives at Blackheath.[5] My youngest daughter was lately married to a M^r. Trench—& is in a fair way I think to a really happy marriage, for he is a fine intelligent, young man who after starting with his relative in the Irish agency line— went cruising over the world for several years: during which he made one journey I think quite new.[6] He started in land from Pernambuco. went straight through the

Indian Country to the Andes & so reached Lima. He has now a lucrative occupation in Westminster;

I enclose you a prospectus of our small Natural History society lately started here, we hope in time to get up a small museum and Lecture Hall. Mr. Mitchell who came here to work the fossil leaf beds helped to set it going and I hope it will excite some interest that may be lasting.[7] Our only Honorary Member yet is Sir C Lyell[8]—and we should be very glad indeed if you will allow us to add your name. I fear we have not much science of the right sort yet— Mr. Austen[9] is our only real geologist I believe, But Mr. Sanders & Mr Baldry[10] are both Cambridge men of good position, and thorough ability

The only one of our old party that I have seen for some time is Hamond[11] who was here at our wedding. I hear from Mellersh & Usborne[12] occasionally. I hope you have followed our movements in Tierra Del Fuego with the Mission, our Missionary—now Bishop of Falklands for the oversight of all South America—(Mr. Stirling) took the bold step of living six months alone with the Natives in Beagle channel (the schooner visiting him twice) to test the safety—and the result decided the Committee to plant a Mission there at once.[13] The Natives that had been at Keppel Island though some were scattered & living near the entrance to Murray Narrows—had influence enough to protect him & to take care of his stores & provisions—divided between four or five huts—without losing a thing. He found on the north shore right opposite to the Murray Narrows a nice little peninsula with a good cove & some clear land. & there is to be the station a clergyman & catechist with their wives and one other assistant are going to live there with some friendly Natives round. They are quite alive to growing potatoes &c and taking care of their goats. and one who with his wife lives at the Murray Narrows has a good garden & goats &c. they keep their house as neat and clean and themselves and children as nicely clothed & clean as Europeans.— Already one shipwrecked crew, though not near there, have been treated kindly by Natives.[14]

My wife joins me in kind regards to Mrs. Darwin and yourself with all your party.

Believe me Dear Darwin | Yours very sincerely | B. J. Sulivan

DAR 177: 293

1 See letter from Robert Cecil, 7 June 1870. The awards of honorary degrees of DCL at Oxford were announced in *The Times*, 20 June 1870, p. 11. A paragraph at the bottom noted, 'The name of Charles Darwin, Esq., F.R.S., would have been included in the foregoing list, but he writes that his health is such that he "could not withstand the fatigue and excitement of receiving an honorary degree."'

2 The admiral whom Sulivan visited may have been Astley Cooper Key, who left Portsmouth probably in June 1870 after being removed from his post in controversial circumstances (*The Times*, 11 June 1870, p. 7). On Sulivan's relationship with Key, see Sulivan ed. 1896.

3 Sulivan attended the State Ball at Buckingham Palace on 17 May 1870 (*The Times*, 18 May 1870, p. 11; Sulivan was incorrectly listed as W. J. Sulivan).

4 Leonard Darwin was at the Royal Military Academy at Woolwich; according to *The Times*, 21 December 1869, p. 12, he had received second prize for mathematics.

5 Henry Norton Sulivan.

6 Sulivan refers to Catherine Sabine Sulivan and William Wallace Trench.

[7] The enclosure has not been found. Sulivan refers to William Stephen Mitchell (see *Correspondence* vol. 14, letter from B. J. Sulivan, 25 December 1866 and nn. 12 and 13).
[8] Charles Lyell.
[9] Presumably Sulivan refers to Robert Alfred Cloyne Godwin-Austen.
[10] Edmund Sandars and Alfred Baldry.
[11] Robert Nicholas Hamond.
[12] Arthur Mellersh and Alexander Burns Usborne.
[13] On Waite Hockin Stirling's activities in South America, and the mission station at Ushuaia, see F. C. Macdonald 1929.
[14] See Hazlewood 2000, pp. 249–53.

From J. J. Weir 27 June 1870

6 Haddo Villas | Blackheath, SE
27th June 1870

My Dear Sir

How I have longed, during my protracted illness, to reply to your enquiries, but this is the first day I have felt myself able, although still very weak & my nerves in a tremble.[1]

In some way or other, I poisoned my hand, the disease spread rapidly, assumed a malignant form, & necessitated a series of acutely painful surgical operations.—

The agony endured during ten days night & day was beyond anything I ever felt, but I will not trouble you further with my own complaints, & have only said this much by way of apology for the delay in answering your kind letter.—

I should say as a general rule that birds when *frightened* closely adpress their feathers & it is astonishing the apparent reduction in bulk, that fear appears to produce in most species, I have noticed it most in the Melopsittacus undulatus, & in the Quail, the latter when frightened looks about half the size when undisturbed. One of the first things a bird does when recovering from a surprise is to shake out the feathers.[2] Some finches when angry, erect the feathers all over the body most species however only spread out the wings and tail, among the former, the Canary, Goldfinch & preeminently, the Goldfinch male, make each "dividual" feather "to stand on end like quills upon the fretful porcupine".

I have a Goldfinch male, often in my dining room, of the most irascible disposition, even if a servant when waiting at table goes too near the cage, he at once appears a ball of erected feathers[3]

The irascibility of most birds, Robin[4] excepted, is greatest during the breeding season, & many then which do not erect the feathers all over the body, erect a ruff of them round the neck, all the buntings & most of the finches do this, the Chaffinch[5] at the same time shewing a small concealed crest at the back of the head, a spot in the same place is not concealed in the female, nor in either sex of the Brambling, "Fringilla montifringilla, nearly all when angry rush at each other with open beak & with threatening gestures.

As to birds exhibiting impatient gestures, the only birds I have observed in my aviary, which shew by signs they are hungry are the Siskins,[6] these if hungry as soon as they see me cling to the wires of the aviary, it may be observed the species mostly seeks its food clinging like a Tit.—

There is one curious habit of captive Ratels[7] which has much interested me, they are in the constant habit of turning summersaults, the head & back being brought with considerable force on to the ground, this appears to me to be the action, which these mellivorus mammals use, to free themselves from the bees, which must attack them in great numbers, when robbing their hives.—

I forgot to observe when speaking of the erection of feathers that the angry robin erects the feathers all over the body, the Whitethroat, C Cinerea,[8] also erects them all around the throat.—

I shall be glad to give at all times any further information in my power.—

I take the opportunity now to mention a few other subjects—

The first bears on the Cytisus Adami.—

My Brother has but two kinds of Laburnum viz Cytisus purpureus, very erect, & Cytisus Alpinus? very pendulous, he had several stocks of the latter grafted with the purple one, & this year, the grafts being two years old, I saw in one, fairly above the stock, about four inches, a raceme of purely yellow flowers, with the usual dark markings, and above them a bunch of purely purple flowers, the branches of the graft in no way shewed an intermediate character, but had the usual rigid growth of Purpureus.—[9] The second case bears on latent gemmules.—[10]

My Brother keeps a large number of Rabbits at Brenchley, not that he is a "Fancier' of them, he has had the breed a long time & they are supposed to be pure bred English lop eared.—

Some time since, in a litter, there appeared one with the very thick coat which distinguishes the variety known as the French, the hair was remarkably long & woolly, but the peculiarity was that it was a grey a color Harrison had never seen in the French variety.—

The specimen was a male & he had from 30 to 40 young from it, without one shewing a trace of the long hair of the male parent, but curiously enough this Spring a litter appeared from a doe, not a descendant of the long hair, with 4 young of the long haired variety, and one of the usual smooth haired.—

The four young were all fawn in color.— The third case bears on the question of hereditary mutilations.—

You will recollect ⟨*two pages missing*⟩

the bones were put by for you, but unfortunately were mislaid & lost.—[11]

The fourth case bears on the question of contagious variegation in plants.—

I have an ivy called the Golden lobe, by Gardeners Hedera helix aurea densa, this has twined into a common Hedera helix[12] which from its close habit I call the clinger; now the clinger has caught the yellow mottling from Aurea Densa, & thus a perfectly new variety has been formed, which was much needed as Aurea Densa was rather too arboreal in its growth.—

Had it been a mere sport I should not have attached any importance to it, but the improbability of the sport being like Aurea Densa, with which its roots & branches are so intimately intertwined, is very great unless by infection or rather perhaps contagion.

I enclose a leaf of Aurea Densa, lobed shaped, of the clinger before sporting & afterwards.—

I am surprised I have had strength to write so much, but I am so interested in your subjects & your enquiry was quite a solace during my illness.—

Believe me | Yours very sincerely | J Jenner Weir

C Darwin Esq[r].

Incomplete
DAR 181: 82

CD ANNOTATIONS
1.1 How ... letter.— 3.3] *crossed pencil*
4.1 I should say] *after opening square bracket pencil*
4.4 One ... feathers. 4.5] *scored blue crayon*
4.6 Some ... tail, 4.7] *underl blue crayon*
6.2 erect ... do this, 6.3] *scored blue crayon*
6.6 "Fringilla ... gestures. 6.7] *double scored red crayon*
6.6 nearly ... gestures. 6.7] *scored blue crayon*
7.1 As ... hives.— 8.5] *crossed pencil*
8.1 There ... hives.— 8.5] 'Ratels' *pencil*
10.1 I shall ... Adami.— 12.1] *crossed pencil*
13.1 My Brother] *after opening square brackets blue crayon, pencil*
13.7 The second case] *after opening square bracket red crayon*
13.7 latent gemmules] *closing square bracket blue crayon, del blue crayon*
14.1 My Brother ... eared.— 14.3] '*Reversion*' *added pencil*
14.1 My Brother] *after opening square bracket red crayon*
14.1 My Brother ... variety.— 15.4] *crossed pencil*
21.1 I have ... growth.— 21.5] '(Contagion of Variegation)' *pencil, square brackets in MS*
21.1 I have] *after opening square bracket pencil*
24.1 I am ... illness.— 24.2] *crossed pencil*
Top of letter: 'J. Jenner Weir | Expression) | (Erection of feathers)' *pencil*

1 For CD's enquiries, see the letter to J. J. Weir, 14 June [1870].

2 CD cited Weir for this information in *Expression*, p. 100. *Melopsittacus undulatus* is the budgerigar. The quail is *Coturnix coturnix*.

3 CD cited Weir for the information about angry birds fluffing out their feathers in *Expression*, p. 99. Weir quotes from Shakespeare's *Hamlet* 1: 5. The canary is *Serinus canaria*; the goldfinch, known to CD as *Carduelis elegans*, is now *C. carduelis*.

4 *Erithacus rubecula*, the European robin.

5 *Fringilla coelebs*.

6 *Carduelis spinus*, the Eurasian siskin.

7 *Mellivora capensis*, also known as the honey badger.

8 Weir presumably refers to the European or common whitethroat, *Sylvia communis*, formerly *Sylvia cinerea* or *Curruca cinerea* (see Montagu 1831, p. 538).

9 Weir's brother was Harrison William Weir. CD discussed *Cytisus adami* in *Variation* 1: 387–90 and 2: 364–5. *Cytisus adami*, a graft hybrid of the common yellow laburnum *C. laburnum* (now *Laburnum anagyroides*) and *C. purpureus*, a dwarf purple broom, is now known as +*Laburnocytisus adamii* (Bean 1970–88, 2: 510–11). See also *Correspondence* vol. 13, letter to J. D. Hooker, [1 May 1865] and n. 3.

10 Latent gemmules were a feature of CD's hypothesis of pangenesis (see *Variation* 2: 357–404), according to which cells threw off minute granules (gemmules) that circulated freely throughout the system, and had the potential to develop into cells like those from which they were derived (*ibid.*, p. 374).

11 CD evidently tore off the half-sheet describing this case, which has not been identified.
12 *Hedera helix* is the English ivy.

From Francis Galton 28 June 1870

42 Rutland Gate SW
June 28/70

My dear Darwin

Of the three things I undertook to do for you I have as yet only done $2\frac{1}{3}^{\text{d}}$.[1]

1. Greg *did* write the article in Fraser. and has no objection at all,—but the contrary,—in being publickly spoken of as the author.[2] He is highly gratified at your appreciation of the article.

2. I have read Duncan's book on Fertility, which seems to me to shew unmistakeably that marriages of young women contribute enormously more to the adult population of the next generation than those of middle aged women.[3]

The number of children he shews to be very much greater He also shews the weight & length of children born to young (not **very** young) mothers to be a maximum & therefore the probability of their being more vigorous As a set off against this he reproduces a table published by a C^{ttee} of the Statistical Society some years ago of which I can make nothing except that the irregularity of its figures shew no great dependance can be put upon it.

($\frac{1}{3}^{\text{d}}$) I have only seen Bartlett *once* out of the 3 promised times, viz: this morning and jogged his memory about the crying monkey.[4] He answered just as you told me he would.

With united kind regards | Francis Galton

We transfused *4* rabbits this morning & I have 1 buck & 1 doe almost quite well after being operated on.[5]

DAR 80: B160–1

CD ANNOTATIONS
1.1 Of the … article. 2.3] *crossed pencil*
5.1 ($\frac{1}{3}^{\text{d}}$) … on. 7.2] *crossed pencil*

1 CD was in London from 24 June to 1 July 1870 (CD's 'Journal' (Appendix II)); no letter to Galton written during this period has been found.
2 William Rathbone Greg had written an anonymous article, 'On the failure of "natural selection" in the case of man', for *Fraser's Magazine* ([Greg] 1868).
3 Galton refers to James Matthews Duncan and Duncan 1866; there is an annotated copy of the second edition (Duncan 1871) in the Darwin Library–CUL (see *Marginalia* 1: 215). CD cited Duncan (but not Duncan 1866) in *Descent* 1: 174 and n. 19.
4 Galton refers to Abraham Dee Bartlett. For an earlier query from CD about crying monkeys, see *Correspondence* vol. 15, letter to Fritz Müller, 15 August [1867] and n. 5. In DAR 189: 138 there is a note in Francis Darwin's hand, possibly taken down at CD's dictation, reading in part, 'The *Macacus inornatus* from Borneo certainly cries, so that the tears roll down its face when in grief or even when pitied. The woman who sold it said so. Bartlett & Sutton have both repeatedly seen it.' Seth Sutton was a keeper at the Zoological Gardens in London; Bartlett was the superintendent.
5 See letter from Francis Galton, 25 June 1870.

From William Robson Scott to E. B. Tylor 28 June 1870

St Leonards, Exeter
June 28th | 1870

My dear Sir,

The question you ask is a difficult one and my knowledge of the history of *these* ⟨ ⟩ good & bad) is not sufficient to enable me to answer it in this manner but I will state what I think may have been the probable origin.[1]

In my chapter on signs I state that there are many signs that are abreviations of natural signs that are now so shortened as to assume all the characters of arbitrary signs and I believe the sign for *bad* to be one of these.[2]

A natural sign for *disaproval*—I will not say *the* one for I think there may be several—is by elevating the hands perpendicularly palms outward as if pushing an object from you & turning aside the face to avoid its sight. Now here the little fingers rise naturally, come nearest the individual disproved, & to lift the hand & present the little finger is only a curtailment of the fuller sign.— Now the sign good i.e the *thumb* elevated may have arisen from its being the digit opposite to the little finger & so taken to express the opposite direction ⟨of⟩ *bad*.— This however I would not like to ⟨suppose⟩ but I think the little finger being used for bad may very probably have arisen in the manner I state & the thumb taken as its opposite for good. *Good* in the North American signs is the hand held horizontally, back upward describing with the arm a horizontal curve outward.

This may probably spring from patting the head or back of a child; but I should rather think myself in the case of the thumb in D & D[3] signs that it arose from its being the opposite of bad—bad being an abreviated form of the natural ⟨sign I⟩ have described, & I can think of no such ⟨sign⟩ as would similarly explain good.—

Such is the best account I can give you of these signs— It is a very busy time with me at present our children are just going away & what with this & the admission of new ones & settlement of a/cs, it keeps me at present very close at work—otherwise I would have looked more into the question for you.—[4]

Believe me | very truly yours | W. R. Scott

P.S. | I see I have neglected to say a word or two of the *opposites* alluded to by Mr Darwin— Opposites we had *greatly* recourse to in teaching the deaf and dumb, and they have a lively sense of these and will use signs accordingly— *Light* for instance ⟨ ⟩ be expressed by opening the eyes wide, & the arms gradually out—while darkness would be expressed by closing the eyes & drawing the arms inwards.

Strength would be represented by doubling the ⟨fists⟩ contracting the muscles of the arms, with a general rigidity of the whole body & erect carriage

weakness—would be just the opposite of this all the muscles relaxed & a depressed appearance &c.

others in the same way—but these ⟨will⟩ perhaps be said to be signs used *[partly]* *[1 word illeg]* being *opposites*, as they are all more or less expressive of the things they are to show I dont know that I have hit the right point Mr Darwin wished, but if he ⟨would⟩ give me a few examples I will then try to answer them;— I shall shortly go to London for a week or so as my holidays are coming on, but any communications ⟨sent here⟩ would follow me.— Do you think there would be anything in my book of interest to M[r] Darwin—I would have great pleasure in sending him one, if he would do me the honour ⟨ ⟩

—Excuse this written amongst message many ⟨ ⟩

DAR 177: 124

CD ANNOTATION
Top of letter: 'Antithes' *pencil*

1 See letter to E. B. Tylor, 25 June 1870.
2 See W. R. Scott 1870, p. 12. CD cited this passage in *Expression*, p. 62. There is an annotated copy of W. R Scott 1870 in the Darwin Library–CUL (see *Marginalia* 1: 749).
3 Deaf and dumb.
4 Scott was principal of the West of England Institution for the Education of the Deaf and Dumb. A/cs: accounts.

From T. H. Farrer 29 June 1870

B[d] of Trade
29 June/70

My dear Mr Darwin

I have at last got a flower of Passiflora Princeps.[1] You were quite right, as you seem always to be The lower or inner coronas, or whatever they are, which cover the nectary, neither fit close to the style, nor lap over one another so as to prevent the entrance of a stiff object. There are wide gaps by which a pin *does* get to the bottom and comes out covered with nectar. But the style is very long: the flower sub-erect: and, so far as I can see, the anthers and stigmas do not droop down as in P. Caerulea. The consequence is that the distance between these organs and the nectary & their relative position is such that a bee on the top of the outer coronas would not touch them. Add to this that the two outer coronas are very little developed: and that the third closes round the style so as to make it an inch from the entrance to the bottom of the nectary— I think with all this we have a correlation of parts suited for Humming birds & not for Bees.—though not exactly the correlation I noted in the others.[2]

I will certainly come and see you as you so kindly suggest—but am very busy till the Session is over—which I find the best thing. It is the leisure & the tired hours which are difficult to bear.[3]

Sincerely yours | T H Farrer

We have been deeply interested in watching the wonderful motions of Passifloras in climbing. They seek & find & hold on & pull up like an animal

DAR 164: 65

1 CD had advised Farrer to look at the pollination mechanism of *Passiflora princeps* in October 1869; Farrer had obtained an imperfect specimen in November 1869 (see *Correspondence* vol. 17, letter to T. H. Farrer, 20 October [1869], and letter from T. H. Farrer, 5 November 1869). *Passiflora princeps* is now *P. racemosa*, the red passionflower.

[2] Farrer had been engaged in a discussion with CD and Fritz Müller about *Passiflora* and *Tacsonia*; see letter to T. H. Farrer, 13 [May 1870] and n. 3. See also *Correspondence* vol. 17, Appendix IV.

[3] Farrer's wife, Frances, had died in May 1870 (see letter from T. H. Farrer, 17 May 1870). Farrer was permanent secretary of the Board of Trade.

To T. H. Farrer [29 June 1870][1]

6. Queen Anne St
Wednesday

My dear Farrer

I have just received from Fritz Müller the enclosed seeds "of a third species of Passiflora, the flowers of which are visited fertilised by a large Humble-bee."[2]

Yours most sincerely | Ch. Darwin

We return home early on Friday morning: if you were in this direction on Thursday & would feel the least inclination to come to luncheon at 1 o clock, I am sure I need not say that we sh^d. all be very glad indeed to see you.—

Linnean Society of London

[1] The date is established by the reference to Fritz Müller's sending seeds of *Passiflora* (see letter to T. H. Farrer, 13 May [1870]), and by the dates of CD's visit to London; the Darwins were in London from 24 June to 1 July 1870 (CD's 'Journal' (Appendix II)). The only Wednesday during this period was 29 June.

[2] No letter from Fritz Müller accompanying the seeds has been found. CD, Müller, and Farrer had been discussing adaptations for pollination in different species of *Passiflora* and *Tacsonia*; see letter from T. H. Farrer, 29 June 1870 and n. 2.

To J. D. Hooker [29 June 1870][1]

Down. [6 Queen Anne Street, London]
Wed.

My dear Hooker

I have heard from Durando & he can send me only seeds of the Iberis.[2]

I have seeds of Hibiscus Africanus & of Nolana prostrata from Carters, said to be matured in Essex; now c^d you give me seeds of these 2 plants raised at Kew; for this w^d be some change of conditions, though not so great as I sh^d wish—[3]

It was a gt disappointment not to get to Kew on Sunday. We return home early on Friday m^g—& if you have them you had better send the seeds to Down—[4]

yours affectly | Ch. Darwin

LS
DAR 94: 173

[1] The date is established by the relationship between this letter and the letters to J. D. Hooker, 25 May [1870] and [13 June 1870?], and by the dates of CD's visit to London; the Darwins were in London from 24 June to 1 July 1870 (CD's 'Journal' (Appendix II)). The only Wednesday during this period was 29 June.

2 Gaetano Durando had a business in Paris selling botanical specimens. No correspondence with him has been found. See also the letter to J. D. Hooker, [13 June 1870?] and n. 2, and *Cross and self fertilisation*, p. 105, where CD says that the seeds were sent by Durando from Algiers.

3 See letter to J. D. Hooker, [13 June 1870?] and nn. 3 and 4. *Hibiscus africanus* is a junior synonym of *H. trionum*. *Nolana prostrata* is now *N. humifusa*. CD probably refers to the seed merchants Carter, Dunnett and Beale of London (*Post Office London directory* 1870).

4 Sunday was 26 June 1870; the Darwins went back to Down on Friday 1 July 1870.

From E. C. Rye 29 June 1870

10, Lower Park Field | Putney | SW
29 June 1870

Dear Sir,

As a former note of mine, ⟨with⟩ regard to insect aid in Orchid ⟨fertili⟩zation, happened to be at the ⟨time⟩ new to you, I venture to ⟨send⟩ a reference to a somewhat ⟨simil⟩ar occurrence, from The ⟨Amer⟩ican Naturalist, Vol. III, 1869, ⟨p. 10⟩9—in which the causes of ⟨death⟩ to Honey bees, through ⟨entan⟩glement in pollen masses of ⟨Asc⟩lepias, are treated of. (cf. 2.c.II ⟨p.⟩ 665).[1] I came across it in working up the matter for 1869 ⟨"Zo⟩ological Record".[2]

Yours truly | E. C. Rye

DAR 176: 229

CD ANNOTATIONS
Top of letter: 'Dichogamy' *pencil*

1 See letter from E. C. Rye, 14 February 1870. CD's reply has not been found. In *American Naturalist* 3 (1869): 109, a correspondent described how honey-bees trapped in the cleft gland of silkweed or milk-weed (*Asclepias*) were forced to pull out the gland, with the pollen masses attached to it, in order to escape; bees thus encumbered were often expelled from the hive and died. *American Naturalist* 2 (1868–9): 665 contained the original report of a honey-bee found dead with its legs bound together with silkweed pollen.

2 Rye mentioned the reports in *American Naturalist* in *Zoological Record* 6 (1869): 130. Further observations were made in *American Naturalist* 3 (1869): 388–9.

To J. J. Weir 29 June [1870][1]

Down, Beckenham, Kent. S.E. [6 Queen Anne Street, London]
June 29

My dear Sir

I was very sorry to hear of your illness. From what I have been told poisoning of the blood is most dangerous and often causes extreme suffering. Now that you have begun to recover (and indeed must have a good deal of strength to have so kindly written me so long a letter) I trust and suppose that there is no risk in your complete and early recovery. Your letter is most rich in facts *even for you*; and I can say nothing stronger than this.[2] What you tell me about the erection of the feathers is exactly what I wanted to learn. It seems in all cases, and more especially with Birds, to be more closely connected with anger than fear.[3] I have seen the Ratel turning head over heels; and if it was really established that they thus free themselves from Bees, the case would be capital for me: I am writing this in London and shall go to Z. G. tomorrow morning and will ask Bartlett whether he has ever heard any authentic accounts from Travellers.[4] The *Cytisus* case ought to be well watched: At present, without some intermediate form, I know that Botanists would simply declare that there was some blunder. I am watching a tree in which the bud of C. purpureum died and the stock pushed forth shoots close beneath.[5]

You do not say so, but I supose the *fatter* of the litter of 4 long-haired rabbits had the blood of the first long-haired rabbit in his veins. I think there can be no doubt *now* that variegation is contagious; but yours is rather a new case.

I shall go to press *very* soon with my Man Book and Sexual Selection and keep Expression for a separate Essay.[6]

With cordial thanks | Yours very sincerely | Ch. Darwin.

Copy
DAR 148: 327

[1] The year is established by the relation between this letter and the letter from J. J. Weir, 27 June 1870.
[2] See letter from J. J. Weir, 27 June 1870.
[3] See letter from J. J. Weir, 27 June 1870 and nn. 2 and 3.
[4] Abraham Dee Bartlett was superintendent of the Zoological Gardens, Regent's Park, London. See letter from J. J. Weir, 27 June 1870 and n. 7.
[5] See letter from J. J. Weir, 27 June 1870 and n. 9. 'C. purpureum': *Cytisus purpureus.*
[6] CD refers to *Descent* and *Expression*, published in early 1871 and late 1872 respectively.

To W. W. Reade 30 June [1870][1]

Down. | Beckenham | *Kent. S.E.* [6 Queen Anne Street, London]
June 30th

My dear Sir

I have to thank you for two letters: one with a specimen of the insect which protects itself with dead ants &c; but I could not detect the *protected* insects. I suppose that it must be a larva; the larvæ of some beetles are known to protect themselves by cases made of their own hardened excrement.—[2] Your second letter, June 4th, is **full** of curious observations: I shall be surprised if you do not make a very interesting book with an account of your Travels & observations on your return.[3]

I am very glad to hear confirmed that negros admire very black negros; but I much wish that time & space had permitted you to enter in some little detail on their ideas of beauty.[4] I have to touch on this point & I shall go to press in a few weeks time; but as the chapter in question is nearly my last, I do not suppose that I shall come to it in the proofs for 3 or 4 months.[5] I have resolved to keep my Essay on Expression in man & animals for subsequent & separate publication.[6]

Thanks for fact about the hand & mouth & astonishment.[7] I wrote some little time begging you to observe about Gorillas & chimpanzee strictly frowning.[8]

Your case of stone tools is interesting: until lately none had been found in Africa, but quite recently they are reported from S. Africa & Egypt; & now you fill up the intermediate gap.—[9]

With every good wish, pray believe me | My dear Sir | Yours very faithfully | Ch. Darwin

DAR 96: 79–80

[1] The year is established by the relationship between this letter and the letter from W. W. Reade, 4 June 1870.
[2] See letter from W. W. Reade, [*c.* 8 or 9 April 1870].
[3] See letter from W. W. Reade, 4 June 1870. Reade published *The African sketch-book* in 1873 (Reade 1873).
[4] See letter from W. W. Reade, 4 June 1870 and n. 4.
[5] CD sent the manuscript of *Descent* to the printers in August 1870 (letter to John Murray, 5 [August 1870]).
[6] *Expression* was published in November 1872 (CD's 'Journal' (DAR 158)).
[7] See letters from W. W. Reade, [*c.* 8 or 9 April 1870] and n. 7, and 4 June 1870 and n. 14.
[8] The letter in which CD asked Reade about gorillas and chimpanzees frowning has not been found. CD wrote about frowning in chimpanzees, orang-utans, and gorillas in *Expression*, pp. 143–4.
[9] See letter from W. W. Reade, 4 June 1870 and n. 7.

To B. J. Sulivan 30 June [1870][1]

Down. | Beckenham | *Kent. S.E.* [6 Queen Anne Street, London]
June 30th

My dear Sulivan

It was very good of you to write to me so long a letter, telling me much about yourself & your children which I was extremely glad to hear.—[2] Think what a benighted wretch I am, seeing no one & reading but little in the newspapers, for I did not know (untill seeing the paper of the your Nat. Hist. Socy.) that you were a K.C.B.[3] Most heartily glad I am, that the Government has at last appreciated your most just claim for this high distinction.— On the other hand I am sorry to hear so poor an account of your health; but you were surely very rash to do all that you did & then pass through so exciting a scene, as a ball at the Palace.[4] It was enough to have tired a man in robust health.— Complete rest, will, however, I hope, quite set you up again. As for myself I have been rather better of late, & if nothing disturbs me I can do some hours work every day. I shall this autumn publish another book partly on Man, which I daresay many will decry as very wicked.—[5] I could have travelled to Oxford, but could no more have withstood the excitement of a commemoration than I could a ball at Buckingham Palace.—[6] Many thanks for your kind remarks about my boys. Thank God all give me complete satisfaction: my third stands second at Woolwich & will be an Engineer officer at Christmas.[7] My wife desires to be very kindly remembered to Lady Sulivan, in which I very sincerely join, & in congratulations about your daughter's marriage.[8] We are at present solitary, for all our younger children are gone on tour in Switzerland.[9] I had never heard a word about the success of the T. del Fuego mission.—[10] It is most wonderful, & shames me, as

I always prophecied utter failure. It is a grand success.— I shall feel proud if your Committee think fit to elect me an Honorary member of your Soc^y.—[11]

With all good wishes & affectionate remembrances of ancient days. Believe me, my dear Sulivan | your sincere friend | Ch. Darwin

Sulivan family (private collection)

[1] The year is established by the relationship between this letter and the letter from B. J. Sulivan, 27 June 1870.

[2] See letter from B. J. Sulivan, 27 June 1870.

[3] Sulivan was made a Knight Commander of the Order of the Bath on 2 June 1870 (*ODNB*).

[4] See letter from B. J. Sulivan, 27 June 1870 and n. 3.

[5] CD refers to *Descent.*

[6] CD had been offered an honorary degree by Oxford University; see letter from B. J. Sulivan, 27 June 1870 and n. 1.

[7] CD refers to Leonard Darwin.

[8] CD refers to Emma Darwin, Sophia Sulivan, and Catherine Sabine Sulivan, who had married William Wallace Trench.

[9] A note in Emma Darwin's diary (DAR 242) for 21 June 1870 reads, 'They went to Swiss—'; another note on 28 July reads, 'Party from Swiss.' It is not known exactly which children went.

[10] See letter from B. J. Sulivan, 27 June 1870 and n. 13.

[11] See letter from B. J. Sulivan, 27 June 1870. CD refers to the Bournemouth Natural History Society.

From J. D. Hooker 1 July 1870

Royal Gardens Kew
July 1/70

My dear Darwin

I am sorry to say that we have no seeds of the *Hibiscus* or *Nolana*;[1] they are things of a class which it is not worth our while to harvest the seeds of— we can buy fresh seeds cheaper annually for ourselves & the rare correspondents who ask for them

I herewith send a list of the best Liliums we can give you plants of most if you do not want many— but I suppose that you want bed's full.—[2]

I have lost a note I made about a work on European orchids that we want; & which you mention in your supplementary paper on Orchid: fertilization. What I want to know is when you saw said work— can you pick up my scattered ideas for me? You quote I think some Society's memoirs, for the work[3]

We have been awfully gay of late at dinners & Garden parties. & I am proportionately behindhand & disgusted with myself.

Ever Yours affect | J D Hooker

DAR 103: 51–2

CD ANNOTATION

Top of letter: 'London Inf. | I work like a Horse & am begun to get MS pages ready for Prg | what a good *[illeg]*'[4] *pencil*

[1] See letter to J. D. Hooker, [29 June 1870]. CD had asked for seeds of *Hibiscus africanus* and *Nolana prostrata.*

[2] The list of species of *Lilium* has not been found.

[3] Hooker refers to 'Fertilization of orchids'; he may refer to Hermann Müller's paper on Westphalian orchids (H. Müller 1868), which is referred to by CD as a memoir (see 'Fertilization of orchids', p. 150 (*Collected papers* 2: 147)).

[4] CD's annotation is a note for his letter to Hooker of 2 July [1870].

From John Murray 1 July 1870

50, Albemarle S^{t} | W.
July 1./70

With John Murrays Comps & thanks to Mr Darwin receipt for £60"9"4 enclosed.[1]
Chas. Darwin Esq | &c &c &c
My Dear Sir

Mr Cooke forgot to point to you yesterday that in arranging the sum paid to you for the last Edition of your "Origin" we had to consider the reduction of its price by one shilling below that of the former Ed^{n}.[2]

Of the Titles you have preferred The Descent of Man—& Selection according to Sex— This need not bind you permanently as to your Title. At all events I am glad we have got rid of an objectionable adjective—[3]

My dear Sir | Yours very faithfully | John Murray

Ch Darwin Esq

DAR 171: 376

[1] An entry in CD's Account books–banking account (Down House MS) for 30 June records a payment of £60 9*s.* 4*d.* The payment is further broken down into £30 9*s.* 4*d.* for expenses for the translation of Fritz Müller's book (F. Müller 1864; Dallas trans. 1869) and £30 for presents of CD's books (CD's Account books–cash account and Classed account books (Down House MS)).

[2] CD evidently visited his publisher, John Murray, the previous day. Murray refers to Robert Cooke and to the fourth and fifth editions of *Origin*. CD was in London from 24 June to 1 July 1870 (see 'Journal' (Appendix II)).

[3] The reference is to *Descent*; the 'objectionable adjective' is not specified, but was presumably 'sexual'. 'According to' was later changed to 'in relation to'.

From B. J. Sulivan 1 July 1870

Bournemouth
July 1/70

My dear Darwin

I send by post the five last numbers of the Mission Magazine as they give all the recent information. Do not return them as I have duplicates.[1] You will also be interested in the Araucanian Mission where little has been done yet beyond Gardiner's Labours at Lota, where there is a considerable European population, and where he first fixed himself through difficulty of residing in the Araucanian portion of the

Country: but where he had much communication with the Indian chiefs, & some Indian children at school.[2] He worked there with all classes, Roman Catholics helping to build church & school, though the priests tried to stop it, but gave up in dispair: and three years since when the law for free religious toleration was hotly debated in the chambers at St Jago.[3] several liberal speakers pointed to the altered condition of Lota & its neighbourhood, and the good a protestant minister had done, as the stong argument for toleration. During the last two years a catechist had been settled at Lebu with several sons of chiefs at school—residing with him: but the war with Chili has done much harm to the work. Still you will see by the letters that something is still being done.[4]

The last letter of M^r. Bridges about the station at Keppel Island you will think more of a farming than a mission character.[5] but from the first we have relied much on teaching the Fuegians to plant Potatoes & vegetables, and take care of cattle sheep and goats, and all the accounts of hundred tons of potatoes and wool of 1300 sheep (all sprung from *16* first landed) refer to so much *native work*. We having only had one farm labourer and his wife on that Island, to teach & superintend the natives, who do nearly all the work. Those on the Island at the last accounts being chiefly new hands—the older trained ones having been some time living again in T. D. Fuego in wooden huts build by themselves, & with some cultivation going on, and every desire to increase it. M^r. Bridges went out as a boy with our first missionary, he has taught all the natives that know any thing, speaks the language fluently, and has long since reduced it to writing in the phonectic character; so that many of them read and write.[6] He is more known & looked up to by the natives than any one who has been there, from his more continued & familiar communications with them, & therefore he will be even safer than M^r. Stirling with the natives, besides having two companions & two females with them. I have always opposed beginning the station as I have feared the danger from the Natives more than any of those at the Mission did. & I could not forget the Murder of our whole party—but one—when they rashly trusted them in spite of my warnings—.[7] but because I refused to be a party, as one of the Committee, to a too hasty attempt to settle among them—Stirling determined to give us the proof it was safe by living alone amongst them; & now I hope it is really quite safe.[8] at all events it is justifiable. I am glad to hear so favorable an account of your health. only dont do too much.

Yours very sincerely | B. J. Sulivan

DAR 177: 294

1 Sulivan refers to the *South American Missionary Magazine*, published every other month; the copies sent to CD have not been found.

2 For the description of work in Araucania, see *South American Missionary Magazine* 4 (1870): 65–9 and 101–3. Lota is a coastal city in the Región del Biobío, over 200 miles south of Santiago, Chile. Allen Weare Gardiner took up his missionary post in Chile around 1865 (*Clergy list* 1864, *Crockford's clerical directory* 1865). 'Araucana' was the name the Spanish gave to the native people (who called themselves 'Mapuche') occupying the area between the rivers Biobío and Toltén in southern Chile (Blancpain 1990, p. 20). The Araucanians were not under the control of the Chilean government.

3 St Jago: Santiago, the capital of Chile.

[4] Lebu is a coastal town south of Lota; the letters referred to are in the *South American Missionary Magazine.* For more on the Chilean government's ongoing attempt to subjugate Araucania, which became part of Chile in 1882, see Blancpain 1990, pp. 140–6. The catechist was Christian Keller.

[5] Sulivan refers to a published letter from Thomas Bridges, which included details of wool production at the mission station on Keppel Island, one of the Falkland Islands (see *South American Missionary Magazine* 4 (1870): 95–9).

[6] As a child, Bridges was found abandoned and was adopted by George Pakenham Despard, who became secretary of the Patagonian Missionary Society. Bridges eventually compiled a dictionary of the Yaghan (Yámana) language. (*ODNB.*)

[7] Sulivan refers to the killing in 1859 of a group of missionaries who tried to start a settlement on Tierra del Fuego (see Hazlewood 2000, pp. 243–53).

[8] Waite Hockin Stirling lived alone among native people at Ushuaia for seven months in 1869 (F. C. Macdonald 1929, pp. 83–113).

To John Murray [after 1 July 1870][1]

Down, Beckenham, Kent
Friday

My dear Sir

On thinking it over I feel more inclined to keep to "The Origin of Man" as the first part of the title to my book.[2]

Yours very faithfully | Ch. Darwin

But I will be guided by you.

Copy
DAR 143: 273

[1] The date is established by the relationship between this letter and the letter from John Murray, 1 July 1870.

[2] The reference is to *Descent.* CD vacillated over his choice for a title throughout 1870, finally approving the title in January 1871 (*Correspondence* vol. 19, letter to Robert Cooke, 12 January [1871]).

From Camille Dareste[1] 2 July 1870

Ville de Lille | (Nord) | Museum d'Histoire Naturelle | rue des Fleurs, 1 | Lille
le 2 juillet *1870*

Monsieur

Permettez-moi de recourir encore une fois à l'inépuisable obligeance dont vous m'avez déjà donné beaucoup de preuves.

Il s'agit toujours de la race des boeufs *niata.* La discussion que j'ai soutenue à ce sujet après avoir porté sur l'existence même de cette race en Amérique, a été actuellement portée sur un tout autre terrain. On conteste que les caractères ostéologiques de la race *niata* soient les mêmes que j'ai décrits dans la tête d'un Veau né de race flamande dont je vous ai envoyé la description, il y a trois ans.[2]

Pour ma part, j'avais cru d'abord à l'existence de quelques différences, légeres il est vrai; tant que je n'avais pu connaître la tête du *niata* que vous avez rapportée d'Amérique que par la brève description d'Owen.[3] Mais ayant pu tout récemment, étudier la figure qui en a été donnée par Rütimeyer dans son mémoire sur les races de boeufs, je me suis convaincu que ces différences n'existaient point;[4] et que, du moins en ne considérant que les profils il y a, sauf les différences d'âge, une véritable identité de forme et de structure.

Mais un point reste indécis pour moi. En étudiant de nouveau la tête dont je vous ai adressé la description,[5] je me suis aperçu d'un caractère nouveau et fort intéressant qui m'avait complètement échappé dans ma première étude. Ce caractère consiste en ce que le vomer vient s'interposer entre les deux os intermaxillaires, et qu'il forme ainsi partie de la voûte palatine. Caractère d'autant plus remarquable qu'il ne se rencontre que chez les Dauphins et un certain nombre de Cétacés.

Pourriez-vous me dire si la tête osseuse que Vous avez rapportée, et qui est au Collège des chirurgiens, présente un semblable caractère? S'il en était ainsi, nous aurions bien évidemment un document incontestable sur l'origine de la race *niata*.

Je Vous prie encore une fois d'excuser mon importunité. Mais il s'agit pour moi d'une question fort importante, par suite des contradictions que j'ai rencontrées lorsque j'ai publié ces faits pour la première fois.[6] D'ailleurs il s'agit ici d'une question beaucoup plus vaste encore de Votre belle théorie sur l'origine des espèces que nous aurons je le crains bien, beaucoup de peine à faire accepter en France.

Veuillez agréer, je vous prie, l'expression de mon sincère et profond respect. | Camille Dareste. | Prof. à la fac. des Sc. de Lille.

DAR 162: 47

CD ANNOTATIONS
4.3 Ce caractère … Cétacés. 4.6] *scored blue crayon; enclosed in square brackets, blue crayon*
4.3 Ce caractère … palatine 4.5] *scored pencil*

1 For a translation of this letter, see Appendix I.
2 Dareste refers to his work *Rapport sur un veau monstreaux* (Dareste 1867; see *Correspondence* vol. 15, letter from Camille Dareste, 19 May 1867 and n. 8). Niata cattle are an abnormally small race from South America, whose peculiarities are now seen as the result of a form of dwarfism. CD discussed niata cattle in *Journal of researches* 2d ed., pp. 145–6, and *Variation* 1: 90–2.
3 For Richard Owen's description of the skull that CD had brought back from South America, see [Owen] 1853, 2: 3832. The skull was in the Hunterian museum of the Royal College of Surgeons, London.
4 The illustration for Ludwig Rütimeyer's description of the niata was a woodcut made from a photograph of the skull in the Hunterian museum (see Rütimeyer 1866, p. 28).
5 See *Correspondence* vol. 15, letter from Camille Dareste, 19 May 1867.
6 Dareste refers to his exchange of papers with André Sanson in the *Comptes rendus hebdomadaires des séances de l'Académie des sciences* (see *Correspondence* vol. 15, letter from Camille Dareste, 19 May 1867 and n. 9).

To J. D. Hooker 2 July [1870][1]

Down. | Beckenham | *Kent. S.E.*
July 2

My dear Hooker

You are very good about the lilies;[2] we want only a few for pots to be brought into the drawing room when in flower; but we had not the least intention of begging bulbs from you.

I presume the paper about which you enquire is by D[r] H Müller on Platanthera & Epipactis (excellent observations) published in Verhandl. d. Nat. Ver. Jahrg. XXV III folge, v. bd. p. 7 to 38.—[3] I know no more than the man in the moon what all this means when expanded but it is correctly copied.

There is another essay published separately by D[r] P. Rohrbach on Epipogium, Göttingen 1866.[4] I have both these papers & c[d] find them with some little trouble; & when in 3 months time I class all my pamphlets I shall certainly come across them; I will however with the greatest pleasure hunt for them now if wanted.

Our London visit was very prosperous in all ways, except in missing Kew.[5]

I am now getting my M.S. ready to send to the printers[6]

yours affectionately. | Ch. Darwin

I have read with greatest interest Bentham's last address: what a lot of thought he puts into a small space: some of the palæontologists will look green about the Proteaceæ.—[7]

LS(A)
DAR 94: 175–6

1 The year is established by the relationship between this letter and the letter from J. D. Hooker, 1 July 1870.

2 See letter from J. D. Hooker, 1 July 1870.

3 See letter from J. D. Hooker, 1 July 1870 and n. 3. CD refers to Hermann Müller's observations on *Platanthera* and *Epipactis* in an article on Westphalian orchids published in *Verhandlungen des naturhistorischen Vereines der preussischen Rheinlande und Westphalens* (H. Müller 1868). CD's annotated copy is in the Darwin Pamphlet Collection–CUL.

4 CD refers to Paul Rohrbach's monograph on the structure and fertilisation of *Epipogium gmelini* (now *E. aphyllum*, the ghost orchid; Rohrbach 1866). CD's annotated copy is in the Darwin Pamphlet Collection–CUL.

5 CD visited London from 24 June to 1 July 1870, but did not visit Hooker (see 'Journal' (Appendix II)).

6 The reference is to the manuscript of *Descent*.

7 In his presidential address to the Linnean Society, George Bentham opposed the view of many palaeontologists that specimens of the order Proteaceae (now a family in the order Proteales) were numerous in Europe during Eocene times (Bentham 1870, pp. lxxxiv–lxxxix).

To B. J. Sulivan 2 July [1870][1]

Down. | Beckenham | *Kent. S.E.*
July 2[d]

My dear Sulivan

Many thanks for your extremely interesting letter & for the Journal, which I shall be very glad to read[2]

Yours most sincerely | Ch. Darwin

Sulivan family (private collection)

1 The year is established by the relationship between this letter and the letter from B. J. Sulivan, 1 July 1870.

[2] CD refers to the copies of *South American Missionary Magazine* that Sulivan sent with his letter of 1 July 1870.

To Friedrich Hildebrand 4 July 1870

Down. | Beckenham | *Kent. S.E.*
July 4 1870

My dear Sir

I believe that your kindness will lead you to aid me if it is in your power. I much want, for some experiments in crossing, seeds of Nolana prostrata & Hibiscus Africanus, which have been matured in Germany or in the more Southern parts of Europe, & therefore under somewhat different conditions from those of England.[1] You will greatly oblige me if you can send pretty soon a few of these seeds in a letter.

With sincere good wishes believe me my dear Sir | yours very faithfully | Ch. Darwin

LS
Eilo Hildebrand (private collection of facsimiles)

[1] CD had earlier asked Joseph Dalton Hooker for seeds of *Hibiscus africanus* and *Nolana prostrata*, but Hooker was unable to supply any (see letter from J. D. Hooker, 1 July 1870 and n. 1).

From J. J. Weir 4 July 1870

6 Haddo Villas | Blackheath SE
4 July 1870

My Dear Sir

I regret that I was not sufficiently clear about the long haired rabbit, what I intended to convey was that he was the parent of between 30 & 40 short haired young, and then of a litter of five, of which he was parent, 4 were long & one short haired, no young were produced in the slightest degree intermediate in pelage.—[1]

The Cytisus case is a good one, the stock was grafted in my Brothers Garden by his own Gardener.—[2]

I particularly examined the tree to see if there was a chance of a double graft, but am quite certain such was not the case.—

I forgot to mention that birds of prey when enraged, elevate the feathers all over the body and also spread the wings & tail, this I have noticed repeatedly in the three species I have kept viz The Kestrel Sparrowhawk & Merlin.—[3]

They are also the only birds, that if unable to fly throw themselves on their backs & strike with their talons upwards.—

The Proverb says that "Hawks do not pick out Hawks eyes", but I find that they do, the female Sparrow Hawks of a nest of 3 or 4 I tried to rear, killed the males & began to eat the heads.—

I am glad to say my thumb is better & the disease is subdued.—[4]

Believe me | Yours very sincerely | J Jenner Weir

C Darwin Esqr

DAR 181: 83

CD ANNOTATIONS
1.1 I regret … pelage.— 1.4] *crossed pencil*
2.1 The … case.— 3.2] *enclosed in square brackets red crayon & blue crayon*
4.1 prey] *underl blue crayon*
4.2 spread … tail] *underl blue crayon*
6.1 The Proverb … subdued.— 7.1] *crossed blue crayon*

1 See letter from J. J. Weir, 27 June 1870 and letter to J. J. Weir, 29 June [1870].
2 See letter from J. J. Weir, 27 June 1870 and letter to J. J. Weir, 29 June [1870]. Weir's brother was Harrison Weir. His gardener has not been identified.
3 See letter from J. J. Weir, 27 June 1870.
4 Weir had suffered blood poisoning (see letter from J. J. Weir, 27 June 1870).

To W. H. Flower 5 July 1870

Down. | Beckenham | *Kent. S.E.*
July 5 1870

My dear Mr Flower

Will you kindly do me a favour & examine the head of the Niata ox which I formerly gave to y^r Museum.[1] If you will look at the part marked with blue in the 2nd page of the enclosed letter, you will see what is wanted.[2] Please to give me a sentence that I may quote to him on your authority; & he is I think quite worthy of your assistance.

I return by this post the book on teeth which I have been glad to see—[3]

Yours very sincerely | Ch. Darwin | Please to return Dareste's letter

LS
Bonham's, 13 March 2002

1 CD refers to a skull from South America, collected by CD while he was on the *Beagle* voyage, and given to the Hunterian museum of the Royal College of Surgeons, London (abstracts of the minutes of the Museum Committee, Archives, Royal College of Surgeons). Flower was the museum curator.
2 CD evidently enclosed the letter from Camille Dareste, 2 July 1870. CD scored in blue a section referring to the position of the vomer in the skull of a monstrous calf that Dareste had examined.
3 The book CD borrowed has not been identified.

To A. R. Wallace 5 [July 1870][1]

Down | Beckenham | *Kent S.E.*
June 5—

My dear Wallace

As imitation & protection are y^r subjects I have thought that you w^d like to possess the enclosed curious drawing. The note tells all I know about it.[2]

yours very sincerely | Ch. Darwin

P.S | I read not long ago a German article on the colour of *female* birds, & the author leaned rather strongly to your side about nidification.[3] I forget who the author was, but he seemed to know a good deal.— | C.D

LS(A)
British Library Board (Add. 46434 ff. 204–5)

[1] The month and year are established by the relationship between this letter and the letter from A. R. Wallace, 6 July 1870. CD evidently wrote 'June' by mistake.

[2] The drawing and note have not been found; however, see the letter from A. R. Wallace, 6 July 1870.

[3] CD refers to Wallace's theory that the colour of female birds that sat on open nests had been modified as a result of the protection gained by camouflage (see Wallace 1867a, [Wallace] 1867b, Wallace 1867c, Wallace 1868; see also *Correspondence* vol. 16, letter to A. R. Wallace, 15 April [1868] and nn. 3 and 4, and letter from A. R. Wallace, 4 October 1868). The German article has not been identified.

From Ernst Haeckel[1] 6 July 1870

Jena
6 Juli 70

Hochverehrter theurer Freund!

Es freut micht sehr, aus Ihrem gütigen Schreiben zu sehen, dass Sie mit Ihrer Gesundheit jetzt mehr zufrieden sind und dass Sie das Manuscript Ihrer wichtigen, von allen Seiten mit Ungeduld erwarteten Buchs über den Menschen glücklich beendigt haben.[2] Ich bin sehr gespannt auf dasselbe und freue mich im Voraus der reichen Belehrung, welche ich dadurch erhalten werde. In unsern deutschen Zeitschriften streiten sich schon jetzt zwei feindliche Parteien darüber, ob Sie die volle und consequente Anwendung der Descendenz Theorie auf den Menschen durchfüren werden oder nicht.

Herr Quatrefages[3] in Paris und die andern naiven Knaben, welche kühn behauptet hatten, dass die "Darwinsche Theorie" den Menschen gar nicht berühre, werden schlecht überrascht sein, wenn sie finden, dass Charles Darwin selbst die animale Descendenz der Menschen als natürlichste und wichtigste Consequenz seiner Lehre anerkennt.

—Es wird mich sehr freuen, wenn Sie mit der neuen Auflage der "Natürlichen Schöpfungsgeschichte" zufrieden sind. Ich habe den ganzen Winter auf ihre Verbesserung verwendet. Im IV. Vortrag wird Sie der neue Passus über *Treviranno* vielleicht interessieren.[4] Viele Mühe habe ich mir gegeben, die genealogische Classification wesentlich zu verbessern und ich glaube dass die Stammbäume der einzelnen Thier Classen wesentlich vervollkommnet sind. Ganz befriedigend können sie freilich niemals werden.

—Gegenwärtig bin ich ausschliesslich mit der Monographie der *Kalkschwämme* beschäftigt, welche sehr viel Mühe und Zeit kostet.[5] Die kleine Thier-Gruppe ist aber sehr interessant wegen ihrer unendlicher Form-Wandelungen und der endlosen vermittelnden Übergangs-Bildungen zwischen den verschiedensten Formen. Man kann entweder 4 oder 20 oder 200 "Canal Species" von Grantien annehmen, oder auch alle Calcispongien nur als eine einzige Species betrachten.[6] Um die feinere Anatomie und Ontogenie der Kalkschwämme zu vollenden, werde ich im August–October an das Mittelmeer gehen, entweder nach Triest oder nach Genua. Es wird aber wohl noch ein paar Jahre dauern, bis mein Buch über die Grantien ganz fertig ist.

—Die "Natürl. Schöpfungsgeschichte" wird jetzt in das Russische & Italianische übersetzt. Wie mir Mr. *Ray Lankester* mittheilte, wird sie vielleicht auch in das Englische übersetzt.[7] Da ich die englische Literatur so sehr schätze, würde mich diese Ehre ganz besonders freuen.

—Nächsten Winter werde ich wieder einen neuen Cyclus von Vorlesungen über den "Darwinismus" halten.—

Über Gegenbaurs Durchführung der Descendenz Theorie in der "Vergleichenden Anatomie" werden Sie sich gewiss gefreut haben.[8] Meine Collegen Gegenbaur und Preyer[9] senden Ihnen die respectvollsten Grüsse und ich bleibe, wie immer, mit der vorzüglichsten Verehrung, und mit den besten Wünschen für Ihr Wohlbefinden

Ihr getreuer | Ernst Haeckel

DAR 166: 54

1 For a translation of this letter, see Appendix I.
2 The reference is to *Descent*.
3 Armand de Quatrefages.
4 In the second edition of *Natürliche Schöpfungsgeschichte* (Natural history of creation; Haeckel 1870a, pp. 83–6), Haeckel maintained that Ludolph Christian Treviranus had presented a monistic interpretation of the unity of nature and argued for the genealogical connectedness of all organisms. For more on Haeckel's monistic philosophy see Haeckel 1899.
5 Haeckel refers to *Die Kalkschwämme: eine Monographie* (Calcareous sponges: a monograph; Haeckel 1872).
6 *Grantia*, a genus of Calcispongiae (now the class Calcarea), is characterised by both radial and incurrent canals (a syconoid canal system; see Bergquist 1978, p. 151), which allow the circulation of water throughout the organism.
7 Edwin Ray Lankester's translation of Haeckel's book, based on the German edition, appeared in 1876 (Haeckel 1876).
8 Haeckel refers to Carl Gegenbaur and *Grundzüge der vergleichenden Anatomie* (Elements of comparative anatomy; Gegenbaur 1870). CD's copy is in the Darwin Library–Down.
9 William Thierry Preyer.

From J. D. Hooker [6 or 7 July 1870][1]

Royal Gardens Kew

My dear Darwin

Thanks for the names of the Orchid: books; please let me see them when you have arranged your pamphlets 3 mo^{s} hence.[2]

Have you read Claparede?[3] it is not so good as I expected it would have been, & *is* rather windy I think— is it worth translation as an avant courier for the "Origin of Man"?[4] if so I would set some one to do it for Nature or some other periodical.—

You & old Brandt are "en lutte" for the Acad: of Sciences. which will be decided I hear on the 15th. or 20th.— What a farce it is![5]

I am delighted to hear that you are on the eve of printing—[6] I am hard at work on Nepenthes for DC. Prodr:— this genus supports Miquels & Wallace's view of the identity of Bornean & Sumatran Zoology & the differences of Java from either *most marvellously.* Who first published on that curious point? I remember old Blume telling me of it in Leyden in 1845. and I think that Miquel has published on it— can you refer me?[7]

Hodgson has been here for 2 days, & is nettled at you not having alluded in your chapter on dogs to his paper on the Indian dog—[8] I told him that I could not doubt but that it was not it's value that you underrated, but that it illustrated no points in his subject.—

Bastian's paper in Nature is full of curious matter but eminently unsatisfactory in treatment I think, & poorly written.[9]

Lyell was here on Tuesday, looking remarkably well— he does not like Bentham's Address at all— or perhaps only the drift of it. I am so glad you admire it's care & thought.[10]

We spent last Sunday at Mr G. Macleays who has taken Pendell Court near Bletchyngly— he is a pleasant & interesting man, & an ardent admirer of his old curmudgeon of a brother— William.[11] I ascended to the top of the N. Downs (Eastern continuation of the Reigate range) & was struck with the capping of very fine gravel. I wish I knew more of tertiary geology, but suppose I should only disbelieve— as I do the whole theory of the upper & lower gravel levels of the Somme & Seine &c. I never can believe that existing rivers or river basins capped these hills with thick beds of gravel which represent either sea shores or drainage over great areas & an enormous amount of denudation of some still higher land, than the hill tops themselves—[12]

Ever your affectionate & ignorant skeptic | J D Hooker

DAR 103: 55–56

1 The date is established by the relationship between this letter and the letters to J. D. Hooker, 2 July [1870] and 8 July [1870], and by the reference to a visit from Charles Lyell on Tuesday. In 1870, the first Tuesday after 2 July was 5 July.

2 See letter to J. D. Hooker, 2 July [1870] and nn. 3 and 4.

3 Hooker refers to Edouard Claparède and his essay review of Wallace 1870a, in which Claparède focused on Alfred Russel Wallace's views on the development of humans (Claparède 1870).

4 Hooker refers to *Descent*; CD had not yet decided on the title (see letter to John Murray, [after 1 July 1870]).

5 Both Johann Friedrich von Brandt and CD were candidates for corresponding membership in the French Académie des Sciences. Brandt was elected to the anatomical and zoological section of the academy on 4 July 1870; CD was elected to the botanical section on 4 August 1878 (*Comptes Rendus de l'Académie des Sciences* 71 (1870): 41, and 87 (1878): 245; Corsi and Weindling 1985, p. 699).

6 Hooker refers to *Descent.*

7 Hooker refers to his monograph 'Nepenthaceae, Cytinaceae', which appeared in the last part of Candolle and Candolle 1824–73 (17: 90–116). Both Friedrich Anton Wilhelm Miquel and Wallace divided the Malay archipelago into biogeographic zones based on the similarity of the flora and fauna (see Keng 1978, pp. xxx–xxxii). The first naturalist to publish on the biogeographic zones of the area was Heinrich Zollinger, who named the whole region 'Malesia' (Zollinger 1857). Wallace referred to

Zollinger's division in Wallace 1869a, 1: 317–18 (see also Johns 1995 for a discussion of the changes in the division of the Malesian region). Hooker also refers to Carl Ludwig Blume and probably to Miquel 1860.

8 In *Variation* 1: 26, CD referred to Brian Houghton Hodgson having tamed young Indian dogs, and cited a notice of Hodgson's paper 'A description of the wild dog of Nepâl' in *Proceedings of the Zoological Society of London* (1833): 111–12. The paper was published in *Asiatic Researches* (Hodgson 1833).

9 Hooker refers to Henry Charlton Bastian and the first and possibly second part of Bastian 1870, a paper on evolution, published in the 30 June and 7 July 1870 issues of *Nature*.

10 Hooker refers to Charles Lyell and to George Bentham's presidential address to the Linnean Society. See letter to J. D. Hooker, 2 July [1870] and n. 7.

11 Hooker refers to George Macleay and William Sharp Macleay. George Macleay had lived at Pendell Court, Bletchingley, Surrey, since 1860 (*ODNB*).

12 Hooker refers to the fluvial theory of the origin of the upper and lower level gravel deposits in river valleys like the Somme and Seine in France or the valleys of the south of England. For more on the theory, see Prestwich 1862a.

From A. R. Wallace 6 July 1870

Holly House, Barking, E.
July 6th. 1870

Dear Darwin

Many thanks for the drawing. I must say however the resemblance to a snake is not very striking unless to a *Cobra* not found in America.[1] It is also evident that it is *not* Mr Bates Catterpillar as that threw the head *backward* so as to show the *feet above* forming imitations of keeled scales.[2]

Claparede has sent me his critique on my book. You will probably have it too. His arguments in reply to my heresy seem to me of the weakest.[3] I hear you have gone to press, & I look forward with fear & trembling to being crushed under a mountain of facts![4]

I hear you were in town the other day.[5] When you are again I sh^{d}. be glad to come at any convenient hour & give you a call.

Hoping your health is improving & with kind remembrances to Mrs Darwin & all your family

Believe me | Yours very faithfully | Alfred R. Wallace

Charles Darwin Esq.

DAR 106: B92–3

1 CD had sent the drawing with an accompanying note with his letter to Wallace of 5 [July 1870]. Neither has been found, but the drawing was evidently one of a late instar of *Papilio cresphontes* (the giant swallowtail butterfly, native to many parts of North America) or a similar species. The caterpillar's thorax resembles the head of a cobra, seen from above, but is also similar to the head shape of American hog-nose snakes (*Heterodon*). Like other species of swallowtail, it also possesses a defensive gland called an osmeterium, which, when everted from its head, resembles the forked tongue of a snake.

2 Wallace refers to Henry Walter Bates. The caterpillar has not been identified.

3 The reference is to Edouard Claparède and his essay review of Wallace 1870a (Claparède 1870). CD's annotated copy is in the Darwin Pamphlet Collection–CUL. Claparède's review focused primarily on Wallace's views on natural selection and development of humans.

[4] Wallace refers to *Descent*. CD and Wallace had disagreed strongly on topics such as sexual selection and protective coloration (see *Correspondence* vols. 15 and 16). Wallace had also recently disputed the applicability of natural selection to aspects of human evolution (see Wallace 1870a, pp. 332–71).

[5] CD visited London from 24 June to 1 July 1870 (see 'Journal' (Appendix II)).

From George Green Gascoyen 7 July 1870

48, Queen Anne St. | *Cavendish Square. W.*
July 7. 1870.

Dear Sir.

Our friend Mivart asked me if I could give him information on two points respecting the Platysma Myoides which you wished to know;[1] one of them I could at once affirm, the other I had not noticed, and have since been making observations, hence the delay in writing to you which I hope you will pardon.

The Platysma is always in a state of violent contraction when a person is struggling for breath. this I have long noticed & I believe is invariably the case.[2]

In the few patients whom I have had the opportunity of observing during the past week and to whom Chloroform was about to be administered for an operation, the action of the Platysma was not so constant or so marked as in the former instance. in most there was distinct contraction but much more in some than in others— the knowledge that the operation will be painless allays greatly the fear of the patient and altho' there is some distrust of the chloroform it is not extreme. I have no doubt that when a person was expecting a severe operation without chloroform the muscle would contract strongly—in company with those which assist in causing the expression of fear by depressing the angles of the mouth.[3]

I yesterday asked a friend who has administered Chloroform in many hundred cases. he said it was a point that had not occurred to him to notice but he was quite sure he had repeatedly seen it— others of whom I asked the question could give me no satisfactory answer.

I remain, Dear Sir, | faithfully yours | Geo. G. Gascoyen

Chas. Darwin Esqr. *F.R.S.*

DAR 165: 11

CD ANNOTATION

2.1 The Platysma … the case. 2.2] *scored pencil*

[1] Gascoyen refers to St George Jackson Mivart. CD may have seen Mivart during his visit to London from 24 June to 1 July 1870 (see 'Journal' (Appendix II)). CD was interested in the contraction of the platysma muscle in connection with his work on expression.

[2] CD had also asked James Paget about contraction of the platysma in persons who were having difficulty in breathing (see *Correspondence* vol. 17, letter to James Paget, 29 April [1869]).

[3] In *Expression*, pp. 300–1, CD discussed contraction of the platysma in patients before an operation, but cited William Ogle for the information.

To J. D. Hooker 8 July [1870][1]

Down. | Beckenham | *Kent. S.E.*
July 8th

My dear Hooker

It suddenly flashed on me in what part of heap, I I shd. find the 2 Pamphlets on orchids, & have extracted them & send them by this post.—[2]

I thought well of Prof. Claparèdes criticism, & I think it wd. be well worth translating & publishing, partly because he is so capital an observer & naturalist, & chiefly because no sort of answer has yet appeared to Wallace.[3] Bates thinks his heterodox views have already done a good deal of mischief to the cause of evolution.[4] There are so few "*ignorant sceptics*" in the world, who think for themselves. Wallace himself thinks Claparede's article very weak;[5] but I conclude that he thinks so because Claparede has arrived at an unpleasant judgement, very much like Lyell about Bentham's address: I wd lay a wager that L. has lately said something about European Proteaceæ in one of his Editions.[6] I do not remember anyone before Wallace on relations of Sumatra & Java: I *obscurely* fancied that Miquel had demurred, or said there were no materials to judge from; but my attention has now been for a long time turned from the grand subject of Geograph. Distrib.[7]

Thanks for your very pleasant letter with lots of all kinds of gossip: I am sorry to have nettled Hodgson, for whom I have much respect, but it is too absurd to suppose that I possibly could refer to all papers on such a large subject as Dogs.—[8] When you see Huxley ask him about Bastian & a fragment of Sphagnum moss, which he mistook for a spont. gen. organism![9]

Farewell my dear old fellow. C.D.

I shd. think I had no chance of election in France against Brandt.[10]

DAR 94: 177–8

1 The year is established by the relationship between this letter and the letter from J. D. Hooker, 10 July 1870.

2 See letter to J. D. Hooker, 2 July [1870] and nn. 3 and 4.

3 See letter from J. D. Hooker, [6 or 7 July 1870] and n. 3. CD refers to Edouard Claparède and Alfred Russel Wallace.

4 Henry Walter Bates had urged CD to write a critique of Wallace's view of natural selection as applied to humans. See letter from H. W. Bates, 20 May 1870 and n. 1.

5 See letter from A. R. Wallace, 6 July 1870 and n. 3.

6 CD refers to Charles Lyell and George Bentham. See letter to J. D. Hooker, 2 July [1870] and n. 7, and letter from J. D. Hooker, [6 or 7 July 1870].

7 CD refers to Friedrich Anton Wilhelm Miquel. See letter from J. D. Hooker, [6 or 7 July 1870] and n. 7.

8 CD refers to Brian Houghton Hodgson. See letter from J. D. Hooker, [6 or 7 July 1870] and n. 8.

9 Thomas Henry Huxley had evidently told CD about an experiment conducted by Henry Charlton Bastian to show the spontaneous generation of organisms. Huxley had pointed out to Bastian that one of the organisms in his test tube looked like *Sphagnum* (peat moss). Bastian wrote to confirm Huxley's observation in a letter of 2 May 1870 (Huxley papers MS 10.238 (Imperial College of Science, Technology and Medicine Archives)). For more on Bastian's work on spontaneous generation within the context of Darwinian theory, see Strick 1999.

10 CD refers to Johann Friedrich von Brandt. See letter from J. D. Hooker, [6 or 7 July 1870] and n. 5.

From J. D. Hooker 10 July 1870

Royal Gardens Kew
July 10/70

Dear Darwin

Enclosed seeds & note just arrived from Playfair of Algiers, I do not quite understand it. as I asked only for what you told me. & here are a whole lot of others seeds— *all were in duplicate parcels*, so I have kept one set; assuming that he means one for us & one for you[1]

Delpino writes asking me where a Mr Curtis has published physiological observations on Dionea, I cannot guess, do you know.[2] I have an obscure recollection of some leaf observations by W. Curtis in his introduction to Botany, a book I have not— Have you any reference to any such observations on Dionaea Cephalotus or Nepenthes?[3]

Very many thanks for the Orchid papers which I will return when done with.[4]

I had a talk with the D. of Argyll last night, with whom I dined, about origin of man, & found him in a "cleft stick" about Wallace, believing him to be right in the fact about man; but allowing that he must be wrong in his argument! (he had not read that paper of Wallaces)—[5] What a clever little beggar it is!— but I cannot follow his views about man; or quite see what he would have us to believe— His chief quarrell with the "Origin" is that you do not state that the order of evolution is preordained though he believes that you would admit this.— I told him that I did not think this was any business of your's—that you did not pretend to go into the origin of life, only into it's phenomena. I could not, before his wife & children especially,[6] go into this matter, & avow my own (& I suppose your) belief that all speculations on preordination are utterly idle in the absence of better materials than theologies & cosmogonies supply us with—that in fact the whole subject is beyond the range of our conceptions:

Thanks for telling me of Sach's Lehrbuch which I will order.[7]

Ever yr affec | J D Hooker

[Enclosure]

British Consulate General | Algiers.
27 June

My dear Dr Hooker

As soon as I received your letter with the enclosure from Dr Darwin I went about to all the gardeners in the place but I could not hear of anyone having any of the seeds in question save only Iberis umbellata[8]

Dr. Darwin also wrote to Durando on the subject and he has sent direct a packet of that seed—[9]

I hope to take a few days at home this summer & to have the pleasure of seeing you & your family band I can only get one month's leave & as I have to go to Switzerland to look for a school for my eldest boy I shall have very little time to spare in England I must go north where I have a big girl, & innumerable relations, & I trust I may be able to be present at the meeting of the B. A. at L'pool—[10] Where will you be between the 15th August & 15 September?

The only chance I shall have of seeing my brother Lyon[11] will be at the B. A.—

Good bye, Yours faithfully | R L Playfair

P.S. Our *Premier President de la Cour Imperiale*[12] has a wonderful collection of Autographs & he is very anxious to have a letter of Darwin's have you any old one you can spare—

DAR 103: 53–4; Royal Botanic Gardens, Kew (DC 17a: 117)

CD ANNOTATIONS
2.1 Delpino. . . W. Curtis 2.3] *double scored pencil*
4.3 he must. . . cannot 4.4] *double scored pencil*

1 See the enclosure from Robert Lambert Playfair. The seeds have not been found but see the memorandum to J. D. Hooker, [13 June 1870?]. CD had asked for seeds of *Iberis amara*, *I. umbellata*, *Nolana prostrata* (now *N. humifusa*), *Hibiscus africanus* (now *H. trionum*), *Canna warszewiczii* (now *C. indica*), and *Mimulus luteus*. The Inwards Book (Royal Botanic Gardens, Kew) records the receipt of seeds of *Lathyrus tingitanus* from Playfair on 4 July 1870. No other seeds were recorded.

2 The references are to Federico Delpino and Moses Ashley Curtis. Curtis had published a description of *Dionaea muscipula* in M. A. Curtis 1834. In *Insectivorous plants*, p. 301 n., CD noted that Curtis was the first person to describe the secretion of the glands in *Dionaea* and cited M. A. Curtis 1834.

3 Hooker refers to William Curtis and probably to Curtis's *Lectures on botany* (W. Curtis 1805).

4 See letters to J. D. Hooker, 2 July [1870] and nn. 3 and 4, and 8 July [1870].

5 George Douglas Campbell evidently agreed with the view expressed by Alfred Russel Wallace in an essay, 'The limits of natural selection as applied to man', that humans were not subject to natural selection to the same extent as other organisms (see Wallace 1870a, pp. 332–71).

6 Campbell's wife was Elizabeth Georgiana Campbell. They had twelve children (*ODNB* s.v. Campbell, George Douglas).

7 Hooker refers to Julius Sachs and to Sachs 1870. CD's annotated presentation copy of Sachs 1870 is in the Darwin Library–CUL (see *Marginalia* 1: 727).

8 Hooker's letter has not been found; the enclosure may have been the memorandum to J. D. Hooker, [13 June 1870?]. *Iberis umbellata* is globe candytuft.

9 Playfair refers to Gaetano Durando. See letter to J. D. Hooker, [29 June 1870] and n. 2. In *Cross and self fertilisation*, p. 105, CD mentions receiving seeds of *Iberis umbellata* sent from Algiers by Durando.

10 Playfair's eldest son has not been identified. The daughter referred to was probably Agnes Mary Playfair. The meeting of the British Association took place in Liverpool from 14 to 21 September 1870 (*Report of the fortieth meeting of the British Association for the Advancement of Science, held at Liverpool*, p. lxxii).

11 Lyon Playfair.

12 The president of the 'Cour Imperiale' has not been identified.

To J. D. Hooker 12 July [1870][1]

Down. | Beckenham | *Kent. S.E.*
July 12th

My dear Hooker

Two of the kinds of seeds will be very useful to me; but why the others were sent by Col. Playfair, I know not, except from the abundance of his kindness.—[2]

I am sure I never heard of Curtis' observations on Dioneæa; nor have I met with anything more than general statements about this plant or about Nepenthes catching insects &c—[3]

I have always thought the D. of Argyll wonderfully clever; but as for calling him "a little beggar" my inherited, instinctive feelings w^d. declare it was a sin thus to speak of a real old Duke.—[4]

Your conclusion that all speculation about preordination is idle waste of time is the only wise one: but how difficult it is not to speculate. My theology is a simple muddle: I cannot look at the Universe as the result of blind chance, yet I can see no evidence of beneficent design, or indeed of design of any kind in the details.— As for each variation that has ever occurred having been preordained for a special end, I can no believe in it, than that the spot on which each drop of rain falls has been specially ordained.—

Spontaneous generations seems almost as great a puzzle as preordination; I cannot persuade myself that such a multiplicity of organisms can have been produced, like crystals, in Bastian's solutions of the same kind.—[5] I am astonished that as yet I have met with no allusion to Wyman's positive statement that if the solutions are boiled for 5 hours, no organisms appear; yet, if my memory serves me, the solutions when opened to air, immediately became stocked.[6] Against all evidence I cannot avoid suspecting that organic particles (my **gemmules** from the separate cells of the lower creatures!) will keep alive & afterwards multiply under proper conditions.[7] What an interesting problem it is.—

Your affect | C. Darwin

DAR 94: 179–180

[1] The year is established by the relationship between this letter and the letter from J. D. Hooker, 10 July 1870.

[2] The reference is to Robert Lambert Playfair. See letter from J. D. Hooker, 10 July 1870 and n. 1.

[3] See letter from J. D. Hooker, 10 July 1870 and nn. 2 and 3. Moses Ashley Curtis, not William Curtis, had made observations on *Dionaea*.

[4] CD refers to George Douglas Campbell. See letter from J. D. Hooker, 10 July 1870.

[5] CD refers to experiments conducted by Henry Charlton Bastian (see letter to J. D. Hooker, 8 July [1870] and n. 9).

[6] CD refers to experiments performed by Jeffries Wyman (see Wyman 1867).

[7] According to CD's hypothesis of heredity, pangenesis, gemmules were minute particles thrown off by individual cells, and capable of generating new cells when needed (see *Variation* 2: 357–404).

From William Bernhard Tegetmeier 14 July 1870

Finchley | N
July 14/70

My dear Sir.

I enclose you a proof of a very interesting letter on Variation in Wild Ducks by M[r] Teebay on whose accuracy of observation I can fully rely.[1]

I do not know whether you are working at varieties now. If you are would you like to look at Dr Cooper's octavo volume on Game fowls.— It is an American Treatise on Fighting Cocks.[2] with a vast amount of valuable information in an American form.— — If you would like to see it I can lend you my own copy or I could get you one at the publishing price 5 dollars. but mine is very much at your service for as long as you like

I am at work on a new edition of my own large poultry book and will, with your permission, avail myself of your store of information in your work on '*Variation*'.[3]

I have lately been making some observations respecting your theory of analogous variation,[4] & I have thought of collecting the **prototypes**? of our varieties of pigeons. and comparing with the varieties.. and having drawings made of each, the analogy is closer than I could have imagined. After all the breeds of the fancier are but imitations of the colorations, etc. of wild species of the same group

The cases would form a good subject for a paper—

I am breeding from the Crested Turkey but the young are not yet old enough for me to say whether any have crests. but you shall know—[5]

I have not been able to work much lately at subjects of purely scientific interest Fortune holds to many hostages and constrains me to work for her but as my work is congenial I am thankful.

Trusting your health is good | Believe me truly yours | W B Tegetmeier

I forget whether or no I ever sent you a drawing of the crested turkey so forward one[6]

[Enclosure]

White Wild Ducks

I have a pure-bred wild duck and drake that last year had two nests of young ones. In the first hatch were ten of the true wild duck colour, and one (which turned out a duck) with pure white plumage. In the second nest there was[7] *again ten of the true wild colour and one pure white drake, and a duck all white, except a black spot on the top of the head and black tail coverts—this duck is now pure white in plumage. The beaks of the two that were pure white are bright orange; the beak of the other ducks is bright orange, with the saddle on the upper beak much speckled with rich black. The legs of all three are orange, the webs of the feet of a more dusky colour. All three are remarkably slender in form, and have the long graceful shape of the pure wild duck.*

This year the same pair of old birds have eleven youngs ones of the true wild duck colour, and one, pure white. Both the old birds were from eggs taken from separate wild duck nests, are both pinioned, and the duck has had no access to any other drake.

Early this year the white drake was paired with the white duck. She has eight young ones nearly full grown, five of which are pure white in plumage, the other three white with the small black spot on the top of the head and black in the coverts. The black on these three is much less than it was when they were in their first feathers, and will very likely be replaced with white when they make up their plumage.

The other duck that had the black spot on the head and the black tail coverts was paired with a wild drake of the true colour. Her nests having been twice destroyed, she has only four young ones, which are not yet feathered, but from appearance three will be pure white, and one with the spot on the head and some black above the tail.

The white drake and two ducks have not been pinioned, and during last autumn and beginning of winter had full liberty of flying miles round, but never, that I know of, alighting anywhere except at their own ponds. When taking a long flight they would go an immense height, and to my mind there are few more beautiful sights than to watch their manœvres in the air on a windy day, especially when coming down.

I am totally at a loss to account for these white birds, as the remainder of the old duck's young are as near like one another and like their parents as possible, and it is more strange that the nearly white duck put to the grey drake should produce white young ones. It is again singular the the coloured feathers should be all black, not one having had a grey feather. Beautiful as these white birds are, if they would retain the black spot on the top of the head, and the black tail coverts some of them have in their earlier feathers, they would be still more beautiful.

Richard Teebay.

DAR 178: 82, DAR 193: 43

CD ANNOTATIONS
3.1 I am . . . '*Variation*'. 3.2] *scored ink*
Enclosure:
Right margin: 'The Field | July 1870' *ink*; '7' *after del* '6'

1 Tegetmeier enclosed an article, 'White wild ducks', by Richard Teebay, that appeared in the 18 July 1870 issue of the *Field* (the annotated cutting is in DAR 193: 43). Teebay described the offspring of a pair of pinioned wild ducks, among which were three white individuals. He also described the progeny of the white individuals. Tegetmeier was pigeon and poultry editor of the *Field* (*ODNB*).
2 Tegetmeier refers to John W. Cooper and Cooper [1869].
3 Tegetmeier refers to *The poultry book* (Tegetmeier 1867) and to *Variation*. The new edition of *The poultry book* (Tegetmeier 1873) appeared in monthly parts between March 1872 and May 1873. Tegetmeier mentioned CD's work in the preface to the new edition, and quoted a long passage from *Variation* on the peafowl (Tegetmeier 1873, pp. 331–2).
4 In *Variation* 2: 348, CD had discussed analogous variation, or the appearance of similar characteristics in several varieties or races descended from the same, or even distinct, species.
5 Tegetmeier had sent CD an engraving of the crested turkey and enclosed specimens of the crest feathers in his letter of 20 July 1869 (*Correspondence* vol. 17).
6 See n. 5, above. The only extant image of the crested turkey in the Darwin Archive–CUL is the one Tegetmeier had already sent. It is in DAR 193: 39.
7 The word 'was' has been crossed through with pen and 'were' added in the margin.

To Albert Günther 15 July [1870][1]

Down. | Beckenham | *Kent. S.E.*
July 15th

My dear Dr. Günther

I do not know which species of Triton, Mr Ford has drawn for me.—[2] I describe briefly in my M.S. *T. punctatus.* Will you kindly write name on proofs & slip them in enclosed envelope, without troubling yourself to write a word.— My Chapt. on Fishes & Reptiles, wd have been much the worst in my volume, had it not been for your kind & most valuable assistance; & now I think it will be one of the best, & certainly the best illustrated.—[3]

Your's very sincerely | Ch. Darwin

Shrewsbury School

[1] The year is established by the reference to a proof-sheet of an illustration for *Descent*, which was published in February 1871 (see below, n. 2).

[2] CD evidently refers to the illustration by George Henry Ford of *Triton cristatus*; it appeared in *Descent* 2: 24 fig. 31. Ford was making illustrations for *Descent* under Günther's supervision (see *Correspondence* vol. 17, letter to Albert Günther, 21 September 1869).

[3] CD refers to *Descent* 2: 1–37. See *Correspondence* vols. 15–17.

To W. B. Tegetmeier 15 July [1870][1]

Down. | Beckenham | *Kent. S.E.*
July 15th

My dear Sir

It is long since I have heard from you, & I am much obliged for your letter.— It is good news that you are going to bring out a new edit. of your Poultry Book & you are quite at liberty to use all my materials.[2] Thanks for curious case of the Wild-Duck variation: I have heard of other instances of a tendency to vary in *one* out of a large litter or family.—[3]

I have too many things in hand at present to profit by your offer of the loan of the American Poultry Book.—[4]

Pray keep firm to your idea of working out the subject of analogous variation with Pigeons; I really think you might thus make a novel & valuable contribution to science.— I can, however, quite understand how much your time must be occupied with the never ending, always beginning Editorial cares.—[5]

I keep much as usual & crawl on with my work.—

Believe me | my dear Sir | Yours very faithfully | Ch. Darwin

Archives of The New York Botanical Garden (Charles Finney Cox Collection) (ser. 14)

[1] The year is established by the relationship between this letter and the letter from W. B. Tegetmeier, 14 July 1870.

[2] See letter from W. B. Tegetmeier, 14 July 1870 and n. 3.

[3] See letter from W. B. Tegetmeier, 14 July 1870 and n. 1.

[4] See letter from W. B. Tegetmeier, 14 July 1870 and n. 2.

[5] See letter from W. B. Tegetmeier, 14 July 1870. Tegetmeier was pigeon and poultry editor of the *Field*.

From William Farr 16 July 1870

General Register Office | *Somerset House*
1870 July 16

My dear Sir

Will you be good enough to look over the enclosed memorandum—which I regret that I have not had time to get copied.[1]

Could you send it here with any—suggestions—& hints as to the *probable* results of the Inquiry?

I saw Sir John Lubbock—& shall go to the House on Monday—when it is expected the Bill will be read a second time— This amendment can be best moved in Committee.[2]

I am my dear Sir | Yours very faithfully | W. Farr

P.S. I got home quite in time to meet my American Friend—Mr. Ruggles—thanks to your very lively horse.[3]

DAR 164: 32

[1] The memorandum has not been found, but concerned the insertion of a question about cousin marriages into the census. See letter to William Farr, 17 July [1870].

[2] The second reading of the bill was on 22 July 1870 (*Hansard parliamentary debates* 3d ser., vol. 203 (1870), cols. 805–18). The Monday following 16 July was 18 July. The amendment, which was moved in committee by John Lubbock on 26 July 1870, read, 'Amendment proposed, in page 2, line 14, after the word "condition," to insert the words "including whether married to a first cousin."' (*Ibid.*, col. 1007). The proposal was defeated by a vote of ninety-two to forty-five (*ibid.*, col. 1010).

[3] According to Emma Darwin's diary (DAR 242), Farr visited CD on 10 July 1870. Farr refers to Samuel Buckley Ruggles.

From Friedrich Hildebrand 16 July 1870

Freiburg i/B
July 16th 1870.

Dear and honoured Sir

You will excuse me that I have not answered your letter at once, but as I had no seeds of Nolana prostrata I wrote for them to Vienna, where I knew that they cultivate this species.[1] I have been waiting for an answer till today, but in vain, so I will not keep you awaiting any longer and send you the seeds of Hibiscus Trionum L = H. africanus Roth,[2] that are grown in our botanical garden. I hope that this will bee the wanted species. Instead of Nolana prostrata I send you some seeds of N. grandiflora, perhaps you can make some use of them.[3]

Last year I experimentized with Oxalis Valdiviana[4] and found that only the combinations of the sexual organs of equal hight in the three forms were fertile. This year I have seen that the offspring of these combinations belongs partly to the mother partly to the father-form, very few to that form, that has had no part in producing the said offspring. I had not expected that there would be any individual that did not belong to the parent forms, but I think that the reason of this may be that one of the parents has been the offspring of the third form—so the childeren belong partly to the form of the mother, partly to that of the father, and a smaller part to that of the grandfather or grandmother. Allow me to give you an example:

The middle styled form was fertilized with the upper antherae of the long styled and the offspring were: middlestyled 21, longstyled 15, shortstyled 2.[5]

I beg your pardon that I have written this rather in a hurry, but I wished to have the letter ready before perhaps the war interrupts the communications.[6]

Believe me dear Sir with my best wishes | yours | respectfully | Hildebrand

DAR 166: 211

1 See letter to Friedrich Hildebrand, 4 July 1870 and n. 1.
2 *Hibiscus africanus* is a synonym of *H. trionum*.
3 CD did not record any experiments with *Nolana grandiflora* in *Cross and self fertilisation*; *N. prostrata* is the only species mentioned, *ibid.*, pp. 186–7.
4 *Oxalis valdiviana* is now *O. valdiviensis*.
5 Hildebrand reported the results of these experiments in his paper 'Experimente und Beobachtungen an einigen trimorphen Oxalis-Arten' (Experiments and observations on some trimorphic *Oxalis* species; Hildebrand 1871). CD's annotated copy of the paper is in the Darwin Pamphlet Collection–CUL.
6 The Franco-Prussian war broke out on 19 July 1870 (Wawro 2003, p. 65).

From John Lubbock 16 July [1870][1]

16 July

My dear M^r. Darwin

The Census Bill is down on the paper for tomorrow & I have mislaid your note.[2]

Would you give me again the exact words in which you think the question had better be put.

I am, dear Mr Darwin, | Yours most sincerely | John Lubbock

DAR 170: 76

1 The year is established by the reference to the Census Bill (see n. 2, below).
2 The second reading of the bill was on 22 July 1870. See letter from William Farr, 16 July 1870 and n. 2. July 16 1870 was a Saturday; Parliament did not sit at weekends.

To William Farr 17 July [1870][1]

Jul 17

My dear Sir

I am very much obliged for the papers.— I feel certain that your judgment on all parts in regard to the Census is incomparably safer than mine. Firstly My impression is that it w^d be better to confine enquiry to 1st cousins.[2] Secondly, I incline to think that your second form of enquiry is the best, that is if both forms are not included; in as much as second will give the proportion out of the whole population of cousins who marry & who have children.

The object of the enquiry is to ascertain whether with mankind, consanguineous marriage within the degree of cousinship is in any way injurious, & The result of such an enquiry is almost equally important whether negative or affirmative.

In England & any part of Europe the marriages of cousins are objected to by fear their supposed injurious consequences; but this belief rests on no direct evidence. It is therefore *[manifestly]* desirable that this belief sh^d either be proved to be false, or sh^d be confirmed so that in the latter case the marriage of cousins might be discouraged. The returns gained if the proper queries are inserted, would show whether married cousins have in their households on the night of the census as many children, as parents who are not related; & sh^d the number prove fewer, we might safely infer either lessened fertility in the parents, which is more probable lessened vitality in the offspring.

If the proportional number of cousins who marry in the whole population was once ascertained, there c^{d} exist a standard by which to judge whether the proportion, (already tabulated in some cases) of persons in asylums for the dumb & deaf the blind & insane who are the offspring of cousins is in excess of the proportion of cousin-offspring in the whole population.—

I fear I have written at too great length, but I have not felt sure, how fully you wished me to write.—

It is moreover much to be wished that the truth of the often repeated assertion, that consanguineous marriages lead to deafness & dumbness, blindness &c sh^{d} be ascertained; & all such opinion c^{d} be easy tested, by the returns from a single census.—

It is beyond my power of filling out all the detail, to see which of your 2 forms of enquiry w^{d} be best, if both are not included.— I could insert the words “married to—” with a blank, showing that the word cousin ought to be introduced; when this is the case.

Draft
DAR 96: 76–7

1 The year is established by the relationship between this letter and the letter from William Farr, 16 July 1870.

2 CD refers to the memorandum that Farr sent with his letter of 16 July 1870; it has not been found. CD evidently returned it with his suggestions. For the final version of the proposed amendment to the Census Bill, see letter from William Farr, 16 July 1870 and n. 2.

To John Lubbock 17 July 1870

Down. | Bromley. | Kent. S.E.
July 17 1870

My dear Lubbock

As I hear that the Census will be brought before the House tomorrow, I write to say how much I hope that you will express your opinion on the desirability of queries in relation to consanguineous marriages being inserted.[1] As you are aware, I have made experiments on the subject during several years; & *it is my clear conviction that there is now ample evidence of the existence of a great physiological law, rendering an enquiry with reference to mankind of much importance. In England & many parts of Europe the marriages of cousins are objected to from their supposed injurious consequences; but this belief rests on no direct evidence. It is therefore manifestly desirable that the belief should either be proved false, or should be confirmed*, so that in this latter case the marriages of cousins might be discouraged.

If the proper queries are inserted, the returns would show whether married cousins have in their households on the night of the census as many children, as have parents who are not related; & should the number prove fewer, we might safely infer either lessened fertility in the parents, or which is more probable, lessened vitality in the offspring.

It is moreover much to be wished that the truth of the often repeated assertion that consanguineous marriages lead to deafness & dumbness, blindness &c, should be ascertained;[2] & all such assertions could be easily tested by the returns from a single census—

Believe me | yours very sincerely | Charles Darwin

LS
DAR 261.7: 6

[1] See letter from William Farr, 16 July 1870 and n. 2, and letter from John Lubbock, 16 July [1870] and n. 2.

[2] For more on the attitudes to consanguineous marriage around this time, see Anderson 1986 and Kuper 2002. See also Child 1862 and Adam 1865.

From Armand de Quatrefages[1] 18 July 1870

Paris
18 Juillet 70

Cher Monsieur et confrere

Soyez assez bon je vous prie pour me dire si vous avez publié le volume sur la Geologie de l'Amérique du Sud qui devait faire partie de l'Expédition du Beagle.—[2] Je l'ai vainement cherché dans notre bibliothèque du Museum; je le chercherai ailleurs. Mais dans le cas où mes investigations seraient infructueuses, il est très important que je sache s'il a paru.— On a dit à l'Académie, dans le but de nuire à votre candidature que vous n'aviez fait que recueillir des objets d'histoire naturelle et que vous aviez appelé des aides pour les décrire.[3] Je tiens à montrer la part que vous avez eue dans la publication.— Dites moi je vous prie quel est le format, combien il y a de pages, de cartes.—

Vous savez deja sans doute qu'a une premiere nomination Brandt ne l'a emporté sur vous qu'au second tour de scrutin.[4] Mais nous avons trois places de correspondants à donner et j'espère bien que nous réussirons Edwards et moi à convaincre l'Académie.[5] Deja nous avons eu deux séances de discussion. Lundi prochain nous recommencerons et ce sera à mon tour a prendre la parole. Croyez bien que je ferai mon possible pour que la bonne cause l'emporte; mais vous êtes vivement attaqué.[6] Hélas! s'il n'y avait encore dans le monde que les batailles de cette nature nous serions bien heureux!—[7]

Votre bien dévoué | De Quatrefages

DAR 175: 7

[1] For a translation of this letter, see Appendix I.

[2] CD's geological observations related to the voyage of the *Beagle* were published in *Coral reefs*, *Volcanic islands*, and *South America*; all three appeared in one volume as *Geology of the 'Beagle'*. Up to this time, there had been no French translation of any part of the work.

[3] Quatrefages refers to the Académie des Sciences. The zoological material related to the voyage of the *Beagle*, *Zoology*, was edited and overseen by CD, but individual parts were done by different authors. See *Correspondence* vol. 2. For more on the opposition to CD's nomination as corresponding member of the Académie, see n. 6, below.

[4] Johann Friedrich von Brandt was elected on 4 July 1870 (see letter from J. D. Hooker, [6 or 7 July 1870] and n. 5).

[5] According to a report in the *Revue des Cours Scientifiques*, 23 July 1870, p. 529, Henri Milne-Edwards spoke on 18 July 1870 in favour of CD's admission.

[6] Emile Blanchard spoke for over an hour against CD's admission. He argued that on the *Beagle* voyage, CD had merely collected and was not capable of making a scientific study of his collections; he held that CD's theory of coral island formation was mostly derivative. He further claimed that CD's pigeon studies demonstrated a lack of true scientific spirit, that his work on cirripedes contained hardly any new facts, and that his theory of descent was both false and unoriginal. Blanchard apparently argued against transformism with examples from Jean Baptiste de Lamarck (*Revue des Cours Scientifiques*, 23 July 1870, p. 529).

[7] Quatrefages alludes to the Franco-Prussian war; France declared war on 19 July 1870 (Wawro 2003, p. 65).

From Arthur Gardiner Butler 19 July 1870

17 Oxford Road, Ealing
19th. July 1870.

Dear Sir

I am sorry that your note should have reached me whilst I am taking holidays, because, as I am very ignorant of *Heterocerous* Lepidoptera, I cannot, whilst away from the Museum, help you to the name of Trimen's Moth, I am pretty sure that *Iris* is the name I suppose you are sure about the *Gynanisa* being all right?[1]

The two species mentioned by Dr Wallace are *Bombyx Pernyi* & *B. Yama-mai*;[2] there is an interesting note on the two species in the Proceedings of the Ent. Soc. for this year, in which Dr Wallace states that hybrid eggs between these species proved fertile; hybrids in the imago state were exhibited at the Soc. 3 or 4 Meetings back.[3]

I have recently been trying some experiments with *Vanessa Urticæ* (the small Tortoiseshell Butt.),[4] breeding in the dark & in differently coloured bottles, the only effects produced were dwarfed examples of a deeper tint than usual & the smaller the dwarf, the deeper the tint, as though the same amount of colour had been spread over a smaller area & thus had become intensified: The caterpillars showed a marked disinclination to change to pupa in the dark for I tried experiments by shutting out the light first on one side of my breeding cage & then on the other & the caterpillars invariably suspended themselves on the lighter side; when both ends were darkened they suspended themselves in the middle.

Believe me to be | yours very sincerely | A G Butler.

Ch. Darwin Esq | &c &c &c

DAR 160: 387

[1] CD's note to Butler has not been found. Butler refers to Roland Trimen; see *Correspondence* vol. 16, letter from Roland Trimen, 13 January 1868 for Trimen's description of *Gynanisa isis* (now *G. maja*). See also *Descent* 2: 132.

[2] Butler refers to Alexander Wallace. In his letter of 15 July [1869] (*Correspondence* vol. 17), Wallace had sent CD statistics on the proportion of sexes of moths bred from *Bombyx pernyi* (now *Antheraea pernyi*, the Chinese oak silkmoth) and *B. yamamai* (now *Antheraea yamamai*, the Japanese oak silkmoth).

[3] Hybrids of *Bombyx pernyi* and *B. yamamai* were exhibited at the 7 March 1870 meeting of the Entomological Society of London (see *Entomologist's Monthly Magazine* (1870): 267).

[4] *Vanessa urticae* is now *Aglais urticae.*

To Armand de Quatrefages 20 July [1870][1]

Down Beckenham | Kent
July 20th

My dear Sir

A German friend lately asked me for a complete copy of my publications; & I have thought it best to send you a copy, though fuller than you require.—[2] I am sure that I feel very grateful & much honoured by the interest which you have taken about my election.— I saw in one of our newspapers the number of votes which I obtained & which I owe to you, & the number was far higher than I had expected.[3]

With very sincere respect & cordial thanks, I remain | Yours very truly obliged | Ch. Darwin

American Philosophical Society (377)

[1] The year is established by the relationship between this letter and the letter from Armand de Quatrefages, 18 July 1870.

[2] See letter from Armand de Quatrefages, 18 July 1870 and n. 2. CD refers to William Thierry Preyer, who had asked CD for a list of his publications for a biographical sketch (see *Correspondence* vol.17, letter from W. T. Preyer, 21 March 1869 and n. 2).

[3] CD refers to Quatrefages's support for his candidacy for election to the French Académie des Sciences. The newspaper CD refers to has not been identified, but according to the report in the *Comptes Rendus Hébdomadaires des Sciences de l'Académie des Sciences* 71 (1870): 41, CD received sixteen votes, Johann Friedrich von Brandt had nineteen votes, Thomas Henry Huxley had three, and Sven Lovén had one. At the second ballot, Brandt received twenty-two votes and CD sixteen.

To John Lubbock 21 July [1870][1]

Down. | Beckenham | *Kent. S.E.*
July 21

My dear Lubbock

Very many thanks for your Book: I have read 4 or 5 Chapts. with extreme interest, too much interest for the good of my internal viscera. It makes me groan that you had not published 4 or 5 months ago: I shd have so profited & saved so much work.— But on some very important points for me, I shall be able & must modify what I have written.—[2] I see that we have struck on some quotations & points in common; but it has surprised me, as on the ornamentation of savages, how largely we have collected our facts from different sources.—[3]

I see we differ a good deal about "moral sense" or rather approach the subject from different points of view; but hardly two men ever do agree on this perplexing subject.— I expect that your book will be a great success.—[4]

I have not yet even looked at the Thysanura, which I shall always love after your description of their playing like lambs & the sweet coyness of the females.—[5]

Your note of the 16th about the Census did not reach me till 10 oclock on the 18th when it was too late to answer. Moreover, I understood from Dr Farr, that the detailed manner of putting the query wd be left to his Office[6]

My dear Lubbock | Yours most truly | Ch. Darwin

Dr N. Hammond (private collection)

1 The year is established by the relationship between this letter and the letter to John Lubbock, 17 July 1870.

2 CD refers to Lubbock's book *The origin of civilisation and the primitive condition of man* (Lubbock 1870a); CD's annotated copy is in the Darwin Library–CUL (see *Marginalia* 1: 511–12). CD added several citations of Lubbock 1870a to notes in *Descent.*

3 For Lubbock's discussion of ornamentation, see Lubbock 1870a, pp. 25–49; for CD's account, see *Descent* 2: 338–54.

4 For Lubbock's views on the development of the moral sense, see Lubbock 1870a, pp. 257–73; for CD's account, see *Descent* 1: 70–106.

5 In 'Notes on the Thysanura. Part III' (Lubbock 1867b), Lubbock described the courtship behaviour of *Smynthurus luteus* in the insect order Thysanura; CD quoted the description in *Descent* 1: 348. CD's annotated copy of the article is in the unbound journal collection in the Darwin Archive–CUL. *Smynthurus luteus* (now *Deuterosminthurus sulphureus*) is no longer considered an insect, but belongs to the order Collembola (springtails) in the class Entognatha.

6 CD refers to William Farr. See letter from John Lubbock, 16 July [1870], and letter from William Farr, 16 July 1870.

From John Lubbock 23 July [1870][1]

15, Lombard Street, E.C.
23 July

My dear Mr. Darwin

As you may perhaps have seen in the Times I brought forward the Cousin question at about half past one last night, or rather this morning.[2]

It would be desirable to get some members to speak to Mr Bruce on the subject.[3] I read most of your letter to the House.[4]

I am glad you like my book.[5] It is almost all sold.

Hoping to see you tomorrow I remain, dear Mr Darwin, | Yours most sincerely | John Lubbock

DAR 170: 20

1 The year is established by the relationship between this letter and the letter to John Lubbock, 17 July 1870.

2 For the report on the second reading of the Census Bill, including Lubbock's proposal that a question on cousin marriages be added, see *The Times*, 23 July 1870, pp. 7–8.

3 Henry Austin Bruce was the home secretary.
4 See letter to John Lubbock, 17 July 1870.
5 See letter to John Lubbock, 21 July [1870] and n. 2.

From John Lubbock 26 July [1870][1]

House of Commons
26 July

My dear Mr. Darwin

We persuaded Bruce to support your amendment, & some good men spoke for it, but as you will see all in vain.[2]

Do not you think you might get most of what you want by an enquiry at one or two of the largest idiot asylums?[3]

I am, dear Mr Darwin, | Yours most sincerely | John Lubbock

DAR 170: 77

1 The year is established by the relationship between this letter and the letter to John Lubbock, 17 July 1870.
2 Lubbock refers to Henry Austin Bruce and the Census Bill of 1870. See letter from William Farr, 16 July 1870 and n. 2. As well as Bruce, Lyon Playfair, Donald Dalrymple, and William Rathbone spoke in favour of the amendment concerning cousin marriage (*Hansard parliamentary debates* 3d ser. vol. 203 (1870), cols. 1007–10).
3 CD had hoped to gather information on possible negative effects on the progeny of marriages between cousins (see letter to William Farr, 17 July [1870]).

To Richard Kippist 28 July [1870][1]

Down. | Beckenham | *Kent. S.E.*
July 28

My dear Sir

I have received 5 Papers from Mr J. P. Mansel Weale of Brooklyn, near K. William's Town, Kaffraria, which he has asked me to communicate to Linnean Soc.—[2] I send them registered, with figures & specimens of the Plants referred to.— The Socy. must settle whether they are worth reading.[3] There are some new & rather curious facts about Disa & Disperis, & the case of the Bee's nest, I believe to be new. The paper on the Asclepiads wants enormous condensation; but I have no spare strength or time to undertake such work.— There is another difficulty that the plants are not named. As I have said there seems to me enough novel matter in these papers to justify me in sending them to the Society. But as there has been so much now written on the fertilisation of Orchids, their publication is another question, about which I will not venture to express an opinion.—

My dear Sir | Yours very faithfully | Ch. Darwin

Linnean Society of London

1 The year is established by the relationship between this letter and the letter to J. P. M. Weale, 30 July 1870.

[2] James Philip Mansel Weale had enclosed the papers with his letter to CD of [25 May 1870]. Kaffraria was a region on the eastern coast of South Africa, between Natal and Cape Colony; it became part of Cape Colony in 1865 (Stewart 1996).
[3] Four of Weale's papers were later published (Weale 1870a, 1870b, 1870c, and 1870d).

From David Forbes 30 July 1870

11, York Place, | Portman Square, | London. W.
30 July 1870

Dear Mr Darwin

I leave this moment for Norway but write you these few lines at the request of Col Fox and some others of the council who wish me to ask if you could send a note if only a page to put in after the discussion of my paper as they say for the good of the journal—since the ethnological wants all the help it can from great names—[1] You have now I suppose had the proofs of the entire paper as it is some days since I corrected the last proofsheets— the paper is over 100 pages and there are 7 plates, 3 of indians—but these are not finished yet by the lithographer I hope you will do this—as you know that you are really the cause of the paper appearing at all—for I do not pretend to be sufficiently strong on Ethnological matters as to rush into print had you not backed me on— I send you a copy of the sermon I gave some Sundays ago on Volcanos & hope it may interest you[2]

The Secretary—Mr Rudler[3] Royal School of Mines—Jermyn Street will do all you wish if you write during my absence in Norway, I shall be away until the 20 August—

Believe me my dear Sir | Yours very truly | David Forbes

I shall not be offended if you "pitch into" me well

DAR 164: 145

[1] Augustus Henry Lane Fox (later A. H. L. Fox Pitt-Rivers) was the honorary general secretary of the Ethnological Society of London. Forbes's paper 'On the Aymara Indians of Bolivia and Peru' was published in the *Journal of the Ethnological Society of London* (Forbes 1870a). No addition by CD was made to follow the printed paper. CD's annotated offprint of the paper is in the Darwin Pamphlet Collection–CUL. See also letter from David Forbes, 13 June 1870.
[2] No copy has been found in the Darwin Archive–CUL, but the lecture on volcanoes was delivered by Forbes at St George's Hall, Langham Place, on 19 June 1870 (*Nature* 2 (1870): 336–7). The text of the lecture appeared in the *Geological Magazine* (Forbes 1870b).
[3] Frederick William Rudler was the assistant secretary of the Ethnological Society (*WWW*).

To J. P. M. Weale 30 July 1870

Down. | Beckenham | *Kent. S.E.*
July 30 1870

My dear Sir

I received 3 days ago your interesting letter of May 25th, the M.S. with the excellent drawings, & the specimens.[1] I have despatched all the latter to the Linn. Soc.; but the session is over, & there will be no meeting till Nov.[2]

The points which have struck me most in your papers are, firstly, the parallelism of Disa & Ophrys;[3] secondly, the structure & the three nectaries of Disperis;[4] & thirdly, the contraction of the caudicles in certain cases.[5] I have also been glad to read what you say about mares & asses; but I believe scarcely any thing is more new than your observation about the acrid secretion from the Soldier-termes.[6]

Whether the Linn. Soc. will publish your papers I cannot of course say; but I heard not long ago that they were overwhelmed with matter, & I doubt about their funds for engraving.[7]

If you will permit me, I w^d suggest that your papers w^d be improved by much condensation, & by omitting all observations which do not lead to definite results.[8] There are not many persons who are interested about the fertilization of flowers.

I forgot, in specifying what seems to me most curious in y^r notes, the case of the bee's nest.[9]

I am very sorry to hear so bad an account of your colony, the success of which has always interested me.[10]

With my best wishes for your success in science & in all other way believe me | my dear Sir yours very faithfully | Ch. Darwin

LS
American Philosophical Society (380)

1 See letter from J. P. M. Weale, [25 May 1870]. The letter is now incomplete.
2 See letter to Richard Kippist, 28 July [1870] and n. 3.
3 Weale had compared the ease of self-fertilisation in *Disa macrantha* to that of *Ophrys muscifera* (see Weale 1870d, p. 47).
4 In his account of the flower structure of the orchid genus *Disperis*, Weale described the two lateral sepals as each having a short nectary (see Weale 1870b, p. 42). These saclike structures are no longer considered to be nectaries, rather they serve as covers for the rostellar arms during development (see Dressler 1993, p. 150).
5 Weale described the caudicle of *Disperis* as bending backward about a minute after the removal of the pollinium (see Weale 1870b, p. 44).
6 Weale's comments on mares and asses were in a now missing section of his letter to CD of [25 May 1870]; only the first part of his comments on *Termes* are still extant.
7 Weale sent fifty-six sheets of watercolour drawings, none of which appeared in the four published articles (Weale 1870a–d); they are in the archive of the Linnean Society, together with a packet of eleven dried plant specimens (Linnean Society of London, Weale Misc, drawer 39).
8 For CD's report on the papers, see *Correspondence* vol. 19, letter to Richard Kippist, 11 March 1871.
9 Weale's remarks on the bee's nest were in a now missing paper enclosed with his letter to CD of [25 May 1870] (see letter from J. P. M. Weale, [25 May 1870] and n. 3).
10 Weale's account of the region of Kaffraria, where he lived, was in a now missing section of his letter to CD of [25 May 1870].

To David Forbes 31 July [1870][1]

Down. | Beckenham | *Kent. S.E.*
July 31st

My dear Mr Forbes

I write a line to thank you very sincerely for your great kindness in sending me the proof-sheets of your paper on the Aymaras, of which I have received up to p. 272, which I presume is not far from the end.[2] It is a most interesting & valuable memoir, & I have been extremely glad to read it before going to press; but the information which you gave me was so substantially correct, that I have had to alter hardly anything. I shall, however, instead of giving some figures refer in most cases my readers to your full details.—[3]

Whatever share I have had in hurrying you to publish has been a right good deed on my part.—

With very true thanks | Yours sincerely | Ch. Darwin

P.S. This appears to me an important erratum

p. 221 10 lines from bottom for longer read shorter[4]

American Philosophical Society (381)

[1] The year is established by the relationship between this letter and the letter from David Forbes, 30 July 1870.

[2] See letter from David Forbes, 30 July 1870 and n. 1.

[3] In *Expression*, p. 318, CD cites Forbes 1870a on the subject of blushing.

[4] This correction was made in the published version of Forbes 1870a.

From Henry Bence Jones 2 August 1870

5 Albion Villas | Folkestone
Aug 2 1870

My dear M[r] Darwin

Your letter was forwarded to me here.[1]

I am not of opinion that the pins & needles need stop your work. They shew some want of motion. Some stopped molecular action I suppose which is somehow related to your indigestion. Altho' the manner is very dark to me and to every one else I suspect.

Whatever you can do in care as to quality & quantity of food, Plenty of air & gentle exercise without fatigue will help this feeling in the nerves away. I wish you would come here and stay ten days with me & I should soon have you well & fit to work at your new book as much as you require[2] I do not think any medicine will help you. A teaspoonful of Sal volatile an hour after each meal in a small wineglass of water is all I advise you to take.[3]

If any more decided nerve symptoms come on you should stop work & ask what more must be done

I am sorry to hear that your son is not well.[4] I do not expect to return to London until the 15th of October

With very kind regards | Believe me | Ys most truly | H Bence Jones

DAR 168: 79

[1] CD's letter has not been found. Bence Jones lived at 31 Brook Street, London, and had a second home in Folkestone where he went sailing; he was a specialist in stomach and renal disorders (*ODNB*; *Correspondence* vol. 15, letter from H. B. Jones to Emma Darwin, 1 October 1867, and n. 5).

[2] CD sent the manuscript of *Descent* to the printers in August 1870 (letter to John Murray, 5 [August 1870]).
[3] 'Sal volatile' is ammonium carbonate, often used as a smelling salt or internally as an antacid. CD took this on the advice of a former doctor of his, William Jenner (see *Correspondence* vol. 12, letter to J. D. Hooker, 13 April [1864] and n. 6, and Colp 1977).
[4] Jones may refer to Horace Darwin, whom CD regarded as an invalid (see letter to W. D. Fox, 18 February [1870], and letter to J. D. Hooker, 25 May [1870]). According to Emma Darwin's diary (DAR 242), Horace was home from school from 22 July to 10 August 1870, but she makes no mention of any illness for any of the boys at around this time.

To H. B. Jones 3 August [1870][1]

Down. | Beckenham | *Kent. S.E.*
Aug. 3d

My dear Dr Bence Jones

I must thank you cordially for your very kind invitation;[2] but my stomach, not withstanding all your good aid, is too diabolical to allow me to visit anywhere except at a very few of my nearest relations, where I can behave as I like. We shall, also, start soon for a visit to my son.—[3] If it had been possible I shd thoroughily have enjoyed accepting your kindness.— Your letter has been an *immense* relief to me; but I will be careful & try & not work too hard.

Yours very gratefully | Ch. Darwin

DAR 249: 87

[1] The year is established by the relationship between this letter and the letter from H. B. Jones, 2 August 1870.
[2] See letter from H. B. Jones, 2 August 1870.
[3] CD, with Emma Darwin and Leonard Darwin, visited William Erasmus Darwin in Southampton from 13 to 26 August 1870 (Emma Darwin's diary (DAR 242), CD's 'Journal' (Appendix II)).

From Paul Broca[1] 4 August 1870

Paris,
4 août 1870

Monsieur,

Je suis heureux du témoignage de bienveillance que vous avez bien voulu me donner; mais je crains bien que la seconde partie de mon discours, qui a paru le lendemain du jour où vous m'écriviez n'ait eu le malheur de vous déplaire.[2] La première partie etait un exposé historique et j ai dû y parler de vous avec les égards dus à un grand naturaliste; mais la seconde partie était une discussion critique et vous aurez peut être à me pardonner quelques passages un peu vifs contre la Selection naturelle.[3] Je vous prie, malgré cela, de ne pas me confondre avec le troupeau de vos adversaires systématiques, pas plus que Quatrefages et Milne-Edwards, qui soutiennent actuellement votre candidature, et aux efforts des quels je joindrais les miens si j'avais l'honneur d'appartenir à l'Academie des Sciences[4]

Veuillez agreer, Monsieur, avec mes remerciements pour votre aimable lettre, l'expression respectueux de mes sentiments de haute estime | P Broca

DAR 160: 313

[1] For a translation of this letter, see Appendix I.

[2] CD's letter has not been found, but was evidently written on 29 July 1870. Broca refers to his lecture 'Sur le transformisme', delivered at the Société d'Anthropologie de Paris in April 1870 and published in the 23 July and 30 July 1870 issues of the *Revue des Cours Scientifiques* (Broca 1870a). A longer version of Broca's paper appeared later in the *Bulletins de la Société d'Anthropologie de Paris* (Broca 1870b).

[3] See Broca 1870a, pp. 553–8. Broca argued that natural selection could only influence functional characteristics, and that many features in an organism had no functional significance; he concluded that while species were not fixed, the causes of their evolution remained undiscovered.

[4] Broca refers to Armand de Quatrefages and Henri Milne-Edwards, both of whom were supporting CD's candidacy for corresponding membership in the zoological section of the French Académie des Sciences. See letter from Armand de Quatrefages, 18 July 1870.

To John Murray 5 [August 1870][1]

Down. | Beckenham | *Kent. S.E.*
July 5th

My dear Sir

In 5 or 6 days I shall have my M.S (except last chapter which is partly written) ready to send to Printers; & all wood-blocks, except for one of later chapters are complete.[2]

Please inform me whether my servant shall leave the M.S (with letters of instructions) at Mss Clowes[3] or to you.— I shall be ready to begin correcting Slips on the 22d, after a short absence from home for rest.—[4]

My dear Sir | Yours sincerely | Ch. Darwin

Do you know anything about Mr Clowes works in hand—will they be able to keep me steadily at work—it drives me mad to change from job to job.—

I ought to begin, at once, for I am a fearfully slow corrector & always liable to interruptions from illness.—

National Library of Scotland (John Murray Archive) (Ms.42152 ff. 209–10)

[1] The month and year are established by the relationship between this letter, the letter from John Murray, 6 August [1870] and the letter to Mr Dorrell, 9 August 1870. CD evidently wrote 'July' in error.

[2] CD refers to *Descent.*

[3] William Clowes & Sons.

[4] CD visited Southampton from 13 to 26 August 1870 (see 'Journal' (Appendix II)).

From William Farr 6 August 1870

General Register Office | *Somerset House*
1870 Aug 6

My dear Sir

I am glad Sir John Lubbock—made the *fight* he did.[1]

It must be kept up.

The Inquiry might be undertaken—through private channels—& a friend of mine has inserted a notice in the Lancet—of which the Editor has sent you a copy.[2]

I shall at all times support—and endeavor to give effect to any suggestions you throw out. Someday—I want to talk to you about zymotic diseases—*biads*[3] & *monads*—but shall not intrude until your opus majus[4] is out.

I am my dear Sir | ever yours faithfully | W. Farr

DAR 164: 33

[1] Farr refers to an amendment to the Census Bill of 1870 (see letter from William Farr, 16 July 1870 and nn. 1 and 2).

[2] The friend has not been identified. A notice, 'The census of 1871', criticising the failure of the committee of the House of Commons to adopt the amendment, appeared in the 'Medical Annotations' section of the *Lancet*, 6 August 1870, p. 195. CD's copy of the notice has not been found.

[3] 'Biad': evidently a coinage by Farr.

[4] Farr refers to *Descent*. *Opus majus*: greater work (Latin).

From John Murray 6 August [1870][1]

50, Albermarle St. | W.
Aug[t] 6

My Dear Sir

I will write to Messrs Clowes to prepare them for the receipt of your MS. directly from you, & I will desire them to give you a written assurance that the Printing shall proceed uninterruptedly usque ad finem—[2] I hope at least you may not suffer on the score of health during its progress. Alas this horrid War is a sad damper to Literature & Science. Until the Thundercloud has passed I shall publish little.[3]

My Dear Sir | Yours very faithfully | John Murray

Ch. Darwin Esq[r]

DAR 171: 377

[1] The year is established by the relationship between this letter and the letter to Mr Dorrell, 9 August 1870.

[2] Murray refers to the printers William Clowes & Sons and to *Descent*. *Usque ad finem*: up to the end (Latin).

[3] The Franco-Prussian war had begun on 19 July 1870 (Wawro 2003, p. 65). *Descent* was published on 24 February 1871 (see 'Journal' (Appendix II)).

From V. O. Kovalevsky 7 August [1870][1]

Nice
7 August

Dear Sir

Having finished all the business I had in Nice I am going now back to Paris and I hope to London.[2] Will You have the kindness Dear Sir to send me a note *Poste restante*

Paris if I can hope to find You about the 15 August in London.[3] I beg Your pardon for this anoyance, but I should be very sorry to miss You as I am going to London expressly for this object

Yours | Very truly | W Kovalevsky

DAR 169: 64

[1] The year is established by the relationship between this letter and the letter from V. O. Kovalevsky, 28 February [1870].

[2] See letters from V. O. Kovalevsky, 20 February [1870] and 28 February [1870].

[3] CD was away from home at the time of Kovalevsky's visit (see Emma Darwin's diary (DAR 242), and CD's 'Journal' (Appendix II)).

To Mr Dorrell 9 August 1870

Down, Beckenham, Kent
Aug 9th./70/

To Mr Dorrell at Messrs Clowes.

Dear Sir

I send all the M.S for my next book, except the last chapter, which I retain for additions.—[1] All the woodcuts are in your hand except for 2 of later chapters in Vol. 2; & I will hereafter send proofs of them with proper page & Chapt. for insertion.—

The book is to be printed of same type & size as my Origin of Species; & I shd like the foot-notes numbered as in my Dom. Animals.[2]

Please to send 2 Slips, printed on paper which is not bibulous.—[3] The proofs of woodcuts in my M.S will serve, instead of fresh proofs of woodcuts.

I shd esteem it a favour if you would **not** send me any slips, till you can send me about 2 sheets,[4] & till you feel sure that you can afterwards keep me at full work, but I work slowly; for I am incapable of changing the nature of my work.—

I shall return home & be ready to begin correcting on Aug. 22d.;[5] or if I shd be delayed a few days I would inform you in good time.—

Will you *oblige* me by a single line telling me, as nearly as you can, when I may expect to receive slips in manner requested; so that I may arrange my occupations.

I hope I shall not cause you much trouble. Dear Sir | Yours faithfully | Ch. Darwin

P.S. I shd like the Chapters to run continuously from I to XXII in the two volumes.—

As far as subject is concerned Chapt XII wd be the best for beginning vol. 2. & I think will be nearly equal. The last Chapt XXII (not sent) will be short; but the Index will be rather long.—

I fear Chapt XII may perhaps have to go in Vol I.; but you will be able to decide better, when the first volume is all printed.[6]

The Henry W. and Albert A. Berg Collection of English and American Literature, The New York Public Library, Astor, Lenox and Tilden Foundations

[1] CD refers to *Descent.*
[2] *Descent* was printed with the same type and was the same size as *Origin* 5th ed. (see R. B. Freeman 1977). As in *Variation*, the footnotes were sequentially numbered in each chapter.
[3] Slips: unpaged proofs. Bibulous: absorbent.
[4] A sheet is equivalent to sixteen pages.
[5] CD visited Southampton from 13 to 26 August 1870 (see 'Journal' (Appendix II)).
[6] The printed version of *Descent* had continuous numbering of chapters over the two volumes and the second volume began with chapter 12.

From Lewis Henry Morgan 9 August 1870

London | 15 Bury St. St James.[1]
August 9. 1870

Dear Sir

I send you by mail an advance copy of the last chapter of my work to which reference is made in the article before the Boston Academy which I sent to you some time ago at the suggestion of Prof Asa Gray of Cambridge.[2] The work is not yet issued, although stereotyped: but it will be about Dec 1. Until it is issued by the Smithsonian Institution, I suppose it should be considered as a private communication.

In a general sense the results of my investigations are in harmony with the Darwinian theory so far as man is concerned: and it occurred to me that you might be glad to look over this chapter in advance.[3] Although apart from the remainder of the text, and the Tables, the force of the positions taken cannot be fully appreciated.

It must be the next work of Ethnology to study with more care the ages of barbarism, to fix the great and successive stages of progress through it, and to find the sequence of customs and institutions which created these epochs. The most interesting and the formative portion of mans physical and mental history lies in the barbarous ages. The stages of his progress were numerous, and by successive reformatory movements as I think each resulting in the creation of new—or in the further developement of old institutions

I intended to call upon you to pay my respects—and to that end provided myself with a letter of introduction from Prof Gray: but I fear now I shall be compelled to forego it until my return to England in the Spring

With great respect | Yours truly | L H. Morgan

The first of living Englishmen | Charles Darwin.

As your country has given you no title I take the liberty of adding an American designation

DAR 171: 238

[1] In 1870, 15 Bury Street, St James's, London, was a lodging house (*Post Office London directory* 1870).
[2] Morgan refers to *Systems of consanguinity and affinity of the human family* (Morgan 1870); he referred to it in his article, 'A conjectural solution of the origin of the classificatory system of relationship' (Morgan 1868, p. 477); no copy of Morgan 1868 has been found in the Darwin Archive–CUL; there is a copy of Morgan 1870 in the Darwin Library–Down.

[3] From an extensive study of kinship terms, in particular those in use in tribes of native Americans and Malays, Morgan concluded that social organisation in humans had developed by gradual stages from an original and universal state of complete promiscuity to a family unit based on stable pairs. He identified two different systems for describing kinship: 'classificatory' terms that acknowledge only degree of relationship, so that, for example, a sister's son or daughter is also referred to as 'son' or 'daughter', and 'descriptive' terms that distinguish individual, lineal relationships (Morgan 1868 and 1870). CD referred to Morgan 1868 in *Descent* 1: 182 in the context of his argument that all civilised nations are descendants of barbarians, and in *Descent* 2: 358–61, in relation to the operation of sexual selection among primitive peoples.

To Mr Dorrell 11 August [1870][1]

Down. | Beckenham | *Kent. S.E.*
Aug. 11

Dear Sir

I am much obliged for your note.—[2] Since writing, my plans have been a little altered, & I shall not be ready to begin correcting until Aug. 28 or 29th., but I hope that this will be equally or more convenient to you.—[3]

Dear Sir | Yours faithfully | Ch. Darwin

Jonathan Haile (private collection)

[1] The year is established by the relationship between this letter and the letter to Mr Dorrell, 9 August 1870.

[2] No letter from Dorrell has been found.

[3] See letter to Mr Dorrell, 9 August 1870 and n. 5.

To L. H. Morgan 11 August [1870][1]

Down. | Beckenham | *Kent. S.E.*
Aug. 11th

Dear Sir

I am much obliged for your extremely kind letter & your present of the concluding chapter, which I am sure I shall read with the greatest interest.[2] In one respect it is fortunate for me that you are compelled to leave England so soon, for I shd. have been much disappointed at not being able to receive you here, & this I could not have done, as I am pledged to a visit for 10 days or a fortnight to a friend, & we all leave home very early on Saturday morning.[3]

I fully agree with your remarks on the extreme importance of studying the habits & institutions, if they can be so called, of savages. I have had lately to attend a little to this subject, as I have sent my M.S. to the Printers for a work on the "Descent of man"; but I have chiefly to treat of veritably primeval times before man was fully man.—[4]

With much respect for your admirable investigations, believe me | Dear Sir | Yours very faithfully | Ch. Darwin

Dept of Rare Books, Special Collections and Preservation, University of Rochester

1 The year is established by the relationship between this letter and the letter from L. H. Morgan, 9 August 1870.
2 See letter from L. H. Morgan, 9 August 1870 and n. 2.
3 See letter from L. H. Morgan, 9 August 1870; CD visited William Erasmus Darwin in Southampton from Saturday 13 to 26 August 1870 (see 'Journal' (Appendix II)).
4 CD had sent the bulk of the manuscript of *Descent* to the printers on 9 August (see letter to Mr Dorrell, 9 August 1870).

From V. O. Kovalevsky 15 August [1870][1]

31. Bloomsbury Str. Bedford Square.

Dear Sir

I am once more for a short time in England and should be very happy to see You if You could accept my visit.[2] It is such a long time I had no news from You; after Your letter to Nice in the Spring I wrote You from a small place near Marseille, but having no answer I fear my letter has been lost.[3] My wife is gone with her sister to the sea side and I am left alone to make excursions;[4] if You are not occupied to much to receive visits, I have an intention to go to Bickley[5] Tuesday and walk from the Station to Down.

I hope to find M$^{rs.}$ Dawin and the ladies[6] in good heal.

Yours | truly | W. Kovalevsky

London 15 August

DAR 169: 65

1 The year is established by the relationship between this letter and the letter from V. O. Kovalevsky, 7 August [1870].
2 See letter from V. O. Kovalevsky, 7 August [1870].
3 Neither CD's letter nor Kovalevsky's reply has been found.
4 Kovalevsky refers to Sofia Vasilyevna Kovalevskaya and Anna Vasilyevna Korvin-Krukovskaya.
5 Bickley Station is just outside Bromley, Kent, about six miles from Down.
6 Emma Darwin, Henrietta Emma Darwin, and Elizabeth Darwin.

From Charles Lawford Acland 17 August 1870

5 West Terrace | Folkstone
Aug 17th/70

Dear Sir,

In studying your work on Variation, especially Vol II pp 252–3 and the chapter on Pangenesis I have been frequently reminded of a conversation I had some years back with D$^{r.}$ Lee, the late well known minister of the Grey Friars Church in Edinburgh—[1] It introduces a thought w$^{h.}$ may appear to you to bear upon your hypothesis.

We had got somehow upon the strong personal resemblance which not unfrequently developes itself between husband and wife after they have been many years married. We had discussed various possible reasons for it, as similarity of tastes, of pursuits, of circumstances &c, and at last Dr Lee said that he believed there was also a physiological reason for it viz: that in his own words: the woman frequently has a portion of the man's blood circulating in her veins for nine months at a time" alluding of course to the period of gestation. If this were at all the true solution, the resemblance ought to be most strongly developed in couples who have had large families, and it ought to be that the woman becomes like the man and not vice versâ. I do not know if any observations have ever been made on this head, or whether matters have been found to accord with the conclusions from Dr. Lee's suggestion. If so the case appears to me to furnish strong collateral evidence in favour of your hypothesis. At any rate I thought it worth mentioning to you—

With much respect I remain | Dear Sir | very faithfully yours | C. L. Acland

DAR 159: A7

[1] Acland refers to Robert Lee, and to CD's 'provisional hypothesis of pangenesis', *Variation* 2: 357–404. In *ibid.*, pp. 252–3, CD dicussed the blending of parental characters and the possible causes of variability in offspring. Acland may also have had in mind *Variation* 2: 365–6, in which CD discussed the direct action of the male element on the female.

To J. V. Carus 18 August 1870

Down. | Beckenham | *Kent. S.E.* [Bassett, Southampton]
Aug 18 1870

My dear Sir

I have just sent the MS. of my new book to the printers, & corrected proofs will be ready in about a month's time.[1] I have thought it best to inform you of this, but under present circumstances your publisher will of course have given up all idea of bringing out a German translation.[2] Nevertheless I may inform you that the book will consist of two volumes of the size & type of my Origin of Species, each containing a little over 400 pages.[3]

What a dreadful misfortune for science is the present war! but I cannot express too strongly how I rejoice at the wonderful success of Germany, & I have not hitherto met a single person who has not entirely participated in this feeling.[4]

My dear Sir | yours very sincerely | Ch. Darwin

The title of my book is "The Descent of Man & Selection in relation to sex"—

LS(A)
Staatsbibliothek zu Berlin – Preußischer Kulturbesitz (Slg. Darmstaedter Lc 1859 Darwin, Charles, Bl. 60/61)

[1] CD refers to *Descent*, which had been sent to the printers, William Clowes & Sons, on 9 August 1870 (see letter from John Murray, 6 August [1870], and letter to Mr Dorrell, 9 August 1870; see also letter to Jean Jacques Moulinié, 18 August 1870).

[2] The Franco-Prussian war had begun on 19 July 1870 (Wawro 2003, p. 65). The publishing firm E. Schweizerbart'sche Verlagshandlung, headed by Eduard Koch, published the German translation of *Descent* in 1871 (Carus trans. 1871). See also letter to J. J. Moulinié, 18 August 1870.

[3] See letter to Mr Dorrell, 9 August 1870, and n. 2.

[4] French troops had crossed the border into Prussia on 31 July 1870. Despite suffering heavy casualties, the Prussian armies outnumbered the French sufficiently to win decisive victories at Wissembourg on 4 August, and at Spicheren and Froeschwiller on 6 August (Wawro 2003). News of the Prussian victory at Mars-la-Tour on 16 August did not reach the British press until 19 August (*The Times*, 19 August 1870, p. 10).

To Jean Jacques Moulinié 18 August 1870

Down Beckenham | Kent [Bassett, Southampton]
Aug 18. 1870

My dear Sir

I have thought it best to tell you that I have sent the M.S. of my new book to the printers, & shall receive the first corrected proofs in about a month's time; but all idea of publishing a translation under present circumstances will of course be given up.[1] I fear also that whatever labour you may have bestowed in preparing the new edition of my Origin of species will likewise be thrown away.[2] The title of my new book is "The Descent of Man & Selection in relation to Sex". It will consist of 2 vols of the type & size of my Origin, each containing a little over 400 pages.[3] If you have leisure at any time, I should like to hear how much you have unfortunately done of the translation of the Origin.

We all in England deeply wish to hear of peace being reestablished in Europe.

My dear Sir | yours sincerely | Ch. Darwin

LS
Bibliothèque de Genève (BGE, Ms. suppl. 66, ff. 16–17)

[1] CD refers to *Descent*, which had been sent to the printers, William Clowes & Sons, on 9 August 1870, and to the Franco-Prussian war, which began on 19 July 1870 (see letter from John Murray, 6 August [1870] and n. 3, and letter to Mr Dorrell, 9 August 1870). CD had corresponded with Moulinié about a French translation of *Descent* in 1869; it was published in 1872 (Moulinié trans. 1872; see *Correspondence* vol. 17, letters from J. J. Moulinié, 20 October 1869 and 5 November 1869, and letter to J. J. Moulinié, 23 October [1869]).

[2] CD had asked Moulinié to undertake a new translation of *Origin* 5th ed.; in the event his translation was based on both *Origin* 5th ed. and *Origin* 6th ed. and appeared in 1873 (Moulinié trans. 1873; see *Correspondence* vol. 17, letter to J. J. Moulinié, 23 October [1869], and letter from J. J. Moulinié, 5 November 1869).

[3] See letter to Mr Dorrell, 9 August 1870 and n. 2.

To F. P. Cobbe 20 August [1870][1]

Bassett | Southampton
Aug 20th

My dear Miss Cobbe

My wife is rather poorly & so I write for her.[2] We are both quite delighted with your admirable & most just article. You editors have more power with your strong right arms than the Knights of old, in righting the oppressed.— Will you be so kind as to put my name down for 1£, or—(whichever you think best) my name for 10^{s} & my wife for 10^{s}.[3]

(Charles Darwin of Down Beckenham Kent)

You & Miss Lloyd[4] need not have your faith in inheritance shaken, with respect to Tropæolum, until you have prevented for 6 or 7 generations any crossing between the vars in same garden. I have lately proved that every shade of colour is transmitted by the most fluctuating garden var. if the flowers are carefully self-fertilised during 6 or 7 generations.—[5]

Thank you for telling me about the articles in Fraser, of which I should not probably have heard.[6]

Pray give my very kind compliments to Miss Lloyd: I hope the dear old white cob in Wales is well.—[7] Pray believe me Yours very sincerely obliged | Ch. Darwin

I forgot to say that I wrote as J.P: for Kent to Home Secretary, calling his attention to Holder's case.—[8]

The Huntington Library (CB 385)

1 The year is established by the reference to the case of Stephen Holder. On 14 August 1870, Holder was sentenced to two months' hard labour by Bromley magistrates' court for taking part in an illegal gambling game of pitch and toss on Sunday 13 August (*Bromley Telegraph*, 20 August 1870; Randal Keynes and Richard Milner, personal communication.)

2 There is no reference to illness in Emma Darwin's diary (DAR 242).

3 Cobbe had published a leading article in the *Echo* on 18 August 1870 denouncing the sentence given to Holder. On 25 August, the *Echo* published a letter purporting to be from CD:

> Sir,– | I have read your admirable and most just article on "Even-handed Justice," and beg to say that if anyone who sympathises with the case be disposed to open a subscription for the benefit of Stephen Holder, I should be happy to contribute to it £1.– | I am, Sir, yours &c. | Charles Darwin | Down, Beckenham, Kent.

Emma Darwin wrote to Henrietta Emma Darwin that CD thought the publication of the letter 'unjustifiable' and had been depressed to discover that Holder had been in prison before (DAR 219.9: 91). Henrietta later wrote that this was Cobbe's own reworking of CD's letter to her, published in this form without his consent (*Emma Darwin* (1915), 2: 302). The original article did not announce the setting up of a subscription fund; Cobbe may have suggested one in a missing letter to CD.

4 Mary Charlotte Lloyd.

5 CD published his observations on the relative effect on colour of cross and self-fertilisation in a number of plant species in *Cross and self fertilisation*, pp. 307–11. Although primarily interested in the relative effects on size and health, he had made notes on the colour of the seventh generation of self-fertilised plants of *Ipomœa purpurea* in 1869 (DAR 78: 89).

6 CD refers to the literary journal *Fraser's Magazine*. The articles have not been identified.

7 Lloyd had lent CD a pony when she, Cobbe, and the Darwins had been staying near one another in Caerdeon, Wales, in July 1869 (see *Correspondence* vol. 17, letter to J. D. Hooker, 8 July [1869] and n. 3).

8 No letter from CD to the home secretary, Henry Austin Bruce, has been found. Only one letter on the Holder case is noted in the *Register of papers received by the Home Office*; this was received on 19 August 1870 but is no longer extant. CD was active as a magistrate in Bromley between 1859 and 1862. (Randal Keynes and Richard Milner, personal communication.)

From William J. Raybould to Abraham Dee Bartlett 22 August 1870

6 Gladstone Place | High Park Road | Smethwick | N^r Birmingham
Augt 22 1870,

To the Manager of the | Royal Zoological Gardens | London[1]
Sir/

I have a Tortoise shell Coloured Cat with *6 claws perfect on each Foot* also 2 young ones with the same Number of Toes or claws, one *Cream & white*, & the other *Blk & white*, a month old Are they worth your notice (*If so make me an offer for them*)[2]

Kindly say & oblige | Yours very truly | W J Raybould

DAR 176: 25

[1] Bartlett was superintendent of the Zoological Society of London's gardens in Regent's Park and had supplied CD with information used in *Variation*; presumably Bartlett forwarded this letter believing that it might be of interest to CD (see n. 2, below).

[2] CD had discussed supernumerary digits in animals and humans in *Variation* 2: 12–17.

To Armand de Quatrefages 23 August [1870][1]

Down Beckenham | Kent [Bassett, Southampton]
Aug. 23. 187

My dear Sir

Under the present terrible state of public affairs, whilst private interests are as nothing, I have hesitated to write to you; but I hope that you will allow me to express briefly my cordial & lasting gratitude for the surprising & very kind interest which you have shown about my election to your Academy.[2] It is almost a pity that you sh^d have been led to waste so much of your valuable time on this affair. As I sh^d not like to trouble M. Milne Edwards with a letter, I hope, if you have any fitting opportunity, that you will express to him my very sincere thanks for the honour which he has conferred on me.[3] I see in the last number of the Révue that M. Edwards is inclined to believe that existing species are the modified descendants of extinct species.[4] Such an admission seems to me very much more important than whether natural selection has been a more or less efficient means of change; though for my own part I sh^d never have been able to admit the evolution of species, unless I c^d have partly explained to myself how the innumerable & beautiful adaptations, which we see all around us, had originated.

If I had been present when M Elie de Beaumont called my science frothy,[5] I might have rejoined that my bubbles have not as yet all burst, whilst his two great bubbles of craters of elevation & the direction of mountain chains according to their age, have every where except in France, burst & vanished into thin air.[6]

With the truest thanks & respects believe me my dear Sir | yours very faithfully | Ch Darwin

P.S. I fear my next book, which will be published in 2 or 3 months, will greatly displease you, but I will send you a copy, as I am sure you will give me the credit of publishing only what I believe to be the truth, after mature consideration.[7]

LS
American Philosophical Society (382)

1 The year is established by the relationship between this letter and the letter from Armand de Quatrefages, 18 July 1870.
2 CD refers to Quatrefages's support for his candidacy for election to the French Académie des Sciences (see letter from Armand de Quatrefages, 18 July 1870, and letter to Armand de Quatrefages, 20 July [1870]). CD also alludes to the Franco-Prussian war, which had begun on 19 July 1870 (Wawro 2003, p. 65).
3 Henri Milne-Edwards had spoken in favour of CD's admission to the Académie des Sciences (see letter from Armand de Quatrefages, 18 July 1870 and n. 5).
4 CD refers to the last but one issue of *Revue des Cours Scientifiques*, of 13 August 1870. It contained a resumé of all the points Milne-Edwards had made in favour of CD during the sessions of the Académie des Sciences when the admission of CD as a corresponding member was being debated ('Opinion de M. H. Milne Edwards de l'Institut sur les travaux de Ch. Darwin', *Revue des Cours Scientifiques* 7 (1870): 588–92).
5 According to the report in *Revue des Cours Scientifiques*, 23 July 1870, p. 529, an academician, believed to be Léonce Elie de Beaumont, shouted, 'C'est de la *science mousseuse*', a remark to which Quatrefages objected strongly.
6 CD refers to Elie de Beaumont's support for a theory originally formulated by Christian Leopold von Buch that volcanoes were caused by pressure from below, which arched the strata into a dome-like formation until the centre collapsed and a vent was formed (Elie de Beaumont 1838; see also *Correspondence* vol. 3, letter to Leonard Horner, 29 August [1844]). He also refers to Elie de Beaumont's theory that the orientation of a mountain chain was determined by the age of its formation (Elie de Beaumont 1831). In his *Principles of geology* (Lyell 1830–3), Charles Lyell vigorously opposed Elie de Beaumont on these subjects, and Lyell's view came to be accepted by the majority of geologists. For more on Elie de Beaumont's theories, see Gohau 2006, pp. 151–8.
7 CD refers to *Descent*; Quatrefages's name appears on CD's presentation list for the book (DAR 210:11. 32; *Correspondence* vol. 19, Appendix IV).

To George F. Kittredge[1] 24 August [1870]

Down. | Beckenham | *Kent. S.E.* [Bassett, Southampton]
August 24th

Dear Sir

I thank you for your obliging letter of Augt. 7th.[2] Should I ever require any information with respect to the Paleontology or Nat. Hist. of your vicinity, I will not forget your kind offer of assistance.— My sons have no strong taste for the natural sciences, excepting one of them, & he is devoted solely to comparative anatomy.[3]

With my Thanks | I remain— Dear Sir | yours faithfully | Charles Darwin

Postmark: AU 24 70
American Philosophical Society (B/D25.271)

1 The recipient is identified by the envelope, which is addressed to 'G. F. Kittredge Esq, 137. Clinton St., Buffalo, New York State, United States'.
2 No letter from Kittredge has been found. It may have referred to Kittredge's work on the origin of species, which was privately published the following year (Kittredge 1871).
3 CD refers to his son Francis, who was studying for a natural sciences degree at Cambridge University (*Alum. Cantab.*).

To Fritz Müller 28 August 1870

Down. | Beckenham | Kent. S.E.
Aug 28—70

My dear Sir

I have to thank you very sincerely for two letters: one of April 25th—containing a very curious account of the structure & morphology of Bonatea.—[1] I feel that it is quite a sin that your letters shd. not all be published; but in truth I have no spare strength to undertake any extra work, which though slight would follow from seeing your letters in English through the press—not but that you write almost as clearly as any Englishman. This same letter also contained some seeds for Mr Farrer which he was very glad to receive.[2]

Your second letter of July 3d was chiefly devoted to mimickry in Lepidoptera: many of your remarks seem to me so good, that I have forwarded your letter to Mr. Bates; but he is out of London, having his summer holiday, & I have not yet heard from him.[3] Your remark about imitators & imitated being of such different sizes, & the lower surface of wings not being altered in colour strike me as the most curious points. I shd. not be at all surprised if your suggestion about sexual selection were to prove true; but it seems rather too speculative to be introduced in my book more especially as my book is already far too speculative.[4] The very same difficulty about brightly coloured caterpillars had occurred to me, & you will see in my Book, what I believe is the true explanation from Wallace.[5] The same view probably applies in part to gaudy Butterflies.— My M.S. is sent to Printers, & I suppose will be published in about 3 months: of course I will send you a copy.[6] By the way I settled with Murray recently with respect to your Book, & had to pay him only 21£"2s"3d which I consider a very small price for the dissemination of your views: he has 547 copies as yet unsold.[7] This most terrible war will stop all science in France & Germany, for a long time: I have heard from nobody in Germany, & know not whether your Brother, Häckel, Gegenbaur Victor Carus or my other friends are serving in the army.[8] Dohrn has joined a cavalry regiment.[9] I have not yet met a soul in England, who does not rejoice in the splendid triumph of Germany over France: it is a most just retribution against that vain-glorious, war-loving nation. As the Posts are all in confusion, I will not send this letter through France.— The Editor has sent me duplicate copies of the Revue des Cours Scientifiques, which contain several articels about my views; so I send you copies, for the chance of your liking to see them.—[10]

My dear Sir | Ever Yours very sincerely | Ch. Darwin

British Library Board (Loan 10:33)

1 The letters referred to have not been found. CD had discussed the structure of *Bonatea speciosa* in *Orchids*, pp. 87, 302–5, 330. In *Orchids* 2d ed., p. 76, CD added a reference to Roland Trimen's description of *Bonatea* (Trimen 1864), but did not mention Müller's account.

2 The seeds were probably of *Passiflora*; Müller had sent seeds of *Passiflora* for Thomas Henry Farrer with his letter of 16 February 1870.

3 Henry Walter Bates had developed a theory about mimicry in butterflies that proposed that a palatable species would benefit from imitating one that was unpalatable (Bates 1861). Müller became interested

in mimicry after reading Wallace 1869c, and began observing resemblances in endemic butterflies (see letter from Fritz Müller, 29 March 1870 and n. 3).

4 Müller also discussed butterflies in two letters to his brother, Hermann Müller, written on 5 May 1870 and 14 June 1870 (Möller ed. 1915–21, 2: 175–9). Müller suggested two points in favour of mimicry being the result of sexual selection. Firstly, he argued, the imitated species tended to be very brightly coloured, so the coloration was more attractive than protective. Secondly, only the upper surface tended to be imitated, so that the imitator was still protected from predators while not flying, since most butterflies rested with their wings closed (see Möller ed. 1915–21, 2: 179). CD refers to *Descent*.

5 In *Descent* 1: 416–17, CD gave Alfred Russel Wallace's explanation that bright colours served as a warning to predators that a caterpillar was unpalatable.

6 *Descent* was published on 24 February 1871 (see *Correspondence* vol. 19, Appendix II). Müller's name appears on CD's presentation list for *Descent* (DAR 210:11. 32; *Correspondence* vol. 19, Appendix IV).

7 CD refers to the English translation of F. Müller 1864, *Facts and arguments for Darwin* (Dallas trans. 1869). According to John Murray's ledger book (John Murray Archive, National Library of Scotland), 1000 copies were printed. Murray received a ten per cent commission for sales and warehousing. After one year, a loss of £21 2*s.* 3*d.* was recorded, but over the next ten years small profits were recorded each year.

8 France had declared war on Prussia on 19 July 1870 (Wawro 2003, p. 65). Prussia had the support of the North German Confederation and the southern states of Baden, Württemberg, and Bavaria. CD refers to Hermann Müller, Ernst Haeckel, Carl Gegenbaur, and Julius Victor Carus. See letter from J. V. Carus, 2 October 1870.

9 For more on Anton Dohrn's service in the Prussian army during the Franco-Prussian war, see Heuss 1991, pp. 104–9.

10 Emile Alglave was the editor of the weekly *Revue des Cours Scientifiques*; CD's copies of some issues of the journal are in the Darwin Pamphlet Collection–CUL. In July and August 1870, a number of issues focused on the debate over CD's candidacy in the zoological section of the Académie des Sciences (see letter from Armand de Quatrefages, 18 July 1870). A paper by Paul Broca on transformism was also published in the journal (Broca 1870a; see letter from Paul Broca, 4 August 1870 and n. 2).

To Philip Lutley Sclater 30 August [1870][1]

Down. | Beckenham | *Kent. S.E.*
August 30th

My dear Mr Sclater,

I hope the Society will permit the enclosed note to be read & printed, as some palliation of my remarks on the woodpecker of the plains, versus Mr. Hudson.—[2]

Yours' very sincerely | Ch. Darwin

[Enclosure]

In the last of Mr. Hudson's valuable articles on the Ornithology of Buenos Ayres, he remarks, with respect to my observations on the *Colaptes campestris*, that it is not possible for a naturalist 'to know much of a species from seeing perhaps one or two individuals in the course of a rapid ride across the Pampas.'[3] My observations were made in Banda Oriental, on the northern bank of the Plata, where, thirty-seven years ago, this bird was common; and during my successive visits, especially near Maldonado, I repeatedly saw many specimens living on the open and undulating plains, at the distance of many miles from a tree.[4] I was confirmed in my belief, that these birds do not frequent trees, by the beaks of some which I shot being muddy, by their tails being but little abraded, and by their alighting on posts or branches of trees (where such grew) horizontally and crosswise, in the manner of ordinary birds, though, as I have stated, they sometimes alighted vertically. When I wrote these notes, I knew nothing of the works of Azara,[5] who lived for many years in Paraguay, and is generally esteemed as an accurate observer. Now Azara calls this bird the Woodpecker of the plains, and remarks that the name is highly appropriate; for, as he asserts, it never visits woods, or climbs up trees, or searches for insects under the bark.[6] He describes its manner of feeding on the open ground, and of alighting, sometimes horizontally and sometimes vertically, on trunks, rocks, &c., exactly as I have done. He states that the legs are longer than those of other species of Woodpeckers. The beak, however, is not so straight and strong, nor the tail-feathers so stiff, as in the typical members of the group. Therefore this species appears to have been to a slight extent modified, in accordance with its less arboreal habits. Azara further states that it builds its nest in holes, excavated in old mud walls or in the banks of streams. I may add that the *Colaptes pitius*, which in Chile represents the Pampas species, likewise frequents dry stony hills, where only a few bushes or trees grow, and may be continually seen feeding on the ground. According to Molina, this *Colaptes* also builds its nest in holes in banks.[7]

Mr. Hudson, on the other hand, states that near Buenos Ayres, where there are some woods, the *Colaptes campestris* climbs trees and bores into the bark like other Woodpeckers. He says, 'it is sometimes found several miles distant from any trees. This, however, is rare, and it is on such occasions always apparently on its way to some tree in the distance. It here builds its nest in holes in trees.'[8] I have not the least doubt that Mr. Hudson's account is perfectly accurate, and that I have committed an error in stating that this species never climbs trees.[9] But is it not possible that this bird may have somewhat different habits in different districts, and that I may not be quite so inaccurate as Mr. Hudson supposes? I cannot doubt, from what I saw in Banda Oriental, that this species there habitually frequents the open plains, and lives exclusively on the food thus obtained. Still less can I doubt the account given by Azara of its general habits of life, and of its manner of nidification. Finally, I trust that Mr. Hudson is mistaken when he says that any one acquainted with the habits of this bird might be induced to believe that I 'had purposely wrested the truth in order to prove' my theory. He exonerates me from this charge; but I should be loath to think that there are many naturalists who, without any evidence, would accuse a fellow worker of telling a deliberate falsehood to prove his theory.

American Philosophical Society (383)

1 The year is established by the date the letters referred to in the enclosure were read (see n. 2, below).

2 Sclater was secretary of the Zoological Society of London. At the 24 February and 24 March 1870 meetings of the society, letters from William Henry Hudson were read in which Hudson claimed that CD was mistaken in his description of the pampas woodpecker, *Colaptes campestris* (now known as the Campo flicker), as a 'woodpecker which never climbs a tree' (*Origin*, p. 184; see Hudson 1870, p. 112, and also n. 4, below).

3 See Hudson 1870, p. 158.

4 For CD's observations of woodpeckers in South America, see *Zoology* 3: 113–14, and Keynes ed. 2000, p. 156.

5 Félix d'Azara.

6 See Azara 1802–5, 2: 311–15.

7 *Colaptes pitius* is now known as the Chilean flicker. Juan Ignacio (Giovanni Ignazio) Molina described the nesting behaviour of the bird in Molina 1788–95, 1: 262. The passage is scored in CD's copy of the book, which is in the Darwin Library–CUL (see *Marginalia* 1: 595).

8 Hudson 1870, p. 159.

9 CD added a reference to Hudson's account in *Origin* 6th ed., p. 142.

From D. Appleton & Co. to Asa Gray 1 September 1870

Sept 1st, 1870

Statement of Sales of Darwin Origin of Species[1] *to* Sept *1st, 1870 By D. Appleton & Co. for account of* Asa Gray

On hand last account	433	*On hand this date*	521
Printed since	500	*Given away*	
		Sold to date	412
	933		933

Sold 412 *Copies.* $\$2^{50}$ *Rate* 5%	*$* 51.50
Feby 1/70 By Balance	298.99
Balance Sept/70	350.49
Value in Sterling	
Exch @ $110\frac{5}{8}$ Gold $117\frac{1}{2}$	£62.5.2

Document
DAR 159: 87

1 This statement gives sales figures for the US edition of *Origin*. Gray had negotiated the original terms of publication of the book with Appleton & Co. (see *Correspondence* vol. 8, Appendix IV).

From Sofia Vasilyevna Kovalevskaya and V. O. Kovalevsky 1 September 1870

London.
1 September. 1870.

Sir!

I am exceedingly thankfull for your kind offer to procure me some books from the library of the Royal Society;[1] and though I really fear that I am misusing your kindness, but the book is so indispensable to me, that I decide myself to profit by it. We ascertained today, that the Royal Society is opened the whole month of September, and the book for which I should ask you an order is *Jacobi. Fundamenta nova theoriae Functionum ellipticarum.*[2] It is not a rare one and could be easily replaced in the improbable case of some misfortune happening to it.

I hope, that you will allow me in the course of the autumn to express you personally my thanks for your kindness. Remember me kindly to M^rss^. Darwin and the Ladies.[3] I am glad to hear, that your excursion to the country has done your health a great deal of good.[4]

Believe me Sir | Your's truly | Sophie Kowalewsky

P.S. May I ask You Dear Sir if You know something of the dredging expedition of Mr Carpenter, I know he is at Gibraltar with the Porcupine, but will he enter the Meditteranean or confine himself only to the Atlantic,[5] I received to day a letter from my brother asking me where M^r^. Carpenter is; he would not shun a trip from Naples even to Tunis or Algeria if he has a chance o meeting him there.—[6]

Yours truly | W Kovalevsky[7]

DAR 169: 53

1 No letter from CD offering to borrow books from the Royal Society of London on Kovalevskaya's behalf has been found.

2 Jacobi 1829.

3 The reference is to Emma, Henrietta Emma, and Elizabeth Darwin. The Kovalevskys visited the Darwins at some point between 15 August and 1 September 1870 (see letter from V. O. Kovalevsky, 15 August [1870]).

4 CD stayed with William Erasmus Darwin in Southampton from 13 to 26 August (see 'Journal' (Appendix II)).

5 The dredging expedition of HMS *Porcupine* began in the Bay of Biscay in July 1870 and took in parts of the Mediterranean off the north African coast and Sicily, concluding at the end of September 1870. William Benjamin Carpenter was in charge of the scientific operations. See C. W. Thomson 1873, pp. 178–96.

6 The reference is to Alexander Onufrievich Kovalevsky. On his plans for marine research in the Mediterranean, see the letter from V. O. Kovalevsky, 28 February [1870].

7 Vladimir Kovalevsky; the 'W' is the German-style transliteration.

From W. W. Reade 3 September 1870

Conservative Club | S^t^. James' Street
Sep^er^. 3. '70

My dear Sir

I have been back a fortnight & sh^d^. have written to you before, but have been much out of sorts. I was disappointed of a passage up the Niger.[1] The trading steamers would take no passengers suddenly smitten with the brilliant idea of the Phenicians & medieval Portugese &c—& are trying to keep their trade in the dark.[2] Not good policy nowadays.

I have seen several albinoes in Africa but have never detected a blush. In fact one avoids these disagreeable objects.

I have often thought I have observed a kind of brown flush in the negro cheek— A German missionary declared to me he had seen a dark mahogany coloured blush— Bates[3] a better observer than the missionary or myself said the same to me the other day.

I am also almost positive that on that part of the cheek where colour is most abundant with us who have colour there is a glowing brown tint in most negroes especially in children corresponding to it. The neck especially the nape appears more pigmentary than the face. I sh[d] like this to be investigated by an experienced observer. I mistrust my own eye.

I c[d] have answered your inquiry about the chimpanzee a few months earlier.[4] My friend Heddle at S[a] Leone had a tame one—[5] I have not see one since. I think chimpanzees often come to Liverpool, live a short time but die before they can be taken to London. Liverpool is the African port & if you have a correspondent there he w[d] be the best person to look out & observe for you. There is no one on that wretched coast who can be relied on in such matters— Trade & red tape & a low form of Christianity—such are the elements of West African Society.[6]

I hope that you will continue to honour me with further inquiries. I have amassed much new material any of which is at your disposal for any fact gathered by me & used by you will acquire a special value on that account. So my offer is perfectly selfish.

Will you allow me to offer one objection to your belief (if I understand it right) that there is no natural standard of beauty?[7] How is it that the Africans men & women always admire long hair? I can positively assert that they do so. Both sexes wear false hair: & they say on the Gold Coast the Europeans would be good looking if they had better teeth; but their hair is beautifully long.[8] I also know an instance of a young Mandingo[9] having fallen in love with a European lady in S[a] Leone.

Yours very truly | Winwood Reade

DAR 176: 39

CD ANNOTATIONS
1.1 I have ... objects. 2.2] *crossed pencil*
2.1 I have ... objects. 2.2] *double scored pencil*
5.1 I c[d]. ... Society. 5.7] *crossed pencil*
7.4 & they say ... long. 7.5] *triple scored pencil*; 'Length of Hair)' *added pencil*
Top of letter: 'Negros blushing.' *pencil*

1 Reade had just returned from an expedition to Africa. He had hoped to discover the source of the Niger river (see letter from W. W. Reade, 24 April 1870, and Driver 2001, pp. 103–6).

2 Phoenicia was an ancient region, situated along the Mediterranean in what is now Lebanon, Syria, Israel, and Palestine; it was known for its thriving maritime trade, as was Portugal (*EB* (1970)).

3 Henry Walter Bates.

4 See letter to W. W. Reade, 30 June [1870] and n. 8.

5 Reade stayed with Charles Heddle in Sierra Leone (see *Correspondence* vol. 17, letter from W. W. Reade, 28 June [1869]).

6 For more on Reade's views of African society, see Reade 1873.

7 CD had alluded to different standards of beauty in his letter to W. W. Reade, 21 May [1868] (*Correspondence* vol. 16).

8 CD cited Reade's views on the similarity between European and African standards of beauty in *Descent* 2: 350.

9 The Mandingo (also known as Mandinka or Malinke) are an ethnic group in west Africa.

To John Murray 4 September [1870][1]

Down. | Beckenham | *Kent. S.E.*
Sept 4th

My dear Sir

Some members of my family have remarked that the old Origin of Species has a very poor appearance compared with my Domestic Animals; & I think this is due to the quality of the paper; so that I hope you will attend to this point for present work. The smooth Binding of Dom: Animals, also, looks much better than Origin.—[2] The Printers have sent me splendid lot of proofs.—[3]

Yours very sincerely | Ch. Darwin

National Library of Scotland (John Murray Archive) (Ms.42152 f. 211)

1 The year is established by the relationship between this letter and the letter from John Murray, 6 August [1870].

2 The first five editions of *Origin* were bound in textured cloth, whereas *Variation* had smooth covers (see R. B. Freeman 1977, p. 123) and was printed on heavier paper. The sixth edition of *Origin* (published in 1872) differed from previous editions in its printing and binding (see R. B. Freeman 1977, p. 79). *Descent* was bound in smooth covers and printed on heavier paper, like *Variation*.

3 CD refers to the proofs of *Descent* and to the printing firm William Clowes & Sons.

From V. O. Kovalevsky 5 September [1870][1]

31 Bloomsbury Str. Bedf. Sq.
Sept. 5.

Dear Sir

I received to day from the Library of the R. Soc. the book my wife wanted and we are both very thankfull for Your kindness.[2] As the price charged by Murray for the first set of casts was very reasonable I think it would be better simply to order him a set and he will charge for it no doubt about the same price as he did for the "Domesticated Animals".—[3] My brother will be certainly delighted by Your mentioning his investigation in Your book, at all event I'll write him today, perhaps he will have something more to communicate to You.[4] I ascertained that Dr. Carpenter will dredge between Malta and Gibraltar[5]

Your very truly | W. Kovalevsky

DAR 169: 80

1 The year is established by the relationship between this letter and the letter from S. V. Kovalevskaya and V. O. Kovalevsky, 1 September 1870.

2 CD had offered to borrow books from the Royal Society of London's library for Sofia Vasilyevna Kovalevskaya (see letter from S. V. Kovalevskaya and V. O. Kovalevsky, 1 September 1870). The book was Jacobi 1829.

3 Kovalevsky was undertaking the Russian translation of *Descent*. He had previously translated *Variation* (V. O. Kovalevsky trans. 1868–9), obtaining stereotypes of the illustrations from John Murray (see *Correspondence* vol. 15, letter to V. O. Kovalevsky, 2 May [1867] and letter from V. O. Kovalevsky, [after 24? May 1867]).

4 The reference is to Alexander Onufrievich Kovalevsky's work on the embryology of ascidians, which suggested an ancestral link between invertebrates and vertebrates. Vladimir Kovalevsky had sent CD

a paper by his brother on this subject in 1867 (A. O. Kovalevsky 1866; *Correspondence* vol. 15, letter from V. O. Kovalevsky, 15 March 1867). CD cited the paper and described this 'discovery of extraordinary interest' in *Descent*, 1: 205–6.

5 The reference is to William Benjamin Carpenter (see letter from S. V. Kovalevskaya and V. O. Kovalevsky, 1 September 1870 and n. 5).

From John Tyndall 7 September 1870

Folkestone
7th Sept.r 1870

My dear Darwin

I have asked Spottiswood & Co to send you the proof of a lecture which I purpose giving in Liverpool on the 16th[1]

I have had no books beside me in writing it; and I would therefore thank you much if you would glance at that portion of the proof where your name is introduced, and kindly inform me whether it is rightly introduced.[2] Prior to page 31 I do not think your name occurs: so that you will not have much to read.

Would you be good enough to send the proof to me with your remarks, if any, to the Royal Institution some time this week?[3] I am very anxious to get it wholly off my hands.

I hope you are well. When I saw you last I was far from well—indeed I was very ill— this accounts for my unsociability.[4] I thank the gods I am now in a fair way. | Yours, ever | John Tyndall

I have been staying with Bence Jones for the last few days,[5]

That remark of Elie de Beaumont where he requires you to recant prior to being admitted into the Academy is simply intolerable—[6] This spirit has much to do with the present condition of France.[7]

DAR 106: C3–4

1 Tyndall refers to the printing firm of William Spottiswood and to the lecture 'Scientific use of the imagination', which he delivered to the British Association for the Advancement of Science in Liverpool on 16 September 1870 (Tyndall 1870).

2 See Tyndall 1870, pp. 31–2.

3 Tyndall was superintendent of the Royal Institution of Great Britain (*ODNB*).

4 Tyndall may have met CD during one of his visits to London in 1870.

5 CD had sought medical advice from Henry Bence Jones in 1865 (see *Correspondence* vol. 13), and still consulted him on occasion (see letter from H. B. Jones, 2 August 1870).

6 The reference is to remarks by Léonce Elie de Beaumont in the dispute surrounding CD's proposed election to the Paris Académie des Sciences: 'M. Darwin a fait de bons travaux qu'il a gâtés par des idées dangereuses et sans fondements: il faut attendre pour l'élire qu'il ait renoncé à ces idées' ('Mr. Darwin has done good work that is tainted by dangerous and unfounded ideas: it is necessary to wait until he has renounced these ideas before electing him'; *Revue des Cours Scientifiques*, 23 July 1870, p. 529. See also the letter to Armand de Quatrefages, 23 August [1870] and nn. 5 and 6.

7 Tyndall alludes to the revolution of 1870 which brought an end to the Second Empire in France. A provisional government was constituted by republican deputies in Paris on 4 September after the defeat of the French army by German forces (see *EB* (1970)).

To John Tyndall 8 September 1870

Down. | Beckenham | *Kent. S.E.*
Sep 8th. 1870

My dear Tyndall

Your whole discourse strikes me as grand & most interesting.[1] What you say about Pangenesis is quite correct, & your expression, "not only is the organism as a whole wrapt up" &c is most happy.[2] What you say about me, coming as it does from you, has pleased me extremely, so much that there must be clouds of vanity in my mind to bring out such pleasant sensations as your remarks passed through them— you are a rash man to say a word for Pangenesis, for it has hardly a friend amongst naturalists, yet after long pondering (how true your remarks are on pondering) I feel a deep conviction that Pangenesis will some day be generally accepted. I have been particularly delighted by your "*as if*" argument. These words have lately been fired at me, & if I had been forced to answer, I should not have known what to say, now I cd answer by a round of your artillery.[3] I have ventured to mark (p. 12. & 40) two sentences which seem to me not so perfectly lucid as all the rest. I may also mention that in the discussion on "the particles of our sky" I read 2 or 3 pages thinking that you referred to the atoms of the air: wd it not be well for blunderers to shew in the early part, that you refer to foreign particles?[4]

With hearty admiration & thanks, yours very truly | Ch. Darwin

P.S. I return the proofs by this post.

When in Brazil I was much struck with the frequent blue haze & I quote (p. 32 of my travels) Humboldt's remark on "the thin vapour which, without changing the transparency of the air, renders its tints more harmonious & softens its effects".[5]

It seemed to me that this phenomenon was much more strongly marked in the tropics than in our temperate regions—

On one day when the beautiful pale blue haze had come on very strong, the difference between the dew point & temperature by Daniel's hygrometer[6] had increased from 7°.5. to 17°—

LS(A)
The Michael Faraday Museum at the Royal Institution of Great Britain, London, reference RI MS JT/2/10/458, spine title: Journal VIIIA 1858–71

1 CD refers to Tyndall 1870. See letter from John Tyndall, 7 September 1870.

2 CD's hypothesis of pangenesis was presented in *Variation* 2: 357–404. Describing pangenesis, Tyndall had written: 'Not only is the organism as a whole wrapped up in the germ, but every organ of the organism has there its special seed' (Tyndall 1870, p. 32).

3 Tyndall used the expression 'as if' to describe the basis on which certain theoretical conceptions were believed. Thus the phenomena of light occur as if the medium of the ether existed. He argued that many assumptions of everyday life were based on the same 'as if' principle (Tyndall 1870, p. 10). For Tyndall's discussion of pondering in relation to scientific discovery, see Tyndall 1867, p. 655, and *Correspondence* vol. 15, letter to J. D. Hooker, 2 September [1867] and n. 5.

4 The phrase 'particles of our sky' appears in Tyndall 1870, p. 26.

5 The quotation from Alexander von Humboldt appears on page 32 of *Journal of researches* 2d ed.

6 CD refers to the dew-point hygrometer designed by John Frederic Daniell (*ODNB*).

From V. O. Kovalevsky 15 September 1870

London
15 Sept. 1870.

Dear Sir!

I have just received a letter from my brother[1] in which he says that he set himself to verify Your old discovery about the supplementary males, but unhappily in whole Naples there is not a single copy of Your monography of the Cirripedia, so I make use of Your kind proposition to lend me for a short time Your own copy which I shall immediately forward to Naples, as soon as my brother has done with the book he schall send it back to You.[2]

At the time when he send me the letter, he was able to see only some of the Scalpellum preserved in Spirit in the museum, but has a good hope of getting a plenty supply of living ones[3]

Your very truly | W. Kowalevsky

31 Bloomsb. Str. | Bedf. Square.

DAR 169: 81

[1] Kovalevsky's brother was Alexander Onufrievich Kovalevsky.

[2] For CD's discussion of complemental males in the barnacle genera *Scalpellum* and *Ibla* see *Living Cirripedia* (1851), pp. 281–93 (see also *Correspondence* vol. 4, Appendix II, and Newman 1993, pp. 377–82, for a discussion of CD's work on complemental males). On CD's concern to have his observations verified, see *Correspondence* vol. 13, letter to Fritz Müller, 10 August [1865].

[3] Kovalevsky refers to the Zoology Museum of Naples.

From George Cupples 17 September 1870

The Cottage, | Guard Bridge | Fifeshire. N.B.
Septr. 17/70—

My Dear Mr Darwin,

I write a single line to say that your Deerhound puppy will be ready at the end of this month. He is the best of the litter—and all are capital specimens.[1]

I am sorry he is not *grey*— he is fawn or cream colour, with black muzzle and ears, similar to one of mine in Kent (Longfield Court, Dartford),[2] which has turned out a splendid animal. The mother of yours is a fine bitch—the father is descended from Sir Walter Scott's celebrated "Maida", and is son of the 1st. prize dog at two successive National Shows (Birmingham.)[3]

He would come in a hamper by rail—and I would send full directions beforehand for management &c.

I hope you are tolerably well and in full ability to go on, also to take interest in what is astir scientifically & otherwise

Ever truly yours George Cupples

To | Charles Darwin, Esqr.

DAR 161: 293

[1] Cupples had previously written about the deer-hound puppy that he promised to breed and rear for CD in his letter of 29 April 1870.
[2] Longfield is a village about three miles south-east of Dartford, in Kent.
[3] On the Birmingham dog shows, see F. Jackson 1990, pp. 41–8. Scott's dog, Maida, was etched by the famous animal painter Edward Landseer (Royal Academy of Arts, London).

From V. O. Kovalevsky 20 September 1870

Dear Sir!

I received yesterday the books You were so kind as to present to my brother;[1] but really I am very sorry as to have given You an occasion to put Yourself in the inconvenient position of buying Your own works for presents.— I must be very careful in the numerous applications I generally adress to Yous, because Your awewhelming kindness overturn all my calculations. I send the books by Post to day to Naples and shall confort myself with the hope that perhaps a successfull revision and confirmation of Your discovery about the males of the Scalpellum will be some sort of repayement for Your kindness.[2]

Your very truly | W. Kovalevsky

20 Sept. 1870.

DAR 169: 82

[1] Kovalevsky had asked to borrow a copy of CD's work on Cirripedia for his brother, Alexander Onufrievich Kovalevsky (see letter from V. O. Kovalevsky, 15 September 1870). CD may have sent the two volumes, *Living Cirripedia* (1851) and (1854).
[2] See letter from V. O. Kovalevsky, 15 September 1870 and n. 2.

From J. D. Hooker 24 September 1870

Royal Gardens Kew
Sept 24/70

Dear old Darwin

I have absolutely nothing whatever to say to you, & so feel quite justified in inflicting a letter on you— If George be still with you please tell him that I hunted up & down the train for him at Willesden, but in vain, & that I therefore sent on his Pall Mall by post:[1]

The Brit. Assoc. meeting was a good one.— very good I suppose, though there was I think a terrible deficiency of good sectional matter—[2] Huxley's address was admirable, but a few such would do up the popularity of President's addresses; for the general public could not follow the subject, & were profoundly ignorant of what he was *driving at*—[3] After the "Lay Sermons", all the Liverpudlians expected brilliant sallies, & spiced heresy—& were woefully disappointed.[4] In a scientific point of view I thought it quite excellent & the proper sort of thing for the occasion. Tyndall's was a wonderful outburst of cracking & sparkling eloquence as delivered, but I was woefully disappointed in the reading of it:[5] it reads ornate, frothy & all but flashy, with a soupcon of *bosh* here & there— I cannot bear to think this of any thing dear old Tyndall has said or sung, & if you think better of it so will I, & to any amount— I will sell my soul for him in such a matter.

Rolleston's opening address was a wonderful affair— full of science, classics, metaphysics, logic, the Fathers, mathematical phraseology (if not method), & what not— delivered in a somewhat jerky & sententious style.[6] A local paper stigmatized it as "Rococo"— a capital hit, if I remember what Rococo is as applied to art—(a jumble of periods & styles each *pronounced* unmistakeably; & not over well matched)— I see that in an after-dinner speech he professed not to have understood the meaning of the term, from which I guess that he feels what he can't understand & pretty sharply too— He made a capital President.— Old Murchison surrounded himself with fireworks: he will never die— he will go up like Elijah/Elisha? in a fiery chariot; & heaven will open to receive him;—whence he will bless the earth & especially the nobles thereof—and we will canonize him.[7] I did not wait over Saturday & so I missed Lubbock's Lecture—which read well.[8] I saw many old faces, & no new ones, but Van Beneden's, a very pleasant one.[9] Lyell[10] looked wondrous grim in a nascent beard, which did not suit him; in fact he looked rather disreputable in it— because of it's shortness & raggedness, & his feebleness & failing sight; which I see with real sorrow— for I never can bear to think of losing Lyell. I do so intensely love & esteem him.— how few men have been so always the same for so many years to us!

I was delighted with the Liverpool Museum, it is a lesson for our Brit. Mus.—[11] the typical series of articulate & lower orders with appended popular descriptions of the tribes were admirable in all respects.

I have polished off an immense lot of small work within the last 3 months, & am now at "Genera plantarum",[12] & hope to be thinking of some general works soon on distribution or Classification— I feel marvellously lightened, more so than since I took the Direction here.[13] God knows how short such feelings of relief usually are, & I do not dare to think how soon they may end. Willy is doing nicely at Mr La Touches,—[14] My wife[15] is at St. Albans, where she has a nice House. I am alone here, this house being under repair, I wish your William or George[16] would come & see me here, & take a run to St. Albans with me any day.

I suppose you have nearly finished your book, I am sorry that you dropped the simple title of "Origin of Man"—[17] I forbear to talk of the war or the loss of the Captain— what harrowing episodes in a life-time, not to talk of a month's time.[18]

Now I shall shut up— when I survey the calligraphy of this epistle I am lost in wonder & admiration—& am tempted to ask you to return it—& to enclose 1d. to make sure you do!

Ever Yrs affect | J D Hooker

DAR 103: 57–9

[1] Hooker refers to George Howard Darwin and the *Pall Mall Gazette.* Willesden Junction was a station in north-west London run by London & North Western Railway.

2 The British Association for the Advancement of Science met in Liverpool from 14 to 21 September 1870 (see *Report of the 40th meeting of the British Association for the Advancement of Science, held at Liverpool*, p. lxxii).

3 Thomas Henry Huxley gave the presidential address on 14 September 1870 (T. H. Huxley 1870c). The topic of Huxley's address was spontaneous generation, and included a summary of recent experiments by Louis Pasteur and John Tyndall that demonstrated the existence of germs. Huxley was generally critical of the existing evidence for spontaneous generation; however he added that the sciences of organic matter were 'yet in their infancy' and that the conditions under which matter becomes living might some day 'be artificially brought together' (T. H. Huxley 1870c, p. lxxxiii). See letters to J. D. Hooker, 8 July [1870] and n. 9, and 12 July [1870].

4 The reference is to *Lay sermons, addresses, and reviews* (T. H. Huxley 1870a). It included one of Huxley's most controversial essays, 'On the physical basis of life' (T. H. Huxley 1869a).

5 Hooker refers to John Tyndall's address, 'On the scientific use of the imagination' (Tyndall 1870). CD had read proofs of the address (see letter to John Tyndall, 8 September 1870).

6 George Rolleston was president of the biology section of the British Association (see Rolleston 1870a).

7 Roderick Impey Murchison was president of the geography and ethnology section (see Murchison 1870). Hooker refers to the Hebrew prophet Elijah and to the story of his departure in II Kings 2: 8.

8 John Lubbock delivered a 'lecture to the operative classes' on 'Savages' (*Report of the 40th meeting of the British Association for the Advancement of Science, held at Liverpool*, p. xxxvii). A summary and lengthy extract from the lecture appeared in *The Times*, 19 September 1870, p. 12; the title was given as 'The social and religious condition of the lower races of man'.

9 Edouard van Beneden became a corresponding member of the British Association in 1870 (*Report of the 40th meeting of the British Association for the Advancement of Science, held at Liverpool*, p. 80).

10 Charles Lyell.

11 British Museum.

12 Bentham and Hooker 1862–83.

13 Hooker was appointed director of the Royal Botanic Garden, Kew, in 1865 (*ODNB*).

14 Hooker's son, William Jackson Hooker, was being tutored by James Digues La Touche (see letter from J. D. Hooker, [31 May 1870]).

15 Frances Harriet Hooker.

16 William Erasmus Darwin and George Howard Darwin.

17 Hooker refers to *Descent.* On the title of the book, see the letter to John Murray, [after 1 July 1870] and n. 2.

18 The Franco-Prussian War had broken out in July 1870 (see Wawro 2003). The *Captain*, a turret-ship with a crew of 500, was lost at sea on 7 September 1870; there were only 20 survivors (see *Annual Register* 1870).

From Alexander Onufrievich Kovalevsky[1] 25 September 1870

Naples.
25 Septem 70.

Monsieur

Mon frère me fait savoir, que vous desiréz inserer dans l'ouvrage que vous publiez une citation, que mes nouvelles etudes sur le developpement des Ascidres constatent mes premieres assertions que les larves des Ascidres simples sont des animaux vertebrés.[2]

Les resultats de mes nouvelles rescherches sur le developpement des Ascidres sont maintenant sous presse, dans l'Archiv fur Mikroscopische Anatomie de *Max Schultze.*[3] Dans cet article, je constate, a l'aide d'une serie de nouvelles observations, que les larves des Ascidres sont de vraies vertebrés, que leurs systeme nerveux se forme a laide d'un sillon dorsal qui se ferme dans tube—système nerveux des Ascidres; je constate encore que la chorde dorsale des larves des Acidres se developpe de la même maniere que chez les vertebrés; Je donne des dessins de la coupe transversale de la larve (fig.) qui presente le systeme nerveux $^{(n)}$, la chorde $^{(ch)}$, l'intestin $^{(d)}$ et les cellules musculaires $^{(m)}$

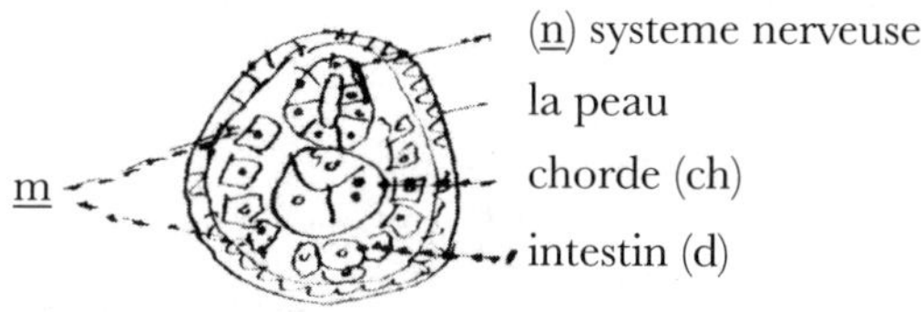

J'espère que j'aurais le plaisir de vous faire parvenir un exemplaire de mes nouvelles études sur le developpement des Ascidres dans le courant du moi Novembre ou Decembre.—

Pendant l'anné passé (1869) j'ai étudié l'embryologie des annelides et je trouvé que la chorde dorsale existe chez beaucoup des Annelides.— Peût–être vous vous rapeléz Monsieur, du grande tube nerveux(?)—*riesiege Rohren Nervenfasern* que Leydig a decrit chez les Lumbricus;[4] *Claparéde* (Zeitschrift fur Wissenschaftliche Zoologie Bd. XIX p. 588, 89–90) a demontrés que cet enorme quasi Nerv se trouve à lexterieur de la Nevrilemme.[5] Mes etudes sur l'embryologie de ces *tubes* m'ont demontré qu'ils se forment de la même maniere que la chorde des vertebrés. Je trouvai chez les embryons des Lumbricidae trois feuilles embryonaire—correspondant aux trois feuilles embryonnaire des vertebrés; du feuille exterieure se forme le systeme nerveux et l'epitehlium exterieur, du feuille interieur (Darmdrüsenblatt)[6] l'epithelium du canal instestinal, du feuille mediane les muscles, la nevrileme et *les chordes ou "riesiege Röhrenfasern"* de Claparède.

Mon Etude sur l'embryologie de Lombric est sous presse dans le Memoires de l'Acad. de St. Petersbourg.[7]

Mon frère m'ecrit qu'il vous serait agreable si j'etudie l'organisation du mal des cirripèdes (Scalpellum) ainsi que l'histoire du developpement et de formation du mal et de ses organes genitaux en comparaison avec la femelle.[8] Je ferais tout mon possible pour ressoudre ces ?? d'autant plus que je rencontrait un fait pareile chez des autre forme, chez la Bonelia viridis.[9] Pressisement, je trouvé, que le mal de Bonelia est une petite Planaire, ou mieux Turbellaire, d'un millimetre.[10] Les males se trouvent dans l'uterus de la femelle, on trouve dans chaque Bonelia femelle 4 ou 5 males— Dans l'organisation du male on trouve plusieur cas de ressemblanse avec la femelle; l'organisation des organe genitaux du male et de la femelle se ressemble beaucoup.[11] Il est bien possible que dans la Bonelia nous trouverons un exemple, que seulement un sexe (femelle) changeat graduellement sa forme primitive, tendit que l'autre sexe (le mal) resta dans sa forme primitive de Turbellaire. J'espère de pouvoir étudier le developpement de la Bonelia, cela devra resoudre la question.

Je resterais a Naples encore toute une anné et je serais extrément heureux si je pourais vous être utile ou en vous envoyant quelque forme de la faune du golfe de Naples, ou en faisant quelques études dont les resultats vous interesserais exclusivement

En vous envoyant cette lettre, Monsieur, je profite de l'ocasion pour exprimér tout mon devouement aux idées dont vous avez enrichit la science et de vous priéer d'agréer ma salutation le plus distingué | Prof. A Kowalevsky

DAR 169: 52

1 For a translation of this letter, see Appendix I.

2 Alexander's brother was Vladimir Onufrievich Kovalevsky. See the letter from V. O. Kovalevsky, 5 September [1870] and n. 4.

3 Kovalevsky refers to A. O. Kovalevsky 1871 and to Max Johann Sigismund Schultze. In *Descent* 1: 205, CD referred to Kovalevsky's earlier paper on the embryology of ascidians (A. O. Kovalevsky 1866b) and to the fact that Kovalevsky had done further research on the subject, but did not cite the later article. No copy of A. O. Kovalevsky 1871 has been found in the Darwin Archive.

4 'Riesiege Rohren Nervenfasern': giant tube of nerve fibres (German; a reference to the median giant fibre). Franz von Leydig described this organ in *Lumbricus* in Leydig 1864, pp. 154–5.

5 The reference is to Edouard Claparède and Claparède 1869. Claparède noted the similarity between the nerve sheath of the earthworm and the myelin sheath of vertebrates (Claparède 1869, p. 590).

6 'Darmdrüsenblatt': gut-gland-layer (German; a reference to the endoderm).

7 Kovalevsky refers to his paper on the embryology of worms and arthropods (A. O. Kovalevsky 1869). It appeared in the 1871 volume of *Mémoires de l'Académie Impériale des Sciences de St.-Pétersbourg.*

8 CD's letter to Vladimir Onufrievich Kovalevsky has not been found, but see the letter from V. O. Kovalevsky, 20 September 1870.

9 *Bonelia viridis* is an incorrect subsequent spelling of *Bonellia viridis*, the green spoonworm.

10 *Bonellia viridis* is no longer placed within the family Planariidae of the class Turbellaria. It is in a separate phylum, Echiura.

11 *Bonellia viridis* is sexually undifferentiated in larval stages, and differentiation is determined by environmental factors alone. For more on the development of sexes in this species, see Berec *et al.* 2005.

From Edward Livingston Youmans 25 September 1870

New York
Sep. 25 1870

Mr Charles Darwin
Sir

I have been long interested in pushing into notice in this country the books on Biological literature which appear in England, and prompted the Appletons to reprint the last edition of the "Origin of Species"[1] I believe they have an arrangement with you to republish your forthcoming book[2] and I now write in their behalf to call your attention to one or two practical business points respecting it.

As the stereotype plates are to be cast in London I desire to say first that it is very important for success here that the page should not be large. Macmillan's American edition of Wallace's late book on Natural Selection is an excellent model.[3] Galton's

Hereditary Genius is too large.[4] Huxley's Lay Sermons is too large and so for the edition we are now printing Mr Macmillan extracted the leads and made the page smaller.[5] Should the page of your new book be large I would urge a similar course with it. Again, it is very important for us that the book should come out simultaneously in the two countries.[6] I will not trouble you with details but it works very badly when the book is just back here. I could show you how Macmillans neglect in this matter cost Mr Galton the success of his book on this side. Mr Longman treated us in the same way in regard to Mr Mills book on Women and although we were bound to pay Mr Mill a "copyright" on the sales the plates came so late that a Philadelphia House got the book out in advance of us, and undersold us hence they had nothing to pay to the author.[7] Of course the English Publisher doesnt care for this and is interested to delay the American reprint that he may send a few copies to the American market but it is the authors interest to prevent this. We should also be greatly helped in handling a new book if we could have the sheets sent us as just as they go through the English press. There is an art in handling a new book so as to give it the fullest and fairest chance with the public, but the publisher has no chance if kept in ignorance of the character and contents of the work.[8]

Very respectfully | E. L. Youmans

DAR 183: 2

1 *Origin* 2d US ed. was published by the New York firm of D. Appleton & Co..

2 On the arrangement with D. Appleton & Co. to publish *Descent,* see the letter from Asa Gray, 14 February 1870 and n. 5.

3 The American edition of Alfred Russel Wallace's *Contributions to the theory of natural selection* (Wallace 1870a) was published by Macmillan and Co. of New York.

4 The American edition of Francis Galton's *Hereditary genius: an inquiry into its laws and consequences* (Galton 1869) was published by D. Appleton & Co. in 1870.

5 The American edition of Thomas Henry Huxley's *Lay sermons, addresses, and reviews* (Huxley 1870a) was published by D. Appleton & Co. in 1870. The reference is to Alexander Macmillan.

6 *Descent* was published in England on 24 February 1871. The page length of *Descent* was 185mm, whereas *Variation* had been 215mm.

7 John Stuart Mill's *The subjection of women* was published by Longmans, Green, Reader and Dyer (Mill 1869). An American edition was published in Philadelphia by Lippincott in 1869; D. Appleton & Co. also published an edition in 1870.

8 In 1871, Youmans founded the International Scientific Series, which provided for publication of books in North America and Europe, with payment to the authors from sales in all countries (see Howsam 2000).

From E. L. Youmans 25 September 1870

New York
Sep 25 1870

Mr C. Darwin
Sir

I have just written you a letter in the interest of D. Appleton & Co Publishers of this city and being not sure of your country address I directed it to the Athenæum—club of which I believe you are a member and to increase the chance of your getting it soon I send this note to what may be your country place[1]

Very respectfully | E. L. Youmans

—As to *my* relation to this matter I may refer you to Huxley Spencer Tyndall Bain and Mill in &c &c—publication of whose works I have been interested though not pecuniarily.[2]

E.L.Y.

DAR 183: 3

1 See first letter from E. L. Youmans, 25 September 1870. CD had been a member of the Athenaeum Club since 1838 (see *Correspondence* vol. 2).

2 The references are to Thomas Henry Huxley, Herbert Spencer, John Tyndall, Alexander Bain, and John Stuart Mill.

To John Murray 26 September 1870

Down. | Beckenham | *Kent. S.E.*
Sept. 26th 1870

My dear Sir,

I must trouble you about some business. As the sheets are printed off I want three clean copies sent me; one for Russian translation, a second for a French translation, which Col. Moulinié wishes to prepare ready for the future; and a third for Messrs Appleton in America.[1] Hereafter I shall want a fourth set for Germany.[2] Will you have the kindness, to order the three sets to be sent me as they are **successively** printed off.

There are 76 woodcuts in my book, 14 of these are from Brehm, & copies cannot be supplied for foreign editions.[3] I have told my correspondents that you would supply stereotypes of the remain 62, at a reasonable price a little above the cost of casting.[4] Please determine & inform me as soon as you conveniently can *[*what the cost will*]* be. Will you be so good as to give an order to Messrs Clowes[5] to send these stereotypes as I may hereafter direct, when I will indicate the 14 of which copies must not be sent. Mr Cooke seemed to think that I had made a mistake in promising stereotypes for America; I regret this much but did not anticipate any difficulty, from the experience of my last book, & I have now **pledged** my word.[6]

I paid Jan 1. 1869 £10.10.9—to Messrs Cassell & Co, for Brehm's beautiful stereotypes, & this sum you can repay me whenever convenient; I enclose the account. (N.B. I shd like if no objection the account returned, as it gives list of Pages & volume of Brehm's Book)[7]

It will be soon time to think about the index, which I have not the heart to make. It will be long. Mr Dallas (Sec. Geolog. Socy. Somerset House) would be excellent for the purpose, as nothing could be better than his last.[8] Will you see him & enter into an agreement; but pray come to a special understanding that he has no other work on hand, so that he will do the clean sheets, which should be sent to him as printed off. The last time he caused nearly a month's delay. Civilly make him understand that we will put the work at once in someone else's hands if there is delay.[9]

I hope to complete the book by the end of November. You will soon have to decide the number of copies to be printed off, about which I should like to hear.

My dear Sir | Yours very sincerely | Ch. Darwin

P.S I enclose mem: about Index.—

Index for Mr Darwin's Book

To be made on the same plan & principle as that of the "Variation of Domestic Animals". Namely the names of all persons and Animals mentioned in text & *foot notes* to be given. Also a fair number of references & cross references to subjects.—

LS(A)
National Library of Scotland (John Murray Archive) (Ms. 42152 ff. 214–17)

1 CD refers to the sheets of *Descent*. A Russian translation was evidently started by Vladimir Onufrievich Kovalevsky, but never completed (see letter from V. O. Kovalevsky, 12 December [1870], and *Correspondence* vol. 19, letter from V. O. Kovalevsky, 14 March 1871). Three different Russian editions were published in 1871 and 1872 (see R. B. Freeman 1977). The French translation was by Jean Jacques Moulinié (Moulinié trans. 1872). The American edition was published by D. Appleton & Co (*Origin* 2d US ed.; see the two letters from E. L. Youmans, 25 September 1870).

2 The German edition of *Descent* was Carus trans. 1871.

3 The reference is to Brehm *et al.* 1864–9. For a list of the plates from Brehm used in *Descent*, see *Correspondence* vol. 16, letter from Cassell, Petter & Galpin, 31 December 1868 and n. 1.

4 On CD's arrangement regarding stereotypes of the illustrations for *Descent*, see *Correspondence* vol. 17, letter from Charles Layton, 23 December 1869. No other correspondence on this matter has been found.

5 William Clowes & Sons, printers to John Murray.

6 CD evidently refers to objections made by Robert Francis Cooke, John Murray's publishing partner, regarding the stereotypes of illustrations for the US edition of *Descent*.

7 CD refers to the letter from Cassell, Petter & Galpin, 31 December 1868 (*Correspondence* vol. 16).

8 William Sweetland Dallas had prepared the index for *Variation*. See *Correspondence* vols. 15 and 16.

9 Dallas's index had taken longer than expected, causing delay to the publication of *Variation* (see *Correspondence* vol. 16, p. xx).

To J. D. Hooker 27 September [1870][1]

Down. | Beckenham, | *Kent. S.E.*
Sept. 27th.

My dear Hooker

Yours is a splendid letter, & I was very curious to hear something about the Liverpool meeting, which I much wished to be successful for Huxley's sake. I am surprised that you think his Address w^d. not have been clear to the public: it seemed to me as clear as water. The general line of his argument might have been answered by the case of spontaneous combustion; tens of thousands of cases of things having been seen to be set on fire w^d be no true argument against any one who maintained that flames sometimes spontaneously burst forth.—[2] I am delighted at the apotheosis of Sir Roderick: I can fancy what neat & appropriate speeches he would make to each nobleman as he entered the gates of Heaven.[3] You ask what I think about Tyndall's Lecture: it seemed to me grand & very interesting, though I could not from ignorance quite follow some parts, & I longed to tell him how immensely it w^d. have been improved, if all the first part had been made very much less egotistical.[4] George[5] independently arrived at the same conclusion & liked all the latter part *extremely*. He thought first part not only egotistical, but rather clap-trap. How well Tyndall puts the "as if" manner of philosophising & shows that it is justifiable; some of those confounded Frenchmen have lately been pitching in me for using this form of proof or argument.—[6]

I have just read Rollestons address in Nature: his style is quite unparalleled![7] I see he quotes you about seed, so yesterday I went & observed more carefully the case given in enclosed paper which you might perhaps like to read & burn.—[8] How true & good what you say about Lyell:[9] he is always the same: Dohrn was here yesterday, & was remarking that no one stood higher in the public estimation of Germany than Lyell.[10]

I am truly & profoundly glad that you are thinking of some general work on Geograph. Distrib. or so forth:[11] I hope to God that your incessant occupations may not interrupt this intention.— As for my book,[12] I shall not have done the accursed proofs till end of November: good Lord what a muddled head I have got on my wretched old shoulders.

My Boys are very much obliged for your invitation for a Sunday; but they cannot accept it at present.—[13]

I wish I could boast of my calligraphy as you do with justice of yours.—

Ever dear old Friend | Yours affec^y. | C. Darwin

DAR 94: 181–3

1 The year is established by the relationship between this letter and the letter from J. D. Hooker, 24 September 1870.

2 See letter from J. D. Hooker, 24 September 1870 and n. 3. The annual meeting of the British Association for the Advancement of Science was held at Liverpool from 14 to 21 September, with Thomas Henry Huxley as president (*Report of the 40th meeting of the British Association for the Advancement of Science*, p. lxxii). Huxley's address (T. H. Huxley 1870c), on spontaneous generation, was published in *Nature* 2 (1870): 400–6.

3 CD refers to Roderick Impey Murchison. See letter from J. D. Hooker, 24 September 1870 and n. 7.

4 The reference is to John Tyndall's address, 'On the scientific use of the imagination' (Tyndall 1870). See letter from J. D. Hooker, 24 September 1870 and n. 5.

5 George Howard Darwin.

6 In a debate over his proposed election to the Paris Académie des Sciences, CD had been criticised for being a scientific amateur and for advancing conclusions that were not demonstrable. See *Revue des Cours Scientifiques*, 23 July 1870, p. 529, and the summary of the French debate in *Nature* 2 (1870): 298 and 309. See also letter from Armand de Quatrefages, 18 July 1870 and n. 6, and letter to John Tyndall, 8 September 1870 and n. 3.

7 CD refers to George Rolleston and to Rolleston 1870. The first portion of Rolleston's address was published in *Nature* 2 (1870): 423–8. See letter from J. D. Hooker, 24 September 1870 and n. 6.

[8] Hooker had tried to discover whether seed from a plant uncovered during the excavation of a lake at the Royal Botanic Gardens, Kew, was present in the surrounding soil. The case was mentioned in connection with the subject of spontaneous generation in Rolleston's address (see Rolleston 1870, p. 98). The enclosure has not been found.

[9] CD refers to Charles Lyell. See letter from J. D. Hooker, 24 September 1870.

[10] CD refers to Anton Dohrn. Dohrn's reminiscence of the visit is given in Heuss 1991, pp. 108–9.

[11] Hooker had been planning to write a general work on the geographical distribution of plants for some years (see, for example, *Correspondence* vol. 14, letter from J. D. Hooker, 4 December 1866 and n. 9).

[12] *Descent.*

[13] Hooker had invited William Erasmus Darwin and George Howard Darwin to visit Kew. See letter from J. D. Hooker, 24 September 1870.

From John Murray 28 September [1870][1]

50, Albemarle St. | *W.*
Sept 28—

Dear Mr Darwin

I will cause 4 sets of fair sheets of your new book to be sent to you when printed off[2]

Clowes has orders to attend to your directions about clichés from the woodcuts.[3] If I understand you aright you have promised Appleton fair sheets & clichés of cuts—(not stereotypes of the *text* of the book)[4] In this I see no great harm

I return Cassells Bill, with a cheque for the amount £10"10"9.[5] I have written to Dallas about Index[6]

You have produced a book wch will caused men to prick up what little has been left them of *ears*—& to elevate their eye brows However it may be eventually judged it cannot fail, I think to be much read—& I think I may venture on an Edition of 2500—copies—

It is with a view to remove any impediments to its general perusal that I wd call your attention to the passage respecting the proportion of advances made by the two Sex in Animals.[7] I wd suggest that it might be toned down—as well as any other sentences liable to the imputation of indelicacy if there be any. I cannot help thinking that in the passages above referred to you scarcely do justice to the females witness the advances of the she cat & mare. I know not whether further on in the book you refer to Parrots & the power not only of speaking, but of applying words to circumstances—showing that they understand the meaning of words—[8] I have a Parrot of my acquaintance whose accomplishments wd amuse if they did not astonish you

I remain My Dear Sir | Yours very faithfully | John Murray

Ch. Darwin Esqr.

DAR 171: 378

[1] The year is established by the relationship between this letter and the letter to John Murray, 26 September 1870.

[2] See letter to John Murray, 26 September 1870 and nn. 1 and 2.

3 The reference is to William Clowes & Sons. See letter to John Murray, 26 September 1870.
4 The reference is to D. Appleton & Co. See letter to John Murray, 26 September 1870 and n. 6.
5 Murray refers to the publishing firm Cassell, Petter & Galpin. See letter to John Murray, 26 September 1870 and nn. 3 and 7.
6 Murray refers to William Sweetland Dallas. See letter to John Murray, 26 September 1870.
7 Murray may refer to the discussion, 'On preference or choice in pairing, as shewn by either sex of quadrupeds' in *Descent* 2: 268–73. See the letter from John Murray, 10 October [1870] and n. 3.
8 CD briefly remarked on the imitative power of parrots (*Descent* 1: 44, 236); but he maintained that only humans had a large power of 'connecting definite sounds with definite ideas' (*ibid.*, p. 54).

To John Murray 29 September [1870][1]

Down. | Beckenham | *Kent. S.E.*
Sept. 29th.

My dear Sir

Many thanks for all your assistance & for the cheque for 10£"10s·9d[2]

I did **not** promise Mess. Appleton stereotypes of text, only of cuts.—[3] If you could manage to let me hear (& you will have hereafter to determine) what to charge for the 62 cliches, I shd like to hear soon & inform my correspondents, so that there may no disappointment hereafter.—

I shd. be particularly glad to hear what passage you thought coarse: I cannot remember any except a *quotation* from Hunter about the female requiring to be courted "to give her desires" or some such words.[4] I somehow fancied that a quotation rendered the sentence less coarse; I felt it was so, & I hope it is almost the only one. How to alter it I do not yet see; but please, if you can, inform me which sentence it is that you object to

My dear Sir | Your's very faithfully | Ch. Darwin

To | J Murray Esqre

National Library of Scotland (John Murray Archive) (Ms. 42152 ff. 212–13)

1 The year is established by the relationship between this letter and the letter from John Murray, 28 September [1870].
2 See letter from John Murray, 28 September [1870] and n. 5.
3 See letter from John Murray, 28 September [1870] and n. 4.
4 See letter from John Murray, 28 September [1870] and n. 7. CD quotes John Hunter's observation that the female generally 'requires to be courted' in *Descent* 1: 273; see also *Descent* 2: 295–6.

To Hermann Settegast 29 September 1870

Down. | Beckenham | *Kent. S.E.*
Sep 29. 1870

Dear Sir

I am very much obliged for your kind letter & present of your beautiful volume. Your work is not new to me, for I heard it so highly spoken of that I procured a copy of the first edition.[1] It was a great gratification to me to find a man who had long studied with a philosophical spirit our domesticated animals, & who was highly competent to judge, agreeing to a large extent with my views. I regretted much that I had not known your work when I published my last volumes.[2]

I am surprized & pleased to hear that science is not quite forgotten under the present exciting state of affairs. Every one whom I know in England is an enthusiastic wisher for the full & complete success of Germany—[3]

With my best thanks & sincere respect I remain | dear Sir | yours faithfully | ⟨Charles Darwin⟩[4]

P.S. I will give one of my two copies of your work to some public scientific library in London

Staatsbibliothek zu Berlin – Preußischer Kulturbesitz (Autogr. I/101, Bl. 1/2)

1 The reference is to *Die Thierzucht* (Settegast 1868). CD's lightly annotated copy is in the Darwin Library–Down (see *Marginalia* 1: 754). CD had ordered a copy in 1868 (see *Correspondence* vol. 16, letter from Williams & Norgate, 2 June 1868). A second edition was published in 1869.

2 CD refers to *Variation.* He cited Settegast 1868 in *Variation* 2d ed., 2: 331 n. 41.

3 CD alludes to the Franco-Prussian war.

4 The signature has been excised.

To [Edward William Blore] [October 1870 or later][1]

Dear Sir

I hope that you will excuse the l. which I take in troubling you.— My son H.[2] has had bad health for several years, & partly in cause of this is extremely backward in Classics & many other subjects. He is now much better but Dr Bence Jones[3] not very long ago said that he must not work hard. My son has a strong taste & some capacity for mathematics. I have /therefore much/ fear that he will not be able to pass his little-go[4] & have advised him to get a private tutor for this sole object; but as he has a strong taste & some cap. for math. I have allowed him to attend Mr Routh's l but I doubt whether he will he will have **strength** or time for both Private tutor & the College Lectures.[5] Under these peculiar circumstances w^d^ it be possible for you to excuse him at present attendance at College Lectures.— I hope that you will forgive me for calling your attention to my son's case & that you will not think of answering this note; but inform my son what you may think right to do.— Dear Sir

Draft
DAR 96: 82

1 The date is established by the fact that Horace Darwin matriculated at Trinity College, Cambridge, in the Michaelmas term of 1870 (*Alum. Cantab.*); the Michaelmas term begins in October. There is a payment to Blore of £25 'for Horace (last year)' recorded in CD's Classed account books (Down House MS) on 31 January 1871, which suggests that Blore was Horace's tutor at Trinity.

2 Horace Darwin.

3 Henry Bence Jones.

4 The little go, or previous examination, was an examination in Latin, Greek, mathematics, and theology usually taken in the second year of residence (Winstanley 1947, pp. 144–5).

[5] Edward John Routh, a highly successful mathematics tutor at Cambridge, had been private tutor to George Howard Darwin (see *Correspondence* vol. 13, letter to Edward Cresy, 19 October [1865]; for Routh's career, see Warwick 2003, pp. 229–40).

From J. V. Carus 2 October 1870

Leipzig,
Oct. 2, 1870.

My dear Sir,

This most horrid war put me entirely out of my regular work and in fact out of order. The very day when your letter (Aug 18.) reached Leipzig I was on the outset to France, not being able to spare one minute to answer it.[1] I took three railway-carriages full of hospital materials with me to distribute amongst the hospitals round Metz and up to Sedan.[2] Then I conducted three hundred wounded to Saxony and did not stop here but a few hours to return again to Douzy with new things.[3] The impressions I got there were so woeful and melancholy that I really wanted some time to get them over and to force myself to the daily work. Pray don't be angry, that I left your kind letter unanswered for such a length of time; but in fact I could not write.

When I returned the second time I found a letter of the publisher asking me to begin the translation as soon as possible.[4] I perfectly agree with him, that the war will not interfere the least with the success of your new book in a german form. People are anxious to find a safe refuge in science, they begin to get tired of the constant excitements brought about by telegrams and shocking war-tales. The results of your work will become a permanent part of the scientific conscience and not be looked over and forgotten as the infallibility will be. The worst of it could only be that it might sell at a less quick rate. But this is a matter of the publisher's. And as he insists on bringing the book out as soon as possible (—and he hopes to be able to bring the first volume at the same time with the original—) I cannot but submit to his judgement.[5]

You would therefore oblige me and the publisher exceedingly if you would be so kind as to *send me the corrected sheets as soon as possible* in the same way as you kindly sent me the sheets of the "Variation under Domestication."[6]

The success of the German armies are indeed wonderful. But after all I may ask if the struggle for superiority between the Romanic and Teutonic race, both taken in the widest sense, could not be fought out in some other form more appropriate to the high standing of their respective culture and civilisation. It is a most dreaful 'struggle for existence' and the only philosophical comfort is that whatever happens, *must* have happened, for without being a necessary consequence of natural conditions it could not happen But for a sensible mind this "nil admirari" is a very hard thing![7] I hope with all my heart that we will soon have peace.

Scientific work is just like oil smoothing down the rolling waves of political and national excitement. I trust you will kindly help me to steady my thoughts again.

Believe me | My dear Sir | Yours very sincerely. | J. Victor Carus

DAR 161: 76

1 Carus refers to the Franco-Prussian War of 1870 to 1871, and the letter to J. V. Carus, 18 August 1870.

2 Metz, a city in north-eastern France, was the headquarters of the Third French Army Corps, and was besieged by the Prussian Second Army from 3 September to 23 October 1870 (Howard 1981, pp. 257–72). The French town of Sedan, in northern France, had been under German control since the battle of Sedan on 2 September 1870 (*ibid.*, pp. 204–23). Carus earned his doctorate in medicine, and had earlier practised as a physician (*NDB*).

3 The kingdom of Saxony was an army-corps area during the Franco-Prussian war; see Howard 1981, p. 22. Douzy is a commune in the department of the Ardennes in northern France, near Sedan.

4 Carus was to translate *Descent* into German; see the letter to J. V. Carus, 18 August 1870. E. Schweizerbart'sche Verlagshandlung, headed by Eduard Koch, published the translation in 1871 (Carus trans. 1871).

5 The first printing of *Descent* was published on 24 February 1871, though presentation copies were released in late 1870 (Freeman 1977, pp. 129–30). The first volume of Carus trans. 1871 was published separately in Germany in April 1871 (*Correspondence* vol. 19, letter from J. V. Carus, 22 April 1871), with the second volume appearing later in the year, probably in June 1871 (*ibid.*, letter from J. V. Carus, 28 May 1871).

6 For Carus's translation of *Variation* (Carus trans. 1868), see *Correspondence* vols. 15 and 16.

7 'Nil admirari': to be astonished at nothing.

To J. V. Carus 5 October [1870][1]

Down. | Beckenham | *Kent. S.E.*
Oct. 5th

My dear Sir

We have all been deeply interested by your letter, just received, & sympathise with every word which you say.[2] I am astonished & pleased to hear that you are prepared to begin translating & the publisher to publish.—[3]

I have corrected 208 pages in **second** Proofs, but no sheets have yet been struck off: I have written again to day for clean sheets of even 2 or 3 first sheets to be sent on as soon as ready; but I know that they wait generally, & do not print any off, till *several* sheets are finally corrected.—[4] As soon as I hear what Murray will charge for stereotypes of 64 wood-blocks, I will let you know: I asked him to charge only a little more than costs[5] You understand that I give *you* the right of Translation, & you must make your own agreement with publisher.—

With the most cordial feelings | Believe me | Yours very sincerely | Ch. Darwin

P.S. You will be pleased to hear that Vol. I consists of only 411 pages.—[6]

Staatsbibliothek zu Berlin – Preußischer Kulturbesitz (Slg. Darmstaedter Lc 1859 Darwin, Charles, Bl. 16/17)

1 The year is established by the relationship between this letter and the letter from J. V. Carus, 2 October 1870.

2 See letter from J. V. Carus, 2 October 1870.

3 Carus was to translate *Descent* into German; the publisher of Carus trans. 1871 was E. Schweizerbart'sche Verlagshandlung (see letter from J. V. Carus, 2 October 1870 and n. 4).

4 CD's letter to William Clowes, the printer of *Descent*, has not been found.

5 See letter to John Murray, 26 September 1870.

6 The first volume of *Descent* was 423 pages long not including preliminaries (8 pages).

From V. O. Kovalevsky 5 October 1870

Sandown
5 Octob. 1870.

Dear Sir

I communicated the wish You expressed to put in Your new work something about the vertebrate character of the Ascidian larva, to my brother and some days ago he sent me a letter to You with some particulars about this question.[1] Till then he has not received Your work on the Cirripedia and so could not thank You for it.[2] I return in three days to London and after a week more we both intent to go to Leipsic for the winter; I hope to see You once more before my departure.[3]

Yours | Very truly | W. Kovalevsky

DAR 169: 83

1 No letter from CD to Kovalevsky regarding the addition to *Descent* has been found; however, Kovalevsky visited the Darwins at some point between 15 August and 1 September 1870 (see letter from V. O. Kovalevsky, 15 August [1870], and letter from S. V. Kovalevskaya and V. O. Kovalevsky, 1 September 1870). For Alexander Onufrievich Kovalevsky's account of ascidian larva forwarded through his brother, see the letter from A. O. Kovalevsky, 25 September 1870.

2 CD evidently gave copies of his monographs on living barnacles (*Living Cirripedia* (1851) and (1854)) to V. O. Kovalevsky who then sent them to his brother in Naples (see letter from V. O. Kovalevsky, 20 September 1870).

3 Kovalevsky had been in England since mid-August (see letter from V. O. Kovalevsky, 15 August [1870]). He and his wife, Sofia Vasilyevna Kovalevskaya, intended to spend the winter in Leipzig, Germany. For Kovalevsky's previous visit to Down, see n. 1, above.

To Robert Main 6 October 1870

Beckenham | Kent S.C.[1]
Octr 6. 1870

Dear Sir

I regret much that I am unable to correct my little Manual.[2] I have lately been unwell from over-work, & I am going to leave home immediately for a complete rest.[3]

On my return I must go on correcting the proofs of a work now on hand,[4] & I really feel incapable of undertaking any other task. Could you ask Professor Phillips to give you a list of books in addition to the few mentioned by me, & perhaps you would persuade him to read over my essay & append to it his own name, some additional suggestions.[5]

As I have not attended much to geology lately, such suggestions would be immeasurably better than any which I could add.

Pray believe me dear Sir | Yours very faithfully | Ch. Darwin

Copy
American Philosophical Society (384)

1 S.C.: copyist's error for 'S.E.'

2 CD refers to his essay 'Geology', first published in *A manual of scientific enquiry; prepared for the use of Her Majesty's Navy: and adapted for travellers in general* (Herschel ed. 1849, pp. 156–95). John William Frederick Herschel published a second edition (Herschel ed. 1851), and Robert Main had superintended a third edition (Main ed. 1859; see *Correspondence* vol. 7, letter to Robert Main, 27 July [1859]).

3 CD was away from home from 13 to 20 October 1870 (see 'Journal' (Appendix II)).

4 CD refers to *Descent.*

5 John Phillips, professor of geology at Oxford University, revised CD's article in Main ed. 1871, pp. 248–75 (the fourth edition of the *Manual of scientific enquiry*; see Main ed. 1871, p. 248).

From Charles Wyville Thomson 7 October 1870

Strandtown | *Belfast*
October 7th 1870.

My dear Sir,

As you are probably aware Dr. Allman has been obliged to resign the Edinburgh Chair—[1] I am a candidate for it and although I have not the honor of your personal acquaintance I cannot refrain from asking you for a few words of recommendation if you think that I am justified in applying for this post in my native town—

I am the more anxious for your sanction of my candidature as my later work has so direct a bearing on the illustration of your views.[2]

Believe me | My dear Sir | Yours with great respect | Wyville Thomson

DAR 178: 113

1 George James Allman resigned from his position as professor of natural history at the University of Edinburgh and keeper of the natural history collection of the Edinburgh Museum of Science and Art due to poor health (Swinney 1999, pp. 55–6).

2 See, for example, Thomson's work on the embryology and development of echinoderms (C. W. Thomson 1861, 1862, and 1863). See also C. W. Thomson 1869; an inscribed and annotated copy is in the Darwin Pamphlet Collection–CUL.

To C. W. Thomson [after 7 October 1870][1]

My dear Sir

I am very glad to hear that you are a candidate for the Chair of N. History in the U. of E.[2] You have my sincere good wishes for your success, for to the best of my judgment, you have proved yourself well fitted for the Post by what you have done in Natural Science & by the course of your recent investigations.[3] Should you succeed you will have an admirable opportunity for bringing forward & encouraging new recruits & as I believe you will do your utmost, I cordially wish for your success.—

I remain | my dear Sir | Yours very faithfully | Ch. Darwin

Draft
DAR 96: 81

[1] The date is established by the relationship between this letter and the letter from C. W. Thomson, 7 October 1870.

[2] See letter from C. W. Thomson, 7 October 1870.

[3] Thomson participated in the cruises of HMS *Lightning* and HMS *Porcupine*; these gathered physical data and biological specimens from the sea floor (*ODNB*; see also Carpenter and Thomson 1868). For some recent publications of Thomson's, see the letter from C. W. Thomson, 7 October 1870 and n. 2. Thomson was appointed to the position on 7 November 1870 (see Swinney 1999, p. 57).

From George Cupples 8 October 1870

The Cottage, | Guard Bridge | Fifeshire. N.B.
Oct. 8/70.

My Dear Mr Darwin,

I was at any rate keeping the puppy for a little longer, in order to harden and acclimatize him (after weaning) to the change of the season, which is here cold at nights.[1] Every day I thought of writing to mention this cause for delay, but did not like to trouble you with such a matter.

I now propose that he shall stay till about the end of the month, after which I will write to tell the day and hour of his arrival at King's Cross station—giving plenty of time beforehand for any note to the contrary.[2] Perhaps we can manage to see him off from Edinburgh ourselves—in which case the proper guard could be seen, otherwise he would go through too many different lines of railway for that. Indeed there is no fear of him in that respect— nor would it be necessary for you to send to meet the hamper, except for the possibility of delay in London.

It is desirable that he should come to his new quarters when very young, in order to get into the habits of the house.[3] All he requires is to have warmth at night, plenty of play—and a companion if possible, which I suppose he would find in Miss Darwin's "Polly", if Polly is goodnatured.[4]

I shall send a brief card of dietary &c—to fix up for the servants. He thrives well, in spite of the season.

I am glad to hear of proofs—though truly sorry that they fatigue you. The excursion, I hope, will set this to rights.[5]

Believe me | Dear Mr Darwin | ever most truly yours | George Cupples

Charles Darwin, Esqr. | Down.

DAR 161: 294

1 Cupples had promised to send CD a Scottish deer-hound puppy that he had bred (see letter from George Cupples, 17 September 1870).
2 King's Cross railway station is a major terminus in north central London.
3 The puppy was evidently born close to 15 August 1870 (see letter from George Cupples, 20 June 1870).
4 'Polly' was a white rough-haired fox-terrier belonging to Henrietta Emma Darwin (R. B. Freeman 1978).
5 CD was working on the proof-sheets for *Descent*; he was to be away from 13 to 20 October 1870 (see 'Journal' (Appendix II)). No letter to Cupples with this information has been found.

From Hermann Hoffmann 8 October 1870

[Enclosure][1]

Phasian. gall. dom. mas[2]
8. X .70

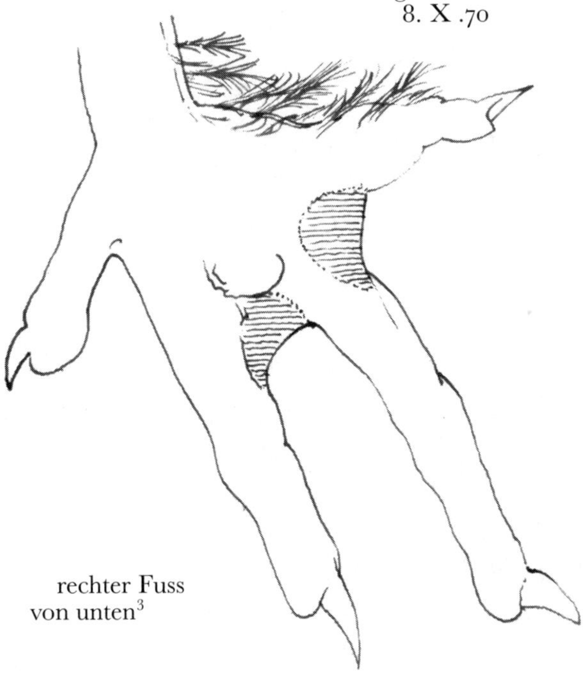

rechter Fuss
von unten[3]

H. H.

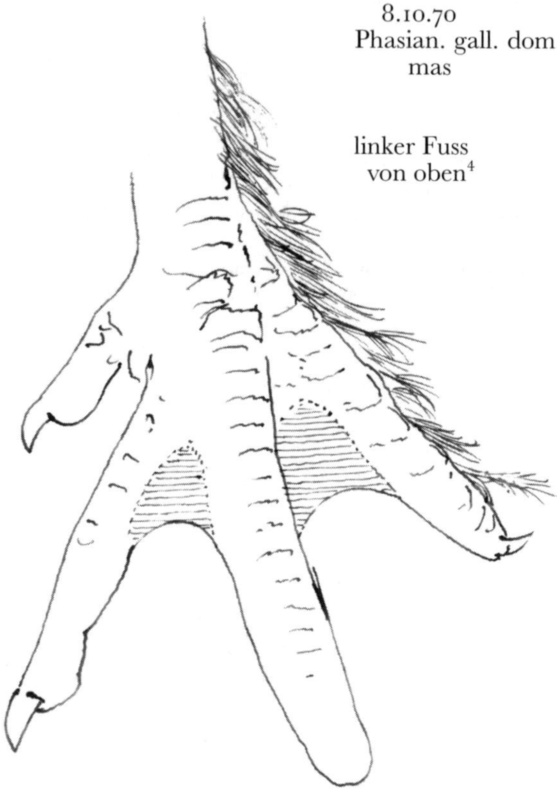

DAR 193: 114

[1] The letter that may have accompanied these sketches has not been found. However, see letter from Hermann Hoffmann, 23 June 1870 and nn. 1 and 2.

[2] *Phasianus gallus domesticus* is now *Gallus gallus domesticus*, the domestic fowl. *Mas*: male.

[3] *Rechter Fuss von unten*: right foot from below (German).

[4] *Linker Fuss von oben*: left foot from above (German).

From John Murray 10 October [1870][1]

50, Albemarle St. | W.
Octr 10th

Dear Mr Darwin

I enclose you Mr Dallas reply respecting your Index—wch reads satisfactorily—so I hope no delay will arise from that quarter.[2]

It is odd that the passage to wch I referred you in your proofs—now eludes my search but it occurs very near to the note, wch you have very properly veiled in Latin.[3]

M[r] Clowes has kept me waiting for his estimate of the cost of Electrotypes— I can now inform you that I will supply the set of 62 cuts for £14" They will cost £7" & the surplus will go to diminish our expenses for Illustrations[4]

You shall have 4 sets of the sheets as fast as they are worked off— I hope you have arranged to receive something handsome from the Americans—in return for this advantage, wch I have given over to you with the desire that you sh[d] benefit. The Yankee publisher cannot fail to gain.[5] I have also a request from Holland for a set for *a Dutch* Translation—[6] What do you say to this?

I remain My Dear Sir | Yours very faithfully | John Murray

Charles Darwin Esq

DAR 171: 379

[1] The year is established by the relationship between this letter and the letter from John Murray, 28 September [1870].

[2] Murray had written to William Sweetland Dallas after CD suggested that Dallas be commissioned to write the index for *Descent* (see letter to John Murray, 26 September 1870, and letter from John Murray, 28 September [1870]). The enclosure has not been found.

[3] For Murray's reference to a passage in the proof-sheets of *Descent*, see the letter from John Murray, 28 September [1870] and n. 7; see also the letter to John Murray, 29 September [1870]. CD wrote *Descent* 2: 345 n. 53 in Latin, but Murray could be referring to the quotation in Latin in *Descent* 2: 273.

[4] Murray refers to illustrations for *Descent* and to one of the family members of the printers William Clowes & Sons. See letter to John Murray, 26 September 1870.

[5] CD had promised the American publisher of *Descent*, D. Appleton & Co., sheets of the text and stereotypes of the woodcuts (see letter to John Murray, 29 September [1870] and n. 3).

[6] Hendrik Adrianus Kramers asked Murray whether he could have early sheets for a translation of *Descent* in a letter of 13 September 1870 (National Library of Scotland, John Murray Archive, Ms. 42153 f. 53). The first Dutch translation of *Descent* was published by Ykema & Van Gyn in Delft (Hartogh Heijs van Zouteveen trans. 1871–2).

To J. V. Carus 11 October [1870][1]

Down. | Beckenham | *Kent. S.E.*
Oct 11th.

My dear Sir

I send by this post the four first sheets of my book.[2]

M[r]. Murray charges £14 for 64 stereotypes, & this I think is very reasonable as the cost of casting is £7 & the profit £7. Your publisher if he thinks fit, had better communicate direct with M[r]. Murray of Albemarle St.[3] Please to observe that I give 14 handsome woodcuts from Brehm of which I cannot supply copies.[4] These are not indispensable but it would be a pity if they were omitted. I enclose a list.[5] Your publisher might negociate for stereotypes of these from the publisher of Brehm's Thierleben

My dear Sir, yours very sincerely | Ch. Darwin

LS
Staatsbibliothek zu Berlin – Preußischer Kulturbesitz (Slg. Darmstaedter Lc 1859 Darwin, Charles, Bl. 18)

[1] The year is established by the relationship between this letter and the letter to J. V. Carus, 5 October [1870].

[2] John Murray had written that he would soon send four sets of sheets of *Descent* (see letter from John Murray, 10 October [1870]); Carus was to translate *Descent* into German (see letter from J. V. Carus, 2 October 1870, and Carus trans. 1871).

[3] See letter from John Murray, 10 October [1870]. Carus's publisher was E. Schweizerbart'sche Verlagshandlung, headed by Eduard Koch.

[4] CD planned to include in *Descent* woodcuts from Alfred Edmund Brehm's multi-volume work on natural history (Brehm *et al.* 1864–9).

[5] The enclosure has not been found. For CD's initial list, and the published woodcuts, see *Correspondence* vol. 16, letter to Cassell, Petter & Galpin, 31 December 1868 and n. 1.

From V. O. Kovalevsky 11 October [1870][1]

Octob. 11 Evening.

Dear Sir

I hope this note may reach You in time before You leave Down,[2] if such should be the case I will pray You to be so kind and send me the four sheets to my adress in London; we leave England only Saturday afternoon. I shall see M^r. Murray an arrange the payment out sending of casts.[3]

I hope Miss Darwin is going better[4]

Yours truly | W. Kovalevsky

P.S. In case this note reaches You to late, that is after Saturday, I shall write another time from Berlin, but the safest thing in this case will be to send the sheets under band *poste-restante* Berlin where I am sure to receive them.—

I received Your letter at ten in the evening and having no stamps I commit the offence (considered impardonable in England) to send it unpaid, that it may go to morrow by the first train.—

Yours | W. K.

DAR 169: 84

CD ANNOTATION

Top of letter: '31 Bloomby St | Bedford Sq^r.' *pencil*[5]

[1] The year is established by the relationship between this letter and the letter from V. O. Kovalevsky, 22 November 1870.

[2] CD was to leave Down on 13 October 1870 (see 'Journal' (Appendix II)).

[3] The reference is to corrected proof-sheets of *Descent*; Kovalevsky intended to translate *Descent* into Russian, but was ultimately not able to complete the work (see *Correspondence* vol. 19, letter from V. O. Kovalevsky, 14 March 1871). Kovalevsky also evidently refers to casts (stereotypes) of woodcuts for *Descent*, and to John Murray. On Kovalevsky's travels, see the letter from V. O. Kovalevsky, 5 October 1870.

[4] Kovalevsky refers to Henrietta Emma or Elizabeth Darwin.

[5] Thirty-one Bloomsbury Street, Bedford Square, was Kovalevsky's London address (letter from V. O. Kovalevsky, 5 September [1870].

To John Murray 11 October [1870][1]

Down. | Beckenham | *Kent. S.E.*
Oct 11th

My dear Sir

I am very much obliged for your note; & I am glad to hear about Mr Dallas, but he will want keeping up to the mark.[2]

I shall be pleased to have a dutch edition of my book, so pray send clean sheets & supply stereotypes of the cuts. But you must inform the publisher that we cannot supply stereotypes of 14 figures from Brehm of which I enclose a list.[3] Could you get any payment to me from the dutch gentleman?

My dear Sir, | Yours very sincerely | Ch. Darwin

I am so knocked up that I am going to have a week's rest from correcting.[4]

LS(A)
National Library of Scotland (John Murray archive) (Ms.42152 ff. 41–2)

1 The year is established by the relationship between this letter and the letter from John Murray, 10 October [1870].

2 The references are to the letter from John Murray, 10 October [1870], and to William Sweetland Dallas, who had taken longer than expected to complete the index for *Variation* (see *Correspondence* vols. 15 and 16).

3 On the Dutch edition of *Descent*, and its publisher, see the letter from John Murray, 10 October [1870] and n. 6. CD's enclosed list of woodcuts from Brehm *et al.* 1864–9 has not been found, but for its probable contents, see *Correspondence* vol. 16, letter to Cassell, Petter & Galpin, 31 December 1868 and n. 1.

4 CD left Down from 13 to 20 October (see 'Journal' (Appendix II)).

From J. D. Hooker 12 October 1870

Kew.
Oct.r 12/70

Dear Darwin

Bentham has translated the Introduction to Miguel's Sumatran supplement to his Flora Ned. Ind—[1] It contains much curious matter regarding the affinities of the Bornean & Sumatran Flora (to the exclusion of the Javan) & should be published. What do you think would be the best vehicle? "Nature" is wretched & too ephemeral— What about Popular Science Review?[2]

Ever yrs | J D Hooker

DAR 103: 60

1 The references are to George Bentham, Friedrich Anton Wilhelm Miquel, and Miquel 1860.

2 No publication of Bentham's translation has been found; however, his manuscript translation of Miquel's 'On the physical geography and vegetation of Sumatra' is in the George Bentham papers, Royal Botanic Gardens, Kew. For more on Hooker's low opinion of *Nature,* which began publication in November 1869, see *Correspondence* vol. 17, letter from J. D. Hooker, 14 November 1869.

To J. D. Hooker 14 October [1870][1]

Leith Hill Place | Dorking | Surrey
Oct. 14.

My dear Hooker

I have come here for a few days idleness & I am doing the job thoroughily.[2] Your question is puzzler: I do not think so poorly of Nature as you do by any means.— I fear the Pop. Sc. R. is rather ephemeral, but more durable, I presume, than Nature.[3] I suppose it w^{d} be against precedent to publish it in Linn. Journal;[4] otherwise that w^{d} be best place. What do you say to Annals & Mag. of N. History?[5] I suppose circulation is on smallest scale.— Anyhow I hope it will be published somewhere.—

What a wretch you have been not to say, as you were writing, whether the case of the charlock impressed you at all;[6] or did you, my chief in the art of wriggling, get out with ease.—[7]

Your lazy friend | C. Darwin

DAR 94: 184–5

[1] The year is established by the relationship between this letter and the letter from J. D. Hooker, 12 October 1870.

[2] CD was visiting the home of his sister Caroline Wedgwood and her husband Josiah Wedgwood III (see 'Journal' (Appendix II)).

[3] Hooker had wondered where George Bentham's translation of the introduction to Miquel 1860 should be published (see letter from J. D. Hooker, 12 October 1870). Pop. Sc. R.: *Popular Science Review.*

[4] *Journal of the Proceedings of the Linnean Society of London.*

[5] *Annals and Magazine of Natural History.*

[6] No recent correspondence with Hooker regarding charlock, or wild mustard (*Sinapis arvensis*), has been found. In 1855, CD had investigated the length of dormancy and the vitality of charlock seeds, and he and Hooker had debated the issue of general seed viability at length (see *Correspondence* vol. 5).

[7] On CD and Hooker's ongoing joke about 'wriggling', see the letter to Asa Gray, 15 March [1870].

To Francis Darwin 18 October [1870][1]

The Moat | Tunbridge
Oct 18.

My dear Frank

I enclose cheque for 115£:[2] I expected of course to have to aid you as you were so much longer than usual at Cambridge, but the above is rather more than I expected; & what is worse I do not see how you can hold out for this quarter with only £46·5·0.—[3] But never, for God's sake, conceal debts from me, & tell me now, whether you owe any more: tell me this & think deliberately when you acknowledge the cheque to Down, where we return on Thursday morning.[4]

As you have not kept accounts it is of course impossible for you know how you have overspent your income.— Let me urge you to make a point of conscience (& then I know it will be done, i.e. if I can persuade you that it is a duty) to keep accounts.— You cannot be sure about paying your debts if you do not, nor can you tell how to economise & spend your income to best advantage. I have never known a man who was too idle to attend to his affairs & accounts, who did not get into difficulties; & he who habitually is in money difficulties, very rarely keeps scrupulously honourable, & God forbid that this sh^{d}. ever be your fate.— If you once got into habit of attending to your money, & this implies keeping accounts, you w^{d}. feel it very little trouble. My father,[5] who was the wisest man I ever knew, thought it the duty of every man, young & old, to keep an account of his money; & I very unwillingly obeyed him; for I was not always so bothersome an old fellow as I daresay I appear to you.—

We have been to Leith Hill & came here yesterday & have enjoyed ourselves; though mamma is not very well to day. This is a wonderfully curious & pretty place.[6]

Your affectionate Father | Ch. Darwin

DAR 211: 6

1 The year is established by the references to the Darwins' holiday at Leith Hill Place and Ightham Mote (see nn. 4 and 6, below).

2 CD listed the amount for 'F[rank]' under 'Boys' in his account book (CD's Classed account books (Down House MS)).

3 Francis had been at Trinity College, Cambridge since 1866; he graduated in 1870 (*ODNB*), taking third place in first class honours in the natural science tripos in December (*The Times*, 19 December 1870, p. 6).

4 The Darwins returned to Down on Thursday 20 October 1870 ('Journal' (Appendix II)). No letter from Francis acknowledging the cheque has been found.

5 Robert Waring Darwin.

6 The Darwins were staying at Ightham Mote, a moated medieval manor house, which is in fact closer to Sevenoaks than to Tunbridge; they had left Down on 13 October, going first to Leith Hill Place ('Journal' (Appendix II)).

From John Gage 19 October 1870

Vineland New Jersey
Oct 19/70.

Charles Darwin.
My friend.

Two years sinse I read with much interest your origin of species.[1] The evidences produced were so clear to my mind that when I had finished the reading, I accepted the theory and felt the better for it, & it has been a text book for me sinse. Two months ago I was in the office of my friend Henry T Child MD, of Philadelphia,[2] talking about your work Origin of Species, & he says I have another work of Darwins that you will be pleased to read, his voyage in the Beagle.[3] I borrowed it & have read it with much interest, & I want to say to you that I found many things in both works, that made me wish I was where I could call on you & ask questions; & there is one on page 250 second Vol. of Voyage, in Harpers publication,[4] that has caused me to write this communication. It is as follows.

"After dinner we stayed to see a curious half superstitious scene, acted by the Malay women. A large wooden spoon dressed in garments, and which had been carried to the grave of a dead man, they pretend becomes inspired at the full of the moon, and will dance & jump about. After the proper preperations, the spoon, held by two women, became convulsed, and danced in good time to the song of the surrounding children & women. It was a most foolish spectacle; but Mr Liesk maintained that many of the Malays believed in its spiritual movements."[5]

How I wish you had written out all the minutiæ of this scene, & any others that you may have witnessed amongst this race of men,[6] whose common parent with our white race, or the original Adam if there was one who was father to us all, must have been far back & low down in the scale of being, & possibly so low as not to have partaken of the fruit of the tree of knowledge of good & evil. I want to know what this race of men *know*, or even what they *believe*, of Spirit manifestations

I have been a student of this spirit intercourse for twenty two years, & have allowed no good opportunity of examining its facts to escape me; & have been well situated for observing the facts amongst the Anglo Saxon race.

Will you do me the favor to write to me all the particulars of this seance, relative to its claims to spirit origin, both for & against.

Did the spoon stand erect?

Did not the women make it dance?

How did the women hold the spoon?

Did the women & children form a circle around the spoon?

Were you so situated as to detect any deception?

If you saw any other manifestations amongst these races that gave evidence of spirit power, or presence; & will communicate them to me you will much oblige. | Your Friend. | John Gage

DAR 165: 2

1 Gage probably read the US edition of *Origin*.

2 Henry T. Child was a physician in Philadelphia, and a spiritualist (US Federal Census, 1870, and Britten 1870, pp. 273–5).

3 Prior to 1870, the American edition of *Journal of researches* US ed. (1846) was reprinted, unaltered, in 1855, 1859, and 1864 (R. B. Freeman 1977, p. 40).

4 The American editions of the *Journal of researches* were published by Harper & Brothers in New York.

5 Gage's transcription of the passage in *Journal of researches* US ed., 2: 250, contains minor punctuation errors. The passage, written 3 April 1836, was part of CD's account of the Cocos (now Cocos-Keeling) Islands. For more on the ceremony described, see Armstrong 1991, p. 59. William C. Liesk was the leader of the Cocos (Keeling) Islanders when the nominal king was away, as during CD's visit (Hughes [1950], p. 41, and Jones 1910, p. 22).

6 CD wrote that the non-British people on the Keeling Islands were 'Malays' from different islands in the 'East Indian Archipelago' (*Journal of researches* US ed., 2: 249). However, the residents then also included people from the Cape (South Africa), New Guinea, and China (Jones 1910, pp. 16–17, 21).

From Alexander Agassiz 22 October 1870

London,
Oct. 22, 1870.

My dear Mr. Darwin:—

Hearing you are in town, I write to ask you if you will allow me to come and bid you good-bye before we sail.[1] I go to Ireland on Thursday next, in the evening, to see the plunder of the last Porcupine Expedition,[2] but any time you can receive me before that it would be a great pleasure for me to see you again before I leave this side of the Atlantic, where I have had such a charming reception and have I trust learned a good deal which will not come amiss in America.

With the kindest remembrances of Mrs. Agassiz and myself to Mrs. Darwin and your family, believe me always | Yours very truly | Alex Agassiz.

G. R. Agassiz ed. 1913, p. 113

1 Alexander Agassiz and his family had been travelling in Europe for over a year and would soon return to Boston (G. R. Agassiz ed. 1913, pp. 97–114); he and his wife, Anna Russell Agassiz, had already visited Down on or shortly before 1 December 1869, and would visit Down again before returning (see *Correspondence* vol. 17, letter to Fritz Müller, 1 December [1869]), and Dobbs 2005, p. 105).

2 During the summer of 1870, HMS *Porcupine* dredged off the coasts of Spain and Portugal, and in the Mediterranean (C. W. Thomson 1873, pp. 82–196). The collections from the *Porcupine* were evidently with Charles Wyville Thomson, at Queen's College, Belfast (see C. W. Thomson 1873, G. R. Agassiz ed. 1913, p. 113, and *ODNB*).

To John Murray 22 October [1870][1]

Down. | Beckenham | *Kent. S.E.*
Oct 22

My dear Sir

My family think, & I quite agree with them, that the binding of Domestic Animals with plain lettering, no golden ornaments & smooth cover, looks much better than the Origin.—[2]

I always give list on back of Title of my books— Can you without much trouble tell me *soon* how many copies you have printed of my Naturalist's Voyage: some years ago I made out, by adding 1500 copies printed by Colburn, that the number was 10,000; & it seems absurd to go on printing same number.—[3]

My dear Sir | Yours very sincerely | Ch. Darwin

What is date of your sale?[4]

National Library of Scotland (John Murray archive) (Ms.42152 ff. 218–19)

1 The year is established by the relationship between this letter and the letter from John Murray, 24 October [1870].

2 CD refers to the binding of *Variation* and *Origin*; see also Peckham ed. 1959, p. 17, and R. B. Freeman 1977, p. 123.

3 The reference is to the editions of *Journal of researches* (1839, 1845, 1852, and 1860). The spine of the 1845 edition read *Naturalist's voyage*, and the spine of the 1860 edition read *Naturalist's voyage round the world* (R. B. Freeman 1977, p. 32). *Journal of researches* was first published by Henry Colburn as the third volume of *The narrative of the surveying voyages of His Majesty's ships Adventure and Beagle* in 1839 (*Journal and remarks*); it was reissued by Colburn as an independent volume (*Journal of researches*) in 1839 and 1840. The second edition was published by John Murray in 1845, and reissued in 1852 and 1860, with the tenth thousand issued in 1860. See R. B. Freeman 1977, pp. 31–40.

4 On Murray's November 'sale dinners' for the principal booksellers, see J. Murray 1908–9, p. 540.

To Alexander Agassiz [23 October 1870?][1]

Down. | Beckenham. Kent.
Feb. 2.

My dear Mr. Agassiz.

It is very good of you to propose calling here, and I shall be delighted to see you— Will you not come to luncheon at 1. o'clock? as that is generally my best time for strength; but do not trouble yourself to answer. & come at any time.[2]

Yours very sincerely. | Ch: Darwin.

Copy
DAR 143: 7

[1] The date is conjectured from the relationship between this letter and the letter from Alexander Agassiz, 22 October 1870, and from the fact that if the letter pertained to Agassiz's earlier visit in December 1869, CD would probably have included train information for travel to Down. The copyist's date is in error as Agassiz was in Paris in February (G. R. Agassiz ed. 1913, pp, 97, 99, 114).

[2] See letter from Alexander Agassiz, 22 October 1870.

From John Murray 24 October [1870][1]

50, Albemarle St. | W.
Oct 24

My Dear Sir

Your wishes regarding a smooth cloth for the binding of your book shall be attended to—[2]

I have resorted to my Ledgers & find you have exaggerated slightly the sale of the Natlt Voyages by about 500 copies Since it has been in my hands—these are the copies sold—[3]

1845	4000.	Col Liby.[4]
1848	1000	"[5]
1852	1000	"
1860	2000	"—[6]

these are now all sold & I have just printed a New Edition, of 1000[7]—wch may go off in 7 or 8 years; that of 1860 having taken ten years to exhaust. I hope to have my sale on the 11th. Novr.[8]

The origin of man is sent to press 2500 copies—& Clowes will send you 5 sets of sheets

Have you instructed him about the sets of clichés?[9]

My Dear Sir | Yours very sincerely | John Murray

Chas Darwin Esq

DAR 171: 380

[1] The year is established by the reference to *Descent*, which first went to press in 1870 (see n. 8, below).

[2] See letter to John Murray, 22 October [1870].

[3] Murray refers to the different editions of CD's *Journal of researches* published by himself. Henry Colburn published earlier editions (see letter to John Murray, 22 October [1870] and n. 3).

[4] *Journal of researches* 2d ed. was published as part of Murray's series, the Colonial and Home Library (Freeman 1977, pp. 35–6).

[5] Murray evidently published another issue of *Journal of researches* 2d ed. in 1848 (see Freeman 1977, pp. 36, 40).
[6] *Journal of researches* (1852), and *Journal of researches* (1860).
[7] *Journal of researches* (1870).
[8] See letter to John Murray, 22 October [1870].
[9] The references are to the first printing of *Descent*, and to the printer William Clowes & Sons. The five sets of sheets were for the Russian, French, German, and Dutch translations, and the US edition (letters to John Murray, 26 September 1870 and nn. 1 and 2, and 11 October [1870] and n. 3). Murray had agreed to supply stereotypes of most of the illustrations in *Descent* (see letter from John Murray, 10 October [1870]).

From T. H. Farrer 27 October 1870

Abinger Hall, | Reigate. (Post Town) | Gomshall (Station) S.E.R.
27 Oct 1870

My dear Mr Darwin

I am sending you back those of your books which I have here—vizt four of Delpinos—[1] The rest shall come back when we get to London where they are—[2] One of the "Atti" of the Italian Società des Scienze Naturali which you have marked as an useless duplicate I keep.[3]

Alas! when will time come for these things. I find every day full and human beings little & big require the time and attention one would willingly give to flowers. Not that the former are less interesting in their way. but one grudges the infinite number of small things to be done, & feels sympathy with womens lot in life. Surely the fact that they have all the little daily things to do is enough to account for their not succeeding in work requiring devotion & abstraction

Sincerly yours | T H Farrer

C Darwin Esq[re]

DAR 164: 66

[1] The books and pamphlets of Federico Delpino's that Farrer borrowed from CD may have included Delpino 1867b, 1867c, 1869a, and the first part of Delpino 1868–74 (see *Marginalia* 1: 191–2). The first three are in the Darwin Pamphlet Collection–CUL, and the last is in the Darwin Library–CUL.
[2] Farrer also maintained a house at Gloucester Terrace, Regent's Park (see letter from T. H. Farrer, 5 June 1870).
[3] Farrer may be referring to Delpino 1867a, published in *Atti della Società Italiana delle Scienze Naturali in Milano.* A copy is in the Darwin Archive–CUL.

From W. D. Fox 28 October [1870][1]

Delamere Rectory | Northwich | Cheshire
Oct 28

My dear Darwin

What an age it is since we have met, or even corresponded.[2] I have often thought of you and wished to write, but knowing how much you are engaged, I have not done so. D[r] Hooker told me a few weeks since, that you are going to astonish us all very shortly by a new work.[3] I shall anxiously look for it. There is such a freshness about your Books that I never tire of them, & when tired & worn out, take down a Volume, with the certainty that I shall soon be relieved from present troubles.

I now write in consequence of *our* having lost the direction of a Governess, which we believe M[rs] Darwin applied to us to help. If we are right in M[rs] Darwin being the Applicant—She will fill up the enclosed for 3 Votes at the proper place. We applied to several others I believe, & I hope they have kept the name & properly applied. Should we be wrong in imagining M[rs] Darwin to be the applicant—you will kindly return the paper.[4]

We go next week to our Winter Quarters at Sandown in Isle of Wight—staying some days with M[rs] Fox's Father at Hampstead.[5]

Should you, by any accident be in London about that time I should greatly enjoy five minutes intercourse with you.

Tell M[rs] Darwin, with our kindest regards, to reply to this letter, and Believe my old & much valued Friend to remain Yours always | W. D Fox

DAR 164: 191

1 The year is established by the relationship between this letter and the letter to W. D. Fox, 15 November [1870].
2 The most recent extant letter between Fox and CD is the letter to W. D. Fox, 18 February [1870].
3 The references are to Joseph Dalton Hooker and *Descent*.
4 Correspondence between Emma Darwin and W. D. Fox or Ellen Sophia Fox regarding a governess and her address has not been found.
5 Ellen Sophia Fox's father was Basil George Woodd of Hillfield House, Hampstead (*Darwin pedigree*).

From S. A. Merrell 28 October 1870

US | Council Bluffs Iowa
Oct 28/1870

Mr Darwin
Dear Sir

Your kind note of reply to the one I sent you regarding the American Cuckoo, came duly to hand—[1] I did not deem it needful to carry the correspondence farther at the time, as I supposed you had no time to spend on irrelavant matters—and have not.

Since that time however I have prepared a short article on the "Unity of Generation" designed to show that the principle is one and the same throughout the entire natural world—also as bearing upon physical progress—[2]

It is in direct relation and support of the Development Theory which I had in view in preparing the paper—

If it does not seem to be an act of presumption on my part to offer you for your inspection the crude reflections of a homunculus[3]—one who has devoted his time mainly to other pursuits, I should be happy to send you the article at some future time

In conversing with an old friend not long since, Major Hoffman,[4] he stated to me he had been observant of the peculiar habits of our cuckoo on two or three distinct occasions

Have not had the pleasure of seeing your new edition of the Origin of Species[5]

With profound respect I remain | Most Truly Yours | S A Merrell MD | Box 325. Council Bluffs | Iowa | US

DAR 171: 158

1 For Merrell's letter on the American cuckoo, see *Correspondence* vol. 17, letter from S. A. Merrell, 1 February 1869. CD's reply has not been found.

2 No published paper by Merrell has been found. No manuscript by Merrell has been found in the Darwin Archive–CUL.

3 *Homunculus*: a little or weak man (Latin).

4 Major Hoffman has not been identified.

5 *Origin* 2d US ed.

From Benjamin Clarke[1] 1 November [1870][2]

November 1st.

I wrote the foregoing letter before I received yours of Oct. 20 containing the Post Office Order for 10s.[3] & I have now only to add that if the Zoological part of the work appears to you of any value you will be receiving back a part of that which you gave when you subscribed to the Relief Fund of the Royal Society.[4]

I enclose to you a few seeds of a plant of Indian Corn which was 3-stemmed without a trace of a male flower, the central stem terminating in a soft slightly hooked spine & one if not both the lateral terminating in cobs. It was artificially fertilised & ripened 10 cobs growing in somewhat unfavourable circumstances being shaded & sparingly watered.[5] It belongs to the continental 6-weeks variety procured by Messrs Gibbs & Co.[6] & was raised from seed produced last year by cutting off the entire of the male flowers of one plant about a week before flowering & fertilising the plant with the flowers of another. This was done with a view of producing varieties by pruning or mutilation as it may be called & it appears that *the extreme effect can be produced in one year* on some of the plants although on a very few; while some of the progeny show no diminution of male flowers;—in these latter there was apparently increased vigour of growth. I will only add that if you grow this seed you will I expect have an opportunity of witnessing this phenomenon & by repeating the experiment increase the number of female plants in the progeny.

You will observe at the back of the title page that science does not in any degree diminish my faith in the bible & that in fact science & the New Testament are according to my views quite in harmony—[7] My reason for placing *Man* in the system of the Mammalia is given at page 44.[8]

AL incomplete
DAR 261.11: 26

[1] The correspondent is identified from the handwriting, and from the relationship between this letter and the letter from Benjamin Clarke, 12 March 1867 (*Correspondence* vol. 15).

[2] The year is established by the reference to the Post Office order (see n. 3, below).

[3] The letters have not been found. CD recorded a payment in his Account books–cash account (Down House MS) of 10*s.* under the heading 'B. Clarke. Syst. Bot. & Zooly.' for 20 October 1870.

[4] Clarke refers to his *On systematic botany and zoology* (Clarke 1870). The botanical part of the work had already been published as *A new arrangement of phanerogamous plants* (Clarke 1866); the zoological part was ready for the press in early 1867 (see *Correspondence* vol. 15, letter from Benjamin Clarke, 12 March 1867) but not published until 1870 (Clarke 1870). CD's donation to the Relief Fund of the Royal Society of London is referred to in the letter from Benjamin Clarke, 25 March 1867 (*Correspondence* vol. 15). For a brief history of the Royal Society's Scientific Relief Fund, see *Record of the Royal Society of London*, p. 111.

[5] In his letter of 25 March 1867 (*Correspondence* vol. 15), Clarke had outlined a procedure for breeding a predominantly female variety of maize by emasculating the breeding stock repeatedly for several years.

[6] Thomas Gibbs and Company were seedsmen; they had premises on the corner of Piccadilly and Half-moon Street, London (*Post Office London directory* 1870).

[7] On the verso of the title page of Clarke 1866 is a poem on the splendour of God as manifest in tropical vegetation. A further poem referring to the 'power of God' was added to the 1870 publication. Both poems were signed B.C.

[8] Clarke qualified his classification of humans among the Mammalia as follows (Clarke 1870, p. 44):

> By including Bimana in this classification it is not intended to say that man was derived from Quadrumana as the other orders of Mammalia it is understood were derived from the Orders below them; Jeremiah, Chap. I, v. 5, is clearly opposed to such a supposition; the angelic host would, if incarnate, of course be classed with Bimana.

To George Cupples 3 November [1870][1]

Down. | Beckenham | *Kent. S.E.*
Nov. 3[d]

My dear M[r] Cupples

A servant shall be at King's Cross on Tuesday 8th. at 9 A.M. ready to receive the precious animal.[2] He shall then take him to some inn & give him water & let him loose for a short time in a stable, & then bring him here at once. So please direct C. Darwin "to be called for".— We are all infinitely obliged for this valuable present & for your most clear instructions which we will endeavour to keep to the letter.

You have been most kind to take so much trouble in giving us such full particulars.

With many thanks | Yours very sincerely | Ch. Darwin

Is the Hamper worth returning? or will you not let me pay for it?

Thanks about the Perch: I cannot but suspect some error; though Serranus is an ally of the Perch & has often been said on good authority to be an Hermaphrodite, but Dr Günther does not believe even in this case, except as an abnormality—[3]

Postmark: 'Beckenham NO 3 70'
Smithsonian Institution Libraries: Dibner Library of History of Science and Technology

[1] The year is established by the relationship between this letter and the letter from George Cupples, 8 October 1870.

[2] Cupples was sending a Scottish deer-hound puppy; see letter from George Cupples, 8 October 1870. Cupples's letter announcing the date and time of the puppy's arrival in London has not been found.

[3] Cupples's letter about perch has not been found. CD mentioned Albert Günther's view of *Serranus* in *Descent* 1: 208 n. 25. The Percidae (perch) and the Serranidae (of which *Serranus* is a member) are both families in the order Perciformes.

From Henry Holland 4 November [1870][1]

72 Brook Street
Nov 4th

My dear Charles,

A few lines, merely to say that a month this autumn in Jamaica, & other W. Indian Islands, has brought me still-more closely within the circle of your doctrine, both as regards the vegetable & the human world. I had no adequate idea before of the luxuriant struggle of tropical vegetation, & of that wonderful parasitic life which plays so large a part in it.

I am interrupted here, but had little indeed more to say than this.

Ever your's affecy | H Holland

DAR 166: 251

[1] The year is established by Holland's reference to the West Indies: he visited the West Indies in 1870 (Holland 1872, pp. 43, 313).

To P. L. Sclater 4 November [1870][1]

Down. | Beckenham | *Kent. S.E.*
Nov. 4th

My dear M[r] Sclater

I am going to beg a favour of you which will cost you little trouble proportionally to the *great* service it will be to me.— I have a most unfortunate weakness, though I strive against it, to copy proper names incorrectly. In my present book on sexual difference I have to give a host of names of Birds.[2] Will you run your eye over the enclosed list & see that names are spelt rightly?— I enclose envelope addressed to save you as much trouble as possible.— You can return the list without a word; except indeed, if you will add that I may send another list for another one or two chapters.—[3] I have checked all names occurring in books which I possess (except one or two just now mislaid) but I have not Jerdon or the Ibis from which many names are taken.[4]

I hope & think this will not cause you much trouble.

Yours very sincerely | Ch. Darwin

Zoological Society of London

[1] The year is established by the relationship between this letter and the letter to P. L. Sclater, 5 December 1870.

[2] CD refers to *Descent*; the chapters on birds are in *Descent* 2: 36–238. CD thanked Sclater for his help in *Descent* 2: 238 n. 59.

[3] Sclater also checked the two chapters on mammals (*Descent* 2: 239–315).

4 CD refers to Thomas Claverhill Jerdon's *Birds of India* (Jerdon 1862–4), and to *Ibis*, the journal of the British Ornithologists' Union.

From Worthington George Smith 4 November 1870

12 Northgrove West | Mildmay Park | London N.
4 Nov. 1870

Dear Sir,

I venture to send you the follo⟨wing⟩ note, thinking it may have some interest for you in connexion with your works.—[1] I have never read of anything *precisely* like it

I have a fair complexion *& light hair*, with very dark eyes & I am married ⟨to a⟩ wife[2] with dark complexion *dark hair* & light eyes—

We have three children they all resemble me in having the fair complexion with *very light hair*; & the mother only, in *all* having very light eyes

The following is the curious point which strikes me as being somewhat new.

The children all have an abundance of *very light hair*, but on very close examination I find about 1 hair in 100 (or perhaps, 500) is almost *jet black*, the black hairs always ⟨ ⟩ **singly** & are *[irregularly]* distributed *[½ line illeg]* has per *[½ line illeg]* of course they cannot be seen unless searched for, we recently saw *one* by accident & on searching found the others growing "sporadically" as I have stated.

It is quite different from the colored pate⟨ ⟩ sometimes seen.

Apologizing for troubling you, with what is perhaps only a piece of absurdity on my part, I remain

Yours faithfully | Worthington G. Smith | F.L.S. &c

DAR 177: 200

CD ANNOTATIONS
1.1 I venture … like it 1.2] *crossed pencil*
Top of letter: 'Inheritance | (not blending) | Black & white Hairs' *pencil*

1 CD had discussed inheritance of differently coloured locks of hair in humans in *Variation* 2: 5–6.
2 Henrietta Smith.

From W. W. Reade 6 November 1870

11 St Mary Abbot's Terrace | Kensington
Nov. 6. '70

My dear Sir

I trust you got the note I sent you just after my return in August.[1] Will you permit me to congratulate you on having completed a work which must have cost you so much labour? I see it announced.[2] I do not know whether you considered the case of "rogueing" among the Jollofs worth quoting.[3] If you require independent evidence as to the beauty of that tribe you will find it in Prichard; though he goes too far when he speaks of their European features.[4]

I am looking forward most eagerly to the appearance of your book because I anticipate that it will suggest points of research. I do not intend to write a narrative of my travels except as part of a work on Africa generally.[5] I am ambitious of doing for negro Africa what Tennent has done for Ceylon;[6] & I intend to read every book of travels relating to W & Central Africa & East Africa. I am now going through the Portuguese authorities. In wading through these chronicles of the old voyagers, which as you are aware, are very rich in detail I might hit upon facts which would be of value to men of science; the danger is that I might pass over some custom or incident insignificant in itself but most important for the solution of a problem with which I am unacquainted. I remember that in the Origin of Species you allude to something wh. you had not met with in the old voyagers, but that you had not completed the examination of them:[7] I need scarcely say that if there is any class of facts in relation to savage people of which you require instances I would note them down, as I go on, & send them to you; and I should be very grateful to you for advice of any kind though I appreciate the value of your time, & the immense mass of correspondence which you have to dispose of.

I am | yours very truly | Winwood Reade

I see Forbes in his Aryamese Indians has spoken of natural selection taking place among the infants of the Indians who are exposed to all weathers.[8] If you remember I suggested that possibly the thickness of the negro scull mt. be accounted for in that way as the infant is always bareheaded in Africa.[9] By the bye the negro does not enjoy *perfect* immunity from malarious fever. Respecting yellow fever I cannot speak from experience: but I have had my own men, pure negroes, down with fever. It is rather curious though, that both the men who suffered from fever in my expedition were light coloured—as many pure Africans are—and the fact that negroes do suffer from fever more or less might indicate, not that Dr. Wells' theory was baseless, but that Nat. Sel: is still going on.[10] We want, not the mere passing notes of travellers, but a series of medical observations extended over years; but how get them in West Africa?

DAR 176: 40

1 See letter from W. W. Reade, 3 September 1870.

2 For an announcement of *Descent*, see the *Academy*, 22 October 1870, p. 8.

3 See letter from W. W. Reade, 4 June 1870 and n. 5. In *Descent* 2: 357, CD quoted Reade's account of the Jollofs' selection of fine-looking slaves.

4 An annotated copy of the third edition of James Cowles Prichard's *Researches into the physical history of mankind* (Prichard 1836–7) is in the Darwin Library–CUL (*Marginalia* 1: 680–3); see *ibid.* 2: 78–9 for the comments on the beauty of the 'Iolofs', and their similarity to Europeans.

5 An annotated copy of Reade 1873 is in the Darwin Library–CUL (see *Marginalia* 1: 697–8).

6 James Emerson Tennent wrote on the natural history and peoples of Ceylon (Sri Lanka); see Tennent 1859 and 1861.

7 In *Origin*, p. 393, CD said that he had not found in the oldest voyages an unequivocal instance of a terrestrial mammal (except domesticated ones) inhabiting an island more than 300 miles from an continent or a continental island, but that he had not finished his search. In *Origin* 5th ed., p. 473, he omitted the phrase about not finishing his search.

[8] See David Forbes's 'On the Aymara Indians of Bolivia and Peru' (Forbes 1870a), p. 224. An inscribed and lightly annotated offprint, with a different pagination, is in the Darwin Pamphlet Collection–CUL.

[9] See letter from W. W. Reade, 4 June 1870.

[10] See letter from W. W. Reade, 4 June 1870 and n. 6. Reade refers to William Charles Wells and Wells 1818. Nat. Sel.: natural selection.

To W. G. Smith 6 November 1870

Down. | Beckenham | *Kent. S.E.*
Nov. 6th

Dear Sir

I am much obliged to you for your kindness in sending me the interesting fact communicated in your letter.[1]

I have not heard of any such case in the human family, but have long suspected that many grey coloured animals have thus originated.

With my best thanks, pray believe me | Dear Sir, yours faithfully | Charles Darwin

LS
British Library Board (Add.54224 f. 196).

[1] See letter from W. G. Smith, 4 November 1870.

From Thompson Forster 8 November 1870

Daventry.
Nov 8th 1870

Dear Sir

I send you the outline of a case I think may be interesting to you if it is, it can be proved or tested anytime as the people live close to this town—

The father when a baby had disease of the hip joints, with dislocation & shortening, & the discharge of a large quantity of pus— in due course he married, & had children all sound but the second one wh had shortening of the leg of the same side at the hip joints, the whole side of the pelvis seemed smaller & the tuber ischii lower, the femur shorter & smaller, but *never* any indication of disease

In the father you feel a thickening see a scar & find by touch & measurement dislocation upwards of the femur, with general thickening, the leg shortened 3 inches. The Son without any evidences of disease with leg shortened, shortening entirely in Femur & pelvis into only general diminution of size[1] should you think this case of any use as, proving the transmission of deformity from disease to offspring, I shall be happy to give you any further information you may require.

believe me | Yours Truly | Thompson Forster.

DAR 164: 160

CD ANNOTATIONS
Top of letter: 'Pan. Inherited Mutilations'[2] *red crayon*

1 CD discussed the inheritance of deformities, disease, and mutilations in *Variation* 2: 7–24, and 72–80.
2 CD discussed mutilations in regard to his hypothesis of pangenesis in *Variation* 2: 398.

From William Ogle [before 9 November 1870][1]

34 Clarges St. | Piccadilly.

Dear Sir,

I have taken the liberty of sending you a copy of a paper which I read at the beginning of the year before the Royal Med. Chir. Society, of which you are an honorary Fellow.[2]

I thought that you might perhaps find some interest in the speculation as to the use of pigment in animals, and its distribution, in which I have indulged in the latter half of the article. (p. 14–28)[3]

I have ventured in one place (p. 19–20) to suggest hypothetically a different account of the relation between the white colour of animals and their liability to be poisoned by noxious plants, than that advanced by yourself, in your work on Domesticated animals and plants.[4]

If it be not asking too great a favour of you, I should much like to hear, whether any facts personally observed by yourself are inconsistent with the hypothesis I have suggested. Whether, that is, you are yourself assured from personal knowledge that coloured animals do eat actually the poisonous plants, without suffering from them, or without suffering so much as do the white.

Again apologising for thus troubling you | Believe me | Yours sincerely | William Ogle.

DAR 173: 3

1 The date is established by the relationship between this letter and the letter to William Ogle, 9 November 1870.
2 Ogle sent an offprint of 'Anosmia; or, cases illustrating the physiology and pathology of the sense of smell' (Ogle 1870a); an annotated copy is in the Darwin Pamphlet Collection–CUL. Ogle read the paper on 25 January 1870. CD was elected an honorary fellow of the Royal Medical and Chirurgical Society of London on 9 June 1868 (see *Correspondence* vol. 16, Appendix III).
3 Ogle refers to the pagination in the offprint he sent CD; the corresponding page numbers in Ogle 1870a are pages 276–90.
4 Ogle argued that animals with less pigment in the olfactory area and with less pigment extending elsewhere had a worse sense of smell, so were more likely to eat poisonous plants and not survive (Ogle 1870a, pp. 281–2). CD had suggested that both lighter and darker animals ate the same poisonous foods, but that the lighter animals were more vulnerable to their damaging effects (*Variation* 2: 335–8).

To William Ogle 9 November 1870

Down, Beckenham, Kent. S.E.
Nov. 9. 1870.

My dear Dr. Ogle,

I have read your paper with the greatest possible interest.[1] It will be extremely curious if your views are confirmed and you are able to explain (not to mention more important points) so many odd little details about the colouring of animals. I have no distinct evidence of the white pigs, sheep, etc. eating the poisonous herbs; but if you will look to the facts given in Vol. 2 p. 337 of my Var. of Dom. animals, you will see that the white parts alone are affected, and this does not seem to agree with your view.[2] By the way, you speak as if the Himalayan rabbit had to provide for its own food when adult, though no doubt you know that it is an artificial Var:[3] I read many years ago Dr Stark's paper, and often wondered over the subject, and this has enhanced my interest in your paper.[4] I hope you will continue your investigation. Would it not be possible to hear of an Albino dog (or some animal at the Zoological Gardens) and it would not be difficult to test a dog's power of smell.

And now I want to beg a little favour of you: I have received conflicting statements about the platysma myoides being brought into strong action in persons suffering from severe dyspnœa;[5] would you kindly observe this point for me? I believe that the contraction of this muscle is easily perceived by the transverse wrinkles on the neck and by the skin near the corners of the mouth being drawn down. I can reconcile the discrepancies between good authorities on this point only by supposing that this muscle acts only during certain forms of dyspnœa.[6]

I have had no communication with you since hearing, about a year and a half ago, that you had been most dangerously ill;[7] and I rejoiced sincerely at your recovery. Since then I have read one or two capital papers by you on the Fertility of flowers.[8] Yours very Sincerely. | Charles Darwin.

P.S. The contraction of the Platysma interests me in relation to expression, which is a hobby-horse of mine,— this muscle is said to contract under great terror. I cannot even positively ascertain whether this is true.— One Doctor declared he had seen it violently contracted in a man with injured brain, who was unconscious and screamed incessantly.[9] In fact this muscle is the bane of existence!

Copy
DAR 147: 193

1 CD refers to Ogle 1870a. See letter from William Ogle, [before 9 November 1870] and n. 2.

2 See *Variation* 2: 337, and letter from William Ogle, [before 9 November 1870] and n. 4.

3 Ogle discussed the foraging of Himalayan rabbits as if they lived and foraged in the wild (Ogle 1870a, pp. 280–1, 283–4).

4 James Stark researched the odour-holding properties of differently coloured substances (Stark 1834). Ogle referred to the paper in Ogle 1870a, p. 286.

5 For the conflicting statements regarding the relation of the platysma myoides, a sheet of muscle in the neck extending from the collar bone to the lower part of the cheek, and dypsnoea (difficulty in breathing), see *Correspondence* vol. 17, letter to James Crichton-Browne, 22 May 1869 and n. 6.

6 See *Expression*, pp. 301–3, for CD's discussion and conclusion about the relationship between the platysma myoides and dypsnoea.

7 The last extant letter from Ogle prior to his letter of [before 9 November 1870] is his letter of [after 7 July 1869] (*Correspondence* vol. 17).

8 Copies of Ogle 1870b and 1870c are in the Darwin Pamphlet Collection–CUL.

9 The doctor CD refers to may have been James Paget, who had offered to make observations of the platysma in patients under the influence of chloroform (see *Correspondence* vol. 15, letter from James Paget, 9 July 1867). See also *Correspondence* vol. 17, letter to James Crichton-Browne, 22 May 1869 and n. 6.

From W. W. Reade 9 November 1870

11 S[t] Mary Abbot's Terrace | Kensington
Nov. 9. '70

My dear Sir

I wrote to you at once on my return at the latter end of August, having come home two months earlier than I anticipated owing to the Phonician jealousy of the merchants who trade up the Niger, & who w[d]. not give me a passage.[1] In that letter I replied to your queries about the frown of the chimpanzee by recommending you if you have a correspondent at Liverpool to make him look out for chimpanzees & there make the experiment.[2] These apes are often brought to Liverpool, & die after a few days or weeks. I had seen one at S[a] Leone—but it was gone before I rec[d] your query. (2) I have seen albinoes but not the blush. By the way I think I must have answered these queries from Africa. I remember asking everybody I met at Accra whether they had seen an albino blush—[3] There was an albino girl there I believe: & it was proposed to make the experiment: but how to raise a blush on an African countenance was the difficulty. My letter must have miscarried:[4] I attributed your silence to your being so much occupied. I thank you much for your kind and cordial letter.[5] The honour of being quoted in your work will be my *first* reward for toils undergone. I assure you I am very proud of it, and also to receive a copy of your work from you.[6] (This address is likely to be permanent).

I do intend to give a full account of my travels: but I think of first bringing out a volume on the history of Africa generally, and then my travels interspersed with essays on parts where I have not been & on scientific questions connected with Africa—the ape & man question for instance—so as to make it a tolerably complete work on Negroland.[7] I have observed that books of travel do not live unless there is something in them besides travels. Can there be a more charming narrative than that of Mungo Park?[8] But it is only a narrative, & is forgotten though he is not. I c[d] easily dash off an amusing work à la Hepworth Dixon in a few months,[9] but I see there is an opening for a comprehensive work on Africa, and I shall try to unite the solidity of a general work with the attraction of a narrative. The task may be beyond my strength; but industry can do much.

I am glad the fever statement caught you in time.[10] I shall bring forward instances of it in my work, but having doctored negros with my own hand given them quinine between the paroxysms &c. I am tolerably confident of the fact. Respecting the beauty question I will express myself more clearly than I have hitherto done. I have examined the question very carefully, and have been assisted in my inquiries not only by intelligent natives speaking English well, but by German missionaries who had lived years among the natives on the Gold Coast & understood their language.[11]

1. I do not venture to assert, nor do I think it probable that the Africans w[d] ever prefer the most beautiful European on mere grounds of physical admiration to a good looking negress.[12] But I assert this: that if Africans had to choose from a number of European women they would certainly select those whom we should select as the best looking. They often talk about the Europeans they see among them, saying so &

so is handsome so & so is ugly. It has been observed that their ideas on these people were the same as ours would be.[13]

Again nothing is more common in Africa than for Europeans to talk with their interpreters & servants about the native women: it has invariably been found that the man's idea of beauty & his master's correspond. To speak not only of the coast but of the interior I have noticed this over & over again: when chatting with the natives one of them has remarked "Dont you think so & so has a very handsome wife" or "look at that pretty girl over there" just as we do in England.[14] I have always found that the faces which are much admired are those which are the farthest removed from the prognathous type—from the regular Guinea negro—a type as exceptional in Africa as that of the Pontina Marsh people in Italy—[15] What are the "points" wh. they admire in a woman I asked a lady who had spent her life among these people. Small hands & small feet she replied: large eyes: a small nose: a well shaped *nape*, or back of the neck. "Why do you tie a string round your upper arm? a native chief was asked. Because that gives one a fine round plump arm he replied. (I have among my notes taken years ago a statement that American women do the same whether Red Indians or Anglo-Americans I dont know: probably the latter who have thin arms).

The admiration of broad hips is carried to extravagance in Central Africa. Tight lacing with us is used not only to compress the waist, but to broaden the shoulders & the hips. "The Europeans" say the Gold Coast negroes w^{d}. be good looking if they had better teeth. But they have fine hair". They themselves use false hair chignons every possible contrivance to make their hair look plentiful just as we do. The *nose* is the only doubtful point to my mind. But girls have been heard to say "I dont want to marry him he's got no nose":[16] this shows that the *very* flat nose is not admired. The large nose, the long nose is also disliked: but the negro nose when large is usually like what we call a bottle-nose though some of them have admirably shaped noses. No doubt habit has something to do with their ideas about the nose—

The black skin is most admired certainly, but that is because the black skin is really the most beautiful. It is not the force of habit, because black skins are the exception not the rule except among a few tribes such as Jollof, the Serracoulis &^{c}.[17] I sh^{d} sum up as my belief & as the belief of those who have lived long among these people that their ideas on the subject of beauty are *on the whole* the same as ours.[18]

Respecting thickness of sculls.[19] The Pagan negroes wear no covering on the head & expose it to a vertical sun. It has been asserted I believe by all anatomists that the negro scull is abnormally thick, & hard.[20] I have seen a bullet wh. had flattened against the frontal bone of a French negro soldier inflicting a slight wound wh I inspected. This thickness is probably intended as a protection against the sun, & may perhaps have been acquired by the practice they have of carrying their infants soon after birth (a week) bare headed in the sun. There is a most curious passage in Herod: comparing the sculls of Egyptians & Persians. The thick sculls of the former he attributes to their being always in the sun, & always bare-headed.[21] I venture to think that this passage which I read on the banks of the Niger of w^{h} river he was the first to speak, may be applied to nat: sel:[22] Any further point you may do me the honour to refer to me upon will be answered to the best of my ability by return of post and I sh[d] be happy now or after the book is published to run down to Beckenham for a couple of hours if there sh[d] happen to be a variety of points to discuss. If you write to me again pray do not fatigue yourself to write autographically but talk to me through your secretary.

Hoping that you will get better health now that your present work is approaching its consummation which so many are eagerly looking forward to

I remain | Yours very truly | Winwood Reade

Jollof is so written.[23]

The people of the Gold Coast have been accustomed to Europeans. But my ideas on this subject are not based on my observations there alone— Much of the apparent horror of the white man among savage negroes is owing to their ideas not that the Devil is white (they have no Devil corresponding to ours) but that all demons, spirits or angels are white.[24] I know the case of one traveller in the far interior who having been carefully surveyed by a woman for some time she said— "He is a fine young man". Again the beard is usually much admired by negro women, though negroes scarcely ever have any.[25]

I may be able to supply you with additional particulars about love-making exogamy &[c] for your 2nd edition: & possibly on other points w[h] will be suggested by your work.[26]

DAR 85: 109–112

CD ANNOTATIONS
1.1 I wrote ... language. 3.7] *crossed pencil*
4.1 1. . . . negress. 4.3] *scored red crayon*
5.3 To speak ... prognathous type— 5.8] *scored red crayon*
5.8 a type ... Italy— 5.9] *double scored red crayon*
6.1 The admiration ... Africa.] *scored red crayon*
6.4 But ... hair".] *double scored red crayon*
6.8 but the ... noses. 6.9] *double scored red crayon*
7.3 I sh[d] ... as ours. 7.5] *double scored red crayon*
8.1 Respecting ... Reade 10.1] *crossed pencil*
12.3 the Devil is white 12.4] *before closing square bracket pencil*
12.4 (they ... man". 12.7] *crossed pencil*
12.7 Again ... any. 12.8] *double scored pencil*
Top of letter: '(Beauty)' *added pencil*

1 See letter from W. W. Reade, 3 September 1870 and n. 2. On Reade's difficulty in attaining passage up the Niger, see Reade 1873, 2: 508–9; see also letter from W. W. Reade, 4 June 1870 and n. 15.

2 CD's query was evidently in a letter that has not been found; see letter to W. W. Reade, 30 June [1870] and n. 8.

3 S[a] Leone: Sierra Leone. Accra is a port in Ghana.

4 In his letter of 3 September 1870, Reade reported that he had seen albinos in Africa but had not observed them blushing.

5 The letter from CD to Reade has not been found.

6 CD cited Reade in *Descent* 1: 289 and 2: 247, 344, 346, 350, 357–8, 374, and included quotations in *Descent* 2: 351, 357. Reade was on CD's presentation list for *Descent* (see DAR 210.11: 32, and *Correspondence* vol. 19, Appendix IV).

7 On Reade's *The martyrdom of man* (Reade 1872), and the unusual structure of Reade's *African sketch-book* (Reade 1873), see Driver 2001.

8 CD read an edition of Mungo Park's *Travels in the interior districts of Africa* (Park 1799) in 1838 (see CD's Reading notebooks, *Correspondence* vol. 4, Appendix IV).

9 William Hepworth Dixon had published books on his travels in the Middle East, the United States, and Russia (Dixon 1865, 1867, and 1870).

10 See the postscript to the letter from W. W. Reade, 6 November 1870.

11 For CD's first query to Reade regarding perceptions of beauty among different peoples, see *Correspondence* vol. 16, letter to W. W. Reade, 21 May [1868]. Reade addressed the question in his letters of 4 June 1870, 3 September 1870, and 6 November 1870. The Gold Coast was the name of the former British colony that now makes up the larger part of the country of Ghana (*Columbia gazetteer of the world* s.v. Ghana).

12 CD quoted this statement of Reade's in *Descent* 2: 351.

13 CD cited Reade on this point in *Descent* 2: 350.

14 CD cited Reade for this observation in *Descent* 2: 344.

15 The Pontine Marshes were in the Lazio region of central Italy, and were famous for malarial fever. (*EB*). See also Reade 1872, p. 272.

16 CD quoted this statement in *Descent* 2: 350.

17 Europeans called some of the peoples living south of the Senegal River the Serrawoolli or Saracolets (Prichard 1836–7, 2: 80).

18 CD cited Reade on this point in *Descent* 2: 350.

19 See the postscript to the letter from W. W. Reade, 6 November 1870.

20 James Cowles Prichard discussed the supposition that Africans had thick skulls and noted authors who had written of it in Prichard 1836–7, 1: 282–3.

21 See Herodotus 3.12.

22 Herodotus 3.32. Nat. sel.: natural selection.

23 CD may have asked for the correct spelling of 'Jollof', as the word was then also sometimes spelled 'Iolof' or 'Whalof' (see Prichard 1836–7, 2: 78).

24 CD cited Reade on this point in *Descent* 2: 346.

25 CD cited Reade on this point in *Descent* 2: 350.

26 CD did cite Reade on additional points in the second edition of *Descent*, though not on the practice of exogamy. For CD's discussion of exogamy, see *Descent* 2: 360, 364.

To P. L. Sclater 9 November 1870

Down. | Beckenham | *Kent. S.E.*
Nov 9. 1870

My dear Mr Sclater

Your offer is *most* generous & w^{d} be an immense benefit for my book.[1] But you do not know what you are offering; there are about 200 pages, of the size of my origin, on birds. I have corrected half of them in slips, & when in sheets they w^{d} be much better for you to read, though there w^{d} be the great disadvantage that I c^{d} not without strong cause, make great alterations. There w^{d} be hardly any thing new to you in any part, & you c^{d} in most cases be able give much better illustrations than I have been able to do. Therefore considering every thing, unless you renew yr offer, I will assume that you have not time to read through so many pages; but in this case I will most gladly accept yr offer of correcting another list of names.

With very sincere thanks | yours very truly | Ch. Darwin

P.S. I have rec[d] M[r] Hudson's proof sheets but have not yet had time to read them—[2]

The only possible interest my 4 chapters on Birds c[d]. have for you is the collection of facts on some trifling points, all relating more or less to sex.

LS(A)
Zoological Society of London

[1] Sclater's letter has not been found, but see the letter to P. L. Sclater, 4 November [1870]. Sclater had evidently offered to check the page proofs of the sections on birds in *Descent*, rather than just lists of names, as CD had suggested.

[2] CD presumably refers to the later parts of William Henry Hudson's letters to the Zoological Society of London on the ornithology of Buenos Aires (Hudson 1870). See letter to P. L. Sclater, 30 August [1870].

From William Ogle [10–17 November 1870][1]

34 Clarges Street | Piccadilly.

Dear M[r]. Darwin

Many thanks to you for your kind letter and good wishes.[2]

I am afraid I must plead guilty to a very stupid blunder in having attributed to the Himalayan rabbits the necessities which only beset wild breeds.[3] As regards the animal with white spots which alone were affected while the colored parts escaped I had considered the case, and thought there did not seem any actual evidence that poison had anything to do with the affection, or had indeed been taken.[4] I find cases recorded, in which white spots suffer, and the cause is supposed to be poison—others, in which the same effect is attributed to poison *and* to exposure to sunlight; and a third group, more numerous, in which the same effect is supposed to have been produced by lightning.

In none of them is there any positive evidence either of poison, or of lightning, while in all, the animals appear to have been most probably, and in some certainly, exposed to great sunlight. I fancy therefore that it is more likely that this one cause produced the similar ill effects in all the cases, than that the causes should have been so heterogeneous. In the case of the animals struck, as it was supposed, with lightning, it is expressly noted (Phil. Trans. 1776) that only the white parts on the back and flanks were affected, the equally white patches on the belly escaping as also even did white parts on the upper side if the hair in them was very long and curly.[5] This is just what might be expected, if sunlight caused the ill effects; but seems unintelligible on the supposition that either poison or lightning was the origin of the mischief.

I will most certainly make a point of observing the condition of the Platysma in any cases of severe dyspnœa which I may find— I have already seen two in the Hospital since I received your letter—one a case of emphysema—the other of cardiac dyspnœa— In neither could I detect any action of this muscle. Certainly there was no transverse wrinkling of the neck, nor drawing of the skin about the corners of the mouth.[6]

It appears however very doubtful whether the contraction of this muscle does, as the anatomical books assert, wrinkle the skin *transversely*. The statement[7]

Incomplete
DAR 173: 4

CD ANNOTATIONS
1.1 Many thanks ... considered the case, 2.4] *crossed pencil*
4.1 I will ... *transversely*. 5.2] *crossed pencil; after opening square bracket blue crayon*
Top of letter: 'Dom. Animals[8] | (Wyman of Boston)[9] | (Platysma) | I suppose correct | Dyspnœa' *pencil*

1 The date range is established by the relationship between this letter and the letters to William Ogle, 9 November 1870 and 17 November [1870].
2 Letter to William Ogle, 9 November 1870
3 See letter to William Ogle, 9 November 1870 and n. 3.
4 See letter to William Ogle, 9 November 1870, and Ogle 1870a, p. 283 n. 1.
5 The reference is to an article about the effects of lightning on a bullock in *Philosophical Transactions of the Royal Society* (Lambert and Green 1776).
6 CD included this information from Ogle, as well as some of his later observations of those experiencing difficulty breathing in *Expression*, p. 301.
7 For CD's account of the platysma myoides, including several of Ogle's observations, see *Expression*, pp. 298–303.
8 *Variation*.
9 Jeffries Wyman. See letter to William Ogle, 17 November [1870].

From W. W. Reade 11 November 1870

Conservative Club
Nov. 11. '70

My dear Sir

I am very glad you think my statement worth quoting:[1] of course I do not deny that the eye is a creature of habit to a great extent.

I am very sorry to hear that you are in such weak health. How great must be your love of knowledge to make you go through such labours, when many other men would be merely torpid or fretful invalids.

When I have had the pleasure of reading your book I will note down such comments as I am able to make on the parts referring to the Africans.[2] By that time I shall have read many books on Africa.

By the bye I forget whether I mentioned a most important case, of a young Mandingo Chief near Sierra Leone falling in love with an English lady in that colony.[3] I believe it is well authenticated; I heard a mulatto lady telling the story but unluckily forgot to get the particulars. I think I shall write to my friend Heddle[4] & get a statement of the facts.

Hoping you will get stronger I remain | my dear sir | yours very truly | Winwood Reade

DAR 176: 41

1 CD's response to Reade's letter of 9 November 1870 has not been found.

[2] Reade refers to *Descent.*
[3] Reade also wrote of the Malinke (Mandingo) chief in his letter of 3 September 1870.
[4] Charles Heddle.

To P. L. Sclater 11 November [1870][1]

Down. | Beckenham | Kent. S.E.
Nov. 11

My dear Mr. Sclater

I will most gladly accept your kindness.—[2] I look at the delay caused as nothing comparatively to the great benefit.— I never expected or hoped for many criticisms, but I still hope you will point out any serious error,—whatever trouble this may cause to my Printers.— I suppose I shall soon receive Revises, but Messrs Clowes[3] sometimes delay the 2d proofs till $\frac{2}{3}$ of a whole vol. is corrected in first proof.—

Mr Hudson's paper is very interesting & it pleases me to see so staunch a hater of evolution a little staggered at the end of his paper.—[4]

Yours very truly obliged | Ch. Darwin

I will not now waste quite so much time in trying to find every name quoted in some book; so you will doubly help me.

Edward Ford (private collection)

[1] The year is established by the relationship between this letter and the letter to P. L. Sclater, 9 November 1870.
[2] Sclater's letter to CD has not been found, but he had evidently reiterated his offer to read the proof-sheets of the chapters on birds and mammals in *Descent* (see letter to P. L. Sclater, 9 November 1870).
[3] William Clowes & Sons were printing *Descent.*
[4] Sclater had sent CD proofs of William Henry Hudson's letters to the Zoological Society of London (Hudson 1870; see letter to P. L. Sclater, 9 November 1870). CD may refer to a passage at the end of a letter read at the Zoological Society on 23 June 1870 (Hudson 1870, p. 550), on the waste of other birds' eggs caused by *Molothrus bonariensis*:

> I often wonder that the little birds in whose nests they lay do not become extinct, or all but extinct, by their means. . . . How strange that it should be so disorderly in the midst of the general order of nature! Or must we come to consider these habits of the *Molothrus bonariensis* 'not as especially endowed or created instincts, but as small consequences of one general law,' namely, transition?

The quotation is from *Origin*, p. 244, where CD had said that it was more satisfactory to him to see such instincts as the cuckoo ejecting other birds from the nest, ants making slaves, and ichneumonidae larvae feeding within the bodies of live caterpillars as the consequences of one general law, 'namely, multiply, vary, let the strongest live and the weakest die', rather than as specially endowed or created instincts. *Molothrus bonariensis* is the shiny cowbird.

From Jacob Heinrich Schmick 12 November 1870

Sir,

The undersigned begs for the honour of being allowed to submit to your judgment two small books of his, in which he has developped and supported by facts a new theory on the great changes of the earth's surface.[1]

This theory, the author hopes, will attract your kind attention so much the more, as it does entirely away with one objection made to your incomparable 'Mutation of Species', viz. the interruptions and want of continuance observable in the fossil remains of anti-diluvian organic life.[2] According to this new theory stoppages of many thousands of years must necessarily have occurred in the animal and vegetable life of the same localities, must be observable in the sudden transitions of petrefacts and so lead to the opinion, as if new creations had taken place; whereas an uninterrupted line of development must exist, and will be seen as soon as we are able to follow up the alternation of strata of both the hemispheres of the earth.

Hoping you will in due time take a little notice of these two books and have the kindness to communicate to the author your opinion about them, he signs as | Your | most ob. serv. | Dr. Schmick, | Professor in the Realschule.[3]

Cologne
Nov. 12th 1870.
(Trutzenberg 68)

DAR 177: 55

1 Copies of Schmick 1869 and 1871 are in the Darwin Pamphlet Collection–CUL; both are annotated.
2 CD discussed the imperfection of the fossil record in *Origin*, pp. 279–311.
3 The Realschule was a secondary school that developed in the mid eighteenth century in Germany, with a greater emphasis on the natural sciences (*EB*).

From George Cupples 14 November 1870

The Cottage, | Guard Bridge | Fifeshire. N.B.
Novr. 14/70—

My Dear Mr Darwin,

Best thanks for your very kind letter, informing me of little "Bran's" safe arrival and hospitable reception.[1] I ought to have acknowleged this trouble on your part before—but was accidentally prevented.

The hamper is not worth returning. I had no expense whatever in connection with the puppy's dispatch to you.[2]

With best regards from Mrs Cupples to Mrs and Miss Darwin,[3] and to you, I remain | Yours ever truly | George Cupples

Charles Darwin, Esqr. | Down— | Kent—

DAR 161: 295

1 CD's letter to Cupples has not been found, but see the letter to George Cupples, 3 November [1870]. Bran was a deer-hound puppy.
2 See letter to George Cupples, 3 November [1870].

[3] Cupples refers to Anne Jane Cupples, and to Emma and Henrietta Emma Darwin.

To W. D. Fox 15 November [1870][1]

Down. | Beckenham | *Kent. S.E.*
Nov. 15

My dear Fox

I suppose that you are now settled for the winter (which has come with a vengeance today) in the Isle of Wight.— My wife wrote about Governesses when I got your letter of Oct. 28th & asked you whether you could come down here, but I suppose that was impossible.—[2] You are a good-for-nothing man, not to have said something about your health & strength. I am a good deal used up with the excessive labour to me of correcting proofs of my present book in 2 Vols; & now it will not be finished till end of year.[3] I owe many facts to you in the larger part viz on sexual differences of Birds.[4] I shall be delighted to send you a copy, whenever it is published, though I have sometimes had misgivings that I ought not to do so, as I fear you will disapprove of my main conclusion on the origin of man; but I can most truly say that I have written nothing without deliberate consideration & acquiring all the knowledge which I possibly could.— I will send it, as I know well that you are a charitable man & do not without good evidence believe in bad motives in others.—[5] It is very delightful to me to hear that you, my very old friend, like my other books, & you were one of my earliest masters in Nat. History.[6]

I wish I had got a little more strength. I feel that each job as finished must be my last.—

Farewell | Your's affectionately | Charles Darwin

American Philosophical Society (385)

[1] The year is established by CD's reference to correcting the proofs of *Descent* (see n. 3, below).
[2] See letter from W. D. Fox, 28 October [1870].
[3] CD was correcting the proof-sheets of *Descent*, which was published early in 1871.
[4] See *Descent* 1: 270, 2: 46, 102–3, 104, 107 n. 7, 114, and *Correspondence* vol. 16.
[5] Fox's name appears on CD's presentation list for *Descent* (see DAR 210.11: 32, and *Correspondence* vol. 19, Appendix IV).
[6] See letter from W. D. Fox, 28 October [1870], and *Correspondence* vol. 1.

To John Murray 15 November [1870][1]

Down. | Beckenham | *Kent. S.E.*
Nov. 15th

My dear Sir

When at leisure I shd much like to hear about your book-sale, as far as I am concerned.—[2]

In the enclosed letter I cannot read the signature— is it Rotterdam sufficient address? I will send him clean sheets as I sh[d] like a translation & he will no doubt buy stereotypes.[3]

I cannot conceive what the reason is, but I have as yet received clean sets of only 3 first sheets from M[essrs] Clowes.—[4]

yours very sincerely | Ch. Darwin

National Library of Scotland (John Murray archive) (Ms.42152 f. 43)

1 The year is established by CD's reference to the Dutch translation of *Descent* (see n. 3, below).
2 On Murray's November 'sale dinners' for the principal booksellers, see J. Murray 1908–9, p. 540.
3 CD refers to a letter from Hendrik Adrianus Kramers, a bookseller and agent for translation rights in Rotterdam; the letter has not been found. It concerned a Dutch translation of *Descent*. See letter from Hermanus Hartogh Heijs van Zouteveen, 27 November 1870. There is a letter from Kramers to Murray requesting proof-sheets of *Descent* for a Dutch translation in the National Library of Scotland, John Murray Archive (42153 f. 53).
4 CD was waiting for clean proof-sheets of *Descent* from the printers, William Clowes & Sons.

From John Murray 16 November [1870][1]

50, Albemarle S[t]. | W.
Nov[r]. 16

My Dear Sir

I think you will not be disappointed when I tell you that 1900 copies were taken of your new Book at my sale on Friday also 340 of Origin of Species.[2]

Clowes will send you immediatel the fair sheets to within 2 of the end.[3] It is prudent to keep back a few sheets in both volumes least any check occurring here—the foreigner sh[d] supplant you at home or anticipate you.

Rotterdam attached to Kramers name will be sufficient address[4]

I am Dear Sir | yours very faithfully | John Murray

Chas Darwin Esq | &c &c &c

DAR 171: 381

1 The year is established by the relationship between this letter and the letter to John Murray, 15 November [1870].
2 See letter to John Murray, 15 November [1870] and n. 2. Murray refers to *Origin* 5th ed. (published in 1869) and *Descent* (published early in 1871).
3 See letter to John Murray, 15 November [1870] and n. 4.
4 See letter to John Murray, 15 November [1870] and n. 3. The reference is to Hendrik Adrianus Kramers.

From John Dean Caton 17 November 1870

Ottawa Illinois USA
Nov 17. 70

Dear Sir.

In my former answer to your inquiry about the difference in color of the male & female elk I expressed the opinion that the male is no darker than the female—[1] I have now to make a partial correction of that statement. That report was made on observations made in the latter part of winter, and spring before the winter coat had been shed. Observations taken in the fall of the year, show that the darker portions, as the neck belly and legs of the male are very much darker than the female while on the sides and back the color on both is substantially the same— During the winter season these dark portions on the male fade out till in the spring they justify my former statement.[2]

I have also to correct a very important error into which I fell when writing "American Cervus". I there stated that the elk sheds its coat but once a year.[3] I have only discovered within the last few months, that in this I was mistaken, they certainly have a second peelage in the latter part of summer— The shades of color and the length of the hairs of the summer coat are so nearly identical with the new winter coat and the former disappears & is replaced by the latter so gradually, that it is exceedingly difficult to detect the change, even when attention is directly called to it—[4] In only two specimens of my herd of 46 was it sufficiently distinct to arrest attention & then it required a months careful study to assure me of my former error

I have during 1869 & '70 added to my collection live specimens both male & female, of the Mule deer (*C. Macrotis*) & Columbia Black Foot deer (*C. Columbianus*) Both upon which I am making careful observations which I shall embody in another paper[5]

It may interest you to state a very important physiological difference I have observed between the elk and the three smaller species which I have of that family—[6] In the elk the *theca* extends forward along the abdomen to within three inches of the umbilicus & has no pendent prepuse almost precisely as is observed in the common bull,[7] while with the three smaller species, the *theca* is suspended from a point so near the scrotum that when the animal is standing it occupies a vertical position not more than half an inch from that member. Its posterior measurement it is from 3 to $4\frac{1}{2}$ inches while anteriorly it is say two thirds that length. This lower portion may be considered an exagerated *prepuse* which as before stated is entirely wanting in the elk.[8] I may attach undue importance to this marked distinction but it really seems to me to be quite worthy of observation

Upon fawns of the virginia deer I observe *invariably*, both male & female, a black spot at the place where the future horn is to appear on the male

On about five percent of the virginia deer when the winter coat first appears, a line of light colored spots appears on each side of the back, extending from the shoulder to the tail about five inches apart on the back but approaching each other towards the tail— Of these there are 16 between the shoulder & hips, and thence to the tail 5. These are more distinct on some specimens than on others and on but few can they be seen all the way back. Similar lines are always seen on the spotted fawn which although brighter in color are not so uniform in outline.[9] Their exact uniformity in number whenever observed fixes them as a true characteristic of the animal

What gradual change may this animal be undergoing in this regard? Is there such a line of spots on the spotted deer of Europe? These spots entirely disappear in a month or six weeks. On one specimen I observed that the summer coat remained on the places of these spots when entirely gone on the regions surrounding them—

My virginia deer have so deteriorated in fertility that the number has diminished one half in four years, confirming your observations on the effect of domestication of many wild animals on their reproductive powers— I think however that a few of the original stock have constantly reproduced vigorous & fertile proginy which has continued through several generations, and I am not without hope of eventually rearing a stock from them, which will bare domestication, & remain vigorous & fertile—[10] The gradual diminution in the growth of the horns through succeeding generations is quite observable.— The female virginia deer has four Mammae equally active, like the common cow. My observations of the other species, in this regard are not completed—

Audubons observations of the tuft of long course hares appearing on the brest of the aged female wild turkis will hardly hold good in this latitude. In all my own observations which embrace perhapes a thousand specimens. I have only seen it on the bird in my parks, on which it appeared when she was one year old. I have enquired of many hunters and of the game dealers of Chicago, who handle very many of these birds anually, and altogether have heard of but one other instance. I am persuaded therefore that this appendage is much more rare here than in Kentuckey 5 or 6 degrees further south where Audubon made his observations—[11] My wild & domestic turkeys cross freely and the hybrids are very fertile[12]

When may we expect your new work?[13] I am looking for it anxously

Yours very truly | J D Caton

Chas Darwin Esq

DAR 83: 172–4

CD ANNOTATIONS
1.5 Observations ... same— 1.7] *scored blue crayon*
5.1 Upon ... male 5.2] 'p. 359' *pencil, circled pencil*
9.2 In all ... fertile 9.9] *scored blue crayon*
Top of letter: 'Colours of ♂ & ♀ elk | Spots on Deer | Hair on Turkey—' *blue crayon*

1 For the discussion regarding the North American elk or wapiti (*Cervus canadensis*), see *Correspondence* vol. 17, letter to J. D. Caton, 20 March 1869, and letter from J. D. Caton, 5 May 1869.

2 See Caton 1877a, pp. 136 and 144–6; CD cited Caton for this information in *Descent* 2: 289. Caton had kept elk (*Cervus canadensis*), and other members of the deer family, in an enclosure since the early 1860s; see Caton 1877a, p. 78.

3 Caton's paper on North American *Cervus* (Caton 1868) was also published in Caton 1880, pp. 146–75; for his discussion of the coat, see pages 148–9 and 152. An annotated offprint of Caton 1868 is in the Darwin Pamphlet Collection–CUL.

4 Caton described the discovery of his error in Caton 1877a, pp. 125–6.

5 For Caton's description of the mule deer and the black-tailed deer, see Caton 1877a, pp. 93–6 and 97–9. Both deer (Caton's *Cervus macrotis* and *C. columbianus*) are now *Odocoileus hemionus*; *C. columbianus* is sometimes classified as a subspecies, *O. hemionus columbianus*.

6 Caton refers to the mule deer, the black-tailed deer, and the common, Virginia, or white-tailed deer (*Cervus virginianus*, now *Odocoileus virginianus*).

7 Caton refers to the sheath, or theca, enclosing the penis, and the prepuce, or foreskin; see Caton 1877a, p. 269.

8 Caton 1877a, p. 270.

9 CD cited Caton for information on spots in adult Virginia deer in *Descent* 2: 303–4. When Caton described them in Caton 1877a, pp. 156–7, he noted that they had been overlooked by naturalists before *Descent* was published.

10 See Caton 1877a, pp. 304–8. For a reference by CD to decreased fertility in plants and animals removed from their natural habitat, see *Origin*, p. 9. CD also discussed the 'evil effects of close interbreeding' in *Variation* 2: 114–22, and discussed 'The advantages and disadvantages of changed conditions of life; sterility from various causes' in *Variation* 2: 145–72. See also letter from J. D. Caton, 5 May 1869 (*Correspondence* vol. 17).

11 See *Correspondence* vol. 17, letter from J. D. Caton, 5 May 1869, and letter to J. D. Caton, 24 May 1869 and n. 3. The references are to John James Audubon, and Audubon 1831–[9], 1: 15. In *Descent* 2: 180, n. 29, CD cited Audubon but also included Caton's observation that females rarely acquired the tuft of hair.

12 See Caton 1877b.

13 *Descent.*

To John Murray 17 November [1870][1]

Down. | Beckenham | *Kent. S.E.*
Nov. 17th

My dear Sir

I am much pleased at your sale. I suggest, though suggestions to you on such a point are superfluous, that 3000 copies be printed.[2] With my distribution & copies & for Review & Libraries, not much more than 500 will now be left on hand.[3]

I fear I shall not have done correcting till end of year; the corrections are frightful & I think I must have another week's rest

Yours very sincerely | Ch. Darwin

National Library of Scotland (John Murray archive) (Ms.42152 f. 242)

1 The year is established by the relationship between this letter and the letter from John Murray, 16 November [1870].

2 See letter from John Murray, 16 November [1870]. Murray had proposed to print 2500 copies of *Descent* (letter from John Murray, 28 September [1870]). In February 1871, 2500 copies were printed, with a second issue of 2000 in March (R. B. Freeman 1977, p. 129).

3 There are eighty names on CD's presentation list for *Descent* (see DAR 210.11: 32 and *Correspondence* vol. 19, Appendix IV).

To William Ogle 17 November [1870][1]

Down. | Beckenham | *Kent. S.E.*
Nov. 17.

My dear Dr Ogle

I am very much obliged about the platysma; & hope that you will keep the subject a little before your mind, more especially as I see that you are quite au fait about expression. It is new to me that this muscle can ever be brought into voluntary action.[2] I have been coming to the same conclusion as you have, viz that this muscle has nothing to do with expression.[3] Dr C-Brown tells me that it does not contract with insane patients under extreme terror.[4] Yet I was unwilling to think that the belief was quite a delusion, owing to Duchenne's striking photographs & some older statements to the same effect.[5] (possibly the source of whole belief) If Gratiolet's statement cd be trusted that this muscle contracts under dyspnoea, it is conceivable that it might act under terror from association with panting breath.[6] One of my sons in sounding certain notes on the flute draws the corners of his mouth much backwards & downwards, & then I can see radiating longitudinal furrows on each side of his neck.[7] A clever surgeon, who has attended to these subjects, tells me that this is produced by another muscle (name forgotten) which is attached below to the clavicles.[8] You will know all about this, & at some future time I shd be grateful to hear whether there is such a muscle, the contraction of which wd produce the longitudinal furrows.

I cannot yet give up the ghost about white colour & vegetable poisons. If you cd prove that white animals were deficient in the power of smell I shd be more staggered.[9] I cannot as yet think that so many observers have been deceived. I wd suggest, (if you are willing, as I hope, to continue the subject,) that you shd send yr paper to Prof. Wyman of Boston, U. States, with a letter asking him if he cd get it observed in Florida, by some careful man whether if the paint-root is given to black & white pigs, both will eat it.[10]

With many thanks for yr very interesting letter yours very sincerely | Ch. Darwin

I wonder whether white pigs find truffles & pig-nuts as well as black pigs.

LS(A)
DAR 261.5: 4 (English Heritage 88205902)

1 The year is established by the relationship between this letter and the letter to William Ogle, 9 November 1870.

2 CD had enquired about the platysma myoides muscle in his letter of 9 November 1870. Some of what Ogle wrote in reply was in the now missing portion of his letter of [10–17 November 1870].

3 See *Expression*, pp. 298–303.

4 See letter from James Crichton-Browne, 15 March 1870, and letter to James Crichton-Browne, 18 March 1870.

5 CD refers to Guillaume Benjamin Amand Duchenne, and photographs in Duchenne 1862. There is an annotated copy of Duchenne 1862 in the Darwin Library–CUL (see *Marginalia* 1: 209–10). See also *Correspondence* vol. 17, letter to James Crichton-Browne, 22 May 1869. CD included a photograph of Duchenne's illustrating the platysma myoides in *Expression*, plate VII, figure 2, as well as a drawing copied from another photograph (*Expression.*, p. 299).

6 See letter to William Ogle, 9 November 1870 and n. 5. CD refers to Louis Pierre Gratiolet and Gratiolet [1865], p. 167. There is an annotated copy of Gratiolet [1865] in the Darwin Library–CUL (see *Marginalia* 1: 345–7).

7 In *Expression*, p. 302, CD mentioned the action of the muscle in 'a young man' playing the flute. See also *Correspondence* vol. 17, letter to James Crichton-Browne, 8 June 1869. The flutist was CD's son Francis (see *Correspondence* vol. 19, letter to Francis Darwin, 25 March [1871]).

8 The surgeon and muscle have not been identified. CD had earlier mentioned the muscle in his letter to James Crichton-Browne, 8 June 1869.

9 See letter from William Ogle, [before 9 November 1870] and n. 4, and letter to William Ogle, 9 November 1870.

[10] CD had given information from Jeffries Wyman about the resistance of black pigs to a poison in the southern United States in both *Origin* 3d ed., p. 12, and *Variation* 2: 227 and 336. Wyman travelled to Florida eight times between 1852 and 1874 (*ANB*).

To ? 17 November [1870][1]

Down. | Beckenham | *Kent. S.E.*
Nov 17

Dear Sir

I am much obliged for your note to Mr Murray;[2] but I have to inform you that I have already agreed with Prof. Victor Carus to translate my next book—

Dear Sir | yours faithfully | Ch. Darwin

LS
Staatsbibliothek zu Berlin – Preußischer Kulturbesitz (Slg. Autogr. Darwin, Charles Robert, Bl. 3/4)

[1] The year is established by the style of the address, with 'Bromley' deleted and 'Beckenham' added by hand, which CD used from April 1869 until May 1871, and from the reference to an agreement with Julius Victor Carus. Carus confirmed his willingness to translate *Descent* into German in his letter of 2 October 1870. In November 1869, CD seems still to have been uncertain whether Carus would accept the task and was asking him to consider other candidates (*Correspondence* vol. 17, letters to J. V. Carus, 9 November [1869] and 20 November 1869).

[2] John Murray may have enclosed the note with his letter of 16 November [1870].

From W. D. Fox 18 [November 1870][1]

Broadlands | Sandown | I. Wight
18

My dear Darwin

The sight of your hand writing did me much good. It would have greatly rejoiced me to have been able to run down to Down for a day, as Mrs Darwin kindly asked me to do, but I am obliged to be careful at this time of year, for fear of being laid up, and I felt I ought not to delay getting to Winter Quarters.[2] As you ask me how I am, I am glad to say much better than I have been for some months, & quite hope to be able to get about thro the winter. I was very ill a few months ago, and hardly thought I should get over it.

I get very stiff and old in my feelings of body, and childish in my mind I think. I do not think I was ever more *young* in mind— in fact I have a very enjoyable existence, & I know few I w^d exchange with. There now.

I wish you could give yourself a little rest, but I know you cannot. In cælo quies, in terra nulla.[3] I hear *sad* tales about your Book about to come forth.[4] I suppose you are about to prove man is a descendant from Monkeys &c &c Well, Well!— I shall much enjoy reading it. I have given up that point now. The three main points of difference to my mind—were that Men drink, smoke & thrash their wives—& Beasts do not.

But alas my faith is overthrown entirely. The Lady Monkey from the Andamans— drinks & smokes like a Christian; & evidently the Gentleman w[d] thrash, if not kill the Lady, if he had an opportunity.[5]

I always look at Books as I do Newspapers. I am not bound to tye my mind to that of the writer. There are points in your unrivaled Book "The Origin of Species"— which I do not *come up to*— but with these few expressions omitted, I go with it completely. I do not think even *you* will persuade me that my ancestors ever were Apes— but we shall see.

I have no *religious* scruples about any of these matters. I see my own way clearly thro them— —but I see many points I cannot get over, which prevent my going "the whole Hog" with you.

In a few years—if not sooner—we shall know a great deal more than we do now.— We are sadly cribbed *here*, and ones mind feels the impossibility of grasping what one longs to do. Well, well! (as a friend of mine always says in a difficulty) let us do *our* best, & hope for better things. I must run over & see you some day. Why not you & M[rs] Darwin run over here, when you have finished your Book— you can study my little Apes & Apesses—

Kindest regards to M[rs] Darwin & thanks for her note—

Always yours Affec[ly] | W D Fox

DAR 164: 192

[1] The month and year are established by the relationship between this letter and the letter to W. D. Fox, 15 November [1870].

[2] See letter to W. D. Fox, 15 November [1870]. Fox's summer home was Delamere Rectory in Cheshire.

[3] *In caelo quies, in terra nulla*: in heaven there is peace, on earth, none (Latin).

[4] Fox refers to *Descent*.

[5] A female 'Andaman monkey' was presented to the Zoological Society of London in July 1869. She was described in *Land and Water*, 24 July 1870, p. 57, by Abraham Dee Bartlett, the superintendent of the Zoological Gardens, who thought she represented a new species, which he proposed to call *Macacus andamanensis*. In April 1870 a letter was received from the Andaman Islands pointing out that there were no monkeys indigenous to the Andaman Islands, and that the few that existed there had been brought from Burma (*Proceedings of the Zoological Society of London* (1870): 220). In June 1870, Bartlett secured a male 'Andaman monkey' for the society; it was identified by Edward Blyth as identical with a species that he had named *Inuus leoninus* (now *Macaca leonina*, the northern pigtail macaque). The female, 'Andaman Jenny', had become well-known to visitors to the Zoological Gardens for smoking and performing other tricks. (*Proceedings of the Zoological Society of London* (1870): 663–4.) The male is illustrated in the plate in *Proceedings of the Zoological Society of London* (1870): facing p. 664. The macaque in the background is presumably Andaman Jenny.

From Asa Gray 21 November 1870

Botanic Garden, | Cambridge, Mass.
Nov. 21. 1870

My Dear Mr. Darwin

I have been a long while studying a family of plants, the *Polemoniaceæ*, in the flowers of which I see, now and then, what ought to be dimorphism of the *Primula* sort, yet not complete or reciprocal.[1]

One of the best-marked cases is that of Gilia (Leptosiphon) nudicaulis (= Collomia nudicaulis Hook. & Arn.)[2] The style in some individuals is very long and more or less exserted; in others short and deeply included in the tube of the corolla. But the stamens are the same in both, i.e. the anthers sessile in the throat. G. (Leptosiphon), androsacea[3]—common in cultivation has generally long style,—sometimes a short one, And I think I have seen the short style in some others of the group.

In the *Ipomopsis* section of *Gilia* (as I limit it) there are 3 species.[4]

1. *G. coronopifolia*—with no dimorphism that I can see, but with stamens & style exserted.

2. *G. aggregata* (pulchella), with stamens in some included in the throat and the style much shorter, i.e. the stigmas lower; others, and more of them, with the stamens exserted, and the style also,—as long or longer.[5]

3. *G. subnuda*, a n. sp. of few known specimens—all uniform, with included stamens & more included style.—like the one form of G. aggregata.

Most Gilias show no attempt at dimorphism.

In one group of 2 species, the stamens (inserted, as is common, just under the sinuses of the corolla,) are so remarkably long that I looked for & expected to see a counterpart form with short filaments or lower-inserted ones.[6] Nothing of the sort is seen. The styles are neither particularly long nor short.

In Polemonium cæruleum—hardly in any other, the stamens, & I think the style with them, varies more or less in proportion to the corolla—but not in any decided way so that I can sort the dried specimens into two kinds.[7]

Finally in *Phlox*, the stamens are on one plan in all,—included, inserted as you know at unequal heights, or the uppermost 2 or even 3 anthers just protruding their tips from the orifice. But the style in each species is either *long* (the stigmas exserted or barely included) or *short*, & when so generally very short, so as to leave the stigmas low down in the tube, below or on a level with the lowest anthers.[8]

As I went on with the study of the species I took with me a foregone conclusion that this was a char. of dimorphism, and therefore would not use it as *specific* or sectional.

And one common species, *P. subulata*, seemed to confirm this view of it. I am obliged to regard as the same species *P. nivalis* (*P. Hentzii*), which has no obvious difference that I can rest on except a short, I think rather variably short, style.—[9] *But also generally a pair of ovules in each cell*, while ordinary *P. subulata* has only one. Yet the differences are not absolute, some short-styled ovaries having solitary ovules in one or two (I am not sure if in all 3) of the cells. Here, you see, if the char. of style and ovules corresponded completely, I should have recognised two species. Is dimorphic variation here going on to set up two independent species.—not interbreeding?

And then, all through, what is the good of the variation in length of style the stamens not reciprocating.? or when the stamens vary being long and short *with* the style? All this is something for you to consider when you are done with *Man* and such comparatively uninteresting subjects.[10]

These either long or short styles are strongly marked in all the far-western species, and only one way in each species in all the specimens (very many in some cases) I have examined. So that I am obliged here to take the style as a specific character.[11] The annual species, all have short style, & 3 or 4 of the Western perennial ones. The long style is the case in the greater number of the species.

If more of the species were decidedly dimorphic, or were so in an orthodox way (with regular longs and shorts), as in Primula, I should conclude that we had here cases of *dimorphism manqué*—like what *Scott* notes of Primula, a few species of which are in the same case.—[12] Unless, by an odd combination of chances one sort only of each has been collected. Or, has one sort been so much more prolific than the other, that the less fertile has died out, and the other got insects to fertilise it perfectly from individual to individual of the same sort?

Well, there are the main facts. We are all agog now, waiting for your book on Man. Being out of my line, I shall not see early sheets, and get a perusal in advance.[13]

We have now been at home a year,—[14] a very busy year it has been with me, and not an idle day, hardly. Yet I look back with dismay upon the small amount of work I have accomplished. Mrs. Gray is always delicate, of late very much so. She desires most affectionate regards to you and all your kind family,— it seems a long, long while ago our most delightful visit two years ago.[15]

Did you see M^r. Morgan of Rochester, N.Y.? He asked for an introduction to you, which I gave, telling you were not always well enough to see any one.[16]

Ever Yours faithfully | Asa Gray

DAR 110: B70–3

CD ANNOTATIONS
1.1 I have … longer. 5.3] *crossed pencil*
2.1 One … throat. 2.4] *scored red crayon*
5.1 2. … longer. 5.3] *scored red crayon*
5.1 *G. aggregata*] *underl red crayon*
6.1 3. … aggregata. 6.2] *crossed red crayon*
7.1 Most … kinds. 9.3] *crossed pencil*
9.1 Polemonium cæruleum] *underl red crayon*
10.2 or the … anthers. 10.5] *scored red crayon*
11.1 As … subjects. 13.4] *crossed blue crayon*
12.1 *P. subulata*,] *underl red crayon*; 'X' *added red crayon*
12.1 I am … only one. 12.4] *scored red crayon*
12.4 Yet … interbreeding? 12.8] *scored blue crayon*
14.4 annual species,] *underl red crayon*
14.4 The long … species. 14.5] *scored red crayon*
15.1 If more … Man. 16.1] *crossed blue crayon*
17.1 We have … any one. 18.2] *crossed red crayon*
Top of letter: 'Variable Pistils. | Also case of Variability' *ink*; '(Kew)' *red crayon*; '*[*cross fertilising*]* | of stamens & Pistils' *blue crayon*; 'See *paper* Proc. Amer. Soc. of Arts & Sciences' *pencil*
Top of 2d sheet: 'Variability' *blue crayon*; 'Variable length of style only point certain' *pencil*; 'Ask for *[illeg]*' *red crayon*

1 *Primula* has one flower form with a short style and long stamens, and another with a long style and short stamens; see CD's paper 'Dimorphic condition in *Primula*'. Gray's paper, 'Revision of the North American Polemoniaceae' (Gray 1870) was read at the 14 June meeting of the American Academy of

Arts and Sciences; he gave a general discussion of 'incipient dimorphism', including genera mentioned in this letter, on page 248. For a modern classification of the family Polemoniaceae (phlox or Jacob's ladder), see Grant 1998.

2 Gray divided *Gilia* species into groups or sections, one of which was Leptosiphon (Gray 1870, p. 265); Leptosiphon is now the genus *Linanthus*. *Gilia* and *Collomia* are now placed in different tribes of the subfamily Polemonioideae (Grant 1998, p. 742). CD referred to Gray's description of *Gilia nudicaulis* (now *Linanthus nucidaule*) in *Forms of flowers*, p. 119. *Collomia nudicaulis* is now *Gymnosteris nudicaulis*.

3 See Gray 1870, p. 265. *Gilia androsacea* is now *Linanthus androsaceus*.

4 See Gray 1870, pp. 275–6.

5 In Gray 1870, p. 275, Gray wrote: 'the tendency to dimorphism, of which there are traces, or perhaps rather incipient manifestations in various portions of the genus, is most marked in *G[ilia]. aggregata*'. CD quoted him in *Forms of flowers*, p. 118. *Gilia aggregata*, now *Ipomopsis aggregata* (scarlet gilia), includes a number of subspecies. *Gilia coronopifolia* is now *Ipomopsis rubra*.

6 Gray refers to his group Giliandra, which included *Gilia stenothrysa* (narrowstem or Uinta Basin gilia) and *G. piniitifida* (sticky gilia) (Gray 1870, p. 276).

7 See Gray 1870, p. 281.

8 For Gray's description of *Phlox* species, see Gray 1870, pp. 248–58.

9 See Gray 1870, pp. 248, 252. CD referred to Gray's conclusions on *Phlox subulata* and *P. nivalis* in *Forms of flowers*, p. 120. The two are now considered separate species.

10 CD was finishing *Descent*.

11 Gray refers to the westernmost areas of the United States; most members of Polemoniaceae are in western North America.

12 In his discussion of dimorphic *Primula* species, John Scott noted several species that appeared to be short-styled and others that appeared to be long-styled, as well as six that he considered non-dimorphic but that sometimes presented variable style and stamen lengths (J. Scott 1864, pp. 80–3).

13 Gray had read proof-sheets of *Variation* (see *Correspondence* vol. 15, letter to Asa Gray, 16 October [1867]).

14 Gray and his wife, Jane Loring Gray, visited England from September 1868 to November 1869, also travelling on the Continent (J. L. Gray ed. 1893, 2: 565; see also *Correspondence* vols. 16 and 17).

15 The Grays and others visited Down House between 24 and 30 October 1868 (Emma Darwin's diary (DAR 242)).

16 Gray refers to Lewis Henry Morgan. See letter from L. H. Morgan, 9 August 1870, and letter to L. H. Morgan, 11 August [1870].

To P. L. Sclater 21 November [1870][1]

Down. | Beckenham | *Kent. S.E.*
Nov. 21

My dear Mr Sclater

I send today two sheets, but the first few pages do not relate to Birds. I hope the corrections will not much interfere with the facity of reading.[2]

Kindly look to footnotes, as they often include names. If you find only names to correct (which is almost too good to be hoped for) please return the sheets to M^essrs^. Clowes, whose address is given at foot of first page, as this will save time.[3] If there are more serious corrections please return the sheets here

You will save me from making many disgraceful misspellings, which would have been widely spread, for 5 foreign editions are now preparing.[4]

Yours very gratefully | Ch. Darwin

Zoological Society of London

1 The year is established by the relationship between this letter and the letter to P. L. Sclater, 9 November 1870.

2 Sclater had offered to read the proof-sheets of the sections of *Descent* concerning birds and mammals (see letter to P. L. Sclater, 11 November [1870] and n. 2.

3 William Clowes & Sons were printing *Descent*.

4 *Descent* was being translated into German by Julius Victor Carus (Carus trans. 1871; see letter from J. V. Carus, 2 October 1870); into French by Jean Jacques Moulinié (Moulinié trans. 1872; see letter to John Murray, 26 September 1870); into Russian by Vladimir Onufrievich Kovalevsky (letter from V. O. Kovalevsky, 22 November 1870; this translation was never published); and into Dutch by Hermanus Hartogh Heijs van Zouteveen (Hartogh Heijs van Zouteveen trans. 1871–2; see letter from Hermanus Hartogh Heijs van Zouteveen, 27 November 1870). A US edition was also being prepared (*Descent* US ed.; see letter from Asa Gray, 14 February 1870 and n. 5; letters from E. L. Youmans, 25 September 1870).

From V. O. Kovalevsky 22 November 1870

Berlin 7. Georgienstrasse.
22 Nov. 1870.

Dear Sir

I did not write You such a long Time not beeing quite sure shall I remain at Berlin or go to Leipsic, and although my wife was not admitted to the University here we resolved to remain at Berlin there beeing here so many means of pursuing studies privately.[1] I hope Your intended visit did You good and delayed a little the printing of Your book what I most fervently wish as I am a little backward with the translation; still, the four sheets You kindly sent me are soon finished and I pray You dear Sir to send me some more if there is any printed.—[2] Could You not inform me when do You hope that both volumes will appear in England.—[3]

From my brother I had news before yesterday, he could not procure himself live Scalpellum's and hopes to get them in the spring, when as the pescatori assured him they are much more abundant.[4] He received an offer from the Kiev University to send him for 5 months to the Red Sea and started the 8 Nov. from Naples, I have letters from Alexandria, where he arrived safe the 11 of November and got the same day to Suez; I expect he will remain the whole time at Tor (Tûr) on the west side of the Sinai peninsula and work there on the coral reefs.[5]

It seems dear Sir that we are going to have a fight with You, that comes all from Your Prussian sympaties; now You allowed Your best and truest ally to be murdered, who shall prevent Russia & Prussia to divide Austria and trottle the Sick man.[6] We are very anxious about the state of Paris, as my sister in Law is shut up there; the siege is drawing to the close as the provisions become scanty, so Paris will surrender without bombardement, one is happy even at this.—[7] My wife send You her compliments and prays to remember her to Mrs Darwin and the ladies

Yours | Very truly | W. Kowalevsky.—

P.S. The proof sheets can be sent under cover by post, it is generally a safer way.—

DAR 169: 85

1 Kovalevsky's last letter to CD is dated 11 October [1870]. Sofia Vasilyevna Kovalevskaya was studying privately in Berlin with Karl Theodor Wilhelm Weierstrass (Tee 1976–7, pp. 122–3).

2 CD had been at Leith Hill Place, the home of his brother-in-law Josiah Wedgwood III, and Ightam Mote, from 13 to 20 October 1870 (see 'Journal' (Appendix II)). Kovalevsky was translating *Descent* from the corrected proof-sheets.

3 *Descent* was published on 24 February 1871 (see R. B. Freeman 1977).

4 Kovalevsky refers to Alexander Onufrievich Kovalevsky; see letter from V. O. Kovalevsky, 15 September 1870 and n. 2. *Pescatori*: fishermen (Italian). *Scalpellum* is a barnacle genus.

5 For more on A. O. Kovalevsky's plans for work in Egypt, see letter from V. O. Kovalevsky, 28 February [1870].

6 The French government declared war on Prussia in July 1870. French hopes that other German states would seize the opportunity to turn on Prussia proved unfounded, and most supported the Prussians; Austria seemed divided in its hostility to both parties and maintained neutrality. The Russian government maintained friendly relations with Prussia, and, in September, announced that it repudiated an article in the Paris Treaty of 1856 referring to the neutrality of the Black Sea. This deeply alarmed the Turkish government, who recognised it as a prelude to the establishment of a Russian fleet in the Black Sea. The Ottoman (Turkish) empire was at the time sometimes referred to as the 'sick man of Europe'. Collusion between the Russians and the Prussians at a time of French weakness was suspected, and the British Government rejected any adjustment of the terms of the treaty. (*Annual Register* (1870): 109 (United Kingdom); 154–221 (France); 233 (Germany); 251 (Austria); 284–5 (Russia); 288 (Turkey).)

7 Paris came under siege by the Prussians in September 1870. In October, Otto von Bismarck had announced his intention of reducing the city by famine rather than bombardment (*Annual Register* (1870): 187). Sofia's sister was Anna Korvin-Krukovskaya.

To A. R. Wallace 22 November [1870][1]

Down. | Beckenham | *Kent. S.E.*
Nov 22d

My dear Wallace

I must ease myself by writing a few words to say how much I & all others in this house admire your article in Nature.[2] You are certainly an unparalleled master in lucidly stating a case & in arguing. Nothing ever was better done than your argument about term "origin of species" & the consequences about much being gained, even if we know nothing about precise cause of each variation.—[3] By chance I have given a few words in my 1st. Vol. now some time printed off about mimetic butterflies & have touched on two of your points, viz on species already widely dissimilar not being made to resemble each other, & about the variations in Lepidoptera being often well-pronounced.[4]

How strange it is that Mr Bennett or anyone else shd. bring in the action of the mind as a leading cause of variation seeing the beautiful & complex adaptations & modifications of structure in plants, which I do not suppose they wd say had minds.—[5]

I have finished 1st Vol. & am half-way through first proofs of 2d Vol. of my confounded book, which half-kills me by fatigue & which I much fear will quite kill me in your good estimation.—[6]

If you have leisure I shd much like a little news of you & your doings & your family.—

Ever yours very sincerely | Ch Darwin

British Library Board (Add 46434 ff. 207–8)

[1] The year is established by the relationship between this letter and the letter from A. R. Wallace, 24 November 1870.
[2] CD refers to Wallace's response to an article by Alfred William Bennett on natural selection (Wallace 1870d; Bennett 1870); Wallace 1870d was published in the 17 November 1870 number of *Nature*. There is an annotated copy in the Darwin Pamphlet Collection–CUL.
[3] In Wallace 1870d, Wallace had responded to Bennett's calculation of the number of generations he thought was necessary before a favourable change in colour was acquired by a mimicking butterfly; as a result of his calculation, Bennett argued that mimicry in Lepidoptera could not occur through natural selection (see Bennett 1870, pp. 30–1, and Wallace 1870d, p. 49). For CD on variations in Lepidoptera, see *Descent* 1: 402, 412.
[4] See *Descent* 1: 412. CD wrote about mimicry in Lepidoptera in *ibid.*, pp. 411–15.
[5] Bennett wrote (Bennett 1870, p. 32): 'I cannot but believe in the existence of an unconscious Organising Intelligence, … And if this inherent innate power of change is admitted, it at once harmonises the tendency to variation which exists in all created beings'.
[6] CD refers to *Descent.*

From A. R. Wallace 24 November 1870

Holly House, Barking.E.
Novr. 24th. 1870

Dear Darwin

Your letter gave me very great pleasure. We still agree I am sure on nineteen points out of twenty, and on the twentieth I am not unconvinceable.[1] But then I must be convinced by facts & arguments, not by high-handed ridicule such as Claparèdes.[2]

I hope you see the difference between such criticisms as his, & that in the last number of the *N. American Review*, where my last chapter is really *criticised*, point by point;[3]—& though I think some of it very weak I admit that some is very strong, & almost converts me from the error of my ways.

As to your new book I am sure it will not make me think less highly of you than I do, unless you do, what you have never done yet,—ignore facts & arguments that go against you.

I am doing nothing just now but writing articles and putting down anti-Darwinians, being dreadfully ridden upon by a horrid old-man-of-the-sea, who has agreed to let me have the piece of land I have set my heart on,[4] & which I have been trying to get of him since last February, but who will not answer letters, will not sigh an agreement, & keeps me week after week in anxiety though I have accepted his own terms unconditionally, one of which is that I pay rent from *last Michaelmas*! And now the finest weather for planting is going by. It is a bit of wilderness that can be made into a splended imitation of a Welsh valley in little, & will enable me to gather round me all the beauties of the temperate flora which I so much admire,[5]—or I would not put up with the old fellow's ways. The fixing on a residence for the rest of your life is an important event, and I am not likely to be in a very settled frame of mind for some time.

I am answering A. Murray's Geog Dist of Coleoptera for my Ent. Soc Pres. Address,[6] and am printing a 2nd. Edition of my 'Essays' with a few notes and additions.[7]

Very glad to see (by your writing yourself) that you are better, & with kind regards to all your family

Believe me Dear Darwin | yours very faithfully | Alfred R. Wallace

My wife[8] is suffering from face-neuralgia, otherwise we are pretty well.

DAR 106: B94–5

1 See letter to A. R. Wallace, 22 November [1870]. CD and Wallace had long been discussing mimicry, sexual selection, and natural selection in Lepidoptera (see *Correspondence* vols. 15–17, and *Descent* 1: 414–15 n. 31).

2 Wallace refers to Edouard Claparède and his essay review of A. R. Wallace 1870a (Claparède 1870); see letter from A. R. Wallace, 6 July 1870, and letter to J. D. Hooker, 8 July [1870] and n. 3.

3 Wallace refers to Chauncey Wright's review of his *Contributions to the theory of natural selection* (Wright 1870; Wallace 1870a). The last chapter was 'The limits of natural selection as applied to man'. There is an annotated copy of Wright 1870 in the Darwin Pamphlet Collection–CUL.

4 Wallace was negotiating with James Theobald for the purchase of four acres of land on the Thames near Grays, Essex (Slotten 2004, p. 298). Theobald was probably a retired farmer, then residing in Chelmsford, Essex (Census returns 1871 (Public Record Office RG10/93/15)).

5 For Wallace's work on his property in Essex, see Raby 2001, pp. 209–11, and Slotten 2004, pp. 298–9.

6 Wallace gave the presidential address to the Entomological Society of London on 23 January 1871 (Wallace 1871a). He responded to Andrew Murray's 'On the geographical relations of the chief coleopterous faunæ' (Murray 1868).

7 Wallace refers to the second edition of *Contributions to the theory of natural selection* (Wallace 1871b).

8 Annie Wallace.

To George Cupples 27 November [1870]

Bran is thriving & growing at a wonderful rate—coat sleek, & not too fat. Plays much with Polly & enjoys English life.[1]

C. Darwin

Nov. 27th—

Endorsement: 'Recd Nov. 30/70'
Postcard
American Philosophical Society (Getz 7383)

1 Bran was a deer-hound puppy that Cupples had sent to CD from Scotland (see letter from George Cupples, 14 November 1870). For Polly, see the letter from George Cupples, 8 October 1870 and n. 4.

From Hermanus Hartogh Heijs van Zouteveen 27 November 1870

Delft
27 November 1870

Sir!

Mr Kramers of Rotterdam told me the contents of your kind letter to him of the 19th inst.[1]

Hearing you would be glad to know, who he would employ for the translation of your new work: the descent of man and selection in relation to sex,[2] I take the liberty to introduce myself— I am naturalist, had my education at the university of Leiden of Prof. I. van der Hoeven and was promoved there in 1866 to Mathereas magister et philosophiae naturalis Doctor[3]

M[r] Kramers wrote to Mr Murray in the name of M[r] Ykema, bookseller and editor in this town.[4]

M[r] Ykema and I, we are happy with your approbation to see your work translated in the Dutch language and were glad to receive the clean sheets of the greatest part of Vol. I.

To obtain now the right of translation we want to show and to depose the title of your book, please send me or M[r] Ykema such as soon as possible with which you wil oblige very much

Sir | Yours faithfully | D[r] H. Hartogh Heijs van Zouteveen

DAR 184: 15

[1] CD's letter to Hendrik Adrianus Kramers has not been found; however, see the letter to John Murray, 15 November [1870], and the letter from John Murray, 16 November [1870].

[2] *Descent.*

[3] Hartogh Heijs van Zouteveen earned two advanced degrees, one of which was in the natural sciences (*NNBW*). His professor was the comparative anatomist and anthropologist Jan van der Hoeven.

[4] See letter from John Murray, 10 October [1870]. Johan Ykema was to publish the Dutch translation of *Descent* (Hartogh Heijs van Zouteveen trans. 1871–2).

From Anne Jane Cupples [28 November 1870][1]

56 Ludgate Hill | EC
Monday morning.

My dear M[r] Darwin,

I have come up to London at the request of Mr Strahan (publisher)[2] who has kindly volunteered to help me to get a pension from Government for my husband. The Duke of Argyle[3] has agreed to present the memorial but as there is no time now to get one up in the usual form, Mr Strahan thinks if I could get a few letters from friends, stating that Mr Cupples was worthy of it. Mr S has asked me to write to you, but as my husband does not know about it I must ask you to excuse the trouble As you do not know him so intimately as the others, the only thing you might be able to say, that he shows powers of some kind and that were his mind more at ease he might yet do something worthy of himself & the world Mr S thinks you may be able to do this judging from his letters to you. If you *can* do this for me, will you be so good as to address the letter to M[r] Strahan. The others I have got begun their letters My dear Mr Strahan I hope you will succeed in this undertaking for my friend Mr Cupples who &c &c" These letters are to be fixed to the memorial.[4] M[r] Cupples has not been very strong these last six months, and I am anxious to get home immediately else I would have been so glad to have run down to see M[rs] Darwin before leaving

please give her and Miss Darwin[5] my very kindest regards. and with the same to yourself I remain | yours very sincerely | A J Cupples

C Darwin Esq | Downe Beckenham Kent,

PS I hope the puppy is getting on nicely.[6] I write this in Mr S's office with the fog so thick it is like night, so I hope you will excuse the writing

DAR 161: 280

1 The date is established by the relationship between this letter and the letter to A. S. Strahan, 29 November 1870. The Monday before 29 November 1870 was 28 November.
2 Alexander Stuart Strahan.
3 George Douglas Campbell.
4 There is no record of a petition for a pension for George Cupples being presented before Parliament.
5 Henrietta Emma Darwin.
6 Anne refers to Bran, a deer-hound puppy that George had sent to CD (see letter from George Cupples, 14 November 1870).

To Alexander Stuart Strahan 29 November 1870[1]

Dear Sir,

I have been informed that an application is to be made to G. for a pension for Mr G. C.[2] Althoug I am not personally acquainted with this gentleman, I have correspondence with him on scientific subjects during several years. On some very intricate points, he has been so kind as not only to collect at the cost of much trouble information from various source, but has likewise made valuable observations for me.—[3] Therefore he has my sincere good wishes for his success.

I have confined myself to the above points relating to Science, and consider Literature beyond my Province.—

I beg leave to remain | Dear sir | Tr. yrs | C.D.

Draft
DAR 161: 280v

1 The transcription is of a draft of the letter in CD's hand. The original letter, signed by CD but in the hand of an amanuensis, is known from a catalogue from Sotheby's New York for a sale on 6 April 1982. The date of the letter is from the sale catalogue, and presumably is a clear date from the letter itself. The catalogue quotes the original letter as follows: 'I have corresponded with him on scientific subjects during several years. On some very intricate points he has been so kind as not only to collect, at the cost of much trouble, information from various sources, but has likewise made for me valuable observations.' Since the draft is more extensive than the catalogue description, the draft has been transcribed here.
2 See letter from A. J. Cupples, [28 November 1870]. CD refers to the Government ('G') and to George Cupples ('Mr G. C.').
3 See *Correspondence* vols. 16–18.

From Charles Anthoni Johnson Brooke 30 November 1870

Sarawak
Nov[br] 30th /70

Dear Sir/

Altho' three years have passed since I rec[d] yr note accompanied by the 17 queries about expressions[1]—since then I have been a year in Europe or I should have sent a reply before—as I have never lost sight of observing the countenances of the people—more particularly the Dyaks of Sakarang and Saribus tribes—[2] From Malays I think little originality can be expected as they are early tutored in conducting themselves in an orthodox Mahomedan Code— for instances—when seated on a mat receiving, or visiting a stranger, the feet are not to be shown—they are doubled up underneath—the hands clasped one on the other—not to show the palms—body slightly stooping & head inclined downwards—eyes looking down— in being surprised, they slowly move their heads to & fro with out expressing any remark— Their customs are brought from other parts more than derived from their fore fathers— I enclose a few answers to yr queries—and regret they are not more complete—

I have lately forwarded a female Maias to the Taunton Museum—[3] I believe a perfect specimen— tho' not large & found in the Rejang river some little distance from Si Munjau & Batang Lupar where Mr Wallace & Signor Brecari obtained their specimens[4]

If I can be of any service in obtaining a few natural specimens I shall be happy to do my best—

I am | Dear Sir | Yrs faithfully | C Brooke

[Enclosure]

Peculiarities of Dyaks

1. Astonishment is expressed by open eyes, general swinging of body & beating of hands on the Chest[5] (this beating of Chest is particularly done by women, sometimes when pleased), calling out "Ake Indai"(oh! Mother.)

2. Shame has rather the effect of producing paleness, the same as anger. Natives say under such feelings they feel the blood drawn from their face.[6] Red eyes are supposed to be a sign of fierceness or rage when they speak of the face as hot as fire— Young men show rage by blustering & talking loud, move their bodies and swing their arms about. I can't say I have discovered perceptible blushes—

4. They show thought by contraction of muscle about the face with a solemn look as when interpreting dreams or omens by the sounds of good or bad birds.

5. When in low spirits, I could never see the corners of their eyes or mouths were altered— they feel acutely, and in most instances let their feelings have full play in grief, even to the enjoyment of it.

6. In good spirits, the eyes perceptibly lighten,[7] and their expressions brighten; women put their hands before their mouths to hide them, when laughing.

7,8, 9 & 10. Dogged expression is fixity or rigidity of countenance & look of sternness (sneering or snarling is not often resorted to). Contempt is shown by a slight smile & in silence—[8] disgust is expressed (particularly by Malays) by *spitting* out the word "Pōēh" or Po-he. Dyaks often utter a gutteral sound and say 'Baka Jelū' (like a beast). They also utter a sound like 'Esh', showing contempt, and the word 'Cheh' or 'Eh' from the throat expresses disbelief or equivalent to nonsense, humbug, &c. These sounds I think a stranger would find it impossible to pronounce as natives, as the 'h' is so thrown away from the mouth—

11. Extreme fear is plainly discernable by the general contortions of face, moving of hands, quivering of eyelids, downcast look & muscular irritability.

12. Laughter sometimes brings tears, the eyes nearly closed, they say they nearly *make water* from laughter, a common expression among the female sex.[9]

13. A man considering he could not do something or prevent anything being done, would not trouble himself & would say he was not clever or lucky.

14. Children protrude their lips when sulky—[10] & put their fists in their mouths & stare as if in rage before generally crying.

15. The guilty, cunning or jealous show marks of their feelings in their countenances to quite as great an extent as Europeans.

16. The Dyak word for Silence is 'Diau'— They do not 'hiss' to express it.

17. Affirmation & negation are not expressed by movements of head, a look from their expressive eyes & slight contraction of brow denotes negation, lifting the brow that of affirmation.[11]

To | Charles Darwin, Esqre | &c &c &c.

DAR 160: 322, 322/1

CD ANNOTATIONS
1.1 Altho' ... particularly 1.4] *crossed blue crayon*
1.4 the Dyaks ... shown 1.7] *crossed pencil*
1.10 slowly ... remark] *scored blue crayon*
1.11 I enclose ... specimens 2.4] *crossed blue crayon*
Enclosure:
2.3 Mother.)] 'In surprise slowly move their heads to & fro and say nothing.' *ink*
4.1 5. When ... enjoyment of it. 4.3] *crossed blue crayon*
8.1 12. Laughter ... sex. 8.2] *crossed red & blue crayon*
Top of enclosure: '36' *red crayon, circled red crayon*

1 CD sent his list of queries on expression to Brooke in March 1867, with a letter that has not been found (see *Correspondence* vol. 15, letter to A. R. Wallace, 7 March [1867] and n. 2). For a list of CD's queries (he sent slightly different versions over time), see Appendix IV; see also *Expression*, pp. 15–16.

2 During the nineteenth century, the term Dyak was often applied to any indigenous people of Borneo (Roth 1896, 1: 39–43). The Dyak people presently comprise Sarawak's largest group of indigenous people, and include the Iban (formerly called Sea Dyaks; Kaur 2001). Roth 1896, 1: 38, listed the 'Skerang Dyaks' as living on the (Batang) Lupar River, and the 'Sarebas' people as living on the Sarebas and Rejang rivers. On the Saribas Iban, see D. Freeman 1981, pp. 5–10; the Skrang Iban are mentioned on p. 8.

3 'Maias' or 'Mias' is the Dyak word for orang-utan (R. B. Freeman and Gautrey 1972, p. 216). The Somerset Archaeological and Natural History Society (SANHS) was then based in Taunton, and in 1874 would move into Taunton Castle, Somerset (http://www.sanhs.org). Brooke sent several zoological specimens to the society (Simon Jones, SANHS, personal communication). Brooke was born in Somerset.

4 Alfred Russel Wallace referred to the Simunjon (or Simunjan) River as a small branch of the Sadong River; these lay south-south-west of the Batang-lupar, or Lupar River (see Wallace 1869a, 1: 46). The area is in the southern, or south-western portion of Sarawak. The Rejang (or Rajang) River to the east is the longest river of Sarawak (*Columbia gazetteer of the world*). Brooke also refers to the Italian botanist Odoardo Beccari.

5 CD cited Brooke for this information in *Expression*, p. 279.

6 CD quoted Brooke's account in *Expression*, p. 318, n. 12.

7 See *Expression*, p. 213.

8 CD cited Brooke for this information in *Expression*, p. 255.

9 CD wrote: 'I hear from the Rajah C. Brooke that it is a common expression with [Dyak women] to say "we nearly made tears from laughter"' (*Expression*, p. 209).

[10] See *Expression*, p. 233.
[11] CD cited Brooke for this information in *Expression*, p. 275; see also p. 277.

To Hermanus Hartogh Heijs van Zouteveen 30 November [1870][1]

Down. | Beckenham | *Kent. S.E.*
Nov. 30th

Dear Sir

I am pleased to hear that the Translation is to be done by one who I have no doubt is fitted so well for the work.—[2]

I enclose a facsimile of the Title-page.—[3]

The whole of Vol. I. &, a small part of Vol. II is printed off, but Mr. Murray objects to *all* the sheets being sent off at once, so that the English edition may not be preceded by *any* foreign edition.—[4] As soon as I receive clean sheets of remainder, I will not fail to send them to you, & I will endeavour not to cause you any delay.—

Dear Sir | Yours faithfully | Ch. Darwin

C. C. Kohler (dealer)

[1] The year is established by the relationship between this letter and the letter from Hermanus Hartogh Heijs van Zouteveen, 27 November 1870.
[2] CD refers to the Dutch translation of *Descent* (Hartogh Heijs van Zouteveen trans. 1871–2). See letter from Hermanus Hartogh Heijs van Zouteveen, 27 November 1870.
[3] The enclosure has not been found.
[4] See letter from John Murray, 16 November [1870].

From Robert Francis Cooke 5 December 1870

50A, Albemarle Street, London, W.
Decr. 5 *1870*

My dear Sir

We have just heard from Rotterdam that the *1st. Volume* of yr new work is announced there for publication on the *10th*.[1]

Are you aware that they intend to publish Vol 1 by itself.?

We have sent them the *Electros*, so they can do so unless they are desired to the contrary.[2]

When do you think we may be able to publish the 2 Vols *here*?[3]

Yours faithfully | Robt. Cooke

Chas. Darwin Esq

DAR 171: 382

[1] Cooke refers to the Dutch translation of *Descent* (Hartogh Heijs van Zouteveen trans. 1871–2). See letter from Hermanus Hartogh Heijs van Zouteveen, 27 November 1870, and letter to Hermanus Hartogh Heijs van Zouteveen, 30 November [1870]. The Dutch translation of *Descent* did not begin to appear until early 1871 (*Correspondence* vol. 19, letter from Hermanus Hartogh Heijs van Zouteveen, 2 March 1871).
[2] Cooke refers to electrotypes of the illustrations to *Descent* (see letter from John Murray, 10 October [1870]).

[3] *Descent* was published on 24 February 1871 (R. B. Freeman 1977).

To Francis Darwin 5 December [1870][1]

Down
Dec. 5th

My dear Frank

I send a cheque for £50 which will clear all the bills mentioned by you.[2] From what you say in your last note, I feel sure that you will be more careful for the future & watch the money as it ouzes out. Altogether it is a rather heavy sum; but I know you & trust you as a good dear affectionate son.—

Ever yours | C. Darwin

DAR 211: 4

[1] The year is established by a reference in CD's Classed account books (Down House MS) dated 5 December 1870 to a payment to Francis of £50.
[2] Francis was still at Cambridge University; see letter to B. J. Sulivan, 20 December [1870] and n. 3.

To P. L. Sclater 5 December 1870

Down. | Beckenham | *Kent. S.E.*
Dec 5th 1870

My dear Mr Sclater,

I send by this post 3 sheets, & tomorrow I will send another. This, with the exception of about $\frac{3}{4}$ of a sheet, completes the part on birds, & I sincerely congratulate you and thank you for your labour will then be completed.[1] Some of the sheets are terribly corrected, but they are chiefly parts in which few names are included. I am so tired with my work that I shall leave this place before post time on Thursday the 8th, & shall stay at "6 Queen Anne Street" for a week.[2] I will call some early day to see you.

Yours very truly obliged | Ch. Darwin

LS(A)
Dr W. N. Livingstone (private collection)

[1] Sclater had agreed to check lists of birds' names in *Descent* for CD (see letter to P. L. Sclater, 4 November [1870]). The chapters on birds are in *Descent* 2: 36–238. Sclater also checked the two chapters on mammals (*Descent* 2: 239–315; see *Descent* 2: 238 n. 59).
[2] CD stayed at his brother Erasmus Alvey Darwin's house in London from 8 to 14 December 1870 (CD's 'Journal' (Appendix II)).

To Asa Gray 7 December 1870

Down. | Beckenham | *Kent. S.E.*
Dec 7. 1870

My dear Gray

I have been very glad to receive yr letter this mg—[1] I have for some time been wishing to write to you, but have been half worked to death in correcting my uncouth English for my new book—[2] I have been glad to hear of yr cases appearing like incipient dimorphism. I believe that they are due to mere variability, & have no significance. I found a good instance in Nolana prostrata, & experimented on it, but the forms did not differ in fertility.[3] So it was with Amsinckia of which you told me.[4] I have long thought that such variations afforded the basis for the development of dimorphism. I was not aware of such cases in Phlox, but have often admired the arrangement of the anthers, causing them to be all raked by an inserted proboscis. I am glad also to hear of your curious case of variability in ovules &c.

I said that I had been wishing to write to you, & this was about yr Drosera, which after many fluctuations between life & death, at last made a shoot, which I c[d] observe.[5] The case is rather interesting; but I must first remind you that the filament of Dionæa is not sensitive to very light prolonged pressure, or to nitrogenous matter, but is exquisitely sensitive to the slightest touch. In our Droseræ the filaments are not sensitive to a slight touch, but are sensitive to prolonged pressure from the smallest object of *any nature*; they are also sensitive to solid or fluid nitrogenous matter. Now in y[r] Drosera the filaments are not sensitive to a rough touch or to any pressure from non-nitrogenous matter, but are sensitive to solid or fluid nitrogenous matter. Is it not curious that there sh[d] be such diversified sensitiveness in allied plants?[6]

I rec[d] a very obliging letter from Mr Morgan, but did not see him, as I think he said he was going to start at once for the continent.[7] I am sorry to hear rather a poor account of M[rs] Gray to whom my wife I both beg to be very kindly remembered—[8]

My dear Gray | Yours affectionately | Ch Darwin

LS(A)
Gray Herbarium of Harvard University (90)

1 See letter from Asa Gray, 21 November 1870.
2 CD was correcting the proof-sheets of *Descent*.
3 See letter to J. D. Hooker, [13 June 1870?] and n. 3. *Nolana prostrata* is now *N. humifusa*.
4 See *Correspondence* vol. 11, letter to Asa Gray, 20 March [1863] and n. 19; CD refers to *Amsinckia spectabilis*.
5 Gray had sent CD bulbs of *Drosera filiformis* (the threadleaf sundew) from Kew in 1869, advising him to see whether he could detect any sensitivity in the glandular hairs; see *Correspondence* vol. 17, letters from Asa Gray, 20 October 1869 and 26 and 29 October [1869].
6 See *Insectivorous plants*, pp. 19–37.
7 CD refers to Lewis Henry Morgan. See letter from L. H. Morgan, 9 August 1870, and letter to L. H. Morgan, 11 August [1870].
8 CD refers to Jane Loring Gray.

From V. O. Kovalevsky 12 December [1870][1]

Berlin 7. Georgienstrasse.—
12 Dec.—

Dear Sir

Excuse me for not writing You immediately to thank for the large parcel of clean sheets You had the kindness to send me, I am busy translating them, the first are already in the press, so that I hope to be not very much behind the original;[2] the reason why I did not write immediately lay in the hope I had to have a letter from You, but either You did not send it (and in this case it is all right) or it is lost in transmission which will be a pity. Mr. Murray sent the casts directly to Russia and made a very moderate charge 14£; though I do not know what quantity of copper cast there is for the 2 volumes.—[3] I wrote You in my last that my brother is gone to the Red Sea, I have letters from him weekly, Suez did not satisfy him much so he is gone a little farther to Tor, a place wher Ehrenberg has been in 1827.—[4] I hope he will return about the 25 of February, when I'll go to meet him at Naples and we shall try together to procure ourselves scalpellums for the verification of Your old discovery.[5] I hope Your health got better from Your autumnal visits.[6]

Pray remember me and Mrs. Kovalevksy to Mrs. Darwin and the ladies.—[7]

Your very truly | W. Kowalevsky

P.S. Have You seen the new book of Häckel "Biologische Studien" it is very interesting, especially his discovery that the Radiolaria contain in their body 40% amylum; such a fact may overtrow much of our prejudices about the dependence of animal life from vegetation an explain the economy of the deep sea inhabitants.—[8]

W. K.

DAR 169: 68

[1] The year is established by the relationship between this letter and the letter from V. O. Kovalevsky, 22 November 1870.

[2] Kovalevsky was preparing a Russian translation of *Descent.* See letter to John Murray, 26 September 1870 and n. 1.

[3] John Murray charged £14 for electrotypes of the illustrations to *Descent*, not including the ones from Brehm *et al.* 1864–9 (letter from John Murray, 10 October [1870]).

[4] Kovalevsky refers to Alexander Onufrievich Kovalevsky and Christian Gottfried Ehrenberg. See letter from V. O. Kovalevsky, 22 November 1870. Ehrenberg took part in an expedition to Egypt between 1820 and 1825 (*DSB*).

[5] See letter from V. O. Kovalevsky, 15 September 1870 and n. 2.

[6] CD had visited Leith Hill Place, Surrey, and Ightham Mote, a fourteenth-century moated manor house in Kent, in October 1870 (CD's 'Journal' (Appendix II)).

[7] Kovalevsky refers to Sofia Vasilyevna Kovalevskaya, and Emma, Henrietta Emma, and Elizabeth Darwin. Vladimir and Sofia had visited Down in late August and early October 1870 (see Appendix II).

[8] The first part of Ernst Haeckel's *Biologische Studien* (Haeckel 1870–7) was a collection of earlier works. Kovalevsky refers to Haeckel 1870–7, 1: 119–27; amylum is starch.

From T. H. Farrer 15 December 1870

3, Gloucester Terrace, | Regent's Park.
15 Dec/70

My dear Mr Darwin,

I humbly cry Peccavi. The fact is I sent off from Abinger your books which I had there, and quite forgot to get out and send others which I had been keeping here.

Sprengel with several others shall go this morning to Cumberland Place.—[1]

I saw Miss Darwin there and told her she ought to be ashamed of being seen about, whilst the book is not yet to be seen.[2]

Sincerely yours | T H Farrer

C Darwin Esq FRS.

DAR 164: 67

[1] Farrer had borrowed, among other books, CD's copy of Christian Konrad Sprengel's *Das entdeckte Geheimniss der Natur im Bau und in der Befruchtung der Blumen* (The secret of nature discovered in the structure and fertilisation of flowers; Sprengel 1793); see *Correspondence* vol. 17, letter from T. H. Farrer, 13 October 1869. Farrer's country residence was Abinger Hall near Dorking. One Cumberland Place, London, was the home of Hensleigh Wedgwood, CD's cousin and brother-in-law.

[2] Henrietta Emma Darwin had been involved in reading the manuscript of *Descent* (see letter to H. E. Darwin, [8 February 1870]).

From Lambert Adolphe Jacques Quetelet[1] 15 December 1870

Académie Royale
des Sciences, des Lettres et des Beaux-Arts
de Belgique.
Art. 14 des Statuts organiques.— La correspondance
de l'Académie se tient par le Secrétaire perpétuel,
organe et interprète de cette Compagnie.
Indicateur, N° 6779.

Bruxelles,
le 15 décembre *1870*.

Monsieur,

J'ai l'honneur de vous informer que la classe des sciences de l'Académie royale de Belgique, voulant vous donner un témoignage de l'estime qu'elle professe pour vos ouvrages, vous a élu *associé* dans la séance de ce jour.[2]

En vous conférant cette distinction, la classe espère vous voir prendre part à ses travaux, en lui communiquant ce qui pourrait l'intéresser dans la branche des sciences que vous cultivez.

Je suis heureux, Monsieur, d'être en cette circonstance l'organe et l'interprête de la Compagnie, et de pouvoir vous offrir mes félicitations en même temps que l'assurance de mes sentiments les plus distingués.

Le Secrétaire perpétuel, | Quetelet

à Monsieur Darwin | Associé de l'Académie, | à Down, Beckenham.

DAR 175: 12

[1] For a translation of this letter, see Appendix I.

[2] The diploma was dated 16 December 1870, but not sent until January 1871 (see *Correspondence* vol. 19, letter from L. A. J. Quetelet, 11 January 1871, and *ibid.*, Appendix III).

To L. A. J. Quetelet [after 15 December 1870][1]

I beg leave to acknowledge your very obliging letter, & to return very sincere thanks for the distinguished honour conferred on me by my election as an associate of your the Royal Academy of Science[2]

I beg the leave to remain | your most obedient | & obliged servant Ch. Darwin

ADraftS
DAR 175: 12v

[1] The date is established by the relationship between this letter and the letter from L. A. J. Quetelet, 15 December 1870.

[2] CD refers to the letter from L. A. J. Quetelet, 15 December 1870, and his election as an associate of the science section of the Académie Royale des Sciences, des Lettres et des Beaux-Arts of Belgium.

From Edouard van Beneden[1] 17 December 1870

Liège
17 X^{bre}. 1870

Monsieur et illustre Confrère,

J'ai éprouvé un véritable joie en apprenant votre nomination comme membre Associé de l'Académie royale de Belgique et je ne puis m'empêcher de vous dire combien je suis heureux de voir qu'en cette circonstance les Belges aient fait la leçon aux membres de l'Institut de France.—[2] Puissent tous les corps savants du monde protester comme l'Académie de Belgique contre la discussion qu'ils ont osé entamer sur la question de vos mérites scientifiques. Votre nom résume tout le mouvement scientifique de ces dernières années et vos immortals travaux sont au dessus de toutes les attaques du cléricalisme et de l'ignorance.—J'ai été indigné comme doivent l'être tous ceux qui ont connaissance de vos travaux, en lisant cette discussion qui est une flétrissure pour cette institution, à ajouter à toutes celles que la science française a éprouvé dans ces dernières années.—

En Belgique la jeune génération a arboré le drapeau de l'indépendance intellectuelle et a jeté au loin les préjugés et les idées préconcues. Et si parmi les hommes d'un âge bien mur il en est qui n'ont pas accepté vos idées, au moins ils ont su mettre vos mérites au dessus de leurs convictions personnelles.—

Je suis heureux et fier d'avoir été nommé correspondant de l'Académie, le jour où vous avez été Élu comme associé.—

Je vous envoie un exemplaire de quelques uns de mes travaux.[3] Veuillez les recevoir comme un témoignage de mon profond respect. Vous verrez que je travaille la même question que Dohrn: l'embryogénie des Crustacés.[4] Je crois être arrivé à un joli résultat: c'est la démonstration par l'embryogénie de la dérivation des articulés d'un forme de Vers. (Embr. des Sacculins.—id des Lernéens)[5]

Je continue ces recherches et les publications et j'espère arriver bientôt à construire l'arbre généalogique de ce groupe.— Je ne suis pas arrivé aux mêmes conclusions que Dohrn.[6] J'espère pouvoir vous envoyer très prochainement le développement des Nebalia.—[7]

J'ai eu l'occasion d'étudier récemment l'organisation du *Anelasma squalicola.* Je crois avec Dohrn que nous avons là le passage aux Peltogaster.[8] J'ai retrouvé parfaitement les racines plongées dans les chairs, telles que vous les avez figurées.—[9] Les pièces de la bouche existent à l'état rudimentaire.

Excusez la liberté que j'ai pris de vous écrire et reçevez, Monsieur et illustre confrère l'expression de mon profond respect. | Prof. D^r^. Edouard Van Beneden | rue Vinave d'Ile à *Liége*

DAR 160: 132

1 For a translation of this letter, see Appendix I.

2 See letter from L. A. J. Quetelet, 15 December 1870. CD had been nominated for the zoological section of the Académie des Sciences on 27 June 1870 (*Comptes Rendus de l'Académie des Sciences* 70 (1870): 1418). The Institut de France includes five academies, one of which is the Académie de Sciences. CD was nominated a further five times without success, until he was finally elected to the botanical section in 1878 (Corsi and Weindling 1985, p. 699). For more on the controversy over CD's election, see the letter from Armand de Quatrefages, 18 July 1870 and n. 6.

3 There are offprints of Beneden 1868, 1869a, 1869b, 1870a, 1870b, 1870c, and 1870d, and Beneden and Bessels 1868, in the Darwin Pamphlet Collection–CUL; Beneden 1868 and 1870b are lightly annotated.

4 Anton Dohrn had written several articles, later collected into a book (Dohrn 1870), on the morphology and development of crustaceans.

5 See Beneden 1870a, pp. 105–6. Beneden suggested that the early development in embryos of *Sacculina* (a genus of Rhizocephala) resembled that of many annelid worms. Beneden had earlier described the embryonic development of *Lernaea*, a genus of parasitic copepods, commonly known as anchor worms (Beneden and Bessels 1868, pp. 440–1).

6 Dohrn outlined some of his differences with Beneden concerning the genealogy of Crustacea in Dohrn 1870, pp. 159–61.

7 Beneden did not publish on *Nebalia*, a genus of sea fleas, now in the order Leptostraca. *Nebalia* had been considered by some to be a divergent form of Phyllopoda, but by 1872 Carl Friedrich Claus had reclassified it as a primitive form of Malacostraca (see Claus 1872, pp. 329–30).

8 The barnacle species *Anelasma squalicola* lives on some species of sharks with its peduncle embedded in the shark's skin. *Peltogaster* is a genus of Rhizocephala that parasitises hermit crabs. CD had suggested *A. squalicola* might be a link to Rhizocephala after reading F. Müller 1864 (see *Correspondence* vol. 13, letter to Fritz Müller, 10 August 1865 and n. 7). For Dohrn's views on the relationship between *A. squalicola* and *Peltogaster*, see Dohrn 1875, p. 72.

9 See *Living Cirripedia* (1851): 380–1 and plate IV, figure 3.

To William Ogle 17 December [1870]

Down | Beckenham | *Kent S.E.*
Dec. 17th

My dear D^r^. Ogle

Absence from home has prevented me from thanking you before for your last to me very valuable letter. I think I need not trouble myself much more about the

Platysma.[1] If you shd. ever see any one suffering great fear & can remember the point do look at the neck.— I will enquire what muscle my informant named when he spoke about the longitudinal furrows.[2] I *think* Duchenne always speaks of transverse furrows.—[3]

I have just been in London for 6 days for some rest, & intended calling on you; but days so short, & not much strength for what I was forced to do, that I failed. I shall be up again in some 6 weeks or 2 months, & much hope then to see you.—[4] If at any time you write again, please tell me whether I may call early & how early—

With very sincere thanks | Yours truly | C. Darwin

Postmark: DE 17 70
DAR 261.5: 5 (English Heritage 88205903)

1 CD stayed at his brother Erasmus Alvey Darwin's house in London from 8 to 14 December 1870 (CD's 'Journal' (Appendix II)). Ogle's letter has not been found, but evidently related to their discussions of the platysma myoides in November; see letters to William Ogle, 9 November 1870 and 17 November [1870], and letter from William Ogle, [10–17 November 1870]. See *Expression*, pp. 298–303.

2 See letter to William Ogle, 17 November [1870], and *Expression*, p. 298.

3 CD refers to Guillaume Benjamim Amand Duchenne and Duchenne 1862.

4 CD was next in London from 23 February to 2 March 1871 (CD's 'Journal' (*Correspondence* vol. 19, Appendix II)). CD and Ogle evidently did meet then; see *Correspondence* vol. 19, letter to William Ogle, 7 March [1871].

From William Ogle [after 17 December 1870][1]

34 Clarges St.

My dear Mr. Darwin

I shall be much pleased if you are able as you kindly propose to pay me a visit when you are next in London.[2] I am always at home from ten to one in the morning excepting on Mondays and Fridays when my Hospital duties take me away earlier than usual.[3] Should I however, hear from you that any hour whatsoever is more convenient to you I will most certainly take care to be then at home, as I should much regret losing the pleasure of seeing you.

I have heard from Professor Wyman, who chances to be just now going to pay a visit to Florida. He has promised to get what information he can as to the question of white and black pigs, and if possible to make some experiments himself on the matter.[4]

Incidentally he mentioned that by an inadvertence in your work, you have spoken of Virginia instead of Florida, as the region in which his observations were made. I do not know whether this is the case; but I thought that if it be so you might like to have your attention called to the unimportant slip, so that it might in after editions be corrected.[5] With kind regards

Believe me | Yours sincerely | William Ogle.

C. Darwin Esqr.

DAR 173: 5

1 The date is established by the relationship between this letter and the letter to William Ogle, 17 December [1870].
2 See letter to William Ogle, 17 December [1870] and n. 4.
3 Ogle was assistant physician at St George's Hospital in Tooting, London (*The Times*, 15 April 1912, p. 9.)
4 Ogle refers to Jeffries Wyman. See letter to William Ogle, 17 November [1870].
5 In *Variation* 2: 227, CD gave a case on the authority of Wyman of all the pigs in a particular region of Virginia being black, since they fed on a root that made the hooves fall off all but the black varieties. He did not make the correction in *Variation* 2d ed., 2: 212.

From B. J. Sulivan 17 December 1870

Bournemouth
Decr. 17./70

My dear Darwin

I must tell you how glad I was today to see that your son had come out second for Engineers at Woolwich, thus rivalling his brother at Cambridge.[1] It must be a great comfort to you and Mrs. Darwin to see one after the other doing themselves such credit. I hope it is an earnest of a splendid career for him in that fine Corps, which leads to so many opportunities of distinction both in the military and Civil Services.

I hope you are still pretty well, and all your party quite well. I went to Southampton a few weeks since to see a family off by steamer, and I called at the Bank to see your son[2] but he was not in.

We were very glad to make the acquaintance of Mr. & Mrs. Langton, and of his father also;[3] they have had the worst weather I ever saw in Bournemouth but they keep very well.

I saw our old shipmate Usborne[4] when passing through Plymouth lately, he has been getting up small steamers to run up & down the harbour and I crossed with him in the "Beagle"— He desired to be kindly remembered to you.

Did you see the account of Lt Musters—(a nephew of Hamonds & our poor little shipmate) having made a years journey with the Patagonians from South to North, going round by the Andes.[5] At the Geographical they thought he was the first European that had been there: but our German catechist of S. A. Mission spent a year going over the same ground about nine years since.[6] I hope Musters had the means with him of fixing positions; he had been out in the River Plate looking after some sheep farm that he started when serving on the station

Our second son[7] who came home so ill last year with fever from Mediterrean went out some months since to join a ship on *Coast* of *Africa*. I suppose because the Doctors who invalided him reported that *he ought not to be sent to a station where he would be liable to fever for three years*. As he wanted a sea going ship of course he did not like to refuse it, but it showed utter indifference on the part of our Naval head-man. The eldest[8] last week was in a harbour ship at Plymouth when a Lieut was wanted at an hours notice to join a ship just starting for China—so he volunteered for the sake of getting to sea

With our kind regards to M^rs. Darwin and all your party believe me | very sin^ly yours | B. J. Sulivan

DAR 177: 295

[1] Leonard Darwin's commission in the Royal Engineers was announced in *The Times*, 17 December 1870, p. 12. He had the second highest marks among candidates commissioned into the Royal Engineers. In 1868, George Howard Darwin had come second in the final examination for the mathematical tripos at Cambridge University (see *Correspondence* vol. 16).
[2] William Erasmus Darwin.
[3] Sulivan refers to Edmund, Emily Caroline, and Charles Langton.
[4] Alexander Burns Usborne.
[5] George Chaworth Musters's paper on his travels in Patagonia, read at the Royal Geographical Society on 15 November 1870, was reported in *The Times*, 17 November 1870, p. 12. He was the nephew of Charles Musters, volunteer 1st class on the *Beagle*, and of Robert Nicholas Hamond, midshipman on the *Beagle* (see *Burke's landed gentry* and *Correspondence* vol. 1, Appendix III). Charles Musters died of fever at Rio de Janeiro in 1832. For G. C. Musters's account of his journey, see Musters 1871.
[6] The South American Missionary Society of London was founded in January 1868; before that it had been the Patagonian Mission of Clifton (F. C. Macdonald 1929, p. 67). The German catechist was Theophilus Schmid.
[7] Thomas Edward Sulivan.
[8] James Young Falkland Sulivan.

From E. A. Darwin 19 [December 1870][1]

19th

Dear Charles

I have received the enclosed which I suppose is all right tho' I never read the resolution referred to.

I see I have 2 packets of Certificates both £12-10-0 shares and an awful heap they are.

Please to return Sec^y. letter & let me know if it is all right as far as you know.[2]

What an escape George has had and it is very well if the instruments have escaped uninjured[3]

ED

DAR 105: B69–70

[1] The month and year are established by the reference to George Howard Darwin's escape (see n. 3, below).
[2] The Maryport and Carlisle Railway had passed a resolution to convert their 4 per cent and $4\frac{1}{2}$ per cent shares into stock (*The Times*, 4 November 1870, p. 7). The 'two packets of certificates' were part of Emma Darwin's trust (Erasmus was a trustee); see letter from E. A. Darwin, 21 [December 1870], and CD's Investment book (Down House MS), p. 124. The secretary was John Addison (*Bradshaw's Railway Manual, Shareholders' Guide and Official Directory* 1870).
[3] George was a member of the expedition to observe the solar eclipse from Sicily (*The Times*, 9 December 1870, p. 7). Their ship, *Psyche*, struck a rock off Catania, Italy, on 15 December 1870; the instruments and the expedition party were saved, but their personal effects lost (see *The Times*, 19 December 1870, p. 5, and 28 December 1870, p. 3). There is a letter from Henrietta Emma Darwin to George, written shortly after she received news of the wreck, in DAR 245: 297.

The eclipse expedition, December 1870, Catania, Sicily. DAR 238.7: 29.
Photograph by Hermann Carl Vogel.
Standing, left to right: Alexander Pedler, Joseph Norman Lockyer, Winifred Lockyer, Thomas Edward Thorpe; seated, left to right: Linnaeus Cumming, George Mitchell Seabroke, Charles Blacker Vignoles, Hutton Vignoles, Miss Thorpe(?).
By permission of the Syndics of Cambridge University Library.

The wreck of the *Psyche* off Catania. DAR 238.7: 30.
Photograph by Hermann Carl Vogel.
By permission of the Syndics of Cambridge University Library.

From Laurent-Guillaume De Koninck[1] 19 December 1870

Liège
le 19 X^bre 1870

Mon cher Confrère,

Vous aurez sans doute déjà reçu la nouvelle officielle de votre nomination en qualité d'associé étranger de l'Académie royale des sciences de Belgique.[2] Malgré cela, je ne puis resister un désir de vous feliciter de cette nomination que j'aurais été heureux de vous apprendre le premier et à laquelle j'ai pu contribuer par mon vote. L'Académie s'honore en s'associant des hommes qui comme vous, mon cher Confrère travaillant sans relâche au progrès de la science et marchent droits vers le but qu'ils se sont proposés, sans s'arrêter aux obstacles qu'ils recontrent sur leur chemin pour l'atteindre, parce qu'ils savent que là est la vérité.

Recevez, mon cher Confrère, avec mes félicitations, l'expression de ma haute estime et de mes sentiments dévoués | L. de Koninck

A Monsieur Darwin, Membre de l'Académie royale de Belgique

DAR 169: 51

1 For a translation of this letter, see Appendix I.
2 See letter from Edouard van Beneden, 17 December 1870.

To L.-G. De Koninck [after 19 December 1870][1]

[Down]

My dear Sir

I thank you sincerely for the kind sympathy which has caused you to write to congratulate me on my e. as Ass of your Academy & for your assistance in conferring this high honour on me.—[2] I assure you that the reception of your letter, from one as so distinguished as a geologist & palæontologist, has been no ordinary *[gratification]* to me.— With the most *[sincere]* respects & my best thanks I am dear Sir Yours very faithfully & obliged | Ch. D.

ADraftS
DAR 96: 83

1 The year is established by the relationship between this letter and the letter from L.-G. De Koninck, 19 December 1870.
2 See letter from L.-G. De Koninck, 19 December 1870.

From W. W. Reade 20 December 1870

11 St. Mary Abbot's Terrace | Kensington
Dec. 20.–70

My dear Sir

It is for me to apologise for my carelessness in not perceiving you had mistaken the spelling, and in the first place for not writing it more clearly—or let us unite in blaming the want of demarcation between T & J. in written character.

Your friend is quite right. Jollof is the word—sometimes spelt Djollof by the French: I believe Wollof is the most correct: but Jollof is the most common & is classical.[1] It is in the Lusiad Canto V. x.

> A Provincia *Jalofo* que reparte
> Por diversas nações a negra gente—[2]

I am looking anxiously for announcements of your date of publication[3] and remain | My dear sir | Yours very truly | Winwood Reade

DAR 176: 42

[1] CD's letter to Reade enquiring about the spelling of 'Jollofs' has not been found. See letters from W. W. Reade, [*c.* 8 or 9 April 1870] and n. 6, 6 November 1870, and 9 November 1870. See also *Descent* 2: 357, where CD discussed whether the beauty of the Jollofs was brought about by selective breeding.

[2] The Lusíad or Lusíads is an epic poem by the Portuguese writer Luiz Vaz de Camões describing Vasco da Gama's voyage to India. Reade's quotation may be translated: 'Leaving in our wake the Jalof province / With its various nations of black people' (see Camões 2001, a modern translation).

[3] Reade refers to *Descent*, published in February 1871.

To B. J. Sulivan 20 December [1870][1]

Down | Beckenham | *Kent S.E.*
Dec. 20th—

My dear Sulivan

I thank you cordially for your very kind letter of congratulation & for your interesting news.[2] All my sons are an infinite satisfaction & none of them have ever given me an hour's uneasiness.— My third son has just passed his examination at Cambridge for Natural Sciences, but he was unfortunate & tells me that he has done very badly & will be very low.—[3] I had not heard of Lieut Musters's surprising Journey: no doubt I shall see some account in Journal of Geograph. Socy.[4] As for myself I keep much as usual, always ailing & grumbling, but able to do some hours work daily. I have lately been working rather too hard in the intolerably tedious labour of correcting the proofs & making uncouth English rather less uncouth for my new book on man, which will disgust you & many others.—[5] It is now nearly done, otherwise it wd. have done for me; for my strength is at a very low ebb.—

It is enough to make one boil over with indignation to hear of the negligence & indifference of the Admiralty in sending your son to the coast of Africa.—[6] If you go to Southampton again, do call on my son William: :[7] he said he intended to call on you, if he went to see the Langtons at Bournemth.[8]

We are all fairly well. With kind remembrances to Lady Sulivan.[9] Ever | My dear Sulivan | Yours very sincerely | Ch. Darwin

Sulivan family (private collection)

[1] The year is established by the relationship between this letter and the letter from B. J. Sulivan, 17 December 1870.

[2] See letter from B. J. Sulivan, 17 December 1870.
[3] Francis Darwin took third place in first class honours in the natural sciences tripos at Cambridge University in 1870 (*The Times*, 19 December 1870, p. 6).
[4] See letter from B. J. Sulivan, 17 December 1870 and n. 5. CD refers to George Chaworth Musters. His paper was not published in the *Journal of the Royal Geographical Society.*
[5] CD refers to *Descent.*
[6] See letter from B. J. Sulivan, 17 December 1870. CD refers to Thomas Edward Sulivan.
[7] William Erasmus Darwin was a banker in Southampton.
[8] CD refers to Charles, Edmund, and Emily Caroline Langton.
[9] Sophia Sulivan.

From E. A. Darwin 21 [December 1870][1]

21st

Dear Charles.

The shares all right as you say

292 £12.10.0 @ 4%
80 D°. $4\frac{1}{2}$%

I find they weigh 4lbs 6oz and I suppose they will go all right by regd Book Post[2]

Sir H came & gave me a great wigging for not having been invited to luncheon the moment you came to town. He was rather in favour of Ld Salisbury being PRS— Sylvester in smoking room last night was saying you or D. of Devonre. ought to be Prest.—[3] If Ld Sy were made he should withdraw his name at once. He got so excited over Kant[4] & Geometry that he almost danced

You must be having jolly time with the Boys[5]

Yours affec. | ED

DAR 105: B71–2

[1] The month and year are established by the relationship between this letter and the letter from E. A. Darwin, 19 [December 1870].
[2] See letter from E. A. Darwin, 19 [December 1870] and n. 2.
[3] Erasmus probably refers to Henry Holland (Sir H), Robert Cecil (the marquess of Salisbury), James Joseph Sylvester, and William Cavendish (the duke of Devonshire). CD had stayed with Erasmus in London from 8 to 14 December 1870 (CD's 'Journal' (Appendix II)). Edward Sabine was president of the Royal Society of London until 30 November 1871 (*Record of the Royal Society of London*, p. 338). He announced his forthcoming resignation at the anniversary meeting of the Royal Society on 30 November 1870 (*Proceedings of the Royal Society* 19 (1870–1): 127).
[4] The reference is to Immanuel Kant's argument that the truths of geometry are synthetic *a priori* truths (Kant 1781).
[5] Erasmus presumably refers to Francis, Leonard, and Horace Darwin.

To Charles Lyell 25 December [1870]

Down. | Beckenham | *Kent. S.E.*
Dec. 25th

My dear Lyell

I am very much obliged for the new book, received yesterday.[1] It seems wonderfully full of matter, & I shall soon look over all the parts likely to contain any new conclusions; but at present I can literally do nothing & think about nothing, except the sickening work of getting my last sheets corrected & recorrected for the press.—[2] I have first proofs to correct of only one more short chapter, & then thank all the powers above & below, I shall be a man again & not a horrid grinding machine.— I wish you joy heartily that your labour is over.—

Ever yours. | C. Darwin

Postmark: DE 26 70
American Philosophical Society (386)

[1] Lyell had evidently sent a presentation copy of *The student's elements of geology* (Lyell 1871); CD's annotated copy is in the Darwin Archive–CUL (see *Marginalia* 1: 545–6).

[2] CD was working on the proof-sheets of *Descent*.

To P. L. Sclater 26 December [1870][1]

The Printers have been slack; but I now send two sheets, & 1½, or thereabouts, more will complete your very kind assistance.[2]

C. Darwin

Dec. 26th.

Zoological Society of London

[1] The year is established by the relationship between this letter and the letter to P. L. Sclater, 5 December 1870.

[2] Sclater was checking the names of the birds and mammals in the proof-sheets of *Descent* (see letter to P. L. Sclater, 5 December 1870 and n. 1).

To P. L. Sclater 29 December 1870

Down. | Beckenham. | *Kent. S.E.*
Dec. 29th

My dear Mr Sclater

I am infinitely obliged for all your corrections, but it is a horrid shame to spoil even an hour of your holidays, for holidays must, I am sure, be rare with you.—[1]

I had some misgivings about the Moschidæ.—[2]

It is rather a bad job about the woodcut. It wd. drive Murray mad, if I proposed to delay for a new woodcut, though it is very kind in you to propose to help me in this manner.—

I will append, beneath the cut, as follows, in parenthesis

(I find that this drawing represents the head of a female, but it serves to show on a reduced scale the characters, of the male.)[3]

If I do **not** hear from you, I will assume that this parenthesis is correct & will insert it— I was at the time rather surprised that the tusks were not larger, but supposed the drawing was of rather young male. One & $\frac{1}{2}$ or $\frac{3}{4}$ sheets will I believe complete your labours.—

Yours most heartily obliged | Ch. Darwin

I will send next sheets, when received, to Hanover Sq^{e}, marked "to be forwarded".—[4]

What a rich & capital volume the last in your Proceedings—[5]

Zoological Society of London

[1] See letters to P. L. Sclater, 5 December 1870 and n. 1, and 26 December [1870]. Sclater presumably made his corrections directly on the proof-sheets of *Descent.*

[2] Moschidae is the family of musk deer. CD mentioned *Moschus moschiferus* in *Descent* 2: 280.

[3] CD refers to his publisher, John Murray, and to the woodcut of the head of an Ethiopian warthog (*Descent* 2: 265 fig. 65). The text in parentheses was inserted as CD suggested.

[4] The headquarters of the Zoological Society of London were in Hanover Square.

[5] CD refers to the *Proceedings of the Zoological Society of London.*

SUPPLEMENT, 1835–1869

This supplement includes all the letters from or before the year 1869 that have been located or re-dated since the publication of *Correspondence* volumes 13–17; earlier supplements were included in volumes 7 and 13. It also includes letters with a wide date range that seem to have been written before 1870. The letters are arranged in chronological order according to the usual practice of the edition. A record of CD's alterations, information about the correspondents and other individuals mentioned in the letters, and bibliographical details, are included in the Manuscript alterations and comments, the Name register, and the Bibliography for volume 18.

From James Busby to Alexander Berry 29 December 1835

Waitangi—Bay of Islands
29th. December 1835

My Dear Sir

I take great pleasure in reminding you that such a person as myself is still in existence by introducing to you the bearer of this, Mr. Darwin: I am sure it will give you pleasure to become acquainted with a Geologist who has spent several years in examining the Mountains of South America and I dont know any person to whom I could introduce him better qualified than yourself to give him an idea of what little is known of the Geology of New Holland.[1]

The Ship I understand will only be a fortnight at Port Jackson and of course Mr. Darwin will have little opportunity of making observations for himself— But what little may be within his power may be greatly improved by judicious advice—

I ought to have mentioned sooner that Mr. Darwin is a passenger with Capt. FitzRoy[2] of H.M.S Beagle.

I see from an occasional glance at the Sydney Newspapers—that the Colony is as much in a ferment as ever— I think the line must now be becoming more & more marked between men of honorable principles and feelings and those who would fall down before the many headed Monster—(*Monstrum horrendum*[3] as it is in New South Wales) of Public opinion to whatever *Supreme head* the many may be supported— How refreshing it is to honest men, to see the individual views of its members defeat the object which a union of rogues might effect—

The fate of the proposal for a Municipal Corporation in Sydney is a proof of what I stated in writing 8 years ago that there was not virtue enough in the Colony for any free Institution which idea I see all but stated in a late charge of Judge Burtons—[4] I have not had time to read the whole of that address but that sentiment caught my eye— And I was delighted to see something like virtue and honesty in high places

I have heard from my brother of your exertions in favour of my father.[5] I would give you my hearty thanks, if I thought that would add thing to the satisfaction which you must have felt in doing what you could to shield an honest man from oppression— We have *heard* of Tory Tyranny— If this is what we are to expect from the Whigs— Every honest man may devoutly pray for a change.— A change which must be at the door for I think the people of England were beginning to awake from their dream of Utopian improvement to the real Evils which have threatened them by the ascendancy of those who are "given to change"— I hope you will excuse this rambling letter— M[rs] Busby joins me in kind regards to M[rs] Berry[6] | And I remain | My Dear Sir | Very Faithfully Yours | James Busby

Alex[r] Berry Esq

State Library of New South Wales (MLMSS 315/51 Item 6, pp. 1–4)

[1] CD travelled in South America between February 1832 and September 1835 (see *Correspondence* vol. 1, Appendix I). New Holland: Australia.

[2] Robert FitzRoy.

[3] *Monstrum horrendum*: 'fearful monster' (Latin). Busby quotes Virgil's description of the cyclops Polyphemus (*Aeneid* 3.658).

[4] William Westbrooke Burton.

[5] James Busby had four brothers; he also refers to his father, John. (*Aust. dict. biog.* s.v. Busby, John).

[6] Busby refers to Agnes Busby and Elizabeth Berry.

To Thomas Spring Rice [before 7 July 1838][1]

To the Right Honourable the Chancellor of the Exchequer.

The Memorial of your Petitioners humbly Showeth,

That your Petitioners have heard with great concern that a recent offer for sale to the British Museum, by Mr. Mansell and Mr. Hawkins, of two valuable collections, illustrating the geology of an important portion of England, has been declined, in consequence of the deficiency of the national revenue for the present year.[2]

That many of your Petitioners who are conversant with the subject have examined these collections, and consider them not only to be of peculiar value as demonstrating the subterraneous structure of this country, but also of general interest to the scientific world, as throwing new and important lights on many branches of research that are now conducted with unexampled activity in all countries, for the purpose of illustrating the physical history of the globe, and applying such knowledge also to the useful purposes of life.

That with a view to the advancement of such objects, societies and museums have been established in nearly all the large provincial towns of this kingdom, and still more extensive museums provided at the national expense by all the Governments in the civilized world. And as your Petitioners would consider it to be conducive to the honour and scientific reputation of England if these collections were placed in the British Museum, so it would be, in their opinion, both detrimental to science and injurious to the honour of the nation if they were allowed to be broken up and dispersed by public sale, or purchased by any foreign Government.

That the getting together of these collections has resulted from a concurrence of such rare opportunities and extraordinary qualifications in the individuals by whom they have been made, that if the occasion which now offers of securing to the nation the fruits of such rare coincidences be not made use of it must be lost for ever.

And your Petitioners will ever pray, &c.

(signed)

Wm. Buckland, Professor of Geology, Oxon.
A. Sedgwick, Woodwardian Professor, Cambridge.
John Phillips, Professor of Geology in King's College, London.
W. Whewill,[3] Pres. G. S.
Rod. Murchison, Vice Pres. Geol. Soc^{y}.
Charles Lyell, Vice Pres. Geol. Soc^{y}.
Charles Stokes, F.R.S., F.G.S., and L.S.
William J. Hamilton, Sec. Geol. Soc^{y}.
Edward Norwich,[4] President of the Linnean Society, F.G.S.
Richard Owen, Hunterian Professor Royal College of Surgeons, London.
William Clift, Conservator of the Museum, ditto.
Charles Babbage.
J. Bostock, M.D., F.R.S., and F.G.S.
P. M. Roget, M.D., Sec. R.S., and F.G.S.
John Taylor, F.R.S., F.L.S., F.G.S., Treasurer to the British Association.
Northampton,[5] F.G.S.
Charles Darwin, Sec. of Geol. Soc^{y}.
W. T. Broderip, F.R.S., F.G.S., F.L.S., &c.

Printed document
House of Commons papers; accounts and papers, 1837/38, XXXVI, 307

1 The date is established by the fact that the petition was presented to the House of Commons on 7, 26, and 27 July 1838 (*Journals of the House of Commons* 93: 683, 755, and 764). A similar version of this petition was presented on 15 and 25 March 1839 (*Journals of the House of Commons* 94: 113, 145).

2 The reference is to Gideon Algernon Mantell and Benjamin Waterhouse Hawkins. Mantell made collections of some 20,000 rocks and fossils from Kent, Surrey, and Sussex. They had previously been displayed at the Sussex Scientific and Literary Institution at Brighton (House of Commons papers, accounts and papers, 1837/8, XXXVI, 305).

3 William Whewell.

4 Edward Stanley, bishop of Norwich.

5 Spencer Joshua Alwyne Compton, second marquis of Northampton.

From Charles Maclaren to Thomas Walker 30 March 1840

Scotsman Office
30 March | 1840

Sir

I have a letter from Mr Charles Darwin Naturalist to the late expedition to the Southern hemisphere stating that he has felt great interest in Mr Kemps notice of Terraces on the Eildon Hills and desiring me to procure him Mr Kemps Address that he may send him a copy of his paper on Glenroy and put some queries to him.[1] Not knowing whether Mr Kemp resides in Galashiels I send this note to you. Will you be so good as furnish him with Mr Kemps Address. Mr Darwin resides at N[o] 12 Upper Gower Street London.

Let me know if the Galashiels Society has a copy of my Geol[y] of Fife and the Lothians.[2] If not I shall be glad to send them a copy.

I am Sir | Your Most Ob Servant | Cha Maclaren

Ruth Cramond and David Cramond (private collection)

1 CD's letter to Maclaren has not been found; Maclaren refers to CD's travels on the *Beagle* from 1831 to 1836 (see *Correspondence* vol. 1). The 15 February 1840 issue of the *Scotsman* contained a report of an essay read by William Kemp to the Galashiels Geological Society, in which Kemp described how CD's Glen Roy theory had encouraged him to look for similar terraces. See *Correspondence* vol.2, letter to Charles Lyell, [19 February 1840] and n. 1. CD's paper was 'Parallel roads of Glen Roy'. The Eildon Hills are in Melrose, Scotland.

2 Maclaren 1839.

To Thomas Walker 5 May [1840][1]

Sir

I have been prevented by ill-health, from which I am still suffering, from sooner acknowledging your letter of the 1st of April.—[2] I hope to be able in a short time to write to M[r] Kemp on the subject of his important discovery of ancient water-lines on your Eildon Hills; and I am very much obliged to your kindness in affording me the means to do so by giving me his address.—[3] I am much gratified at the impression, you have been pleased to form of my geological work in the southern hemisphere.—[4] I fully concur with you in estimating most highly the pleasure derived from investigating the ancient history of the world, and every geologist must hail with joy, so auspicious a fact, as the concurrence of several warm followers of the science in the formation of a country Society.—[5]

With every wish for the prosperity of your Society, Believe me | Dear Sir | Yours respectfully | Chas. Darwin

12 Upper Gower St | Tuesday May 5th

Jim Somerville (private collection)

1 The year is established by the relationship between this letter and the letter from Charles Maclaren to Thomas Walker, 30 March 1840 (this volume, Supplement).

2 This letter has not been found.

3 See this volume, Supplement, letter from Charles Maclaren to Thomas Walker, 30 March 1840 and n. 1. CD refers to William Kemp.

4 CD had made observations on South American geology while on the *Beagle* voyage. See *Journal of researches* and *Geology of the 'Beagle'*.

5 Thomas Walker was secretary of the Galashiels Geological Society ('The Galashiels geological society', *Scotsman*, 21 March 1840, p. 3).

To William Kemp 1[8] May [1840][1]

12 Upper Gower St
Monday May 1⟨8⟩

Sir

I was very much interested by an abstract of your paper which appeared some time since in the Scotsman.[2] Having learnt your address, through the kindness of Mr. Walker,[3] I trust you will excuse the liberty I take in addressing you, and that will do me the pleasure of accepting a copy of my paper o⟨n⟩ Glen Roy.—[4] I wish very much to know, whether your paper is printed anywhere at greater length than in the Scotsman? I should feel much obliged if you would answer me this question, & at the same time perhaps you will be good enough to inform me, whether any more facts have been discovered, & whether your views on the origin of ⟨th⟩e terraces round the summit of the Eildons have been confirmed, or the contrary?—[5] I should much like to know by what means you have estimated the height of the terraces above the sea, and their corresponding level on the different hills?[6] Over how *many miles* of country have traces of these marks at the same level been observed? Should you feel so inclined, any observations on th⟨is⟩ subject, which you would have the kindness to communicate with me, I should receive with the greatest interest.— I had intended to have visited your part of Scotland this summer, but the state of my health renders it very doubtful, whether I shall be able. I should have much enjoy⟨ed⟩

Incomplete
Ruth Cramond and David Cramond (private collection)

1 The year is established by the relationship between this letter and the letter from Charles Maclaren to Thomas Walker, 30 March 1840 (this volume, Supplement). The date is partially obscured, but reads May 10, 16, or 18. In 1840, 18 May fell on a Monday.

2 See this volume, Supplement, letter from Charles Maclaren to Thomas Walker, 30 March 1840 and n. 1.

3 Thomas Walker.

4 CD refers to 'Parallel roads of Glen Roy'.

5 Kemp wrote that the terraces in the Eildon Hills were raised sea beaches, in accordance with CD's theory about the terraces at Glen Roy in 'Parallel roads of Glen Roy', which Kemp had already read (*Scotsman*, 15 February 1840, p. 3). He did not publish further on the terraces.

6 CD measured the terraces at Glen Roy by taking measurements at either end of each terrace with a mountain barometer (see 'Parallel roads of Glen Roy', pp. 44–5). The terraces are the 'parallel roads' referred to in 'Parallel roads of Glen Roy'.

To William Allport Leighton 25 November [1840–6][1]

I trouble you with a line to thank you for the trouble you have kindly taken in procuring me cuttings of the weeping yew, which my Father's gardner will take great care of & I hope in future years to have a fine tree to remind me of Shropshire.[2] I have often thought of that curious case of the heath in the stone quarry & the more I think, the greater the pity appears, that it shd not be well made out & published. Have you ever written about it?[3] [...] bit of clay to the end of the cutting, & send it direct in a letter to me, & I w^{d} not lose a day in forwarding it to Hayes Place.—[4] Pray do not think a moment more about this, if your relation to the individual sh^{d} make it appear to you a liberty.

Believe me | yours vy faithfully | C. Darwin

P.S. It w^{d} be necessary of course to know whether the Heath has hitherto lived out of doors through the *winters*

Incomplete[5]
Unknown dealer

[1] The year range is established by the reference to the heath in the stone quarry and by CD's assumption that it had not been published; the heath was found in 1840 or 1841 and published in 1847 (Sidebotham 1847). See n. 3, below. Sidebotham 1847 was dated 15 February 1847; according to the editor of the journal it was published in, articles were normally published less than two weeks after their receipt (*Phytologist* 2 (pt. 3) (1847): vi).

[2] Leighton was a botanist and curate living in Shrewsbury, where CD had grown up. CD later cited observations of the weeping yew from Leighton's *The flora of Shropshire* (Leighton 1841) in *Variation* 2: 18. CD also refers to Robert Waring Darwin and John Abberley.

[3] CD probably refers to a sample of heath sent to Leighton by Joseph Sidebotham; the heath was found growing out of a fossil tree in a quarry at Tintwistle, Derbyshire. Leighton identified the sample as *Erica arborea* (tree heath; see Sidebotham 1847, p. 779). In Sidebotham 1847, Sidebotham said the heath was found 'six or seven years ago'.

[4] CD probably refers to Hayes Place, Hayes, Kent, the house of Wilhelmina Barbra Traill. Hayes was three miles away from CD's house in Down.

[5] The original letter is complete and is described in the sale catalogue as being four pages long. The first paragraph is the dealer's transcription, and the rest was transcribed from a photograph.

From Alexander Bridport Becher 1 September 1842

Sept 1st 1842

C Darwin Esqr
Sir—

When Capt Beaufort[1] went away he told me to order three copies of your new work on Corals—[2] Please to send them as soon as convenient & you will oblige

ABB

Contemporary copy
United Kingdom Hydrographic Office: Letter Book 10 (1841–2), p. 510

[1] Francis Beaufort.

[2] Becher refers to *Coral reefs*.

To Charles Edward and Mary Kingford Mudie 10 December [1842–5 or 1855–68][1]

Down Bromley Kent
Dec. 10th.

Mr Darwin presents his compliments to Mr & Mrs. Mudie & is much obliged for the honour of their invitation. Mr. Darwin fears that he cannot come up to London on that day, otherwise he would have had much pleasure in accepting Mr & Mrs. Mudies' kind invitation.—

The Rare Book & Manuscript Library of the University of Illinois at Urbana-Champaign (Charles E. Mudie Collection, 1816–1897: Correspondence, B29)

1 The date range is established by the form of the address, which CD used from October 1842 to April 1846, and from October 1855 until March 1869.

To ? 23 January [1843–6 or 1856–69][1]

Down Bromley Kent
Jan. 23d

Dear Sir

I am much obliged to you for sending me a copy of your memoir, with its wonderfully beautiful illustrations, on the most interesting scientific point, which has occurred for many years.—[2]

With my best thanks I beg leave to remain | Dear Sir | Yours faithfully | Ch. Darwin

Edward Ford (private collection)

1 The date range is established by the form of the address; CD used this address from October 1842 to April 1846, and from October 1855 to March 1869.

2 The memoir has not been identified.

To William Kemp 7 April [1843][1]

Down Bromley | Kent
April 7th.—

Dear Sir

Owing to my change of residence your communication has been slow in arriving here.[2] I am much obliged for it, as it promises to be very wonderful.— I will today send one or two of the seeds to Mr. Robert Brown (confessedly the first Botanist in Europe) & find out whether he knows them or their family; & I will send the remainder to Professor Lindley with a request to have them planted at the Horticultural Garden of London & taken care of.—[3] Let me urge on you to watch your plants & if they do not prosper to vary their treatment. I think you do not know whether they inhabited a cold or hot, a wet or dry climate. As any plant dies, I earnestly recommend you to dry it between paper most carefully even if it has only the germinal leaves, & preserve if they flower some in Spirits of Wine & I will get (if you choose) the best botanists to examine them.— Should the Plants turn out known ones, your case no doubt is deserving of record, but sh[d] they turn out unknown species your case would be magnificent.—[4] D[r] Daubeny of Oxford is attending to the subject of germination of old seeds, but he unfortunately has just started for Spain.—[5] With respect to your paper for Taylor's Journal, I will do as you please, but if it were my own nothing sh[d] induce me to publish it, till the seeds had fully grown & had been examined by Botanists with *known* names.—[6] In present state some of the Public w[d] say he has been imposed upon by seeds having been placed there, as a joke; & others w[d] say, no doubt they are some common British Plants. It is well to make every paper to tell with its whole effect *at once.* If the seeds turn into unknown plants, *perhaps* you w[d] prefer your Paper being read before the Geological Society; sh[d] you prefer this, I w[d] (when the plants have been examined) with pleasure transmit it to the Soc.— In that case the latter part of the Paper, if you will allow me to suggest it, had better be a little simplified:— If you follow my advice, & retain your paper till the plants have flowered or have died & have been examined—Could you not in the ⟨in⟩terval reexamine the situation & try to get more or hunt for shells or bones &c in the gravel.—

I will wait till I hear from you again.— I will inform you if M[r] Brown has anything particular to say about the seeds. & on the ultimate fate of those seeds, which I send to the Horticultural Gardens.—

Believe me | Yours very faithfully | C. Darwin

Ruth Cramond and David Cramond (private collection)

1 The year is established by the relationship between this letter and the letter from William Kemp, 2 October 1843 (*Correspondence* vol. 2).

2 CD moved to Down House in Down, Kent, on 14 September 1842 (see *Correspondence* vol. 2, Appendix II). Kemp's letter has not been found.

3 Kemp had found seeds in a sandpit near Melrose, Scotland; see *Correspondence* vol. 2, letter from William Kemp, 2 October 1843. CD refers to John Lindley (see *Correspondence* vol. 2, letter to John Lindley, 8 [April 1843]). CD's letter to Robert Brown has not been found.

4 Kemp wished to argue that the seeds had been deposited in the sand thousands of years previously when the pit would have been a river bank or lake shore (see *Correspondence* vol. 2, letter to John Lindley, 8 [April 1843]).

5 From 1841 to 1857, Charles Giles Bridle Daubeny served on a committee of the British Association for the Advancement of Science experimenting on the growth and vitality of seeds. In the committee's report to the British Association meeting in 1842, Daubeny and Hugh Edwin Strickland requested old seeds to be submitted for experiment (Strickland and Daubeny 1842; see also *Correspondence* vol. 2, letter from J. S. Henslow, 2 November 1840).

6 CD refers to *Annals and Magazine of Natural History*. Kemp's original paper has not been found, but he later published a much-edited account in the *Annals* (Kemp 1844). The account was drafted by CD based on Kemp's letters (see *Correspondence* vol. 2, Appendix VI).

To William Kemp 24 [April 1843][1]

Down | Bromley | Kent
24th

Dear Sir

I am glad you are willing to follow my advice about deferring your account till the plants have been ascertained.— I return the copy of the Paper you sent me.—[2] I am sorry to say I have not yet heard from M^r. R. Brown & I fear my letter has miscarried or perhaps he waits till I see him in town: I will not fail to write if he has anything particular to observe.—[3]

I have heard from Prof. Lindley, who says one of the seeds is certainly a Polygonum & the other probably a Chenopodium: he anticipates they will turn out common species— He has had seeds carefully planted & will communicate the result to me if they germinate.—[4]

I hope you will carefully dry some of your specimens in their flowering & seeding state | Yours very faithfully | C. Darwin

Ruth Cramond and David Cramond (private collection)

[1] The month and year are established by the relationship between this letter, the letter from William Kemp, 2 October 1843 (*Correspondence* vol. 2), and the letter to John Lindley, 8 [April 1843] (*Correspondence* vol. 2).

[2] See this volume, Supplement, letter to William Kemp, 7 April [1843] and n. 6.

[3] See this volume, Supplement, letter to William Kemp, 7 April [1843]. CD refers to Robert Brown.

[4] *Polygonum* and *Chenopodium* are both in the order Caryophyllales. See *Correspondence* vol. 2, letter from John Lindley, [after 8 April 1843].

To William Kemp [8 September 1843]

Down. Bromley, | Kent

My dear Sir

The seeds you sent me from the sand-pit & which I forwarded to Prof: Lindley, have germinated & turn out to be Rumex acetosella & an Atriplex, which Lindley does not know.— He does not think it can be a variety of *A. patula.*[1] He has sent specimens to Prof. Henslow who writes to me that he is carefully examining it— He does not recognise it as a British Plant,—but think it may *perhaps* turn out to be A. hastata, a French species.—[2] He seems to have doubted whether it might not be a monstrous variety of *A. patula.*—

If you have any specimens in flower, w^d it not be worth your while to send him some by Post in a Fresh condition; it w^d, I daresay aid his investigations.— His direction is Prof. Henslow

Hitcham Rectory
Hadleigh
Suffolk.—

If you send any, you must write with them, else he will not know, from whom they come.—

This is really a very curious case, & when the Botanists have made up their minds about the plant, you will I hope publish an account of it, with a careful account of the Quarry.—[3] It is a pity, that some of the earth & rubbish, in which the seeds were found, was not preserved, perhaps minute seeds in it might have germinated.—

I did not write some months ago to say that R. Brown w[d] not offer any opinion on the seeds; the specimens, which I sent him, were too imperfect.—[4]

Believe me | Very truly yours | C. Darwin

Endorsement: 'Septr 8 1843'
Ruth Cramond and David Cramond (private collection)

[1] John Lindley had thought one of the seeds to be in the genus *Polygonum* and the other in the genus *Chenopodium*. *Rumex acetosella*, sheep's sorrel, like *Polygonum*, is a member of the *Polygonaceae*. *Atriplex*, like *Chenopodium*, is a member of the *Chenopodioideae*. See *Correspondence* vol. 2, letter from John Lindley, [before 2 September 1843]. *Atriplex patula* is the spear saltbush.

[2] The letter from John Stevens Henslow has not been found. See *Correspondence* vol. 2, letter to J. S. Henslow, [2 September 1843], and letter from William Kemp, 2 October 1843. *Atriplex hastata* is now *A. prostata*, the triangle orache.

[3] Kemp replied to CD with an account and map of the situation in which he had found the seeds (see *Correspondence* vol. 2, letter from William Kemp, 2 October 1843). An account was later published in *Annals and Magazine of Natural History* as Kemp 1844.

[4] Letters to and from Robert Brown have not been found. See this volume, Supplement, letter to William Kemp, 7 April [1843].

To William Kemp 9 October [1843][1]

Down. near Bromley Kent
October 9th

Dear Sir

I write a line to inform you that your paper has reached me safely.— I will consult with M[r] Lyell, (who has taken much interest in your discovery) what will be the best place of publishing it.—[2]

I am flattered by the compliment you make me in addressing your communication in a letter to me.—

I think it w[d] be a thousand pities to publish the Paper before Prof Henslow has definitely made out whether the Atriplex is the *A. hastata* of France or not.— I will write by today's post to urge him to complete his examination.—[3]

Will you inform me, *soon* about how many feet the layer with the seeds was *above the present level of the nearest part of the* present *river.*— I think this had better be clearly introduced, & I do not quite understand your description in reference to this point.—[4]

As you permit me to shorten some parts of your paper, I will do so— I assure you the retrenchment of redundancies, is a reform that I every month learn to value more & more in my own writings—

Your letter must have a heading & the account of Henslow's conclusions introduced.— No species, of Atriplex, can properly be considered, I believe, as a climbing plant, at most they can be called only *trailing* plants.— I will make the alterations which appear to me adviseable & then send you a copy for your approval.— But we must wait for Henslow's decisi⟨on⟩ as the whole *force* of the Paper turns on this.[5]
Yours very truly | C. Darwin

P.S. | Can you not inform me (& I w[d] incorporate it with your paper) what relation the bed of sand bears to the period of the dispersal of boulders whether by floating ice or glaciers— Are there any *large* or *angular* boulders of foreign rocks on top of sand or embedded in it, or in the gravel beneath it? Are the rocks scored, polished, or hummocked beneath it— Do endeavour to throw some light on these points, as they w^d^ be very interesting.—[6]

I will enquire from Prof. Lindley how he raised the seeds, whether in open ⟨½ *line missing*⟩ or in the Green house.[7]

I think your speculation about the former dampness of climate too bold— What reason have you for supposing that there were more lakes then than now? Is it not also doubtful, how far a few more or less lakes, w^d^ affect the surrounding climate?—[8]

What a pity that no fresh-water shells were in the sand.—

Ruth Cramond and David Cramond (private collection)

[1] The year is established by the relationship between this letter and the letter from William Kemp, 2 October 1843 (*Correspondence* vol. 2).

[2] CD refers to the letter from William Kemp, 2 October 1843 (*Correspondence* vol. 2). No correspondence on the subject has been found. On the publication of Kemp 1844, based on Kemp's letters to CD, see *Correspondence* vol. 2, Appendix VI.

[3] No letter from CD to John Stevens Henslow has been found written on 9 October 1943, but see *Correspondence* vol. 2, letter from J. S. Henslow, 9 October 1843, in which Henslow said he inclined to believe that what had been thought to be *Atriplex hastata* was in fact a variety of *A. patula*, a common wasteland plant.

[4] Kemp had sent CD a description of the location of the seeds in his letter of 2 October 1843, but omitted the depth at which the sands were buried. CD was preparing an account of the discovery for *Annals and Magazine of Natural History* which he constructed mostly of excerpts from Kemp's letters. See *Correspondence* vol. 2, Appendix VI, and Kemp 1844.

[5] The plant identifications from Henslow, John Lindley, and Charles Cardale Babington were introduced at the beginning of the paper CD compiled from his correspondence with Kemp. See Appendix VI, *Correspondence* vol. 2, and Kemp 1844.

[6] CD and Kemp believed that identification of the seeds could be considered along with geological arguments to demonstrate that the valley where they were found had been an ancient sea beach (see this volume, *Supplement*, letter to William Kemp 1[8] May [1840] and n. 5).

[7] CD had sent some of Kemp's seeds to John Lindley for identification. See *Correspondence* vol. 2, letter from John Lindley, [before 2 September 1843] and n. 3. CD's letter to Lindley has not been found.

[8] In his previous letter Kemp had argued that the comparative health of the plants grown from his seeds in the damp garden over those grown in the dry house demonstrated that the sandpit they were retrieved from must once also have been wet (*Correspondence* vol. 2, letter from William Kemp, 2 October 1843).

To William Kemp [14 October 1843][1]

Down near Bromley | Kent [Shrewsbury][2]
Saturday

Dear Sir

I have heard from Prof. Henslow, who expresses a doubt, whether some of the seeds may not have been contained in the soil, in which you planted the seeds from the sand-pit.—[3] In the case of the Atriplex, I think, the circumstance of the same strange form having come up from the seeds planted by you & at the Hort. Soc. shows that there could have been no mistake in this case.[4] But will you be so good, as to inform me, whether you took any precaution against this source of error.—

I am very sorry to say that Prof. Henslow, now suspects that the Atriplex is merely a variety of the common *A. patula.*[5] in this case, I fear, no editor would wish to insert more than a brief notice of the germination of seeds buried at the depth &c &c under which you found them. But we had better pause and hear what the Botanists determine. I will send a specimen to Mr Babington, who is an accurate discriminator of British species & hear what he says.[6]

Believe me | Yours very faithfully | C. Darwin

Postmark: Shrewsbury OC 15 1843
Ruth Cramond and David Cramond (private collection)

1 The date is established by the postmark. In 1843, the Saturday before 15 October was 14 October.
2 On 12 October 1843 CD went to Shrewsbury for twelve days (*Correspondence* vol. 2, Appendix II).
3 See *Correspondence* vol. 2, letter from J. S. Henslow, 9 October 1843.
4 One of the seeds sent to John Lindley turned out to be an *Atriplex*, as well as one of the plants grown by Kemp. See *Correspondence* vol. 2, letter from John Lindley, [before 2 September 1843], and letter from J. S. Henslow, 9 October 1843.
5 CD refers to John Stevens Henslow. See *Correspondence* vol. 2, letter from J. S. Henslow, 9 October [1843].
6 CD's letter to Charles Cardale Babington has not been found. See *Correspondence* vol. 2, letter from C. C. Babington, 26 October 1843. Babington identified the sample as *Atriplex angustifolia* (now *A. patula*, the common orach).

To William Kemp 1 November [1843]

Down, near Bro⟨m⟩ley, | Kent
Nov. 1st.

Dear Sir

I have been detained from home, otherwise I sh^d^ have sooner answered your last communication.—[1]

I am sorry to say that Mr Babington (who from the te⟨s⟩timony of several Botanists, is a first-rate judge in the genus Atriplex) pronounces the plant reared at the Hort. Soc. to be A. angustifolium; I enclose his le⟨t⟩ter, as you might like to see it.—[2] I am quite p⟨u⟩zzled what to recommend you about publication,—b⟨u⟩t first let me say, that, from what I have seen ⟨of⟩ your ardent love of natural science, I would as soon believe myself capable of playing a hoax o⟨n⟩ the public, as I would believe you to be so.— T⟨h⟩e most sceptical of the public w^d^ only su⟨s⟩pect an accidental error. For myself I believe the seeds had been deposited, as you describe & that they produced the plants, you sent to Prof. H⟨e⟩nslow.— and I think I have shown by my ⟨co⟩rrespondence, that I do not underrate the interest of the[3] ⟨*most of line missing*⟩ seeds were sown, & with you, two species of Polygonum. P. aviculare & P. convolvulus (as Henslow informs me) & one of Atriplex came up,[4] whereas at the Hort. Soc. another species or variety of Atriplex (for I sh^d^ have told you, that the Atriplex reared by you is undoubtedly according to Henslow the A. patula) & a Rumex R. acetosella:[5] therefore 3. genera, and 4 or 5 species have come up from

2 kind of seeds.— Moreover unfortunately the plants are just those, which might be expected to come up from garden soil. If the Atriplex had turned out a new species that alone w[d] have been good evidence; or even if the same variety of this genus had come up with you and at the Hort. Soc, there w[d] have been some evidence. I daresay the Rumex at the Hort. Soc was from a seed in the soil, but the Public (when all the facts are stated, which of c⟨our⟩se you would, wou⟨ld⟩ I fear, say it is more probable that the seeds of *all* these plants were contained in the soil, than that such ancient seeds sh[d] retain their vitality.— I believe differently myself; but I declare I do not see, how you can make your case sufficiently clear, to justify its publication.— I take great blame to myself that I did not urge Prof. Lindley to plant the seeds in a circle or in a perfectly straight line, for in that case, or had you done so, the evidence w[d] have been good for the majority of the plants which came up, though even then not for every individual one.

Let me hear what you think of what I have said.— I have all your papers safe & w[d] return them to you at anytime—[6] I assure you, I feel extremely disappointed at the prospect ⟨of⟩ your trouble being thrown away, & of so curious (& as I believe real) ⟨a⟩ fact being lost— I trust you will credit me, when I say, that I have done as much in endeavouring to obtain information for you, as I could if I had discovered the seeds myself; & that I have advised you to the best of my judgment.—

Believe me, dear Sir | Yours very faithfully | C. Darwin

Postmark: 1 NO 1843
Ruth Cramond and David Cramond (private collection)

1 See this volume, Supplement, letter to William Kemp, [14 October 1843], n. 2, and *Correspondence* vol. 2, letter from William Kemp, 17 October 1843.

2 See *Correspondence* vol. 2, letter from C. C. Babington, 26 October 1843. The specimen of *Atriplex angustifolia* had been sent to John Lindley, who had not been able to identify it (see *Correspondence* vol. 2, letter from John Lindley, [before 2 September 1843]). CD refers to Charles Cardale Babington.

3 Kemp had written to CD to ask whether or not to publish an account of his discovery of seeds twenty-five feet deep in a sandpit (see *Correspondence* vol. 2, letter from William Kemp, 17 October 1843). CD refers to John Stevens Henslow.

4 The plants grown by Kemp were identified by Henslow as *Polygonum convolvulus*, *P. aviculare*, and *Atriplex patula* (see *Correspondence* vol. 2, letters from J. S. Henslow, 9 October 1843 and 17 October 1843).

5 John Lindley had identified one of the seeds sent to him as *Rumex acetosella*. See *Correspondence* vol. 2, letter from John Lindley, [before 2 September 1843]. The *Atriplex* was identified by Babington as *Atriplex angustifolia* (see above n. 3).

6 Kemp's accounts of the discovery of the seeds were included in his letters to CD. See *Correspondence* vol. 2, letters from William Kemp, 2 October 1843 and 17 October 1843.

To William Kemp [9 November 1843][1]

Down. Bromley Kent
Thursday

Dear Sir

Allow me to express my respect at the pleasant tone of your last letter, after what would have been to me, in your case a considerable mortification.[2]

Since writing to you, I have been reflecting on Prof. Lindleys opinion on the seed, before it was planted—& I think after all, a pretty good case may be made out, fairly stating every circumstance[3]

With your permission I will draw up a **short** statement with as little theory as possible & will send it you for your approval.[4]

Would you please return me Mr Babington's letter,[5] that I may quote it.— I suppose you did not actually mark the spot, where you sowed each seed in the pot—but I presume you knew very nearly where each seed lay.—[6]

In Haste, | Yours faithfully | C. Darwin

P.S. | You will perceive, whe⟨n⟩ you view my abstract of your several documents how Prof. Lindleys opinion on the seeds is valuable evidence.

Ruth Cramond and David Cramond (private collection)

1 The date is established by the relationship between this letter and the letters from William Kemp of 4 November 1843 and 10 November 1843 (*Correspondence* vol. 2). The first Thursday after 4 November 1843 was 9 November.

2 See *Correspondence* vol. 2, letter from William Kemp, 4 November 1843.

3 John Lindley originally had written that the seeds would prove to be common varieties of *Polygonum* and *Chenopodium* (see *Correspondence*, vol. 2, letter from John Lindley, [after 8 April 1843]).

4 CD had advised against publication concerning Kemp's seeds; see this volume, Supplement, letter to William Kemp, 1 November [1843]. The short statement was published as Kemp 1844; see *Correspondence* vol. 2, Appendix VI.

5 CD refers to Charles Cardale Babington. In his letter to William Kemp of 1 November [1843] (this volume, Supplement), CD had enclosed the letter from C. C. Babington, 26 October 1843 (*Correspondence* vol. 2).

6 As all the plants were common in garden soil, CD thought it necessary to be able to prove that each plant had in fact grown from one of the seeds Kemp had found buried in a sand pit (see this volume, supplement, letter to William Kemp, 1 November [1843]).

To William Kemp 22 November [1843]

Down near Bromley Kent
Nov 22d

Dear Sir

I am pleased how good a case your facts have make.— Your last letter with an account of your manner of planting &c &c was very useful.—[1] You will perceive the bearing of some of your facts by reading the article on Agyptian Wheat in the accompanying Gardeners Chroni⟨cle⟩ which please to return to me.—[2] You will think I have greatly curtailed your paper, I am **certain** that a **short** paper with **no theory** is desirable in this case. Please to make any corrections o⟨r⟩ remarks or improvements on a separate piece of paper, & I will introduce them.[3] I have written to Professor Henslow to le⟨arn⟩ what your doubtful plant was, & will make ⟨any⟩ alteration which may be desirable—[4] I propose to send your paper to the Annals of Natural Histo⟨ry⟩ with a request for its early insertion (proba⟨bl⟩y it can not be admitted for 5 or 6 weeks)[5] & for a few copies for yourself— I hope you will be satisfied with my abstract, which I have tak⟨en⟩ pains with & will correct, when in type.— The paper is not geological enough for the Geological Soc.—

Believe me | With good wishes | Yours very faithfully | C. Darwin

P.S. You will observe that I have introduced a few remarks not in your papers, which I thought desirable, which I do not doubt you will excuse[6]

Postmark: 22NO22 | 1843
Ruth Cramond and David Cramond (private collection)

[1] See *Correspondence* vol. 2, letter from William Kemp, 10 November 1843. Kemp explained in this letter that he had planted his seeds in rows.
[2] An article about wheat grown from seeds found in Egyptian mummies' tombs was printed in *Gardeners' Chronicle* on 11 November 1843, pp. 787–8.
[3] CD's draft of the paper (Kemp 1844) is in DAR 50: 1–6. It is primarily made up of passages extracted from the letters from William Kemp of 2 October 1843, 17 October 1843, 4 November 1843, and 10 November 1843 (*Correspondence* vol. 2).
[4] CD refers to John Stevens Henslow. See *Correspondence* vol. 2, letter to J. S. Henslow, [4 November 1843].
[5] The paper was published in *Annals and Magazine of Natural History* in February 1844 (Kemp 1844).
[6] For CD's draft, with alterations, see *Correspondence* vol. 2, Appendix VI.

To William Kemp 1 December [1843][1]

Down Bromley Kent
Nov. 31.

Dear Sir

You will be weary of my interminable letters.— Conversing yesterday with Mr. R. Brown (the great Botanist) I found him very skeptical, & to my great surprise he threw some shadow of doubt on the Agyptian Wheat case, & the doubt rested (& likewise in the case of the Raspberry seed) on the evidence not being clear that Sir G. Wilkinson did really get the seed himself or personally knew the person who with his own hands got the seed out of the Mummy Case.[2] Will you, therefore, please to inform me, whether you yourself extracted any of the seeds.—or whether Mr — Bell (please inform me of his *Christian name*) actually saw the seeds disinterred?[3] or whether any other *middle aged* inhabitent (giving ⟨*1 line missing*⟩ witnesses.— Were ⟨th⟩ere many work-people present who were witnesses? Could any of them have buried the seed in hopes of being rewarded for the discovery (it has been suspected that this was the case with the Raspberry seeds & *known* to be the case with *some* Mummy Wheat).—

Will you please to give me a **brief** & *correct* account of this part of the evidence.[4]

& believe me | dear Sir | Yours very faithfully | C. Darwin

Is Mr Bell of Melrose the actual land-owner of the Sand-pit? or does he rent it?

Ruth Cramond and David Cramond (private collection)

[1] The year is established by the relationship between this letter and the letter from William Kemp, 4 December 1843 (*Correspondence* vol. 2). CD evidently wrote 'Nov. 31.' in error.

[2] Robert Brown had been consulted about Kemp's seeds, but had not previously given any opinion on them (see letter to William Kemp, [2 September – 1 October 1843] and n. 5. CD refers to John Gardner Wilkinson's attempt to grow wheat grains found in an Egyptian tomb (see *Correspondence* vol. 2, letter from J. S. Henslow, 2 November 1840 and nn. 3 and 4; see also this volume, Supplement, letter to William Kemp, 22 November [1843] and n. 2). CD also refers to John Lindley's interest in raspberry seeds found in Dorsetshire in a coffin estimated to be sixteen or seventeen hundred years old (*Correspondence* vol. 2, letter to John Lindley, 8 [April 1843] and n. 3, and Lindley 1840, p. 185).

[3] John Bell owned the sand pit in which Kemp's seeds had been collected by Thomas Welsh and was present when they were unearthed. See *Correspondence* vol. 2, letter from William Kemp, 24 November 1843 and n. 2.

[4] For Kemp's description of the discoverer and witness, see *Correspondence* vol. 2, letter from William Kemp, 4 December 1843.

To William Kemp 7 December [1843]

Down.
Dec. 7th.—

Dear Sir

I have the pleasure to inform you that to day I have sent your paper to the Annals[1]—with the particulars last sent added.—[2] I hope you do not think, that I have been needlessly scrupulous,—I have had no motive but adding to the value of your discovery & establishing an interesting fact. I am sorry you thought it adviseable to send me a testimonial of yourself—[3] Your correspondence & pursuits rendered that quite superfluous.

Yours faithfully | with best wishes | C. Darwin

Postmark: 7 DE 7 | 1843
Ruth Cramond and David Cramond (private collection)

[1] A paper CD had compiled from letters written to him by William Kemp on 2 October, 17 October, 4 November, and 10 November 1843 describing the discovery of seeds twenty-five feet deep in a sandpit near Galashiels, Scotland, was published in *Annals and Magazine of Natural History* in February 1844 (Kemp 1844). See *Correspondence* vol. 2.

[2] Kemp had sent further details about the discovery of his seeds with his letter of 4 December 1843 (*Correspondence* vol. 2).

[3] Kemp enclosed testimonials for himself, Thomas Welsh, who had found the seeds, and John Bell, who was present at the discovery, with his letter of 4 December 1843 (*Correspondence* vol. 2).

To Henry Denny 5 February [1844][1]

Down Bromley | Kent
Feb 5

Dear Sir

The specimen ticketed on green 153 is from the *Condor*,[2] Chile

Red 395 is from a Hawk, killed at Maldonado on banks of Rio Plata.—[3] I believe the Hawk is the Circus megaspilus (v. Zoolog of Beagle's Voyage. Part Birds).[4]

But at some time M^r. Waterhouse could probably tell you from the catalogue of the Zoological Soc. what hawk has *my* number 1396[5]

Red 3⟨35⟩ from Procellaria glacialoides *H. Smith.* killed in Lat 43^0 S. coast of Patagonia (v Zoolog. Beagle's Voyage Birds p. 140)[6]

Green 561. from Man.[7]

I send you some **rough** (which must not be quoted verbatim) notes on subject, which please return sometime. I have specimens in Spirits, of this Louse, which I could sometime get out, if you wished it— Would you be so obliging as sometime to inform me, whether you believe in Mr Martiall's story, as given in my notes.[8]

With my best wishes | Believe me | dear Sir | Yours very faithfully | C. Darwin

P.S. Would you be so kind as also to inform me sometime whether the Human Pediculus is the *Europæan species*[9]

Jeremy and Helen Evans (private collection)

1 The year has been conjectured from the relationship between this letter and the letters to Henry Denny, 20 January [1844] and 3 June [1844] (*Correspondence* vol. 3).

2 CD refers to the specimens of lice he had collected while travelling on HMS *Beagle*. Numbered tags on each specimen corresponded to specimen lists in CD's notebooks. This list is in Down House Notebooks 63.4 and 63.5 (see R. D. Keynes ed. 2000, pp. 370–415). Specimen 153 is of *Ricinus* sp. (family Ricinidae) from a condor, collected in Valparaiso in August 1834. For CD's observations of lice on dying condors, see Barlow ed. 1963, p. 243, and DAR 29.3.

3 Red 395 (specimen 1395) consists of lice taken from a falcon in Maldonado, Uruguay, in July 1833 (R. D. Keynes ed. 2000, p. 389).

4 *Circus megaspilus* is now *Circus buffoni*, the long-winged harrier (K. G. V. Smith 1987). CD refers to *Birds*.

5 CD refers to George Robert Waterhouse. Specimen 1396 is the falcon from which CD took specimen 1395 (see nn. 3 and 4, above).

6 *Procellaria glacialoides* is now *Fulmarus glacialoides*, the southern fulmar (Creuwels *et al.* 2007, p. 1084). CD refers to *Birds*.

7 Specimen green 561 (2561) was lice and fleas collected from the inhabitants of Chiloé Island, Chile, in January 1835 (DAR 31: 315 and R. D. Keynes ed. 2000, p. 410).

8 CD probably refers to his zoological notebooks from the *Beagle* voyage, in which he gave more detailed accounts of the collection of his specimens (DAR 30 and 31). In DAR 31: 315 CD reported that the lice from Chiloe were quite different from those found in England, and repeated a story from a Mr Martial that they were different also from those found on the Sandwich Islands.

9 CD refers to specimen 2561; see n. 7, above. The human *Pediculus* was probably *Pediculus corporis*, the body louse. Only two other species of louse are found on humans: *Pediculus capitis*, the head louse, and *Phthirus pubis*, the pubic louse. However, these species inhabit different parts of the body, rather than different parts of the world. See *Correspondence* vol. 13, letter to Henry Denny, 17 January [1865] and n. 5.

To William Kemp 23 February [1844][1]

Down Bromley Kent
Feb. 23^d

Dear Sir

I much regret there has been so much delay in sending you your copies of your Paper—it was caused by a mistake of the Publishers.—[2]

I have taken the liberty of sending a copy to Prof. Lindley of the Gardener's Chronicle & another to Mr Babington.—[3]

Yours very faithfully | C. Darwin

Ruth Cramond and David Cramond (private collection)

[1] The year is determined by the relationship between this letter and the letter to William Kemp, 7 December [1843] (this volume, Supplement).

[2] CD refers to Kemp 1844. CD had compiled it and sent it to the editors of *Annals and Magazine of Natural History*. See this volume, Supplement, letter to William Kemp, 7 December [1843] and *Correspondence* vol. 2, Appendix VI.

[3] John Lindley and Charles Cardale Babington had both helped identify the plants discussed in Kemp 1844. See *Correspondence* vol. 2, letter from John Lindley, [before 2 September 1843] and letter from C. C. Babington, 26 October 1843.

To Josiah Wedgwood III 31 July 1844

Down
Wednesday | July 31st. 44

Dear Jos

I write merely to acknowledge the order for £48"4^s"8^d & to thank you.—

It sounds very dangerous the 5 000£ being deposited without any security, but if you like to keep this note as security, I here engage to give to you full indemnification.[1]

Ever yours | Charles R. Darwin

Alan Wedgwood (private collection)

[1] The £5000 referred to was part of a £20,000 sum made over to Robert Waring Darwin by CD and Emma Darwin; the £20,000 formed the basis of a mortgage loan to Edward Herbert, second earl of Powis (R. W. Darwin's Investment book (Down House MS), pp. 57, 70). See *Correspondence* vol. 3, letter to Josiah Wedgwood III and E. A. Darwin, 25 July 1844. The other transaction has not been identified.

To Thomas Gold Appleton 31 March [1846][1]

Down Bromley Kent
March 31

Dear M^r. Appleton

I was much surprised a few days since by receiving your kind present of Fremont's Expedition,[2]—a work of which I had seen extracts & had been for some time wishing to read—so that it is a most desirable present in itself & I feel very much gratified at your kind remembrance of us.— M^rs. Darwin & the children are all very well & we are leading our usual quiet life in our quiet village of Down.— Some not long time since we had a visit from M^rs. Mackintosh & Robert, with only the Baby, as the other children were a little unwell.[3] Your sister looked decidedly better than when I last saw her before, but yet sadly delicate— I hope to hear of her paying you a visit at Boston this summer.— Robert has at last finished his collected works of Sir James, &

I am delighted to hear that they are likely to sell well; it always has appeared to me rather a hazardous undertaking in a money point of view.[4]

With Emma's kind remembrances pray believe me, with thanks | yours very sincerely | C. Darwin

Private collection

[1] The year is established by the reference to Mackintosh 1846.

[2] CD refers to John Charles Frémont and Frémont 1845. CD's annotated copy of Frémont 1845 is in the Darwin Library–CUL; he read it in 1846 (see CD's Reading notebooks, *Correspondence* vol. 4, Appendix IV, 119: 17a).

[3] CD refers to Mary and Robert James Mackintosh and their children Angus, Eva, and Ronald. Mary was Appleton's sister.

[4] CD refers to Mackintosh 1846. James Mackintosh was Robert Mackintosh's father.

To Georgina Tollet 1 October [1846?–71][1]

Down.—
Oct 1

My dear Georgina

This morning there was plenty of cobwebs on the K. Garden Hedge,— so many that the children remarked on them; though I have occasionally seen more.— I marked 13 with red worsted. At 12 I passed down the Hedge, & having forgotten all about it, I did not notice *one*. Then I remembered & returned with Emma as witness; & walking slowly & looking pretty carefully I could see hardly any & began to put faith in your account. When I put my head close into hedge & examined each square foot I could find a good many. But when I came to my marked webs, they were *all* in statu quo. I tried same thing a week or two ago, with same result.—

I do not know what conclusions you will draw; but I know what audacious conclusions I draw.— Emma, who has so good a heart, says that as when I looked really closely I could see webs everywhere, she thinks you, who are so sharp-sighted, could not have overlooked them before, so that they must have disappeared. I, alas, have not nearly so good a heart.— Anyhow in the 13 webs marked neither spiders or anything else had caused a line to disappear.—

I am very ungrateful not at least to wish to see (,or rather not see) what you saw; as you cheered me so in my work on Species.—[2]

Pray believe me, yours affectionately | Ch. Darwin

DAR 185: 140

[1] The starting year of the date range is conjectured from its being the earliest date at which CD had two children aged 5 or over, on the assumption that younger children would not have 'remarked' on the number of cobwebs. The end year is established from the fact that Georgina Tollet died in the third quarter of 1872 (*BMD* (*Death index*)). The reference to 'the children' suggests that this letter was probably written earlier rather than later in this range.

[2] Georgina read the manuscript of *Origin* before it was published (see *Correspondence* vol. 7, letter to John Murray, 5 April [1859]), and may have given CD encouragement earlier. Emma Darwin commented to William Erasmus Darwin in an undated letter, 'She read a great deal of Papas M.S. & is very useful to him in making him explain things that are not quite clear.' (DAR 219.1: 33.) No further reference to the disappearance of cobwebs has been found in the Darwin Archive–CUL or in CD's published works.

To David Thomas Ansted 25 November [1846][1]

Down Farnborough Kent
Nov. 25th

My dear Ansted

Enclosed are a set proofs of the fossil shells in my American vol, which Mr. Horner wished to have to cut up, for his illustrated catalogue.—[2] Also some specimens, as an advertisement, from Sowerby, which I have no means of disseminating, but which you perhaps will be so kind as to scatter abroad.[3]

I hope you are better.— I send copies of some papers which you once expressed a wish to have.[4] | Ever yours | C. Darwin

P.S. | If you have the means wd you kindly add to the list of Mr Price's work on the Invertebrata my Brother's name[5]

E. Darwin Esq
7. Park St
Grosvenor Sqre

Hans A. O. Baumann (private collection)

[1] The year is established by the reference to *South America*, which was published in October 1846; see *Correspondence* vol. 3, letter to Smith, Elder & Co., [19 October 1846]. CD was in London from 18 November for business at the Geological Society of London; see *Correspondence* vol. 3, letter to J. D. Hooker, [6 November 1846]. Ansted was assistant secretary at the Geological Society until January 1847 (*ODNB*).

[2] CD refers to *South America*. Leonard Horner was preparing an illustrated copy of John Morris's *Catalogue of British fossils* (Morris 1843) for visitors to the Geological Society's collections (Woodward 1907, p. 159).

[3] George Brettingham Sowerby drew the illustrations of fossil shells for *South America*.

[4] The papers have not been identified.

[5] Mr Price and his work on invertebrates have not been identified, although John Price, who had an interest in natural history, had been a friend since schooldays both of CD and of his brother, Erasmus Alvey Darwin.

To Josiah Wedgwood III 14 August [1847][1]

Aug 14th

My dear Jos.

Many thanks for your letter. I enclose my note for Stokes, which will you kindly take the trouble to read & his previous note to me.—[2] The briefest note from you to him, authorising the purchase & explaining the payment & giving your address will suffice.— There is, however, one point will you be so good as to decide.— Very stupidly I have quite forgot, what was the sum you have advanced for me in paying the calls &c.—but I judge from the sum you mention you have to pay to Elizabeth,[3] that the money now in hand will nearly cover that & that there is an additional 400£ to be received presently: in this latter case, that money will do for the 5£ calls, (which **if** I understand Stokes) note will amount to 200£ on the 15th of October.— Will you be so kind as to decide this point for me, which I hope won't much trouble you.—

The 50£ interest is delightful.[4]

Remember sometime you have to repay me 15£.[5] I wonder whether we cd pay off all the London & N Western calls.—[6] I long, almost as much as you can, to have done with all this eternal trouble about calls.— You must, I imagine just mention to Stokes, whether he must deduct or whether you will pay the Brokerage: I have no idea what that is per 1000£.—

Do not execrate me much for this trouble.— | Yours affectionately | C. Darwin

If you do not approve for any reason of my note to Stokes, do not send it, but write yourself without enclosing mine & I would write independently to thank him.—

Alan Wedgwood (private collection)

1 The year is established by the relationship between this letter and the letter to Josiah Wedgwood III, [20? August 1847] (*Correspondence* vol. 4).
2 CD refers to Charles Stokes, his long-time friend and stockbroker. Neither the letter from Josiah Wedgwood III, nor any letter to Stokes, has been found. CD wanted him to buy Leeds & Bradford railway shares for Emma's trust fund (*Correspondence* vol. 4, letter to Josiah Wedgwood III, [22 June – 17 August 1847], which should now be dated [22 June – 13 August 1847], and letter to Josiah Wedgwood III, [20? August 1847] and n. 3).
3 Sarah Elizabeth Wedgwood.
4 CD recorded a £50 payment on 14 August 1847 for interest on Maer, the former home of Josiah Wedgwood II (CD's Classed account books (Down House MS)).
5 CD had advanced £15 to Emma's trust fund on 17 June 1847 and was repaid 31 August 1847 (see *Correspondence* vol. 4, letter to Josiah Wedgwood III, [22 June – 17 August 1847] and n. 4).
6 Shares in the London and North West Railway were purchased with money from Emma Darwin's trust property on 30 June 1846 (CD's Investment book (Down House MS)).

To Josiah Wedgwood III [22 August? 1847][1]

Sunday

Dear Jos.

As I do not know whether Mr Stokes writes to you, I send the enclosed.[2] I have acknowledged the receipt & asked him to pay the calls in advance of 200£., if he has received no instructions to the contrary from you.—[3]

It appears there is now no dividend.

	£	
Paid	3742.	10
Call	200	
	3942"	10[4]

Will you undertake that Mr Stokes is paid.—

I have got Erasmus to write to N. W. Railway to know whether I can pay off all the little calls in that Railway.— There is no hurry whatever; but whenever I know what money is over, I think I will get one or more shares (as it may be) in this same Railway.—[5]

I cannot tell you how heartily I congratulate you both on having secured Leith Hall Place & being fixed for life.—[6]

Yours affect | C. Darwin

Alan Wedgwood (private collection)

1 The date is conjectured from the relationship between this letter and an entry in CD's Investment book (Down House MS) recording Charles Stokes's purchase of stock in the Leeds and Bradford Guaranteed Railway on 21 August 1847. The Sunday following 21 August 1847 was 22 August.

2 The enclosure has not been found.

3 CD's letter to Stokes has not been found. See n. 1, above.

4 This sum is recorded in CD's Investment book (Down House MS).

5 CD had purchased shares in the London and North West Railway on 30 June 1846 (CD's Investment book (Down House MS)). There is no record of further stock purchases in the London and North West Railway. CD refers to his brother, Erasmus Alvey Darwin.

6 Josiah Wedgwood III and his wife Caroline Sarah Wedgwood purchased Leith Hill Place, near Dorking, Surrey, in 1847 (Bright 1884).

To John Russell [10 July 1848][1]

To the Right Honorable Lord John Russell, M.P., First Lord of the Treasury, &c. &c.

The Memorial of the undersigned Graduates and former Members of the Universities of Oxford and Cambridge with some of the Fellows of the Royal Society.

Sheweth,

That the present system of the ancient English Universities has not advanced, and is not calculated to advance the interests of religious and useful learning to an extent commensurate with the great resources and high position of these bodies.[2]

That the constitution of the Universities of Oxford and Cambridge and of the Colleges (now inseparably connected with their academical system) is such as in a great measure to preclude them from introducing those changes which are necessary for increasing their usefulness and efficacy.

That under these circumstances, believing that the aid of the Crown is the only available remedy for the above-mentioned defects, your Memorialists pray that your Lordship will advise Her Majesty to issue Her Royal Commission of Inquiry into the best methods of securing the improvement of the Universities of Oxford and Cambridge.[3]

N. W. Senior, M.A., late Fellow of Magdalen College, Oxford (Master in Chancery).

J. S. Henslow, (Rev.,) M.A., St John's College, Cambridge (Professor of Botany).

B. Powell, (Rev.,) F.R.S., M.A., Oriel College, Oxford (Savilian Professor of Geometry).

B. Price, M.A., Worcester College, Oxford.
T. J. Phillips, M.A., late Fellow of Trinity College, Cambridge.
James Heywood, M.P., F.R.S., Trinity College, Cambridge.
Edmund Head, (Bart.,) M.A., late Fellow of Merton College, Oxford (Governor of New Brunswick).
T. J. Agar Robartes, M.P., B.A., Christ Church, Oxford.
Ph. Le Breton, (Rev.,) M.A., Exeter College, Oxford.
Nugent, (Lord,)[4] M.P., Brasenose College, Oxford.
Chas. Lyell, F.R.S., M.A., Exeter College, Oxford.
Harry Verney, (Bart.,) M.P., Downing College, Cambridge.
P. J. Locke King, (Hon.,) M.P., M.A., Trinity College, Cambridge.
H. R. Yorke,[5] M.P., Christ's College, Cambridge.
Joseph Kay, B.A., Trinity College, Cambridge.
E. F. Percival, M.A., Brasenose College, Oxford.
E. Horsman, M.P., Trinity College, Cambridge.
Erasmus Darwin,[6] M.B., Christ's College, Cambridge.
Charles Darwin, M.A., Christ's College, Cambridge.
H. Wedgwood, M.A., late Fellow of Christ's College, Cambridge.
Thos. H. Farrer, B.A., Balliol College, Oxford.
[*and 203 others.*]

Printed document
Cambridge Pamphlets, Folio Series, vol. 4: CUL Cam.a.500.5/124

1 The date is written in pencil on the document.
2 Concern over the state of education in the natural sciences at Oxford and Cambridge led to reforms in both universities; in 1848 Cambridge instituted a tripos examination in the natural sciences, and in 1850 Oxford created a School of Natural Science (Sanderson ed. 1975, p. 75).
3 Russell set up Royal Commissions to investigate science teaching at Oxford and Cambridge in 1850 (Sanderson ed. 1975, p. 76).
4 George Nugent Grenville, second Baron Nugent of Carlanstown.
5 Henry Galgacus Redhead Yorke.
6 Erasmus Alvey Darwin, CD's brother.

To Henri Milne-Edwards 2 March [1849][1]

Down Farnborough Kent
March 2[d]

My Dear [Sir]

My health [has] been so bad all this winter that I [have been] advised to go [to] Malvern [for one] or two months & consequently [cannot come to] Paris this spring.—[2] I am however not in the least less anxious to accept your truly generous offer of assistance in [identific]ations.[3] The delay in my coming will [at least be] advantageous in one respect, f[or I fin]d the [species are] most difficult to [distinguish] by only [external] characters & [therefore] the more perfectly I [know the class the more*]* I will profit by my visit to the Jardin des *[*Plantes*]*.[4] I suppose from *[illeg]* delayed that I shall not come *[*for six*]* or seven months. I have already lost four to five months *[*of my time *[illeg]* donate*]* some *[*new*] [*genus*]* or species to the Museum.—[5]

*[*As you thought*]* it most useful to *[*do, depositing*]* one species of each genus. You would confer a great favour on me if you could send me a single specimen of the Alepas parasite of which a large group is figured in the Voyages of the Astrolabe & of the Coquille.[6] If you care to thus far oblige me, you might send it (to avoid custom-house opening it) by some friend in a very little bottle, directed to me *either* at "7 Park S.t Grosvenor Sq.r"[7] or to the "Geological Soc: Somerset House." Pray do not waste your time in answering this letter. By working every day on which I could, I have made some progress & I hope that my work may meet hereafter with your approval, which I should prize more than that of any man's in Europe.—

With my thanks | I remain, with much respect | Yours very faithfully | C. Darwin

Piasa SA, Paris, 2008

1 The year is established by the relationship between this letter, the letter to Henri Milne-Edwards, 1 September [1848], and the letter to William Thompson, [1 March 1849] (*Correspondence* vol. 4).

2 CD went to Malvern for the water cure from 10 March to 30 June 1849 (see *Correspondence* vol. 4, Appendix I).

3 The letter containing Milne-Edwards's offer has not been found; however, see *Correspondence* vol. 4, letter to Henri Milne-Edwards, 1 September [1848]. CD was working on his monograph on cirripedes (Lepadidae; *Living Cirripedia* (1851)). Milne-Edwards was in charge of the collection of Crustacea at the Muséum d'Histoire Naturelle, Paris, and had written a monograph on Crustacea (Milne-Edwards 1834–40).

4 CD did not visit the Jardin des Plantes in Paris.

5 For cirripede species newly described by CD, see *Living Cirripedia* (1854), pp. 611–40.

6 CD refers to *Alepas parasita*, to Quoy and Gaimard 1830[–4] (*Voyage de découvertes de l'Astrolabe*: *Zoologie*), and to Duperrey, 1825–30 (*Voyage autour du monde … sur la corvette … La Coquille*). See also *Correspondence* vol. 4, letter to Henri Milne-Edwards, 1 September [1848]. For CD's diagnosis of *A. parasita*, see *Living Cirripedia* (1851), pp. 163–5.

7 Seven Park Street was the house of CD's brother, Erasmus Alvey Darwin (see *Correspondence* vol. 3, letter to J. D. Forbes, 13 [November 1844]).

To Lovell Augustus Reeve [before 14 March 1849][1]

I have heard an account with what uncommon zeal you have pursued natural history and on this ground I shall be happy to append my signature to your paper, and to wish you all success.[2] I hope for the honour of that Society your being in the publishing trade cannot be the smallest objection to you.

Incomplete
Melvill 1900, p. 352

1 The date is established by Reeve's reference in his diary to his receiving CD's letter on 14 March 1849 (Melvill 1900, p. 352).

[2] Reeve was applying for membership in the Royal Society of London; he did not succeed (*ODNB*). Reeve's publishing company, Reeve Brothers, had supplied a coloured plate for *South America* (see *Correspondence* vol. 3, letter to Reeve Brothers, [August 1846]).

To Rowland Hill 6 January 1851[1]

Here is the apteryx in three positions, resting on his beak, wading in a great pool, and running forth in quest of cockroaches or bugs or anything else he can catch. I send it for your amusement[2]

Incomplete[3]
Sotheby's, New York, 2006

[1] The date is taken from the sale catalogue. According to the catalogue, the letter is written on paper with the letterhead of the Zoological Society of London.

[2] Rowland Hill collected both living and dead specimens of animals and birds. He lived in Shropshire, and was acquainted with CD's friend Thomas Campbell Eyton (*Correspondence* vol. 1, letter from T. C. Eyton, 12 November 1833). *Apteryx* is the kiwi bird; CD had referred to *Apteryx* in *Journal of researches* 2d ed., p. 200. Three small sketches appear at the head of the letter; however, it is not possible to reproduce them.

[3] The letter advertised in the sale catalogue is incomplete; only the first leaf was for sale.

To John Richardson 4 November [1851][1]

Down Farnboro' Kent
Nov.r 4th.

My dear Sir John

I have received the Cirripedes all safely, & am very much obliged for your kindness.—[2]

They shall in a few weeks be returned with their names.— They will be valuable to me, as giving Ranges.—[3] I am ashamed to trouble you, but will you kindly sometime tell me (if you can) where "Assistance Bay" *[1 word illeg]* *[*whence*]* Mr Sutherland's specimens come.—[4] I can find no such place in my Gazeteer.— The species *[2 lines illeg]*

Incomplete
J. A. Stargardt 26–7 June 2007

[1] The year is established by the relationship between this letter and the letter to John Richardson, 30 December [1851] (this volume, Supplement), and by the reference to Peter Cormack Sutherland's specimens, which he brought back from his Arctic expedition in 1851 (see Sutherland 1852).

[2] CD thanked Richardson for specimens and information in *Living Cirripedia* (1851), p. 10. Richardson returned from a search for John Franklin's ill-fated expedition in 1849 (*ODNB*).

[3] In *Living Cirripedia* (1854), CD referred to Richardson's specimens of Cirripedia to demonstrate the habitat range of *Balanus porcatus* (p. 258); he referred to specimens found by Sutherland to show the range of *B. crenatus* (p. 261), *B. porcatus* (p. 257), and *B. balanoides* (p. 272). *Balanus porcatus* is now *B. balanus*; *B. balanoides* is now *Semibalanus balanoides*.

[4] Assistance Bay is off Barrow Straight, Nunavut, Canada. Sutherland made collections there of zoological, botanical, and geological specimens (Sutherland 1852, 2: 284).

To John Richardson 30 December [1851][1]

Down Farnborough Kent
Dec.[r] 30th.

My dear Sir John.

Tomorrow I will send your Box (carriage paid) by the Carrier to London to be booked by Railway to you on Thursday, & I suppose you will receive it, (I hope safely) on Friday or Saturday. I enclose according to your desire the names of the cirripedes. They are all British species; but I have been *particularly* glad to see them on account of their high range, & am very truly obliged to you for the loan of them.—[2]

I have lately been reading with much interest the account of your last expedition.[3]

Pray believe me | your's truly obliged | C. Darwin

To | Sir J. Richardson C. B.

James E. Murdoch Papers, Mss. 667, Louisiana and Lower Mississippi Valley Collections, Louisiana State University Libraries

1 The year is established by the relationship between this letter and the letter to John Richardson, 4 November [1851] (this volume, Supplement), the days of the carrier, and *Living Cirripedia* (1854). Peter Cormack Sutherland's specimens, referred to in *Living Cirripedia* (1854) and in the letter to John Richardson, 4 November [1851], were brought back from North America in 1851. The package carrier from Down travelled to London early every Thursday (*Post Office directory of the six home counties* 1851). Between 1851 and 1854, 30 December fell on a Tuesday only in 1851.

2 See letter to John Richardson, 4 November [1851] and nn. 2 and 3.

3 CD refers to Richardson 1851.

To George Ransome 5 August [1852][1]

Down Farnborough | Kent
Aug 5th.

Dear Sir

In answer to the Circular from the Ipswich Museum, I have the pleasure of forwarding an order for £1. [s]1. [d]0 made payable to yourself.[2]

My dear Sir | yours sincerely | C. Darwin

George Ransome Esq[re]

Yale University, Cushing/Whitney Medical Library

1 The year is established by the relationship between this letter and the letter to J. S. Henslow, 8 March [1853] (*Correspondence* vol. 5).

2 CD recorded in his Classed account book (Down House MS) that on 5 August 1852 he made a payment to Ipswich Museum of £1 1*s.* George Ransome was secretary of Ipswich Museum.

To Francis Galton 22 February [1855][1]

Down Farnborough Kent
Feb[y]. 22

My dear Galton

Many thanks for your note & your trouble in searching out about the Pigeons.—[2]

I am most particularly obliged to you for having obtained M^r. Andersson very kind offer of information about the several Breeds of cattle in S. Africa:[3] the information will be of *real value* to me. Please to give to Mr Andersson my *cordial* thanks. If he will take the trouble I sh^d. be much interested to hear about any difference in the breeds of any other domestic animals, sheep. goats dogs &c &c, kept by the several tribes. I do not at all want long descriptions; a *very* brief notice w^d suffice for my purpose. Perhaps you would forward this note. i e if my horrid handwriting would be legible by a foreigner. If M^r. Andersson by an extraordinary chance sh^d. know anything about the origins of the breeds, such information w^d. be eminently valuable to me. Do the natives ever procure a choice animal to cross their breeds with? Or do they ever make any selection of any peculiar colour, or any good useful character, & & breed from these? or is the breeding *quite* promiscuous? In periods of famine, w^d. they make any selection & keep any particular animals as more valuable for breeding up their stock again?— These queries are not altogether vague, but are founded, more or less, on statements which I have heard.— I am really deeply interested on this subject, & shall be greatly indebted to you & Mr Andersson for any information, however slight.

Your's very truly | C. Darwin

Photocopy
National Library of South Africa, Cape Town

1 The year is established by the relationship between this letter, the letter to Francis Galton, 24 July [1853] (*Correspondence* vol. 5), and the letter from C. J. Andersson to Francis Galton, 18 May 1854 (Galton 195, University College, London, Special Collections), and by the form of the address, which CD used until August 1855. CD's first letter to Galton after Galton's return from Africa was dated 24 July [1853] (*Correspondence* vol. 5). Galton wrote to Andersson on 30 July and 30 November 1853, but Andersson did not reply to any letters from this period until May 1854 (letter from C. J. Andersson to Francis Galton, 18 May 1854 (Galton 195, University College, London, Special Collections)); accordingly, Andersson could not have offered to send observations to CD before that date.

2 Galton's letter has not been found.

3 CD refers to Charles John Andersson, who travelled in South Africa with Galton between 1850 and 1852. CD cited Andersson's observations of cattle-breeding in South Africa in *Variation* 1: 87–8 and 2: 207.

From Francis Galton to Charles John Andersson [after 22 February 1855][1]

Cattle

Number of distinct breeds viz (Cape?) Namaqua Damara Bechuana. (Ovampo buy & import)[2]

Characteristics of these; height, colour: horns (*[*Back*]* *[*Kopfs*]*— some Hottentot Ian *[*Ross*]* or someonse else is noted for a breed of these) feet, temper. (sicknesses?) (If Damara cattle *die* in the Colony?. They do in St Helena)[3]

Price of bulls compared to oxen— The Namaquas set store upon good Damara bulls do the Bechuanas? Admiration in which fine cattle are held by the Damaras

&c—do they take any *particular* means to weed out the bad ones? . . . . . & to breed good ones?

Colored teams of oxen. to match— In describing, compare as much as possible to the different well known English breeds.

Goats—large ones, like what Hans had.[4] small, ordinary breed—the very small ones are rare in Ovampo land.—color &c.

Sheep. Wool hair—tails—colour—size—horns.[5]

Dogs native breeds, are there any. Good dogs especially prized.[6]

Dear Andersson,

I wrote to Charles Darwin saying that you would be glad, before long, to give him any information you possessed about native *breeds of cattle*, and that I was sure you would be a most reliable authority, as you had had so much to do with them.[7] He sends me the enclosed note to forward to you; & on the top of this bit of paper I have jotted down a few heads of enquiry as they come into my mind— You will doubtless add more.[8]

Ever yours | Francis Galton

Photocopy
National Library of South Africa, Cape Town

1 The date is established by the relationship between this letter and the letter to Francis Galton, 22 February [1855] (this volume, Supplement).

2 CD had asked Galton several questions about the breeding of domestic animals in South Africa. See this volume, Supplement, letter to Francis Galton, 22 February [1855]. Damara, Bechuana, Namaqua, and Ovampo were the names of peoples, and by association, regions of southern Africa. Bechuanaland is now Botswana. Damaraland, Namaqualand, and Ovampo are regions of Namibia. CD described Namaqua, Damara, and Bechuana cattle in *Variation* 1: 88.

3 No letter from Andersson concerning cattle has been found, but he describes Damara, Bechuana, and Namaqua cattle in Andersson 1856, pp. 315–36. The Colony is Cape Colony, South Africa.

4 Hans Larsen was in charge of Galton's livestock in Africa, including goats (Galton 1853, p. 75).

5 For Galton and Andersson's descriptions of the Damara sheep they kept in Africa, see Galton 1853, pp. 242–3, and Andersson 1856, p. 232.

6 For Andersson's descriptions of Namaqua dogs, see Andersson 1856, pp. 284–5.

7 Galton's letter has not been found. Andersson traded cattle in Africa during his expeditions between 1850 and 1854 (Andersson 1856).

8 Galton enclosed CD's letter to him of 22 February [1855] (this volume, Supplement).

To C. J. Andersson 25 March [1856][1]

Down Bromley Kent
March 25th.

Dear Sir

I am very much obliged to you for the Proof sheets, received this morning.[2] But before I refer to the cattle, let me congratulate you on producing a work, which judging from the 32 pages promises to be so interesting, and how wonderfully you have mastered English! I declare you write a better style, forgive me for saying, than half our English authors.—

The facts you mention about the cattle, & about the Damara prizing the great horns & colours, are *exactly* the sort of facts, which I wanted to know.—[3] Will you think me very unreasonable if I ask you a few other questions?— Do you suppose that the natives ever unite a particular Bull & particular cow together?[4] as to breed any particular colour or excellence. I presume you saw no other breeds (*or even sub-breeds*) than the Damara, Bechuana & Namaqua breeds?—[5]

Will you have the kindness to tell me whether you noticed any different breeds of Dogs (or of Sheep) in the same or in different tribes of natives? Do the natives keep Fowls or other poultry? But I fear you will think me very unreasonable in asking so many questions; but I am most deeply interested about all domesticated animals. I can only trust to your friendship towards my relation F. Galton to make you forgive me being so troublesome.[6]

With very sincere thanks, pray believe me Yours very sincerely | Ch. Darwin

I cannot resist one other very odd question,— do you suppose that the natives select their wives for their personal appearance or beauty; & does a native & European at all agree in their ideas of the beauty of the female sex; or have they a different beau ideal?[7]

Photocopy
National Library of South Africa, Cape Town

1 The year is established by the relationship between this letter and the letter from C. J. Andersson, [6 April 1856] (*Correspondence* vol. 6).

2 CD refers to the proof-sheets of *Lake Ngami: or, explorations and discoveries during four years' wanderings in the wilds of South Western Africa* (Andersson 1856). They have not been found in the Darwin Archive–CUL.

3 CD cited Andersson's observations of cattle in Andersson 1856 (under the title *Travels in South Africa*) in *Variation* 1: 88 and n. 62, 2: 207 and n. 59, and 2: 300 and n. 27.

4 The part of Andersson's reply relating to cattle has not been found (see *Correspondence* vol. 6, letter from C. J. Andersson, [6 April 1856] for the surviving part); but CD wrote in *Variation* 2: 207, 'Mr. Andersson further informs me that the natives frequently match a particular bull with a particular cow.'

5 Damara, Bechuana, and Namaqua were the names of peoples, and by association, regions of southern Africa. Bechuanaland is now Botswana. Damaraland and Namaqualand are regions of Namibia.

6 Francis Galton and Andersson travelled together in southern Africa from 1850 to 1852 (Andersson 1856).

7 For Andersson's reply to this question, see *Correspondence* vol. 6, letter from C. J. Andersson, [6 April 1856].

To Walter Baldock Durrant Mantell [before 10 April 1856][1]

Have you any idea whether the New Zealanders ideas of Beauty either in face or whole person, in their women, is like ours? i.e. would the handsomest woman in a tribe in the eyes of a New Zealander, be the handsomest woman in our eyes?

Yes

I am aware it is a mere chance whether you can answer this odd question; & certainly no one less familiar with the natives than you are, could answer it.— Lastly do the chief men generally succeed in getting for their wives the handsomest women, or do they care more for a good work-woman?

Both— In the old times almost every girl pretty or promising to be so, whom one might see in a pa was sure to be taken to some chief—[2] But the aristō's took care to have wives skilled in watu-ing mats[3] and in cookery— In the straits Ngatiruanui ladies were highly prized from their skill in roasting potatos—[4]

Another charm was rank in their own or a neighb^{g}. tribe Te Hapuku of Ahuriri,[5] has succeeded thus in greatly increasing his importance & doubling that of his children—

No room to add more

AL incomplete
DAR 85: A99

CD ANNOTATIONS
4.1 Both ... potatos— 4.4] 'chiefs seized on prettiest women' *in margin, pencil*
5.1 Another charm ... children— 5.3] *crossed pencil*

1 The date is established by the relationship between this letter and the letter to W. B. D. Mantell, 10 April [1856] (*Correspondence* vol. 6). In his letter to Mantell of 10 April [1856], CD wrote, 'I fear you cannot answer my question whether the beau ideal of beauty amongst the less civilised natives ... would agree with ours'. The assumption is that he had sent this letter to Mantell before 10 April and that Mantell, once his memory had been jogged, sent his replies, written on the original letter, after 10 April. See also *Correspondence* vol. 6, letter to W. B. D. Mantell, 3 April [1856], for CD's request to Mantell to answer queries on the paper they were written on, and this volume, Supplement, letter to C. J. Andersson, 25 March [1856], for a similar query. In his letter to Mantell of 5 June [1856–9] (*Correspondence* vol. 6), CD wrote, 'I think that you were so kind as to answer fully all my previous questions.—'. '*Yes*', and the section 'Both ... more' are in Mantell's hand.

2 CD cited Mantell on this point in *Descent* 2: 369. *Pa*: fortified village (Maori).

3 *Watu*, or *whatu*: weave (Maori: Williams 1971). The mats in question were called *whariki*, and were of great cultural significance.

4 The Ngāti Ruanui are a Maori people traditionally based in the Taranaki region of the North Island of New Zealand, north of the Cook Strait. The potatoes would traditionally have been sweet potatoes.

5 Ahuriri is now a suburb of the modern city of Napier, in Hawke's Bay. Te Hapuku, an influential Maori leader in the Hawke's Bay region, had taken up residence in 1859 at Poukawa, inland and south of Napier (*DNZB*). He may have been born at Ahuriri (*ENZ* s.v. Te Hapuku).

To the Royal Society 8 October [1856?][1]

Royal Soc^{y} | Somerset House

The Bearer has called for the Books

C. Darwin

Down | Oct. 8th—

Will you please also send me, Isis Band XXI Heft II. and Isis 1828. H. 2. (I daresay this is the same volume.)—[2]

Also Isis 1829 VI p. 639 | Brehm, Abstammung der Hauskatke.[3]

Photocopy
DAR 249: 111

[1] The year is conjectured from CD's interest in research on cats (see n. 3, below), and from the day the Down carrier visited London. The Down carrier service, operated by George Snow, went to London early on Thursdays (*Post Office directory of the six home counties* 1851), so it is likely that CD wrote this note on a Wednesday. Between 1842, when the Darwins moved to Down, and end of 1856, when the library of the Royal Society commenced its move from Somerset House (*Record of the Royal Society of London*, p. 65), 8 October fell on a Wednesday in 1845 (when, however, CD was away from Down from 15 September to 26 October; see *Correspondence* vol. 3, Appendix II), 1851, and 1856. In 1851 CD was working on the proofs of *Fossil Cirripedia* between August and November (*Correspondence* vol. 5, Appendix I), and there are no references to cats in his extant correspondence. In mid-October 1856, CD finished the second chapter of the manuscript of his 'big book' on species (*Correspondence* vol. 6, Appendix III). According to CD's table of contents for his manuscript of his 'big book', cats were discussed in this chapter (*Natural selection*, p. 25; the first two chapters of the manuscript are no longer extant). Cats are mentioned frequently in his 1856 correspondence (see *Correspondence* vol. 6).

[2] Volume 21 of *Isis* was the volume for 1828; CD may have been interested in Christian Ludwig Brehm's article 'Ueber *Columba domestica, livia et Amalia*', in part 2, pp. 135–41.

[3] CD refers to Brehm's article 'Wo stammt die Hauskatze her?' (Where does the domestic cat originate?), *Isis* 22 (1829): 639.

To James Dwight Dana 21 December [1856][1]

Down Bromley Kent
Dec. 21st

My dear Sir

I am very much obliged for all the trouble you have so kindly taken for me, & for your letter of Nov. 28th.—[2] I have been these few last days comparing & thinking over the many definitions of species; & puzzling over the different variability of different forms; & I think I will venture to quote your remark on the "libration" of species.—[3] But first let me thank you for your paper on Geological Development:[4] everything which you have ever written interests me; but I own this last paper is rather too grand for my digestion; not but that I am very much inclined, from different reasons, to agree with you on antiquity of continents & oceans; & I have lately had some written discussion on subject with Lyell, who differs toto coelo[5] from me. In my coral volume,[6] I remember coming to same conclusion of continents now & lately having risen, ocean-beds having sunk, & great archipelagoes having oscillated. But I have never satisfied myself on secular refrigeration: I forget whether you have discussed subject, but I never could understand how refrigeration could cause in same area, the prodigious oscillations of level, which we must believe in, during the successive sedimentary formations of Europe.—

What you say about the importance of the infertility of species, when crossed, I quite agree with; & I think those who attribute our domestic varieties to the crossing of ma[n]y primordial forms, *greatly* underrate the evidence on this question— Nevertheless after a very careful study of all the facts, these facts do not seem to me to make so important a distinction between species & varieties, as I at first thought: I have to a certain extent satisfied my own mind on this head, but whether I shall satisfy anyone else, I know not.— My subject as I treat it, is almost ridiculous in one way; that I never come to any conclusion on any point, without having to advance a heap of difficulties & arguments on the other side. How I wish I had some of your knowledge & powers of work, & my work would be a very different affair.— But I am loth to throw away 19 years collection of facts & observations of my own, & I will do my best.

Dr. A. Gray has sent me his paper,[7] & I have had a good deal of correspondence with him on the subject before & since he published.— I admire extremely all that he writes & it seems to me his only fault is too much modesty about his results. A rare fault with a vengeance amongst men of science.

J. Lubbock is going to publish a nice paper on Daphnia— he has made out the spermatozoa, the structure of ephippium, & some facts on double means of procreation, nicely.—[8] Prof. Huxley has been at work lately on Crustacea, & believes he has made out in the higher Crustacea the homologues of the cementing apparatus in my old enemies the Cirripedes—[9]

I have just been looking over your letter again; & your doctrine of species being definite like chemical elements, rubs against me; but I am terribly in want of facts *whether variable forms are variable in all* **places** (can you give me any facts?) **& all times**. Protean genera, like Rubus Rosa &c[10] puzzle me greatly. But I am running on at an unreasonable length, so will conclude with thanks & all good wishes, yours sincerely | C. Darwin

P.S. | This morning I have received your note of the 8th,[11] & *very much* obliged I am for it. I am particularly glad to hear about the Cave Rat, for though the case does not differ in any essential respect in its relations from, for instance, an alpine Fauna to the continent in which the mountain stands; the cave animals have always seemed to me an eminently curious instance.

I am, also, very glad to hear your opinion on the parallelism of embryological & palæontological development: I think I am *nearly* in your frame of mind, strongly inclined to believe in it; only I feel that I want more numerous & striking instances to convince me than those as yet given. I adduced the fact of Chthamalus, *merely* as case of negative evidence not being trustworthy; as far as development goes it is quite right that the Chthamalinæ shd. appear before the Balaninæ.—[12] As my work very much consists in the comparison of the opinions & results of good workers in different branches, I am very glad to hear the opinion of one like yourself, in whom I have so much faith.—

My principle of work is to consider every subject which I can at all master, under this one point of view,—would such & such be the case, if species are descended, like varieties, from a common parent. And according as larger or smaller classes of facts can be thus explained, so should I conclude that species really were, or were not, lineal descendants of other species.—

I forget whether I ever mentioned that I have taken up domestic Pigeons,—thinking it better to study well one group than slightly several.— I am collecting specimens from all quarters of world, & have got all breeds known in Britain alive; & am making skeletons, crossing breeds &c &c.— I mention this on the mere chance of your ever stumbling on any Fancier: if so I shd. much like to hear whether he believes that any breed (or *sub*-breed rather) has become *at all* modified or different from those of Europe. It is not likely, but yet worth enquiry. The old Pigeon literature, together with Persian & Indian specimens, has thrown some light on the origin of domestic varieties, which really differ in an astonishing degree.—[13]

Catherine Barnes, Historical Autographs & Documents

1 The year is established by the reference to the letter from J. D. Dana, 8 December 1856 (*Correspondence* vol. 6).
2 Dana's letter of 28 November 1856 has not been found.
3 In the manuscript of his 'big book' on species, CD wrote, and then crossed out, 'Many authors seem to consider that each species was created with a certain fixed amount of variability, or to use an expression in a letter of Prof. Dana, with "its system of librations under the influences of nature to which it may be subject", & this would include both recognised varieties & individual variations.' (*Natural selection*, p. 105.)
4 Dana's paper 'On the plan of development in the geological history of North America' was published in the *American Journal of Science and Arts* (Dana 1856); it was also published together with Dana 1855 as a pamphlet, 'On American geological history' (New Haven: Ezekiel Hayes, 1856). There is an annotated copy of the pamphlet in the Darwin Pamphlet Collection–CUL. CD wrote on the back, '(Very poor Pamphlet)'.
5 CD refers to Charles Lyell. *Toto caelo*: completely, utterly (Latin).
6 *Coral reefs*.
7 CD refers to Asa Gray and the first part of A. Gray 1856–7 ('Statistics of the flora of the northern United States'); see *Correspondence* vol. 6, letter to Asa Gray, 12 October [1856]. CD's annotated copy of A. Gray 1856–7 is in DAR 135 (3).
8 CD refers to John Lubbock and Lubbock 1857.
9 CD refers to Thomas Henry Huxley and T. H. Huxley 1856–7, 14: 354. See *Correspondence* vol. 6, letter to T. H. Huxley, 9 December [1856] and n. 7; see also *Fossil Cirripedia* (1851) and (1854), and *Living Cirripedia* (1851) and (1854).
10 See *Origin*, p. 46, where CD gave *Rubus* and *Rosa* as examples of genera in which the species presented 'an inordinate amount of variation'.
11 *Correspondence* vol. 6, letter from J. D. Dana, 8 December 1856.
12 CD had told Dana of a new species of fossil cirripede, in the genus *Chthamalus*, found in a Cretaceous formation, having previously believed that no cirripedes of that type existed so long ago; see *Correspondence* vol. 6, letter to J. D. Dana, 29 September [1856]. See also *Fossil Cirripedia* (1854), p. 5. The fossil cirripede turned out not to be a fossil, however (Southward ed. 1987, p. 47).
13 CD discussed domestic pigeons in *Origin*, pp. 20–9. See also *Correspondence* vol. 6.

From J. Knightly Ince[1] [1857–61?][2]

2

excess of advancement retrograde into mere muddle—

The *Schimmels*—	the grays are—
Eisen—	irongray
apfel—	applegray
Honig	honey gray— we call it nutmeg—& a very bad grey it is & getting unhappily very common

but you know all this as well as I do— Silber, schwarz, roth & grau schimmel I mean— "rothschimmel" tho' I ought to say is a "strawberry roan"—

The "*Falb*"'s

are "*dun*"—*fulvus*, "fallow", '*fallow* dun' &c &c—

Kuhfalb.	red dun
Reh falb—	fallow deer dun
semmelfalb.	cream coloured—

The Braun's

Hellbraun—	bright bay
Kastaninenbraun	Châtaine[3]
Kirschbraun	clarety Brown
Fahl braun	greyish Brown

Fuchs
Chesnuts

Zobel fuchs—	sable Chesnut
Stichel fuchs	dappled slotted Chesnut. The Chesnut with those black smears one sees.

Sir, you deserve well of y^r^ country—& I trust these Germans will not bemuddle you in your own clear headedness & humbug hating soul—[4]

These I am convinced you possess—

Pray give our kindest regards to Mrs Darwin & our little lassies love to the little ones

Y^rs^ ever dear Sir with sincere respect very truly | J Knightly Ince

Incomplete
DAR 261.11: 20 (EH 88206072)

CD ANNOTATIONS
3.1 "*Falb*"'s] 'Falb' *added after, pencil*
4.4 semmelfalb.] 'bread' *added above, pencil*
7.1 *Fuchs*] 'Fuchs' *added after, pencil*
9.3 smears] *underl pencil*
9.3 one] 'one' *added over, pencil*
10.1 Sir ... truly 13.1] *crossed pencil*
Top of letter: 'p12 & p96'[5] *pencil*

1 J. Knightly Ince has not been identified.
2 The date range is conjectured from the subject matter, colours of horses. Most of CD's letters on this subject were written between 1857 and 1861.
3 *Châtaine*: chestnut-brown (French).
4 CD had presumably asked Ince about German words for the colours of horses. He may have been reading Hofacker 1828, a German work on inheritance including a discussion of colour in horses. CD

cited Hofacker 1828 in his discussion of inheritance of colour in horses in *Variation* 1: 59, and in a letter on the subject to the *Field*, [before 25 May 1861] (*Correspondence* vol. 9).

5 The page references are probably to Hofacker 1828; these pages are annotated in CD's copy in the Darwin Library–CUL (see *Marginalia* 1: 382–4).

To William Walmisley Baxter 23 September [1857–9?][1]

Down.
Sept 23

Dear Sir

I am sorry to say that the returned gloves are not at all tougher or stiffer than those which I possess. If you could get me a pair made with stiffer hairs, or bristles, I sh^{d}. be very much obliged.[2]

Yours faithfully | Ch. Darwin.

Lyon & Turnbull, Edinburgh, 4 February 2009

1 The year span is conjectured from the date when W. W. Baxter took over the Bromley pharmacy of his father, William Baxter (1857), and based on the years after then when CD took the most hydropathic treatments (see n. 2, below). CD gave up hydropathy after 1863 (Browne 2002, p. 238), and he was not at home on 23 September in 1860, 1862, and 1863. He underwent a number of hydropathic treatments in 1857, 1858, and 1859. See *Correspondence* vols. 6–11. There is a handwritten note on the letter, reading, 'from Wm Baxter Junr March 18*[88]*. | Bromley Kent', which suggests that W. W. Baxter sent the letter to someone else after CD's death. However, it is possible that the letter was written to his father. CD first took the water cure in 1849.

2 Horsehair gloves were used to apply friction as part of hydropathic treatment, including as part of the treatment for nervous indigestion, from which CD suffered (Gully 1846, p. 162).

To Skeffington Poole 20 October [1858][1]

Down Bromley Kent
Oct 20th

Sir

I write merely to thank you for so kindly taking the trouble to acknowledge my note, and for your promise of any information in your power.[2] By native horses I meant any with which you were acquainted, and more especially the Kattywar Breed.[3]

Sir | Yours faithfully & obliged | Ch.Darwin

P.S | I have seen it stated that the Kattywar horses are stumpy or cob-built; but I notice that Col. Curtis[4] incidentally speaks of them as rather long-legged or lanky & for *Indian Horses* tall. How is this, would you be so good as to tell me.

I would also ask whether you ever happened to notice any dark stripe on the head, or near the eye?

Ronald Levine, Modern 1st Editions

1 The date is established by the relationship between this letter and the letter to Skeffington Poole, 13 October [1858] (*Correspondence* vol. 7).

[2] CD refers to his letter to Poole of 13 October [1858] (*Correspondence* vol. 7); Poole's acknowledgement has not been found. The questions that CD enclosed with his letter to Poole have not been found, but may have been similar to those enclosed with the letter to Hugh Falconer, 23 November 1857 (*Correspondence* vol. 6); see also this volume, Supplement, letter from Skeffington Poole, [21 October 1858]. CD cited Poole for information in *Origin*, pp. 163, 164, 166, and *Variation* 1: 58, 59, 2: 43.

[3] Kathiawar (Kattywar) is a peninsula mostly in Gujarat state, India. Kattywar horses are now generally known as Kathiawari horses.

[4] Possibly James Gray William Curtis.

From Skeffington Poole [21 October 1858][1]

Premising the breed of horses inquired about, to be that of Kattywar.—

I beg to offer the following replies to the questions sent, but at the same time give them from recollection only.[2]

(1) "Dunn" is a colour between a brown and a black, but there are various shades of it in the Katty horse—

"Dunn" has nothing to do with "*Cream Colour*" A "cream-coloured" horse has a *mealy mouth*, and this is never seen in a "*Dun*" horse— Neither Bay nor Chestnut have anything to do with "*Dun*" nor is it so understood in India— "Mouse-Dun" is a colour like a mouse, only tinged with "dun". There are also *Mouse* Coloured horses that you would not apply dun to at all.

(2) "Tiger Marks" are stripes on the cheeks, noses withers, and fore and hind legs, many shades darker than the colour of the horses, and are called so from their resemblance to the stripes on a Tiger.

I have never heard these stripes called "*Zebra Marks*" by the Natives nor do I suppose the people of Kattywar, where these horses are bred ever saw or heard of a Zebra, besides which, the stripes on the Katty horse are not analogous to the Zebra— Colonel Curtis[3] I daresay applies both terms with the same meaning.

(3) There are so many shades of "Dun", "Mouse-dun", & Mouse colour that the colt or filly varies very much in colour when first foaled, and it would be almost impossible to describe what is asked—indeed none but a most experienced hand can tell what the colour of a foal will be—

The stripes are more distinct on the colts being first foaled than at any other time (just the same as the wild Ass, which is found in numbers on the banks of the Runn)[4] they then become less & less distinct till after the first coat is shed when they come out as strong as before, and fade or grow out as the age of the horse increases, but in some horses I have known the stripes perceptible at the age of 25 years

(4) The "Tiger" like bars, or stripes on the legs, are common in "dun", "Mouse-dun" and "Mouse" coloured horses, the stripes are common on both fore and hind legs, but not more so on the fore than the hind legs— These stripes are also often to be seen on the cheeks and sides of the nose, also round the eyes— If this breed of horses of the above colour had not the stripes about them they would not be of the

pure Katty breed—and horses of these colours without the stripes are not considered pure—

There are plenty of these coloured horses in "India" but they are an inferior Animal altogether

In Kutch the adjoining Province these coloured horses are numerous but worth very little in comparison, the pure Katty being valued at as many thousands of rupees as the Kutch is at hundreds—

(5) (1) The shoulder stripe like that of a domesticated ass, is general on all these pure Katty horses of the above colours—

(2) One half of the "Indian dun" horses may have a shoulder stripe, and also have the "*List*"—[5]

(3) I have often seen the shoulder stripe *double* and sometimes **treble**, these are considered extra well bred.

(6) I have seen both Greys & Bays with these stripes when first foaled but they soon fade away. I have never known them in a chestnut. Black horses are very rare in India—

(7) No crossing of other colours will produce a "*dun*" in my opinion, either the dam or sire must be a "*dun*"—[6]

(8) I don't believe that any distinct breeds of horses whether "*dun*" or any other colour when crossed would produce offspring with "*List*" or shoulder "stripes" or tiger like stripes on the legs unless such marks appear in one of the parents.—

Draft or contemporary copy
Ronald Levine, Modern 1st Editions

[1] The date is established by the relationship between this letter and the letters to Skeffington Poole, 20 October [1858] and 22 October [1858] (this volume, Supplement).

[2] See *Correspondence* vol. 7, letter to Skeffington Poole, 13 October [1858] and n. 4, and this volume, Supplement, letter to Skeffington Poole, 20 October [1858]. CD cited information from this letter in *Origin*, pp. 164 and 166, and *Variation* 1: 58 and 59.

[3] See this volume, Supplement, letter to Skeffington Poole, 20 October [1858] and n. 4.

[4] The Runn of Kutch (or Rann of Cutch) is a salt marsh north of Kathiawar (Kattywar).

[5] List: a stripe of colour (*OED*).

[6] CD was interested in whether the dun colour in horses could appear when neither parent was dun, as an instance of reversion to the ancestral colour; see also this volume, Supplement, letter from J. K. Ince, [1857–61?] and n. 4.

To Skeffington Poole 22 October [1858][1]

Down Bromley Kent
Oct 22

Dear Sir

I really do not know how to thank you sufficiently for your great kindness. I never dreamed of receiving so much, & to me such important, information. You *cannot* imagine how much your communication has interested me.[2]

You are so kind as to say that I may ask you any other question; but I will not again (ie. after this note) be so unreasonable as to trouble you again.

My last note crossed your communication[3] you have fully anticipated my questions about stripes on face.

(1) will you inform me whether the Kattywars are cob-built?

(2) I earnestly hope that you will permit me to quote your name as my authority for such facts as I shall have occasion to give.[4]

(3) You refer to the wild Ass on the banks of the **Runn**.[5] I presume that this is the Equus Hemionus of Naturalists; if so, it has list down the back, but **no** shoulder stripe. Is this so? You refer to stripes on the young of this animal: any information on this head would be **particularly** valuable to me. What is general colour of the young? I have seen it stated that sometimes the faintest trace of a shoulder stripe may be detected on the Hemionus.

The reason why I asked about Duns (with stripes) appearing when horses of other colours are crossed, was because I know a case of a Dun (in my sense of word) Cart-Horse of the Belgian breed with double-shoulder stripe; & a case of very small Dun Welch Pony with treble shoulder stripe (these facts made me ask about double shoulder stripe in Kattywar breed, which latter fact is quite new to me) & I thought it improbable that both great Cart-Horse & small Welch Pony should have had a cross with any fine pure Dun Breed, such as that of Kattywar.[6]

I asked about face-stripes, from the singular fact of such appearing in two hybrids from the Hemionus & Common Ass.[7]

I have seen it stated in print that the common domestic ass sometimes has double shoulder stripes, but I know not whether to believe it.[8] But perhaps these details may be troubling you as trifling, though so interesting to me, in trying to make out something on the inheritance of characters & variation in general.—

I shall soon be writing to Col. Curtis[9] & shall tell him how deeply indebted I am to your very great kindness.—

With my cordial thanks, | I beg permission to remain Dear Sir | Yours faithfully & obliged | Charles Darwin

Ronald Levine, Modern 1st Editions

1 The date is established by the relationship between this letter and the letter to Skeffington Poole, 20 October [1858] (this volume, Supplement).

2 See this volume, Supplement, letter from Skeffington Poole, [21 October 1858].

3 See this volume, Supplement, letter to Skeffington Poole, 20 October [1858], and letter from Skeffington Poole, [21 October 1858].

4 CD cited Skeffington Poole by name in *Origin*, pp. 163, 164, and 166, and *Variation* 1: 58, 59, 2: 43.

5 See this volume, Supplement, letter from Skeffington Poole, [21 October 1858] and n. 4.

6 See *Origin*, p. 164, and *Variation* 1: 57. See also *Correspondence* vol. 7, letter to W. E. Darwin, [26 May 1858] and n. 9. CD was interested in collecting examples of stripes appearing in dun-coloured horses as an illustration of reversion to ancestral characters (see *Origin*, pp. 163–7, and *Variation* 1: 55–61). He believed that stripes occurred most often in duns.

7 See *Origin*, p. 165.

8 See *Variation* 1: 63. CD refers to W. C. L. Martin 1845, p. 205. There is an annotated copy of W. C. L. Martin 1845 in the Darwin Library–CUL (*Marginalia* 1: 568–9).

[9] See this volume, Supplement, letter to Skeffington Poole, 20 October [1858] and n. 4.

From Skeffington Poole 23 October 1858

23rd. Oct. 1858

Dear Sir,

I have yours of yesterday, & I am glad that the information I have been able to give is useful to you.[1]

You will observe that what I have stated is "*from recollection only*". I mention this, as you wish to quote my name as authority, this I have no objection to, but as there may be some little inaccuracy, I should not like to assert all I have stated, as positive facts, though to the best of my recollection they are so, under this proviso you may quote my name—

The true Kattywar horse is not a "Stumpy" "Cob-built" animal in any respect, but well proportioned with a fine high crest and withers, standing from 15 to 16 hands high (I have heard them 17 hands high) they have very easy paces and are pleasant to ride from having high action—

There are plenty of animals that are Bull-necked, "*Stumpy*" & "Cob-like"— These are generally called "Tattoos" though they may be the height of a horse—

The true breed of Kattywar horses has much deteriorated in the last twenty years, and will soon become extinct from want of care. Not one twentieth of the horses being bred there now that there were thirty years ago. There are many fine horses still in existence, but are difficult to be seen as they are located in the *[*houses*]* with the people who are very *[*anxious*]* about them— I have seen many **bay** mares of the Katty breed equal in appearance to the best English blood.

The wild Ass on the banks of the Runn, is the same as that found in Northern Persia, but I do not know the scientific name. It stands higher than the domestic ass, is most beautifully proportioned— The head and legs are striped when first foaled, but the mark across the withers is not so distinct, as either that of the tame ass, or the Kattywar horse and soon becomes *[illeg]*.[2]

The colour is much the same as that of the domestic *light* coloured Ass—

They are very fleet, so much so, that it is a rare occurrence to overtake them on the most fleet horse, although run over a country perfectly level— They keep in herds of about 15 or 20 and are very difficult to be got near— They frequent the Runn more about *[illeg]* *[*Jamna*]* and Naggar Parkar[3] than any other point— When the dam is going to bring forth, she leaves the herd and hides herself in any adjoining jungle, where she leaves the young, visiting it each day till it can run strongly—as should the old males get at it, and it prove to be a male, they emasculate it by biting out the testicles, at least this is the way in which the natives account for so many Geldings being found in the herds, which can always be recognised, they being so much larger than the others, and when shot they have always proved to be Geldings— The Villagers at the breeding season watch the dams into the jungle, & by this means obtain their young— The doctor of my Regt. at Deesa[4] had two a male and female, which he brought up & though every care was taken of them, they could not be tamed, so as to be used for riding or any domestic purpose, indeed they are said to be untameable.

I believe it to be well known, that some years ago, some of the Katty breed of horses were sent home to one of our Sovereigns, this may account for the breed appearing in England— It is to be traced to a breed of horses now to be seen at Elvaston Castle (Lord Harrington's) and these are known to have come originally from the Royal Stud.—[5]

Draft or Contemporary Copy
Ronald Levine, Modern 1st Editions

[1] See this volume, Supplement, letter to Skeffington Poole, 22 October [1858].
[2] CD cited Poole for this information in *Origin*, p. 163. The Runn of Kutch (or Rann of Cutch) is a salt marsh north of Kathiawar (Kattywar).
[3] Nagar Parker is a town north of the Rann of Cutch, and is now in Pakistan.
[4] Deesa is a town east of Cutch, in Gujurat, India. Poole's regiment was listed as being stationed there in 1841 and 1848; the surgeon on both occasions was Abraham Israel Montefiore (*India list*).
[5] Elvaston Castle in Derbyshire was the seat of the fifth earl of Harrington, Leicester FitzGerald Charles Stanhope. The story about Kathiawari horses being sent to England has not been substantiated.

To Skeffington Poole [27 October 1858][1]

Moor Park, Farnham | Surrey
Wednesday

Dear Sir

Your letter has been forwarded to me here from my home, and I write merely to thank you for this further proof of your great kindness.[2]

The facts about the young wild ass, which I do not doubt is the Hemionus[3] (& I shall be able to find out positively), are quite new and interesting to me. I am particularly obliged for your permission to use your name & I will be cautious in my manner of quoting you. But the tenor of your communication leaves little doubt in my mind of the entire accuracy of your memory.

With my cordial thanks for your great kindness, I beg leave to remain Dear Sir | yours truly obliged | Ch Darwin

Ronald Levine, Modern 1st Editions

[1] The date is established by the relationship between this letter and the letter from Skeffington Poole, 23 October 1858 (this volume, Supplement), and by the address and day of the week given. CD visited Moor Park in Surrey for hydropathic treatment from 25 October to 1 November 1858 (*Correspondence* vol. 7, Appendix II); there was one Wednesday in this period, 27 October.
[2] See this volume, Supplement, letter from Skeffington Poole, 23 October 1858.
[3] *Equus hemionus* (the onager).

To Benjamin Collins Brodie 26 April [1859][1]

Down Bromley Kent
Ap. 26th

Dear Sir Benjamin Brodie

In answer to your note, I cannot think of any Zoologist who has any especial claim for the Royal Medal.[2] Though not a Botanist myself, it seems to me that Mr Bentham has high claims for this honour. His contributions to De Candolles great Prodromus, almost by itself shows his European reputation as a systematic Botanist:[3] & many of his memoirs, as I can speak from my own knowledge, contain much important information on Classification, Distribution & such general subjects.—

Should the Council think of a geologist, to the best of my judgment, M[r] Prestwich[4] amply deserves a medal, for his remarkable & excellent labours on the British Tertiary formations, & their correlation with those on the continent. I am tempted to add that it would be a facility in thinking over the medals if any list of the names of individuals existed to whom the medals of all kinds have been given: it would aid in apportioning the medals fairly for different subjects.—

With much respect pray believe me | Yours sincerely | C. Darwin

The Royal Society

[1] The year is established by the subject matter. Benjamin Collins Brodie (1783–1862) was president of the Royal Society of London from 30 November 1858 to 30 November 1861; George Bentham was awarded a Royal Medal in 1859 (*Record of the Royal Society of London*).

[2] Brodie's letter has not been found.

[3] CD refers to Alphonse de Candolle and Candolle and Candolle 1824–73 (*Prodromus systematis naturalis regni vegetabilis*). Bentham worked on a monograph on the heaths, and on the order Scrophularineae (which is roughly equivalent to the modern family Scrophulariaceae) and the Labiatae (now the Lamiaceae), for Candolle and Candolle 1824–73 (B. D. Jackson 1906, pp. 115, 137).

[4] Joseph Prestwich.

From E. A. Darwin 15 August [1859 or later][1]

Aug 15th

Dear Charles

I have fallen upon a book in London Lib which may possibly be in your way. The Psychology of Shakespeare by D[r] Bucknell, Editor of the Journal of Mental Science.[2] The book is much narrower than it sounds, as he is pleased to say in his Preface that in modern usage, Psychology is confined to the science of diseases &c of the mind, but as he is a mad doctor, I suppose he has got to use the word technically. I have not read very much as it does not seem really to enter the

Incomplete
DAR 105: B68

[1] The date is established by the date of publication of Bucknill 1859. Bucknill 1859 was published between 31 May and 14 June 1859 (*Publishers' Circular* 1859).

[2] John Charles Bucknill's editorship of the *Journal of Mental Science* is mentioned on the title page of Bucknill 1859. The London Library is a private lending library in St James's Square, Piccadilly, London.

To Charles Kingsley 1 December [1859][1]

Ilkley Wells House | Otley Yorkshire
Dec 1

Dear Sir

Thank you for your second note.[2] I am very glad you wrote the Winter Garden, I did so extremely like it.[3] I felt myself amongst the tall fir-trees.— With respect to

Cirripeds, the Ray Soc. published two awfully big volumes of my work with many Plates[4] They cost together 1.1.0; but would not be in the least worth your buying, as I describe all species of world & go into great detail of anatomy, homologies & metamorphosis. Owen in his second Edition of Comp. Anat gives good abstract.[5] In the new Edit. of Penny Cyclop. I wrote a brief abstract for Dr. Lankester of my work.—[6] If you shd care much for subject, you might borrow the vols. from any of the Scientific societies. The facts given on the *Complemental* (i e accessory males married to a hermaphrodite) males are, I think, really very curious, & these males, almost the most degraded creatures in the world. It took me months before I could believe my own eyes & I have given evidence in great detail.—[7]

I would venture to advise you to read one of the abstracts, & then if you care for subject borrow my volumes.—

I can in truth hardly say how glad I am you have permitted me to quote the sentence in your note.[8]

Pray believe me | Yours very sincerely | C. Darwin

Thomas Fisher Rare Book Library, University of Toronto

1 The year is established by the relationship between this letter and the letter to Charles Kingsley, 30 November [1859] (*Correspondence* vol. 7).

2 Kingsley's second letter has not been found, but see *Correspondence* vol. 7, letter from Charles Kingsley, 18 November 1859.

3 CD refers to Kingsley 1858.

4 CD refers to *Living Cirripedia* (1851) and *Living Cirripedia* (1854).

5 The reference is to Richard Owen and Owen 1855, pp. 277–95.

6 CD refers to Edwin Lankester and the *English cyclopædia*; Lankester was in charge of the natural history articles (*ibid.* 1: viii). In the entry for Cirripedia, Lankester quoted extensively from CD's abstract (see *ibid.* 1: 1096–1118).

7 CD first discovered 'complemental males' in the hermaphrodite barnacle *Scalpellum vulgare* in 1848 (see *Correspondence* vol. 4, letters to J. D. Hooker, 10 May 1848 and n. 12 and 6 October [1848], and *Living Cirripedia* (1851), pp. 281–93). In some species, these complemental males lacked stomach and mouth parts and several had no penis.

8 See *Correspondence* vol. 7, letter from Charles Kingsley, 18 November 1859 and n. 4.

To T. H. Huxley [1860–70?][1]

6, Q. A. St.[2]
Wednesday

My dear Huxley

I must thank you for the extremely pleasant evening you gave Frank;[3] but pleasant is not the right word, delightful, most delightful, was the mildest of a crowd of epithets, which came bubbling out of his mouth in inextricable confusion.

Ever yours | C. Darwin

Copy
DAR 145: 244

1 The date is conjectured on the basis of the change in house numbers in Queen Anne Street in 1860 (see n. 2, below), and the likely age of Francis Darwin at the time the letter was written (see n. 3, below). No visits to Queen Anne Street by CD are recorded in Emma Darwin's diary for 1864 (DAR 242).

2 In the spring of 1860, the house numbers in Queen Anne Street, London, changed; the number of Erasmus Darwin's house, where CD stayed on his visits to London, changed from 57 to 6. See *Correspondence* vol. 8, letter to a bookseller, 13 May 1860, n. 1, and *Post Office London directory* 1861).

3 Francis Darwin was born in 1848. For a similar comment on his enthusiasm, see *Correspondence* vol. 16, letter to J. D. Hooker, 17 [June 1868].

To a bookseller 9 April [1860]

Down Bromley Kent
Ap. 9th

Dear Sir

Please procure & send me by **Post**, Mr Pat. Matthew's work on "Naval Timber & Arboriculture" 1831. Adam Black Edinburgh & Longman London.—[1]

Dear Sir, | Yours faithfully | C. Darwin

Endorsement: '1860'
Uppsala University Library, Waller Manuscript Collection (Waller MS alb-54:068)

1 CD had become aware of Patrick Matthew's book (Matthew 1831) after Matthew published a letter in the *Gardeners' Chronicle*, 7 April 1860, pp. 312–13, in which he claimed to have published a principle of natural selection before the publication of *Origin*. See *Correspondence* vol. 8 and Dempster 1996. There is an annotated copy of Matthew 1831 in the Darwin Library–CUL (see *Marginalia* 1: 571).

To Williams & Norgate 10 April [1860][1]

Down Bromley Kent
Ap 10

Dear Sir

Will you be so kind as to send those two copies of the Translation of my Origin, which I received this morning (I presume through you) & which I now send to you by Post, to

Prof. Van der Hoeven

& to

Prof. Moleschott.—[2]

I believe Van der Hoeven lives at Leyden, but I do not know where Prof. Moleschott lives, but Prof. Huxley[3] tells me that you will probably know.

Kindly charge carriage to me; likewise for a copy of English Edition, which I have asked Mr Murray to send you for my Prof Liebig of Munich.—[4]

Dear Sir | yours faithfully | C. Darwin

Alexander Autographs (dealers), 20 February 2005, Lot 273

1 The year is established by the relationship between this letter and the letter to H. G. Bronn, 10 April [1860] (*Correspondence* vol. 8).

[2] CD refers to the German translation of *Origin*, Bronn trans. 1860, to Jan van der Hoeven, and to Jacob Moleschott. Hoeven and Moleschott's names appear on CD's presentation list for the German translation of *Origin* (see *Correspondence* vol. 8, Appendix III).

[3] Thomas Henry Huxley.

[4] CD refers to John Murray and Justus von Liebig. Liebig's name appears on CD's presentation list for the second edition of *Origin* (*Correspondence* vol. 8, Appendix III).

To ? 14 June [1860][1]

Down Bromley Kent
June 14

My dear Sir

Many thanks for your kind note & offer—I have sent the list to Dr Hooker & have told him to write direct to you if they ⌈desire⌉ any or all at Kew—[2] As Hooker is very much over-worked I have told him that I have told you, that if you did not hear in 2 or 3 or 4 days, the seeds were not wanted at Kew I really have no where else to recommend.—

I am sorry to say my Translator could not agree with Publisher, & I see no chance whatever of French Translation.—[3] There was good Review lately in the Revue des Deux Mondes & in Bib. Univ. of Geneva.—[4]

I am extremely much pleased to hear what you say about young men: I knew well I sh^d. never convert old stagers.—

I am sorry to say that my eldest daughter has long been very ill; & I am much harassed and unwell—[5] I fear there is little chance of my being at Oxford—[6]

In Haste | My dear Sir | Yours truly obliged | C. Darwin

Barlow Humphreys (private collection)

[1] The year is established by the reference to reviews of *Origin* in *Revue des Deux Mondes* and *Bibliotheque Universelle* (see n. 4, below); these reviews were published in April and March 1860, respectively.

[2] Joseph Dalton Hooker was assistant director of the Royal Botanic Gardens, Kew.

[3] CD had engaged Pierre Theodore Alfred Talandier to make a French translation of *Origin* (*Correspondence* vol. 8, letter to Edward Cresy, 20 January [1860]), but Talandier failed to secure a publisher (*ibid.*, letter to J. L. A. de Quatrefages de Bréau, 30 March [1860]).

[4] CD refers to Pictet de la Rive 1860 and Laugel 1860; there are annotated copies in the Darwin Pamphlet Collection–CUL.

[5] Henrietta Emma Darwin was ill from late April 1860 until early 1861 (see *Correspondence* vol. 8, Appendix II, and *Correspondence* vol. 9, p. xix).

[6] In 1860, the meeting of the British Association for the Advancement of Science was held at Oxford. For accounts of the meeting, see *Correspondence* vol. 8.

From W. D. Fox 16 August [1860?][1]

Deepdene | Hythe | Southampton
August 16

My dear Darwin

In answer to your enquiry, I believe I might truly state—not one dozen, but dozens of white cats came under my observation.[2] The first I saw was a half bred Persian at Ryde.[3] I had a kitten of hers, from whom I had a great number, never having less than six at a time for years. In every case, if *pure* white & with blue eyes, they were stone deaf. I used to show this by making all sorts of the loudest noises close to them, which they never in the least perceived.

I have remarked them in various other places—once at the large Inn at Chichester (which I think was a male)—2 Norwegian Cats whose owner I amazed (as I have done several others) by remarking "that it was a pity such fine cats should be deaf". These were females.

Of course the greater part of those we bred & kept were females, but I also had males, as I kept my breed pure for many years. I am *certain* the females were deaf, & I have no doubt about the males being so also, as I must have observed them if they had not been.

I forget where I wrote several letters on the subject of these blue eyed cats—it is now 20 to 30 years since—but they elicited exactly similar facts.

I bred several that had only one eye blue; and we were quite convinced that in those cases, they were deaf only on the blue side. All my elder children remember "Lily" our first cat—as perfectly deaf & blue eyed—& she certainly brought litters of kittens for many years—

So that the sex is undeniable, tho' I once had a Male Tortoishell (or at least so reputed) who had a family—to the great delight of my children.

Ever yours W D Fox

After writing the above I asked my wife & cubs if they remembered the white cats. "Lily" was too far back, but "Glaucops" was well remembered for her azure eyes, and kittens innumerable.

Fanny,[4] who is just returned from Bradwell nr G[t] Yarmouth—added "that there is a cat there perfectly white, but with the usual green eyes, that is deaf—like the blueeyed ones." She remarked this particularly, as opposed to the blue eyed theory.

"But at Caister Rectory close by Mr Steward[5] has a magnificent blue eyed cat which is quite deaf."

I shall probably see both these cats before long, and shall very likely find a Colony about there.

The Caister Cat—being described as *very fine*, *may* probably be a male.

Your Son William[6] was kind enough to call here on Friday, and most favorably impressed us all.

DAR 164: 188

1 The year is conjectured on the supposition that the letter was written before the publication of *Variation* in 1868 and by comparison with W. D. Fox's correspondence and diaries, which make any other year unlikely. See Larkum 2009.

2 CD first asked Fox about the correlation between deafness and the colour of fur and eyes in cats in 1856 (see *Correspondence* vol. 6, letter to W. D. Fox, 8 [June 1856]). The first two chapters of CD's species book, composed in 1856 and dealing with 'Variation under domestication', are not extant, but CD later used much of the material for *Variation*. He wrote: 'The Rev. W. Darwin Fox informs me that he has seen

more than a dozen instances of this correlation in English, Persian, and Danish cats' (*Variation* 2: 329).

3 Ryde is a seaside town on the Isle of Wight.

4 Frances Maria Fox.

5 George William Steward. Caister-next-Yarmouth (now Caister-on-Sea) was just north of Great Yarmouth, Norfolk.

6 William Erasmus Darwin.

To a bookseller 6 September [1860]

Down Bromley Kent
Sept. 6

Dear Sir

Will you oblige me by procuring Fraser's Magazine for July, & send it, or cut out Review on my Origin of Species (according to Postage) & send it to

Prof. Asa Gray
Cambridge
Massachussetts
U. States

Please charge me postage & excuse my asking this favour as Prof. A. Gray wants it immediately[1]

Dear Sir | In Haste | Yours sincerely | C. Darwin

Endorsement: '1860'
Bernard Quaritch (dealers) 2003

1 See *Correspondence* vol. 8, letter to Asa Gray, 10 September [1860]. The review was Hopkins 1860.

To John Medows Rodwell 15 October [1860][1]

15 Marine Parade | Eastbourne
Oct. 15th.

Dear Sir

I am truly obliged for your interesting letter.[2] I am very far from being surprised at anyone not accepting my conclusions on the origin of species; as the argument rests almost solely on the view explaining & grouping phenomena, otherwise inexplicable. I have some confidence that I am in the main right; for I find it as yet a universal rule, that those naturalists who go a little way with me, the more they reflect on the subject the further they go.— I am at work on my larger work, but ill-health & other interruptions make my progress very slow.—[3] Your remarks on language seem to me very striking; & the "struggle for life" with words is quite new to me.— I had often thought that a striking resemblance might be traced in the genesis of words & species; but was much too ignorant to attempt it.—[4] It was done to a certain extent some 4 or 5 or 6 months ago in the Cornhill Magazine by Lewes in one of his Zoological papers.—[5] Could you not publish an essay on the subject?

I have been particularly interested by your case of the Horses. I have somewhere read a nearly parallel case. I am sorry to give you trouble; but I sh[d] very much like to know whether the case was published; & if you can give me any further particulars; such as how many Horses were affected; how soon they recovered &c.—[6] Prof. Wyman has sent me an analogous case with respect to pigs in Florida: he was surprised at seeing them all black; & he found that they eat a certain root, which injures & kills the white pigs, but does not hurt the black; & the farmers added, "we help it by *selection*, for we kill the young white pigs".—[7]

It is not white cats, but white cats with *blue* eyes, which are deaf,—if one eye is blue the cat is deaf on that one side—if your cat has *distinctly* blue eyes & is not deaf, I sh[d] be particularly obliged for the fact, as it will be the sole exception which I have heard of.—[8]

Pray accept my best thanks for your interesting letter & with respect, I beg leave to remain | Dear Sir | yours faithfully & obliged | Charles Darwin | of Down, Bromley, Kent

I shall be here for about 10 days[9]

DAR 185: 149

1 The year is established by the relationship between this letter and the letter from J. M. Rodwell, 31 October 1860 (*Correspondence* vol. 8).

2 Rodwell's letter has not been found.

3 CD was working on the manuscript of *Variation*, the first part of what was intended as a three-part work explaining in greater detail the operation of natural selection; the other two parts were never published. See *Correspondence* vol. 8, Appendix II and n. 1.

4 Rodwell made translations from Hebrew, Arabic, Ethiopic, and Coptic; he planned to write a book on the 'birth, life, and death of words' (see *Correspondence* vol. 8, letter to J. M. Rodwell, 5 November [1860]), but apparently it was never published. On the application of CD's theories to language, see Alter 1999.

5 CD refers to George Henry Lewes and Lewes 1860, pp. 445–7.

6 Rodwell's answer was in the portion of his letter of 31 October 1860 (*Correspondence* vol. 8) that is now missing; see *ibid.*, letter to J. M. Rodwell, 5 November [1860]. CD gave Rodwell's information in *Variation* 2: 337:

> The Rev. J. Rodwell informs me that his father turned out about fifteen cart-horses into a field of tares which in parts swarmed with black aphides, and which no doubt were honeydewed, and probably mildewed; the horses, with two exceptions, were chesnuts and bays with white marks on their faces and pasterns, and the white parts alone swelled and became angry scabs. The two bay horses with no white marks entirely escaped all injury.

7 Asa Gray had sent CD information from Jeffries Wyman on the black pigs of Florida, which was later confirmed by Wyman himself (see letter to Asa Gray, 3 April [1860], and letter from Jeffries Wyman, [*c.* 15] September 1860). The case was given in *Origin* 3d ed., p. 12, and in *Variation* 2: 227, 336, where the pigs were incorrectly said to live in Virginia.

8 CD discussed the correlation between blue eyes and deafness in cats in *Origin*, pp. 12 and 144, where he stated that the connection was invariable. In *Origin* 3d ed., p. 12, he altered the text to state that cats with blue eyes were 'generally' deaf. He discussed the topic again in *Variation* 2: 329. In *Variation* CD mentioned, without citing Rodwell's case, that he had since heard of a few exceptions to the rule. See also *Correspondence* vol. 8, letter from J. M. Rodwell, 31 October 1860.

9 The Darwins stayed at Eastbourne from 22 September to 10 November 1860 in the hope that a stay at the seaside would improve Henrietta Emma Darwin's health (*Correspondence* vol. 8, Appendix II, and

letter to G. V. Reed, 15 September [1860]; their stay was prolonged by Henrietta's attack of sickness on 22 October (*ibid.*, letter to John Innes, 26 October [1860]).

From E. A. Darwin 12 November [1860–8][1]

6. Queen Anne St
12 Nov

Dear Charles.

I send you the Tithes.[2] Langton tells me he has written a letter on Investments which you are to send me in due time[3]

Yours affec. | E D

DAR 105: B11

1 The date range is established by the change of Erasmus's house number from 37 to 6 Queen Anne Street in early 1860 (*Post Office London directory*), and by the dates when CD entered his annual 'Tithes' in his Account books (see n.2, below); after 1868, they were entered in months other than November (except in 1871, when CD entered them on 29 November).

2 In his Account and Investment books (Down House MSS), CD referred to an annual income ranging in the 1860s from £15 to £21 as the 'Castle Morton Tithes'. CD and each of his five siblings inherited one sixth of a 'Castle Morton trust' from their father, Robert Waring Darwin, in 1837. The trust was evidently related to a property of CD's grandfather, Josiah Wedgwood I, in the parish of Castle Morton, Worcestershire. (Robert Waring Darwin's Investment book, Down House MS.)

3 The letter from Charles Langton, CD and Erasmus's brother-in-law, has not been found.

To Henry Fawcett 6 December [1860][1]

Down Bromley Kent
Dec. 6th

My dear Sir

I had intended writing to you before I received your obliging note this morning.[2] I have been struck with admiration at your clear & forcible exposition of my views; & I beg leave to thank you cordially for all your kind & generous expression towards me, & likewise for the sake of the subject.— I have been much pleased with your metaphor or rather simile of Mont Blanc; & as I am now preparing a corrected Edition, I shall give this simile as suggested by you.[3] Many parts of the Article have interested me much.— I am indebted to you for having recalled to my recollection Mr Hopkins passage about Denudation. He is mistaken in supposing that I ever meant to say that when a formation has once been fairly upraised that it has been wholly denuded away. I referred exclusively to matter recently deposited, & not consolidated being denuded during its first elevation.[4] But as the subject has been broached by him, I will touch on it, & ask how he understands the presence of thousands of square miles (as in N.E., S. America, described by Humboldt) of naked metamorphic schists & plutonic rocks. Will anyone believe that these have been metamorphosed in their present uncovered state? on the contrary we must admit that they were once covered & that subsequently *all* the sedimentary strata have in such cases been denuded away over whole regions.—[5]

Nothing has pleased me so much personally as your remarks (with the quotation from Mills) on the proper spirit & method of scientific research.[6] I never could see that I was wrong, however much the inductive system was hurled at my head, & now I feel sure I was not.[7] As you seem so kindly interested in my work, I may mention that I believe that the key of my work was gained by an unusually inductive line of research. For when I began to suspect from geographical distribution &c &c &c, that new species have been formed by descent, I determined to work at domestic productions with not one single idea in my head; & no one can know the years of blind labour I had, before I clearly saw that Selection was man's chief means. When I had got thus far I strongly suspected that this was the key to nature's work; but it was some time before I could conceive how it could be applicable,—not indeed until I chanced to read Malthus's noble work—[8] I may just add, simple as it may now appear, that it took me many years thought before I saw the absolute necessity of (& manner of understanding) the principle as I have called it of divergence; which no one seems to perceive is new. Oh the work I had in tabulating the plants of differently sized areas, & puzzling why everywhere there was so much diversity of forms!— But I really ought to beg your pardon, occupied as you now are with exciting political concerns, for troubling you with these personal details, which I have not mentioned to others.—

I truly sympathise with you in your great calamity of loss of eyesight: The admiration of every right-minded man at your noble & courageous zeal to play a fitting part in life, must be some, though a poor, consolation to you.—[9]

With very sincere thanks, I beg leave to remain | My dear Sir | Yours very faithfully | Charles Darwin

Karpeles Manuscript Library Museums

1 The year is established by the reference to Fawcett's review of *Origin* (Fawcett 1860).

2 Fawcett's letter to CD has not been found.

3 In his review of *Origin* in *Macmillan's Magazine*, Fawcett wrote:

> An individual would excite a smile of ridicule who, having discovered that Mont Blanc three thousand years ago was of the same altitude as it is at the present time, should consider that he had refuted those theories of modern geology which suppose that the stupendous peaks of Switzerland were lifted from off their ocean bed, and that every physical change in this earth's appearance has been produced by the indefinitely prolonged operation of the same physical causes which on every side around us continue in ceaseless activity. The extinction of species and the introduction of new ones are associated with periods which can only be described as geological epochs; and the time which has elapsed since the occurrence of the most remote recorded historical event is but an instant compared with the period which is indicated by the deposition of one of the strata which tell the history of this planet's structure.

(Fawcett 1860, p. 87.) CD referred to this passage in *Origin* 3d ed., p. 138.

4 In *Origin*, pp. 289–91, CD had argued that the geological record was incomplete in part because large periods of time separated successive strata, and gave as an example of the absence of significant deposits

on the west coast of South America, because of the action of the waves as the land was uplifted from the sea. Fawcett, in his review, had quoted William Hopkins's argument that it was extremely unlikely that a whole stratum could be removed by denudation, since when the land was uplifted, denudation tended to be uneven (Fawcett 1860, p. 90).

5 CD put forward this argument in *Origin* 3d ed., pp. 314–15. CD refers to Alexander von Humboldt and Humboldt 1814–29.

6 In Fawcett 1860, p. 84, Fawcett quoted John Stuart Mill:

> The mode of investigation which, from the proved inapplicability of direct methods of observation and experiment, remains to us as the main source of the knowledge we possess, or can acquire, respecting the conditions and laws of recurrence of the more complex phenomena, is called in its most general expression the deductive method, and consists of three operations—the first, one of direct induction; the second, of ratiocination; and the third, of verification.

7 See, for example, *Correspondence* vol. 7, letter from Adam Sedgwick, 24 November 1859: 'You have *deserted*—after a start in that tram-road of all solid physical truth—the the true method of induction'.

8 CD read Thomas Robert Malthus's *Essay on the principle of population* (Malthus 1826) in 1838 (see *Notebooks*, Notebook D, 134e and 135e). For an analysis of CD's thinking and method of work during this period, see Kohn 1980.

9 Fawcett had been blinded in a shooting accident in 1858, but had determined that the misfortune would make no difference to his political ambitions (*ODNB*). He decided to stand for election in the London borough of Southwark in November 1860; he withdrew on 8 December (Stephen 1885, pp. 189–94).

To David Forbes 11 December [1860][1]

Down Bromley Kent.
Dec. 11.

My dear Sir

Your letter has pleased me exceedingly.[2] When I saw how much error you detected in d'Orbigny, I said to myself "Good Heavens my turn will come next".—[3]

I hope that you will be able to send your observations on Chile &c to Geolog. Soc. before your departure (which I am sorry to hear is so very soon); but of course if not written out already that will be quite impossible.[4] On your return if you wish to publish in extenso, your case would be an excellent one for aid from the Government grant to Royal Soc<u>y</u>.[5] Pray bear this in mind.— You would get plenty of backers.— Your paper just sent to G. Soc. will at once put you in that position, that you might boldly demand aid.

I am sorry for my blunder about Andesite.[6] Have you ever reflected on the subject of plutonic rocks being same character throughout extensive areas at same epoch. I remember this subject treated of by (I think) Keilhau: it seems to me very mysterious & indicating some great unknown class of facts.—[7]

How curious I shall be at some future time to read about the Uspallata formation.[8] I know how busy you are & must be; but it would be a *great favour*, if you would write literally one line to me to say whether the great mound of detritus at Val del Yeso has any connection with ice-action; because I am preparing new Edition of my Origin of Species, & if you say "no" I will strike out 3 lines; if you say "yes" I will leave them as they now stand.—[9] I saw in the valley of Cauquenes boulders which I now think must have had some connexion with Glacial action.—[10]

When in S. America, would it trouble you to make a few observations on Mules, as by enclosed paper, & some time write me the results?[11]

With every good wish for your success & prosperity, & with cordial thanks for your letter & kind expressions. Believe me | My dear Sir | Yours very sincerely | C. Darwin

DAR 185: 150

1 The year is established by the references to Forbes's paper on the geology of Bolivia and southern Peru (Forbes 1860), which was read before the Geological Society of London on 21 November 1860. A summary of this letter was published in *Correspondence* vol. 13, Supplement.

2 Forbes's letter has not been found. The letter from David Forbes, [November? 1860] (*Correspondence* vol. 8), which should now be dated [after 11 December 1860], is a reply to this letter.

3 CD refers to Forbes 1860, in which Forbes challenged many of Alcide Charles Victor Dessalines d'Orbigny's findings on the geology of South America (Orbigny 1835–47; see *Correspondence* vol. 8, letter to Charles Lyell, 4 December [1860]). For a summary of the differences between Forbes and Orbigny, see *Proceedings of the Geological Society of London* 33 (1877): 44–6.

4 Forbes 1860 was intended as the first part of a memoir that would include two further parts on the geology of Chile and the Argentine (Forbes 1860, p. 62). However, Forbes read no further paper on the geology of South America to the Geological Society.

5 Forbes did not publish on the geology of Chile and Argentina with the Royal Society of London.

6 CD had described a volcanic rock that he found in the vicinity of the Beagle Channel as andesite (*South America*, pp. 154–5 *et seq.*). Forbes had revised CD's nomenclature, preferring to call it diorite on account of its similarity to European diorites (Forbes 1860, p. 29).

7 CD had earlier been impressed by Baltazar Mathias Keilhau's theory of the origin of rocks by heat-induced crystallisation (see Keilhau 1838–40 and 1844, and *Correspondence* vol. 3, letter to Adolf von Morlot, 10 October [1844] and n. 7).

8 CD crossed the Uspallata pass and the Uspallata range in Argentina in 1835 (see *Journal of researches* 2d ed., pp. 331, 333; see also *South America*, pp. 60, 175).

9 CD described the geological features of the Yeso valley, in Chile, in *South America*, p. 178. This may have been the source for the passage about the 'vast mound of detritus' in an Andean valley in *Origin*, p. 373. After receiving Forbes's reply (*Correspondence* vol. 8, letter from David Forbes, [November? 1860], now redated [after 11 December 1860]), CD restated his conviction that the mound was due to glacial action (*Origin* 3d ed., pp. 403–4).

10 See *South America*, pp. 64–7. While a small number of boulders near Cauquenes in Chile had previously been described by Adolphe Théodore Brongniart as erratics deposited by glaciers, CD argued that river action and uplift could have displaced them from their original strata (*South America*, p. 66 and n.).

11 The enclosure listing questions for Forbes has not been found. By mules, CD probably means hybrids between goats and sheep. For Forbes's reply, see *Correspondence* vol. 8, letter from David Forbes, [November? 1860], now redated [after 11 December 1860]. CD cited Forbes on the sheep of Chile in *Variation* 1: 95, and considered the question of the possible relatedness of sheep to goats in *Variation* 1: 94, 98–9, and 102.

To W. B. Tegetmeier [28 March – 14 April 1861][1]

send them to me by enclosed address. as I rather want to see two of them.—
My dear Sir | yours sincerely | C. Darwin

I am getting together a few points to investigate.—

I think the Half-lop, had better be killed (not knocked on Head) if you get one.—[2]

I have not yet opened the Box with the Skulls; for I have not yet come to Skeleton.—[3]

Incomplete[4]
Max Rambod (dealer), February 2008

[1] The date range is established by the relationship between this letter and the letters to W. B. Tegetmeier, 28 March [1861] and 14 April [1861] (*Correspondence* vol. 9).

[2] CD was interested in how the skulls of domestic rabbits were modified by the lengthening and lopping of their ears; the half-lop rabbit has one ear lopping forward, and a correspondingly asymmetrical skull (see *Variation* 2: 324–5).

[3] Tegetmeier had lent CD his collection of fowl skulls for his work on *Variation* (see *Variation* 1: 260, and *Correspondence* vol. 9, letter to W. B. Tegetmeier, 28 March [1861]).

[4] The letter advertised by the dealer is incomplete; all the available text is transcribed above.

To ? 7 May [1861–8?][1]

Down. | *Bromley.* | *Kent. S.E.*
May 7th

Sir

I am much obliged for your note & offer, but I am sorry to say that I am too much occupied to contribute to any Periodicals.—

Sir | Your obliged & obet servant | Charles Darwin

I need not say that your Review[2] has my *cordial* good wishes.—

Photocopy
DAR 249: 122

[1] The date range is conjectured from the style of the address, which CD used regularly from May 1861 until April 1869.

[2] The review has not been identified.

To Thomas Campbell Eyton 12 [May 1861 – April 1863][1]

Down. | *Bromley.* | *Kent. S.E.*
12th

Dear Eyton

I am particularly obliged to you for telling me about your crossed pigs; but the other day I was wishing to know how it would be.— When your young crossed pigs are pretty well grown up, I should very much like to hear what they are like & whether they resemble each other.—[2]

I doubt from what I have heard, whether your half-bred *G. Sonneratii* will be productive, yet I was assured many years ago that this once occurred—[3]

With many thanks, | Dear Eyton | Believe me | Yours very sincerely | C. Darwin

University of Birmingham (Special collections: EYT45)

[1] The date is established by the style of the printed stationery, which is a type that CD started using in May 1861, and by the fact that CD refers to having heard of fertile hybrids of *Gallus sonneratii* 'many years ago' (see also n. 3, below). In 1862, hybrids of *G. sonneratii* and *G. bankiva* (assumed to be the domestic fowl) were shown to have limited fertility at the Zoological Gardens in London (S. J. A. Salter 1863). CD reported these findings in *Variation* 1: 234, citing S. J. A. Salter 1863, which was published in the April issue of *Natural History Review*. After April 1863, one would expect CD to refer Eyton to John William Salter's publication.

[2] No further information on Eyton's crossed pigs has been found. CD and Eyton had corresponded in 1856 about the offspring of an African pig, belonging to Rowland Hill, and a common pig (see *Correspondence* vol. 6, letter to T. C. Eyton, 21 August [1856]).

[3] See *Correspondence* vol. 10, letter from T. C. Eyton, [after 19 May 1862?]; the 'jungle fowl' mentioned in that letter (to which this letter may be a reply) may have been *Gallus sonneratii*. In *Natural selection*, p. 435 n. 12, CD wrote, 'several years ago I myself saw at the Zoological Gardens young birds, which were the *offspring of hybrids* inter se from a Sonnerat cock & Bantam Hen'. (*Natural selection*, the manuscript of CD's 'big book' on species, was written between 1856 and 1858.) See also *Variation* 1: 234, where CD also cited Edward Blyth for information about *Gallus sonneratii* crossing with domestic hens in India.

To ? 20 June [1861–8][1]

Down. | Bromley. | Kent. S.E.
June 20th

Dear Sir

As it will save you trouble of referring I send by this post a spare copy of the paper about which you enquire.— Permit me to remark that it was never in remote degree intended for publication & was written many years ago. I had no connection with the publication of these extracts beyond giving my permission.—[2]

Dear Sir | Yours faithfully | Ch. Darwin

Mr. Wallace's paper seems to me excellent.—

American Philosophical Society (Getz 13879)

[1] The date range is established by the printed stationery, which is a type that CD used between May 1861 and April 1869.

[2] The paper referred to is probably CD's part of the Darwin–Alfred Russel Wallace papers read to the Linnean Society by Joseph Dalton Hooker and Charles Lyell in 1858 (C. Darwin and Wallace 1858). CD's paper contained extracts from his 1844 manuscript essay on species (see *Foundations*) and from the letter to Asa Gray, 5 September [1857] (*Correspondence* vol. 6); a footnote explained that the manuscript essay was not intended for publication and so was not written with care.

To Frances Julia Wedgwood 11 July [1861][1]

[Torquay][2]
July 11

My dear Snow

Some one has sent us 'Macmillan'; and I must tell you how much I admire your Article;[3] though at the same time I must confess that I could not clearly follow you in some parts, which probably is in main part due to my not being at all accustomed to metaphysical trains of thought.[4] I think that you understand my book perfectly, and that I find a very rare event with my critics. The ideas in the last page have several times vaguely crossed my mind.[5] Owing to several correspondents I have been led lately to think, or rather to try to think over some of the chief points discussed by you. But the result has been with me a maze—something like thinking on the origin of evil, to which you allude.[6] The mind refuses to look at this universe, being what it is, without having been designed; yet, where one would most expect design, viz. in the structure of a sentient being, the more I think on the subject, the less I can see proof of design.[7] Asa Gray and some others look at each variation, or at least at each beneficial variation (which A. Gray would compare with the rain drops which do not fall on the sea, but on to the land to fertilize it) as having been providentially designed.[8] Yet when I ask him whether he looks at each variation in the rock-pigeon, by which man has made by accumulation a pouter or fantail pigeon, as providentially designed for man's amusement, he does not know what to answer;[9] and if he, or any one, admits these variations are accidental, as far as purpose is concerned (of course not accidental as to their cause or origin); then I can see no reason why he should rank the accumulated variations by which the beautifully adapted woodpecker has been formed, as providentially designed. For it would be easy to imagine the enlarged crop of the pouter or tail of the fan-tail as of some use to birds, in a state of nature, having peculiar habits of life. These are the considerations which perplex me about design; but whether you will care to hear them, I know not.— I think that your article will strike & interest many reflecting minds.

We are all very jolly here; & I admire the beautiful scenery more than could be reasonably expected of an acknowledged descendent of an Ape.

My dear Snow | yours affectionately | Charles Darwin

LL I: 313–14; Christie's, 3 March 2004[10]

[1] The year is that given by Francis Darwin in *LL* I: 313. The slightly shorter version of this letter published in *LL* was reproduced in *Correspondence* vol. 9 (down to 'whether you will care to hear them, I know not'). The text from 'providentially designed. For it . . . ' to the valediction appeared in facsimile in a Christie's sale catalogue in 2004, and the transcription has been corrected and added to from the facsimile.

[2] The address is given in the sale catalogue description. The Darwins rented a house in Hesketh Crescent, Torquay, Devon, in July and August 1861 (*Correspondence* vol. 9, Appendix II and n. 14).

[3] CD refers to an unsigned article entitled 'The boundaries of science. A second dialogue', published in *Macmillan's Magazine* in July 1861 (*Wellesley index*). The first part of the dialogue appeared in the June 1860 issue ([F. J. Wedgwood] 1860–1).

[4] The articles take the form of a dialogue between the religiously orthodox Philocalos (lover of beauty) and the defender of CD's views, Philalethes (lover of truth). Topics discussed include the implications of the theory of natural selection for the connection between the material and spiritual realms, the division between reason and faith, and the relationship of the Creator to his creation. Frances had been strongly influenced by the religious teachings of Thomas Erskine and his friend John Frederick Denison Maurice, one of the founders of Christian Socialism. She had spent several months with followers gathered at Erskine's estate at Linlathen, in Forfarshire, Scotland, in 1859. See B. Wedgwood and Wedgwood 1980, pp. 254, 267.

[5] The closing pages of the second part of the dialogue address the question of whether there is design in

nature and, if so, how this informs religious belief and scientific inquiry ([F. J. Wedgwood] 1860–1, 4: 244–7).

6 The problem of the existence of evil in the world forms part of the discussion mentioned in n. 5, above.

7 Philalethes argues that while there are 'unmistakeable evidences of *design*' in nature, any attempt to make the apparent harmony in nature's operations a complete one by bringing humans into consideration is bound to fail ([F. J. Wedgwood] 1860–1, 4: 244–5). The discussion continues:

> *Philoc.* You mean to say, then, that somewhere or other there is a misfit between man and his dwelling-place; that, if the machinery was ever perfect, some evil power has laid its hand on the mainspring, and deranged the working ever since? *Philal.* Yes; and that influence lies wholly without the boundaries of science, which is exclusively occupied with the machinery itself, and can take no account of any influence from without. Go back as far as we will, therefore, science only shows us the working of the present order of things.

8 Gray presented this view in the third part of his three-part series of articles on CD's theory published in 1860 and reprinted in A. Gray 1861a.

9 See *Correspondence* vol. 8, letter to Asa Gray, 26 November [1860].

10 The letter sold at Christie's was complete and was four pages long; only the last page was given in facsimile, and has been transcribed here. The rest of the letter has been transcribed from the version in *LL* (see n. 1, above).

To Henry Fawcett 20 July [1861][1]

2. Hesketh Crescent | Torquay
July 20th

My dear Sir

Nothing would have given me more pleasure than seeing you at Down, but as the above address shows, we are away from home.[2] We shall remain here 5 or 6 weeks longer having come here on account of my daughter's health & my own.—[3]

You could not possibly have told me anything which would have given me more satisfaction than what you say about Mr Mill's opinion.—[4] Until your Review appeared I began to think that perhaps I did not understand at all how to reason scientifically.—[5]

I thank you cordially for your very kind note. Perhaps at some future time you may be in our neighbourhood & we may have the pleasure of meeting.— I am forced to write in a great hurry as I must send some other letters by this Post—

My dear Sir | [Yours sincerely Charles Darwin][6]

AL(S)
Karpeles Manuscript Library Museums

1 The year is established by the address; the Darwins stayed at Hesketh Terrace in Torquay in July and August 1861 (see *Correspondence* vol. 9). A fragment of this letter, from a published source, was published in *Correspondence* vol. 9).

2 Fawcett had written to CD to acknowledge CD's letter thanking him for his review of *Origin* in *Macmillan's Magazine* (see *Correspondence* vol. 9, letter from Henry Fawcett, 16 July [1861], this volume, Supplement, letter to Henry Fawcett, 6 December [1860], and Fawcett 1860).

3 Henrietta Emma Darwin was suffering from the after-effects of what was thought to have been typhus fever (see *Correspondence* vol. 8, letter to J. D. Hooker, 11 May [1860] and n. 5).

[4] Fawcett had told CD about a conversation he had had with John Stuart Mill in which Mill endorsed the logic of CD's reasoning in *Origin* (see *Correspondence* vol. 9, letter from Henry Fawcett, 16 July [1861] and n. 3, and this volume, Supplement, letter to Henry Fawcett, 6 December [1860] and n. 6).

[5] Both Adam Sedgwick and Samuel Wilberforce in their reviews of *Origin* had criticised CD's method of reasoning ([A. Sedgwick] 1860 and [Wilberforce] 1860). Fawcett's review published in *Macmillan's Magazine* (Fawcett 1860) countered these criticisms.

[6] The signature has been excised, and 'Yours sincerely Charles Darwin' has been added in an unknown hand.

To ? 24 August [1861–8][1]

Down. | Bromley. | Kent. S.E.
Aug 24—

Dear Sir

Although you are so kind as to tell me not to acknowledge your letter, I must beg permission to thank you. The case which you relate is extremely interesting, & though I have heard of several yet I am very glad to have heard of this remarkable instance of inheritance[2]

With my thanks pray believe me my dear Sir | yours faithfully | Charles Darwin

LS
Wellcome Library (Wellcome MS.7781/28)

[1] The date range is established by the form of address, which is one that was used by CD between May 1861 and April 1869.

[2] While CD was working on *Variation* and after the book was published he was occasionally sent unusual cases of inheritance in plants, humans, and other animals. See, for example, *Correspondence* vol. 15, letter to William Ogle, 29 March [1867].

To John Rogers 14 October [1861][1]

Down. | Bromley. | Kent. S.E.
Oct 14th

Dear Sir

Mr Veitch has kindly sent me a Catasetum & I think I at last understand its mechanism.[2] I shd. like to quote what you said;[3] but I am anxious to hear (if you will kindly inform me) whether you often touched the "trigger"; & if so whether you did it with your finger & in this case, whether the sticky hair or gland always or generally hit your finger so as to stick to it. I shd. be very much obliged for this little bit of information.

Dear Sir | Yours faithfully | Charles Darwin

Did you ever notice any nectar in the pouch or basin in the labellum of Catasetum? Have you cultivated Stanhopea and you ever notice nectar in the cup of this genus? or in Oncidium?

Rogers family (private collection)

[1] The year is established by the relationship between this letter and the letter to J. D. Hooker, 11 October [1861] (*Correspondence* vol. 9).
[2] CD refers to James Veitch. See *Correspondence* vol. 9, letter to J. D. Hooker, 11 October [1861].
[3] CD cited Rogers (misspelt 'Rodgers') on *Catasetum* in *Orchids*, p. 236 n. *.

To Robert Scot Skirving 16 November [1861–8][1]

Down, Bromley, Kent, S.E.
Nov. 16

My dear Sir

I am sorry to say that I know nothing of the habits of earwigs.[2] Mr. Newman[3] with whom I am slightly acquainted is I believe a good sort of man, and certainly a good entomologist, and he may be trusted I should think on the point in question. The Forficula or Labia minor is a good flier and excepting in size would not be distinguished from the common earwig.[4] It is certainly strange that an insect, having wings, should not fly; but even if this is the truth yet I should not be surprised at the occasional appearance of a flying earwig on the same principle that the bug occasionally has been known to possess wings. In answer to your kind inquiries I am sorry to say that my health still keeps very indifferent.

Believe me my dear Sir | Yours very faithfully | Charles Darwin

Copy
DAR 147: 481

[1] The date range is established by the address, which is a form that CD used between May 1861 and April 1869.
[2] No letter from Skirving on this subject has been found.
[3] Edward Newman.
[4] *Labia minor* was originally named *Forficula minor* by Linnaeus (Linnaeus 1758–9, 1: 423). *Labia* and *Forficula* are both genera of the order Dermaptera, earwigs. *Forficula* is now a monospecific genus; *Forficula auricularia* is the European earwig (referred to by CD as the common earwig). *Labia minor* is the lesser earwig. Newman thought that *F. auricularia* did not fly, but the point was disputed (*Zoologist* 8 (1850): 2695, 2759, 2831; *Entomologist* 3 (1866–7): 154).

To ? 25 November [1861–8][1]

Down. | *Bromley.* | *Kent. S.E.*
[Nov] 25

Dear Sir

I have just read your notice on bent cleavage;[2] & it has occurred to me that you might possibly like to hear that I observed the same fact in Chili. You will find a brief notice of this subject in my Geological Observations on South America 1846 footnote page 160.[3] In case you thought it worth while to refer to the passage, you will find a copy of this book at the Geolog. Soc.[4] I have suggested as a conjectural explanation the action of earthquakes. I presume that in a district lying for instance south of a volcanic focus the vibrations w[d] travel across the country in a nearly regular manner.

Pray do not trouble yourself to acknowledge the receipt of this as I do not know whether it is worth sending, & I beg leave to remain, dear Sir, | yours faithfully | Charles Darwin

LS
M. D. Conybeare (private collection)

[1] The date range is established by the printed stationery, which is a type that CD used between May 1861 and April 1869. The month is ambiguous; it looks as if the amanuensis may have tried to write an 'N' over a 'D'; the final two letters are 'ov'.
[2] The notice has not been identified.
[3] See *South America*, pp. 160–1 n. 23. By 'bent cleavage' CD means the bending of the tops of the laminae of clay-slate.
[4] CD refers to the Geological Society of London.

To Dorothy Fanny Nevill 27 November [1861][1]

Down. | Bromley. | Kent. S.E.
Nov. 27th

Madam

I will not attempt to thank your Ladyship for all your kindness, for it is beyond my power— I am pleased to hear that my Books have at all interested you; but I fear my little Orchid book will be dry.[2] This summer when at the sea, I meant merely to write a paper for some scientific journal, but the subject grew on me till my M.S. got rather too long for a paper.[3] I am convinced that orchids have a wicked power of witchcraft, for I ought all these months to be working at the dry old bones of poultry, pigeons, and rabbits instead of intensely admiring beautiful orchids.— I mention all this, because, though I can hardly bear to write the words, I must beg your Ladyship not to send any more of your treasures; though perhaps at some future period I may indulge myself with the examination of a few more orchids.— I will not forget your Ladyships most generous offer to give me other flowers, if I require them for observation, & I have no doubt that I shall some time be a beggar again *[less than 2 pages missing]* of the beauties of Dangstein, which I shall now read with interest.[4]

If your Ladyship should meet Mr Knox I hope that you will remember me to him: I spent many years ago a very pleasant morning with him & Sir Philip Egerton at the Zoological Gardens.—[5]

I beg leave to remain | with cordial thanks | your Ladyships | sincerely obliged | Charles Darwin

Since writing I have reason to hope that I shall receive a flower of Mormodes from Mr Rucker of Wandsworth[6]

Incomplete[7]
Max Rambod (dealer)

[1] The year is established by the relationship between this letter, the letter to John Lindley, 16 November [1861], and the letter to D. F. Nevill, 19 November [1861] (*Correspondence* vol. 9). See n. 6, below.

RN48524
Lady Dorothy Fanny Nevill (née Walpole)
by Henry Richard Graves, 1855.
Photograph National Portrait Gallery, London; unknown collection.

[2] Nevill's letter to CD has not been found, but was evidently a reply to his letter of 19 November [1861], in which CD had promised to send Nevill a copy of *Orchids* when it was published.
[3] CD was at Torquay from 1 July to 27 August 1861 (*Correspondence* vol. 9, Appendix II). CD decided in September 1861 that the subject of orchids would be worth publishing as a short book (see *Correspondence* vol. 9, letter to John Murray, 21 September [1861]).
[4] Nevill had invited CD to see the orchid collection in her garden at Dangstein, near Midhurst, Sussex, but CD had declined on account of his health (see *Correspondence* vol. 9, letter to D. F. Nevill, 19 November [1861]). CD probably refers to an article, 'The gardens and conservatories at Dangstein', that appeared in two parts in the 26 November and 3 December 1861 issues of the *Journal of Horticulture* (n.s. 2: 168–9, 183–5).
[5] CD refers to Arthur Edward Knox and Philip de Malpas Grey-Egerton.
[6] In his letter to John Lindley of 16 November [1861] (*Correspondence* vol. 9), CD mentioned that he had been promised orchids (*Catasetum* and *Dendrobium*) by Sigismund Rucker, who lived in West Hill, Wandsworth. In *Orchids*, p. 249 n., CD thanked Rucker for lending him a plant of *Mormodes ignea*. Rucker evidently offered to send the plant after CD had made a second request for flowers of *Mormodes* from Nevill (see *Correspondence* vol. 9, letter to D. F. Nevill, 19 November [1861]).
[7] The original letter is complete and is described in the sale catalogue as being four pages long. The first and last pages have been transcribed from a photograph in the sale catalogue; the section 'of witchcraft . . . beggar again' is the dealer's transcription.

To ? 31 December [1861–8][1]

Down. | Bromley. | Kent. S.E.
Dec 31st

Dear Sir

As I have never especially attended to Conchology I am sorry to say I cannot tell you the name of the enclosed shell which I now return—

Yours faithfully | Ch. Darwin

J. David Archibald (private collection)

[1] The date range is established by the printed stationery, which is a type that CD used between May 1861 and April 1869.

To Ladies 24 February [1862–9][1]

Down. | Bromley. | Kent. S.E.
Feb 24th

Ladies!

You express yourselves in very good English, & even if you had expressed yourselves obscurely, I should be a very dull man, if I did not perceive your kind feelings towards me, & did not acknowledge the honour which you have done me.—

I beg leave to remain, with much respect, | Yours truly obliged | Ch. Darwin

R. M. Smythe & Co., New York

[1] The date is established by the form of the address, which CD used from May 1861 to April 1869. The recipients have not been identified.

To John William Salter 28 February [1862][1]

Down Bromley Kent
Feb. 28th

My dear Sir

Mr. Sowerby[2] sent me the enclosed & I see on corner it is to be returned. I have been very glad to see it; for it certainly seems a very striking series; though there is a rather wide gap between the Coal & Tertiary. And I further presume the true affinities of Crustaceans must be very difficult to make out in Fossils.[3] Nevertheless such series delight me.— I wish I could persuade you to publish papers on such subjects.[4]

That series which you showed me of the Brachiopods was very striking.[5] Its publication would counter-balance half-a-dozen of the most spiteful Reviews,[6]

With many thanks | My dear Sir | Yours sincerely | Ch. Darwin

American Philosophical Society (Getz 5019)

1 The year is established by the reference to the series of Crustacea (see n. 3, below) and by the mourning border on the notepaper, which was used in January and February 1862, for Charlotte Langton, Emma Darwin's elder sister.

2 CD probably refers to James de Carle Sowerby, Salter's father-in-law (*ODNB*). Sowerby's brother, George Brettingham Sowerby, had been engaged by CD to do the illustrations for *Orchids*.

3 The enclosure has not been found, but may have been a draft of a series illustrated in Salter 1862, p. 90 (read May 1862). The illustration showed a series of Phyllopoda (a subclass of the Crustacea), going from the Triassic to the Tertiary period without showing any intermediate forms. The Crustacea discussed by Salter have since been reclassified as part of the subclass Phyllocarida.

4 Salter did, in fact, publish the series of Crustacea (see n. 3, above). CD had earlier expressed his regret that he 'could not persuade Mr Salter to publish a little paper on the subject' of the affinities among the brachiopods in the context of his transmutation theory (*Correspondence* vol. 9, letter to T. W. St C. Davidson, 26 April 1861).

5 Salter had shown CD a series of fossil brachiopods at the Museum of Practical Geology in April 1860. Salter had made a chart, following CD's own illustration of divergence facing p. 117 in *Origin* (see *Correspondence* vol. 8, first letter to Andrew Murray, 28 April [1860] and n. 12, and *Correspondence* vol. 9, letter to T. W. St C. Davidson, 26 April 1861 and n. 3).

6 On the reviews of *Origin*, see *Correspondence* vol. 8, Appendix VII and Ellegård 1990.

To ? 11 March [1862–9][1]

Down Bromley Kent
March 11th.

Dear Sir

You are most entirely welcome to insert anything from any of my works in your Magazine. I shd. think my Journal would give better extracts on the habits of animals than the "Origin".[2]

Dear Sir | In Haste | your faithful servant | C. Darwin

Karpeles Manuscript Library Museums

1 The date is established by the form of the address, which CD used from May 1861 to April 1869. The recipient has not been identified.
2 CD refers to *Journal of researches* and *Origin.*

To ? 29 March [1862–9][1]

Down. | Bromley. | Kent. S.E.
Mar 29

Dear Sir

I am extremely sorry that you have had so much trouble in writing to me at such length. The intended publication seems highly promising & valuable & I am sure I wish it all success.[2] But the state of my health conjoined with the quantity of matter which I wish to publish renders it impossible that I sh[d] be a contributor; and this answer I have been compelled unwillingly to make on previous occasions.

I should be proud to be a fellow labourer, but as there is no chance of my writing in the intended Periodical, I do not think it w[d] be right that my name sh[d] appear.

I beg leave to repeat my regret for the trouble which I have inadvertently caused you, & with my best wishes for the success of your undertaking, I remain | Dear Sir | Yours very faithfully | Charles Darwin

LS
Wellcome Library (Wellcome MS.7781/8)

1 The date range is established by the form of address, which is one that was used by CD between May 1861 and April 1869.
2 The periodical has not been identified. CD was often asked to contribute or lend his name to new periodicals during this period, but routinely refused such requests (see, for example, *Correspondence* vol. 15, letter to Charles Kingsley, 30 April [1867]).

To Joseph Bullar 27 June [1862][1]

Down. | Bromley. | Kent. S.E.
June 27th

M[r] Darwin presents his compliments to D[r]. Bullar & is very much obliged to him for so kindly sending him the orchid flowers with Diptera—

M[r] Darwin has just made out by the aid of one of his sons, that orchis maculata is fertilised by the aid of Diptera.—[2]

American Philosophical Society (256)

1 The year is established by the relationship between this letter and CD's notes on *Orchis maculata* dated 20 June 1862 (see n. 2, below).
2 In a note dated 20 June 1862 (DAR 70: 12–14), CD recorded that his son George Howard Darwin had caught six flies (*Empis lucida*) with pollinia of *Orchis maculata* attached to their eyes, as well as one specimen of *E. pennipes* with pollinia attached to its thorax. He also noted that his son William Erasmus Darwin saw a fly remove pollinia of *O. maculata* in Southampton. In the same note CD recorded having

received three flowers of *O. latifolia* with a black fly in the nectary from a doctor in Southampton, evidently a reference to Bullar, who practised there (*Medical directory* 1862).

To ? 21 August [1862][1]

1 Carlton Terrace, Southampton.
Aug. 21

Dear Sir,

Severe illness in my family on a journey has prevented me from sooner thanking you for your very obliging note.[2] I did not receive the specimen, which followed me from place to place, until it was quite withered; but the monstrosity is a common one. I should not expect this fusion of two flowers to be inherited, but it would be worth trial; nor indeed should I expect that the flower would set seed, but both these points would be worth ascertaining. Such facts, as I have collected, lead me quite to disbelieve in "jumps of any size", but I have no objection to them on any other score.[3]

I am pleased that any of my works sh^d. have interested you & with my best thanks, I remain | Dear Sir | Your obliged & obe^t ser^t | Ch. Darwin | (of Down, Bromley Kent)

Lawrences Auctioneers, Crewkerne, Somerset, 2009[4]

1 The year is established by the reference to CD and his family's being detained in Southampton owing to illness (see n. 2, below).

2 The note has not been found. In August 1862, CD and his family set out for Bournemouth for the sake of Leonard Darwin, who had scarlet fever; however, Emma Darwin was herself taken ill with scarlet fever at Southampton, where she, CD, and Leonard remained from 13 August to 1 September (see *Correspondence* vol. 10, Appendix II). One Carlton Terrace was the home of William Erasmus Darwin, CD and Emma's eldest son (*ibid.*, letter to John Lubbock, 21 August [1862]).

3 The specimen has not been identified. The sender evidently speculated that a dramatic monstrosity could engender a new species; CD, however, did not believe in the possiblity of sudden change of this sort (see, for example, *Origin*, p. 206).

4 The text down to 'ascertaining.' is from the dealer's transcription; the rest is from an image of the letter.

To Hugh Falconer 4 October [1862][1]

Down. | Bromley. | Kent. S.E.
Oct 4th

My dear Falconer

I returned home on Tuesday evening & on Thursday I returned your M.S., with my letter, to Park Crescent, & there I have no doubt you will find it safe.—[2] I sh^d. indeed be grieved if it were lost.—

In Haste | Ever yours | C. Darwin

Maggs Bros, Autograph Letters & Historical Documents, Cat. 1345, 2003

1 The year is established by the relationship between this letter and the letter from Hugh Falconer, 3 October 1862 (*Correspondence* vol. 10).

[2] In his letter of 3 October 1862 (*Correspondence* vol. 10), Falconer asked CD whether he had received the manuscript that Falconer had sent him on 27 September 1862. The manuscript was part of a paper on fossil and living species of elephant (Falconer 1863). CD commented favourably on the manuscript in his letter to Hugh Falconer, 1 October [1862] (*Correspondence* vol. 10).

To W. A. Leighton 4 December [1862][1]

Down. | *Bromley.* | *Kent. S.E.*
Dec. 4th

My dear Sir

How extremely kind you have been, but I am ashamed to think of the trouble which I have caused you. The difference in the hair is indeed striking: I see De Jonghe just *alludes* to the patent & [adpressed] hairs in G. Ch. 1858 p. 103.[2] I grieve to find, & the more grieved after all the trouble which you have taken, that I have not knowledge enough to work out the curious fruit *[text missing]*

I had a Verbasum in my garden (which I now know to be a natural hybrid from V. thapsus & lychnit[i]s) & I wished to make some experiments on it; so I asked Mrs. Darwin to come & tie in thread the quite young flowers.[3] To do this she had with forceps to pull off buds on each side. When I came to fertilise them, I cried out 'Why have you chosen old flowers & the corollas are all dropped off?' She declared she had not & tied some more & when I came I found them also all off. So I said we will make sure this time for we will pull each flower and see the corolla is firm; so we again pulled off the buds on each side & tied the thread, *when lo & behold* off dropped all *the corollas.* It is caused by the *movement* of the *sepals.*— So I came *[illeg]* the hair & [revisited] *[illeg]* & at last *[illeg]* Sir [J. S] Smith British *[illeg]* statement on [auth] *[illeg]* [come to same] *[illeg]* to V. *pulverulentum* *[illeg]* *virgatum*,[4] which *[illeg]* plants If I do **not** *[illeg]* I *will rely that it* *[illeg]* plants *virgatum*; for *[text missing]*

My dear Sir, with cordial thanks | Yours sincerely | Ch. Darwin

Incomplete[5]
Unknown dealer

[1] The date is established by the relationship between this letter, the letter to J. D. Hooker, 6 October [1862] (*Correspondence* vol. 10), and an experimental note of CD's dated 21 June 1863 (DAR 108: 5; see *Correspondence* vol. 11, enclosure to letter from W. A. Leighton, [before 21 June 1863] and n. 5. CD discovered that the *Verbascum* that he mentions in this letter was a hybrid in 1862 (*Correspondence* vol. 10, letter to J. D. Hooker, 6 October 1862). If he had already made the experiments he mentions in his note of 21 June 1863, he would not have made the comment he makes in this letter about the movement of the sepals.

[2] In an article 'On the strawberry' by J. De Jonghe in *Gardeners' Chronicle*, 6 March 1858, p. 173, the author mentioned that different varieties of strawberry could have horizontal or ascending hairs on the leaf stalk. J. De Jonghe has not been further identified.

[3] Notes in DAR 108: 2, dated 28 June and 2 July 1862, report CD's observations regarding the dropping off of the flowers of what is described as 'Verbascum Thapsiforma?'. Emma Darwin's participation in the thread-tying is not mentioned. In a later note, dated 4 October 1862 (DAR 108: 3), CD described finding a dozen 'yellow Verbascum, like lychnitis', but sterile. He noted morphological features of the

plants similar to both *Verbascum thapsus* (common mullein) and *V. lychnitis* (white mullein) and concluded the plants were natural hybrids (see also *Correspondence* vol, 10, letter to J. D. Hooker, 6 October [1862] and n. 14).

4 CD refers to James Edward Smith's description of *Verbascum pulverulentum* (yellow hoary mullein) and *V. virgatum* (large-flowered mullein) in *Flora Britannica* (J. E. Smith 1800–4, 1: 251–2).

5 The original letter is complete and according to the sale catalogue is eight pages long. The letterhead, date, salutation, and incomplete first paragraph are transcribed from an image, as is the valediction. The second paragraph, down to 'sepals', is from the dealer's transcription, though some words at the end of this section are visible in the image. The fragmentary lines at the end of the second paragraph are transcribed from a partial image.

From E. A. Darwin [1863–6?][1]

Dear Charles

I have signed for SED[2] also, as I presume the shares stand in our joint names.[3]

I am afraid your good time has not lasted long from Henrietta's acc^{t}[4]

Yours affec. | E D

DAR 105: B34

1 The date range is conjectured by the reference to CD's health and by the date of his sister's death (see n. 2, below). CD's 'good time', which had 'not lasted long', may refer to a period preceding one of the intervals of poor health that he suffered during 1863, 1864, and 1865 (see *Correspondence* vols. 11, 12, and 13, Bowlby 1990, pp. 370–84, and Colp 1998).

2 Susan Elizabeth Darwin, CD and Erasmus's elder sister, died on 3 October 1866 (R. B. Freeman 1978).

3 The shares have not been identified but were probably related to a legacy to Susan, Erasmus, and CD from a Darwin or Wedgwood relation. CD's legacies and investments are recorded in his Investment book (Down House MS).

4 CD's daughter Henrietta Emma Darwin may have visited Erasmus's residence in London on some of her many trips to London from 1863 to 1865; Emma Darwin's diary (DAR 242) records Henrietta travelling to London at least four times in 1863, twelve times in 1864, and nine times in 1865.

From William Bowman 31 January [1863][1]

5 Clifford St
Jan 31

My dear Sir

I will with g^{t} pleasure give you any infor^{n}. you require, so far as my knowledge extends—and if you will give me a little interval—Say a week or two.—[2]

Yrs vy sincerely | W Bowman

When I have done so, you ought to come up & examine me vivâ voce, for the subject is large & branches out & w^{d} need explanation— I could give you a bed at Hampstead for a night.[3]

Yrs vy truly | WB.

Sir John Paget Bowman

[1] The year is established by the relationship between this letter and the letter to William Bowman, 10 March [1863] (this volume, Supplement).

[2] No letter from CD requesting information from Bowman has been found.

[3] Bowman resided at 5 Clifford Street, London, and North End, Hampstead (*Post Office London directory* 1865). CD may have met with Bowman during his visit to London from 4 to 14 February 1863 (*Correspondence* vol. 11, Appendix II).

From Richard Frean 17 February 1863

Plymouth
Feb 17th 1863.

(Copy)
Dr Sir

I am much interested in the subject of your book on the "Origin of Species"[1] & this must be my apology for addressing you At the time of its appearance I was a Medical Student in London, & much attached to the study of Geology, general Natural History &c, & more particularly at that time—Animal and Vegetable Morphology.[2]

It is not surprising therefore that I was obliged to accept your Hypothesis as the only satisfactory explanation of all that is so perplexing & mysterious in these latter subjects. Of course other considerations,—especially the inhabitants of the Galapagos Archipelago & other facts in the Geographical distribution of Animals & Plants contributed to this result.

The remarkable unsymmetrical structure of the Sole, Turbot & other Fishes of the Pleuronectidæ had attracted my attention & appeared a strong argument in its favor.[3]

The Teats & Lacteal glands of the Mammalia present a difficulty I cannot imagine how these can have been developed by 'Natural selection' I am anxious to know the views of others on the subject.

The degeneration of civilized man appears to be a necessary consequence of your hypothesis—if the following reasoning is correct. Man's organs of locomotion, hearing smell, taste & sight were modified & maintained by natural selection when the life of the individual & the existence of the species depended upon their efficientcy. After a time his more perfect hand & superior reasoning powers gave him ascendancy over kindred forms in the struggle for life.

Civilization ensues; the strong protect the weak, machinery is used for locomotion, for defence against natural enemies, for preparing food &c &c, & robust health & acute senses are no longer the conditions of the life of the individual & the perpetuation of the species. They must therefore necessarily decay. The consideration of cultivated plants & domesticated animals will prove that variations of scructure are greatest when the plant or animal is placed under unnatural conditions. Civilized Man is thus placed. His various organs &c. will be liable to great variations & since natural selection cannot act in preserving the good & eliminating the bad, They must all ultimately be injurious for those which at first are beneficial, will by exaggeration become so.

This process of degeneration may have a limit for it may be supposed that natural selection will act when variations of structure interfere with a prolific reproduction, yet it is difficult to see how this principle can do so when variations are large & frequent. If these deductions are correct they are of practical importance. The future object of Medical Science will be the prevention rather than the cure of disease; & this will be accomplished by subjecting the body as far as practicable to its former conditions. A more natural diet, thin clothing, a large amount of exercise &c will be some of the conditions of health & physical enjoyment. Extremes will meet for the highest civilization will approximate the savage state.

I am | Dr Sir &c | R F

C. H. Darwin Esq. M.A

Contemporary copy
Launceston Library (State Library of Tasmania): Local Studies Collection – Manuscripts: LMSS 0020 (Robert Norman Smith Diaries and correspondence)

1 *Origin*.

2 Frean later trained as a surgeon at the Middlesex Hospital, London, and qualified in 1871, but no record of his earlier medical studies has been found.

3 Pleuronectidae is the family of dabs, halibuts, and righteye flounders. During development the left eye migrates to the right side of the head, allowing the fish to lie on the ocean bottom on its left side, with both eyes facing up.

To William Bowman 10 March [1863][1]

Down. | *Bromley.* | *Kent. S.E.*
March 10th

My dear M^r. Bowman

I thank you heartily for the trouble which you have taken. I will venture to give you a little, & but a little, more trouble by sending you at most 3 or 4 pages of M.S, *well copied out*, after I have worked your remarks in.—[2]

You will thus be able to see whether I have made any blunder, & possibly some cases might occur to you worth adding; for instance any inherited deviation in form or function of the iris, or in shape of lens.—[3] The reading over my few pages cannot cause you much trouble & you could strike out anything with a pencil—

With sincere thanks | Yours very sincerely | Ch. Darwin

I shall not get the M.S. ready for a week or so,—

I sh^d. be glad of any case of ill-growth of eye-lashes inherited.—[4]

American Philosophical Society (B/D25.303)

1 The year is established by the relationship between this letter and the letter to James Paget, 11 March [1863] (*Correspondence* vol. 11; see n. 4, below).

2 See this volume, Supplement, letter from William Bowman, 31 January [1863]. Bowman had suggested meeting CD to talk about a subject in which CD was interested. Bowman's remarks have not been found in any letter, but may have been given to CD in person while he was visiting London from 4 to 14 February 1863. CD included information from Bowman on inherited imperfections of the eye in *Variation* 2: 8–10.

[3] There is no specific mention of inherited deviations in the iris or the shape of the lens in the information supplied by Bowman in *Variation*; the particular conditions on which CD cited Bowman were hypermetropia, myopia, squinting, and cataract (*Variation* 2: 8–9).

[4] CD asked the same question of James Paget (see *Correspondence* vol. 11, letter to James Paget, 11 March [1863]). Information from Paget on the inheritance of long hairs in eyebrows, but not eyelashes, is given in *Variation* 2: 8.

To W. H. Flower 25 May [1863][1]

Down. | Bromley. | Kent. S.E.
May 25th

My dear Sir

As a cheque will probably be less troublesome than a P. order I enclose one with very many thanks, for the Photographs.—[2]

M. Quatrefages has received them & is much pleased & surprised at the appearance of the Skull.—[3]

With very sincere thanks for your great kindness, believe me | yours sincerely | C. Darwin

R & R Enterprises Autograph Auctions, November 2006

[1] The year is established by the relationship between this letter and the letter from W. H. Flower, 9 May 1863 (*Correspondence* vol. 11).

[2] According to CD's Classed account book (Down House MS), CD made a payment of 17*s.* for the photographs on 25 May 1863.

[3] Flower had sent CD photographs of the niata ox skull at the Hunterian Museum and CD had sent these to Armand de Quatrefages (see *Correspondence* vol. 11, letter from W. H. Flower, 9 May 1863, and letter to Armand de Quatrefages, 14 May [1863]). Quatrefages acknowledged receipt of the photographs in his letter of 19 May [1863] (*ibid.*).

To Lydia Ernestine Becker [26 or 27 May 1863][1]

The dark purple anthers are a mass of some Cryptogamic plant, allied, I suppose, to the smut of Wheat. In the bud the pollen grains can be distinguished, afterwards they are wholly corrupted.[2] There remains a pretty case of a reversion from a diœcious to a hermaphrodite condition.[3]

Incomplete
Journal of Botany 7: 291–2

[1] The date is established by the relationship between this letter and the letters from L. E. Becker, 23–4 May [1863] and 28 May [1863] (*Correspondence* vol. 11). In 1863, 24 May was a Sunday; Becker must have posted her letter of 23–4 May on 25 May. The letter would have reached CD by 26 May at the earliest.

[2] See *Correspondence* vol. 11, letter from L. E. Becker, 23–4 May [1863] and nn. 5 and 6. Becker had sent specimens of flowers of *Lychnis dioica* (now *Silene dioica*, red campion) with large dark purple anthers rather than the usual small yellow anthers. The anther-smut fungus that affects mostly female flowers has now been identified as *Microbotryum violaceum*; discoloration of the pollen grains is not evident during the earliest stages of development (see Uchida *et al.* 2003, p. 244).

[3] In *Cross and self fertilisation*, pp. 410–11, CD used the rudimentary stamens and pistils found, respectively, in the female and male plants of this species to argue that some dioecious plants evolved from hermaphrodite species.

To Arthur Rawson 6 June [1863][1]

Down. | Bromley. | Kent. S.E.
June 6

Dear Sir

I am very much obliged to you for your kindness in sending me the Cypripedium. It is a plant which I have never seen & therefore have been glad to examine it.[2]

The contrivance by which insects, after they have entered the labellum by the mouth, are forced to crawl out by one of the small lateral passages & thus get smeared with the viscid pollen, is very curious; it is exactly the same principle on which traps are made to catch insects in kitchen, namely the edge of the mouth, or large opening into the labellum being bent inwards so that the insect instead of getting out falls back.[3]

With my best thanks, believe me dear Sir yours faithfully | Charles Darwin

LS
American Philosophical Society (Getz 5563)

[1] The year is established by the relationship between this letter and the letters from Arthur Rawson, 1 April [1863] and [6 April 1863] (*Correspondence* vol. 11).

[2] Rawson had loaned CD a specimen of the orchid *Cypripedium pubescens* (greater yellow lady's slipper; see *Correspondence* vol. 11, letter from Arthur Rawson, [6 April 1863] and n. 2).

[3] In *Orchids* 2d ed., p. 230, CD added a description of his experiment with *Cypripedium pubescens* and compared the working of the labellum to insect traps.

From Henry Farncombe Billinghurst [21 July 1863 – August 1870?][1]

Southampton

Sir

I beg to acknowledge the receipt of your favor of the 20th inst returning 7 transfers duly signed by yourself & Mr. G. H. Darwin[2]

I am Sir | Your obdt Servt | H F Billinghurst | Sub Country Manager

LS(A)
DAR 80: B18v

[1] The date range is conjectured on the basis of George Howard Darwin's birth date and a note by CD on the verso of the letter. George was born on 9 July 1845. The earliest he could have signed a legal document was July 1863. On the verso of this letter is a note by CD mentioning the explanation for the basis of sympathy given in Smith 1759 and Bain 1868. CD notes that neither work can account for the fact that sympathy is stronger when excited by a loved one rather than a stranger. He makes this point in *Descent* 1: 81–2 and refers to these sources. CD was working on the manuscript of *Descent* between 1868 and 1870 (*Correspondence* vols. 16, 17, and this volume). It was sent to the printers in August 1870 (this volume, letter to Mr Dorrell, 9 August 1870).

[2] The reference is to George Howard Darwin. The transfers have not been identified.

To W. B. Tegetmeier 2 February [1864][1]

Down Bromley | Kent.
Feb 2.

My dear Sir

I have found the box & will send it tomorrow night to the Hotel. I am ashamed of myself & thought it had been returned long ago.[2] Perhaps one or two skulls have been taken out & are placed with specimens to be drawn. They shall ultimately be returned, but I really have not strength now to search for them.[3] During the last 5 months I have done actually nothing but I hope I am now slowly recovering.

I am much obliged for your information about the crosses.[4] It w^{d} be no use your calling here as it would be 100 to 1 that I sh^{d} not be able to see you.[5] I am often not able to see even my own children The box will be at the Hotel on Thursday morning

I am Dear Sir | yours faithfully & sincerely | Ch. Darwin

LS(A)
New York Botanical Garden Library (Charles Finney Cox collection) (ser. 14)

[1] The year is established by the relationship between this letter and the letter from W. B. Tegetmeier, 1 February 1864 (*Correspondence* vol. 12).

[2] CD refers to a box of skulls of various poultry breeds that Tegetmeier had sent to CD in April 1861. Tegetmeier wanted to show the skulls at a meeting of the Ethnological Society and asked CD to send them to the Gordon Hotel, Covent Garden (see *Correspondence* vol. 12, letter from W. B. Tegetmeier, 1 February 1864 and nn. 2–4).

[3] Some of the skulls were retained to be engraved for *Variation* and returned in January 1867 (see *Correspondence* vol. 15, letter to W. B. Tegetmeier, 6 January [1867]).

[4] Tegetmeier had given CD an update on some poultry-crossing experiments. His results had so far shown no lessening of fertility in the progeny of crossed fowls (see *Correspondence* vol. 12, letter from W. B. Tegetmeier, 1 February 1864 and nn. 5 and 7).

[5] In his letter of 1 February 1864 (*Correspondence* vol. 12), Tegetmeier had offered to collect the skulls from CD. CD was in poor health during the second half of 1863 and the early months of 1864 (see *Correspondence* vol. 12, letter to Asa Gray, 25 February [1864] and n. 2).

To William Erasmus Darwin [1 May 1864][1]

My dear W^{m}

I have seen about the transfer very many thanks for all you are doing for me.[2]

It was the anthers & not the stigma which latter are variable,[3] but I am now nearly sure about the anthers & almost about the flowers.

But I am greatly interested in this case. Look at Broomfield about Rhamnus if you come across any flowers of the 2 sexes I sh[d] much like to see them[4]

yours affectly | Ch Darwin

I C.D am very well today I have no doubt that the long sty. have largest flowers at least in kitch gar.[5]

Frank & Ruck Lenny & another appeared from Clapham about 2.30 yesterday having walked from school, in 2 parties after the manner of boys.[6]

We stuffed them & croquetted them & sent them home & it did very well.— Alfred is still here & very harmless.[7]

L(S)
DAR 210.6: 122

1 The date is established by the reference to the visit of the boys from school (see n. 6, below).

2 The transfer has not been identified.

3 CD refers to variation in the anthers in different forms of *Pulmonaria angustifolia* (blue cowslip); in his letter of 3 May [1864] (*Correspondence* vol. 12), CD sent flowers of long-styled and short-styled *P. angustifolia* and asked William to outline the size of the anthers of the two forms.

4 CD refers to *Flora Vectensis: being a systematic description of the phænogamous or flowering plants and ferns indigenous to the Isle of Wight* (Bromfield 1856). William later found flowers of both sexes of *Rhamnus catharticus* (now *R. cathartica*, buckthorn) growing on the Isle of Wight; most of CD's work with this species was done in 1866 (see *Correspondence* vol. 14). CD also knew of dimorphic American species of *Rhamnus* (see *Correspondence* vol. 9, letter from Asa Gray, 11 October 1861). CD evidently saw similarities between *Rhamnus* and *Pulmonaria.* In *Forms of flowers*, p. 287, he suggested that from the number of bad pollen grains in the small anthers of the long-styled form of *P. angustifolia*, it appeared that the form was tending to become female.

5 CD refers to plants of *Pulmonaria angustifolia.* He had received seedlings collected on the Isle of Wight by William in 1863 (see *Correspondence* vol. 11, letter to W. E. Darwin, [25 July 1863] and nn. 2 and 3); a note dated 6 May 1864 in DAR 110: A51 begins, 'Pulmonaria I. of Wight in kitchen garden'. In *Forms of flowers* p. 105, CD noted William's observation concerning *P. angustifolia*, that the 'corolla, though variable in size, was generally larger in the long-styled flowers'.

6 Francis Darwin, Arthur Ashley Ruck, and Leonard Darwin were pupils at Clapham Grammar School. Ruck was a 'great friend' of Francis's at school (B. Darwin 1955, p. 100). Emma Darwin wrote in her diary (DAR 242) for 30 April 1864, 'Children from school & Alfred' (see n. 7, below).

7 CD refers to Alfred Allen Wedgwood.

To William Lonsdale 6 May [1864][1]

Down, Bromley, | Kent;
May 6th.

My Dear Lonsdale,

I received your letter with as much surprise as from one dead;[2] for it so happened that 3 or 4 nights ago I was thinking about you and I saw you as plainly as in the old days in your little room at the Geolog. Soc.[3] I was thinking how I could learn any news of you. I am sorry at the poor account that you give, and I know how long and how much you have suffered.[4] I had quite forgotten about the coral, but

I now remember its appearance. Your MS. is arrived and I am fairly astounded at the labour you have bestowed on the subject. It seems a very great pity that such labour should be wasted. Had I not better send the specimen and MS. to the Geolog. Soc. to be printed or kept in the archives?[5] It might be of extreme use to any one working on the subject.

I have myself been ill for the last 9 months, but am slowly recovering and hope still to do a little work in Nat. History.[6]

Believe me, my dear Lonsdale, I shall ever remember your uniform kindness to me in old long past days, and our many pleasant conversations.[7]

I remain yours | Very sincerely, | Charles Darwin.

Murch 1893, pp. 436–7

1 The year is established by the reference to CD's long illness (see n. 6, below).

2 The letter from Lonsdale has not been found.

3 Lonsdale had been curator and librarian, then assistant secretary and librarian, of the Geological Society of London from 1829 to 1842 (*DNB*). He had described CD's fossil coral specimens from Tasmania, collected during the *Beagle* voyage (see *Volcanic islands*, pp. 161–9).

4 Lonsdale had resigned his post at the Geological Society in 1842 because of poor health (*DNB*). Upon his resignation, CD had contributed to a subscription fund to assist Lonsdale in his work on fossil corals (*Correspondence* vol. 2, letter to A. S. Horner, [4 October 1842] and n. 4, and letter to Charles Lyell, [5 and 7 October 1842]).

5 The specimen and manuscript have not been found; however, in a letter to Peter Martin Duncan of 19 September 1876 (*Calendar* no. 10608), CD wrote: 'On turning out a pile of rubbish in one of my rooms I found a parcel with a fossil coral & a long M.S account (beautifully written) by Lonsdale. I believe that this arrived during one of my long illnesses, anyhow it was completely forgotten'. CD said that he would send the specimen and manuscript to Duncan, who was then president of the Geological Society.

6 CD had been seriously ill for the last five months of 1863 and the first four months of 1864; his symptoms including persistent vomiting (see *Correspondence* vols. 11 and 12).

7 CD had been introduced to Lonsdale by Adam Sedgwick before his departure on the *Beagle* voyage (see *Correspondence* vol. 1, letter from Adam Sedgwick, 18 September 1831); he later remarked on the 'most cordial reception' given him by Lonsdale (see *Correspondence* vol. 1, letter to J. S. Henslow, [30–1 October 1836]). CD had considered Lonsdale as a possible editor of his 1844 species sketch, in the event of his death (see *Correspondence* vol. 3, letter to Emma Darwin, 5 July 1844, n. 5).

To W. E. Darwin [19 May 1864][1]

Thursday

My dear W^m^

Y^r^ observation on the 1 big anther with its differently sized pollen grains is a quite new & most extraordinary discovery.[2] I remember distinctly seeing one anther much larger than all the others; but I never dreamed of the pollen-grains differing.

Never mind whether it kills you or not search more flowers & measure pollen & keep memorandum how many times you succeed for I shall lay great stress on this to me most unexpected discovery.

I return your letter (which keep most safely) & it may be of service to you, but more especially because I think you have written one, or 2 words wrong which I have pencilled for correction.[3]

Never mind Menianthes Rhamnus[4] or your life but work this out. Goodbye | what weather | C. D.

I wish you many ever get the little mare.[5] She does beautifully—

L(S)
DAR 210.6: 186

1 The date is established by the relationship between this letter and the letters from W. E. Darwin, 18 May [1864] and [20 May 1864] (*Correspondence* vol. 12).

2 See *Correspondence* vol. 12, letter from W. E. Darwin, 18 May [1864]. William had included sketches of the anthers of long-styled and short-styled flowers of *Pulmonaria angustifolia* (blue cowslip); see also letter from W. E. Darwin, [20 May 1864] and n. 3.

3 For CD's annotations, see *Correspondence* vol. 12, letter from W. E. Darwin, 18 May [1864].

4 For CD's query regarding *Menyanthes* (buckbean or marsh trefoil), see the letter from Emma Darwin to W. E. Darwin, [17 May 1864] (*Correspondence* vol. 12). For CD's interest in *Rhamnus* (buckthorn), see *ibid.*, letter from H. E. Darwin to W. E. Darwin, [18 May 1864] and n. 2. Most of CD's work on *Rhamnus* was done in 1866 (see *Correspondence* vol. 14).

5 The mare was evidently a horse that William planned to buy, but which was kept at Down for some time (see *Correspondence* vol. 12, letter from H. E. Darwin to W. E. Darwin, [18 May 1864]).

To ? 1 August [1864 or 1865][1]

Down. | *Bromley.* | *Kent. S.E.*
Augt 1st

My dear Sir

I enclose with much pleasure my photograph made by my son, for I have no other.[2] I shall be very glad to receive yours, if you will be so good as to send me one

My dear sir | Yours sincerely | Charles Darwin

Muséum National d'Histoire Naturelle: Bibliothèque Cryptogamie (Ms 493 fol. 637)

1 The year is established by the reference to the photograph (see n. 2, below).

2 CD refers to a photograph taken by his eldest son, William Erasmus Darwin in the spring of 1864 (see *Correspondence* vol. 12, letter from W. E. Darwin, [19 May 1864] and n. 8). It is reproduced as the frontispiece to *Correspondence* vol. 12. CD's next photograph was probably taken in November 1865 (see *Correspondence* vol. 13, letter from E. A. Darwin to Emma Darwin 25 [November 1865]).

From Francis Trevelyan Buckland [1865–6][1]

Dear–Mr Darwin.

I was most pleased that you were good enough to speak to me the other day at the museum—[2] ⟨ ⟩ favour.

Please to—sign the enclosed and return it to me as above—I am very glad that our Oxford Chancellor joins the Society.[3]

Yours most obliged | F Buckland—

I should *so like* to show you my museum you have not touched oysters in your book[4] I could show you some curious things thus

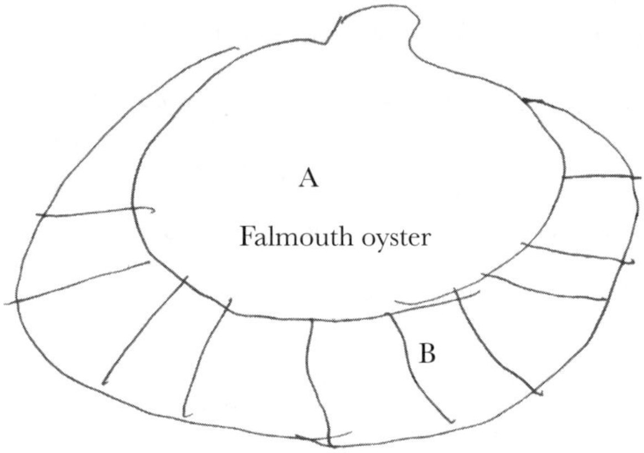

A. Falmouth growth
B. French growth—quite different
Yours ever | F Buckland

DAR 160: 364

1 The date range is established by the reference to the Society for the Acclimatisation of Animals, Birds, Fishes, Insects, and Vegetables within the United Kingdom and by the reference to oyster culture and Buckland's museum (see nn. 3 and 4, below).
2 Buckland probably refers to the British Museum, which CD frequently visited when in London.
3 Buckland refers to the Society for the Acclimatisation of Animals, Birds, Fishes, Insects, and Vegetables within the United Kingdom. The society was founded by Buckland in 1860 and came to an end in 1866 (Ritvo 1990, p. 239). The chancellor of the University of Oxford was Edward George Geoffrey Smith Stanley. No record of his membership of the society has been found, but his son Edward Henry Stanley had been a subscriber since 1862 and was vice-president from 1863 to 1865 (*Annual Report of the Society for the Acclimatisation of Animals, Birds, Fishes, Insects, and Vegetables within the United Kingdom* 1865). The enclosure has not been found.
4 Buckland refers to the South Kensington Museum. After his appointment as scientific referee to the museum in May 1865, Buckland moved his collection illustrating fish and oyster culture there (Bompas 1885, p. 153). Buckland also refers to *Origin*. In *Variation* 2: 280, CD cited Buckland for information on regional variation in oysters.

To Richard Kippist 18 January [1865][1]

Down. | Bromley. | Kent. S.E.
Jan 18th.

My dear Sir

I send by the carrier tonight a long paper for the Society.[2] Will you have the kindness to send me one line to say that it has been received. I will write tomorrow—the post is going.[3]

My dear Sir | Yours very faithfully | Ch. Darwin

LS
Alexander Autographs 2008

[1] A description of this letter was published in *Correspondence* vol. 13. The year is established by the reference to the 'long paper' sent to the Linnean Society (see n. 2, below), and by the relationship between this letter and the letter to J. D. Hooker, 19 January [1865] (*Correspondence* vol. 13).

[2] The reference is to 'Climbing plants'. In the letter to J. D. Hooker, 7 January [1865] (*Correspondence* vol. 13), CD commented on the great length of the manuscript of 'Climbing plants', which he was about to send to the Linnean Society for publication. Richard Kippist was librarian of the Linnean Society (*ODNB*).

[3] Kippist's reply and CD's second letter have not been found.

To John Murray 14 April [1865][1]

Down Bromley Kent
Ap. 14th

My dear Sir

Many thanks for the advertisement.— I am much pleased that you are content with the title.—[2]

Yours sincerely | C. Darwin

John Wilson Manuscripts

[1] The year is established by the relationship between this letter, the letter from John Murray, 1 April 1865, and the letter to John Murray, 4 April [1865] (*Correspondence* vol. 13).

[2] Murray advertised CD's forthcoming book, *Variation*, in the *Reader*, 15 April 1865, p. 427. The notice is under the heading 'Mr. Murray's quarterly list of works in the press', and the title given is 'Domesticated Animals and Cultivated Plants, or the Principles of Variation, Inheritance, Re-union, Crossing, Interbreeding, and Selection under Domestication'. The word 're-union' was a misprint for 'reversion' (see *Correspondence* vol. 13, letter to John Murray, 2 June [1865]). *Variation* was not published until 1868.

From W. D. Fox [before 1 March 1866][1]

[2 pages missing]

Do you know anything ⟨about⟩ Woking. A *grand* man—a Dr Brushfield[2]—wants to become the Superintendant of a new Asylum there. He is a perfect man for the situation & I shd much like to see him there.

⟨*1⅔ pages severely damaged*⟩

AL incomplete
DAR 164: 205

[1] The date is established by the date of appointment of the superintendent of Brookwood Asylum (see n. 2, below).

[2] Thomas Nadauld Brushfield was appointed superintendent of Brookwood Asylum, Woking, on 1 March 1866.

To Robert McLachlan 23 March 1866

Down, Bromley, Kent,
March 23, 1866[1]

I am very much obliged to you for your paper on Sterrha, which I have read with very great interest.[2] Many of your observations are quite new to me. I was particularly glad to read about the varying organs in the Neuroptera;[3] Should you ascertain similar cases in the Trichoptera,[4] I shd. be grateful for information. Would it not be well worthwhile to get some of your friends to breed Eupithecia from eggs of the same female & to feed them on different plants & thus ascertain positively how the mimicry is affected.[5] I have been very glad to see (whether or not you have been influenced by my writings) that you have given up to a great extent the belief in the immutability of species;[6] & I feel sure that as you attend to other subjects besides the discrimination of species you will ultimately go further in your belief; at least I have hitherto found this to be the case with those who have doubted to a limited degree.

LS incomplete[7]
Profiles in History (March 2006)

1 The address and date are given in the sale catalogue description.

2 CD refers to McLachlan's paper 'On some remarkable varieties of *Sterrha sacraria* Linn., with general notes on variation in Lepidoptera' (McLachlan 1865). *Sterrha sacraria* is now *Rhodometra sacraria*, the vestal. It belongs to the family Geometridae.

3 See McLachlan 1865, p. 467. The order Neuroptera is now restricted to lacewings and their kin, but used to include dragonflies and damselflies, which are now in the order Odonata (see also n. 6, below).

4 Trichoptera is the order of caddisflies.

5 *Eupithecia* is a genus of the family Geometridae; it includes most of the moths commonly known as pugs. McLachlan had observed that, among Lepidoptera, while the larva in this genus showed the greatest tendency to variation, the imago was most constant. He suggested that the larval variation was related to a tendency of the larva to mimic the colour of the food plant on which it developed (see McLachlan 1865, p. 465).

6 McLachlan had referred to transitional species of *Gomphus* (club-tailed dragonflies, then in the order Neuroptera) as evidence for development theory (McLachlan 1865, p. 467).

7 The original letter is complete and is described in the sale catalogue as being four pages long. Probably only the salutation and valediction are missing.

To J. V. Carus 26 December [1866][1]

Down. | Bromley. | Kent. S.E.
Dec 26

My dear Sir

Owing to Dr Günther's recent observations will you be so kind as to alter the beginning of the paragraph at page 417 as follows—[2]

"Turning to the sea we find the same law. The marine Faunas on the Eastern & western shores of Southern & Central America are very distinct, with hardly a single shell, Crustacean, or other animal in common, except as lately shewn by Dr Günther, some fish; yet these great" &c

I hope you are making good progress in your laborious undertaking.

Pray believe me | My dear Sir | yours sincerely | Charles Darwin

LS

Staatsbibliothek zu Berlin – Preußischer Kulturbesitz, Slg. Darmstaedter Lc 1859 Darwin, Charles (Carus Nr. 30)

[1] The year is established by the reference to Carus's translation of *Origin* 4th ed. (see n. 2, below).

[2] Carus was translating *Origin* 4th ed. into German. CD had just heard of Albert Günther's paper reporting that about a third of fish species were identical on either side of the Isthmus of Panama (Günther 1864–6; see *Correspondence* vol. 14, letter from J. D. Hooker, 14 December 1866 and n. 10). The change appears in the third German edition (Bronn and Carus trans. 1867, p. 417). CD made a similar change to *Origin* 5th ed., p. 424.

From Ellen Frances Lubbock [1867–8?][1]

Dear Mr. Darwin

When you spoke of studying *expression*, I fancied you meant that of feature— It occurs to me that you may allude to forms of speech: but however, in case my first idea was right, I have turned down a page of "Adam Bede" which may interest you, if you had forgotten it—[2] It is painful to feel that one can only be referred to as a student of works of fiction— but after a dose of Relationships, one requires a "halfpenny worth of bread to all that sack".[3] And at any rate, the line I have taken in literature is thoroughly novel.

It was such a pleasure to see you & Mrs. Darwin this morning. If you knew *how* great, I think you would come oftener.

Believe me yours most truly | Ellen Lubbock

Page 132.[4] I have put a cross.

DAR 104: 227–228

[1] The date range is conjectured from the reference to CD studying expression and by the quotation (see nn. 2 and 4, below). Most of CD's research on expression in children was carried out from 1867 to 1868.

[2] The reference is to *Adam Bede* (Eliot 1859). The book went through several editions during Eliot's lifetime (for more on the editions, see Eliot 2001, pp. liv–lxxxvii).

[3] The quotation is from Shakespeare's *Henry IV* Part 1, 2.4.

[4] The quotation referred to is probably 'Totty showed her teeth and wrinkled her nose in great glee', which occurs on page 131 of the first volume in most editions, near the end of chapter 7. The character 'Totty' is a young child in Eliot 1859.

From Edward Blyth [after 24 February 1867][1]

years ago, and ⟨$\frac{1}{2}$ *line excised*⟩ remarks were quite forgotten by myself.[2] I have not much now to comment upon them. I remember that Gould assured me about the difference of sexes in the common Gallinule, saying that the female is both larger and handsomer—colouring; but a much more decided instance[3] *[missing section of unknown length]*

The fact stated by Gould of the extraordinary development of the trachea in the female only of *Rhynchea australis*, is the more remarkable that nothing of the kind occurs in the female of *R. bengalensis*; yet the birds are scarcely distinguishable externally, except that the Australian species has shorter toes.[4] *[missing section of unknown length]*

though more slowly in the females, the only difference being that the fully mature females are duller-coloured on the back.

In a remarkable hornbill (*Aceros nipalensis*) inhabiting the S.E. Himalaya and the mountains of Burma, the female (or what from analogy I presume to be such) has wholly black plumage with white tail-tip, whereas the (presumed) male has the head, neck, and lower-parts bright ferruginous.[5] Now in a nestling bird of this species in the Calcutta museum, the colouring is that of the (presumed) male, and possibly the sexes may be thus distinguished from the first. They are slightly so in the British blackbird and stonechat;[6] and in a parrakeet very common in Burma (*Palæornis javanicus*), the young from the first are conspicuously distinguishable by the male having the upper mandible coral-red, while in the female it is black; this continues for the first year at least, after the adult plumage has been attained, but finally the upper mandible of the female becomes red as in the male, and I have witnessed the change in a caged bird, besides that I have seen many specimens with the bill in process of change.[7]

By *Ardeola* I meant *Ardetta*, which is the name now given to the little bittern of this country and its immediate congeners.[8] In *Ardeola*, of which the European squacco heron (*A. ralloides*) is characteristic, there is very great difference in the summer and winter plumage, the latter being so similar in some of the races that they cannot be told apart, whilst in summer dress they are strongly distinguished.[9]

The seasonal change in the plovers and sandpipers is completed in India before the several species leave the country for their breeding haunts. We really know very little of the southern

AL incomplete
DAR 84.1: 105–6

CD ANNOTATIONS
1.1 years … instance 1.4] *crossed blue crayon*
2.1 trachea in the female 2.2] *underl blue crayon*
3.1 though … on the back. 3.2] 'P' *top of page, blue crayon* '(Case of Parrots Bill)' *end of para, pencil*
3.1 though … process of change. 4.12] *crossed pencil*
3.7 (*Palæomis javanicus*),] 'P' *in margin, blue crayon*
4.12 change.] '3' *circled blue crayon, end of para*
6.1 The … southern 6.3] *crossed pencil*
6.2 We … southern 6.3] *crossed blue crayon*

1 The date is established by the relationship between this letter and the letter from Edward Blyth, 24 February 1867 (*Correspondence* vol. 15).

2 Blyth refers to two pages of galley proofs of a translation of Georges Cuvier's *Règne animal* with footnotes by Blyth (Cuvier 1840 and 1849, pp. 158–9; these are pages 146 and 147 in Cuvier 1859). The pages,

with annotations by both CD and Blyth, are in DAR 84.1: 179–80. When CD asked some questions about Blyth's editorial remarks on these pages, Blyth asked CD to send the pages because he could not remember them (see *Correspondence* vol. 15, letter to Edward Blyth, 23 February [1867] and letter from Edward Blyth, 24 February 1867).

3 In his footnote, Blyth had referred to the 'typical state' of plumage in birds, which was usually only attained by adult males, but in the common gallinule, it was attained only by the adult female (Cuvier 1840 and 1849, p. 158). Blyth refers to John Gould.

4 *Rhynchaea australis* and *R. bengalensis* are now the sub-species *australis* and *benghalensis* of *Rostratula benghalensis*, the greater painted snipe. Tracheal elongation in females is described by John Gould in Gould 1865, 2: 275. CD scored the passage on the trachea in his copy of Gould 1865 (see *Marginalia* 1: 340). In *Descent* 2: 202–3, CD cited Blyth and Gould on the sexual differences in the trachea of these birds.

5 *Aceros nipalensis* is the rufous-necked hornbill.

6 In *Descent* 2: 219 and 220 n. 49, CD referred to the male and female young resembling, respectively, the male and female adults of the blackbird, '*Turdus merula*', and the stonechat, '*Saxicola rubicola*'.

7 *Palaeornis javanicus* is now *Psittacula alexandri fasciata*, a subspecies of the red-breasted parakeet. In *Descent* 2: 179, CD cited Blyth's description of the differences in the colour of male and female upper mandibles. Modern descriptions differ from Blyth's (see Juniper and Parr 1998, p. 414).

8 *Ardetta minuta*, mentioned on p. 158 of the galley proof (see n. 2, above), is now *Ixobrychus minutus*, the little bittern. CD crossed out 'Ardeola' and added 'Ardetta' in the margin of p. 158 of his copy of the galley proof. CD cited Blyth's notes to Cuvier 1840 and 1849 in *Descent* 2: 180 n. 29.

9 CD discussed the summer and winter plumage of several species of *Ardeola* (pond herons) in *Descent* 2: 190.

From Henrietta Anne Huxley to Emma Darwin 22 March [1867][1]

26 Abbey Place | St John's Wood
22 March.

Dear M[rs] Darwin

Yet another query about bedsteads & bedding— Will it not be more convenient to have no foot board or foot rails to the bedstead if it has a somnier portatif[2] two mattresses— Surely the bed is more easily made & looks better— which have you?— Am I a plague?

I have just got a new nurse—& today she has refused to wheel the perambulator—[3] Is it not provoking. I have not yet got an under nurse—

Baby has not solved the riddle for M[r] Darwin but thus much I noted the other morning when I entered the nursery— She was anxious to come to me yet did not cry only pushed away her nurse— At the same time the corner of her right eyebrow was very elevated.—but there was no drawing down of her mouth corners—[4] Is this observation good for anything?

I am very glad that Horace is better—[5] it is so sad to have one's children ill— It is hard—writing with Baby upon one's lap—& I had better end

With love to all | I am | Yours very affectionately | Nettie Huxley

DAR 166: 287

CD ANNOTATIONS

1.1 Yet . . . nurse— 2.2] *crossed pencil*
3.1 Baby . . . anything? 3.5] 'about 5 or 6 months old'[6] *added pencil*
4.1 I am . . . end 4.2] *crossed pencil*

[1] The year is established by the age of the Huxleys' youngest child, and by the reference to Horace Darwin's illness (see nn. 4 and 5, below).
[2] Sommier portatif: portable spring mattress (French).
[3] The nurse has not been identified.
[4] Ethel Gladys Huxley, the Huxleys' youngest child, was born in May 1866 (Clark 1968). CD was collecting reports of observations of facial expressions in infants for *Expression.*
[5] According to Emma Darwin's diary (DAR 242), on 16 March 1867 Horace Darwin had returned home from school with a fever.
[6] In March 1867, Ethel would have been about ten months old.

From John Murray 2 April [1867]

Albemarle St
Ap 2

My Dear Sir

Will you have the kindness to tell me whether Mr Schweitzerbarth be the real Simon Pure to whom I am to entrust the Clichés of your new work on Domestication[1] I shall of course make him pay some thing above the cost of the mere Electros[2] to go towards diminishing our expenses in the Woodcuts

I remain Dear Sir | Yours very faithfully | John Murray

Chas Darwin Esq

[Enclosure][3]

E. Schweizerbart, | *Verlagsbuchhandlung.* | Stuttgart
29 März | 1867

Herrn John Murray in London.

Wie ich von Herrn Ch. Darwin höre ist dessen neues Werk: The Variation of Animals & Plants usw bereits unter der Presse und da derselbe mir gestattet auch von diesem Büche eine deutsche Ausgabe zu veranstalten, so wende ich mich hiermit an Sie mit der Bitte mir von den Holzschnitten dazu *güte* Clichés machen zu lassen. Bei den Orchideen[4] waren die Herren Williams & Norgate so gefällig das Paket nach Leipzig zu senden, wo selbst ich diesen auch die Kosten für die Clichés bezalte; vielleicht Könnte es wieder so gehalten werden.

Ich danke Ihnen zum Voraus für die gütige Besorgung der Sache und zeichen mit collegialischer Hochachtung | E Schweizerbart[5]

National Library of Scotland (John Murray Archive) (Ms. 42153 ff. 30–1)

[1] Murray refers to Christian Friedrich Schweizerbart and to *Variation.* 'Simon Pure': the real man, the genuine article. The reference is to a character in *A bold stroke for a wife* ([Centlivre] 1718).
[2] Electrotypes.
[3] For a translation of the enclosure to the letter, see Appendix I.
[4] Schweizerbart's firm had published a German translation of *Orchids* (Bronn trans. 1862).
[5] In his business communications, C. F. Schweizerbart continued to use the signature of his uncle, Wilhelm Emanuel Schweizerbart, from whom he had purchased the publishing firm in 1841.

To Francis Darwin 30 August [1867–70][1]

Down
Aug 30

My dear Frank

Will you look over the enclosed Latin sentence & see that it is correct.[2] If any of your Classical friends are up, perhaps you c^{d}. get one to glance over it.—[3] Please return it tolerably soon.

In Haste Yours affect^{ly}. | C. Darwin

DAR 271.3: (1 of 16)

[1] The date range is established by the reference to Francis Darwin's being at Cambridge (see n. 3, below).

[2] The enclosure has not been found. It may have been the text of *Descent* 2: 345, n. 53, an account of steatopygia that CD had translated into Latin in order to disguise content considered indelicate (see letter from John Murray, 10 October [1870] and n. 3).

[3] Francis went up to Cambridge on 12 October 1866 and was admitted as a student at Trinity College, Cambridge on 29 September 1866; he graduated in December 1870 (Emma Darwin's diary (DAR 242); *Alum. Cantab.*; *Cambridge University calendar*).

To W. B. Dawkins 30 [August 1867][1]

Down. | Bromley. | Kent. S.E.
30th

My dear Sir

Excuse me for troubling you with one line as a suggestion, viz to collect hereafter all the facts on the changes in the skulls or skeletons of animals long confined, *but not generated* in menageries, & publish a short separate paper on subject.—[2] I am pretty well read up on such subjects, & I believe the fact to be as new as it is surprising.—

In Haste | My dear Sir | Yours very faithfully | C. Darwin

Harvey Cushing / John Hay Whitney Medical Library, Yale University

[1] The date is established by the relationship between this letter and the letters from W. B. Dawkins, 27 August 1867 and 7 September 1867 (*Correspondence* vol. 15).

[2] In his letter of 27 August 1867 (*Correspondence* vol. 15), Dawkins discussed the effect of captivity in modifying skull shape in hyaenas and remarked that he found the bones of menagerie animals useless for comparison. In his reply to this letter, Dawkins agreed to take up CD's suggestion, but no paper on the subject by Dawkins has been found (see *Correspondence* vol. 15, letter from W. B. Dawkins, 7 September 1867).

To John Murray [24 September – 31 October 1867][1]

Index for Mr Darwin's work on the Variation of animals & Plants under Domestication.[2]

The names of all animals & plants to be given in Index. The names of all the authors, with very brief statement of subject– There need not be very copious references to the subjects discussed, as the Heading of Chapters show this. But when the

same subject is alluded to in several places; it is highly advisable that it shd be given in Index. Thus I remember that /Peloric/ flowers or /Pelorism/ is referred to in 3 or 4 places. My Book certainly requires a good index in an eminent degree.
Ch. Darwin

Memorandum
Heritage Auction Galleries, USA, 2006

[1] The date range is established by the relationship between this letter, and the letters from John Murray, 23 September [1867] and 1 November 1867 (*Correspondence* vol. 15).
[2] The index for *Variation* was prepared by William Sweetland Dallas (see *Correspondence* vol. 15).

From John Lubbock 27 October [1867][1]

High Elms
27 Oct.

My dear Mr. Darwin

Thanks for Tylor's certificate.[2] Can you give me any information as to the exact manner in which the Fuegian harpoons (like the one you gave me sometime ago) were handled & used?[3] I should be very much obliged if you would.

Norgate has sent me word that he has only a dozen copies of Prehistoric Times left, so I must be preparing for a second Edition.[4]

Yours very sincerely | John Lubbock

C Darwin Esq

DAR 170: 59

CD ANNOTATIONS
1.2 the Fuegian harpoons] 'FitzRoys volume'[5] *above, pencil*
2.1 Norgate … left, 2.2] 'Put /Headings/' *before, pencil*
Top of letter: 'glass bottles' *pencil*

[1] The year is established by the reference to Edward Burnett Tylor's certificate (see n. 2, below).
[2] Tylor's certificate as a candidate for fellowship of the Royal Society of London is dated 19 October 1867; it was read to the Society on 21 November 1867, balloted on 4 June 1868, and resubmitted in 1871. Tylor was elected a fellow on 8 June 1871 (Certificates of Election and Candidature, GB 117, ref. no. EC/187/1/12, Royal Society of London). CD's is the first signature in the list of proposers 'from personal knowledge'.
[3] No reply from CD has been found. In 1842, CD donated two Fuegian spears to John Stevens Henslow for his 'Marquee museum', but no record of a similar gift to Lubbock has been found (see *Correspondence* vol. 2, letter to J. S. Henslow, 16 September [1842]). There is an engraving of a Fuegian bone harpoon in both the first and second editions of *Prehistoric times*, but Lubbock made no changes to the text of this section (see Lubbock 1865, p. 436, and Lubbock 1869, p. 529).
[4] Lubbock refers to Frederick Norgate of the publishing firm Williams & Norgate and to his book *Prehistoric times* (see n. 3, above).
[5] Lubbock refers to Robert FitzRoy and to *Narrative* 2: 184, in which FitzRoy describes the use of spears by Fuegians in hunting seals and porpoises. The passage is scored in CD's copy, which is in the Rare Books Department–CUL.

From Osbert Salvin [1868?][1]

[Enclosure]

Richard B. Sharpe informs me that *Carcineutes* is the one form in the whole group in which the sexes differ conspicuously. In the ♂ the back, tail and crown of head are fine blue banded with black with much red about the head. In the female the whole upper surface 'reddish-brown banded with black. Beneath white with black markings'.[2] In the closely allied genus *Dacelo* the distinguishing mark is that in the male, the tail is blue with black bars, in the female brown with bluish-black bars, so that the tail differs in a strictly analogous manner as the whole plumage in the last genus.[3]

Curious modifications are also exhibited in the genus *Ceryle*. In some neotropical species the whole lower surface of the male is red, whilst in the female the breast is crossed by a blackish-grey band.[4] In certain other species of the same section *Chloroceryle* the breast alone of the male is red whilst in the female no part is red, and the breast is crossed by an interrupted dark colored band, in some degree resembling the band in the first mentioned species.[5] , i.e The presence or absence of a pectoral band is also the distinguishing mark of the sexes of the old world sections of the same genus (*Ceryle*).[6] In fact the section *Megaceryle* of the old world is analogous to the section *Streptoceryle* of the New &. the section Ceryle of the Old World is also analogous to the section Chloroceryle of the new.[7]

DAR 84.1: 172–3

CD ANNOTATIONS
1.2 conspicuously] 'in colour' *interl after, pencil*
1.2 back ... blue banded 1.3] *underl blue crayon*
1.3 about the head] *underl blue crayon*; '& not *[illeg]*' *added above, ink*; '& partly faun-coloured below' *interl after, pencil*; 'below' *del ink*, ', when' *added ink*
1.4 Beneath] *del ink*; '& the lower part' *added ink*
1.4 white ... markings 1.5] *underl blue crayon*
1.6 the tail ... with black bars] *underl blue crayon*
1.6 bluish-black bars] *underl blue crayon*
1.8 genus.] 'Here we have cases of Birds nesting in holes, differing | Kingfisher group'[8] *added pencil*
2.2 the male is red] 'brick' *interl before* 'red' *pencil*
2.2 the breast is crossed 2.3] 'the same' *interl before* 'crossed' *pencil*
2.3 *Chloroceryle* 2.4] *parentheses added pencil*

1 The year is conjectured by the reference to Richard Bowdler Sharpe's research on kingfishers (see n. 2, below).

2 Sharpe described two species of *Carcineutes* in his monograph of the Alcedinidae (Sharpe 1868–71). These descriptions appeared in the first part of the monograph, published on 1 July 1868 (Sharpe 1868–71, p. ii). *Carcineutes pulchellus* (now *Lacedo pulchella*, the banded kingfisher) is the species described here by Salvin (see Sharpe 1868–71, pp. 251–2 and pl. 91).

3 Salvin probably refers to Sharpe's description of *Dacelo leachi* (see Sharpe 1868–71, pp. 289–90 and pl. 115). Sharpe divided the family Alcedinidae into two subfamilies; both *Carcineutes* and *Dacelo* were in his subfamily Daceloninae (see Sharpe 1868–71, pp. x–xi).

4 Salvin describes *Ceryle torquata* (now *Megaceryle torquata*, the ringed kingfisher; Sharpe 1868–71, pp. 73–5 and pl. 24).

[5] The reference is to *Ceryle americana* (now *Chloroceryle americana*, the green kingfisher; Sharpe 1868–71, pp. 89–90 and pl. 28).

[6] Salvin probably refers to the African *Ceryle maxima* (now *Megaceryle maxima*, the giant kingfisher; Sharpe 1868–71, pp. 67–9 and pl. 22).

[7] Sharpe did not divide these genera in the way Salvin indicates; rather he placed both old and new world species within *Ceryle* and rejected many of the genera proposed by other authors, retaining only those that he believed were characterised by marked structural differences (Sharpe 1868–71, pp. viii–ix).

[8] CD refers to the fact that although the sexes are dimorphic in these species of kingfisher, they are all birds that build concealed nests. Alfred Russel Wallace had argued in a paper on nidification (Wallace 1868) that differences in male and female plumage often related to the type of nest built. Species with open nests tended to have females with subdued plumage compared to the males, whereas in species with concealed nests (like kingfishers), females had little need for protective coloration. In a letter to Wallace of 5 May [1868] (*Correspondence* vol. 16), CD indicated that he did not entirely accept Wallace's generalisations on bird's nests. In *Descent* 2: 173–4, CD cited Sharpe's description of *Carcineutes pulchellus* as an exception to Wallace's rule.

From Hensleigh Wedgwood [1868–70?][1]

Dear Charles

If I understand the argument in favour of an early civilisation from the complex nature of the rudest languages of the rudest people it is that people in a barbarous state could never have formed such a complex language.[2]

Perhaps the best answer to this assertion would be to point out that language was never formed designedly[3] It grew spontaneously out of the efforts made by the speaker to make his hearer understand, and so every modification of it would arise from the slurring over of what was once more fully expressed. Every change would be an instinctive, not a meditated operation and therefore there is no reason why it should not have been carried to any extent by those who had no thought of their modes of expression at all Practically we see that people in the rudest stage do understand these complex modes of speech. Therefore as it appears to me there can be nothing in the uncivilised condition of life that could have hindered speech from gradually growing to such a stage

But if you suppose speech to have actually grown from the beginning the supposition of an originally civilised condition seems wholly cut away as it is impossible to conceive a civilised state antecedent to the acquisition of speech

HW.

DAR 80: 164–5

CD ANNOTATION
Top of letter: 'Rudiments' *pencil*

[1] The date range is conjectured from the content of the letter, which relates to material used in *Descent* (see n. 2, below). CD began his research for *Descent* in 1868; the manuscript was sent to the printer in August 1870 and it was published in 1871 (see 'Journal' (Appendix II) and R. B. Freeman 1977).

[2] In *Descent* 1: 61, CD discussed an argument about the complexity of language structure and its relationship to civilisation found in Wake 1868. CD scored the passage that he later quoted in his copy of Wake 1868 (see *Marginalia* 1: 832). He evidently asked Wedgwood how he would refute the argument.

[3] In *Descent* 1: 61, CD argued that language was not 'methodically formed', but did not cite Wedgwood on this point.

To A. D. Bartlett 16 February [1868][1]

Down. | *Bromley.* | *Kent. S.E.*
Feb. 16th

My dear Sir

I want to beg two favours of you.— I wish to ascertain whether the Bower-Bird discriminates colours. Will you have all the coloured worsted removed from cage & the bower, & then put, all in a row, at same distance from bower, the enclosed coloured worsted, & mark whether the Bird *at first* makes any selection.[2] Each packet contains equal quantity; the packets had better be separated & each thread put separate, but close together; perhaps it would be fairest, if the several colours were put alternately,—one thread of bright scarlet, one thread of brown, &c &c. There are six colours.— Will you have kindness to tell me whether the birds prefer one colour to another?

Secondly, I very much want several heads of the fancy & long-domesticated rabbits to measure the capacity of skull. I want only *small* kinds; such as Himalaya, small Angora, Silver Grey or any small-sized rabbit which has long been domesticated. The Silver-Grey from *warrens* would be of little use. The animals must be adult. & the smaller the breed the better.[3]

Now when any one dies would you send me the carcase *named*; if the skin is of any value it might be skinned; but it would be rather better with skin & I could make present to any keeper to whom the skin is perquisite.

This would be *great* assistance to me, if you would have kindness thus to aid me.

My dear Sir | Yours sincerely | Ch. Darwin

Address for dead Rabbits

Ch. Darwin Esqre care of Down Postman Bromley | Kent | per Rail

Wellcome Library (Wellcome Western MS. 7781/7)

[1] The year is established by the relationship between this letter and the letter from A. D. Bartlett, 9 December 1867 (*Correspondence* vol. 15, and by its connection with CD's research in 1868 on the perception of colour in birds (see n. 2, below).

[2] Though Bartlett wrote that he would try the experiment with the bowerbird (*Chlamydera*) in his letter of 9 December 1867, no further correspondence on the subject has been found, and CD does not provide results of the experiment in his discussion of the bowerbird's 'taste for the beautiful' in *Descent* 2: 112–13. See also *Descent* 1: 63, and 2: 69–71. CD's ideas regarding beauty, sexual plumage of birds, and natural selection, had recently been challenged by George Douglas Campbell, the duke of Argyll (see *Correspondence* vols. 14 and 15). For more on CD's interest in colour discrimination in birds, see *Correspondence* vol. 16, letter to J. J. Weir, 27 February [1868].

[3] CD had attended to the osteology of different breeds of rabbits from 1855 to 1865, particularly in 1856 and 1861 when he had skeletonised different rabbits (see, for example, *Correspondence* vol. 6, letter to W. D. Fox, 8 [June 1856], and *Correspondence* vol. 9, letter to W. B. Tegetmeier, 22 March [1861]); he usually asked that the skull be preserved when the rabbit was killed. See *Variation* 1: 124–9 for CD's discussion of the brain capacity of domestic rabbits, and the table in *ibid.*, p. 127 for a comparison of skull length and capacity in wild, feral (Porto Santo), and domestic rabbits, including the Himalayan, Angora, and a silver-grey from a warren. CD wanted small specimens of domestic rabbits for comparison with wild rabbits because the former tended to be larger. Two years earlier, CD had asked Bartlett for the body of one of his Porto Santo rabbits, with the skull undamaged (*Correspondence* vol. 13, letter to A. D. Bartlett, 30 January [1865]). There is no further extant correspondence on rabbit skulls; however, CD included remarks about the modification of the skull and brain size of rabbits in the course of domestication in *Descent* 1: 147–8.

To Edward A. Smith [11–31 March 1868][1]

Mr. Smith—

Woodcuts to serve merely to show general difference between sexes: must be neatly executed & finished.—[2]

Chiasognathus Grantii.— fig. of *male*, reduced to half or third nat. size Head & thorax of female equally reduced.[3]

Bledius Taurus— male magnified twice or if necessary thrice: head & thorax of female equally magnified.— Viewed laterally & rather obliquely.—[4]

Please observe, that figures ought to be *reduced* (or magnified) as much as is compatible with perfect distinctly of Horns.

In each case *male & female*, in female the head, thorax & base of elytra sufficient.— Figures lateral, seen a little obliquely from above.—in following order.— about size of figure I must trust to your judgment. Please mark in each case how much reduced.— The Beetles must be neatly shaded; & the legs arranged naturally.

Chalcosoma atlas	♂	&	♀[5]	} reduced[6]
Dynastes Hercules		do		
Dipelicus Cantori		do		
Onthophagus rangifer		do		(magnified to match other figures)
Copris Isidis	♂		♀	} (reduced)[7]
Phanæus hastifer	♂		♀	

Attend particularly to degree of development of rudiments of Horns in the females of the latter species.

(Crabro cribrarius, male & front part of female.)[8]
(Taphroderes distortus Male magnified & Head & thorax of female.)[9]
(Pneumora male & female)[10]

4 Chester Place | Regents Park. N.W.

DAR 81: 94–5

[1] The date range is established by the return address and by the relationship between this letter, the letter to H. W. Bates, 11 February [1868], and the letter from H. W. Bates, 26 February 1868 (*Correspondence* vol. 16). CD stayed at 4 Chester Place, London, the house of Sarah Elizabeth Wedgwood, Emma Darwin's sister, from 11 to 31 March 1868. Sarah moved to Down later in 1868 (*Emma Darwin* (1915) 2: 189.

[2] Smith made several drawings of Coleoptera for *Descent.* CD had asked the entomologist Henry Walter Bates for suggestions of good insect species to engrave (see *Correspondence* vol. 16, letter to H. W. Bates, 11 February [1868]).

[3] Illustrations by Smith of *Chiasognathus grantii* appeared in *Descent* 1: 377. CD had collected specimens of this beetle from the island of Chiloe while on the *Beagle* voyage (see *Transactions of the Entomological Society of London* 2 (1837): xli).

[4] For the illustrations of *Bledius taurus* (now *B. furcatus*), shown from the side, see *Descent* 1: 374.

[5] For illustrations of *Chalcosoma atlas* (the atlas beetle), see *Descent* 1: 368.

[6] *Dynastes hercules* (the Hercules beetle) is not illustrated in *Descent.*

[7] Images of *Dipelicus cantori* and *Onthophagus rangifer* appear in *Descent* 1: 369. *Copris isidis* (now *Heliocopris gigas*) is also figured (*ibid.*). *Phanaeus hastifer* is now *Sulcophanaeus leander*; *P. lancifer* (which might be what CD meant; see CD annotations) is now *Coprophanaeus lancifer. Phanaeus faunus* (now *Sulcophanaeus faunus*) is illustrated (*ibid.*). CD noted that there was little difference in horn development in the sexes of *P. lancifer* (*ibid.*, p. 370).

[8] Illustrations of *Crabro cribrarius* (the slender-bodied digger wasp) appear in *Descent* 1: 343.

[9] Images of *Taphroderes distortus* appear in *Descent* 1: 344. For more on the genus *Taphroderes*, see *Correspondence* vol. 16, letter from C. O. Waterhouse, 19 February 1868.

[10] Illustrations of an unidentified species of *Pneumora* (bladder grasshoppers) appear in *Descent* 1: 358.

From E. A. Darwin [September 1868 – August 1870][1]

I suppose this Bill is Carolines[2] so send it on as I dare say it is paid— The Clock is not come.

to Charles

Sylvester came & said some officers having been speaking so highly of Leonard that I thought you would like to hear it— in their expressive language 'he is all there' & perhaps Lenny can translate the military language.[3] This note is all topsy turvy as I dont know to whom I am writing

DAR 105: B73

[1] The date range is established by the reference to Leonard Darwin's performance at military college (see n. 3, below).

[2] The reference is to Caroline Sarah Wedgwood.

[3] James Joseph Sylvester was professor of mathematics at the Royal Military Academy, Woolwich, where Leonard Darwin was a student from August 1868 until 1871. Sylvester's successor at Woolwich was appointed in August 1870. (RMAS archives, Sandhurst Collection.)

From Lawrence Ruck [after 29 April 1869?][1]

Pantlludw, | Machynlleth.

1 Castration of Lambs is usually so barbarously performed viz by the teeth that the operation is attended more or less by weakness, sometimes death should frost or wet set in soon after consequently the horns are affected in growth according to the strength of the animal left to it after recovery which in some cases is very tardy & perhaps never so you find some Ram-Lambs with strongish horns & some with stunted horns after the operation.[2] There appears to be no uniformity as to size or length of horn I should think the wether horn is reduced to about one third of the matured Ram's horn

2 Lambs at six weeks— Sheep after the rutting *[season]* The growth of the horn would cease after castration should the Lambs be allowed to run free over a year or more

3 According to the mode of castrating Lambs hereabout it would be dangerous to work the Lambs at an earlier age than usual I cant say what a skillful "Vet" might do with proper care & attention.

—The Ewe Lambs develope small horns in this country occasionally consequently it is in the nature of the animal; in the blood.[3] Castration affects animals differently as to kind or sort for instance the horse pig cat ox vary in size after also as to habits or courage etc

I dont suppose the Cock would crow much afterwards. which I leave for you to find out The subject requires a treatise I cant say that I have myself curiosity upon the subject We find it necessary & useful at times. Please extract from this if you like

AL incomplete
DAR 88: 134–5

CD ANNOTATIONS
1.6 to size … Ram's horn 1.8] *scored pencil*

1 The date is conjectured from the relationship between this letter and the letter from M. A. Ruck to Francis Darwin, [29 April – 27 May 1869] (*Correspondence* vol. 17).

2 Mary Anne Ruck also passed on information on the growth of sheep's horns that she had received from a neighbour in Wales (see *Correspondence* vol. 17, letter from Edward Jones to M. A. Ruck, 28 April 1869).

3 CD was gathering information on the age at which horns appeared in sheep breeds in which both sexes had horns in comparison with those breeds in which only the males were horned (see *Correspondence* vol. 17, letter to J. V. Carus, 5 January 1869). In *Descent* 1: 289 n. 26, CD noted that information received from north Wales indicated that horns could be felt in both sexes of sheep at birth.

From W. B. Dawkins 31 July 1869

Norman Road, Rusholme, Manchester.
31 July 1869.

Dear Mr. Darwin,

I returned from Rhagatt yesterday, after having had a most pleasant day at Cave-digging.[1]

M[rs] Lloyd[2] entertained me en Prince, and on Friday organised a party to explore the cave and placed several workmen at my disposal. On reaching the place after a drive of 14 miles, I found that all the bones you sent me were discovered in a space between two limestone beds, where the intervening softer material had been carried away by water. Its maximum heighth was about 4 to 6 inches and its width 20 feet or more, and it extended parallel to the bedding of the rock

Bones.

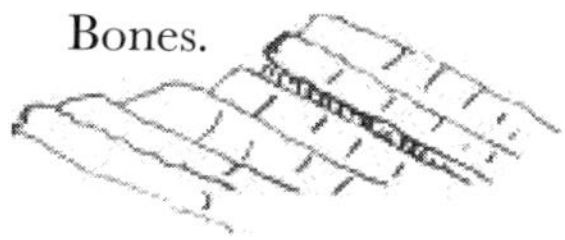

The bones had manifestly been conveyed in by water, and belonged to the Prehistoric epoch (which I have defined in a paper now in the press, that will be sent to you shortly).[3] They were washed out of an old refuse-heap, which has now utterly disappeared.

At a little distance and much lower down in the rock there is a cave, full of red cave-earth, a kind of modern red silt, that I have found invariably to be barren of any prehistoric or postglacial remains. We dug into it and then forsook it for a spot that seemed to me to be likely to contain bones. Here we met with a curious jumble of human bones, and remains of horse, goat, hare, rabbit, badger, enormous birds, wolf, wild cat, fox and Celtic Shorthorn. In the small area excavated there were the remains of at least 5 corpses varying by age from infancy to the decline of life. All the adult teeth were ground smooth like the so called Ancient British teeth. The place that contained them was a small rock-shelter, gradually passing into a hole, which may perhaps be the entrance into a cave. We dug 6 feet into it, and M[rs] Lloyd intends to carry the exploration on for her own satisfaction.[4] The example set by M. Lartet in the Aurignac cave, prevents my hazarding a guess at the age of the human bones, or at the cause by which they were introduced.[5] Most of the long-bones were perfect; the few that were imperfect were broken by the fall of stones from the roof. The perfection of the vertebræ implies also that neither dogs, nor wolves had gained access to the corpses to which they belong, since both those animals invariably eat the centra and most of the neural arches. There were

AL incomplete
The Huntington Library (CB 847)

[1] After the discovery of cave bones on the Rhagatt estate near Corwen in Wales, CD, who was on vacation in Wales at the time, sent the bones to Dawkins for identification (see *Correspondence* vol. 17, letter to W. B. Dawkins, 19 July [1869], and letter to M. C. Lloyd, [24 July 1869]).

[2] Gertrude Jane Mary Lloyd. CD evidently enclosed this letter in his letter to Mary Charlotte Lloyd of 4 August [1869] (*Correspondence* vol. 17).

[3] Dawkins refers to his paper 'The Denbighshire caves' (Dawkins 1869).

[4] For more on Lloyd's exploration of the cave, see *Correspondence* vol. 17, letter from W. B. Dawkins, 19 October 1869. See also Lucas 2007, pp. 326–7.

[5] In 1860, Edouard Lartet had discovered mammoth and cave-bear bones along with implements of stone and bone in a cave near Aurignac in southern France. Lartet was very cautious in dating the human occupancy of the cave, but concluded it probably pre-dated the diluvian period. (See Lartet 1862, p. 71.)

APPENDIX I
Translations of letters

From Federico Delpino[1] [1870][2]

Honourable Sir!

Allow me to offer you as a gift another small piece of work, "l'applicazione della Teoria Darwiniana ai fiori e agl'insetti visitatori dei fiori".[3]

Both author and translator[4] can be counted amongst your disciples, all the more so as they are convinced and persuaded of the great truths exposed in the Doctrine that is named after you. May you in your great kindness accept this slender gift.

Perhaps it would not be the best use of your time, if you resolved to read this pamphlet; still, it would be extremely important to me if you read my annotations on pp. 31. 32.

In this annotation I have briefly set out the reasons that have profoundly convinced me of the infinite variability of beings, though in perfect harmony with the vitalistic and teleological theories propounded over here.

Please accept anew the profession of my highest esteem and gratitude | Your most obedient disciple | Federico Delpino | Professor of botany, Vallombrosa

(Firenze per Vallombrosa)

DAR 162: 141

1 For a transcription of this letter in its original Italian, see p. 1.

2 The date is established by the reference to Delpino trans. 1870a (see n. 3, below).

3 Delpino refers to Delpino trans. 1870a (The application of Darwinian theory to flowers and to the insects that visit flowers), an annotated translation of H. Müller 1869. There is an annotated copy in the Darwin Pamphlet Collection–CUL.

4 Delpino refers to himself and to Hermann Müller.

From Federico Delpino[1] 28 February 1870

Florence
on the 28. February 1870.

Most distinguished man!

I have received, through the signal kindness of Your Honour, the two numbers of the "scientific opinion", where you pleased to publish, in English translation, the essay on the Marantaceae.[2] Please accept my most heartfelt thanks.

You will be pleased to notice, I think, that in all my work, however small its intrinsic merit may be, it is still always my purpose to pay respect to and to develop the doctrine which Your Honour has with such force of genius founded and established. Other may call hypothesis the theory that is named after Your Honour; I, however, shall not hesitate to propound it with full conviction, not just as a theory but as a *doctrine of fact.* All the parts of which the science of phytology is constituted are plain and intelligible, if one allows the doctrine of transformation; if, on the other hand, one one allows for the contrary theory, nothing makes sense any more—

Then there are other parts, such as teratology, which directly make no sense without the doctrine of transformation— The existence of rudimentary organs then is an absolute proof for the truth of the aforesaid doctrine— Thus Your Honour will have a disciple in me who, however little he may be able to follow himself with success the trail that you have blazed, is still resolved, convinced, and in his convictions stubborn and immutable.

Now I share with the whole scientific world the lively desire that the treatises "on man" and "on the variation of beings in a state of nature" will soon see the light of day.[3] It is easy to predict the great success which they will have, and how well they will stand side by side with the immortal books "on the origin of species" and "variation under domestication"

In the anticipation of these, I have the honour to declare my profound esteem and liveliest admiration of Your Honour

your most respectful disciple | Federico Delpino

DAR 162: 146

1 For a transcription of this letter in its original Italian, see p. 51.

2 Delpino refers to the translation of his article on the biology and genealogy of the Marantaceae that appeared in *Scientific Opinion* on 2 and 9 February 1870 (Delpino 1870b).

3 Delpino refers to *Descent*, and to the 'second work' that CD discussed in *Variation* 1: 4–8, but which was never published.

From R. F. Albrecht[1] 16 March 1870

Highly honoured Sir!

I am at the moment in the process of putting together, on the basis of the principles of your theory of the origin of species, an argument on the evolution of the apparatus that enables birds to fly. In order to do so it is absolutely necessary that I include in my brochure a brief outline of the laws of variation. In this, however, I shall be unable to avoid using your own words. Hence I would like to ask you to authorise me to do so, since otherwise it might come to legal proceedings against me for violations of the copyright.— At the same time I also would like to communicate to Your Honour that I intend to publish a second brochure, which as yet exists only in my imagination, "On the intermediate or transitional forms. After completion of my work I shall permit myself to send you copies of it.[2] That you may want to cry out at the German supporters of your theory, as Dr. A. B. Meyer[3] put it: "May God protect me from my friends, & with my enemies I shall deal myself", I can only confirm.

I, however, will conduct myself as such a friend and follower of your theory that you will not need to tell me this.

Lastly I would like to encourage you to assemble and then publish in this form all of your publications to date, and to do likewise with your work thereafter, so that it will be more accessible than is unfortunately now the case, both for the German people and indeed for all other peoples. I look forward to your reply, | and remain most respectfully | R. F. Albrecht.

Berlin, Bahnhofstrasse 3. | Berlin, 16. March 1870.

DAR 159: 33

[1] For a transcription of this letter in its original German, see p. 70.
[2] The publications have not been identified.
[3] Adolf Bernhard Meyer.

From Georg Recht[1] 17 March 1870

Mist. Darvin!

I venture to communicate to you the latest results of my research, which I can summarise in one sentence, since I state that the law of inertia—vis inertiae—the fundamental law of our mechanics, and thus of our engineering, our physics, our astronomy, and of all our knowledge, is false or an error.

According to this law all phenomena of nature are the effect of a primitive impulse, which the astronomers call tangential force and the chemists affinity; and since this impulse is not demonstrable in astronomy, for the hypothesis of Laplace[2] is untenable, the philosophers accept the kinetic condition of the universe as the primitive state, that is, they assume that the present state of the universe is constant.

According to this aforementioned fundamental law, physicists regard all phenomena of nature as merely mechanical without any further activity of other forces, while according to my own fundamental law these phenomena are all dynamic.

For mathematicians have, from arithmetics, geometry and a particular mechanics, constructed a special universe, within which they can compute, and this theoretical or mathematical universe has been accepted as congruent or identical with the physical universe. This is the great error. The fundamental law of analytical or mathematical mechanics is the following:

Once a body has been set in motion by an external cause whether rectilinear or rotational or both at once—it stays in this state on its own, without further action of any force.

This fundamental law is based on the following assumptions:

(a) Every moving object assumes its proper velocity without intermediate stages of velocity;

(b) Every moving object assumes any velocity in each and every one of its points in its entire spatial extension.

Subsequent conclusions from the fundamental law are, however:

1) Every presently acting force produces a uniform rectilinear motion;

2) Every constant force produces an accelerated, indeed a uniformly accelerated motion.

In contrast to this I offer these laws for physical bodies:

1) Every physical body enters into motion only successively bit by bit; and consequently in the state of motion it is *expanded*;— proof of the molecular composition of bodies, but above all of the *bonds* between these molecules among one another, which are broken apart in a physical and chemical way—

2) Every body assumes a particular velocity only successively and by degrees, starting, to be sure, from zero.

As a result of the first law a body in a state of motion is elastically excited; and, since heat not only expands bodies, but the reverse, that expansion also binds heat, so transforms its thermal state— In a vacuum, the rotation of a ball can cause it to glow at its equator, while its poles remain cold or even cool down— This elastic expansion or compression, depending on whether the force is pulling or pushing, operates upon the extinction of the force, it transforms every motion of a physical body into a retarding one and reduces the body to its primitive spatial state in respect to size & form, and thus with respect to its thermal state, by which, however, it also enters the state of rest.

I had the opportunity to perform experiments on the railway, and I found that the retardation time of a train depends solely on the elastic state of the drawbar, whereby the movement itself or the limit of elasticity must be fundamentally distinguished from the size of its elastic force. By *laying* rubber rings *on top of one another* one increases only the limit of elasticity, by *laying them side by side* one increases the elastic force.

According to this account, a force absolutely never occurs in nature by itself alone, but there always are two of them acting together, that is, an external one and an internal one; this combination is simply the state of equilibrium of these, and not an unstable but a stable equilibrium, for with the alteration of one of the two forces changes, the other changes at the same time.

If the onset of operation of two opposing forces that are in balance happens at the same time, then we get an equilibrium at rest, if one of them starts after the other, we get an atom of time and thus an atom of velocity or motion.

Any body set in motion by a force thus has among its inner qualities a tendency towards rest, which becomes apparent only after the force has already taken effect, and as a result of this tendency, the body gradually comes to rest as the force fades, i.e., unless the force is constantly renewed. Therein lies my fundamental law of motion. From this follows:

1) Uniform motion is the result of a constant force or of a continuous *renewal of force.*

2) Accelerated motion is the effect of a continuous *increase of force.*

When one considers nature with these laws of motion in mind, all her riddles appear solved, all her secrets are uncovered before us. The process of individuation from the formation of the world-body down to the genesis of the simplest cell becomes clear if one associates it with the nature of the ether-body, and with the true size, form & motion of its atoms and molecules, by apprehending their physical and chemical association with particular bodies and individuals. This whole matter has become entirely clear to me, and all these secrets lie thus open before me, e.g. the nature or the cause of gravity, which has hitherto been understood only in its effects and not in its causes, like mathematical theorems; only the space of a letter is not sufficient for its exposition. There is just one point, however, for which I venture to claim your patience, namely the concept of elasticity.

What is commonly understood by the elasticity of a body is its tendency to restore its original size & form should these its spatial elements be altered by an external cause.

The size of a body depends on the size of its three dimensions, which may be mutually dependent as e.g. in a ball, or which may be entirely independent of one another, e.g. in an ellipsoid with three axes or in a parallelepiped.

The shape of a body consists in angles that form its dimensions.

Thus we must distinguish above all between: 1) elasticity of size and 2) elasticity of form; the former is either just linear, or doubly linear—surface elasticity—or triply linear—body elasticity—

As every dimension of a body can again be either extended or reduced, we must in each kind of elasticity distinguish between elasticity in tension and elasticity in compression, or between positive and negative elasticity.

However, since it is not merely the body as such which is elastic, but also its molecules, we must distinguish the elasticity of the body from its molecular elasticity.

Regarding the elasticity of form, any angle can be extended or reduced, and this elasticity causes a bending of the body, so long as the angle planes remain the same. Any such plane, however, may turn to either side, and from this results a double rotational elasticity. Together, the bending and the rotational elasticity make up the elasticity of form.

If we now e.g. stretch a body, its elasticity of width may or may not change with its elasticity of length; but in any case, this elongation entails a change in the molecular bonds. These, consequently, must be able to move closer together and further away from one another without becoming disconnected. Such molecular movements can only be rotational, and hence it follows that changes in the body must always be a combination of rotation with translational movement.

Elasticity is latent velocity or latent heat or the result of latent aether, which thereby changes into what is commonly understood by substance or matter. The aether itself consists exclusively of immutable and hence inelastic atoms, which solely through their speed and their connectedness form the foundation of the universe, that is, the so-called spiritual and the material foundation.

The idea of a tendency in matter, a notion that as such is incongruous with the concept of matter, must be understood as a tendency towards speed or motion. Thus, it is brought about by the presence of an atom of velocity without motion setting in. E.g. if within the molecules in their respective elasticity there is an atom of velocity that is unable either to shatter the body or to induce a development in it, then there is a tendency in the body.

When a body is e.g. under the influence of an attraction by another body, it is excited in its molecular elasticity, which brings about a movement of the body as a whole, and if this movement is neutralised by a supporting surface, the excited molecular elasticity continues as a tendency towards movement.

As for the elasticity of natural bodies, unitary bodies are singly elastic, binary bodies doubly elastic, and ternary bodies are triply elastic, i.e. they are spatially linear or linear elastic in all directions. This elasticity relates to the molecules, while the elastic differentiation of the bodies divides them up into solid, fluid and gaseous bodies.

In organic nature solid bodies are represented by linearly elastic plants, the doubly elastic by the animal kingdom, and those that are in all directions linear elastic or gaseous are represented by man.

In animals the muscles of the spine and of the extremities form a right angle and thus the elements of a plane, while the spine, the cerebellum and the anterior brain lie on a single (straight) line, which is why animals are mentally linear elastic only. Man, on the other hand, is physically linear and mentally doubly elastic, and his spine and cerebral line are mutually perpendicular.

Finally I venture to establish only the concept of *general* movement.

The first and basic movement is that of a steady linear motion, the second that of a steady rotation. General linear motion proceeds outward from one point in two opposite directions, and general rotation likewise proceeds in opposite directions. In a helix or a spiral line both movements are simply connected with one another, while in a double helix, in which two spiral lines move through opposite rotations towards the two opposite directions, also in translatory manner, general linear motion is demonstrated.

Consider a body in which every atom is connected with every other atom, and general linear movement is applied to every connection within this body, then one has a body in which from every point, i.e. in every atom, there occurs general linear movement in all directions. Such a body is an aetherbody, which thus contains all the conditions of the so-called mental & material life of the universe, and furthermore all the conditions of the entire process of individuation.

Thus, general linear motion with opposite rotations and opposite translations can be observed in any beam of light, in any electrical or thermal motion, as well as in any chemical process. In every single beam of light we see substance move in opposite directions in combination with two opposite rotations, which are presumed to be undulations.

These movements can however also be observed in the animal kingdom, in the transition from land animals to aquatic animals and vice versa, from aquatic animals to land animals; from land animals to flying animals and from flying animals to land animals; from aquatic to flying animals and from flying to aquatic animals, not to mention the migration of organs from genus to genus. Thus, e.g. the sex life in humans starts with the sense of vision, that of land animals with the sense of smell, while in birds it is based on the sense of hearing. The sense of vision in birds ends in the renal points, etc. etc.

It would make me very happy, sir, if you would devote some time to these lines, and would bring them to the attention of the Royal Society. I would consider myself infinitely fortunate to receive a reply regarding their reception there. I hardly dare to hope that I would be so fortunate as to appear before the Royal Society and to reply in person to any questions that may be put to me.[3]

With my highest esteem and veneration | Your Honour | your most humble servant | Dr. Recht | in Munich.

$\frac{17}{3}$ 70.

DAR 176: 77

[1] For a transcription of this letter in its original German, see pp. 72–6.

[2] In his letter of 1 March 1869 (*Correspondence* vol. 17), Recht had discussed the application of physical principles concerning motion to living systems. Recht refers to the nebular hypothesis formulated by by Pierre Simon, marquis de Laplace, in *Exposition du système du monde* (Laplace 1796).

[3] No papers by Recht appear in the *Proceedings of the Royal Society of London.* Recht had asked CD to communicate an earlier monograph (Recht 1869) to a scientific society (see *Correspondence* vol. 17, letter from Georg Recht, 9 February 1869).

From Armand de Quatrefages[1] 30 March 1870

Paris
30 March 1870

Dear Sir and colleague

For a long time I have been wanting to write to you and let you know of the joy that your last letter gave me. But—we say in French that Hell is paved with good intentions—and I have allowed myself to go and add another stone to the paving of the infernal resting-place. My long absence last year left me such an arrears that I have hardly escaped from it, and, swept along from one thing to another, I have neglected many duties.

Let me repeat that your letter made me very happy by proving to me that our Scientific dissensions would not diminish the cordiality of our relations.[2] I would have trusted so in any case; for the nobility of your character, which shows itself on every page of your books, reassured me in advance. Yes, I dare to say it, we are both pursuing truth and that alone must establish links between us which are stronger than the causes of discord resulting from the difference in our way of seeing things. So I shall not hesitate to send you the new edition of my studies on transformism, which are being printed in one Volume at this moment.[3] You will receive them, I am certain, in the same spirit in which they are sent.

I am also very happy that an opportunity other than words has presented itself for showing all the esteem I have for your work. We are to appoint a correspondent to our section of the Academy of Sciences. A few days ago I was chatting to M. Edwards about it and I saw with joy that the same thought had come to both of us, that of presenting You as our candidate.[4] I hope that we will be supported by our other colleagues; but in any case, rest assured that I shall be a zealous and convinced advocate before the Academy. Our very dissensions will give more weight to my words, if need be, and I do not doubt of your nomination. The pleasure of demonstrating all there is of value in your work, will compensate me a little for the regret that I feel at having been *forced* to combat you on certain points. But matters had come to such a pass that I could no longer keep silent without seeming to abandon my convictions. This consideration alone, you may be assured, put the pen into my hand for I dislike controversy.

Dear Sir and colleague, please accept the expression of the affectionate devotion of your most devoted | De Quatrefages

DAR 175: 6

1 For a transcription of this letter in the original French, see pp. 86–7.

2 CD's letter to Quatrefages has not been found.

3 Quatrefages refers to *Charles Darwin et ses précurseurs français* (Quatrefages 1870). There is a copy in the Darwin Library–CUL.

4 Quatrefages refers to Henri Milne-Edwards, who was the head of the zoology and anatomy section of the French Académie des Sciences. CD was nominated six times between 1870 and 1878 for membership of the zoological section of the Académie, but was not elected until 1878, and then it was to the botanical section (Corsi and Weindling 1985, p. 699). Among the other nominees on this occasion were Thomas Henry Huxley, Rudolf Leuckart, and Carl Vogt; the successful candidate was Johann Friedrich von Brandt (*Comptes Rendus Hebdomadaires des Seances de l'Académie des Sciences* 70 (1870): 1418; 71 (1870): 41). See also letter from Armand de Quatrefages, 18 July 1870.

From Giovanni Canestrini[1] 21 April 1870

Padua
21/4 1870

Dear Sir,

I have delayed my reply to your esteemed letter until now because I wanted to give you news about the Italian translation of your latest valuable work.[2]

There have been some difficulties with Mr Zanichelli[3] in Bologna, regarding the edition of the translation, and the translation has until now remained undone. We have now an agreement with another publisher, namely Mr Vincenzi[4] in Modena, and the printing has already started. The named author of the translation is a pupil of mine, namely Dr Bonizzi,[5] who has indeed translated the last chapters. Mr Vincenzi in Modena will very soon settle the value of the stereotypes with your publisher. I am sorry about the delay.[6]

As soon as the translation appears, I will make certain that a copy is dispatched to you without delay.

In a few weeks I shall take the liberty to send you one of my works, in which I defend the *natural* origin of man, based especially 1. on the abnormal characteristics of man, 2. on the rudimentary organs, 3. on the very old (prehistoric) shapes of the skull, and 4. on embryological characteristics.[7]

Your theory of the origin of species has made great strides here and is winning more supporters every day. Just as the translation of your first work has achieved much, so I hope that that of the second will also do much.[8]

I remain most respectfully | your most devoted | Dr. J. Canestrini | professor at the university of Padua

DAR 161: 35

1 For a transcription of this letter in its original German, see pp. 101–2.
2 CD's letter to Canestrini has not been found. Canestrini refers to *Variation.*
3 Nicola Zanichelli.
4 Mr Vincenzi has not been further identified.
5 Paolo Bonizzi.
6 In the event, Canestrini's translation of *Variation* (Canestrini trans. 1876, based on the second English edition) was published by Unione Typografico-Editrice in Turin.
7 There is a copy of Canestrini's *Origine dell'uomo* (Canestrini 1870) in the Darwin Library–Down.
8 Canestrini and Leonardo Salimbeni had translated *Origin* into Italian in 1864. On the reception of Darwinism in Italy, see Corsi and Weindling 1985 and Pancaldi 1991.

From F. C. Donders[1] 17 May 1870

Utrecht,
17 May | 1870

Dear Sir,

I now feel that I ought to have given you the sad news of the loss of our daughter, our only child, the object and the centre of our life, our happiness and our hope. She died on 5th March, just 5 Days after having given birth to two babies, who, though very weak, have survived her up to now.[2] We are living with a sorrow which it seems can never end, and just like my much-loved son-in-law, Doctor Engelmann,[3] I have found it hard to return to the laboratory and to attend to our occupations. In the last two months of her pregnancy, our daughter was already suffering so much that we lived in constant apprehension. But for this the small piece of work you did me the honour of confiding to me, would have been finished long since. However, I have just turned my hand to it once more, and you will have it in a few days. I have examined the circulation of the eye with some care, and all I am lacking is the inclination to finish my note on this subject, which I want to publish in my Journal: in general I believe I can confirm the views of Sir Charles Bell.[4] Doctor Moore[5] will send you the English translation of this article. As for the other question, I don't think that the supposed relationship exists.[6] In my next letter, I shall formulate my reply. You may be certain of having it in hand within a week of receiving this one, and by then Mr. Moore will have received a printed copy of my article on the first question.

Please accept the assurance of my respect and affection, | Dear Sir, | Your most devoted | Donders

DAR 162: 224

[1] For a transcription of this letter in its original French, see pp. 121–3.
[2] Donders's daughter was Marie Engelmann; his wife was Ernestine J. A. Donders. The two babies have not been identified.
[3] Theodor Wilhelm Engelmann.
[4] CD had asked Donders questions about the contraction of the muscles around the eye, and about Charles Bell's account of the physiology of weeping (see letter to F. C. Donders, 13 May [1870] and n. 1). Donders refers to the *Nederlandsch Lancet*, of which he was editor (*DSB*).
[5] William Daniel Moore.
[6] Donders refers to CD's query regarding the possibility of contraction of the muscles around the eyes producing tears (see letter to F. C. Donders, 13 May [1870], and *Correspondence* vol. 17, letter to F. C. Donders, 6 September 1869 and n. 3).

From Federico Delpino[1] 20 May 1870

Royal Museum | of | Physics and Natural History of Florence
20. May 1870—

Honourable Sir!

I received your esteemed letter of the 15th of the month only today and I hasten to reply.[2]

I am sorry but at the moment we have no seeds of *Canna Warszewiczii* in our botanical gardens;[3] but today I will write a letter to my friends the directors of the botanical gardens at Siena, Modena and Naples, who might be able to supply me; since I see that the seeds of the same plant figure in the inventory or catalogue of seeds published by those establishments for the year of 1870.

Thus I hope to be able to send Your most Illustrious Honour a packet of the desired seeds, from a different source, within a few days.

I thank Your Honour for the communication concerning the leguminous collection your son made at Ravenna. The plant is a *Lotus siliquosus* of Linneus.[4]

I have to date not had occasion to study the dichogamic apparatus of this plant. However, I have studied and described that of *Lotus corniculatus*, of *Lotus hirsutus*, and of *Lotus ornithopodioides*. (*Sugli apparecchi della fecondaz. etc.* p. 25, and *Ulteriori osservazione etc.* p. 45 and 46). Now, by examining the sample sent to me by Your most Illustrious Honour, I was able to learn that the apparatus of *Lotus siliquosus* is in fact similar to that of *Lotus corniculatus*, and of the others.[5]

This dichogamic apparatus, which I have called "*papilionaceous pump apparatus*" is rather marvellous. It is similar to the instrument with which one makes pasta at the vermicelli-makers. To suck up the honey, the fertilizing insects, which are exclusively of the family of bees (*apis*, *eucera*, *anthophora*, and *Bombus*), press down the carina, and this action causes the top of the filaments, which are clubshaped and swollen at the end, to act as a piston and to squirt out the pollen in worm-like shapes which attach themselves to the breast and abdomen of the insects.[6]

This apparatus has also been described by Severin Axell, on p. 29—of his book 'Om anordiningarna etc' "Sätter sig nu en insekt på kölen, så sänkes denna något, den styfva pistillen och filamenterna verka som en tryckpump på pollenmassen, hvilken utjagas i form af ett litet moln genom hålet i kölen spets".[7] The operation of the insect is described well enough, but there is an inexactness in so far as he says that the pollen exit from the end of the carina in the form of a cloudlet (moln), while rather more precisely it exits in the shape of wormlets.

Noble and excellent Man, please accept the sincere and respectful testimony of my consideration—

Your most humble disciple | Federico Delpino

DAR 162: 147

[1] For a transcription of this letter in its original Italian, see pp. 126–7.

[2] CD's letter has not been found.

[3] *Canna warszewiczii* is now *C. indica.* Delpino had described the structure of the flowers of *Canna* in 1867 (Delpino 1867b, p. 23).

[4] It is not known which of CD's sons made the collection, but William Erasmus Darwin was in Italy in April 1870, and according to Henrietta Emma Darwin, visited Florence (letter from H. E. Darwin to Emma Darwin, 14 April [1870]; DAR 219.9: 87). *Lotus siliquosus* is now *L. maritimus.* Delpino refers to Carl von Linné (Carolus Linnaeus).

[5] Delpino refers to two works on dichogamy which he had sent to CD, Delpino 1867b, p. 25 and Delpino 1868–75, part 1, pp. 45–6. *Lotus corniculatus* is bird's-foot trefoil; *L. hirsutus* is hairy canary clover.

[6] See Delpino 1868–75, part 1, pp. 45, 47, 52. *Apis*, *Eucera*, *Anthophora*, and *Bombus* are all genera in the family Apidae, which includes bumble bees, euglossine bees, honey bees, and stingless bees.

[7] Delpino quotes from Johan Severin Axell's monograph *Om anordningarna för de fanerogama växternas befruktning* (On the mechanisms for the fertilisation of phanerogamic plants; Axell 1869, p. 29). The passage reads, 'If now an insect settles on the carina, this is somewhat lowered, the stiff pistil and the filaments work as a pressure pump on the pollen-mass, which is ejected in shape of a small cloud through the hole in the cusp of the carina.'

From Charles Renard[1] 20 May 1870

Société Impériale des Naturalistes de Moscou. | *Moscow,*
$\frac{8}{20}$ May *1870*[2]

Dear Sir

Your extensive and useful work for the advancement of science, to which Your wide-ranging and important research carried out on the circumnavigatory voyage of the Beagle, Your geological observations on volcanic islands, Your monograph on the Cirripedes, Your geological observations on South America &. &. and lastly the work on the origin of species that is so justly famous, bear broad and ample witness, has been admired by the Société Impériale des Naturalistes de Moscou for a long time.—[3] The society has therefore obtained the satisfaction of electing You unanimously as its honorary member at its meeting of 19th February, and it has charged me with sending You its diploma, which will follow these few lines tomorrow.—[4]

The Société Impériale des Naturalistes, which owns a very rich library, the product solely of its members' gifts and of its literary connections (no funds are set aside or employed for its augmentation) would be delighted if You were able to send it some of Your publications which You may still have available.—

Please accept, dear Sir, the expression of my highest and most perfect consideration, with which I have the honour to be | Sir | Your most devoted servant | Dr. Renard | secretary of the I. Société des Naturalistes

My address: His Excellency, the current Councillor of State | Dr. Ch^s. Renard. | Moscow. Miloutinskoe Peréoulok, house of Askarkhanoff.

to Professor Charles Darwin, London.

DAR 176: 120

[1] For a transcription of this letter in its original French, see pp. 127–8.
[2] Renard gives both the Julian (8 May) and Gregorian (20 May) calendar dates.
[3] Renard refers to *Living Cirripedia* (1851), *Living Cirripedia* (1854), *Fossil Cirripedia* (1851), *Fossil Cirripedia* (1854); he also alludes to *Journal of researches*, *Volcanic islands*, *South America*, and *Origin.*
[4] The diploma is in DAR 229: 23. See Appendix III.

From F. C. Donders[1] 27 May 1870

Utrecht,
27 May | 1870

My dear Sir,

The expression of your affectionate sympathy was very dear to me. You, who have suffered the painful loss of a beloved daughter, in some sense have the measure of what it is to lose one's only child.[2] Besides the void, no hope of ever seeing it filled. I think of it all the time, but best not to speak of it. So I have only to thank you for your kind words of sympathy, where every consolation must necessarily fall short. The courage to use the days of life that remain to me well, as far as possible, is gradually returning to me. All my efforts are directed to that end.

I have done the article on the action of the eyelids in cases of ocular congestion by expiratory pressure. I shall send it this very day to Doctor Moore at Dublin, who has kindly agreed to undertake the translation.[3] At the same time, I shall send you the proofs of the article, in Dutch. You will not be satisfied with it: you will find many things that you scarcely need, and the answer to the fundamental question will not seem decisive enough to you. But you will at least see that I have not treated the question lightly. Allow me to give you a brief summary of my work.

First I confirmed the fact of the occlusion of the eyelids in cases of excess expiratory pressure. Then I examined the way in which that pressure produces congestions and the dilation of vessels in general. Next I studied the special conditions applying to the anterior, intra-ocular and retro-ocular vessels, and I described the symptoms produced by expiratory pressure in these three orders of vessels. In the meantime, the opportunity of examining how far palpebral pressure is able to suppress them has presented itself.

I sum up in the following terms, more or less:

"We are convinced that the external vessels of the eye, as well as the intra-ocular and retro-ocular vessels, dilate under the influence of immoderate respiratory pressure. We have seen that the occlusion of the eyelids tends to diminish or to suppress this dilation, firstly through the pressure they exert, and additionally, it appears, by virtue of a certain action associated with that occlusion. We have confirmed that at each immoderate expiratory pressure the eyelids close or at least exhibit a tendency to close. Thus we are justified in concluding that the occlusion of the eyelids protects the vessels and the eye itself against the dangers which *may* flow from an excessive expiratory pressure.

The weak point is—that there are not sufficiently certain indications of the danger that this pressure causes under normal circumstances in the eyes."

I had the intention of adding to this letter the reply to the other question;[4] but I must ask you to wait until tomorrow. It is annoying that I am not master of my own time.— In the meantime, please accept the assurance of my respect & my most perfect consideration.

Your wholly devoted | Donders

DAR 162: 225

[1] For a transcription of this letter in the original French, see pp. 135–6.
[2] Donders refers to Anne Elizabeth Darwin and Marie Engelmann. See letter to F. C. Donders, 19 May 1870 and nn. 1 and 2.
[3] See letter from F. C. Donders, 17 May 1870. Donders refers to his article, 'De beweging van het oog toegelicht met het phaenophthalmotroop' (Donders 1870a), and the English translation by William Daniel Moore, 'On the action of the eyelids in determination of blood from expiratory effort'. (Donders 1870b). CD discussed Donders's conclusions in *Expression*, pp. 160–2.
[4] For CD's second question, see the letter to F. C. Donders, 13 May [1870].

From F. C. Donders[1] 28 May 1870

Utrecht
28 May 1870

My dear Sir,

I am continuing the letter which I had the honour of addressing to you yesterday.[2] You will allow me, I hope, to enter into some preliminary discussion which will facilitate the exposition of my ideas.

Tear secretion takes place under the direct influence of specific secretory nerve fibres included in the portio major nervi trigemini.[3] They branch out from it to form the lachrymal nerve, which originates in the 1st and perhaps also in the 2nd branch of nerve V. The secretory fibres possess a centrifugal action; all the other fibres of these two great branches seem only to possess a centripetal action: they are sensitive. In a way, it is thus an anomaly that the fibres which preside over tear secretion are contained within the same neurilemma.[4] But in fact this arrangement has no consequences, since the function of the nerve depends only on its centre of origin and its peripheral distribution. In the same way, we see some fibres branching off from the facial nerve (the chorda tympani)[5] to preside over the secretion of the submaxillary gland.

It is thus a matter of finding, in the medulla oblongata, the special nucleus—that is to say the centre—of nerve cells, which through its action determines the innervation of the nerve fibres that produce tear secretion, and of studying its connections. From what we know of the nuclei of other nerves corresponding to determinate functions (according to Stilling, Schroeder Van der Kolk, Deiters and other anatomists),[6] such as the function of the electrical organ, the consensual motion of the eyes, the contraction of the pupil, etc. etc., we are justified in accepting a special nucleus for tear secretion, saline secretion, accommodation, and each group of muscles whose action corresponds to a specific function, such as locomotion, expression, etc.

Now, two types of nerve fibre act on the nucleus of the lachrymal nerve: 1st. central fibres, in direct communication with the large hemispheres, 2nd. fibres acting by reflex action, which branch off from the periphery and terminate at the nucleus.

A similar nucleus must correspond with the orbicular muscle, and is likewise put into play 1st. by central fibres which determine voluntary and automatic motion, 2nd. by fibres which, branching off from the periphery, act by reflex.

In cases where the action of the orbicular muscle and that of the lachrymal gland are combined, the action of central and reflex fibres must act simultaneously on the two nuclei. The interaction could be such that action in the muscle provoked by one means or another determines tear secretion, or vice versa, by reflex action. This is another question to be resolved.

Let us first examine if there is communication between the *central fibres* of *both* nuclei:

1st. voluntary innervation of the orbicular muscle can occur, even in a high degree, without associated action by the lachrymal gland. When there are some signs of tears, this phenomenon can be explained either by means of the pressure exerted directly on the gland, which evacuates its contents in some way, or by means of eye irritation, which, as we shall see, stimulates tear secretion by reflex.

2nd. automatic innervation of the lachrymal gland, in turn, can occur independently of that of the orbicular muscle. Everyone knows that a painful memory makes tears flow without contraction of the orbicular muscle, and that children who are frustrated in their desires often turn their eyes to us wide open and filled with tears. If the eyes close later, it is probably a reflex action produced by eye irritation.

3rd. It may occur that the central fibres communicate violent pain to both nuclei simultaneously, e.g. during a great sorrow, or a violent grief. Then children and nervous people, in particular, may be seen to shed tears with closed eyes. At that moment, the reflex action is contributing, but it does not predominate.

As for communication between the fibres which *through reflex action* affect *both* nuclei, this seems to be far more common. Such fibres belong to two different classes of nerves: a. the optical nerve and b. the nerve of the fifth pair.

a. As regards the optical nerve, irritating it with a bright light causes contraction in the orbicular muscle, rather than tear secretion. However, it is not uncommon to find secretion increased at the same time.

Mr. Herkenstein found, in his experiments upon animals, that irritation of the optical nerve of a single eye provokes tear secretion in both eyes; while conversely only the eye on the same side reacts to irritation of the mucosa, which is supplied by the fifth pair.[7]

b. The irritation of certain sensitive branches of the 5th pair acts on both nuclei, or, to put it better, there are points on the periphery where, through reflex action of the nerves, irritation simultaneously provokes orbicular contraction and tear secretion. This is what we find in irritation of the ocular conjunctiva, especially near the cornea and of the cornea itself,—whether from touching alone, or from a permanent foreign body, or from superficial inflammation in phlyctenular or scrophulous opthalmia, etc.— I have also observed some cases of a very curious affection: after a touch that produced neither a wound nor contusion, for example friction from clothing, a spasm of the orbicular muscle occurred, with very profuse weeping, which lasted no longer than an hour, perhaps. But from time to time thereafter, sometimes at large intervals, perhaps of several weeks, there were violent episodes of spasm in the orbicular muscle and tear secretion, with primary or secondary redness of the eye.[8] There are thus points on the eye surface that act on both nuclei.— There are also some in the nasal mucosa. Irritation of this mucosa causes tears to flow & simultaneously produces contraction first in the orbital part of the orbicular muscle alone, quite often as a preliminary to sneezing, which, in causing an explosion, causes the eyelids to be tightly closed and favours tear secretion even more: this is caused by the fact that a secretion that is usually intermittent, once established by nervous irritation, is regularly augmented by sanguinary congestion.— This dual reflex action in irritation of the nasal mucosa seems to correspond to what is required to eliminate the irritation—produced by a foreign body—as much by the violent current of air in sneezing as by the tears descending into the nose.

If during vomiting the orbicular muscle contracts and there is tear flow simultaneously, is there not, most commonly, irritation of the nasal mucosa by vomited substances, or at least of the back of the throat, which also seems to affect both phenomena.

In such cases of irritation of the nasal mucosa, there is no true pain, just a slight prickling or a feeling of constriction, the first of which in particular excites sneezing, and the second of which we easily miss.

Irritation of the lachrymal sac seems only to produce weeping. At least in cases of chronic inflammatory irritation, tears flow when the eye is not closed, and, which is interesting, if one destroys the lachrymal sac in such cases, as soon as scarring takes place, once there is no condition of irritation, tear secretion is often nil, as long as the eye is free of all irritation: no tears are to be seen, despite the complete occlusion of the ducts. This is a good proof that tear secretion is intermittent, like that of saline, etc.—

Ordinary blinking therefore does not seem to be accompanied by tear secretion: It is a reflex action as much from the irritation of the optical nerve as from the nerve of the fifth pair (the surface of the eye). However, I should not like to assert absolutely that tear secretion is totally absent during blinking: besides, we know that the mechanism of tear absorption by the lachrymal points is put into play by such blinking. For the moment, My dear Sir, I could not tell you any more. You have yourself remarked that the voluntary contraction of the orbicular muscle has, so to speak, no effect on tear secretion. Now this ought not to astonish us, for it is very certain that these two actions are not in a cause-and-effect relation to one another: it's only the irritation produced by the contraction that could indirectly give rise to weeping, and the irritation of the eyes and even eyelids by tears that could give rise to a contraction,—in both cases by reflex action.

One must thus fall back on the other nerves, the nuclei, as Stelling has termed them.[9] I am convinced that a very assiduous anatomical examination may allow us to trace the fibres of the lachrymal nerve, as far as the nerve cells where they terminate: this is difficult and laborious, but not impossible. One would expect the same thing for the chorda tympani, if one wanted to: "where there is a will, there is a way". Perhaps I shall be able to engage one of my students to undertake the examination.

Still, it is true that the first question to appear in the sequence of ideas that concerns us consists of the precise determination of the nuclei, taking the peripheral fibres as a point of departure. Later arises the more difficult question of the connections between these nuclei and the reflex or central fibres, and their mutual connections, that is to say the connections from each nucleus to the rest. If there is a manifestation of the cerebral life in facial expressions, it is that there is perfect organic correspondence in the details, every muscle or muscle group having its own nerve cells, on which local and possibly specific actions, linked to psychic manifestation, act in a harmonious manner.

Please excuse me if while writing I have become rather carried away. Perhaps here or there, I may allude to some physiological fact that might allow you to settle your ideas regarding the question that you did me the honour of proposing. You may be very certain that you will make me happy by addressing yourself to me on points that fall within the scope of physiology, not all of the details of that your mind commands. I have boundless admiration and respect for your works.

My dear Sir, I beg you to accept the assurance of my boundless esteem. | Donders

DAR 162: 226

1 For a transcription of this letter in its original French, see pp. 136–9.

2 See letter from F. C. Donders, 27 May 1870. Donders there discussed the first of two questions that CD had put to him in the letter to F. C. Donders, 13 May [1870].

3 For CD's question on tears, see the letter to F. C. Donders, 13 May [1870]. The 'portio major nervi trigemini' is the large sensory root of the trigeminal nerve.

4 Donders evidently uses the term 'neurilemma' to refer to the sheath of connective tissue around a nerve or nerve fascicle (*OED*).

5 'Chorda tympani': a branch of the facial nerve which traverses the mucous membrane of the tympanum (*OED* s.v. 'tympanichord').

6 Donders refers to Benedict Stilling, Jacob Ludwig Conrad Schroeder van der Kolk, and Otto Friedrich Karl Deiters.

7 Ulrich Herzenstein reported these experiments in his article, 'Zur Physiologie der Thränensecretion' (On the physiology of tear secretion; Herzenstein 1867).

8 CD quoted, in translation, the two preceding sentences in *Expression*, p. 166.

9 'Stilling nucleus': a column of large neurons that gives rise to the dorsal spinocerebellar tract (*Stedman's medical eponyms*).

From Joachim Barrande[1] 19 June 1870

Prague | 419-Kleinseite | Choteksgasse
19 June 1870—

Most honoured master,

At the moment when your kind letter of 28 May ult. reached me, I was finishing these pages attached, in reply to a French geologist, who had informed me of his impressions after reading my work on the *distrib^n of Cephalop. Silur.*[2]

Even though these pages are not intended to be published, I would have had doubts about the courtesy of sending them to their destination without communicating them to you. I hope you will excuse my scruples, if the pages contain some circumstances that were not able to be brought thus far to your attention.

In any case, I am happy to take this opportunity, to express to you my sincere and profound admiration for such varied works, through which you have enriched science and which no one would appreciate more highly than me.

Accept, most honoured master, the expression of my most distinguished sentiments. | J Barrande

[Enclosure]

Prague
14 June 1870—

Copy.

Sir and most honoured colleague,

I highly praise your straightforward candour in recognising that the results of my studies on the Cephalop. Silur. upset your convictions concerning the doctrine of filiation and transformation, which you have adopted.

Since you intend to seek an entirely satisfying solution to the problem of successive appearance of organic beings, following from the theoretical principle in question, I call your attention to certain facts which seem unamenable to this theory.

According to the current views of transformists, Eozoon would be the first known type, or the first stock of all the animals.[3]

Now you know, like me, how the primitive type approaches that of the Polypidoms,[4] especially the elements of its structure. One could say, that between Eozoon and the Polypidoms, there is only one step to take, in the route of filiation and of transformation. According to the theoretical ideas, this step would also have had to be the first made along this path. The Polypidoms then would have to have covered the sea floor, after the Eozoic era and before the predominance of other types of higher structure.

In spite of this theoretical verisimilitude, one has not recovered any trace of calcareous Polypidoms, neither in the rocks that contain Eozoon, nor in the earths called anteprimordial, nor even on the horizon of the primordial Silurian fauna,

fauna one tells us is relatively recent in relation to the Eozoic eras. One only perceives the first vestiges of these Polypidoms belatedly, that is to say, in the second silurian fauna and their great development has no place except in the third fauna and subsequent eras.

It would be difficult to believe, that their traces have everywhere disappeared in formations before the second Silurian fauna, for, by reason of their calcareous nature, they would be preserved at least as well as the delicate and micoscopic remnants of Eozoon.

By contrast, the remains of Trilobites appear with frequency incomparable in all regions of the two continents, on in the most distant reaches where a fauna has been found, that is to say on the horizon of the primordial Silurian fauna.

Why did the fragile wrapping of these Crustaceas enjoy the privilege of preservation, if the sturdy wrapping of the Polypidoms, contemporary or previous, actually disappeared, in all regions that have been explored.

If one considers that the primordial fauna is found well developed in a large number of countries geographically very separated, one must ask, why do all the descendants of Eozoon, after a thousand supposed transformations, over an immense sequence of anteprimordial ages, and under the influence of the most diverse circumstances, were then so to speak given the word, to converge at the same time into the same trilobitic type, so well characterised and so uniform around the whole world, despite its local variations. This singular fact well merits explanation by your theories.

Let us also note that, in the primordial fauna the Trilobites are found amongst some other types, providing relatively rare individuals. These are primarily Brachiopods and Pteropods, which represent the Molluscs next to the Articulata.

If all these types are truly derived from a common prototype, their coexistence manifesting uniformly everywhere one encounters primordial fauna, we are led to believe that they are the most related by their structure among the articulata and the Molluscs, since they are the most related by the filiation of their common source. However, do zoologists show us this relative affinity between Trilobites, Brachiopods and Pteropods?

While admitting the common descent of all these types from the same prototype, there still remains the difficulty that the theory encounters at each step, that is, the inexplicable disappearance of all intermediate forms between the Eozoon and each of the primordial types. These forms would have to have been represented by innumerable individuals during the anteprimordial era. Can the theory account for this invariable disappearance in all countries, while it puts forth the preservation of Eozoon?

It is highly doubtful that an abrupt transformation by chance, which would seem to offer a happier solution than slow transformation, could be a more satisfactory explanation of all the facts.

To introduce any accidents whatever as a cause of the appearance of successive types, is to attribute these appearances to chance. However, he who ventures to invoke chance as an organizer of the gradual development of the animal sequence, when, to the eyes of all men of science, this sequence presents the results of a unique plan, marvellously executed, despite all the troubles undergone by the terrestrial crust and by all the material elements, amidst which life had to spread.

I end by wishing you the most ingenious and happy ideas, in your future theory. But you will not find it bad that I repeat to you that which I have already expressed in my last work on the Cephalopods: I do not hope that the human spirit will finally penetrate the mystery that envelops the first appearance and the succession of the forms of life on our planet.

Accept &c | J B

P.S— You will soon receive next my *Défense des Colonies—IV*[5] and you will find that I have established (p. 145–) this remarkable fact: that certain Trilobites of Bohemia, offering various oscillations of form, returned to their primitive form, after a long interval of time. These facts merit attention, since they make us imagine that the modifications of form produced under the influence of man, could not be in truth anything but temporary oscillations.

DAR 160: 44–5

[1] For a transcription of this letter in its original French, see pp. 169–71.

[2] CD's letter to Barrande of 28 May 1870 has not been found. Barrande refers to Barrande 1870. The geologist has not been identified.

[3] Barrande presumably refers to *Eozoon canadense.* In 1864, John William Dawson identified samples taken from pre-Silurian strata in eastern Canada as fossilised Foraminifera, single-celled protists with shells; he named the species *Eozoon canadense*, the 'Dawn animal from Canada' (Dawson 1864). Further samples were sent to William Benjamin Carpenter, an expert on Foraminifera, who confirmed Dawson's interpretation (Carpenter 1864). CD added information on the discovery of *Eozoon canadense* to *Origin* 4th ed., p. 371, as substantiating his claim, made in *Origin*, p. 307, that life existed before the Silurian period. The interpretation of the samples as pre-Silurian fossils remained controversial, however (see, for example, Carpenter 1866, and King and Rowney 1866); and by the end of the century, comparisons with similar, more recent, formations indicated that the samples were mineral in origin (see Schopf 2000).

[4] Polypiers (French) or polypidoms (English) are now (in English) polypariums. A polyparium is 'the common stem or supporting structure of a colonial cnidarian (or other colonial invertebrate), to which the individual polyps are attached', or (an obsolete meaning) a colonial cnidarian (*OED*).

[5] Barrande refers to the fourth volume of Barrande 1861–81.

From Camille Dareste[1] 2 July 1870

Ville de Lille | (Nord) | Museum d'Histoire Naturelle | rue des Fleurs, 1 | Lille
2 July *1870*

Dear Sir

Allow me to appeal once more to the infinite obligingness of which you have already shown me so much evidence.

My subject is still the race of *niata* cattle. The debate I have sustained on this matter, after having turned on the very existence of this race in America, has now been carried into quite a different domain. Some are now questioning whether the osteological characteristics of the *niata* race are the same as those I described in the head of a Calf born of the Flemish race, the description of which I sent you three years ago.[2]

For my part, I had at first believed in the existence of certain differences, though admittedly slight ones, for all I was only able to know the *niata* head which you brought back from America from Owen's brief description.[3] But having very recently been able to study the picture Rütimeyer gives in his memoir on the races of cattle, I have become convinced that these differences do not exist;[4] and that, at least when considering the profiles alone, there is, differences of age excepted, a true identity between form and structure.

But one point remains inconclusive for me. While looking again at the head whose description I sent you,[5] I became aware of a new and most interesting character that had completely escaped me during my first investigation. This character consists of the fact that the vomer is interposed between the two intermaxillary bones, thus forming part of the hard palate. A character the more remarkable in that it is only found among the Dolphins and a certain number of Cetaceans.

Can you tell me whether the skull that You brought back, and which is at the College of surgeons, exhibits a similar character? If it were so, we would very evidently have an incontestable document concerning the origin of the *niata* race.

Once again, I beg You to excuse my importunity. But for me this is a most important question, because of the conflict I encountered when I published these facts for the first time.[6] Besides, there is also a far more extensive question here of Your fine theory on the origin of species, which we shall have, I greatly fear, much trouble in getting accepted in France.

Please allow me to express my sincere and profound respects. | Camille Dareste. | Prof. at the Sc. fac. of Lille.

DAR 162: 47

1 For a transcription of this letter in its original French, see pp. 196–7.

2 Dareste refers to his work *Rapport sur un veau monstreaux* (Dareste 1867; see *Correspondence* vol. 15, letter from Camille Dareste, 19 May 1867 and n. 8). Niata cattle are an abnormally small race from South America, whose peculiarities are now seen as the result of a form of dwarfism. CD discussed niata cattle in *Journal of researches* 2d ed., pp. 145–6, and *Variation* 1: 90–2.

3 For Richard Owen's description of the skull that CD had brought back from South America, see [Owen] 1853, 2: 3832. The skull was in the Hunterian museum of the Royal College of Surgeons, London.

4 The illustration for Ludwig Rütimeyer's description of the niata was a woodcut made from a photograph of the skull in the Hunterian museum (see Rütimeyer 1866, p. 28).

5 See *Correspondence* vol. 15, letter from Camille Dareste, 19 May 1867.

6 Dareste refers to his exchange of papers with André Sanson in the *Comptes rendus hebdomadaires des séances de l'Académie des sciences* (see *Correspondence* vol. 15, letter from Camille Dareste, 19 May 1867 and n. 9).

From Ernst Haeckel[1] 6 July 1870

Jena
6 July 70

Highly esteemed, dear friend!

I am very glad to see from your kind letter that you are now more satisfied with your health and that you have fortunately completed the manuscript of your important book on man, so eagerly awaited on all sides.[2] I am terribly excited about it, and am looking forward already to the rich instruction that I shall derive from it. In our German newspapers two opposed camps are already quarrelling about the question whether you will carry out a full and consistent application of the theory of descent to man.

Mr Quatrefages[3] in Paris, and the other naive fellows who have so boldly claimed that "Darwinian theory" does not apply to man, will get a nasty surprise when they find out that Charles Darwin himself acknowledges man's descent from animals as the most natural and important consequence of his theory.

—I would be delighted if you are pleased with the new edition of "Natürliche Schöpfungsgeschichte". I have devoted the whole winter to its improvement. In lecture IV a new passage on *Treviranus* will perhaps be of interest.[4] I have taken great pains to substantially improve the genealogical classification, and I believe that the genealogical trees of the various classes of animals are now essentially complete. Of course, they can never be wholly satisfactory.

—I am at the moment occupied exclusively with the monograph on *calcareous sponges*, which takes a lot of effort and time.[5] The small animal group is, however, very interesting due to its infinite changes of form and the innumerable transitional stages among the most varied forms. One can accept 4 or 20 or 200 "canal species" of Grantia, or consider all Calcispongiae as just one single species.[6] In order to complete the sections on the precise anatomy and ontogeny of the calcareous sponges, I shall visit the Mediterranean between August and October, either Trieste or Genoa. Still, it will very likely take a couple of years before my book on Grantia is quite ready.

—The "Natürl. Schöpfungsgeschichte" is now going to be translated into Russian and Italian, and Mr. *Ray Lankester* has told me that it may be translated into English.[7] Since I treasure English literature so very much, I would be especially pleased by this honour.

—Next winter I will again give a new cycle of lectures on "Darwinism".—

I'm sure you were very pleased about Gegenbaur's exposition of the theory of descent in his "Vergleichenden Anatomie".[8] My colleagues Gegenbaur and Preyer[9] send their most respectful greetings and I remain, as ever, with the profoundest reverence and best wishes for your health,

Faithfully yours | Ernst Haeckel

DAR 166: 54

1 For a transcription of this letter in its original German, see pp. 201–2.

2 The reference is to *Descent*.

3 Armand de Quatrefages.

4 In the second edition of *Natürliche Schöpfungsgeschichte* (Natural history of creation; Haeckel 1870a, pp. 83–6), Haeckel maintained that Ludolph Christian Treviranus had presented a monistic interpretation of the unity of nature and argued for the genealogical connectedness of all organisms. For more on Haeckel's monistic philosophy see Haeckel 1899.

5 Haeckel refers to *Die Kalkschwämme: eine Monographie* (Calcareous sponges: a monograph; Haeckel 1872).

6 *Grantia*, a genus of Calcispongiae (now the class Calcarea), is characterised by both radial and incurrent canals (a syconoid canal system; see Bergquist 1978, p. 151), which allow the circulation of water throughout the organism.

[7] Edwin Ray Lankester's translation of Haeckel's book, based on the German edition, appeared in 1876 (Haeckel 1876).

[8] Haeckel refers to Carl Gegenbaur and *Grundzüge der vergleichenden Anatomie* (Elements of comparative anatomy; Gegenbaur 1870). CD's copy is in the Darwin Library–Down.

[9] William Thierry Preyer.

From Armand de Quatrefages[1] 18 July 1870

Paris
18 July 1870

Dear Sir and colleague

Be so good, pray, as to tell me whether you have published the volume on the Geology of South America that was supposed to be part of the Expedition of the Beagle.—[2] I have looked for it in vain in our library at the Museum; I shall seek it elsewhere. But in case my investigations should prove fruitless, it is very important for me to know whether it has appeared.— It is being said at the Academy, with a view to harming your candidature, that you had done no more than collect natural history specimens and that you had called on assistants to describe them.[3] I am bent on showing the part you had in the publication.— Tell me if you please what the format is, and how many pages and maps there are—

No doubt you already know that in an earlier nomination Brandt only won out over you at the second ballot.[4] But we have three vacancies for correspondents to award and I very much hope that we will succeed, Edwards[5] and I, in convincing the Academy. We have already had two discussion sessions. Next Monday we shall start again and it will be my turn to speak. Please believe that I shall do what I can to make the good cause win; but you are being strongly attacked.[6] Alas! If there were nothing but battles of this sort in the world, we should be very fortunate!—[7]

Your most devoted | De Quatrefages

DAR 175: 7

[1] For a transcription of this letter in its original French, see p. 216.

[2] CD's geological observations related to the voyage of the *Beagle* were published in *Coral reefs*, *Volcanic islands*, and *South America*; all three appeared in one volume as *Geology of the 'Beagle'*. Up to this time, there had been no French translation of any part of the work.

[3] Quatrefages refers to the Académie des Sciences. The zoological material related to the voyage of the *Beagle*, *Zoology*, was edited and overseen by CD, but individual parts were done by different authors. See *Correspondence* vol. 2. For more on the opposition to CD's nomination as corresponding member of the académie, see n. 6, below.

[4] Johann Friedrich von Brandt was elected on 4 July 1870 (see letter from J. D. Hooker, [6 or 7 July 1870] and n. 5).

[5] According to a report in the *Revue des Cours Scientifiques*, 23 July 1870, p. 529, Henri Milne-Edwards spoke on 18 July 1870 in favour of CD's admission.

[6] Emile Blanchard spoke for over an hour against CD's admission. He argued that on the *Beagle* voyage, CD had merely collected and was not capable of making a scientific study of his collections; he held that CD's theory of coral island formation was mostly derivative. He further claimed that CD's pigeon studies demonstrated a lack of true scientific spirit, that his work on cirripedes contained hardly any new

facts, and that his theory of descent was both false and unoriginal. Blanchard apparently argued against transformism with examples from Jean Baptiste de Lamarck (*Revue des Cours Scientifiques*, 23 July 1870, p. 529).

7 Quatrefages alludes to the Franco-Prussian war; France declared war on 19 July 1870 (Wawro 2003, p. 65).

From Paul Broca[1] 4 August 1870

Paris,
4 August 1870

Dear Sir,

I am pleased by the mark of benevolence you were so good as to give me; but I do fear that the second part of my discourse, which appeared the day after you had written to me, may have had the misfortune of displeasing you.[2] The first part was a historical summary, and here I had to speak of you with the regard that is due to a great naturalist; but the second part was a critical discussion and you may have to excuse me some rather lively passages against Natural Selection.[3] Despite this, I beg you not to confound me with the herd of your systematic adversaries, any more than Quatrefages and Milne-Edwards, who are currently supporting your candidature, and whose efforts I would second if I had the honour of belonging to the Academy of Sciences[4]

Please accept, dear Sir, along with my thanks for your amiable letter, the respectful expression of my high esteem | P Broca

DAR 160: 313

1 For a transcription of this letter in its original French, see p. 224–5.

2 CD's letter has not been found, but was evidently written on 29 July 1870. Broca refers to his lecture 'Sur le transformisme', delivered at the Société d'Anthropologie de Paris in April 1870 and published in the 23 July and 30 July 1870 issues of the *Revue des Cours Scientifiques* (Broca 1870a). A longer version of Broca's paper appeared later in the *Bulletins de la Société d'Anthropologie de Paris* (Broca 1870b).

3 See Broca 1870a, pp. 553–8. Broca argued that natural selection could only influence functional characteristics, and that many features in an organism had no functional significance; he concluded that while species were not fixed, the causes of their evolution remained undiscovered.

4 Broca refers to Armand de Quatrefages and Henri Milne-Edwards, both of whom were supporting CD's candidacy for corresponding membership in the zoological section of the French Académie des Sciences. See letter from Armand de Quatrefages, 18 July 1870.

From Alexander Onufrievich Kovalevsky[1] 25 September 1870

Naples.
25 Septem 70.

Dear Sir

My brother has let me know that you want to insert a citation in the work you are publishing, to the effect that my new studies on the development of the Ascidians confirm my earlier assertions that the larvae of simple Ascidians are vertebrate animals.[2]

The results of my new research on the development of the Ascidians are now in press, in *Max Schultze*'s Archiv fur Mikroscopische Anatomie.[3] In this article, I confirm, with the help of a series of new observations, that the larvae of the Ascidians are true vertebrates, and that their nervous system forms by means of a dorsal sulcus which closes into a tube—the nervous system of the Ascidians; I also confirm that the dorsal chord of the larvae of the Ascidians develops in the same way as in the vertebrates; I give drawings of the transverse section of the larva (fig.) which shows the nervous system $^{(n)}$, the chord $^{(ch)}$, the intestine $^{(d)}$ and the muscle cells $^{(m)}$

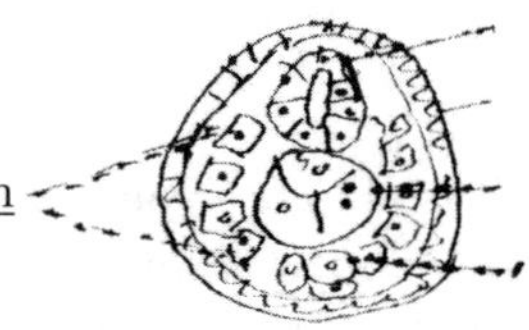

I hope that I shall have the pleasure of sending you a copy of my new studies on Ascidian development in the course of the month of November or December.—

In the last year (1869) I have studied annelid embryology and I have found that the dorsal chord exists in many Annelids.— Perhaps you recall, Sir, the great nerve tube(?)—*riesiege Rohren Nervenfasern* that Leydig describes in Lumbricus;[4] *Claparéde* (Zeitschrift fur Wissenschaftliche Zoologie Bd. XIX p. 588, 89–90) has demonstrated that this enormous quasi Nerve is found outside the Neurilemma.[5] My studies on the embryology of these *tubes* have shown me that they form in the same way as the vertebrate chord. I found three embryonic layers in Lumbricidae embryos—corresponding to the three embryonic layers in vertebrates; the nervous system and the exterior epithelium form from the external layer, the epithelium of the intestinal canal from the internal layer (Darmdrüsenblatt),[6] the muscles, the neurilemma and *the chords or "riesiege Röhrenfasern"* of Claparède from the median layer.

My study on Earthworm embryology is in press in the Memoirs of the Acad. of St. Petersburg.[7]

My brother writes that it would please you if I studied the organisation of the male in cirripedes (Scalpellum) as well as the history of the development and formation of the male and of his genital organs compared with the female.[8] I would do everything I could to resolve these ??, the more so because I have encountered a comparable fact in another form, Bonelia viridis.[9] Namely, I find that the male of Bonelia is a small Planarian, or rather Turbellarian, a millimetre long.[10] The males are found in the female's uterus, and in each Bonelia female one finds 4 or 5 males— In the male's organisation one finds several cases of resemblance with the female; the organisation of the genital organs of the male and female is very similar.[11] It is indeed possible that in Bonelia we find an example in which just one sex (female) gradually changed its primitive form, whilst the other sex (the male) remained in its primitive form, the Turbellarian. I hope to be able to study the development of Bonelia, that ought to resolve the question.

I will remain at Naples for another whole year and I should be most happy if I could be of service to you either by sending you some form of fauna from the gulf of

Naples, or by carrying out some studies whose results would be of interest exclusively to you

In sending you this letter, dear Sir, I shall take the opportunity to express all my attachment to the ideas with which you have enriched science and to offer you my most distinguished greetings | Prof. A Kowalevsky

DAR 169: 52

1 For a transcription of this letter in its original French, see pp. 248–50.

2 Alexander's brother was Vladimir Onufrievich Kovalevsky. See the letter from V. O. Kovalevsky, 5 September [1870] and n. 4.

3 Kovalevsky refers to A. O. Kovalevsky 1871 and to Max Johann Sigismund Schultze. In *Descent* 1: 205, CD referrred to Kovalevsky's earlier paper on the embryology of ascidians (A. O. Kovalevsky 1866b) and to the fact that Kovalevsky had done further research on the subject, but did not cite the later article. No copy of A. O. Kovalevsky 1871 has been found in the Darwin Archive.

4 'Riesiege Rohren Nervenfasern': giant tube of nerve fibres (German; a reference to the median giant fibre). Franz von Leydig described this organ in *Lumbricus* in Leydig 1864, pp. 154–5.

5 The reference is to Edouard Claparède and Claparède 1869. Claparède noted the similarity between the nerve sheath of the earthworm and the myelin sheath of vertebrates (Claparède 1869, p. 590).

6 'Darmdrüsenblatt': gut-gland-layer (German; a reference to the endoderm).

7 Kovalevsky refers to his paper on the embryology of worms and arthropods (Kovalevsky 1869). It appeared in the 1871 volume of *Mémoires de l'Académie Impériale des Sciences de St.-Pétersbourg*.

8 CD's letter to Vladimir Onufrievich Kovalevsky has not been found, but see the letter from V. O. Kovalevsky, 20 September 1870.

9 *Bonelia viridis* is an incorrect subsequent spelling of *Bonellia viridis*, the green spoonworm.

10 *Bonellia viridis* is no longer placed within the family Planariidae of the class Turbellaria. It is in a separate phylum, Echiura.

11 *Bonellia viridis* is sexually undifferentiated in larval stages, and differentiation is determined by environmental factors alone. For more on the development of sexes in this species, see Berec *et al.* 2005.

From L. A. J. Quetelet[1] 15 December 1870

Académie Royale
des Sciences, des Lettres et des Beaux-Arts
de Belgique.
Art. 14 of the organic Statutes.— The Academy's correspondence
will be conducted by the Perpetual Secretary,
organ and interpreter of that Company.
Indicator, N° 6779.

Brussels,
15 December *1870*.

Dear Sir,

I have the honour of informing you that the science section of the Royal Academy of Belgium, wishing to display the esteem that it has for your work, has elected you an *associate* at today's meeting.[2]

In conferring this distinction upon you, the section hopes to see you participate in its work, by communicating whatever might be of interest to it in the branch of science that you cultivate.

I am happy, Sir, to be the Company's organ and interpreter in this matter, and to be able to offer you my congratulations together with my most distinguished respects.

The Perpetual Secretary, | Quetelet

to Mr. Darwin | Associate of the Academy, | at Down, Beckenham.

DAR 175: 12

1 For a transcription of this letter in its original French, see p. 314.

2 The diploma was dated 16 December 1870, but not sent until January 1871 (see *Correspondence* vol. 19, letter from L. A. J. Quetelet, 11 January 1871, and *ibid.*, Appendix III).

From Edouard van Beneden[1] 17 December 1870

Liège
17 December 1870

Dear Sir and illustrious Colleague,

I felt a real joy on learning of your nomination as an Associate member of the Académie royale de Belgique and I cannot prevent myself from telling you how happy I am to see that in this matter the Belgians have taught the members of the Institut de France a lesson.—[2] May all the learned bodies of the world protest, like the Academy of Belgium, against the debate they dared instigate on the question of your scientific merits. Your name summarises the whole scientific movement of recent years and your immortal work is above all attacks by clericalism or ignorance.— I was indignant, as must all be who know your work, on reading this discussion, which is another blot on the record for this institution, to add to all those that French science has experienced in recent years.—

In Belgium the younger generation has hoisted the standard of intellectual independence and has thrown off prejudice and preconceived ideas. And if among men of a riper age there are some who have not accepted your ideas, at least they have been able to place your merits above their personal convictions.—

I am happy and proud to have been named a correspondent of the Academy on the same day that you were Elected as an associate.—

I am sending you copies of some of my work.[3] Please receive them as a demonstration of my profound respect. As you will see, I am working on the same question as Dohrn: the embryogeny of the Crustaceans.[4] I think I have reached a fine result: the demonstration through embryogeny of the derivation of the articulates from a sort of Worm. (Embr. of Sacculina.—id. of the Lernaeans)[5]

I am continuing these researches and publications and I hope soon to be able to construct the genealogical tree for this group.— I have not come to the same conclusions as Dohrn.[6] I hope to be able to send you the development of Nebalia very soon.—[7]

Recently I had the opportunity to study the organisation of *Anelasma squalicola.* I hold, with Dohrn, that this is the transition to the Peltagastes.[8] I was perfectly able to find the roots plunged into the flesh, just as you depicted them.—[9] The mouthparts exist in a rudimentary state.

Excuse the liberty I have taken in writing to you and accept, dear Sir and illustrious colleague, my profoundest respects. | Prof. D^r. Edouard Van Beneden | rue Vinave d'Ile, *Liège*

DAR 160: 132

1 For a transcription of this letter in its original French, see pp. 315–16.

2 See letter from L. A. J. Quetelet, 15 December 1870. CD had been nominated for the zoological section of the Académie des Sciences on 27 June 1870 (*Comptes Rendus de l'Académie des Sciences* 70 (1870): 1418). The Institut de France includes five academies, one of which is the Académie de Sciences. CD was nominated a further five times without success, until he was finally elected to the botanical section in 1878 (Corsi and Weindling 1985, p. 699). For more on the controversy over CD's election, see the letter from Armand de Quatrefages, 18 July 1870 and n. 6.

3 There are offprints of Beneden 1868, 1869a, 1869b, 1870a, 1870b, 1870c, and 1870d, and Beneden and Bessels 1868, in the Darwin Pamphlet Collection–CUL; Beneden 1868 and 1870b are lightly annotated.

4 Anton Dohrn had written several articles, later collected into a book (Dohrn 1870), on the morphology and development of crustaceans.

5 See Beneden 1870a, pp. 105–6. Beneden suggested that the early development in embryos of *Sacculina* (a genus of Rhizocephala) resembled that of many annelid worms. Beneden had earlier described the embryonic development of *Lernaea*, a genus of parasitic copepods, commonly known as anchor worms (Beneden and Bessels 1868, pp. 440–1).

6 Dohrn outlined some of his differences with Beneden concerning the genealogy of Crustacea in Dohrn 1870, pp. 159–61.

7 Beneden did not publish on *Nebalia*, a genus of sea fleas, now in the order Leptostraca. *Nebalia* had been considered by some to be a divergent form of Phyllopoda, but by 1872 Carl Friedrich Claus had reclassified it as a primitive form of Malacostraca (see Claus 1872, pp. 329–30).

8 The barnacle species *Anelasma squalicola* lives on some species of sharks with its peduncle embedded in the shark's skin. *Peltogaster* is a genus of Rhizocephala that parasitises hermit crabs. CD had suggested *A. squalicola* might be a link to Rhizocephala after reading F. Müller 1864 (see *Correspondence* vol. 13, letter to Fritz Müller, 10 August 1865 and n. 7). For Dohrn's views on the relationship between *A. squalicola* and *Peltogaster*, see Dohrn 1875, p. 72.

9 See *Living Cirripedia* (1851): 380–1 and plate IV, figure 3.

From L.-G. De Koninck[1] 19 December 1870

Liège
19 December 1870

My dear Colleague,

You will undoubtedly already have received the official news of your nomination as foreign associate of the Académie royale des sciences of Belgium.[2] In spite of that, I cannot resist the desire to congratulate you on this nomination, about which I would have been happy to have been the first to inform you and to which I was able to contribute with my vote. The Academy does itself honour by associating itself with men like you, my dear Colleague, who work unremittingly for scientific progress and head straight for the goal they have set themselves, without stopping at obstacles they encounter on the way, because they know that truth lies there.

Please accept along with my congratulations, my dear Colleague, the expression of my high esteem and devoted regards | L. de Koninck

To Mr Darwin, Member of the Académie royale of Belgium

DAR 169: 51

1 For a transcription of this letter in its original French, see p. 322.

2 See letter from Edouard van Beneden, 17 December 1870.

From John Murray 2 April [1867][1]

[Enclosure]

E. Schweizerbart, | *Verlagsbuchhandlung.* | Stuttgart
29 March | 1867

Mr John Murray in London.

Having heard from Mr Ch. Darwin that his new work: The Variation of Animals & Plants etc will soon go to press and that he has allowed me to prepare a German edition of this book, I now apply to you with a request to provide me with clichés of the woodcuts. In the case of Orchids,[2] Williams & Norgate kindly sent the package to Leipzig where I myself paid for the cost of the clichés; perhaps this could be handled the same way.

I thank you in advance for your kind attention in this matter and remain with collegial respect | E Schweizerbart[3]

National Library of Scotland (John Murray Archive) (Ms. 42153 ff. 30–1)

1 For a transcription of the enclosure in its original German, and the letter with which it was enclosed see p. 406.

2 Schweizerbart's firm had published a German translation of *Orchids* (Bronn trans. 1862).

3 In his business communications, C. F. Schweizerbart continued to use the signature of his uncle, Wilhelm Emanuel Schweizerbart, from whom he had purchased the publishing firm in 1841.

APPENDIX II
Chronology 1870

This appendix contains a transcription of Darwin's 'Journal' for the year 1870. Darwin commenced his 'Journal' in August 1838 and continued to maintain it until December 1881. In this small notebook, measuring 3 inches by $4\frac{1}{2}$ inches, Darwin recorded the periods he was away from home, the progress and publication of his work, and important events in his family life.

The version published by Sir Gavin de Beer as 'Darwin's Journal' (de Beer ed. 1959) was edited before the original 'Journal' had been found and relied upon a transcription made by an unknown copyist. The original, now in the Darwin Archive in Cambridge University Library (DAR 158), reveals that the copyist did not clearly distinguish between the various types of entries it contains and that the transcription made was incomplete.

From 1845 onward, Darwin recorded all that pertained to his work (including his illnesses, since these accounted for time lost from work) on the left-hand pages of the 'Journal', while the periods he was away from home, and family events, were noted on the right-hand pages. In order to show clearly Darwin's deliberate separation of the types of entries he made in his 'Journal', the transcription has the left- and right-hand pages labelled.

All alterations, interlineations, additions, and the use of a different ink or pencil have been noted. In addition, the editors have inserted additional information relevant to Darwin's correspondence throughout this transcription of the 'Journal' for 1870. These interpolations are enclosed in square brackets to distinguish them from Darwin's own entries, the source of the information being given in the footnotes.

[**Left**]

1870

I find Murray has printed 9000 copies of my Journal of Travels, & Colburn 1500 = 10,500 copies[1]

The whole of the year at work on the Descent of Man & Selection in relation to Sex.[2] I began this work on Febr 4. 1868; but many interruptions. Went to press Aug 30/70/ & finished last proof

[15 June. French translation of *Orchids* published.][3]
[9 August. CD sends manuscript of *Descent* to printer.][4]
[Before 4 September. Proofs of *Descent* sent to CD.][5]
[Before 24 October. *Descent* sent to press.][6]

[Right]

1870

[8–10 January. Francis Galton visits.][7]
[22–4 January. Alfred Newton, Albert Günther, and Robert Swinhoe visit.]
[23–4 January. Joseph Dalton Hooker visits.][8]
March 5–10 Erasmus[9]
[15 April. Rudolf Albert von Kölliker and Albert Günther visit][10]
May 20–24 Cambridge
[13 June. J. D. Hooker visits.][11]
Jun 24 to July 1. Erasmus
[26 and 30 June. CD visits Zoological Gardens, Regent's Park, London.]
[10 July. William Farr visits.][13]
Aug 13–26 Southampton at William[14]
[27 August – 1 September. Vladimir Onufrievich Kovalevsky and Sofia Vasilyevna Kovalevskaya visit during this period.][15]
[26 September. Anton Dohrn visits.][16]
[29 September. Thomas Henry and Henrietta Anne Huxley visit.][17]
[8–12 October. V. O. Kovalevsky and S. V. Kovalevskaya may have visited during this period.][18]
Oct 13–20 Leith Hill & Ightham Mote[19]
Dec. 8–14 Erasmus

1 CD refers to different printings and editions of his *Journal of researches*. See letter to John Murray, 22 October [1870] and n. 3.
2 *Descent.*
3 See letter to Louis Rérolle, 15 June 1870 and n. 1.
4 See letter to Mr Dorrell, 9 August 1870.
5 See letter to John Murray, 4 September [1870].
6 See letter from John Murray, 24 October [1870].
7 Emma Darwin's diary (DAR 242).
8 Emma Darwin's diary (DAR 242).
9 CD's brother, Erasmus Alvey Darwin, lived at 6 Queen Anne Street, London. It is likely that CD actually stayed in London until 12 March, based on Emma Darwin's diary (DAR 242) and the letter to J. D. Hooker, 8 March [1870], where CD writes that he returns home on Saturday. In 1870, 12 March was a Saturday.
10 Letter from Albert Günther, 12 April 1870.
11 Letter to St G. J. Mivart, 13 June [1870].
12 See letter from T. W. Wood, [after 26 June 1870] and n. 1, and letter to J. J. Weir, 29 June [1870]).
13 Emma Darwin's diary (DAR 242).

[14] William Erasmus Darwin.

[15] See letter from V. O. Kovalevsky, 15 August [1870], and letter from S. V. Kovalevskaya and V. O. Kovalevsky, 1 September 1870.

[16] See letter to J. D. Hooker, 27 September [1870].

[17] Emma Darwin's diary (DAR 242).

[18] See letter from V. O. Kovalevsky, 5 October 1870, and letter from Henrietta Emma Darwin to L. M. Forster, 12 October 1870 (DAR 245: 417).

[19] Leith Hill Place, Surrey, was the home of Josiah Wedgwood III and Caroline Wedgwood. Ightham Mote is a fourteenth-century house near Sevenoaks, Kent.

APPENDIX III
Diplomas presented to Charles Darwin

In 1870, Darwin received the following diplomas. Although not letters in the conventional sense, diplomas represent significant communication between Darwin and scientific organisations, and the citations in such diplomas often provide valuable indications of those aspects of Darwin's work that were considered worthy of honour. In view of this, they have been included here, with translations.

From Société Impériale des Naturalistes de Moscou 19 February 1870

AUSPICIIS AUGUSTISSIMI
POTENTISSIMI ATQUE CLEMENTISSIMI PRINCIPIS
ALEXANDRI SECUNDI
OMNIUM RUSSIARUM
IMPERATORIS ET AUTOCRATORIS,
ET CETERA, ET CETERA, ET CETERA
SOCIETAS CÆSAREA NATURÆ CURIOSUM MOSQUENSIS,
CONVENTU DIE XIX FEBRUARII ANNI MDCCCLXX MEMBRIS SUIS
ADSCRIPSIT HONORARIIS
VIRUM NOBILISSIMUM, DOCTISSIMUM
CAROLUM R. DARWIN ESQ. F.R.S.

Praeses Princeps Schizinsky-Schichmotoff
Vice-Praeses Alexander Fischer de Waldheim.
Secretarii Dr. Renard
D[r]. N. *[illeg]*

№. 1036
Diarii generalis Soc.
anni 1870.

[Translation]

UNDER THE MOST VENERABLE AUSPICES
OF THE MOST POWERFUL AND MOST MERCIFUL PRINCE
ALEXANDER THE SECOND

EMPEROR AND SOVEREIGN
OF ALL RUSSIANS
ET CETERA, ET CETERA, ET CETERA·
THE IMPERIAL SOCIETY OF NATURALISTS OF MOSCOW,
CONVENING ON THE NINETEENTH DAY OF FEBRUARY OF THE YEAR 1870,
APPOINTED TO ITS HONORARY MEMBERS
THE MOST NOBLE AND MOST LEARNED
CHARLES R. DARWIN ESQ. F.R.S.

President President Schizinsky-Schichmotoff
Vice-President Alexander Fischer de Waldheim.
Secretary Dr. Renard
D^r. N. *[illeg]*
N°. 1036 General diaries of the Soc. of the year 1870.

DAR 229: 23

[1] Schizinsky-Schichmotoff has not been identified.
[2] Alexander Fischer von Waldheim.
[3] Charles Renard.
[4] N. *[illeg]* has not been identified.

From Società Geografica Italiana 15 May 1870

SOCIETÀ GEOGRAFICA ITALIANA

La Società Geografica Italiana nell'adunanza generale del 14 Marzo 1870, udita la proposta della Presidenza,

Ha deliberato ad unanimità di eleggere a Socio d'onore il Signor Carlo Darwin

La Presidenza ha quindi ordinate ordinato di iscriverlo nell'albo dei Soci d'onore, e d'inviargli il presente diploma.

Firenze, 15 Maggio 1870.

PRESIDENTE
Negri Cristoforo[1]

Il Segretario
Giglioli[2]

Articolo 26 dello Statuto della Societa

Il Consiglio elegge fra gli stranieri dei membri corrispondenti e propone alla Società la nomina di Soci d'onore.

A Soci d'onore sono eleggibili soltanto nomini universalmente acclamati per aver fatto progredire la scienza. Se questi si presentano alla Societè siedono coi membri del Consiglio e si ammettono al voto.

[Translation]

ITALIAN GEOGRAPHICAL SOCIETY

The Italian Geographical Society at its general meeting of the 14 March 1870, having heard the proposal of the President,

Unanimously resolved to elect as an honorary Member Mr Charles Darwin

The President accordingly commanded that he be recorded on the roll of honour of the Society, and that he be sent this diploma.

Florence, 15 May 1870.

PRESIDENT
Negri Cristoforo

Secretary
Giglioli

Article 26 of the Statutes of the Society

The Council elects among foreigners correspondent members and proposes to the Society the nomination of honorary Members.

An honorary Member is only eligible for nomination by universal approval for having advanced science. If they present someone to the Society the members of the Council will sit and put it to the vote.

DAR 229: 24

[1] Cristoforo Negri.
[2] Enrico Hillyer Giglioli.

APPENDIX IV
Darwin's *Queries about expression*

In 1868, Darwin sent out a number of printed copies of a list of queries on human expression (see, for example, enclosure to letter to G. H. K. Thwaites, 31 January [1868]). In the Darwin Archive–CUL, three copies of these queries have been found (DAR 53: B2, DAR 181: 41, DAR 186: 1). There is also what is apparently a proof copy of this printed version (DAR 96: 1) with handwritten corrections on it, some by Darwin, that are incorporated into the three other copies. It is likely that this printed version was not made until late 1867 or early 1868 (see Freeman and Gautrey 1972 and 1975), and that it represents a near final version of the queries; in particular, questions 2, 5, 6, 10, and 13 contain additions that rarely appear in extant early versions of the queries, but that do appear in the version that Darwin reproduced in the introduction to *Expression*. The final version of the printed queries in the Darwin Archive–CUL is transcribed below for its own intrinsic interest and as a convenience to readers wishing to see the printed questions when answers to a now unavailable list of queries appear in the *Correspondence*. At least two additional copies of Darwin's printed list of queries survive in other archives (see Freeman and Gautrey 1975).

QUERIES ABOUT EXPRESSION

(1.) Is astonishment expressed by the eyes and mouth being opened wide, and by the eyebrows being raised?

(2.) Does shame excite a blush when the colour of the skin allows it to be visible? and especially how low down the body does the blush extend?

(3.) When a man is indignant or defiant does he frown, hold his body and head erect, square his shoulders and clench his fists?

(4.) When considering deeply on any subject, or trying to understand any puzzle, does he frown, or wrinkle the skin beneath the lower eyelids.

(5.) When in low spirits, are the corners of the mouth depressed, and the inner corner of the eyebrows raised by that muscle which the French call the "Grief muscle?" The eyebrow in this state becomes slightly oblique, with a little swelling at the inner end; and the forehead is transversely wrinkled in the middle part; but not across the whole breadth, as when the eyebrows are raised in surprise.

(6.) When in good spirits do the eyes sparkle, with the skin a little wrinkled round and under them, and with the mouth a little drawn back at the corners?

(7.) When a man sneers or snarls at another, is the corner of the upper lip over the canine or eye tooth raised on the side facing the man whom he addresses?

(8.) Can a dogged or obstinate expression be recognized, which is chiefly shewn by the mouth being firmly closed, a lowering brow and a slight frown?

(9.) Is contempt expressed by a slight protrusion of the lips and by turning up the nose, with a slight expiration?

(10.) Is disgust shewn by the lower lip being turned down, the upper lip slightly raised, with a sudden expiration, something like incipient vomiting, or like something spat out of the mouth?

(11.) Is extreme fear expressed in the same general manner as with Europeans?

(12.) Is laughter ever carried to such an extreme as to bring tears into the eyes?

(13.) When a man wishes to shew that he cannot prevent something being done, does he shrug his shoulders, turn inwards his elbows, extend outward his hands and open the palms; with the eyebrows raised?

(14.) Do children when sulky, pout or greatly protrude the lips?

(15.) Can guilty, or sly, or jealous expressions be recognized? though I know not how these can be defined.

(16.) As a sign to keep silent, is a gentle hiss uttered?

(17.) Is the head nodded vertically in affirmation, and shaken laterally in negation?

Observations on natives who have had little communication with Europeans would be of course most valuable, though those made on any natives would be of much interest to me.

General remarks on expression are of comparatively little value; and memory is so deceptive that I earnestly beg it may not be trusted.

A definite description of the countenance under any emotion or frame of mind, with a statement of the circumstances under which it occurred, would possess much value. An answer within six or eight months, or even a year, to any *single* one of the foregoing questions would be gratefully accepted. In sending answers, the questions need not be copied, but reference may be made to the numbers of each query.

CHARLES DARWIN,

DOWN, BROMLEY, KENT,
1867.

MANUSCRIPT ALTERATIONS AND COMMENTS

The alteration notes and comments are keyed to the letter texts by paragraph and line numbers. The precise section of the letter text to which the note applies precedes the square bracket. The changes recorded are those made to the manuscript by CD; changes of hand in letters written partly by CD and partly by amanuenses are also recorded. Readers should consult the Note on editorial policy in the front matter for details of editorial practice and intent. The following terms are used in the notes as here defined:

del	deleted
illeg	illegible
interl	interlined, i.e., inserted between existing text lines
omitted	omitted by the editors to clarify the transcription
over	written over, i.e., superimposed

To Anton Dohrn 4 January 1870
0.1 Beckenham] *before del* '*Bromley.*'
6.2 daily] *after del* 'dela'
Enclosure
0.1 Beckenham] *before del* '*Bromley.*'

To Abraham Dee Bartlett 5 January [1870]
0.1 Beckenham] *before del* '*Bromley.*'
2.1 the skin round the 2.2] *above del* 'its'
3.2 or] *over* 'C'

To Richard Kippist 10 January [1870 or 1871]
0.1 Beckenham] *before del* '*Bromley.*'

To Albert Günther 13 January [1870]
0.1 Beckenham] *before del* 'Bromley.'

To Adolf Reuter 15 January [1870]
3.1 My . . . Darwin] *in CD's hand*

To Albert Günther 16 January [1870]
0.1 Beckenham] *before del* '*Bromley.*'
3.1 Many] 'M' *over* 'm'

To Richard Kippist 16 January [1870 or 1871]
0.1 Beckenham] *before del* '*Bromley.*'
1.2 & I sh^d^ . . . as] *interl*
1.2 wishes] *altered from* 'wishing'

To Charles Boner 20 January [1870]
0.1 Beckenham] *before del* '*Bromley.*'

To A. R. Wallace 26 January [1870]
0.1 Beckenham] *before del* '*Bromley.*'
1.2 still] *interl*
2.6 several] *after del illeg*
3.1 Man] 'M' *over* 'm'

To the American Philosophical Society 5 February 1870
0.1 Beckenham] *before del* '*Bromley.*'

To Henrietta Emma Darwin [8 February 1870]
1.2 , also,] *interl*
1.5 marked thus] *interl*
2.1 foot-notes] 'foot-' *interl*
2.4 separate] *interl*

2.4 them] *interl after illeg*
2.5 each] *after del* 'them'

To Alfred Newton 9 February [1870]
0.1 Beckenham] *before del* '*Bromley.*'

To Gustav Jäger 17 February 1870
1.1 I] *interl*
1.2 since] *after del illeg*

To W. T. Preyer 17 February [1870]
0.1 Beckenham] *before del* '*Bromley.*'
1.2 certainly] *interl*
1.10 rate of] *interl*
1.11 of the poison] *interl*
2.1 in . . . optics 2.2] *interl*
3.3 any taste for] *interl*
3.6 person] *interl*
3.10 ever since] *interl*
3.14 benefit] *after del* 'was'
3.16 Near] *after del* 'At'
3.17 always] *interl*
3.18 much] *above del* 'a great'
3.19 by far] *interl*
3.21 made to] 'to' *interl*

To W. D. Fox 18 February [1870]
0.1 Beckenham] *before del* '*Bromley.*'
1.4 here] *over* '&'
1.8 on] *interl*
2.9 has] *over* 'he'
2.10 I fear] *interl*

To Joseph Dalton Hooker 21 February [1870]
0.1 Beckenham] *before del* '*Bromley.*'
1.5 one] *over* 'on'
1.5 living . . . island, 1.6] *interl above del* 'on mainland'
1.6 Land] *interl*]
1.10 island] *above del* 'land'
2.3 Round Island 2.4] *above del* 'Isl^{d}. Mauritius.'
2.4 Mauritius] *above del* 'latter'
2.4 island; 1.6] *over del* 'islands'
2.6 & the] 'the' *interl*
2.6 marine] *after del* 'number of recent'
2.13 & on] 'on' *interl*

To James Orton 24 February [1870]
0.1 Beckenham] *before del* '*Bromley*'
1.4 old] *interl*
3.5 is] *interl*

To John Lubbock 26 February [1870]
0.1 Beckenham] *before del* '*Bromley*'
2.1 you.] *interl*

To J. D. Hooker 8 March [1870]
1.1 on] *over* 'th'
1.5 every day 1.6] *interl*
7.1 large] *interl*

To Thomas Woolner 10 March [1870]
1.1 The] *added*

To Asa Gray 15 March [1870]
0.1 Beckenham] *before del* '*Bromley*'
3.2 very difficult] 'very' *interl*
4.3 your] *interl*

To Edwin Ray Lankester 15 March [1870]
1.8 turn] *after del* 'into'

To J. J. Weir 17 March [1870]
0.1 Beckenham] *before del* '*Bromley.*'
1.3 a] *interl*
1.4 the Pig] *after del caret*
1.4 by Mr Giles.] *interl*
1.7 see . . . p. 55 1.8] *interl*
1.9 Is] *after del* 'Now'
1.9 at all] *interl*
1.10 Vol. . . . 325,] *interl*
1.11 Vol I . . . p 41] *interl*
2.4 view] *interl*
2.8 &c] *interl*
5.5 ass-colts] *after del* 'co'
5.5 apparently] *interl*
5.5 to . . . offspring 5.6] *interl*
5.6 as characters] 'characters' *after del* 'to'
6.2 *subsequent*] *interl*

To Albert Günther 23 March [1870]
4.1 my] *interl*

To ? 23 March [1870–1]
0.1 Beckenham] *before del* '*Bromley.*'
1.3 to Physics] 'to' *over* 'if'
2.1 less] *interl*

To [Henry Hussey Vivian?] [April or May 1870?]
1.1 liberty] *after del* 'great'
1.1 subject 1.2] *pencil above del pencil* 'point'
1.2 to be] *above del* 'is'
1.2 scientific] *above del* 'much'
1.2 , & which] *after del* 'for science'
1.2 greatly] *after del* 'fur'
1.3 you sh^{d} think fit, to give the *[illeg]*] *interl*; 'sh^{d}' *interl pencil*; 'think' *altered from* 'thought' *pencil*; ', to

give the *[illeg]*' *pencil*
1.3 consulting] *after del* 'due'; *altered from* 'consultation' *pencil*
1.3 those competent] *after del pencil* 'with'
1.3 an opinion 1.4] 'opinion' *after interl and del* 'sound'
1.4 I have been] *above del* 'The course of my studies has'
1.4 to study] *after del* 'me'
1.4 experimentally] *after interl and del pencil* '& am still studying'
1.4 theoretically] *interl*
1.5 & the evil] *after del* 'distinct breeds, or in the case of plants, distinct individuals'
1.5 close] *interl*
1.5 ; & the difference 1.6] 'difference' *above del illeg*
1.6 & size] *after del illeg interl*; 'size' *circled pencil*
1.6 & power] *below del illeg*
1.6 & health] *interl pencil*
1.6 the offspr[ing] in the 2 cases is] *above del* 'if cross-bred again'
1.7 in most] *after del illeg*; 'most' *pencil above del* 'many'
1.7 In Chap 17] *after interl & del* 'In many instances the effect on the'
1.7 Chap 17] *below del* 'the 2^{d} volume'
1.7 Chap 17 of my] 'Chap 17 of' *added above* 'my'
1.7 variation of] *interl*
1.7 Domestic] *interl after del* 'Domestic'
1.8 I have given] *after del interl illeg*
1.8 & these ... subject. 1.9] *added*; 'these might' *after del* 'which'; 'at present' *after del* '*[*known*]*'
1.9 various] *pencil below del* 'all' *below del pencil* 'many'
1.9 European] 'an' *added pencil*; *after del pencil* 'parts of'
1.10 countries] *interl pencil*
1.10 owing to ... data 1.11] *interl*; '*[*continued*]*' *interl*
1.11 found] *below del* 'formed'
1.11 near relatives] *above del* 'cousins'
2.1 Some] *after del* '& no data have *as yet [*interl*] ever been collected. by which *[8 words illeg]* convinced that this is an important subject for *[illeg]* of mankind. &'; 'S' *pencil over* 's'
2.1 went to the] *above del* 'suggested'
2.1 of the] *after del pencil* 'office'
2.1 suggesting] *interl*
2.1 during] *after del* 'if'
2.2 a return] *after del* 'th'
2.2 should be demanded] *above del* 'had to be made'
2.2 in each family the parents were] *above del* 'each man was a'
2.3 related] *altered from* 'relative'
2.3 in blood] 'in' *after del* 'of'
2.3 for it w^{d}] *after del* 'to his wife'; *above del* '&'
2.3 by] *over* 'to'
2.3 average] *interl*
2.4 the *[*surviving*]*] *above del* 'their'
2.4 of related & not related] *above del* 'with the average number of'
2.4 to judge of the] *interl after del* 'which were related with the number of *[illeg]* parents not related; & this w^{d} serve as a guide to the'
2.4 vitality] *after interl & del illeg*; *after del* 'to the'
2.5 , or of the fertility of the parents] *added pencil*
3.1 soon] *interl pencil*
3.2 highly] *above del* 'an'
3.2 for the welfare] *after del* 'point'
3.2 the subject 3.3] *below del* 'it'
3.3 *[8 words illeg]*] *pencil above del pencil illeg*

To John Murray 9 April [1870]
1.2 &] *over* '—'

To Rudolf Albert von Kölliker 18 April [1870]
0.1 Beckenham] *before del* '*Bromley.*'
6.1 old] *interl*
6.1 ( )] *square brackets in MS*

To Tord Tamerlan Teodor Thorell 20 April 1870
0.1 Beckenham] *before del* '*Bromley.*'

To St G. J. Mivart 21 April [1870]
0.1 Beckenham] *before del* '*Bromley.*'
2.2 the terms used in] *interl*
2.3 & considering whole organisation 2.4] *interl*
2.8 (if] *parenthesis over comma*
2.9 as] *above del* 'the'
3.2 on rudiments—variability &c] *added below* 'much very interesting to me'
4.1 a fossil] 'a' *over del* 'an'; 'fossil' *interl*
5.2 believe] *after del* 'pray'

To St G. J. Mivart 23 April [1870]
0.1 Beckenham] *before del* '*Bromley.*'

To Francis Galton [27 April 1870?]
1.1 much] *after del* 'very'
1.1 & very sorry] *interl*
1.1 glad] *after del* 'very'
1.2 spirits.—] *below del* 'determination'

To Fritz Müller 12 May 1870
4.1 (Pangenesis ... day!)] *added in CD's hand*

To Alfred Newton [22 May 1870]
1.2 as ... today] *interl*

To J. D. Hooker 25 May [1870]
0.1 Beckenham] *before del* '*Bromley*'
1.12 residence] *interl*
1.14 who was] *interl*
1.15 that] *interl*
1.15 was] *interl*
1.21 dreamed] *above del* 'dreamed'
3.2 always] *interl*
3.2 partly] *interl*
3.6 first] *interl*
4.1 or any time,] *interl*
5.1 success] *interl*
5.1 is] *after del illeg*

To T. H. Farrer 28 May [1870]
0.1 Beckenham] *before del* '*Bromley.*'
3.3 what it] 'it' *interl*
5.2 I . . . for] *interl*
7.2 almost liquid] *interl*

To Adam Sedgwick 1 June [1870]
0.1 Beckenham] *before del* '*Bromley.*'
2.4 was] *interl*
2.5 (as . . . prove)] *parentheses over commas*

To James Paget 4 June [1870]
0.1 Beckenham] *before del* '*Bromley.*'
1.5 with,] *added above* 'or'
1.6 spasmodic] *interl*
1.9 with nothing . . . stomach 1.10] *interl*
1.10 strongly] *interl*

To St G. J. Mivart 13 June [1870]
0.1 Beckenham] *before del* '*Bromley.*'
1.3 served] *above del* 'send'
1.3 might] *above del* 'my'
2.1 (& . . . me)] *parentheses over commas*
2.4 fly Ophrys] 'Ophrys' *interl*
2.4 that] *interl*
2.5 the backs of 2.6] *interl*
3.1 difficult:] *after del* 'complex:'
3.6 colour or] *interl*
3.7 the cheeks of] *interl*
3.9 being] *interl*
3.11 recently discovered] *interl*
3.11 this] *over* 'the'
3.11 note] *above del* 'note'
3.13 on] *over illeg*
6.1 artificial] *after del* 'organic'
6.3 any] *interl*
6.3 such] *interl*
6.4 fact] *interl*

To J. J. Weir 14 June [1870]
0.1 Beckenham] *before del* '*Bromley.*'
1.4 only] *interl*

To T. H. Huxley 20 June [1870]
0.1 Beckenham] *before del* '*Bromley.*'
1.1 Icones Phy.] *interl*
1.1 (Part] *interl*
1.5 23^{d}] *above del* '16th'
3.1 with] *above del* 'about'

To Ernst Haeckel 23 June [1870]
0.1 Beckenham] *before del* '*Bromley*'
2.2 from you] *interl*

To Edward Burnett Tylor 25 June 1870
3.2 any or many] *interl in CD's hand*

To Frederick Bates 26 June [1870?]
1.5 soon] *interl*

To T. H. Farrer [29 June 1870]
1.2 the flowers of] *interl*

To W. W. Reade 30 June [1870]
0.1 Beckenham] *before del* '*Bromley.*'
1.3 larva] 'a' *over* 'æ'
2.5 in the proofs] *interl*

To B. J. Sulivan 30 June [1870]
0.1 Beckenham] *before del* '*Bromley.*'

To J. D. Hooker 2 July [1870]
6.1 Ch. Darwin . . . Proteaceæ— 7.3] *in CD's hand*

To B. J. Sulivan 2 July [1870]
0.1 Beckenham] *before del* '*Bromley.*'

To A. R. Wallace 5 [July 1870]
2.1 Ch. Darwin . . . C.D 3.3] *in CD's hand*

To J. D. Hooker 8 July [1870]
0.1 Beckenham] *before del* '*Bromley.*'
2.1 Prof.] *interl*
2.2 chiefly] *after del* 'partly'
2.6 Claparede has arrived 2.7] *above del* 'he comes to'
2.7 an unpleasant judgement,] *above del* 'different conclusion'
2.7 about] *after del* 'with'
2.8 L.] *over* 'he'
2.9 in] *after del* 'in'

To J. D. Hooker 12 July [1870]

0.1 Beckenham] *before del 'Bromley.'*
2.2 or about] 'about' *interl*
5.2 can] *above del* 'could'
5.4 with] *interl*
5.7 my] *interl*
5.8 afterwards] *interl*

To Albert Günther 15 July [1870]
0.1 Beckenham] *before del 'Bromley.'*
1.2 briefly in my M.S.] *interl*
1.2 them] *interl*
1.5 will be] *interl*

To W. B. Tegetmeier 15 July [1870]
0.1 Beckenham] *before del 'Bromley.'*
2.2 American] *interl*

To William Farr 17 July [1870]
1.2 parts] *after del* 'the'
1.2 in regard to] *interl after del* 'of'
1.2 Census is] 'is' *above del* 'on'
1.2 Firstly] *interl*
1.3 Secondly ... children. 1.6] *bracket in margin, ink*
1.3 I incline to think] *above del* 'It seems to me'
1.4 that is] *interl*
1.4 forms] *interl*
1.5 in as much as] *above del* 'as It'
1.5 out of the whole population] *interl*
1.6 have children.] *after del interl* 'happen to'
1.6 have children.] *altered from* 'have children at home on the night of the census.'
2.1 The object ... injurious, & 2.2] *crossed ink*
2.1 object] *above del* 'importance'
2.2 The] 'T' *over* 't'
2.2 of such an enquiry 2.3] *interl*
2.3 is] *above del* 'are'
2.3 whether negative or affirmative] *circled ink, line drawn to caret after* 'importance'; *before del* 'The census returns will show whether cousins ['have' *del*] produce on an average as many children'
3.1 Europe] *above del* 'the continent'
3.1 cousins are] 'are' *above del* 'is'
3.1 to by] 'by' *interl, before del interl* 'some persons'
3.2 consequences] *above del* 'results'
3.2 belief] *interl*
3.2 direct] *interl*
3.2 evidence.] *point over comma*
3.2 It is] *after del* 'only on a rather far *widely stretched [*interl*] drawn analogy.'
3.3 therefore] *interl*
3.3 belief] *interl*
3.3 either] *above del* 'either'
3.3 proved to be] 'to be' *after del interl* 'either'
3.3 false] *above del* 'empty an idle proposition'
3.3 sh^{d} be 3.4] *interl*
3.4 in the latter case] *interl*
3.4 might] *after del* 'may'
3.5 if the] *after del* 'by'
3.5 proper queries are inserted] *interl after del* 'query' *above del* 'supposed query inserted into the census'
3.5 show whether] *after del* 'indirectly'
3.5 married] *interl after del* 'in the household'
3.6 have ... census] *above del* 'produce on an average'
3.7 sh^{d} the number] *after del* 'which'
3.7 we might] *after del* 'sh^{d} be [*interl*] may'
3.8 either] *interl*
3.8 in the parents,] *interl*; 'parents' *after del* 'either', *before del* 'when related'
3.8 probable] 'e' *over* 'y'
3.8 in the offspring. 3.9] *before del* 'of such consanguineous marriages.'
4.1 If the proportional ... offspring in the whole population. 4.5] *crossed ink*
4.1 If] *interl after del* 'Lastly'; 'I' *over* 'i'
4.1 number] *interl before del illeg*
4.1 of cousins] *after del* 'out of the whole population'
4.1 in the whole population] *interl*
4.1 was] *above del* 'by'
4.2 exist] *above del* 'be'
4.3 the dumb] 'the' *above del* 'blind,'
4.3 the blind & 4.4] *interl above del* '&'
4.4 of cousin-offspring in 4.5] *interl*; 'cousin' *after del* 'fertility prod cousin-marriage In offspring children producing'
4.5 the whole population.] *after del* 'fer'
5.1 I fear ... to write.— 5.2] *crossed ink*
5.1 but I have] 'have' *after del* 'suppose your enquiry is far from done rather'
6.1 much to be wished] *above del* 'highly highly desirable'
6.1 often repeated] *above del* 'often asserted'
6.2 consanguineous] *after del* 'the'
6.3 opinion] *below del* 'conclusion'
6.3 by the returns] 'by' *over del* 'in'
7.1 It is beyond ... case. 7.4] *crossed ink*
7.2 could insert] *after del* 'enclose here the first form; & I suppose you c^{d} somehow call especial attention to'
7.3 the word] *after del illeg*

To John Lubbock 21 July [1870]
0.1 Beckenham] *before del 'Bromley.'*
1.6 largely] *interl*

To Richard Kippist 28 July [1870]
0.1 Beckenham] *before del 'Bromley.'*

To David Forbes 31 July [1870]
0.1 Beckenham] *before del* 'Bromley.'
1.4 but] 'b' *over* 'w'

To H. B. Jones 3 August [1870]
0.1 Beckenham] *before del* 'Bromley.'
1.3 , also,] *interl*

To John Murray 5 [August 1870]
0.1 Beckenham] *before del* 'Bromley.'
1.1 5] *over* '4'
2.2 Slips] *above del* 'Revises'

To Mr Dorrell 9 August 1870
3.1 of] *after del* 'in my'
3.1 proofs ... my 3.2] *interl*; 'of woodcuts' *after del* 'in my'
3.2 instead] *after del illeg*
4.1 **not**] *interl*
10.1 may perhaps] *above del* 'will'

To Mr Dorrell 11 August [1870]
0.1 Beckenham] *before del* 'Bromley.'
1.2 28 or] *interl*

To L. H. Morgan 11 August [1870]
0.1 Beckenham] *before del* 'Bromley.'
2.4 chiefly] *interl*

To J. V. Carus 18 August 1870
4.1 sex"—] *before del* 'Ch.' *in CD's hand*

To F. P. Cobbe 20 August [1870]
3.4 if the] 'the' *after del* 'crossing'

To George F. Kittredge 24 August [1870]
0.1 Beckenham] *before del* 'Bromley.'

To Fritz Müller 28 August 1870
0.1 Beckenham] *before del* 'Bromley.'
2.5 in colour] *interl*
2.7 in my book] *interl*
2.9 you] *after del* '&'
2.13 with ... Book,] *interl*
2.15 as yet] *interl*
2.21 letter] *interl*

To Philip Lutley Sclater 30 August [1870]
0.1 Beckenham] *before del* 'Bromley.'
1.1 as] *above del* 'in'
1.2 on] *over* 'of'

To John Murray 4 September [1870]
0.1 Beckenham] *before del* 'Bromley.'
1.3 point] *interl*

To John Tyndall 8 September 1870
6.1 day] *in CD's hand above del* 'occasion'

To John Murray 26 September 1870
2.4 Please ... be. 2.5] *interl in CD's hand*
2.5 stereotypes] *interl in CD's hand*
3.1 Jan 1. 1869] *interl in CD's hand*
3.3 (N.B.... Book) 3.4] *in CD's hand*
3.3 if no objection] *interl*
4.2 (Sec ... House)] *interl in CD's hand*
7.1 P.S.... Index.—] *in CD's hand*

To J. D. Hooker 27 September [1870]
0.1 Beckenham,] *before del* 'Bromley.'
1.5 having ... to be 1.6] *interl*
1.6 against] *after del* 'to anyone'
1.6 any one who] *above del* 'someone'
1.6 maintained] 'ed' *over* 'ing'
1.9 gates] *after del* 'Kingdo'
2.2 more] *interl*
3.3 shall not] 'not' *interl*

To John Murray 29 September [1870]
0.1 Beckenham] *before del* 'Bromley.'
2.1 M[ess]. Appleton] *interl*
2.2 (& ... determine)] *brackets over commas*
3.2 about] *after del* '&'
3.4 so,] *interl*
3.5 sentence ... that] *interl*
5.1 Esqre] *over illeg*

To [Edward William Blore] [October 1870 or later]
1.1 I hope] *after del* 'My'
1.4 said] *above del* 'thought'
1.4 My son ... mathematics. 1.5] *circled ink*
1.5 I have] *after del* 'Under these circumstances'; 'have' *above del* 'am'
1.5 [therefore much]] *interl*
1.5 that] *above del* 'very doubtful whether'
1.7 I have allowed ... 1 but] *interl*
1.7 attend] *altered from* 'attending'; *after del* 'he is'
1.10 excuse] *above del* 'permit'

To J. V. Carus 5 October [1870]
0.1 Beckenham] *before del* 'Bromley.'
2.1 pages] *interl*
2.2 first] *interl*
2.4 finally corrected] *above del* 'ready'
2.5 I asked ... costs 2.6] *added*

To C. W. Thomson [after 7 October 1870]
1.3 have ... yourself] *above del* 'are [exceptionally]'

1.5 admirable] *after del illeg*
1.5 opportunity] *above del* 'field'
1.6 new] *above del* 'fresh'
1.6 &] *after del* ', in natural science,'
1.6 cordially] *after del* 'wish'
1.6 for] *interl*
2.1 I remain] *after del* 'With my sincere respects'

To J. D. Hooker 14 October [1870]
2.1 , as . . . writing,] *interl*
2.2 my] *interl*

To Francis Darwin 18 October [1870]
1.5 & think deliberately] *interl*
1.6 to] *above del* 'at'
2.2 overspent] *above del* 'overpent'
2.11 an] *added*

To John Murray 22 October [1870]
0.1 Beckenham] *before del* '*Bromley.*'

To George Cupples 3 November [1870]
0.1 Beckenham] *interl before del* '*Bromley.*'
1.4 C. Darwin] *interl*

To P. L. Sclater 4 November [1870]
0.1 Beckenham] *before del* '*Bromley.*'
1.6 list] *interl*
1.7 send] *interl*

To P. L. Sclater 9 November 1870
1.6 be able] *interl in CD's hand*
4.1 The only . . . sex. 4.2] *in CD's hand*

To P. L. Sclater 11 November [1870]
0.1 Beckenham] *before del* '*Bromley.*'

To W. D. Fox 15 November [1870]
0.1 Beckenham] *before del* '*Bromley.*'
1.3 that] *altered from* 'what'
1.6 it] *after del* '&'
1.7 owe] 'e' *over* 'ed'
1.7 differences of] 'of' *over* 'in'
1.10 truly] *above del* 'truy'

To John Murray 15 November [1870]
0.1 Beckenham] *before del* '*Bromley.*'
3.1 as yet] *interl above del illeg*
3.1 clean] *interl*
3.2 first] *interl*

To John Murray 17 November [1870]
0.1 Beckenham] *before del* '*Bromley.*'

To William Ogle 17 November [1870]
1.4 viz] *interl in CD's hand*
1.8 (possibly . . . belief)] *interl in CD's hand*
1.12 clever] *interl in CD's hand*
2.4 (if . . . subject,)] *parentheses in CD's hand*
4.1 I wonder . . . pigs.] *in CD's hand*

To P. L. Sclater 21 November [1870]
0.1 Beckenham] *before del* 'Bromley.'
2.1 names.] *above del* 'notes.'
2.3 , as] *comma over point*
2.3 as . . . time.] *added*
2.4 If . . . here] *added*

To A. R. Wallace 22 November [1870]
0.1 Beckenham] *interl before del* '*Bromley.*'
1.7 already] *interl*
2.1 it is that] *interl before del* 'that'
2.1 action] *after del illeg*

To Alexander Stuart Strahan 29 November 1870
1.2 M^r] *after del* 'the'
1.2 G.] *above* 'M^r'
1.2 correspondence 1.3] *after del* 'had much'
1.3 during . . . years.] *interl*
1.4 at the cost . . . trouble] *interl*
2.2 beyond] *after del* 'as'
3.1 I beg leave to remain] *above del* 'Pray believe me'

To Hermanus Hartogh Heijs van Zouteveen 30 November [1870]
0.1 Beckenham] *before del* '*Bromley.*'
3.1 a small] *interl*
3.3 preceded] *above del* 'preceed'

To P. L. Sclater 5 December 1870
1.3 of the] *in CD's hand above del* 'other'

To Asa Gray 7 December 1870
2.7 smallest] *above del* '*slightest*'
4.1 My . . . Darwin] *in CD's hand*

To L. J. A. Quetelet [15 December 1870 – 11 January 1871]
1.2 my . . . your 1.3] *interl*
1.3 Science] *above del* 'Belgium'

To William Ogle 17 December [1870]
0.1 Beckenham] *interl before del* '*Bromley.*'

To L.-G. De Koninck [after 19 December 1870]
1.3 high] *interl*
1.5 respects] *above del* 'esteem'

To B. J. Sulivan 20 December [1870]

0.1 Beckenham] *interl before del* '*Bromley.*'
1.8 the] *after del* 't'
2.4 the] *interl*

To Charles Lyell 25 December [1870]
0.1 Beckenham] *interl before del* '*Bromley.*'
1.5 first . . . of] *interl*

To P. L. Sclater 26 December [1870]
1.2 more] *interl*

To P. L. Sclater 29 December 1870
0.1 Beckenham] *interl before del* '*Bromley*'
2.1 some] *interl*
5.1 represents] *above del* 'is of'
6.3 the drawing] *above del* 'it'

SUPPLEMENT

To William Kemp 1[8] May [1840]
1.3 that] *added*
1.6 I should] *after del* '& how I could'
1.10 above the sea, 1.11] *interl*

To William Allport Leighton 25 November [1840–6]
1.6 the end] 'the' *interl*
1.6 send] 'd' *over* 't'

To William Kemp 7 April [1843]
1.9 dry it] *caret before 'it'*
1.9 between paper] *interl*
1.10 & preserve . . . Wine] *interl*
1.14 Spain.—] *after del* 'Oxford.—'
1.16 publish it,] 'it,' *interl*
1.19 some common] 'some' *interl*
1.22 transmit it] 'it' *interl*
2.2 , which] *added*

To William Kemp 24 [April 1843]
1.4 he waits] 'e' over 'is'; 'waits' *after del* 'customary'
2.3 result to] *interl*

To William Kemp [8 September 1843]
1.1 Prof:] *interl*
2.1 flower, w^{d}] 'w^{d}' *after del* 'it'
2.2 investigations] *above del* 'suggestion'

To William Kemp 9 October [1843]
1.3 of publishing it] *interl*
3.1 Prof] *interl*
4.1 *soon*] *added*
4.2 present] *interl*
4.2 better] *after del* 'clearly been'
7.2 floating] *interl*
7.3 *large* or *angular*] *interl*
7.3 of foreign rocks] *interl, circled*
7.4 the gravel] 'the' *interl*
8.1 Lindley] 'L' *over* 'H'
9.2 Is it] *over illeg*
9.2 not also 9.3] *interl*
10.1 fresh-water] *interl*

To William Kemp [14 October 1843]
0.1 Down] *after del* 'Shrewsbury'
1.6 against] *after del* 'aga'

To William Kemp 1 November [1843]
2.2 is a first-rate] 'is' *after del* 'I'
2.7 su⟨s⟩pect] *after del* 'believe'
2.8 had] *after del* 'were'
2.11 Polygonum] 'num' *over illeg*
2.11 P. aviculare & P. convolvulus 2.12] *interl*
2.12 came] *after del* 'wherea'
2.13 or variety] 'or' *over illeg*
2.14 according to Henslow] *interl*
2.15 R. acetosella] *interl*
2.16 plants] *after del illeg*
2.17 a new] 'a' *interl*
2.17 species] *interl*

To William Kemp [9 November 1843]
1.2 in your case] *interl*

To William Kemp 22 November [1843]
1.6 or improvements 1.7] *interl*
1.7 piece of] *interl*

To William Kemp 1 December [1843]
1.11 being] *after del* 'disc'
1.12 & *known* . . . Wheat 1.13] *above* 'Raspberry seeds'
4.1 Is] *added*
4.1 the actual] 'the' *after del* 'is'

To Henry Denny 5 February [1844]
1.1 from the *Condor*] 'from' *after smudged* 'p.', *&* *over* '15'
2.1 395] *after del* '1'
5.1 (which . . . verbatim)] *interl*
5.4 Martiall's] 'tiall' *over* 'shall'

To Georgina Tollet 1 October [1846?–71]
1.4 with Emma as witness] *interl*

To Josiah Wedgwood III [22 August? 1847]
1.2 , if he . . . you.— 1.3] *added*
2.1 there is] 'is' *interl, below del interl* 'are'
5.1 cannot] 'not' *added*

To Henri Milne-Edwards 2 March [1849]
2.3 parasite of] *interl*

To John Richardson 30 December [1851]
1.3 Friday or Saturday] *above del* 'Thursday'
2.1 the] *over* 'your'

To Francis Galton 22 February [1855]
2.7 i e] *interl*

To C. J. Andersson 25 March [1856]
2.4 unite] *interl*
3.2 (or of Sheep)] *interl*
3.2 in different] 'in' *interl*
3.3 Fowls] *after del* 'Poultry'

To Walter Baldock Durrant Mantell [before 10 April 1856]
1.1 either . . . person, 1.2] *interl*

To James Dwight Dana 21 December [1856]
1.1 obliged] *interl*
1.6 this] *above del* 'your'
1.7 but] *interl*
1.11 -beds] *interl*
2.6 mind] *after del illeg*
5.3 (can ... facts?)] *interl; square brackets in ms*
6.4 mountain] *interl*
6.4 cave animals] *interl*
7.5 case] *after del* 'add'
7.6 right] *above del* 'write'
8.3 larger] *after del illeg*

To Skeffington Poole 20 October [1858]
3.2 incidentally] *interl*

To Skeffington Poole 22 October [1858]
5.2 give.] *after del* 'quote'
6.5 shoulder] *interl*
7.1 (with stripes)] *interl*
9.1 domestic] *interl*

To Benjamin Collins Brodie 26 April [1859]
2.4 of the names . . . to whom 2.5] *interl above del* 'existed of all'

To Charles Kingsley 1 December [1859]
1.3 with many Plates 1.4] *interl*
1.7 a] 'an' *altered to* 'a'
1.7 brief] *interl*
1.7 for] *over* 'of'
1.9 (i e . . . hermaphrodite) 1.10] *interl*

To a bookseller 9 April [1860]
1.1 &] *interl*

To John Medows Rodwell 15 October [1860]
1.3 almost] *interl*
1.3 explaining] *after del* 'given'
1.10 much] *interl*

To Henry Fawcett 6 December [1860]
1.6 Article] 'A' *over* 'a'
1.11 being denuded] *interl*
1.13 naked] *interl*
1.16 they were . . . subsequently] *interl*
2.1 (with] *after del* 'on'
2.4 so] *interl*
2.6 research] *above del* 'work'
2.7 new] *interl*
2.8 single] *interl*
2.13 (& manner 2.14] *parenthesis over comma*
2.15 perceive] *after del* 'know'

To David Forbes 11 December [1860]
2.4 one] *above del* 'case'
4.6 they now] 'now' *interl*

To Frances Julia Wedgwood 11 July [1861]
1.25 many] *interl*

To Henry Fawcett 20 July [1861]
3.3 must] *above del* 'have to'

To ? 21 August [1862]
1.6 quite] *interl*

To W. A. Leighton 4 December [1862]
2.12 plants *virgatum*] 'plants' *interl*
2.12 *will*] *interl*

To W. B. Tegetmeier 2 February [1864]
3.1 & sincerely] *in CD's hand*

To Francis Darwin 30 August [1867–70]
1.2 c^{d}.] *interl*

To W. B. Dawkins 30 [August 1867]
1.1 hereafter] *interl*
1.2 or skeletons] *interl*
1.3 *generated*] *interl above del* 'bred'
1.4 it is] *interl*

To A. D. Bartlett 16 February [1868]
1.2 coloured] *interl*
1.3 all] *interl after del* 'in separate packets'
1.5 packets] *above del* 'bits'
1.6 several] *above del* 'the'

2.4 & the . . . better. 2.5] *added*
3.1 *named*;] *interl*

To Edward A. Smith [11–31 March 1868]
2.1 Grantii] *interl*
2.1 or third] *interl*
2.1 nat. size] '(?)' *added pencil*
4.1 Please . . . Horns. 4.2] *scored blue crayon*
5.1 in female] *interl*
5.1 .— Figures 5.2] *after del* 'for female'
5.2 of figure 5.3] *interl*
5.10 Copris] *altered in pencil from* 'Heliocopris'
5.11 reduced] *below del* 'increas'
5.12 hastifer] *pencil, below ink del pencil* 'lancifer', *triple underl red crayon*
7.1 (Crabro . . . female) 9.1] *square brackets in MS*

BIOGRAPHICAL REGISTER

This list includes all correspondents and all persons mentioned in the letters and notes that the editors have been able to identify. Dates of letters to and from correspondents are given in chronological order. Letters to correspondents are listed in roman type; letters from correspondents in italic type; third-party letters are listed with the name of the recipient or sender given in parentheses.

Abberley, John (b. 1813/14). Gardener. Robert Waring Darwin's gardener at The Mount, *circa* 1841. (Census returns 1841 (Public Record Office HO107/926/8).)

Acland, Charles Lawford (1833–1903). East-Indian born Anglican cleric, educator, and author. Ordained priest, 1860. Curate, Holy Trinity, Gray's Inn Road, London, 1859; St Olave's, Hart Street, 1863. Fellow, Radley College, Berkshire, 1863–7; mathematical tutor, 1866–7. Headmaster, Colchester Grammar School, 1870–92. Vicar, All Saints', Cambridge, 1892–1903. Wrote on the antiquities of the Hebrides. (*Alum. Cantab.*)
17 August 1870

Adanson, Michel (1727–1806). French botanist. Studied the flora, fauna, geography, and peoples of Senegal, 1748–54; published *L'histoire naturelle du Sénégal* (1757). Spent much of the rest of his life publishing on botany and working on a comprehensive taxonomical system. (*DBF.*)

Addison, John. Secretary and general manager of the Maryport & Carlisle Railway Company. Of Castle Hill, Maryport, Cumberland. (*Post Office directory of the county of Durham* 1873.)

Agassiz, Alexander (1835–1910). Swiss-born zoologist, oceanographer, and mining engineer. Son of Louis Agassiz. Emigrated to the United States in 1849. Joined the US survey of Washington Territory boundaries in 1859. Superintendent, Calumet copper mine, Michigan, 1867; later president of the corporation. Curator of the Harvard Museum of Comparative Zoology, 1874. (*DAB.*)
22 October 1870, [23 October 1870?]

Agassiz, Anna Russell (1839/40–73). Daughter of George Russell, a Boston textile merchant. Married Alexander Agassiz in 1860. Gave birth to three sons in 1862, 1866, and 1871. (Dobbs 2005.)

Agassiz, Elizabeth Cabot Cary (Elizabeth) (1822–1907). Educator. A founder of the educational establishment for women that later became Radcliffe College, Cambridge, Massachusetts. President of Radcliffe College, 1894–9; honorary president, 1900–3. Married Louis Agassiz in 1850. (*ANB.*)

Agassiz, Jean Louis Rodolphe (Louis) (1807–73). Swiss-born zoologist and geologist. Professor of natural history, Neuchâtel, 1832–46. Emigrated to the

Agassiz, Jean Louis Rodolphe (Louis), cont.
United States in 1846. Professor of zoology and geology, Harvard University, 1847–73. Established the Museum of Comparative Zoology at Harvard in 1859. Foreign member, Royal Society of London, 1838. (*ANB*, *DAB*, *DSB*, *Record of the Royal Society of London.*)

Albert Edward, prince of Wales. *See* Edward VII.

Albrecht, R. F.
16 March 1870

Alglave, Emile (1842–1928). French economist and editor. Studied palaeography and law at the Ecole des Chartes, Paris, and received a doctorate in law, 1868. Professor, faculty of law, Douai, 1870; taught political economy, Lille, 1873; professor, scientific finance, faculty of law, Paris, 1878. Co-founder and director of the journals *Revue des cours scientifiques* and *Revue des cours littéraires*, 1864–80. Worked mainly in the area of tax reform. (*DBF.*)

Allen, Frederick. Riding-school master. Of 70 Seymour Place, Bryanston Square, London. (*Post Office London directory* 1869.)

Allman, George James (1812–98). Botanist and zoologist. Professor of botany, Dublin University, 1844; regius professor of natural history, Edinburgh University, 1855–70. FRS 1854. (*ODNB.*)

American Philosophical Society.
5 February 1870

Andersson, Charles John (Carl Johann) (1827–67). Swedish-born ornithologist, hunter, and explorer. Accompanied Francis Galton to South Africa in 1849. Organised a successful expedition to Lake Ngami and published an account of the expedition in 1856. Continued to explore the South African interior and to carry out ornithological studies. (*DSAB*, *Modern English biography*, *SBL.*)
[after 22 February 1855] (from Francis Galton), 25 March [1856]

Anon.
23 January [1843–6 or 1856–69], 14 June [1860], 7 May [1861–8?], 20 June [1861–8], 24 August [1861–8], 25 November [1861–8], 31 December [1861–8], 11 March [1862–9], 29 March [1862–9], 21 August [1862], 1 August [1864 or 1865], 23 March [1870–1], 14 June [1870], 17 November [1870]

Ansted, David Thomas (1814–80). Geologist. Professor of geology, King's College, London, 1840–5. Assistant secretary to the Geological Society of London, 1844–7; editor of the society's journal from 1844. Lecturer on geology at the military college at Addiscombe, Surrey, and professor of geology at the College of Civil Engineers, Putney, from 1845. Consulting geologist and mining engineer from *circa* 1850. FRS 1844. (*ODNB*, *Geological Magazine* n.s. 7 (1880): 336, *Record of the Royal Society of London*, Sarjeant 1980–96.)
25 November [1846]

Appleton, Charles Edward Cutts Birchall (1841–79). Philosopher and journal editor. Lecturer in philosophy at St John's College, Oxford, from 1867. Founded the *Academy: a Monthly Record of Literature, Learning, Science, and Art*; editor, 1869–79. (*ODNB.*)

2 March [1870], 18 May [1870]

Appleton, Thomas Gold (1812–84). American essayist, poet, and artist. Promoted the growth and improvement of Boston; a trustee of the Athenæum and the Public Library; founder and benefactor of the Museum of Fine Arts. (*ANB.*)

31 March [1846]

Appleton, William Henry (1814–99). American publisher. Partner in the New York firm D. Appleton & Co., 1838–48. Became head of the firm in 1848 on the retirement of his father, Daniel Appleton. D. Appleton & Co. published works by British scientists such as CD and Herbert Spencer. (*ANB.*)

D. Appleton & Co. New York publishing house. Founded by Daniel Appleton (1785–1849) in 1831. His son William Henry Appleton (1814–99) was taken into partnership in 1838. American publishers of works by CD and Herbert Spencer. (*ANB.*)

1 February 1870 (to Asa Gray), *1 September 1870* (to Asa Gray)

Axell, Johan Severin (1843–92). Swedish botanist. Studied at Uppsala University, 1861–9; DPhil., 1869. Lecturer in botany, 1869–70. Wholesale dealer, 1870–91. (*Taxonomic literature. Supplement.*)

Azara, Félix d' (1746–1811). Spanish explorer and army officer. Sent to South America to survey Spanish and Portuguese territories. Published works on the fauna of Paraguay. (*NBU.*)

Babbage, Charles (1791–1871). Mathematician and pioneer in the design of mechanical computers. FRS 1816. (*DSB, ODNB.*)

Babington, Charles Cardale (1808–95). Botanist, entomologist, and archaeologist. Involved in natural history activities at Cambridge for more than forty years; an expert on plant taxonomy. A founding member of the Cambridge Entomological Society and the Cambridge Antiquarian Society. Editor of *Annals and Magazine of Natural History* from 1842. Chairman, Cambrian Archaeological Association, 1855–85. Professor of botany, Cambridge University, 1861–95. FRS 1851. (*DNB, DSB.*)

Baillon, Ernest Henri (1827–95). French botanist. Professor of medical natural history, and of hygiene and natural history applied to industry, Ecole Centrale des Arts et Manufactures, Paris, 1863–95. Editor of the botanical journal, *Adansonia*, 1860–70. (*DBF.*)

Bain, Alexander (1818–1903). Scottish philosopher and psychologist. Graduated from Marischal College, Aberdeen, 1840; stayed as assistant and teacher in philosophy. Assistant secretary, metropolitan sanitary commission of the Board of Health, London, 1846. Lecturer at Bedford College, 1851, with other posts following. Professor of logic, University of Aberdeen, 1860. Founded the journal *Mind* in 1876; editor until 1892. Considered a founder of modern psychology. (*ODNB.*)

Baldry, Alfred (1828–1915). Entered Trinity College, Cambridge, 1847; BA 1851, MA 1854. Rowed at Cambridge. Moved to Bournemouth, 1865. (*Alum. Cantab.*)

Barkly, Henry (1815–98). Colonial administrator. Served as governor of British Guiana, 1849–53; Jamaica, 1853–6; Victoria, 1856–63; Mauritius, 1863–70; and Cape Colony, 1870–77. Sent plants to the Royal Botanic Gardens, Kew. Investigated the ferns of Jamaica and Mauritius and its dependencies. Knighted, 1853. FRS 1864. (Gunn and Codd 1981, *ODNB*, *Proceedings of the Royal Society of London* 75 (1905): 23–5.)

Barrande, Joachim (1799–1883). French-born palaeontologist and stratigrapher. Employed as a tutor by the French royal family until their exile in 1830. Railway engineer in Prague, where he became interested in fossils. Collected and described the fossils of the Bohemian basin, 1840–83. The results of his studies were published in thirty volumes entitled *Système Silurien du centre de la Bohême* (1852–1902). (*DSB*.)

19 June 1870

Bartlett, Abraham Dee (1812–97). Taxidermist and zoo superintendent. Taxidermist, *circa* 1834–52. Superintendent of the natural history department, Crystal Palace, 1852–9; of the Zoological Society's gardens, Regent's Park, 1859–97. (*Modern English biography*, *ODNB*.)

16 February [1868], 5 January [1870], 22 August 1870 (from W. J. Raybould)

Bastian, Henry Charlton (1837–1915). Physician and specialist in clinical neurology. Assistant physician and lecturer on pathology, St Mary's Hospital, London, *circa* 1863–6. Professor of pathological anatomy, University College, London, 1867; physician, University College Hospital, 1878. Best known for his work on spontaneous generation. FRS 1868. (*DSB*.)

Bates, Frederick (1829–1903). Brewer and naturalist. Younger brother of Henry Walter Bates. Managing director of a brewing company in Leicester. A keen entomologist, particularly interested in the Heteromera, publishing many papers and forming an extensive collection. Also formed collections of tiger beetles and British Coleoptera. Cultivated orchids for some years. (*Entomologist's Monthly Magazine* 39 (1903): 286–7; Gilbert 1977.)

26 June [1870?]

Bates, Henry Walter (1825–92). Entomologist. Undertook a joint expedition to the Amazon with Alfred Russel Wallace, 1848–9; continued to explore the area, after Wallace returned to England, until 1859. Provided the first comprehensive scientific explanation of the phenomenon subsequently known as Batesian mimicry. Published an account of his travels, *The naturalist on the River Amazons*, in 1863. Assistant secretary, Royal Geographical Society of London, 1864–92. President, Entomological Society of London, 1868, 1869, and 1878. FRS 1881. (*DSB*, *ODNB*.)

17 January 1870, *20 May 1870*, [22 May 1870]

Baxter, William (d. 1857). Chemist and druggist. 'Dresser' to the surgeon James

Scott. A founding member of the Pharmaceutical Society of Great Britain and a founder of the Bromley Literary Institute. Opened a chemist's shop in Bromley in 1818, and later branches in Dartford and Greenwich. (Horsburgh 1980, p. 406.)

Baxter, William Walmisley (1829/30–1900). Chemist. Succeeded his father in business at 40 High Street, Bromley, Kent, with branches in Dartford and Greenwich, 1857; opened another branch in Beckenham; retired in 1897. Churchwarden, manager of the National Schools, Trustee of the Philanthropic Society, and honorary secretary of Bromley Literary Institute. (*BMD* (*Death index*), Horsburgh 1980, pp. 406–7, *Post Office directory of the six home counties* 1859, Watkins 1977.)

23 September [1857–9?]

Beale, Lionel Smith (1828–1906). Physician and microscopist. Professor of physiology, King's College, London, 1853–69; physician, 1869; professor of medicine, 1876–96. Published on the clinical application of microscopy. A vitalist. FRS 1857. (R. Desmond 1994, *ODNB*.)

Beaufort, Francis (1774–1857). Naval officer and hydrographer. Retired as rear-admiral in 1846. Hydrographer to the navy, 1829–55. Originator of the Beaufort scale for wind force. FRS 1814. (*DNB*, *Record of the Royal Society of London*.)

Beccari, Odoardo (1843–1920). Italian naturalist. Collected and described botanical specimens primarily. Took part in several expeditions to South-east Asia (including Borneo), New Guinea, and Australia. (*DBI*.)

Becher, Alexander Bridport (1796–1876). Naval officer and hydrographer. Catalogued holdings of the Hydrographic Office, 1823–6, and was subsequently appointed chief assistant to the hydrographer. Founding editor of *Nautical Magazine*, 1832–71. Gained the rank of rear admiral the year before his retirement in 1865. (Dunkin 1879.)

1 September 1842

Becker, Lydia Ernestine (1827–90). Leading member of the women's suffrage movement, botanist, and astronomer. Published *Botany for novices* (1864); awarded a Horticultural Society Gold Medal, 1865. Founder and president of the Manchester Ladies' Literary Society, 1867. Secretary to the Manchester National Society for Women's Suffrage from 1867. Member of the Manchester School Board, 1870. Editor of and regular contributor to the *Women's Suffrage Journal* from 1870. Secretary to the London central committee of the National Society for Women's Suffrage from 1881. (Blackburn 1902, *Journal of Botany* 3 (1865): 164, *Macmillan dictionary of women's biography*, *ODNB*.)

[26 or 27 May 1863]

Bell, Charles (1774–1842). Anatomist and surgeon. Best known for his investigations of the nervous system and the expression of emotions in humans. Illustrated his own works. Co-owner of and principal lecturer at the Great Windmill Street School of Anatomy, London, 1814–25. Surgeon at the Middlesex Hospital, 1812–36. Professor of surgery, Edinburgh University, 1836. Knighted, 1831. FRS 1826. (*DSB*, *ODNB*, *Record of the Royal Society of London*.)

Bell, John (b. 1793/4). Scottish stocking-maker in Melrose. Owned a sandpit near Galashiels. (1851 Scotland Census, Melrose Parish, 2A/8; *Correspondence* vol. 2, letter from William Kemp, 4 December 1843.)

Beneden, Edouard Joseph Louis Marie (Edouard) van (1846–1910). Zoologist. Specialised in the comparative anatomy and embryology of invertebrates. Professor of zoology, University of Liège, from 1870. (*DSB*.)

17 December 1870

Bennett, Alfred William (1833–1902). Botanist, bookseller, and publisher. Proprietor and editor of the *Friend*, the monthly journal of the Society of Friends. Lecturer in botany, Bedford College, London, 1868; Westminster Hospital, London, 1869–73. Author of several papers on pollination, 1871–3, and other botanical works. Biological sub-editor for *Nature*, 1871–4. Vice-president of the Microscopical Society, 1892, 1899, and 1900; editor of the society's journal, 1897–1902. Vice-president of the Linnean Society of London, 1891–2. (R. Desmond 1994; *Journal of the Royal Microscopical Society* 1879–1902; *List of the Linnean Society of London*, 1891–2; *Medical directory* 1869–73; *ODNB*.)

Benson, Berry (1843–1923). American soldier and accountant. Fought on the side of the Confederacy in the American Civil War, 1861–5. Served as the model for a statue of an anonymous enlisted man erected in Augusta, Georgia, in 1878. After 1868, worked as a cotton broker and then an accountant. Achieved national recognition with his defence of Leo Frank, a Jewish factory manager accused of murdering a young worker. (Edward J. Cashin, 'Berry Benson (1843–1923)', 2 October 2006, http://www.georgiaencyclopedia.org (accessed 12 March 2008).)

9 February 1870

Bentham, George (1800–84). Botanist. Moved his botanical library and collections to the Royal Botanic Gardens, Kew, in 1854, and was provided with facilities there for his research from 1861. President of the Linnean Society of London, 1861–74. Published *Genera plantarum* (1862–83) with Joseph Dalton Hooker. FRS 1862. (*DSB, ODNB*.)

Berry, Alexander (1781–1873). Scottish-born merchant. Studied medicine at Edinburgh University. Worked as a surgeon's mate. Gave up medicine for commerce in 1806. Travelled to Australia for trade, 1807–12. Returned to Australia in 1819. Nominee in the Legislative Council, 1828–61. (*Aust. dict. biog.*)

29 December 1835 (from James Busby)

Berry, Elizabeth (1782–1845). Colonist. Née Wollstonecraft. Married Alexander Berry in 1827. (*Aust. dict. biog.* s.v. Berry, Alexander.)

Billinghurst, Henry Farncombe (1831/2–1912). Banker. Country manager at the London & Westminster Bank, 1875 to late 1890s. (1861 England Census (RG9/400/176), *BMD* (*Death index*), Royal Bank of Scotland Archives.)

[21 July 1863 – August 1870?]

Bischoff, Theodor Ludwig Wilhelm (1807–82). Physician. MD 1832. Professor of anatomy, Munich, 1854. Opposed CD's theory of evolution and its adaptation by Ernst Haeckel, emphasising anatomical differences between humans and monkeys. (*DBE.*)

Bismarck, Otto Eduard Leopold (Otto) von (1815–98). German politician. Prime minister of Prussia, 1862–73. Chancellor of Germany, 1871–90. (*EB.*)

Bissell, George P. (b. 1825/6.) Physician. Emigrated to the United States, residing in western Oregon by 1855. Practised in Salem, Oregon, *circa* 1870. (Salem Public Library Reference Staff, Salem, Oregon; 1870 United States Federal Census (M593_1287/14/28b).)

10 February 1870

Blanchard, Charles Emile (Emile) (1819/20–1900). French naturalist. Worked at the Muséum National d'Histoire Naturelle, Paris, from the age of 14; accompanied Henri Milne-Edwards on an expedition to the coast of Sicily. Assistant to Milne-Edwards and Armand de Quatrefages at the Muséum and at the Ecole Normale Supérieure, 1847. Member, Académie des Sciences, from 1862. Administrative professor at the Muséum National d'Histoire Naturelle, 1862–96. Professor, Institut agronomique, 1876–88. Published extensively on entomology, agriculture, and geology, and against evolution. (Conry 1974, *DBF*, Tort 1996.)

Blore, Edward William (1828–85). Clergyman and scholar. Tutor at Trinity College, Cambridge. BA Cambridge (Trinity College), 1851; fellow, 1853; tutor, 1857–75; vice-master, 1880–5. Ordained priest, 1855. (*Alum. Cantab.*)

[October 1870 or later]

Blume, Carl Ludwig (1796–1862). Dutch botanist. (*NNBW.*)

Blyth, Edward (1810–73). Zoologist. Druggist in Tooting, London, *circa* 1832–7. Wrote and edited zoological works under the pseudonym Zoophilus. Curator of the museum of the Asiatic Society of Bengal, Calcutta, India, 1841–62. Provided CD with information on the plants and animals of India in correspondence between 1855 and 1858 and later. Returned to Britain in 1863, and continued to write on zoology and on the question of the origin of species. (*Correspondence* vols. 5–7, *DSB*, *ODNB.*)

[after 24 February 1867]

Boner, Charles (1815–70). Journalist, translator, and poet. Tutored the two eldest sons of John Constable, 1831–7. Employed by the family of Prince Thurn und Taxis in Regensburg, Germany, 1840–60. Translated several German works, wrote verse, articles, and books; travelled as a naturalist and huntsman. Settled in Munich in 1860. (*ODNB.*)

[before 8 January 1870], *8 January 1870*, 20 January [1870]

Bonham-Carter, Alice (1828/9–1912). Daughter of Joanna Maria Bonham-Carter, a family friend of the Darwins. The family lived at Keston, Kent, from 1853. (Bonham-Carter 1960, *Burke's landed gentry* 1965.)

25 January [1870] (to Emma Darwin)

Bonham-Carter, Elinor Mary (1837–1923). Daughter of Joanna Maria Bonham-Carter, a family friend of the Darwins. The family lived at Keston, Kent, from 1853. Married Albert Venn Dicey (*ODNB*) in 1872. (Bonham-Carter 1960, *Burke's landed gentry* 1965.)

Bonizzi, Paolo (1840–89). Italian naturalist. Of Modena. Studied with Giovanni Canestrini. (*Catalogue of the library of tbe British Museum* (*Natural History*); letter from Giovanni Canestrini, 21 April 1870.)

A bookseller.

9 April [1860], 6 September [1860]

Boott, Mary (1794/5–1871). Daughter of the botanist Lucy Hardcastle of Derby. Married Francis Boott in 1820. (Census returns 1861 (Public Record Office RG9/173: 123); R. Desmond 1994 s.v. Hardcastle, Lucy; *ODNB* s.v. Boott, Francis; *BMD* (*Death index*); *Journal of the Proceedings of the Linnean Society* 8 (1865): xxiii–xxvii.)

Bostock, John (1772–1846). Physician and medicinal chemist. Attended Joseph Priestley's chemical lectures at Hackney College, Middlesex, in 1792. Studied as an apothecary in Liverpool until 1794. MD, University of Edinburgh, 1798. Physician to the General Dispensary, Liverpool, 1798–1817. Wrote medical and scientific articles for the *Monthly Review* and the *Edinburgh Medical and Surgical Journal*. Moved to London in 1817 and gave up medicine. Chemical lecturer at Guy's Hospital. Co-edited *Annals of Physiology*. Wrote the first recorded description of what became known as hay fever in 1819. FRS 1818, council member from 1829, vice-president from 1832. (*ODNB*, *Record of the Royal Society of London*.)

Bowman, William (1816–92). Ophthalmic surgeon. Author of works on physiology. Assistant surgeon, King's College Hospital, London, 1839–56; surgeon, 1856–62; elected joint professor of physiology and of general and morbid anatomy, King's College, London, 1848; assistant surgeon at the Royal London Ophthalmic Hospital, Moorfields, 1846–51; surgeon, 1851–76. Created baronet, 1884. FRS 1841. (*DSB*, *ODNB*.)

31 January [1863], 10 March [1863]

Brandt, Johann Friedrich von (1802–79). German zoologist and palaeontologist. Studied medicine at the University of Berlin, 1821–6. Taught medical botany and pharmacology there, 1829–31. Appointed to the Academy of Sciences, St Petersburg, 1831; director of the Zoological Museum until 1879. Professor of zoology and comparative anatomy at the Medical Surgical Academy, St Petersburg, 1851–69. (*DBE*.)

Brehm, Alfred Edmund (1829–84). German zoologist and traveller. Travelled in Egypt and the Sudan, Spain, Norway and Lapland, Siberia, and Turkestan. Studied natural history in Jena and Vienna, 1853–6. Director of the zoological gardens, Hamburg, 1862–7. Founded the aquarium in Berlin, 1867. Published books and articles on zoology. (*ADB*, *BHGW*.)

Brehm, Christian Ludwig (1787–1864). German clergyman and ornithologist.

His bird collection formed the basis of the Rothschild museum in Tring, Hertfordshire. (*NDB.*)

Breton, Philip le (1779–60). Clergyman and educator. BA (Exeter College), Oxford, 1800; MA 1803. Curate of Farringdon. Rector of St Saviour's, Jersey. Left the church and established a school for boys in Westminster. Served on several Unitarian councils. (*Alum. Oxon.*, *Christian reformer, or Unitarian magazine and review* n.s. 16 (1860): 756–8.)

Bridges, Thomas (b. *c.* 1842 d. 1898). Dictionary compiler, missionary, and sheep farmer. Missionary on Keppel Island, West Falklands, 1856–68; established a mission at Ushuaia in the Beagle Channel, 1867. Settled permanently in Ushuaia with his wife, Mary Ann Varder, and daughter in 1871. Abandoned missionary work in 1884 and became a sheep farmer near Ushuaia from 1887. Author of a Yámana dictionary. (E. L. Bridges 1948, *ODNB.*)

Bristowe, Henry Fox (1824–93). Barrister. Called to the bar, 1847; incorporated at Lincoln's Inn, 1854. Nephew of William Darwin Fox. Married Selina Bridgeman in 1850. Knighted, 1887. (*Alum. Cantab.*, *Darwin pedigree.*)

Broca, Pierre Paul (Paul) (1824–80). French surgeon and anthropologist. MD, Paris, 1849. Assistant professor of the medical faculty in Paris and surgeon of the Central Bureau, 1853. Pioneer in the field of anthropology. Elected professor of pathology at Paris, 1867; of clinical surgery, 1868. Vice-president, French Academy of Medicine. (*DBF*, *DSB.*)
4 August 1870

Broderip, William John (1789–1859). Magistrate and naturalist. BA (Oriel College), Oxford, 1812. Called to the bar, 1817. Magistrate in London, 1822–56. A founder of the Zoological Society of London, 1826. Possessed an unrivalled conchological cabinet. FRS 1828. (*ODNB*, Sarjeant 1980–96.)

Brodie, Benjamin Collins (1783–1862). Surgeon. Professor of comparative anatomy and physiology, Royal College of Surgeons, 1819–23. Sergeant-surgeon to Queen Victoria. President of the Royal Society of London, 1858–61. Created baronet, 1834. FRS 1810. (*DSB*, *ODNB.*)
26 April [1859]

Brodie, Benjamin Collins (1817–80). Chemist. Awarded a doctorate at Giessen, Germany, in 1850 for his analysis of beeswax. Professor of chemistry, Oxford, 1855–72. Devised an alternative approach to chemistry from that based on atomic theory, and developed a calculus of chemical operations. Succeeded to the baronetcy, 1862. FRS 1849. (*Alum. Oxon.*, *DSB*, *ODNB.*)
9 June 1870

Brongniart, Adolphe Théodore (1801–76). French palaeobotanist, plant anatomist, and taxonomist. One of the founders of the *Annales des Sciences Naturelles*, 1824. Professor of botany, Muséum d'Histoire Naturelle, from 1833. (*DSB.*)

Brooke, Charles Anthoni Johnson (1829–1917). Naval officer, colonial officer, and second raja of Sarawak. Served in the Royal Navy, 1842–52. Joined his uncle

Brooke, Charles Anthoni Johnson, cont.
James Brooke, raja of Sarawak, in 1852. Named as heir, 1867; became second raja on his uncle's death in 1868. In later life, spent part of each year in England. Knighted, 1888. (*ODNB*.)
30 November 1870

Brougham, Henry Peter, 1st Baron Brougham and Vaux (1778–1868). Scottish-born statesman, lawyer, reformer, and educationalist. A founder of the *Edinburgh Review*, 1802, the Society for the Diffusion of Useful Knowledge, 1826, and London University, 1826. Prolific author. Lord chancellor, 1830–4. A founder of the National Association for the Promotion of Social Science, 1857. Elected chancellor of Edinburgh University, 1859. Created Baron Brougham and Vaux, 1830. FRS 1803. (*ODNB*, *Record of the Royal Society of London*.)

Brown, Robert (1773–1858). Scottish botanist. Naturalist to the expedition surveying the coast of Australia, 1801–5; published descriptions of the plants he collected. Librarian to the Linnean Society of London, 1805–22; to Sir Joseph Banks, 1810–20. Continued as curator of Banks's collections after his death in 1820 and negotiated their transfer to the British Museum in 1827. Keeper of the botanical collections, British Museum, 1827–58. Discoverer of 'Brownian motion'. FRS 1811. (*DSB*, *ODNB*.)

Browne, John Hutton Balfour (1845–1921). Barrister. Studied at Edinburgh University. Called to the bar, 1870; bencher, Middle Temple, 1890. Registrar and secretary to the Railway Commission, 1874. Son of William Alexander Francis Browne; brother of James Crichton-Browne. (*ODNB* s.v. Browne, William Alexander Francis; *WWW*.)

Browne, Vincent de Paul (1847/8–70). Son of William Alexander Francis Browne and Magdelene Howden Browne (née Balfour), and brother of James Crichton-Browne. (*ODNB*.)

Browne, William Alexander Francis (1805–85). Physician. President of the Plinian Society, Edinburgh, when CD was a member. Physician at Stirling, 1830. Superintendent of Montrose lunatic asylum; superintendent of Crichton Institute at Dumfries, 1839. Commissioner in lunacy for Scotland, 1857–70. (*Modern English biography*.)

Bruce, Henry Austin, Baron Aberdare (1815–95). Politician. Liberal MP for Merthyr Tudful, 1852–68. Under-secretary at the Home Office, 1862–4. Privy councillor, 1864–6. Home secretary, 1868–73. MP for Renfrewshire, 1869–73. Interested in education. Created Baron Aberdare, 1873. FRS 1876. (*ODNB*.)

Brushfield, Thomas Nadauld (1828–1910). Physician and antiquary. Studied medicine and surgery, London Hospital, 1845–9. Matriculated with honours at London University, 1848. MD, St Andrews, 1862. Superintendent, Brookwood Asylum, Woking, 1865–82. As an antiquary, studied the career of Walter Ralegh and published a series of papers on him. (*ODNB*.)

Buch, Christian Leopold von (1774–1853). German geologist and palaeontologist. Travelled extensively in Europe and became a supporter of the geological catastrophe theory of Cuvier. After 1826, devoted himself mainly to palaeontology. Recognised the chronological significance of fossils. Foreign member, Royal Society of London, 1828. (*DBE*, *DSB*.)

Büchner, Friedrich Karl Christian Ludwig (Ludwig) (1824–99). German materialist philosopher and physician. Lecturer in medicine, especially forensic medicine, at Tübingen University, 1854–5. Following the publication of his first work, *Kraft und Stoff* (1855), he was debarred from academic teaching and returned to general medicine. (*DBE*, *NDB*.)

Buckland, Francis Trevelyan (1826–80). Naturalist, popular science-writer, and surgeon. Son of William Buckland. Trained and practised medicine at St George's hospital, London, 1848–53. Assistant surgeon in the second Life Guards, 1854–63. Staff writer for the *Field*, 1856–65. In 1865, appointed scientific referee to the South Kensington Museum, where he established an exhibit on pisciculture. Launched a weekly journal, *Land and Water*, in 1866. Inspector of salmon fisheries from 1867. (Bompas 1885, *ODNB*.)

[1865–6]

Buckland, William (1784–1856). Geologist and clergyman. Reader in mineralogy, Oxford University, 1813; reader in geology, 1819–49. President of the Geological Society of London, 1824–5 and 1840–1. Dean of Westminster from 1845. FRS 1818. (*DSB*, *ODNB*, *Record of the Royal Society of London*.)

Bucknill, John Charles (1817–97). Psychiatrist and writer. Superintendent, Devon County Asylum, Exminster, 1844–62. Co-founder, 1853, *Asylum Journal*, later *Journal of Mental Science*; editor 1853–62. Wrote extensively on insanity. Lord Chancellor's visitor in lunacy, 1862–76. Co-founder of *Brain*, 1878. FRCP 1859. FRS 1866. Knighted 1894. (*ODNB*.)

Bullar, Joseph (1808–69). Physician and author. Born in Southampton, the son of a schoolmaster. Trained as a surgeon in Dorset, 1824–8. Entered St Bartholomew's Hospital, London, 1828. Studied medicine in Paris and Dublin, 1830–2. MD, Edinburgh 1833. In general medical practice in Southampton, 1833–60. With his brother William founded the Royal South Hampshire Infirmary in 1835 and was surgeon there. Physician to the infirmary, 1852–67. Researched and wrote on botanical subjects, and on the relationship of science and Christian belief. First president, Southampton Microscopical Society, 1861. (Dayman 1869.)

27 June [1862]

Burton, Richard Francis (1821–90). Soldier, diplomat, explorer, and writer. Joined the Indian army in 1842. Studied Asiatic languages and culture. Assistant on the survey of Sind, 1843. Adopting an assumed identity, went on pilgrimage to Mecca in 1853. Explored Somalia in 1853. Served in the Crimea in 1855. Participated in the search for the source of the Nile, 1856–9. British consul, Fernando Po, West Africa, 1861–5; Santos, Brazil, 1865–9; Damascus, 1869–71; Trieste, 1872. In later years, in England, devoted himself to literary scholarship. (*DNB*.)

Burton, William Westbrooke (1794–1888). Judge. Entered the navy as a midshipman in 1807. Served in the Mediterranean, East and West Indies, China, and New Orleans. Called to the bar, 1824. Recorder of Daventry, 1826–7. Accepted a seat on the bench of the newly constituted Supreme Court at the Cape of Good Hope in 1827; at New South Wales in 1832. Puisne judge, Madras, 1844–57. Member of the Australian Legislative Council, 1857; president, 1858–61. Retired to England in 1861. Knighted, 1844. (*Aust. dict. biog.*, *ODNB*.)

Busby, Agnes (1800/1–89). Colonist. Née Dow. Married James Busby in 1832. (*Aust. dict. biog.* s.v. Busby, James, *Friends of West Norwood Cemetery*, http://www.fownc.org/newsletters/no58.shtml (accessed 3 December 2008).)

Busby, James (1801–71). Scottish-born viticulturalist and civil servant. Son of John Busby. Studied viticulture in France. Emigrated with his parents to New South Wales in 1824. Superintendent at the Male Orphan School near Liverpool, 1824–7. Collector of internal revenue, 1827–9. Toured vineyards in France and Spain in 1831. British resident in New Zealand, 1833–40. Represented the Bay of Islands in the Auckland Provincial Council, 1835–5 and 1857–63. Published on viticulture. (*Aust. dict. biog.*, *DNZB*.)

29 December 1835 (to Alexander Berry)

Busby, John (1765–1857). Surveyor and civil engineer. Worked as a coal-mine manager and mineral surveyor from 1784. Engaged on various public works in Scotland. Emigrated to Australia in 1824. Planned and oversaw construction of a regular water supply to Sydney, 1825–37. (*Aust. dict. biog.*)

Busk, George (1807–86). Russian-born naval surgeon and naturalist. Served on the hospital ship at Greenwich, 1832–55. Retired from medical practice in 1855. Member of several scientific societies. President of the Microscopical Society, 1848–9; of the Anthropological Institute, 1873–4. Zoological secretary of the Linnean Society of London, 1857–68. Hunterian Professor of comparative anatomy, Royal College of Surgeons of England, 1856–9; council member, 1863; member of board of examiners, 1868; president, 1871. Specialised in palaeontology and in the study of Bryozoa. FRS 1850. (*DNB*, *DSB*, Plarr 1930.)

Butler, Arthur Gardiner (1844–1925). Entomologist. Assistant, zoological department, British Museum, 1863; assistant keeper, 1879–1901. Worked mostly on Lepidoptera. (*Entomologist* 58 (1925): 175–6, *Proceedings of the Linnean Society of London* (1926): 75–6.)

19 July 1870

Camões, Luiz Vaz de (1524–80). Portuguese poet. (*EB* s.v. Camoens, Luis Vas de.)

Campbell, Elizabeth Georgiana (d. 1878). Daughter of George Granville Leveson-Gower, second duke of Sutherland. Married George Douglas Campbell in 1844. (*Burke's peerage* 1980; *ODNB* s.v. Campbell, George Douglas.)

Campbell, George Douglas, 8th duke of Argyll (1823–1900). Scottish statesman and author of works on science, religion, and politics. A defender of the concept of design in nature. Chancellor of St Andrews University, 1851–1900.

President of the Royal Society of Edinburgh, 1860–4. Privy seal, 1852–5, 1859–60, 1860–6, and 1880–1; postmaster-general, 1855–8 and 1860; secretary of state for India, 1868–74. Succeeded to the dukedom in 1847. FRS 1851. (*ODNB*.)

Candolle, Alphonse de (1806–93). Swiss botanist, lawyer, and politician. Active in the administration of the city of Geneva until 1860. Responsible for the introduction of postage stamps to Switzerland. Professor of botany and director of the botanic gardens, Geneva, from 1835. Concentrated on his own research after 1850. Foreign member, Royal Society of London, 1869. (*DSB*, *Record of the Royal Society of London*.)

Canestrini, Giovanni (1835–1900). Italian zoologist. Studied philosophy and natural sciences at Vienna, receiving his degree in 1861. Professor of natural history, Modena, 1862–9. Professor of zoology and comparative anatomy and physiology, Padua, 1869–1900. Translated the first Italian edition of *Origin* (1864) in association with Leonardo Salimbeni. Worked mainly in ichthyology and later on human origins. (*DBI*, Pancaldi 1991.)

21 April 1870

Carpenter, William Benjamin (1813–85). Naturalist. Fullerian Professor of physiology at the Royal Institution of Great Britain, 1844–8; physiology lecturer, London Hospital, 1845–56; professor of forensic medicine, University College, London, 1849–59. Registrar of the University of London, 1856–79. President of the British Association for the Advancement of Science, 1872. Founding member of the Marine Biological Association. FRS 1844. (*DNB*, *DSB*, *Modern English biography*, Royal Institution of Great Britain, http://www.rigb.org.)

Carus, Julius Victor (1823–1903). German comparative anatomist. Conservator of the Museum of Comparative Anatomy, Oxford University, 1849–51. Professor extraordinarius of comparative anatomy and director of the zoological museum, University of Leipzig, 1853. Translated the third German edition of *Origin* (1867) and, subsequently, twelve other works by CD. (*DSB*, *NDB*.)

26 December [1866], *16 April [1870]*, *18 April 1870*, 18 August 1870, *2 October 1870*, 5 October [1870], 11 October [1870]

Cassell, Petter, & Galpin. Publishing and printing firm with premises at La Belle Sauvage printing works, Ludgate Hill, London. (*Post Office London directory* 1866.)

Caton, John Dean (1812–95). American jurist, author, and naturalist. Opened Chicago's first law office, 1833. Admitted to the Illinois bar. Appointed associate justice of the supreme court of Illinois, 1842, eventually serving as chief justice in 1855, 1857–64. Owned a large property in Ottawa, Illinois. Travelled widely in the United States, Europe, and Asia. Wrote several books, including *The antelope and deer of America* (1877). (*DAB*.)

17 November 1870

Cavendish, William, 7th duke of Devonshire and 2d earl of Burlington (1808–91). Statesman and benefactor of scientific and industrial enterprises. Whig MP for Cambridge University, 1829–31; for Malton, Yorkshire, 1831; for Derbyshire, 1831; for North Derbyshire, 1832–4. Succeeded to the earldom, 1834;

Cavendish, William, 7th duke of Devonshire and 2d earl of Burlington, cont.
to the dukedom, 1858. Chancellor of London University, 1836–56; of Cambridge University, 1861–91. President of the Royal Agricultural Society, 1870. Privy councillor, 1876. Knighted, 1858. FRS 1829. (*DNB*, *Record of the Royal Society of London*.)

Cecil, Robert Arthur Talbot Gascoyne- (Robert), 3d marquess of Salisbury (1830–1903). Statesman. Attended Christ Church, Oxford, 1847–9. Conservative MP for Stamford, 1853–68. Published many political articles in the *Saturday Review*, Bentley's *Quarterly Review*, and John Murray's *Quarterly Review* from 1856. Became Viscount Cranborne when his elder brother died in 1865; marquess of Salisbury on his father's death in 1868. Secretary of state for India, 1874–8. Foreign secretary, 1878. Prime minister, 1885, 1886–92, 1895–1902. FRS 1869. (*ODNB*.)
7 June 1870

Child, Gilbert William (1832–96). Physician. BA, Exeter College, Oxford, 1854. MD, 1859. Examiner in medicine, Oxford University, 1863; in natural science, 1867. Practised in Oxford until *circa* 1882. Wrote on sanitation, amongst other subjects. Fellow of the Linnean Society of London, 1868. (*List of the Linnean Society*, *Modern English biography*.)
26 January [1870], 16 February [1870]

Child, Henry T. (b. 1816/17). American physician. Born and resided in Philadelphia. Treasurer of the First Association of Spritualists (of Philadelphia), 1864. (1870 United States Federal Census (M593_1391/219/441); Britten 1870, pp. 273–5.)

Claparède, Jean Louis René Antoine Edouard (Edouard) (1832–71). Swiss naturalist and invertebrate zoologist. Professor of zoology and comparative anatomy, Academy of Geneva, 1862. Specialised in invertebrate anatomy, histology, embryology, and evolution. One of the first Swiss naturalists to endorse CD's theory. (*Dictionnaire historique & biographique de la Suisse*, Gilbert 1977.)

Clarke, Benjamin (1813–90). Systematic botanist. Trained as a medical practitioner, but rarely practised. Devised his own system of classification. Fellow of the Linnean Society of London, 1845. (R. Desmond 1994, *Journal of Botany* 28 (1890): 84–6.)
1 November [1870]

Clift, William (1775–1849). Naturalist. Curator of the Hunterian Museum at the Royal College of Surgeons, 1793–1842. In 1835, Clift's daughter Caroline married Richard Owen, who succeeded Clift as curator in 1844. FRS 1823. (Dobson 1954, p. 103; *DSB*; *ODNB*.)

William Clowes & Sons. Printers. William Clowes (1807–83), eldest son of William Clowes (1779–1847), joined his father's printing business in 1823; the name of the firm was changed to William Clowes & Sons in 1839. Printed the official catalogue of the Great Exhibition of 1851. Introduced improvements in music printing. Printers to John Murray. (*ODNB*.)

Cobbe, Frances Power (1822–1904). Writer and philanthropist. Wrote extensively on religious and ethical subjects. Leading campaigner for women's rights and against animal vivisection. (*ODNB*.)

23 March [1870?], *28 March [1870?]*, 20 August [1870]

Colburn, Henry (1784/5–1855). Publisher. Proprietor of the English and Foreign Circulating Library in London, where he also began his publishing operation, 1806–24. Co-founded the *New Monthly Magazine and Universal Register*; founded the *Literary Gazette* and other periodicals. Ran publishing operations in New Burlington Street, 1824–32. Set up premises at Great Marlborough Street in 1836. (*ODNB*.)

Colenso, Francis Ernest (Frank) (1852–1910). Barrister and campaigner for Zulu rights. Younger son of John William Colenso. BA, Cambridge, 1874. Admitted to the Inner Temple, 1874; called to the bar, 1877. Practised as a barrister in Natal, 1877–9. Assistant secretary and assistant actuary to the Norwich Union, Norwich, from 1879; actuary and secretary to the English and Scottish Law Life Assurance Association, London, 1886–94; Eagle Insurance Company, London, 1894. Supported his father's campaign for the rights of the African populations of Natal and Zululand. (*ODNB*.)

Colenso, Robert John (1850–1926). Physician and clergyman. Elder son of John William Colenso. BA, MA (Trinity College), Oxford, 1877; BMed, 1879. (*Alum. Oxon*, *BMD* (*Death index*), *IGI*.)

Compton, Spencer Joshua Alwyne, 2d marquess of Northampton (1790–1851). Patron of science and the arts. MP for Northampton, 1812–20. President of the Royal Society, 1838–48. Succeeded his father as second marquess of Northampton, 1828. FRS 1830. (*ODNB*, *Record of the Royal Society of London*.)

Cooke, Robert Francis (1816–91). Publisher. Cousin of John Murray and partner in his publishing company, which published many of CD's books. (*Correspondence* vol. 13, letter from John Murray, 1 April 1865, *Modern English biography*.)

5 December 1870

Cooper, John W. (b. 1802/3). American physician. Resident in West Chester, Pennsylvania. (1870 United States Federal Census (M593_1325/170).)

Cope, Edward Drinker (1840–97). American biologist and palaeontologist. Worked at the Smithsonian Institution on the reptile collection, 1859. Studied the collections of Harvard Museum of Comparative Zoology and the Academy of Natural Sciences, Philadelphia, in the 1860s. Taught zoology at Haverford College, Pennsylvania, 1864–7. Made several palaeontological expeditions to the western states from the early 1870s. Professor of geology and later zoology, University of Pennsylvania, from 1889. (*ANB*.)

Coquand, Henri-Jean-Baptiste (Henri) (1813–81). French geologist. Carried out geological exploration in Spain, Algeria, and Morocco. Professor of geology, Besançon, 1852; Poitiers, then Marseilles, 1859–64. Prospected in Albania, the Ionian Islands, and Sicily. Settled in Marseilles. (*DBF*.)

Crichton, Elizabeth (1779–1862). Scottish philanthropist. A daughter of Robert Grierson, fifth baronet of Lag and Rockhall. Married James Crichton (1765–1823), a distinguished physician, in 1810. Used the money left in her husband's

Crichton, Elizabeth, cont.
will for charitable purposes to found the Crichton Institution, a mental hospital in Dumfries. Godmother of James Crichton-Browne, the son of William Alexander Francis Browne, first medical superintendent of the institution. (*ODNB.*)

Crichton-Browne, James (1840–1938). Scottish physician, psychiatrist, and author. Graduated from the College of Surgeons, Edinburgh, in 1862. Medical director, West Riding Asylum, Wakefield, 1866–75. Lord chancellor's visitor in lunacy, 1875–1922. Pioneering researcher into the causes and management of mental illness. Editor, *West Riding Lunatic Asylum medical reports*, 1871–5. One of the founders of *Brain*, 1878; co-editor, 1878–85. Author of a number of popular works on mental health and social policy. Knighted, 1886. FRS 1883. (*DNB.*)
31 January [1870], *15 March 1870*, 18 March 1870, 2 April [1870], *6 June 1870*, *[6 June 1870]*, 8 June [1870]

Crompton, Joseph (1813/14–78). Clergyman. MA, University of Glasgow, 1834. Unitarian minister at the Octagon in Norwich, 1839–52. Founder and minister of the Free Christian Church. President of the Norfolk and Norwich Naturalists' Society, 1869–71. Joined the Church of England, 1875. Curate of St Mary-at-Coslany with St Michael-at-Pleas, Norwich, 1872–6; rector of St Lawrence, Norwich, from 1876. (*BMD* (*Death index*); *Crockford's clerical directory* 1878; *Roll of the graduates of the University of Glasgow*; *Transactions of the Norfolk and Norwich Naturalists' Society* 1 (1869–74).)

Cumming, Linnaeus (1843–1927). Schoolteacher. BA (Trinity College), Cambridge, 1865. Assistant master, Cheltenham College, 1869–75; Rugby, 1875–1911. Member of the eclipse expedition to Sicily, 1870 and 1871. Studied botany and geology. (*Alum. Cantab.*)

Cupples, Anne Jane (1839–98). Scottish author. Second daughter of Archibald Douglas. Married George Cupples in 1858. Wrote children's books. Lived in New Zealand from 1891. (*Modern English biography.*)
[28 November 1870]

Cupples, George (1822–91). Scottish writer and dog breeder. Served as an apprentice on an eighteen-month voyage to India and back on the *Patriot King*, *circa* 1838; had his indentures cancelled on his return. Studied arts and theology at Edinburgh University for eight years. Published a number of novels and other books, and wrote many articles and stories for journals. Bred Scottish deer-hounds. (*Correspondence* vol. 16, letter from George Cupples, 1 May 1868, Cupples 1894, *Modern English biography.*)
29 April 1870, *20 June 1870*, *17 September 1870*, *8 October 1870*, 3 November [1870], *14 November 1870*, 27 November [1870]

Curtis, James Gray William (d. 1870). Army officer in India. Deputy assistant commissary general, 1849–50. Retired with the rank of colonel, 1854. (*Modern English biography.*)

Curtis, Moses Ashley (1808–72). American botanist and minister. Ordained in the Episcopal Church, 1834. Carried out missionary work in North Carolina for most of his life. Studied and collected North American lichens and fungi. (*DAB*.)

Curtis, William (1746–99). Botanist and entomologist. Demonstrator of plants and *praefectus horti* at the Chelsea Physic Garden, 1772–7. Established botanic gardens in Bermondsey in 1773, Lambeth in 1779, Brompton in 1789. Founded the *Botanical Magazine* in 1781. Published on entomology and botany. (*ODNB*.)

Cuvier, Jean Léopold Nicolas Frédéric (Georges) (1769–1832). French systematist, comparative anatomist, palaeontologist, and administrator. Professor of natural history, Collège de France, 1800–32; professor of comparative anatomy, Muséum d'Histoire Naturelle, 1802–32. Permanent secretary to the Académie des Sciences from 1803. Foreign member, Royal Society of London, 1806. (*DBF*, *DSB*.)

Dallas, William Sweetland (1824–90). Entomologist, author, and translator. Prepared lists of insects for the British Museum, 1847–58. Curator of the Yorkshire Philosophical Society's museum, 1858–68. Assistant secretary to the Geological Society of London, 1868–90. Translated Fritz Müller's *Für Darwin* (1869); prepared the index for *Variation* and the glossary for *Origin* 6th ed. Editor, *Annals and Magazine of Natural History*, 1868–90, *Popular Science Review*, 1877–80. (R. B. Freeman 1978, *Geological Magazine* n.s. decade 3, vol. 7 (1890): 333–6, *Modern English biography*, Sarjeant 1980–96.)

Dalrymple, Donald (b. 1814 d. *c.* 1873). Surgeon and member of parliament. Senior surgeon at the Hospital for Sick Children; proprieter and surgeon of the Heigham Retreat Lunatic Asylum. Retired from the profession in 1862. Magistrate and deputy lieutenant for Norfolk. Liberal MP for Bath from 1868. A local secretary of the British Association for the Advancement of Science meeting at Norwich in 1868. (*Medical directory* 1868; *Report of the 38th meeting of the British Association for the Advancement of Science, held at Norwich in August 1868*, List of members; Stenton 1976.)

Dana, James Dwight (1813–95). American geologist and zoologist. Geologist and mineralogist with Charles Wilkes's expedition to the South Seas, 1838–42; wrote reports on the geology, zoophytes, and Crustacea. An associate editor of the *American Journal of Science and Arts* from 1846. Professor of natural history, Yale University, 1855–64; professor of geology and mineralogy, 1864–90. Foreign member, Royal Society of London, 1884. (*ANB*, *DSB*, *Record of the Royal Society of London*.)

21 December [1856]

Daniell, John Frederic (1790–1845). Experimental philosopher. Worked for a sugar refining establishment, 1808–21. Designed meteorological instruments. Helped found the Society for the Diffusion of Useful Knowledge in 1827. First professor of chemistry, King's College, London, from 1831. FRS 1814. (*ODNB*.)

Dareste, Gabriel-Madeleine-Camille (Camille) (1822–99). French zoologist. A specialist in experimental embryology. Doctor of medicine, 1847. Doctor of

Dareste, Gabriel-Madeleine-Camille (Camille), cont.

science, 1851. Taught natural history at various provincial institutions. Professor of zoology, University of Lille, 1864–72. Professor of ichthyology and herpetology, Muséum d'Histoire Naturelle, Paris, 1872. Director of the laboratory of teratology, later attached to the Ecole des Hautes-Etudes, 1875. Awarded the grand prize in physiology by the Académie des Sciences for *Recherches sur la production artificielle de monstruosités* (1877). (*DBF*, *Dictionnaire universel des contemporains.*)

2 July 1870

Darwin, Anne Elizabeth (Annie) (1841–51). CD's daughter. (*Darwin pedigree.*)

Darwin, Elizabeth (Bessy/Lizzy) (1847–1926). CD's daughter. (*Darwin pedigree*, R. B. Freeman 1978.)

Darwin, Emma (1808–96). Youngest daughter of Josiah Wedgwood II. Married CD, her cousin, in 1839. (*Emma Darwin* (1904) and (1915).)

22 March [1867] (from H. A. Huxley)

Darwin, Erasmus (1731–1802). CD's grandfather. Physician, botanist, and poet. Advanced a theory of transmutation similar to that subsequently propounded by Jean Baptiste de Lamarck. FRS 1761. (*DSB*, King-Hele 1999, *ODNB.*)

Darwin, Erasmus Alvey (1804–81). CD's brother. Attended Shrewsbury School, 1815–22. Matriculated at Christ's College, Cambridge, 1822; Edinburgh University, 1825–6. Qualified in medicine but never practised. Lived in London from 1829. (*Alum. Cantab.*, R. B. Freeman 1978.)

15 August [1859 or later], 12 November [1860–8], [1863–6?], [September 1868 – August 1870], 1 April [1870?], 19 [December 1870], 21 [December 1870]

Darwin, Francis (1848–1925). CD's son. Botanist. BA (Trinity College) Cambridge, 1870. Qualified as a physician but did not practise. Collaborated with CD on several botanical projects, 1875–82. Lecturer in botany, Cambridge University, 1884; reader, 1888–1904. Published *LL* and *ML*. President of the British Association for the Advancement of Science, 1908. Knighted, 1913. FRS 1882. (*DSB*, *ODNB.*)

30 August [1867–70], 18 October [1870], 5 December [1870]

Darwin, George Howard (1845–1912). CD's son. Mathematician. BA (Trinity College), Cambridge, 1868; fellow, 1868–78. Studied law in London, 1869–72; called to the bar, 1872, but did not practise. Plumian Professor of astronomy and experimental philosophy, Cambridge University, 1883–1912. President of the British Association for the Advancement of Science, 1905. Knighted, 1905. FRS 1879. (*DSB*, *Men-at-the-bar*, *ODNB.*)

[1 February 1870 or earlier], [3 February 1870 or earlier], [4 February 1870], [20 April 1870]

Darwin, Henrietta Emma (1843–1927). CD's daughter. Married Richard Buckley Litchfield (*Alum. Cantab.*) in 1871. Assisted CD with some of his work. Edited *Emma Darwin* (1904) and (1915). (*Burke's landed gentry* 1952, *Correspondence* vol. 11, R. B. Freeman 1978.)

[8 February 1870], *[after 8 February 1870]*, [March?] 1870, *[30 March 1870]*

Darwin, Horace (1851–1928). CD's son. Civil engineer. BA, Trinity College, Cambridge, 1874. Apprenticed to an engineering firm in Kent; returned to Cambridge in 1877 to design and make scientific instruments. Founder and director of the Cambridge Scientific Instrument Company. Mayor of Cambridge, 1896–7. Knighted, 1918. FRS 1903. (*Alum. Cantab.*, *ODNB*.)

Darwin, Leonard (1850–1943). CD's son. Military engineer. Attended the Royal Military Academy, Woolwich. Commissioned in the Royal Engineers, 1871; major, 1889. Served on several scientific expeditions, including those for the observation of the transit of Venus in 1874 and 1882. Instructor in chemistry and photography, School of Military Engineering, Chatham, 1877–82. Intelligence service, War Office, 1885–90. Liberal Unionist MP, Lichfield division of Staffordshire, 1892–5. President, Royal Geographical Society of London, 1908–11; Eugenics Education Society, 1911–28. Chairman, Bedford College, London University, 1913–20. (Margaret Keynes 1943, RMA Gentleman Cadet Register (RMAS archives, Sandhurst collection), Sarjeant 1980–96, *WWW*.)

Darwin, Robert Waring (1766–1848). CD's father. Physician. Had a large practice in Shrewsbury and resided at The Mount. Son of Erasmus Darwin (*ODNB*) and his first wife, Mary Howard. Married Susannah, daughter of Josiah Wedgwood I (*ODNB*), in 1796. FRS 1788. (R. B. Freeman 1978.)

Darwin, Susan Elizabeth (1803–66). CD's sister. Lived at The Mount, Shrewsbury, the family home, until her death. (*Darwin pedigree*, R. B. Freeman 1978.)

Darwin, William Erasmus (1839–1914). CD's eldest son. Banker. BA, Christ's College, Cambridge, 1862. Partner in the Southampton and Hampshire Bank, Southampton, 1861. Chairman of the Southampton Water Company. Amateur photographer. (*Alum. Cantab.*, F. Darwin 1914.)

[1 May 1864], [19 May 1864]

Daubeny, Charles Giles Bridle (1795–1867). Chemist and botanist. Professor of chemistry, Oxford University, 1822–55; professor of botany, 1834–67; and of rural economy, from 1840. An active supporter of the British Association for the Advancement of Science; vice-president, 1847, president, 1856. FRS 1822. (*Alum. Oxon.*, *ODNB*, *Record of the Royal Society of London*.)

Dawkins, William Boyd (1837–1929). Geologist and palaeontologist. Member of the Geological Survey of Great Britain, 1861–9. Curator of natural history, Manchester Museum, 1869. Professor of geology, Owens College, Manchester, 1874–1908. Specialised in fossil mammals. FRS 1867. (*ODNB*.)

30 [August 1867], *31 July 1869*, *29 January [1870]*

Dawson, George Mercer (1849–1901). Canadian scholar and civil servant. Studied at the Royal School of Mines, London, 1869–72; graduated with distinction, winning the Forbes Medal and Prize in natural history and palaeontology, and the Murchison Medal in geology. Naturalist and geologist on the international boundary survey, 1872–4. Geologist, Geological Survey of Canada, 1875–83; assistant director, 1883–5; acting director, 1885–6. Wrote on the ethnology and anthropology of western Canadian native peoples. FRS 1891. (*DCB*.)

Dawson, John William (1820–99). Canadian geologist and educationalist. Investigated the geology of the maritime provinces with Charles Lyell in 1842 and 1852. Superintendent of education for common schools in Nova Scotia, 1850. Appointed principal and professor of geology at McGill University, 1855. Knighted, 1884. FRS 1862. (*DNB, DSB.*)

Dawson, Margaret Ann Young (1825/6–1913). Scottish born. Née Mercer; married John William Dawson in 1847. (1841 Scotland Census (4C/11/860); *Standard* (Montreal), 27 September 1913.)

Deiters, Otto Friedrich Karl (1834–63). German anatomist and histologist. After early training in Bonn, was an assistant in Rudolf Virchow's institute of pathology, Berlin, 1857. Habilitated at Bonn, 1858. Developed a technique to isolate cells later called neurons under the microscope. (*BLA.*)

De Koninck, Laurent-Guillaume (1809–87). Belgian palaeontologist and medical doctor. Professor of chemistry at the University of Liège. Member, Académie Royale de Belgique, 1842. Worked on Belgian fossils of the Carboniferous period. (*Nouveau dictionnaire des Belges.*)

19 December 1870, [after 19 December 1870]

Delpino, Federico (1833–1905). Italian botanist. Travelled extensively for botanical purposes as a youth and in 1873. Civil servant, ministry of finances, Turin, 1852–6; assistant in the botanic garden and museum, Florence, 1867; lecturer, Vallombrosa school of forestry, 1871; professor of botany and director of the botanic garden, Genoa, 1875–84; professor, University of Bologna, 1884; professor of botany and head of the botanic garden, Naples, 1894. (*DBI*, Mayerhöfer 1959–70, Penzig 1905.)

[1870], 28 February 1870, 20 May 1870, 1 June [1870]

Demosthenes (384–322 BCE). Athenian orator. (*Oxford classical dictionary.*)

Denny, Henry (1803–71). Zoologist and botanist. Specialised in entomology, studying parasitic insects. Curator and assistant secretary, museum of the Literary and Philosophical Society of Leeds, 1826–71. Secretary to the West-Riding Geological and Polytechnic Society. (*DNB*, Gilbert 1977, *Proceedings of the Linnean Society of London* (1870–1): lxxxiv–lxxxv.)

5 February [1844]

Despine, Prosper (b. 1812). French doctor and author. Practised medicine in Marseilles. Member of the Société médico-psychologique, Paris. Published on psychology and criminality. (*DBF.*)

Dixon, William Hepworth (1821–79). Historian, traveller, man of letters. Published lives of William Penn (1851), Robert Blake (1852), and Francis Bacon (1862). Editor of the *Athenaeum*, 1853–69. Justice of the peace for Middlesex and Westminster, 1869. Member of the London School Board, 1870. Fellow of the Society of Antiquaries and the Royal Geographical Society of London. (*DNB.*)

Dohrn, Felix Anton (Anton) (1840–1909). German zoologist. Studied medicine and zoology at various German universities. PhD, Breslau, 1865. Studied with Ernst Haeckel and became Haeckel's first assistant at Jena, where he habilitated in 1868. Founded the Zoological Station at Naples, built between 1872 and 1874. The station was the first marine laboratory, and served as a model for other similar institutions throughout the world. (*DBE*, *DSB*, Heuss 1991.)

4 January 1870, *13 January 1870*

Donders, Ernestine J. A. (d. 1887). Née Zimmerman, daughter of a Lutheran minister. Married Frans Cornelis Donders in 1845. (*DSB* s.v. Donders, Franciscus Cornelis.)

Donders, Frans Cornelis (Franciscus Cornelius) (1818–89). Dutch physiologist and ophthalmologist. MD, Leiden University, 1840. Professor of anatomy at Utrecht military school; professor of physiology at University of Utrecht, 1862. President of the science section of the Royal Amsterdam academy of sciences, 1865–83. Founder and director of a physiology laboratory at the University of Utrecht, 1866–88. FRS 1866 (Tort 1996; *NNBW*, *Proceedings of the Royal Society of London* 49 (1890–1): vii–xxiv.)

13 May [1870], *17 May 1870*, 19 May 1870, *27 May 1870*, *28 May 1870*, 3 June [1870], 21 June 1870

Dorrell, Mr. Head of typesetting at William Clowes and Sons, printers to CD's publisher, John Murray. (*Correspondence* vol. 14, letter from ?, 10 May [1866?]; this volume, letter to Mr Dorrell, 9 August 1870.)

9 August 1870, 11 August [1870]

Douglas, George Sholto, 17th earl of Morton (1789–1858). Diplomat and statesman. Retired from service, 1825. Succeeded to earldom, 1827; a representative peer of Scotland, 1830–58. Lieutenant-colonel, Midlothian yeomanry cavalry, 1843–4; vice-lieutenant of Midlothian, 1854–8. (*Modern English biography* s.v. Morton.)

Du Bois-Reymond, Emil Heinrich (1818–96). German physiologist. A founder of the Physikalische Gesellschaft in Berlin, 1845. Instructor in anatomy, Berlin Academy of Art, 1848–53. Elected member of the Prussian Academy of Sciences, 1851; permanent secretary, 1876. Appointed professor of physiology at the University of Berlin, 1858. Worked on animal electricity, and in 1850 invented the nerve galvanometer to measure the electrical impulses in nerve and muscle tissue. (*ADB*, *DSB*, *NDB*.)

Duchartre, Pierre Etienne Simon (1811–94). French botanist. Specialised in plant physiology. Professor of botany, Agricultural Institute, Versailles, 1849–52; professor of botany, Faculty of Sciences, Paris, 1861. (Barnhart comp. 1965, *DBF*.)

Duchenne, Guillaume Benjamin Amand (1806–75). French physician. Studied medicine at Paris, then practised in his native town of Boulogne-sur-Mer. Returned to Paris in 1855 and carried out experiments on the therapeutic use of electricity. One of the founders of neurology. (*DBF*.)

Duncan, James Matthews (1826–90). Scottish physician. Physician, Edinburgh Royal Infirmary, 1861. President of the Obstetrical Society of Edinburgh, 1873–5. Lecturer on midwifery and obstetric physician, St Bartholomew's Hospital, London, 1877–90. FRS 1883. (*ODNB*.)

Duncan, Peter Martin (1824–91). Physician, zoologist, and geologist. Physician, Essex and Colchester Hospital, 1848–59; consultant physician, county asylum and Oldham Club. Practised at Blackheath from 1860. Professor of geology, King's College, London, 1870. Secretary of the Geological Society of London, 1864–70; president, 1876–8. Specialist on living and fossil corals and Mesozoic echinoids. FRS 1868. (*Medical directory* 1849–76, *ODNB*.)

Durando, Gaetano Leone (Gaetano) (1811–92). Italian botanist. Served in the Sardinian army. Later collected plants in Europe and North Africa, and sold and exchanged specimens from his business in Paris. (*Aluka* http://www.aluka.org, accessed 10 November 2008; *London Journal of Botany* 5 (1846): 660.)

Ecker, Alexander (1816–87). Physiologist. MD Freiburg, 1837. Professor of physiology and anatomy, Freiburg, 1850. (*DBE*.)

Edward VII (1841–1910). Eldest son of Queen Victoria and Prince Albert. King of Great Britain and Ireland and of the British dominions beyond the seas, and emperor of India, from 1901. As Albert Edward, prince of Wales, married Princess Alexandra of Schleswig-Holstein-Sonderburg-Glücksburg on 10 March 1863. FRS 1863. (*ODNB*.)

Egerton, Philip de Malpas Grey-, 10th baronet (1806–81). Palaeontologist. Of Oulton Park, Cheshire. Tory MP for South Cheshire, 1835–68; for West Cheshire, 1868–81. Specialised in fossil fish. FRS 1831. (*ODNB*, Sarjeant 1980–96.)

Ehrenberg, Christian Gottfried (1795–1876). German zoologist, comparative anatomist, and microscopist. Wrote extensively on infusoria and the development of coral reefs. Professor of medicine at Berlin University, 1839. Foreign member, Royal Society of London, 1837. (*DSB*, *NDB*.)

Elie de Beaumont, Jean-Baptiste-Armand-Louis-Léonce (Léonce) (1798–1874). French mining engineer and geologist. Participated in the drawing-up of a geological map of France, 1825–41. Professor of geology, Ecole des Mines, 1827; professor, Collège de France, 1832. Foreign member, Royal Society of London, 1835. (*DBF*, *DSB*, *Record of the Royal Society of London*.)

Ellis, Joseph (1805/6–70). Coal miner. Resided in Sheffield. Admitted to the West Riding Asylum, Wakefield, Yorkshire, on 17 November 1866 after attempting to commit suicide. Died in the asylum, of dysentery, 12 January 1870. (Wakefield Headquarters (GB 201) C85/3/6/113/p212A–212Bi/2).)

Emerson, Ralph Waldo (1803–82). American clergyman, essayist, and poet. (*ANB*.)

Engelmann, Marie (d. 1870). Daughter of Frans Cornelis Donders. Married Theodor Wilhelm Engelmann in 1869. Died in childbirth. (*DSB* s.v. Engelmann, Theodor Wilhelm; *NNBW* 1: 823 s.v. Engelmann, Theodoor Wilhelm.)

Engelmann, Theodor Wilhelm (1843–1909). German physiologist. Son of Wilhelm Engelmann. Studied in Jena, Leipzig, Heidelberg, and Göttingen. Professor of biology and histology, Utrecht, 1871; professor of physiology, 1888. Professor of physiology, Berlin, 1897. (*NDB*.)

Erskine, Thomas, of Linlathen (1788–1870). Scottish theologian and advocate. Practised as an advocate until 1816, when he inherited the family estate. Wrote on theology. Part of a cosmopolitan literary circle, including Thomas Carlyle, Charles Kingsley, F. D. Maurice, and Dean Stanley. (*ODNB*.)

Evans, Robert Speke (1811–74). Clerk to the Admiralty. Of Temple Mount, Lesness Heath, Erith, Kent. His daughter Frances (d. 1921) married William Boyd Dawkins in 1866. (*BMD* (*Death index*); *DNB*; *IGI* (accessed 1 March 2008); letter from W. B. Dawkins, 29 January [1870].)

Evrainova, Anna Mikhailovna (Zhanna) (1844–1919). Russian student of law and journal editor. Escaped from Russia to Heidelberg, 1869; moved to Leipzig, 1870. Doctorate in jurisprudence, Leipzig, 1873. Edited the journal *Severnyi vestnik* (*Northern Messenger*). Cousin of Sofia Vasilyevna Kovalevskaya. (Koblitz 1983.)

Eyton, Thomas Campbell (1809–80). Shropshire naturalist. Friend and Cambridge contemporary of CD. Author of several works on natural history. On coming into possession of the family estate at Eyton, Shropshire, in 1855, he built a museum for which he formed a collection of skins and skeletons of European birds. An active magistrate throughout his adult life. (*ODNB*.)

12 [May 1861 – April 1863]

Falconer, Hugh (1808–65). Palaeontologist and botanist. Superintendent of the botanic garden, Saharanpur, India, 1832–42. Superintended the arrangement of Indian fossils for the British Museum in 1844. Superintendent of the Calcutta botanic garden and professor of botany, Calcutta Medical College, 1848–55. Retired owing to ill health and returned to Britain in 1855; pursued palaeontological research while travelling in southern Europe. Vice-president of the Royal Society of London and foreign secretary of the Geological Society of London, 1865. FRS 1845. (*DSB*, *ODNB*.)

4 October [1862]

Farr, William (1807–83). Statistician and epidemiologist. Studied medicine in Shrewsbury, Paris, and London. Worked as a medical practitioner, public lecturer, and medical journalist in London from 1833. Worked for the General Register Office from 1837, from 1839 as compiler of abstracts; superintendent of the Statistical Department, 1842–80. From the mid-1850s to the mid-1860s, worked closely with Florence Nightingale on hospital and army sanitary reform. FRS 1855. (*ODNB*.)

16 July 1870, 17 July [1870], *6 August 1870*

Farrer, Frances (1833–70). Daughter of Maitland and William Erskine. Her mother Maitland, a daughter of Sir James Mackintosh, was a half-sister of Frances Emma Elizabeth Wedgwood (1800–89). Married Thomas Henry Farrer in 1854. (*ODNB* s.v. Farrer, Thomas Henry; B. Wedgwood and Wedgwood 1980 (family tree).)

Farrer, Thomas Henry, 1st Baron Farrer (1819–99). Civil servant. BA, Oxford (Balliol College), 1840. Called to the bar, 1844; ceased to practise in 1848. Secretary of the marine department, Board of Trade, 1850, rising to sole

Farrer, Thomas Henry, 1st Baron Farrer, cont.

permanent secretary of the Board of Trade, 1867–86. In 1854, married Frances Erskine, whose mother, Maitland, was the half-sister of Frances Emma Elizabeth Wedgwood; in 1873, married Katherine Euphemia Wedgwood, daughter of Frances and Hensleigh Wedgwood. Made a baronet, 1883; created Baron Farrer of Abinger, 1893. (*ODNB*, B. Wedgwood and Wedgwood 1980.)

13 [May 1870], *17 May 1870*, *26 May 1870*, 28 May [1870], *5 June 1870*, *29 June 1870*, [29 June 1870], *27 October 1870*, *15 December 1870*

Fawcett, Henry (1833–84). Economist and politician. BA, Cambridge, 1856. Blinded in a shooting accident in 1858. Appointed professor of political economy, Cambridge University, 1863. MP for Brighton, 1865–74; Hackney, 1874–84. Post-master general, 1880. FRS 1882. (*ODNB*.)

6 December [1860], 20 July [1861]

Fernow, Carl Ludwig (1763–1808). German art historian and librarian. Studied philosophy in Jena, 1791–3. Lived in Rome, 1794–1803. Associate professor of aesthetics, Jena. Librarian at the court in Weimar, 1804. (*DBE*.)

Fischer von Waldheim, Alexander (1803–84). German botanist in Russia. MD, Moscow, 1825. Professor extraordinarius, botany and pharmacology, Medical–Surgical Academy, Moscow; professor of botany, Moscow University, 1830–65. On the council of the Société Impériale des Naturalistes de Moscou. (*DBE*.)

FitzRoy, Robert (1805–65). Naval officer, hydrographer, and meteorologist. Commander of HMS *Beagle*, 1828–36. Tory MP for Durham, 1841–3. Governor of New Zealand, 1843–5. Superintendent of the dockyard at Woolwich, 1848–50. Chief of the meteorological department at the Board of Trade, 1854; chief of the Meteorological Office from 1855. Rear-admiral, 1857; vice-admiral, 1863. FRS 1851. (*DSB*, *ODNB*.)

Flower, William Henry (1831–99). Anatomist and zoologist. Curator of the Hunterian Museum, Royal College of Surgeons of England, 1861–84; Hunterian Professor of comparative anatomy, Royal College of Surgeons, 1870–84. Director of the Natural History Museum, London, 1884–98. President of the Zoological Society of London, 1879–99. President of the British Association for the Advancement of Science, 1889. Knighted, 1892. FRS 1864. (*DNB*.)

25 May [1863], 25 March [1870], 5 July 1870

Forbes, David (1828–76). Geologist and mining engineer. Brother of Edward Forbes. Supervised mining and metallurgical works in Norway, 1848–57. Travelled in South America, 1857–60, in search of ores of nickel and cobalt for the firm of Evans & Askins, nickel–cobalt refiners of Birmingham. After more years of travelling, he settled in England and became foreign secretary to the Iron and Steel Institute in 1869. FRS 1858. (*ODNB*.)

11 December [1860], *13 June 1870*, *30 July 1870*, 31 July [1870]

Ford, George Henry (1809–76). South African artist. Assistant to Andrew Smith, 1821–4. Artist, Cape Town Museum, 1825; expedition for exploring Central

Africa, 1834–6. Moved to England in 1837. Artist, British Museum, 1837 until *circa* 1875. Friend of Albert Günther and illustrator of several of his works; provided illustrations for the second volume of *Descent.* (Gunther 1972, Gunther 1975.)

Forster, Thompson (1835–1901). Surgeon. Received medical training in Manchester, Dublin, London, Brussels, and Paris. MRCS and LSA, 1856. House-surgeon, Manchester Royal Infirmary. Practised in Daventry, Northamptonshire, from 1858. (*British Medical Journal* (1901), pt. 1: 1593.)
8 November 1870

Fox, Ellen Sophia (1820–87). Daughter of Basil George Woodd of Hillfield, Hampstead. Married William Darwin Fox in 1846. (*Darwin pedigree.*)

Fox, Frances Maria (b. 1848). Daughter of William Darwin Fox and Ellen Sophia Fox. Married Alexander Pearce in 1878. (*Darwin pedigree.*)

Fox, William Darwin (1805–80). Clergyman. CD's second cousin. A friend of CD's at Cambridge; introduced CD to entomology. Maintained an active interest in natural history throughout his life and provided CD with much information. Rector of Delamere, Cheshire, 1838–73. Spent the last years of his life at Sandown, Isle of Wight. (*Alum. Cantab.*, *Autobiography*, *Correspondence.*)
16 August [1860?], [before 1 March 1866], 15 February [1870], 18 February [1870], *28 October [1870]*, 15 November [1870], *18 [November 1870]*

Franklin, John (1786–1847). Naval officer and explorer. Lieutenant-governor of Van Diemen's Land (Tasmania), 1837–43. Leader of the 1845 expedition in search of a north-west passage during which all hands perished. Knighted, 1829. FRS 1823. (*ODNB*, *Record of the Royal Society of London.*)

Fraser, Oscar Louis (b. 1849). Osteologist. Assistant prosector to the Zoological Society of London; osteologist to the museum at Calcutta from *circa* 1871. (*BMD* (Birth index); Census returns 1871 (Public Record Office RG10/209/116/44); Galton 1871, p. 395.)

Frean, Richard. (b. 1839). Physician and surgeon. Trained at the Middlesex Hospital, London; governor's prizeman in clinical surgery and medicine, 1871. Resident obstetrical assistant at the Middlesex Hospital. Practiced in Plymouth, 1872. Emigrated to Australia, where he became surgeon at Hay Hospital, New South Wales, 1874–6. Resident meical officer, Sydney Infirmary, from 1877. (*BMD* (*Birth index*), *Medical directory*.)
17 February 1863

Frémont, John Charles (1813–90). American explorer, politician, and soldier. Served with the United States Topographical Corps, and in the early 1840s, led a series of surveying expeditions to the American west on behalf of the War Department. A popular hero for his role in the conquest of California in the war against Mexico. Elected to the United States Senate, 1850; nominated to run for president, 1856 and 1864. Served as a major-general in the Union army during the American Civil War. Governor of Arizona, 1878–83. (*DAB.*)

Gage, John (b. 1801/2). American farmer. Born in New York State. Resided in Vineland, New Jersey. (1870 United States Federal Census (M593_859/420/626).)

19 October 1870

Galton, Frances Anne Violetta (Violetta) (1783–1874). Daughter of Erasmus and Elizabeth Collier Darwin. Married Samuel Tertius Galton in 1807. (*Darwin pedigree*.)

Galton, Francis (1822–1911). Traveller, statistician, and scientific writer. CD's cousin. Explored in south-western Africa, 1850–2. Carried out various researches on heredity. Founder of the eugenics movement. FRS 1860. (*DSB*, *ODNB*.)

22 February [1855], *[after 22 February 1855]* (to Charles John Andersson), *15 March 1870*, *17 March 1870*, *22 March 1870*, *31 March 1870*, *8 April 1870*, *26 April 1870*, [27 April 1870?], *12 May 1870*, *1 June 1870*, *25 June 1870*, *28 June 1870*

Gama, Vasco da (b. *c.* 1460 d. 1524). Portuguese navigator and discoverer of the sea-route to India. (*EB*.)

Gardiner, Allen Weare (1831/2–78). Clergyman and missionary. Son of Allen Francis Gardiner. BA, Oxford, 1855; MA, 1857. Minister at Nailsworth, Gloucester, 1860–5. Chaplain of Lota, Chile, from 1869 to 1873 or 1874. Chaplain to the bishop of Bathurst, New South Wales, Australia, 1874–8. (*Alum. Oxon.*, *Clergy list*, *Crockford's clerical directory*.)

Garner, Robert (1808–90). Naturalist and surgeon in Stoke-upon-Trent. Surgeon to the North Staffordshire Infirmary, 1834. First president of the Staffordshire Branch of the British Medical Association. Fellow of the Linnean Society of London, 1856. President of the North Staffordshire Naturalists' Field Club, 1878. Specialised in anatomical work on invertebrates and vertebrates. (*Modern English biography*, Plarr 1930.)

22 February [1870–1]

Gascoyen, George Green (1829–75). Surgeon. House surgeon, St Mary's Hospital, London, 1855–6; demonstrator of anatomy, 1856–62; dean of the school, 1860–4; surgeon from 1870; lecturer in anatomy, 1871. Assistant surgeon, Lock Hospital, London, 1859–60; surgeon from 1860. Secretary of the Royal Medical and Chirurgical Society, 1866–9. (*Medical directory* 1871, *Proceedings of the Royal Medical and Chirurgical Society of London* 8 (1875): 77–8.)

7 July 1870

Gaudry, Albert-Jean (Albert) (1827–1908). French palaeontologist. Assistant to his brother-in-law, Alcide d'Orbigny, professor of palaeontology at the Muséum d'Histoire Naturelle. Carried out excavations at Pikermi, Attica, in 1855 and 1860, and published *Animaux fossiles et géologie de l'Attique* (1862–7). Studied the fossils of small reptiles and batrachians, 1866–92. Taught a course in palaeontology at the Sorbonne, 1868–71; appointed professor of palaeontology at the Muséum, 1872. (*DBF*, *DSB*.)

Gegenbaur, Carl (or Karl) (1826–1903). German anatomist and zoologist. A supporter of CD; emphasised the importance of comparative anatomy in evolutionary reconstruction. Professor extraordinarius of zoology, Jena, 1855–8; professor of anatomy and zoology, 1858–62; of anatomy, 1862–73. Professor of anatomy and comparative anatomy, Heidelberg, 1873–1901. Elected to the Deutsche Akademie der Naturforscher Leopoldina, 1857. (*DBE*, *DSB*, *NDB*.)

Giglioli, Enrico Hillyer (1845–1909). Italian zoologist. Born in London to an Italian father and an English mother. Studied at Genoa and Pavia, and at the School of Mines in London. Became friends with Filippo De Filippi, whom he accompanied on the *Magenta* expedition, a circumnavigation of the globe. Professor of vertebrate zoology, Florence, 1869. (*DBI*.)

Gilchrist, James (d. 1885). Scottish physician. Apprenticed to a draper, then trained for the Presbyterian ministry, but had to abandon his studies due to ill health. MD Edinburgh, 1850. Assistant, Crichton Royal Institution, Dumfries; medical superintendent, Royal Lunatic Asylum, Montrose, 1853–7; medical superintendent, Crichton Royal Institution and Southern Counties Asylum, 1857–79. Member of the Botanical Society of Edinburgh, and president of the Dumfries Field Club. (*Medical directory*, *Transactions of the Botanical Society of Edinburgh* 17 (1887–9): 2–11.)

Giles, Daniel (1761–1831). Landowner and member of parliament. MP for East Grinstead, 1802–7; St Albans, 1809–12. Practised as a barrister until *circa* 1813. Inherited his father's estate at Youngsbury, near Ware, Hertfordshire. (Thorne 1986 4: 23–4.

Girard, Maurice Jean Auguste (Maurice) (1822–86). French entomologist. Held an official position on the Commission for Public Instruction in Natural History. President of the French entomological society, 1867. Author of *Traité élémentaire d'Entomologie*. (*Entomologist's Monthly Magazine* 23 (1886): 113; Gilbert 1977.)

Godron, Dominique Alexandre (1807–80). French botanist, zoologist, and ethnologist. In 1854, became head of the science faculty at Nancy, where he established a natural history museum and a botanic garden. (*DBF*.)

Godwin-Austen, Robert Alfred Cloyne (1808–84). Geologist. Noted for his work on the stratigraphy of southern England. Predicted the existence of coal-bearing strata in the south-east of England. Secretary of the Geological Society of London, 1843–4 and 1853–4. Assumed the additional name Godwin from 1854. FRS 1849. (*DSB*, *ODNB*.)

Goethe, Johann Wolfgang von (1749–1832). German poet and naturalist. (*DSB*, *NDB*.)

Gould, John (1804–81). Ornithologist and artist. Taxidermist to the Zoological Society of London, 1828–81. Described the birds collected by CD on the *Beagle* expedition. FRS 1843. (*ODNB*.)

Graham, George (1801–88). Civil servant. Military secretary in Bombay, 1828–30. Private secretary to his brother, Sir James Robert George Graham, 1831–4. Registrar general, 1838–79. (*Modern English biography*.)

Grant, Robert Edmond (1793–1874). Scottish physician and zoologist. An early

Grant, Robert Edmond, cont.

supporter of the theory of the transmutation of species. Befriended CD in Edinburgh. Professor of comparative anatomy and zoology, University College, London, 1827–74. FRS 1836. (*ODNB.*)

Gratiolet, Louis Pierre (1815–65). French anatomist and anthropologist. Laboratory assistant, Muséum d'Histoire Naturelle, Paris, 1842–53; lectured on anatomy, 1844–50; director of anatomical studies, 1853–62. Deputy to the professor of zoology, Faculty of Science, Paris, 1862–3; professor, 1863–5. (*DSB.*)

Gray, Asa (1810–88). American botanist. Fisher Professor of natural history, Harvard University, 1842–88. Wrote numerous botanical textbooks and works on North American flora. President of the American Academy of Arts and Sciences, 1863–73; of the American Association for the Advancement of Science, 1872; a regent of the Smithsonian Institution, 1874–88. Foreign member, Royal Society of London, 1873. (*DAB, DSB, Record of the Royal Society of London.*)

1 February 1870 (from D. Appleton & Co.), *14 February 1870*, *27 February and 1 March 1870*, 15 March [1870], 1 September 1870 (from D. Appleton & Co.), *21 November 1870*, 7 December 1870

Gray, Jane Loring (1821–1909). Daughter of Charles Greely Loring, Boston lawyer and politician, and Anna Pierce Brace. Married Asa Gray in 1848. Edited the *Letters of Asa Gray* (1893). (Barnhart comp. 1965; Dupree 1959, pp. 177–84.)

14 February 1870

Greg, William Rathbone (1809–81). Essayist. Mill-owner, 1832–50. Author of the *Creed of Christendom* (1851). Wrote articles for the leading quarterlies, and books, mostly on politics and economics. Comptroller of the Stationery Office, 1864–77. (*ODNB*, Tort 1996.)

Grenville, George Nugent, 2d Baron Nugent of Carlanstown (1788–1850). Irish-born politician and writer. Matriculated at Brasenose College, Oxford, in 1804. Succeeded as Baron Nugent of Carlanstown in 1812. MP for Buckingham from 1810; for Aylesbury from 1812 and 1847. Lord of the Treasury, 1830–2. Lord high commissioner of the Ionian Islands 1832–5. Advocated parliamentary reform and religious liberty. (*ODNB.*)

Grierson, Robert, 1st baronet (1655/6–1733). Scottish landowner. Laird of Lag, 1666. Created baronet of Nova Scotia, 1685. Principally known for his persecution of the covenanters. (*ODNB.*)

Groom Napier, Charles Ottley (1839–94). Tobago-born imposter. Wrote books and articles on natural history topics from a Christian viewpoint. Adopted the additional name Napier in 1865. Adopted the title prince of Mantua and Montferrat, and claimed descent from, among others, King David of Israel. Advocate of vegetarianism and teetotalism. (*ODNB.*)

January 1870

Grove, William Robert (1811–96). Lawyer, judge, and natural philosopher. Professor of experimental philosophy, London Institution, 1841. An active member of the Royal Society of London; founder member of the Philosophical Club,

1847; Royal Medallist, 1847. Appointed judge, Court of Common Pleas, 1871. Knighted, 1872. FRS 1840. (*DSB*, *ODNB*.)

Gunn, John (1801–90). Geologist, palaeontologist, and clergyman. Rural dean, vicar of Barton Turf, and rector of Irstead, Norfolk. Leading member of the Norfolk and Norwich Museum; one of the founders of the Norfolk and Norwich Archaeological Society, 1846; of the Norwich Geological Society, 1864; president, Norwich Geological Society, 1864–77 and 1883–8. (L. Huxley ed. 1918, Woodward ed. 1891.)

Günther, Albrecht Carl Ludwig Gotthilf (Albert) (1830–1914). German-born zoologist. Began his association with the British Museum in 1857; made catalogues of the museum's specimens of amphibia, reptiles, and fish; officially joined the staff in 1862. Assistant keeper of the zoological department, 1872–5; keeper, 1875–95. Edited the *Record of Zoological Literature*, 1864–9. FRS 1867. (*NDB*, *ODNB*.)

13 January [1870], *[after 13 January 1870]*, 16 January [1870], 23 March [1870], *[*c. *9 April 1870]*, *12 April 1870*, 15 July [1870]

Günther, Eleonora Louise (1806–94). Daughter of Ludwig Friedrich Nagel, a Lutheran pastor in Germany. Married Friedrich Gotthilf Günther in 1829. Mother of Albert Günther. Language teacher in Brighton. (Gunther 1975, p. 214, *ODNB* s.v. Günther, Albert Charles Lewis Gotthilf.)

Gunther, Robert William Theodore (1869–1940). Historian of science. Founded the Museum of the History of Science, Oxford, in 1924. Son of Albert and Roberta Günther. (*ODNB*.)

Günther, Roberta (1843–69). Natural history artist. Daughter of the baillie John M'Intosh of St Andrews and Elizabeth Mitchell. Met Albert Günther at the British Association for the Advancement of Science meeting in Dundee in 1867, and married him in 1868. Died after giving birth to her son, Robert William Theodore Gunther. (*IGI* (accessed 27 August 2008); *ODNB* s.v. Günther, Albert Charles Lewis Gotthilf, and Gunther, Robert William Theodore.)

Haeckel, Ernst Philipp August (Ernst) (1834–1919). German zoologist. MD, Berlin, 1857. Lecturer in comparative anatomy, University of Jena, 1861–2; professor extraordinarius of zoology, 1862–5; professor of zoology and director of the Zoological Institute, 1865–1909. Specialist in marine invertebrates. Leading populariser of evolutionary theory. His *Generelle Morphologie der Organismen* (1866) linked morphology to the study of the phylogenetic evolution of organisms. (*DSB*, *NDB*.)

23 June [1870], *6 July 1870*

Hamilton, William John (1805–67). Geologist. MP for Newport, Isle of Wight, 1841–7. Secretary of the Geological Society of London, 1832–54; president, 1854 and 1865. (*DNB*, Sarjeant 1980–96.)

Hamond, Robert Nicholas (1809–83). Naval officer. Lieutenant, 1827. Midshipman on HMS *Beagle*, 1832–3. (*Burke's landed gentry* 1952, *Correspondence* vol. 1, Appendix III, O'Byrne 1849.)

Hartogh Heijs van Zouteveen, Hermanus (1841–91). Dutch naturalist and geologist. Translator and champion of Darwin's work in the Netherlands. (*NNBW.*)
27 November 1870, 30 November [1870]

Hauer, Franz von (1822–99). Austrian geologist and geographer. Research assistant to Wilhelm von Haidinger and lectured on palaeontology from 1846. Director of the Imperial Geological Institute, 1866–85. Taught at the School for Soil Cultivation from 1874. Head of the Natural History Museum in Vienna, 1885–96. (*DBE.*)

Hauxwell, John. Collector of birds and insects. Lived in the Amazon region for more than thirty years. (Orton 1870, p. 227; *Proceedings of the American Philosophical Society* 23 (1886): 156.)

Hawkins, Benjamin Waterhouse (1807–89). Artist, sculptor, geologist, and popular natural history lecturer and author. Assistant superintendent of the Great Exhibition, 1851. Commissioned by the Crystal Palace Company to construct life-size replicas of extinct animals for the park in 1852. Illustrated *Fish* (part 4) and *Reptiles* (part 5) for the *Zoology of the voyage of HMS Beagle.* (R. B. Freeman 1978, *Men and women of the time* 1865.)

Hayden, Ferdinand Vandeveer (1828?–87). American geologist and naturalist. Explored the western territories of the United States, 1853–60. Lecturer on geology in the Auxiliary Department of Medicine at the University of Pennsylvania, 1865–72. Directed several surveys for the Department of the Interior, 1867–78. Worked for the US Geological Survey from its creation in 1879 until his retirement in 1886. (*ANB.*)
2 May [1870], *14 June 1870*

Head, Edmund Walker, eighth baronet (1804–68). Statesman and scholar. BA (Oriel College), Oxford, 1827; fellow, Merton College, 1830–8. Travelled in Italy, Germany, and Spain, 1828–35. Entered Lincoln's Inn in 1835. Appointed assistant poor-law commissioner, 1836; chief commissioner, 1841–7. Succeeded to the baronetcy in 1838. Lieutenant-governor of New Brunswick, 1848–54. Governor-in-chief of British North America, 1854–61. Returned to England in 1861. Civil service commissioner, 1862–8. Governor of the Hudson's Bay Company, 1863–8. FRS 1863. (*ODNB*, *Record of the Royal Society of London.*)

Hector, James (1834–1907). Scottish geologist. Surgeon and geologist on the government expedition to the western parts of British North America, 1857–60. Geologist to the provincial government of Otago, New Zealand, 1861–5. Director of the Geological Survey of New Zealand, 1865. Director of the meteorological department of the New Zealand Institute, of the Colonial Museum, and of the botanical garden, Wellington, 1866–1903. Knighted, 1887. FRS 1866. (*DNZB.*)

Heddle, Charles William Maxwell (Charles) (1811/12–99). Senegal-born merchant. Born in Senegal of a Scottish father and Senegalese mother. Educated in Scotland. Worked for the London firm Forster and Smith, importers and exporters for west Africa. Returned to west Africa in 1834; set up as an import and export trader in Freetown, Sierra Leone, in 1838. Appointed member of the governor's council (later the legislative council), 1845. Retired to France in 1871. One of the richest traders in west Africa. (*ODNB.*)

Henderson, Andrew (b. *c.* 1823 d. 1906). Nurseryman. Proprietor of the family business, Wellington Road Nurseries, at 3 Wellington Road, St John's Wood, London. Took charge of the nursery at his father's death, with Pineapple Nursery, Maida Vale. Later emigrated to Florida. (R. Desmond 1994, *Post Office London directory* 1869.)

Hennell, Sara Sophia (1812–99). Author. Educated at home. Worked as a governess, 1832–41. Corresponded with George Eliot. Wrote on religious subjects. (*ODNB.*)

Henslow, George (1835–1925). Clergyman, teacher, and botanist. BA, Christ's College, Cambridge, 1858. Curate of Steyning, 1859–61; of St John's Wood Chapel, 1868–70; of St James's Marylebone, 1870–87. Headmaster at Hampton Lucy Grammar School, Warwick, 1861–4; at the Grammar School, Store Street, London, 1865–72. Lecturer in botany at St Bartholomew's Hospital, 1866–80. Honorary professor of botany at the Royal Horticultural Society, 1880–1918. Author of a number of religious books, including *Plants of the Bible* (1907), and of children's books on natural history. Younger son of John Stevens Henslow. (*Alum. Cantab.*, *Crockford's clerical directory*, R. Desmond 1994, Lightman ed. 2004, *ODNB* s.v. Henslow, John Stevens.)
3 March 1870

Henslow, John Stevens (1796–1861). Clergyman, botanist, and mineralogist. CD's teacher and friend. Professor of mineralogy, Cambridge University, 1822–7; professor of botany, 1825–61. Extended and remodelled the Cambridge botanic garden. Curate of Little St Mary's Church, Cambridge, 1824–32; vicar of Cholsey-cum-Moulsford, Berkshire, 1832–7; rector of Hitcham, Suffolk, 1837–61. (*DSB*, *Historical register of the University of Cambridge*, *ODNB.*)

Herbert, Edward, 2d earl of Powis (1785–1848). Politician. Tory MP for Ludlow, 1806–39. Assumed the surname Herbert, in lieu of Clive, in 1807. Succeeded to the earldom in 1839. (*ODNB.*)

Herodotus (5th century BCE). Greek historian. Author of the first great narrative history of the ancient world (*Oxford classical dictionary.*)

Herschel, John Frederick William, 1st baronet (1792–1871). Astronomer, mathematician, chemist, and philosopher. Member of many learned societies. Carried out astronomical observations at the Cape of Good Hope, 1834–8. Master of the Royal Mint, 1850–5. Created baronet, 1838. FRS 1813. (*DSB*, *ODNB.*)

Herzenstein, Ulrich (*fl.* 1860s). Physician and opthalmologist. Practised in Odessa from 1862. (*Archiv für Ophthalmologie* 12 (1866): 76.)

Heywood, James (1810–97). University reformer and philanthropist. Attended universities of Edinburgh and Geneva. Worked as a banker in Manchester until inheriting a fortune. Passed the mathematical tripos at Cambridge University, 1833, but was debarred from taking his degree because he was Unitarian.

Heywood, James, cont.
Called to the bar, Inner Temple, 1838. Trustee of Cross Street Chapel in Manchester. Helped found the Manchester Athenaeum in 1839. Compiled and published statistics on the state of English universities in the 1840s. MP for North Lancashire, 1847–57. Carried an amendment removing religious tests for those matriculating or taking BA degrees at Oxford University, 1854. A supporter of women's suffrage and education. President of New College, Manchester, 1853–8. FRS 1839. (*ODNB*.)

Hildebrand, Friedrich Hermann Gustav (Friedrich) (1835–1915). German botanist. After studying mineralogy, geology, and agriculture at Berlin, he took up botany, studying at Bonn, then from 1855 to 1858 at Berlin, where he received his doctorate. Habilitated at Bonn, becoming privat-dozent there, in 1859. Professor of botany, Freiburg im Breisgau, 1868–1907. Worked mainly on hybridity, dimorphism, and generation. (Correns 1916, Junker 1989, Tort 1996.)
4 July 1870, *16 July 1870*

Hill, Rowland, 2d Viscount (1800–75). Politician. MP for Shropshire, 1821–32; North Shropshire, 1832–42. Lord lieutenant of Shropshire, 1845–75. (*Alum. Oxon.*, *Complete peerage*.)
6 January 1851

Hodgson, Brian Houghton (1801?–94). Diplomat and Nepalese scholar. In the service of the East India Company from 1816; assistant resident in Nepal, 1820; resident, 1833–43. Acquired an important collection of Sanskrit and Tibetan manuscripts. Wrote extensively on the geography, ethnography, and natural history of India and the Himalayas. Lived in Darjeeling, 1845–58; thereafter settled in England. FRS 1877. (*ODNB*.)

Hoeven, Jan van der (1801–68). Dutch comparative anatomist and anthropologist. Lecturer on zoology, Leiden, 1826–68. An adherent of Cuvier's theory of catastrophism; believed in creation rather than evolution, attempting in his works to show how the infinite wisdom of the Creator is reflected in the harmony of the world's creatures. (*BWN*, *DSB*.)

Hoffmann, Hermann (1819–91). German botanist and physician. Studied medicine and biology at the universities of Giessen and Berlin, 1837–41. GP and lecturer in medicine, Giessen, from 1842. A pioneer in bacteriology. Also worked on variability, propagation, migration, and dissemination in plants. (*DBE*.)
23 June 1870, 8 October 1870

Holder, Stephen (*fl.* 1870s). Carter at Orpington, Kent. Sentenced to two months' imprisonment for gambling, 1870. (*Echo*, 18 August 1870, p. 1.)

Holland, Henry, 1st baronet (1788–1873). Physician. Related to Josiah Wedgwood I. Physician-in-ordinary to Prince Albert, 1840; to Queen Victoria, 1852. President of the Royal Institution of Great Britain, 1865–73. Created baronet, 1853. FRS 1815. (Caroe 1985, *DNB*, *Emma Darwin* (1904), *Record of the Royal Society of London*.)
4 November [1870]

Hooker, Charles Paget (1855–1933). Physician and surgeon. Third child of Frances Harriet and Joseph Dalton Hooker. Trained at St Bartholomew's Hospital, London; made a licentiate of the Royal College of Physicians of London and the Royal College of Surgeons, Edinburgh, 1879, before being appointed to the staff of the Hertfordshire General Infirmary. Cottishall Cottage Hospital, Norfolk, 1880–5; Cirencester Cottage Hospital, Gloucestershire, 1885–1912. (Allan 1967, *Medical directory* 1881–1933, *Medical who's who* 1914.)

Hooker, Joseph Dalton (1817–1911). Botanist. Worked chiefly on taxonomy and plant geography. Son of William Jackson Hooker. Friend and confidant of CD. Accompanied James Clark Ross on his Antarctic expedition, 1839–43, and published the botanical results of the voyage. Appointed palaeobotanist to the Geological Survey of Great Britain, 1846. Travelled in the Himalayas, 1847–9. Assistant director, Royal Botanic Gardens, Kew, 1855–65; director, 1865–85. Knighted, 1877. FRS 1847. (*DSB*, *ODNB*.)

21 February [1870], *[7 March 1870]*, 8 March [1870], *[22 May 1870]*, 25 May [1870], *[31 May 1870]*, 2 [June 1870], [13 June 1870?], [29 June 1870], *1 July 1870*, 2 July [1870], *[6 or 7 July 1870]*, 8 July [1870], *10 July 1870*, 12 July [1870], *24 September 1870*, 27 September [1870], *12 October 1870*, 14 October [1870]

Hooker, William Henslow (1853–1942). Eldest child of Frances Harriet and Joseph Dalton Hooker. Civil servant, India Office, 1877–1904. Encouraged imperial ties between metropolitan institutions (particularly the Royal Botanic Gardens, Kew) and British East Africa, *circa* 1896–1906. (Allan 1967; *India list* 1904–5; *Zanzibar Gazette*, 5 February 1896, p. 6, and 28 November 1900, p. 5.)

Hooker, William Jackson (1785–1865). Botanist. Father of Joseph Dalton Hooker. Regius professor of botany, Glasgow University, 1820–41. Appointed first director of the Royal Botanic Gardens, Kew, 1841. Knighted, 1836. FRS 1812. (*DSB*, *ODNB*.)

Hopkins, William (1793–1866). Mathematician and geologist. Tutor in mathematics at Cambridge University. President of the Geological Society of London, 1851–3. Specialised in quantitative studies of geological and geophysical questions. FRS 1837. (*DSB*, *ODNB*, *Record of the Royal Society of London*.)

Horner, Leonard (1785–1864). Scottish geologist and educationalist. Founded the Edinburgh School of Arts in 1821. Warden of University College, London, 1827–31. Inspector of factories, 1833–59. A promoter of science-based education at all social levels. President of the Geological Society of London, 1845–7 and 1860–2. Father-in-law of Charles Lyell. FRS 1813. (*DSB*, *ODNB*.)

Horsman, Edward (1807–76). Scottish politician. Attended Trinity College, Cambridge, from 1823. Admitted advocate of the Scottish bar, 1832. Commissioner of inquiry into the Church of Scotland. MP for Cockermouth, 1836–52; for Stroud, 1853–68; for Liskeard, 1869–76. Chief secretary for Ireland, 1855–7. (*Alum. Cantab.*, *ODNB*.)

Hudson, William Henry (1841–1922). Argentinian-born author and naturalist. Collected bird skins for the Smithsonian Institution, 1866–9. Corresponding

Hudson, William Henry, cont.
member, Zoological Society of London, 1870. Moved to England in 1874. Wrote both popular natural history and works of fiction. (*ODNB.*)

Humboldt, Friedrich Wilhelm Heinrich Alexander (Alexander) von (1769–1859). Prussian naturalist, geographer, and traveller. Official in the Prussian mining service, 1792–6. Explored northern South America, Cuba, Mexico, and the United States, 1799–1804. Travelled in Siberia in 1829. Foreign member, Royal Society of London, 1815. (*DBE, DSB, NDB.*)

Hunter, John (1728–93). Surgeon and anatomist. His collection of zoological specimens formed the basis of the Hunterian Museum of the Royal College of Surgeons. FRS 1767. (*DSB, ODNB.*)

Huxley, Ethel Gladys (1866–1941). Daughter and youngest child of Henrietta Anne and Thomas Henry Huxley. Married artist–author John Collier in 1889. Because of the British law against marrying a deceased wife's sister they married in Norway. (Clark 1968, p. 110, A. Desmond 1994–7, 1: 348.)

Huxley, Henrietta Anne (1825–1915). Born Henrietta Anne Heathorn. Emigrated to Australia in 1843. Met Thomas Henry Huxley in Sydney, Australia, in 1847, and married him in 1855. (A. Desmond 1994–7, Freeman 1978.)
22 March [1867] (to Emma Darwin)

Huxley, Thomas Henry (1825–95). Zoologist. Assistant-surgeon on HMS *Rattlesnake*, 1846–50, during which time he investigated Hydrozoa and other marine invertebrates. Lecturer in natural history, Royal School of Mines, 1854; professor, 1857. Appointed naturalist to the Geological Survey of Great Britain, 1854. Hunterian Professor, Royal College of Surgeons of England, 1862–9. Fullerian Professor of physiology, Royal Institution of Great Britain, 1855–8, 1866–9. President of the Royal Society of London, 1883–5. FRS 1851. (Clark 1968, A. Desmond 1994–7, *DSB, ODNB.*)
[1860–70?], *21 January 1870*, 20 June [1870], *22 June 1870*

Ince, J. Knightly
[1857–61?]

Innes, John Brodie (1817–94). Clergyman. Perpetual curate of Down, 1846–68; vicar, 1868–9. Left Down in 1862 after inheriting an entailed estate at Milton Brodie, near Forres, Scotland; changed his name to Brodie Innes in 1861 as required by the entail. Priest in charge of Milton Brodie Mission and general licentiate of the diocese of Moray, 1861. Chaplain to the Bishop of Moray, 1861–80 and 1886–94. (*Clergy list, County families* 1864, *Crockford's clerical directory*, R. B. Freeman 1978, J. R. Moore 1985.)
28 January 1870

Jäger, Gustav (1832–1917). German physician and zoologist. Co-founder and director of the Vienna zoological garden, 1858–66. Professor of zoology and anthropology, Hohenheim Academy, from 1867; Stuttgart Polytechnic, from 1870. Taught physiology and histology at the Veterinary School, Stuttgart, from 1874. Co-editor of *Kosmos*, 1877–9. Returned to medical practice in 1884. (*DBE*, R. B.

Freeman 1978, *NDB*, Weinreich 1993.)

17 February 1870

Jenner, William, 1st baronet (1815–98). Physician. Professor of pathological anatomy at University College, London, 1849; Holme Professor of clinical medicine, 1860. Assistant, then full, physician to University College Hospital, 1849–76; consulting physician, 1879. Physician extraordinary to Queen Victoria, 1861; physician-in-ordinary, 1862. Established, through clinical studies, definitive evidence that typhus and typhoid were different diseases. FRS 1864. (*DNB*.)

Jerdon, Thomas Claverhill (1811–72). Zoologist. Joined the East India Company as assistant surgeon in the Madras service in 1835; later civil surgeon at Telicherry and major. Resigned from government service in 1868 and returned to England in 1872. Author of works on the birds and mammals of India. (Lightman ed. 2004, *ODNB*.)

Jones, Henry Bence (1814–73). Physician and chemist. Physician to St George's Hospital, 1846–62. Secretary of the Royal Institution, 1860–72. In his chemical studies, devoted himself especially to the application of chemistry to pathology and medicine. FRS 1846. (*ODNB*.)

2 August 1870, 3 August [1870]

Kant, Immanuel (1724–1804). German philosopher. (*DSB*, *NDB*.)

Kay, Joseph (1821–78). Economist and lawyer. BA (Trinity College), Cambridge, 1845; MA, 1849. Admitted at the Inner Temple, 1844; called to the bar, 1848. Travelled in Europe as a travelling bachelor of Trinity College from 1845. Judge of the Salford hundred court from 1862. QC, 1869. Solicitor-general, county palatine of Durham from 1872. Wrote works in favour of free trade in land. (*ODNB*.)

Keilhau, Baltazar Mathias (1797–1858). Norwegian geologist and petrologist. Lecturer, University of Christiania, 1821; professor, 1834. (*NBL*, Sarjeant 1980–96.)

Keller, Christian. Missionary at Lebu, Araucania (Chile). (*South American Missionary Magazine* 4 (1870): 65–9, 101–3.)

Kemp, William (1788–1864). Engineer and businessman. Turner and cooper in Galashiels from 1821. Manager of the Gas Company, Galashiels, 1834–60. Amateur geologist and contributor to *Chambers's Journal*. Member of the Glasgow Geological Society from 1840, and lecturer to the Galashiels Mechanics' Institute. Honorary curator of the local museum. Contributed to the *London, Edinburgh and Dublin Philosophical Magazine*; corresponded with Robert Chambers about terraces in the Tweed river valley. (David Cramond, personal information, R. Hall 1898, pp. 558–60.)

1[8] May [1840], 7 April [1843], 24 [April 1843], [8 September 1843], 9 October [1843], [14 October 1843], 1 November [1843], [9 November 1843], 22 November [1843], 1 December [1843], 7 December [1843], 23 February [1844]

Kettle, Rosa Mackenzie (d. 1895). Novelist and poet. Born Mary Rosa Stuart

Kettle, Rosa Mackenzie, cont.
Kettle, but adopted Mackenzie, the maiden name of her mother. (*Modern English biography.*)

Key, Astley Cooper (1821–88). Naval officer. Entered the navy, 1833; commissioned 1842. Royal Commission on national defence, 1859–60. Captain of the steam reserve, Devonport, 1860–3. Captain of the *Excellent* and superintendent of the naval college, 1864. Rear admiral and first director of naval ordnance, 1866. Admiral superintendent, Portsmouth, 1869; Malta, 1870. Established the Royal Naval College, Greenwich, 1872–6. Vice-admiral, 1873. Commanded the North American station, 1876. Commanded the Baltic fleet, 1878. Principal naval aide-de-camp to the queen, 1879. First naval lord, 1879–85. Knighted, 1873. (*ODNB*; *The Times*, 11 June 1870, p. 7.)

Kindermann, Adolph Diedrich (1823–92). German painter and photographer. Of Hamburg. CD used some of his photographs in *Expression*. (Hannavy 2008, 1: 384.)

King, Peter John Locke (1811–85). Politician. BA (Trinity College), Cambridge, 1831; MA 1833. MP for East Surrey, 1847–74. (*ODNB*.)

Kingsley, Charles (1819–75). Author and clergyman. Appointed professor of English, Queen's College for Women, London, 1848. Regius professor of modern history, Cambridge University, 1860–9. Rector of Eversley, Hampshire, 1844–75. Chaplain to the queen, 1859–75. (*ODNB*.)
1 December [1859]

Kippist, Richard (1812–82). Botanist. Librarian of the Linnean Society of London, 1842–80. Specialist in Australian plants. (R. Desmond 1994, *ODNB*.)
18 January [1865], 10 January [1870 or 1871], 16 January [1870 or 1871], 28 July [1870]

Kirchhoff, Gustav Robert (1824–87). German physicist. Professor extraordinarius, Breslau, 1850–4; Heidelberg, 1854–75. Professor of theoretical physics, Berlin, 1875–86. Made major contributions to the study of electromagnetic radiation and electrical currents. A close friend of Robert Wilhelm Eberhard Bunsen; the two collaborated in laying the foundation of the method of spectral analysis, 1857–63. (*DSB*, *NDB*.)

Kittredge, George F. (*fl.* 1870s). American naturalist and amateur author. (Kittredge 1871; letter to G. F. Kittredge, 24 August [1870].)
24 August [1870]

Knox, Arthur Edward (1808–86). Army officer and ornithologist. An authority on falconry. Author of ornithological works. (*Modern English biography.*)

Koch, Eduard Friedrich (Eduard) (1838–97). German publisher. Took over E. Schweizerbart'sche Verlagsbuchhandlung in 1867, after which the firm published mostly scientific titles. Published a multi-volume edition of CD's works, translated by Julius Victor Carus. (*Biographisches Jahrbuch und deutscher Nekrolog* 2 (1898): 227.)

Kölliker, Rudolf Albert von (1817–1905). Swiss anatomist and physiologist. Professor of physiology and comparative anatomy, University of Würzburg,

1847–64; professor of anatomy, 1849–97. Foreign member, Royal Society of London, 1860. (*DSB.*)

18 April [1870], 14 June 1870

Korvin-Krukovskaya, Anna Vasilyevna (1843–87). Russian writer and revolutionary. Sister of Sofia Vasilyevna Kovalevskaya. One of her novels was published by Dostoyevsky. Went to Paris in 1867, where she worked for a publisher. Fled to Switzerland with the French communist Victor Jaclard in 1870; married him in 1871. Joined the Russian section of the Internationale. Returned to Paris and worked in the Commune in education and medical services. Co-founded the journal *La Social.* Fled to Switzerland in 1872 with her husband, after they were both condemned to forced labour. Went to Russia in 1874. Returned to Paris in 1880 after being pardoned. (*DBF,* Koblitz 1983.)

Kovalevskaya, Sofia Vasilyevna (1850–91). Russian mathematician and author. Wife of Vladimir Onufrievich Kovalevsky. Studied mathematics with private tutors, in Heidelberg from 1869, then in Berlin under Karl Weierstrass from 1871. Doctorate in absentia from Göttingen, 1874. Lecturer in mathematics, University of Stockholm, 1883; professor, 1884. Life professorship, 1889. Wrote mathematical papers and autobiographical novels. (*BDWS* s.v. Kovalevskaia, Sofia Vasilyevna.)

1 September 1870 (and V. O. Kovalevsky)

Kovalevsky, Alexander Onufrievich (1840–1901). Russian embryologist. Brother of Vladimir Onufrievich Kovalevsky. Held academic posts at various Russian universities; professor of histology, St Petersburg, 1890–4. His studies of ascidian embryology revealed that tunicates were chordates and gave strong support to Darwinian transmutation theory. Foreign member of the Royal Society of London, 1885. (*DSB, GSE* s.v. Tunicata.)

25 September 1870

Kovalevsky, Vladimir Onufrievich (1842–83). Russian palaeontologist. Graduated from the School of Jurisprudence in 1861. Thereafter published, translated, and edited works by CD, Charles Lyell, Louis Agassiz, and others. Studied natural science and palaeontology, travelling throughout Europe, 1869–74. Submitted doctoral thesis on the palaeontology of horses at the university of Jena in 1872. Associate professor, Moscow University, 1880–3. (*DSB.*)

20 February [1870], 22 February [1870], *28 February [1870], 7 August [1870], 15 August [1870], 1 September 1870* (and S. V. Kovalevskaya), *5 September [1870], 15 September 1870, 20 September 1870, 5 October 1870, 11 October [1870], 22 November 1870, 12 December [1870]*

Kramers, Hendrik Adrianus (1816–98). Dutch bookseller. Of Rotterdam. Imported French and German books from 1837. In 1850, visited publishing houses in London in order to deal directly with them. An agent for translation rights. (*NNBW,* Weel 2003.)

Kundt, August Adolph Eduard Eberhard (August) (1839–94). German physicist. Studied physics and mathematics at Leipzig and Berlin, 1859–64; received a doctorate in 1864. Privat-dozent, Berlin, 1867. Professor of physics, Zurich

Kundt, August Adolph Eduard Eberhard (August), cont.
polytechnic, 1868; Würzburg, 1870; Strasbourg 1872. Professor of experimental physics, Berlin, 1888–94. (*DBE.*)

Lacaze-Duthiers, Félix Joseph Henri de (1821–1901). French invertebrate zoologist. Assistant to Henri Milne-Edwards. Professor of zoology in Lille, 1854–64; at the Muséum d'Histoire Naturelle, Paris, 1856–9; at the Faculté des Sciences, Paris, from 1869. (*DSB.*)

Ladies
24 February [1862–9]

Lamarck, Jean Baptiste Pierre Antoine de Monet (Jean Baptiste) de (1744–1829). French naturalist. Held various botanical positions at the Jardin du Roi, 1788–93. Appointed professor of zoology, Muséum d'Histoire Naturelle, 1793. Believed in spontaneous generation and the progressive development of animal types; propounded a theory of transmutation. (*DSB.*)

Langer, Karl von (1819–87). Austrian anatomist. Professor of zoology, Budapest; professor of anatomy, Vienna. Discovered and mapped lines of normal elasticity in skin (Langer's lines). (*OBL*, *WWWS.*)

Langton, Charles (1801–86). Rector of Onibury, Shropshire, 1832–41. Left the Church of England in 1841. Resided at Maer, Staffordshire, 1841–7, and at Hartfield Grove, Hartfield, Sussex, 1847–63. Married Emma Darwin's sister, Charlotte Wedgwood, in 1832. After her death, married CD's sister, Emily Catherine Darwin, in 1863. (*Alum. Oxon.*, *Emma Darwin* (1915), R. B. Freeman 1978.)

Langton, Charlotte (1797–1862). Emma Darwin's sister. Married Charles Langton in 1832. Resided at Maer, Staffordshire, 1840–6, and at Hartfield Grove, Hartfield, Sussex, 1847–62. (*Emma Darwin* (1915), R. B. Freeman 1978.)

Langton, Charlotte Mildred (Mildred) (1868–1940). Daughter of Edmund and Caroline Langton. Assumed the surname Massingberd in 1887. Married Leonard Darwin in 1900. (R. B. Freeman 1978, *Darwin pedigree.*)

Langton, Edmund (1841–75). Son of Charles and Charlotte Langton. BA (Trinity College), Cambridge, 1864. Admitted at Lincoln's Inn, 1864. (*Alum. Cantab.*, *Emma Darwin* (1915).)

Langton, Emily Caroline (Lena). Daughter of Charles Langton Massingberd. Married Edmund Langton in 1867. Assumed the surname Massingberd in 1887. (R. B. Freeman 1978, *Darwin pedigree.*)

Lankester, Edwin (1814–74). Surgeon, physician, medical reformer, anatomist, botanist, and popular author and lecturer. Founder and editor of the *Journal of Social Science*, 1865–74. Secretary for the biological section of the British Association for the Advancement of Science, 1839–64; of the Ray Society from 1844. President of the Microscopical Society of London, 1859. FRS 1845. (*DNB*, *Medical directory* 1845–75.)

Lankester, Edwin Ray (1847–1929). Zoologist. Studied natural sciences at Oxford under George Rolleston, 1866–8; physiology at Leipzig and Vienna; morphology under Ernst Haeckel at Jena; marine zoology with Anton Dohrn in Naples,

1871–2. Fellow and tutor, Exeter College, Oxford, 1872–5; professor of zoology, University College, London, 1875–91; Linacre Professor of comparative anatomy, Oxford, 1891–8; director of the natural history departments and keeper of zoology, British Museum, 1898–1907. Knighted, 1907. FRS 1875. (*ODNB*.)
15 March [1870]

Laplace, Pierre Simon, marquis de (1749–1827). French mathematician, physicist, and cosmologist. (*DSB*.)

Larsen, Hans (*fl.* 1850s). Danish-born cattle drover. Worked as a sailor. Left ship at Walvis Bay, Namibia, *circa* 1843. Worked as a cattle drover. Head servant to Galton's expedition in southern Africa, 1850–1. Emigrated to Australia in 1853. (Andersson 1856, p. 347; Galton 1853, pp. 57 and 75; Wallis 1936, p. 53.)

Lartet, Edouard Amant Isidore Hippolyte (Edouard) (1801–71). French palaeontologist and prehistorian. Trained as a lawyer. Began palaeontological research in the 1830s, and became interested in the question of the antiquity of humans; carried out excavations in 1860 that provided proof of the contemporaneity of humans with extinct animal species. Named professor of palaeontology, Muséum d'Histoire Naturelle, 1869, but ill health prevented him from fulfilling his duties. (*DSB*, Sarjeant 1980–96.)

La Touche, James Digues (1824–99). Irish cleric, tutor, geologist, and palaeontologist. Also interested in entomology, botany, meteorology, and astronomy. Vicar of Stokesay, Shropshire, 1855–99. Tutored the sons of Joseph Dalton Hooker, John William Colenso, and Friedrich Max Müller, and the nephews of Charles Lyell. Published *A hand-book of the geology of Shropshire* (1884). (*Geological Magazine* n.s. 6 (1899): 235–7.)

Lecoq, Henri (1802–71). French naturalist and vulcanologist. Professor of natural history, University of Clermont-Ferrand, and director of the city's botanic garden, 1826–54. Taught at the Preparatory School of Medicine and Pharmacy, Clermont-Ferrand, from 1840; professor, science faculty, from 1854. Published widely on botany, agriculture, and meteorology. (*Grande encyclopédie*, Sarjeant 1980–96.)

Lee, Robert (1804–68). Church of Scotland minister. Minister, Arbroath, Forfarshire, 1833; Campsie, Stirlingshire, 1836; Old Greyfriars, Edinburgh, 1843–68. First professor of biblical criticism, University of Edinburgh, and dean of the Chapel Royal, 1847. Undertook controversial reform of the liturgy to include music and written prayer. (*ODNB*.)

Leighton, William Allport (1805–89). Botanist, clergyman, and antiquary. Went to school with CD in Shrewsbury, 1817. Priest at Lichfield, 1843–6. Curate of St Giles's, Shrewsbury, 1846–8. Edited the *Transactions of the Shropshire Archaeological Society* for many years. Published *The flora of Shropshire* (1841). (*Clergy list* 1846–8, R. Desmond 1994, R. B. Freeman 1978, *ODNB*.)
25 November [1840–6], 4 December [1862]

Lesseps, Ferdinand Marie (Ferdinand) de (1805–94). French engineer. Constructed a canal across the Isthmus of Suez, 1859–69; commissioned to attempt

Lesseps, Ferdinand Marie (Ferdinand) de, cont.
a similar project at Panama, 1879–89, but the attempt failed and he was charged with fraud. (*EB* (1970).)

Lethbridge, Christopher (1812/13–1910). Solicitor. Admitted to practise, 1866. Worked with his father, John Christopher Lethbridge, at 25 Abingdon Street, Westminster. They both lived in Granville House, Blackheath. (*BMD* (*Death index*); Census returns 1861 (Public Record Office RG9/413/80); *Law list* 1870; *Post Office London directory* 1869.)

Lethbridge, John Christopher (b. 1816/17). Solicitor. Admitted to practise, 1837. Worked with his son, Christopher Lethbridge, at 25 Abingdon Street, Westminster. They both lived in Granville House, Blackheath. (Census returns 1861 (Public Records Office RG9/413/80); *Law list* 1870; *Post Office London directory* 1869.)

Leuckart, Karl Georg Friedrich Rudolf (Rudolf) (1822–98). German zoologist. Lecturer in zoology, University of Göttingen, 1847–50. Professor extraordinarius of zoology, University of Giessen, 1850–5; professor, 1855–69. Appointed professor of zoology in Leipzig, 1869; dean of the philosophical faculty, 1873; rector, 1877–8. Carried out important work in morphology and parasitology. (*DBE*, *DSB*, *NDB*.)

Lewes, George Henry (1817–78). Writer. Author of a biography of Goethe (1855). Contributed articles on literary and philosophical subjects to numerous journals. Editor, *Fortnightly Review*, 1865–6. Published on physiology and on the nervous system in the 1860s and 1870s. Lived with Mary Ann Evans (George Eliot) from 1854. (Ashton 1991, *ODNB*.)

Leydig, Franz von (1821–1908). German zoologist. Lecturer in physiology, University of Würzburg, 1849–55; professor extraordinarius, 1855–7. Professor of zoology, Tübingen, 1857–75; of comparative anatomy, Bonn, 1875–87. Foreign member, Royal Society of London, 1901. (*DSB*, *NDB*.)

Liddon, Henry Parry (1829–90). Theological scholar and clergyman. A leader of the Oxford Movement. Canon of St Paul's Cathedral from 1870. Declined participation in CD's memorial committee. (*ODNB*.)

Liebig, Justus von (1803–73). German organic chemist. Professor of chemistry, University of Giessen, 1825–51; University of Munich, 1852–73. Played a major role in the development of organic chemistry. Developed artificial agricultural fertilisers. President of the Bavarian Academy of Science, 1859–73. Foreign member, Royal Society of London, 1840. (*ADB*, *DBE*, *DSB*, *NDB*.)

Liesk, William C. (*fl.* 1820s–30s). Ship's mate and colonist. Mate on *Borneo*, one of the first ships to carry European colonists to the Cocos (Keeling) Islands, in 1827. Second in command of the Cocos (Keeling) Islands when the nominal leader was away. Left for Batavia (Jakarta) *circa* December 1837. (Armstrong 1991, pp. 50–1; Hughes [1950], pp. 38, 41, 48–51; *Journal of researches* 2d ed., p. 452.)

Lindley, John (1799–1865). Botanist and horticulturalist. Assistant in Joseph Banks's library and herbarium, 1819. Garden assistant secretary, Horticultural Society of London, 1822–6; assistant secretary, 1826–41; vice-secretary, 1841–58;

honorary secretary, 1858–62. Lecturer on botany, Apothecaries' Company, from 1836. Professor of botany, London University (later University College, London), 1829–60. Horticultural editor of the *Gardeners' Chronicle* from 1841. FRS 1828. (R. Desmond 1994, *DSB*, *ODNB*.)

Linnaeus, Carolus. *See* Linné, Carl von.

Linné, Carl von (Carolus Linnaeus) (1707–78). Swedish botanist and zoologist. Professor of practical medicine, University of Uppsala, 1741; professor of botany, diatetics, and materia medica, 1742; court physician, 1747. Proposed a system for the classification of the natural world, and reformed scientific nomenclature. FRS 1753. (*DSB*, *Record of the Royal Society of London*.)

Lloyd, Gertrude Jane Mary (b. *c.* 1815). Daughter of Philip Lake Godsal. Married John Lloyd in 1847. Sister-in-law of Mary Charlotte Lloyd. Superintended the collection of fossil remains found in caves on the estate at Rhagatt near Corwen, Wales. (*IGI* s.v. Godsal, Gertrude Jane Mary; Nicholas 1872, 2: 705.)

Lloyd, Mary Charlotte (1819–96). Welsh landowner and artist. Studied in Italy with the sculptor John Gibson. Met Frances Power Cobbe there and became her lifelong companion. Neighbour of the Darwin family during their stay at Caerdeon, Barmouth, Wales, in summer 1869. (Lucas 2007.)

Lloyd-Mostyn, Edward Mostyn, 2d Baron Mostyn (1795–1884). Politician and horse-breeder. MP for Flintshire, 1831–7, 1841–2, 1847–54; Lichfield, 1846–7. Lord lieutenant of Merioneth, 1840. Colonel of the Merioneth county militia, 1847–52. Interested in horse-racing and horse-breeding. (*Modern English biography*.)

Lockyer, Joseph Norman (1836–1920). Astronomer. Civil servant in the War Office, 1857–75; published papers on solar physics. Secretary to the Royal Commission on Scientific Instruction and the Advancement of Science, 1870–5; employee of the Science and Art Department at South Kensington from 1875; first director of the Solar Physics Observatory, and professor of astronomical physics, Royal College of Science, South Kensington, 1890–1911. Established the journal *Nature* in 1869. Knighted, 1897. FRS 1869. (*DNB*, *DSB*.)

Lockyer, Winifred (d. 1879). Translator and assistant astronomer. Daughter of William James, a railway promoter. Married Joseph Norman Lockyer in 1858. Translated French science books into English. Assisted her husband on the eclipse expedition to Sicily, 1870. (*ODNB* s.v. Lockyer, Joseph Norman, *The Times*, 9 December 1870, p. 7.)

Lonsdale, William (1794–1871). Geologist. Served the Geological Society of London, 1829–42, first as curator and librarian, and after 1838 as assistant secretary and librarian. Suggested that fossils, alongside other factors, were a reliable basis for estimating the age of geological strata. (*DSB*, *ODNB*.)

6 May [1864]

Lovén, Sven (1809–95). Swedish marine biologist. Travelled to Spitsbergen and northern Norway, 1836–7. Curator of the invertebrate section of the Museum of Natural History, Stockholm, 1841. Studied shell banks on the west coast of Sweden, providing evidence that an Arctic sea had once covered much of the present

Lovén, Sven, cont.
Scandinavian land-mass. (*Catalogue of the library of the British Museum (Natural History)*, *DSB*.)

Lubbock, Ellen Frances (1834/5–79). Daughter of Peter Hordern, clergyman, of Chorlton-cum-Hardy, Lancashire. Married John Lubbock in 1856. (*Burke's peerage* 1970, Census returns 1861 (Public Record Office RG9/462: 75).)
[1867–8?]

Lubbock, John, 4th baronet and 1st Baron Avebury (1834–1913). Banker, politician, and naturalist. Son of John William Lubbock and a neighbour of CD's in Down. Studied entomology and anthropology. Worked at the family bank from 1849; head of the bank from 1865. Liberal MP for Maidstone, Kent, 1870–80; for London University, 1880–1900. Succeeded to the baronetcy in 1865. Created Baron Avebury, 1900. FRS 1858. (*DSB*, Hutchinson 1914, *ODNB*, *Record of the Royal Society of London*.)
27 October [1867], 26 February [1870], *27 February [1870]*, *16 July [1870]*, 17 July 1870, 21 July [1870], *23 July [1870]*, *26 July [1870]*

Lucas, Eduard (1816–82). German gardener. Trained as a gardener at Dessau; gardening assistant in Frankfurt an der Oder, Greifswald, and Erfurt. From 1838, at the botanic gardens in Münich, where he succeeded in cultivating exotic plants. Head of the botanic gardens in Regensburg, 1841, of the institute gardens in Hohenheim (now part of Stuttgart), 1843. Studied and taught the cultivation of fruit and vegetables. Founded the institute of pomiculture at Reutlingen, 1860. (*DBE*.)

Lyell, Charles, 1st baronet (1797–1875). Scottish geologist. Uniformitarian geologist whose *Principles of geology* (1830–3), *Elements of geology* (1838), and *Antiquity of man* (1863) appeared in many editions. Professor of geology, King's College, London, 1831. President of the Geological Society of London, 1835–7 and 1849–51; of the British Association for the Advancement of Science, 1864. Travelled widely and published accounts of his trips to the United States. CD's scientific mentor and friend. Knighted, 1848; created baronet, 1864. FRS 1826. (*DSB*, *ODNB*.)
25 December [1870]

Lyell, Francis Horner (1852–1934). Son of Henry Lyell. (*Burke's peerage*.)

Lyell, Mary Elizabeth (1808–73). Eldest child of Leonard Horner. Married Charles Lyell in 1832. (R. B. Freeman 1978.)

M'Cann, James. Clergyman. University of Giessen, PhD 1863; D.Theol. 1868. Deacon, 1863; priest, 1864. Curate of Holmfirth, 1863; of St Paul's, Huddersfield. Incumbent of St Paul's, Glasgow. Wrote on science and religion. (*Crockford's clerical directory* 1879.)

Mackintosh, Angus (b. 1845). Son of Robert James Mackintosh. (Census returns 1861 (Public Record Office RG9/578/81/15), *Eton school lists*.)

Mackintosh, Eva (b. 1843/4). Daughter of Robert James and Mary Mackintosh. (Census returns 1861 (Public Record Office RG9/578/81/61), B. Wedgwood and Wedgwood 1980.)

Mackintosh, James (1765–1832). Philosopher and historian. Professor of law and general politics at the East India Company College, Haileybury, 1818–24. Married Catherine Allen, his second wife, sister-in-law of Josiah Wedgwood, in 1798. Knighted, 1803 (*DNB*, *ODNB*.)

Mackintosh, Mary. American-born wife of Emma Darwin's cousin, Robert James Mackintosh. Sister of Thomas Gold Appleton. (B. Wedgwood and Wedgwood 1980.)

Mackintosh, Robert James (1806–64). Emma Darwin's cousin. Barrister-at-law, Lincoln's Inn, 1833. (*Alum. Oxon.*, O'Leary 1989, B. Wedgwood and Wedgwood 1980.)

Mackintosh, Ronald (b. 1839/40). Son of Robert James Mackintosh. (Census returns 1841 (Public Record Office HO107/684/18), B. Wedgwood and Wedgwood 1980.)

McLachlan, Robert (1837–1904). Entomologist. Travelled to China and Australia, 1855. Worked on the Neuroptera and Trichoptera, and wrote monographs on both these groups. Editor, *Entomologist's Monthly Magazine*, from 1864; proprietor, from 1902. FRS 1877. (*ODNB*.)
23 March 1866

Maclaren, Charles (1782–1866). Scottish newspaper editor. Established the *Scotsman*, 1817; Editor, 1820–45. Wrote on geological subjects. Fellow of the Royal Society of Edinburgh, 1837; Geological Society of London, 1846. (*ODNB*, Sarjeant 1980–96.)
30 March 1840 (to Thomas Walker)

McLean, S. Scottish veterinary surgeon. Of Mill Burn Cottage, Inverness. (*Correspondence* vol. 16, letter from J. B. Innes, 7 December 1868, and letter from S. McLean to J. B. Innes, 15 December 1868.)

Macleay, George (1809–91). Traveller and statesman. Went to Australia in 1827 and accompanied Charles Sturt in an exploring expedition down the Murray River. Member of the legislative council, 1851–6; legislative assembly, 1856–9. Travelled widely in Europe and western Asia from 1859. Knighted, 1875. (*ODNB* s.v. Macleay, Alexander.)

Macleay, William Sharp (1792–1865). Naturalist and diplomat. Brother of George Macleay. Originated the circular or quinary system of classification. Diplomat in France and then Havana; retired in 1837. In 1839, emigrated to Sydney, Australia, where he devoted himself to the study of Australian natural history. Made large collections of Australian insects and marine fauna, and supervised the cultivation of exotic plants in his private gardens. Trustee of the Australian Museum, 1841–62. (*Aust. dict. biog.*, *ODNB*.)

Macmillan, Alexander (1818–96). Publisher. Co-founder with his brother Daniel of the firm Macmillan & Co. in 1843. After moving the firm's headquarters to London in the early 1860s, Macmillan began commissioning textbooks on botany, chemistry, and geology for children in elementary schools. (*ODNB*.)

Main, Robert (1808–78). Astronomer. Chief assistant at the Royal Observatory,

Main, Robert, cont.
1835–60. An active member of the Royal Astronomical Society. Appointed Radcliffe observer, Oxford, in 1860. FRS 1860. (*ODNB*.)
6 October 1870

Malthus, Thomas Robert (1766–1834). Clergyman and political economist. First professor of history and political economy at the East India Company College, Haileybury, 1805–34. Quantified the relationship between growth in population and food supplies in *An essay on the principle of population* (1798). FRS 1818. (*DSB*, *ODNB*.)

Mann, Gustav (1836–1916). German-born botanist. Gardener at the Royal Botanic Gardens, Kew, 1859. Botanical collector to the Niger expedition, 1859–62. Worked for the Indian Forest Service, 1863–91. (R. Desmond 1994, R. Desmond 1995.)

Mantell, Gideon Algernon (1790–1852). Surgeon, geologist, and palaeontologist. Scientific lecturer and author. Made noted collections of invertebrate and dinosaur fossils. Secretary, Geological Society of London, 1841–2; vice-president, 1848–9. FRS 1825. (*DNB*, *DSB*.)

Mantell, Walter Baldock Durrant (1820–95). English-born New Zealand public servant and naturalist. Son of Gideon Algernon Mantell. Emigrated to New Zealand, arriving in 1840. Commissioner for extinguishing native titles, Middle Island (South Island), 1848–51; commission of Crown lands for the Southern District of the Province of New Munster, 1851–6. Visited England, 1855–9; returned to New Zealand and was elected to the House of Representatives for Wallace, 1861–6. Attempted unsuccessfully to secure the fulfilment of promises made to the Ngai Tahu at the time of the original land purchases. Retired from the House and accepted a seat on the Legislative Council, 1866. Worked with Charles Lyell and Richard Owen. (*DNZB*.)
[before 10 April 1856]

Marshall (or Martial), Mr. Surgeon on an English whaling vessel whom CD met in 1834 during the *Beagle* voyage. (*Correspondence* vol. 3, letter to Henry Denny, 3 June [1844], *Correspondence* vol. 13, letter to Henry Denny, 17 January [1865].)

Martins, Charles Frédéric (1806–89). French botanist. Qualified as a doctor in Paris in 1834. Professor of botany, Faculté de Montpellier, 1846. Also published on geology and meteorology. (*Dictionnaire universel des contemporains*, *NBU*.)

Matthew, Patrick (1790–1874). Scottish gentleman farmer. Author of works on political and agricultural subjects. Advanced a theory of natural selection in the 1830s. (Dempster 1996, R. Desmond 1994.)

Maudsley, Henry (1835–1918). Medical doctor, psychologist, psychiatrist, and author. Studied medicine at University College Hospital, London. MD 1857. Assistant medical officer, Wakefield Asylum, 1857–8; medical superintendent, Manchester Royal Lunatic Asylum, 1859–62. Appointed physician, West London Hospital, 1864; lecturer on insanity, St Mary's Hospital, London, 1868–81; professor, medical jurisprudence, University College, London, 1869–79. Joint editor,

Journal of Mental Science, 1863–78. Developed a lucrative private practice. Researched heredity and degeneration, and promoted belief in the physical causes of mental illness. Wrote extensively on medical jurisprudence. Elected president, Medico-Psychological Association, 1870. (*ODNB*.)

Maurice, John Frederick Denison (1805–72). Clergyman and theologian. Chaplain of Guy's Hospital, London, 1836–46. Professor of modern history and English literature, King's College, London, 1840–6; of ecclesiastical history, 1846–53. Chaplain of Lincoln's Inn, 1846–60. Perpetual curate of the chapel of St Peter's, Vere Street, London, 1860–6. Knightbridge professor of moral philosophy, Cambridge, 1866–72. Vicar of St Edward's, Cambridge, 1870–2. (*Alum. Cantab.*, *ODNB*.)

Mellersh, Arthur (1812–94). Naval officer. Midshipman and mate on HMS *Beagle*, 1825–36. Served off the coast of Syria, then in command of HMS *Rattler* in the Burma campaign in 1852. Served off the coast of China in the 1850s, suppressing piracy, and in the Caribbean and South America before retiring in 1864. (*Modern English biography*; *The Times*, 28 September 1894, p. 4.)

Melville, Elizabeth Helen. Author. Daughter of Randal Callender of Craigforth, Stirling. Married Michael Linning Melville, later judge in the Courts of Mixed Commission at Sierra Leone, in 1840. (*County families* 1874, s.v. Melville, Michael Linning.)

Merrell, S. A. (b. 1828). American homeopathic doctor. Graduated from Illinois College, Jacksonville, 1853; Eclectic Medical School of Cincinnati. Worked at the United States Military Hospital, Kansas City, Missouri, 1863–5. Moved to Council Bluffs, Iowa, in 1865. (Cleave 1873.)

28 October 1870

Meyer, Adolf Bernhard (1840–1911). German zoologist, anthropologist, and museum director. Trained as a physician in Berlin. Travelled extensively in the Eastern Archipelago, publishing on the birds of the Celebes. Director, Royal Museum of Zoology, Anthropology, and Ethnography in Dresden, from 1874. (*Auk* 28 (1911): 519.)

Michels, John (*fl.* 1870–94). Microscopist. Member of the Quekett Microscopical Club, 1870. Emigrated to the United States. Member of the American Society of Microscopists, 1891–4. Chief microscopist, Bureau of Animal Industry, Chicago, by 1891. (*Journal of the Quekett Microscopical Club* 2 (1870): 89; *Proceedings of the American Society of Microscopists* 13 (1891): 59, 216; 15 (1894): 254.)

3 May 1870

Mill, John Stuart (1806–73). Philosopher and political economist. (*DSB*, *ODNB*.)

Milne-Edwards, Henri (1800–85). French zoologist. Professor of hygiene and natural history, Ecole centrale des arts et manufactures, 1832. Professor of entomology, Muséum d'histoire naturelle, 1841, with responsibility for the collections of crustaceans, myriapods, and arachnids as well as insects; professor of mammalogy, 1861. Foreign member, Royal Society, 1848. (*DSB*.)

2 March [1849]

Miquel, Friedrich Anton Wilhelm (1811–71). Dutch botanist. Described the flora of the Dutch East Indies. Director, Rotterdam botanic garden, 1835–46. Professor of botany, Amsterdam, 1846–59; Utrecht, 1859–71. (*DSB.*)

Mitchell, William Stephen (1840–92). Botanist and businessman. BA, Cambridge, 1866. Did some work on fossil leaves. Went into business as a wine merchant. (*Alum. Cantab.*, R. Desmond 1994.)

Mitford, Mary Russell (1787–1855). Playwright and writer. Published *Our village* (1824–32). Corresponded with many other writers. Gave Elizabeth Barrett Browning the spaniel Flush in 1841. (*ODNB.*)

Mivart, Caroline Georgina (1784/5–1870). Née Cunningham. Of Scottish/Irish descent. Mother of St George Jackson Mivart by her second marriage, to James Edward Mivart. (*BMD* (*Death index*); Gruber 1960.)

Mivart, Mary Anne (b. 1835/6). Of London. Wife of St George Jackson Mivart. (Census returns 1861 (Public Record Office RG9/87/107/13).)

Mivart, St George Jackson (1827–1900). Comparative anatomist. Converted to Catholicism, 1844. Called to the bar, 1851, but never practised. Established his reputation as an anatomist by his studies on primates. Lecturer in comparative anatomy, St Mary's Hospital Medical School, London, 1862–84. Secretary, Linnean Society of London, 1874–80; vice-president, 1892. Professor of the philosophy of natural history, University of Louvain, 1890–3. Excommunicated, 1900. Vigorous critic of Darwinism. Attempted to reconcile evolutionary theory and Catholicism. FRS 1869. (*DNB.*)

8 March [1870], 21 April [1870], *22 April 1870*, 23 April [1870], *25 April 1870*, *11 June 1870*, 13 June [1870]

Moleschott, Jacobus Albertus Willebrordus (Jacob) (1822–93). Physician and physiologist. General practitioner, Utrecht, 1845–7. Lecturer in physiology and anthropology, Heidelberg, 1847–54. Appointed professor of physiology, Zürich, 1856; of experimental physiology and physiological chemistry, Turin, 1861. Noted as an advocate of materialist philosophy. (*ADB*, *BHGW*, *DSB.*)

Molina, Juan Ignacio (Giovanni Ignazio) (1740–1829). Chilean priest and natural historian. Educated at the Jesuit school at Concepción. Left Chile in 1768, when the Jesuit order was expelled, and settled in Bologna, where he became a professor of natural history. Wrote on the natural history of Chile. (Medina 1906.)

Montefiore, Abraham Israel (d. 1852). Surgeon in the Bengal Military Establishment. (*Bombay calendar and almanac for 1853* (Bengal: n.p.), p. 123; *India list* 1849.)

Moore, William Daniel (1813–71). Irish physician and scholar of Dutch and Scandinavian languages. MB, Dublin, 1843; MD, comitatis causa, Oxford University, 1862; MA, Cambridge University (incorporated from Dublin), 1866. Fellow of the Swedish Society of Physicians. Translated several medical works. (*Alum. Cantab.*)

Morley, John (1838–1923). Politician and writer. Lost his faith while at Lincoln College, Oxford, and left for London, where he earned a living from 1860 as a teacher and journalist. Editor of the *Fortnightly Review*, 1867–82; of the *Pall Mall*

Gazette, 1880–3; of *Macmillan's Magazine*, 1883–5. Liberal MP for Newcastle, 1883–95; for Montrose Burghs, 1896–1908. Chief secretary for Ireland, 1886 and 1892–5; secretary of state for India, 1905–10. Created Viscount Morley of Blackburn, 1908. (*ODNB*.)

Morgan, Lewis Henry (1818–81). Anthropologist. Practised law in Rochester, New York, from 1844 until the mid 1860s. Studied the Iroquois, and kinship in the Iroquois and other Native-American peoples. Published *League of the Ho-de'no-sau-nee, or Iroquois* (1851); *Systems of consanguinity and affinity of the human family* (1871); and *Ancient society* (1877). (*ANB*.)
9 August 1870, 11 August [1870]

Morris, John (1810–86). Geologist. Originally a pharmaceutical chemist in Kensington. Published *Catalogue of British fossils* (1845). Professor of geology, University College London, 1854–77. (*ODNB*, Sarjeant 1980–96.)

Morton, Lord. *See* Douglas, George Sholto.

Moulinié, Jean Jacques (1830–73). Swiss zoologist and militia inspector. Translated *Variation*, *Descent*, and the sixth edition of *Origin* into French. (Tort 1996.)
18 August 1870

Mudie, Charles Edward (1818–90). Founder of a lending library. In 1840, opened a shop in London selling newspapers and loaning books. Within two years, the book-lending side of the business was his main concern. Had over 25,000 subscribers and was influential in publishing because of his purchasing power (*ODNB*.)
10 December [1842–5 or 1855–68] (and M. K. Mudie)

Mudie, Mary Kingsford (1820/1–98). Married Charles Edward Mudie in 1847. Daughter of the Reverend Henry Carpenter Pawling. (*BMD* (*Death index*), *ODNB* s.v. Mudie, Charles Edward.)
10 December [1842–5 or 1855–68] (and C. E. Mudie)

Müller, Heinrich Ludwig Hermann (Hermann) (1829–83). German botanist and entomologist. Brother of Johann Friedrich Theodor (Fritz) Müller. Schoolteacher in Schwerin, 1854–5. Studied blind cave insects in Krain, 1855. Teacher of natural sciences at the Realschule in Lippstadt, 1855–83; became director of the school. After settling in Lippstadt, studied the local flora, in particular the mosses. CD's *Orchids* directed Müller's attention to the pollination and fertilisation of flowers, on which he published several papers and books. (Gilbert 1977, Krause 1883, *Science* 2 (1883): 487–8.)
8 March 1870, 14 March 1870

Müller, Johann Friedrich Theodor (Fritz) (1822–97). German naturalist. Emigrated to the German colony in Blumenau, Brazil, in 1852. Taught mathematics at the Lyceum in Destêrro (now Florianópolis), 1856–67. *Naturalista Viajante* of the National Museum, Rio de Janeiro, 1876–92. His anatomical studies on invertebrates and work on mimicry provided important support for CD's theories. (*ADB*, *DBE*, Möller ed. 1915–21, *NDB*, West 2003.)
16 February 1870, *29 March 1870*, 12 May 1870, 28 August 1870

Müller, Johannes Peter (1801–58). German comparative anatomist, physiologist, and zoologist. Became professor of anatomy and physiology at Berlin University in 1833. Foreign member, Royal Society, 1840. (*DSB*, *Record of the Royal Society of London*.)

Murchison, Roderick Impey, 1st baronet (1792–1871). Geologist and army officer. Served in the British army, 1807–15. Noted for his work on the Silurian system. President of the Geological Society of London, 1831–3 and 1841–3; of the British Association for the Advancement of Science, 1846; of the Royal Geographical Society of London, 1843–4, 1851, 1857–8, 1862–70. Director-general of the Geological Survey of Great Britain, 1855. Knighted, 1863; created baronet, 1866. FRS 1826. (*DSB*, *Journal of the Royal Geographical Society*, *ODNB*.)

Murie, James (1832–1925). Physician and naturalist. MD, Glasgow, 1857; appointed pathologist to the Glasgow Royal Infirmary, 1857. Naturalist and medical officer on John Petherick's expedition to the upper White Nile, 1861–3. Prosector to the Zoological Society, 1865–70. Secretary, Linnean Society of London, 1876–80; librarian, 1880–8. (R. Desmond 1994; *Proceedings of the Linnean Society of London* (1925–6): 92–4.)

Murphy, Joseph John (1827–94). Irish mill-owner, merchant, and writer. Published poetry and works on natural theology. Originally a Quaker, but became a member of Church of Ireland. Resided in Belfast. (*Modern English biography*.)

Murray, Andrew Dickson (Andrew) (1812–78). Lawyer, entomologist, and botanist. Practised law in Edinburgh; later moved to London. Assistant secretary to the Royal Horticultural Society, 1861; member of the council from 1865; member of the scientific committee from 1868. An expert on insects harmful to crops. In entomology, specialised in the Coleoptera; in botany, in the Coniferae. (*ODNB*.)

Murray, Arthur (1856–70). Youngest son of John Murray (1808–92). Died at Eton of congestion of the lungs. (Paston 1932, pp. 149, 308.)

Murray, John (1808–92). Publisher, and author of guide-books. CD's publisher from 1845. (R. B. Freeman 1978, *ODNB* s.v. Murray family, publishers.)

14 April [1865], *2 April [1867]*, [24 September – 31 October 1867], 9 April [1870], *10 June [1870]*, *1 July 1870*, [after 1 July 1870], 5 [August 1870], *6 August [1870]*, 4 September [1870], 26 September 1870, *28 September [1870]*, 29 September [1870], *10 October [1870]*, 11 October [1870], 22 October [1870], *24 October [1870]*, 15 November [1870], *16 November [1870]*, 17 November [1870]

Musters, Charles (d. 1832). Fourth son of John Musters of Colwick Hall, Nottinghamshire. Volunteer 1st Class in the *Beagle*. Died of fever at Rio de Janeiro. (*Burke's landed gentry 1879*.)

Musters, George Chaworth (1841–79). Naval officer and explorer. Served in the Royal Navy from 1854. Travelled in South America, 1869–70, living with Patagonian Indians, and crossed northern Patagonia from east to west. Visited Vancouver Island in 1872, and travelled in Bolivia, 1874–6. (*ODNB*.)

Naudin, Charles Victor (1815–99). French botanist. Joined the herbarium staff at the Muséum d'Histoire Naturelle and became professor of zoology at the Collège

Chaptal, Paris, in 1846. Resigned his professorship almost immediately owing to a severe nervous disorder. Appointed aide-naturaliste at the Muséum d'Histoire Naturelle, 1854. Established a private experimental garden at Collioure in 1869, earning his living by selling seeds and specimens. First director of the state-run experimental garden at Antibes, 1878. Experimented widely on plants, particularly on acclimatisation and hybridity. Published a theory of transmutation based on hybridisation. (*DSB*, *Taxonomic literature.*)

Negri, Cristoforo (1809–96). Italian geographer and political scientist. Founder and president of the Società Geografica Italiana. (*Rivista Geografica Italiana* 3 (1896): 125.)

Nevill, Dorothy Fanny (1826–1913). Society hostess and horticulturalist. Daughter of Horatio Walpole, third earl of Orford; married Reginald Henry Nevill in 1847. Developed a notable garden at Dangstein, near Petersfield, Hampshire, where she cultivated orchids, pitcher plants, and other tropical plants; employed thirty-four gardeners. (*ODNB.*)

27 November [1861]

Newman, Edward (1801–76). Naturalist. Founder of the Entomological Club, the precursor of the Entomological Society, 1826. Natural history editor of the *Field*, 1858–76. (R. Desmond 1994, Gilbert 1977, *ODNB.*)

Newton, Alfred (1829–1907). Zoologist and ornithologist. Travelled extensively throughout northern Europe and North America on ornithological expeditions, 1854–63. Editor of *Ibis*, the journal of the British Ornithologists' Union, 1865–70. Professor of zoology and comparative anatomy, Cambridge University, 1866–1907. FRS 1870. (*DNB.*)

9 February [1870], *11 February 1870*, [22 May 1870]

Nicol, Patrick (1847–73). Scottish-born physician and mental health practitioner. Studied at the University of Aberdeen; MA 1866; MB CM 1869; MD 1871. Clinical assistant, West Riding Lunatic Asylum, Wakefield, Yorkshire; assistant medical officer, Sussex County Asylum, 1870; physician, Bradford Infirmary. Forced to resign from his post at Sussex County Asylum for insulting the medical superintendent. Published papers on mental illness. (Gardner 1999; *West Riding Lunatic Asylum Medical Reports* 1 (1871): 178–208; *Roll of the graduates of the University of Aberdeen.*)

13 May 1870

Norgate, Frederick. Publisher. Formed the science publishing house Williams & Norgate of Covent Garden, London, and South Frederick Street, Edinburgh, in partnership with Edmund Sydney Williams. (*Modern English biography* s.v. Williams, Edmund Sydney.)

Ogle, William (1827–1912). Physician and naturalist. MD, 1861. Lecturer on physiology at the medical school, and assistant physician, St George's Hospital, 1869–72. Superintendent of statistics, General Register Office, 1880–1903. Translated Aristotle's *De partibus animalium* into English in 1882. Published on flower structure and mechanisms for pollination. (*Alum. Oxon.*, *The Times*, 15 April 1912, p. 9.)

Ogle, William, cont.

[before 9 November 1870], 9 November 1870, *[10–17 November 1870]*, 17 November [1870], 17 December [1870], *[after 17 December 1870]*

Oldham, Thomas (1816–78). Irish geologist. Chief geological assistant to the Geological Survey of Ireland, 1839–50. Professor of geology, Trinity College, Dublin, 1845–51. In 1849, discovered in the rocks of Bray Head, County Wicklow, fossils or organic marks, which were named *Oldhamia*, after him. Director, Geological Survey of India, 1850–76. FRS 1848. (*Modern English biography*, *ODNB*, Sarjeant 1980–96.)

Oliver, John Ryder (1834–1909). Army officer. Lieutenant, Royal Artillery, 1855; served in the Indian Mutiny campaigns, 1857–9. Lieutenant, Royal Horse Artillery, 1860–4; captain, 1864; at St Helena, 1866–9; major, 1874. Brigade major, Royal Artillery, Aldershot, 1869–74; lieutenant-colonel, 1882; brevet colonel, 1885. Senior professor, Royal Military College, Canada, 1877–86; commandant, 1886–8. Honorary major-general, 1887. Inventor of a sundial that showed true mean time. (*Alum. Cantab.*, *WWW* 1897–1915.)

Orbigny, Alcide Charles Victor Dessalines d' (1802–57). French palaeontologist. Travelled widely in South America, 1826–34, on a commission for the Muséum d'Histoire Naturelle. Professor of palaeontology, Muséum d'Histoire Naturelle, 1853. (*DSB*.)

Orton, James (1830–77). American natural historian, explorer, and educator. AB, Williams College, 1855; graduated from Andover Theological Seminary, 1858. Ordained in the Presbyterian ministry, 1860. Instructor in natural history, University of Rochester, 1866. Leader, Andes and Amazon expedition, 1867–9, 1873, 1876. Professor of natural history, Vassar College, 1869–77. Wrote a textbook on comparative zoology and also published on women's education. (*ANB*, Elliott 1979.)

24 February [1870], *31 March 1870*

Owen, Mary Susan (b. 1836). CD's niece. Daughter of Henry Parker and CD's sister Marianne. Married Edward Mostyn Owen in 1866. (*Darwin pedigree*.)

2 February [1870], *7 February [1870]*

Owen, Maud (1869–87). Daughter of Mary Susan and Edward Mostyn Owen. CD's great-niece and god-daughter. (*Darwin pedigree*; letter from M. S. Owen, 7 February [1870].)

Owen, Richard (1804–92). Comparative anatomist. Assistant conservator of the Hunterian Museum, Royal College of Surgeons of England, 1827; Hunterian Professor of comparative anatomy and physiology, 1836–56. Superintendent of the natural history departments, British Museum, 1856–84; prime mover in establishing the Natural History Museum, South Kensington, 1881. President of the British Association for the Advancement of Science, 1858. Described the *Beagle* fossil mammal specimens. Knighted, 1884. FRS 1834. (*DSB*, *ODNB*.)

Packard, Alpheus Spring, Jr (1839–1905). American entomologist. Studied under Louis Agassiz. Qualified as a physician but practised only for a few months.

In 1867, co-founder, with Edward Sylvester Morse, Alpheus Hyatt, and Frederic Ward Putnam, of the *American Naturalist*, which he edited for approximately twenty years. Professor of zoology and geology at Brown University, 1878–1905. (*ANB*, *DSB*.)

Paget, James, 1st baronet (1814–99). Surgeon. Assistant surgeon at St Bartholomew's Hospital, London, 1847; surgeon, 1861–71. Arris and Gale Professor of anatomy and surgery at the Royal College of Surgeons of England, 1847–52. Lectured on physiology in the medical school, St Bartholomew's, 1859–61; on surgery, 1865–9. Appointed surgeon-extraordinary to Queen Victoria, 1858; serjeant-surgeon, 1877. Created baronet, 1871. FRS 1851. (*ODNB*.)

4 June [1870]

Park, Mungo (1771–1806). Scottish surgeon and explorer. Conducted early explorations of the River Niger, and disappeared during his second trip. (R. Desmond 1994, *ODNB*.)

Parker, Mary Susan. *See* Owen, Mary Susan.

Pasteur, Louis (1822–95). French chemist and microbiologist. Professor of chemistry, Strasbourg University, 1849–54. Professor of chemistry and dean of the science faculty, Lille University, 1854–7. Administrator and director of scientific studies, Ecole Normale, Paris, 1857–67; director of the laboratory of physiological chemistry, 1867–88. Professor of chemistry, the Sorbonne, Paris, 1867–74. Director of the Institut Pasteur, Paris, 1888–95. Renowned for his work on fermentation and for experiments providing evidence against the theory of spontaneous generation. Foreign member, Royal Society of London, 1869. (*Dictionnaire universel des contemporains*, *DSB*, *Record of the Royal Society of London*.)

Pedler, Alexander (1849–1918). Chemist. Chemical assistant at the Royal Institution of Great Britain, 1867. Member of the eclipse expedition to Sicily, 1870. Professor of chemistry, Presidency College of Bengal, 1873–96; principal, 1896–1904. Helped establish the Indian Association for the Cultivation of Science. FRS 1892. Knighted, 1906. (*Nature* 2 (1870): 409–10; *ODNB*.)

Pender, Emma (1816/17–90). Daughter of Henry Denison of Daybrook, Arnold, Nottinghamshire. Second wife of John Pender, 1851, with whom she had two daughters and two sons. (Census returns 1861 (RG9/2972/6/5); *ODNB* s.v. Pender, John.)

[before 23 June 1870] (and John Pender)

Pender, John (1816–96). Scottish textile merchant and telegraph entrepreneur. Textile trader in Glasgow and Manchester, trading with China and India. In 1864, founded the Telegraph Construction and Maintenance Company, which successfully laid cable between Ireland and Newfoundland in 1866. Reorganised the four companies involved in the link to India; the line was completed in June 1870. Formed the Eastern Telegraph Company in 1872. Liberal MP for Totnes, 1862–6; for the Wick burghs, 1872–85; and, as a Liberal Unionist, 1892–6. Knighted, 1888. (*ODNB*.)

[before 23 June 1870] (and Emma Pender)

Percival, Edward France (1821/2–49). Lawyer. BA (Brasenose College), Oxford, 1844; MA 1847. Barrister-at-law, Inner Temple, 1847. (*Alum. Oxon.*)

Perkin, William Henry (1838–1907). Chemist. Studied at the Royal College of Chemistry, 1853–6. In 1856, discovered the first aniline dye, aniline purple, by accident while attempting to synthesise quinine. Opened a chemical factory in 1858; pooled resources with BASF, 1869–70. After selling the business in 1873, devoted himself to pure chemical research at home. FRS 1866; royal medallist, 1879. Knighted, 1906. (*ODNB.*)

Peschel, Oscar Ferdinand (1826–75). German geographer. Studied law and history at Heidelberg and Leipzig, 1845–8. Editor, *Augsburger Allgemeine Zeitung*, 1850. Chief editor at J. G. Cotta'schen Buchhandlung in Augsburg, 1854. Director of the weekly geographical newspaper *Das Ausland*, 1854. First chair of the geography department, University of Leipzig, 1871. Introduced Darwinian concepts of change and distribution as analytical tools in geography. (*ADB, NDB.*)

Phillips, John (1800–74). Geologist. Keeper of the Yorkshire Museum, 1825–40. Assistant secretary, British Association for the Advancement of Science, 1832–62. Professor of geology, King's College, London, 1834–9. Palaeontologist to the Geological Survey of Great Britain, 1841–4. Deputy reader in geology, Oxford University, 1853; professor, 1860–74. FRS 1834. (*DSB, ODNB.*)

Phillips, Thomas Jodrell (b. 1803/4). Lawyer. BA (Trinity College), Cambridge, 1829; MA 1832; fellow, 1830. Called to the bar, Inner Temple, 1835. Assumed surname and arms of Jodrell, 1868. (*Alum. Cantab.*)

Planchon, Jules Emile (1823–88). French botanist. Herbarium assistant, Royal Botanic Gardens, Kew, 1844–8. Professor of botany, Institute of Horticulture, Ghent, Belgium, 1849–51. Professor, School of Medicine and Pharmacy, Nancy, 1851–3. Professor of botany, faculty of sciences, Montpellier, 1853–81; director of the school of pharmacy, 1856. Became professor of botany in the faculty of medicine and director of the botanic garden, Montpellier, in 1881. (R. Desmond 1994, *Grande encyclopédie.*)

Playfair, Agnes Mary (b. 1853). Eldest daughter of Robert Lambert Playfair. (*IGI.*)

Playfair, Lyon, 1st Baron Playfair of St Andrews (1818–98). Statesman and chemist. Chemist to the Geological Survey of Great Britain and professor of chemistry at the School of Mines, 1845. Secretary at the Department of Science and Art, 1853–8. President of the Chemical Society, 1857–9. Professor of chemistry, Edinburgh University, 1858–69. Liberal MP for the Universities of Edinburgh and St Andrews, 1868–85; for South Leeds, 1885–92. Postmaster-general, 1873; chairman and deputy speaker of the House of Commons, 1880–3. Knighted, 1883; created Baron Playfair, 1892. FRS 1848. (*DNB, DSB.*)

Playfair, Robert Lambert (1828–99). Colonial administrator and author. Royal (Madras) Artillery, 1846–61; Madras staff corps, 1861–3. Consul and lieutenant-colonel, Zanzibar, 1863–7; consul-general, Algeria, 1867; territory of Algeria and northern Africa, 1889. Wrote books for travellers on Algeria and other parts of the region, as well as extensive bibliographies of Algeria and Morocco. (*ODNB.*)

Poole, Skeffington (1803–76). Soldier. Entered the East India Company's service in 1820. Served in the first regiment light cavalry of the Bombay Military Establishment. Retired with the rank of lieutenant-colonel in 1850. (*India list* 1853; http://www.familysearch.org, Pedigree Resource File.)
20 October [1858], *[21 October 1858]*, 22 October [1858], *23 October 1858*, [27 October 1858]

Powell, Baden (1796–1860). Clergyman and writer. Wrote on optics and theological topics; active in university reform. Vicar of Plumstead, Kent, 1821–7. Savilian Professor of geometry at Oxford University, 1827–60. Author of an article on the study of the evidences of Christianity in *Essays and reviews* (1860). FRS 1824. (*DSB*, *ODNB*.)

Prestwich, Joseph (1812–96). Geologist and businessman. Entered the family wine business in London in 1830; became proprietor in 1842. Professor of geology, Oxford University, 1874–88. President of the Geological Society of London, 1870–2. An expert on the Tertiary geology of Europe. Prominent in studies of human prehistory. Knighted, 1896. FRS 1853. (*DSB*, *ODNB*.)

Preyer, William Thierry (1841–97). English-born German physiologist. Studied medicine and natural science at Bonn, Heidelberg, Berlin, and Vienna. PhD, Heidelberg, 1862. Habilitated at Bonn, 1865. MD, Jena, 1866. Professor of physiology, Jena, 1869; Berlin, 1888–94. Worked mainly on developmental physiology. A proponent of science teaching in schools. (*DBE*.)
[before 17 February 1870], 17 February [1870], *9 May 1870*, 15 May 1870

Price, Bonamy (1807–88). Economist. BA (Worcester College), Oxford, 1825; MA, 1832. Appointed mathematical master, Rugby School, 1830–32; classical master, 1832–50. Lived in London, 1850–68. Drummond Professor of political economy, Oxford, 1868–88. Wrote articles for *Fraser's magazine*, *Blackwood's*, and the *Contemporary Review*. (*ODNB*.)

Price, John (1803–87). Scholar, schoolteacher, and naturalist. Educated at Shrewsbury School, 1818–22, and St John's College, Cambridge. Assistant master, Shrewsbury, 1826–7. Headmaster of the junior department at Bristol College, then classics principal at the Liverpool High School, before settling in Chester. A founding member of the Chester Natural Science Society. (*Alum. Cantab.*, *Eagle* (St John's College, Cambridge) 15 (1888): 169–72, *Modern English biography*.)

Prichard, James Cowles (1786–1848). Physician and ethnologist. MD, Edinburgh, 1808. Studied at Trinity College, Cambridge, and St John's and Trinity Colleges, Oxford. Physician, Bristol Infirmary, until 1843. In *Researches into the physical history of man* (1813), maintained that human races were varieties of one species. A commissioner of lunacy in London, 1845. FRS 1827. (*DSB*, *ODNB*.)

Pusey, Edward Bouverie (1800–82). Clergyman and university professor. Regius professor of Hebrew, Oxford, from 1828. Ordained and installed as a canon of Christ Church, 1828. Member of the Oxford Movement. (*ODNB*.)

Quatrefages de Bréau, Jean Louis Armand de (Armand de Quatrefages) (1810–92). French zoologist and anthropologist. Doctorate in the physical

Quatrefages de Bréau, Jean Louis Armand de (Armand de Quatrefages), cont.
sciences, University of Strasbourg, 1830; doctorate in medicine, 1832. Founded *Journal de médicine et de chirurgie de Toulouse*, 1836. Moved to Paris and took a doctorate in the natural sciences, 1840. Professor of natural history at the Lycée Henri IV, Paris, 1850; professor of anthropology, Muséum d'Histoire Naturelle, 1855. Foreign member, Royal Society of London, 1879. (*DSB*, *Record of the Royal Society of London*.)
30 March 1870, 28 May [1870], *18 July 1870*, 20 July [1870], 23 August [1870]

Quetelet, Lambert Adolphe Jacques (1796–1874). Belgian statistician. Astronomer, Brussels Royal Observatory, 1828–74. Secretary, Académie Royale des Sciences et Belles-Lettres, Brussels, 1834–74. Foreign member, Royal Society, 1839. (*BNB*, *DSB*.)
15 December 1870, [15 December 1870 – 11 January 1871]

Ransome, George. Chemist and druggist in Ipswich. Secretary of the Ipswich Museum until 1852. (*Post Office directory of Cambridge, Norfolk and Suffolk* 1853, Russell-Gebbett 1977.)
5 August [1852]

Rathbone, William (1819–1902). Merchant and philanthropist. Partner in his father's firm, Rathbone Brothers & Co., from 1841. Involved in philanthropic activities in Liverpool; in the establishment of district nursing; and in education. Liberal MP for Liverpool, 1868–80; Caernarvonshire, 1881–5; North Caernarvonshire, 1885–95. (*ODNB*.)

Rawson, Arthur (1818–91). Clergyman. Ordained deacon, 1841; priest, 1842. Curate, Bromsgrove, Worcestershire, 1841–3; perpetual curate, Trinity Church, Bromley, Kent, 1843–82. Cultivated orchids and florists' flowers, particularly pelargoniums. (*Alum. Cantab.*, R. Desmond 1994.)
6 June [1863]

Raybould, William J. (*fl.* 1870s). Beer retailer in Smethwick, near Birmingham. (*Post Office directory of Birmingham* 1872.)
22 August 1870 (to A. D. Bartlett)

Reade, William Winwood (1838–75). Traveller, novelist, and journalist. Travelled in West Africa, 1861–3. Studied at St Mary's Hospital, London, in 1865; worked at a cholera hospital in Southampton, 1866. Travelled in West Africa, 1868–70, and again in 1873 as correspondent for *The Times*. Wrote novels as well as travel observations. (Letter from W. W. Reade, 3 September 1870, *ODNB*.)
*[*c. *8 or 9 April 1870]*, *24 April 1870*, *4 June 1870*, 30 June [1870], *3 September 1870*, *6 November 1870*, *9 November 1870*, *11 November 1870*, *20 December 1870*

Recht, Georg (b. *c.* 1816 d. 1873). German mathematician. Studied mathematics at Munich; PhD, 1840; habilitated, 1842; professor extraordinarius, 1849–67. Wrote on mathematics and philosophy. (Bosl ed. 1983.)
17 March 1870

Reeve, Lovell Augustus (1814–65). Conchologist, publisher, and bookseller. Set up a shop in King William Street, Strand, for the sale of natural history specimens and the publication of conchological works; moved to Henrietta Street, Covent Garden, in 1848. Editor and proprietor of the *Literary Gazette*, 1850–6. (*ODNB*, Sarjeant 1980–96.)

[before 14 March 1849]

Reinwald, Charles-Ferdinand (1812–91). German-born bookseller and editor. Founded a business exporting French books in Paris, 1849. Editor in particular of foreign scientific works, and of the *Dictionnaire universel de la langue française*, by M. P. Poitevin. Published the *Catalogue annuel de la librairie française.* (*Dictionnaire universel des contemporains.*)

Rejlander, Oscar Gustaf (1813–75). Swedish?-born photographer. Studied art in Rome. Opened a painter's studio in Wolverhampton in 1845. Took up photography in 1853. Moved to London in 1862. Famous for genre photographs. CD used some of his photographs in *Expression.* (*ODNB.*)

Renard, Charles (Carl Ivanovich) (1809–86). Russian physician. MD, Giessen, 1832. Secretary of the imperial society of naturalists of Moscow, 1840–72; vice-president, 1872–84; president, 1884–6. Editor, *Bulletin de la Société des Naturalistes de Moscou*, 1845–63. (*Bol'shaya entsiklopediya.*)

20 May 1870, 28 May 1870

Rérolle, Louis (1849–1928). French naturalist and museum director. Completed a degree in natural science, Lyon, 1876. Travelled in South America for the museum of Lyon, 1877–8. Teacher, agricultural school, Ecully, 1880–2; Société d'ensignement professionnel du Rhône,1881–5. Director, natural history museum of Grenoble, 1887–1919. (Personal communication from the Museum d'Histoire Naturelle, Grenoble.)

15 June 1870, *22 June 1870*

Reuter, Adolf (1825–1901). German horticulturalist. Studied horticulture in the Prussian royal botanic gardens at Sanssouci and Berlin from 1843; Liege, Ghent, Paris, and Kew, 1850–4. Planteur in the royal nursery, Sanssouci, 1854; inspector, 1865; head gardener, 1866. Head court gardener, 1894. (*Gartenflora* 51 (1902): 64–8.)

11 January 1870, 15 January [1870]

Reynolds, Joshua (1723–92). Portrait-painter. (*ODNB.*)

Rice, Thomas Spring. *See* Spring Rice, Thomas.

Richardson, John (1787–1865). Arctic explorer and naturalist. Assistant surgeon and surgeon on various ships, then surgeon and naturalist on John Franklin's polar expeditions, 1819–22 and 1826–7. Surgeon to the Chatham division of marines, 1824–38. Appointed physician to the Royal Hospital, Haslar, 1838. Conducted a search expedition for John Franklin, 1847–9. Retired to Grasmere, where he accomplished much literary work. Knighted, 1846. FRS 1825. (Johnson 1976, *ODNB.*)

4 November [1851], 30 December [1851]

Rinz, Sebastian (1782–1861). German gardener and landscape architect. City gardener for Frankfurt am Main, 1808–61. Together with his son, ran a nursery in the Westend district of Frankfurt. (*Frankfurter Biographie.*)

Rivers, Augustus Henry Lane Fox Pitt- (1827–1900). Archaeologist and anthropologist. Educated at the Royal Military College, Sandhurst, and commissioned into the Grenadier Guards in 1845. Became interested in archaeology during his military career and after 1867, concentrated on field excavation. Inspector of ancient monuments, 1882. FRS 1876. (*ODNB.*)

Robartes, Thomas James Agar (1807/8–82). Politician. BA (Christ Church), Oxford, 1830. MP for Cornwall, 1847–68. Assumed surname and arms of Robartes by royal license in 1822. Created Baron Robartes, 1869. (*Alum. Oxon.*)

Robertson, Charles Alexander Lockhart (d. 1897). Physician. Assistant staff surgeon, Military lunatic asylum, Great Yarmouth, 1845–50. Appointed Linacre demonstrator in anatomy, Oxford, in 1860. Medical superintendent, Sussex County lunatic asylum, Hayward's Heath, 1859–70. Lord chancellor's visitor in lunacy, 1870–96. (*Alum. Cantab.*, *Modern English biography*.)

Robertson, John (1811/12–75). Writer and editor. Educated at the universities of Glasgow and Aberdeen. Admitted to Lincoln's Inn, 1834. Reporter on the *Morning Chronicle*; editor of the *London and Westminster Review*, 1837–40. (*Modern English Biography*, *Records of Lincoln's Inn.*)

Rodwell, John Medows (1808–1900). Clergyman and orientalist. Friend and contemporary of CD's at Cambridge. Rector of St Peter's, Saffron Hill, London, 1836–43; of St Ethelburga's, Bishopsgate, 1843–1900. Commenced oriental studies as a young man. Published an English version of the Koran in 1861. (*DNB.*)
15 October [1860]

Rogers, John (1807–67). Barrister. Resided at River Hill House, Sevenoaks Weald, Kent. Keen orchid grower; supplied CD with information on orchids. FRS 1839. (*Modern English biography*; *Orchids*, p. 236 n.; *Post Office directory of the six home counties* 1862.)
14 October [1861]

Roget, Peter Mark (1779–1869). Physician and philologist. First Fullerian Professor of physiology at the Royal Institution of Great Britain, 1834–7. Secretary of the Royal Society of London, 1827–48. Compiler of the *Thesaurus of English words and phrases* (1852). FRS 1815. (*ODNB.*)

Rohlfs, Gerhard Friedrich (1831–96). German explorer. Studied medicine in Göttingen, Heidelberg, and Würzburg. Served in the Foreign Legion, Algeria, 1855–60. Travelled to Morocco, crossed North Africa from Tripoli via Bornu to the Gulf of Guinea, 1865–7, and explored the Libyan desert, 1869, 1873–4. First European to reach Kufra, 1878. Took part in an English expedition to Abyssinia, 1880–1. German consul-general, Zanzibar, 1884–5. (*NDB.*)

Rohrbach, Paul (1846–71). German botanist. (*ADB*, Barnhart comp. 1965.)

Rolleston, George (1829–81). Physician and physiologist. Appointed physician to the British Civil Hospital at Smyrna (Izmir), Turkey, 1855, during the Crimean War. Physician to Radcliffe Infirmary, Oxford, and Lee's Reader in anatomy at Christ Church, Oxford, 1857. Linacre Professor of anatomy and physiology, Oxford University, 1860–81. FRS 1862. (*ODNB.*)

20 April 1870

Rouse, Rolla Charles Meadows (1832–1904). Clergyman. BA, Cambridge, 1856; MA, 1859. Ordained deacon 1856, priest 1857. Vicar of Southwold, Suffolk, 1867–70; rector, Woodbridge, Suffolk, 1870–87; Rayleigh, Essex, 1887–1904. (*Alum. Cantab.*)

Routh, Edward John (1831–1907). Mathematician. BA Cambridge, 1854; senior wrangler, James Clerk-Maxwell coming second. From 1862 until his retirement from coaching in 1888 he was the best-known mathematics tutor at Cambridge, having twenty-eight senior wranglers and forty-three Smith prizemen among his pupils. (*Alum. Cantab.*, *DNB.*)

Royal Society of London

8 October [1856?]

Royds, Ernest Edmund Molyneux (1848–92). Banker, magistrate, and rabbit and poultry fancier. Of Greenhill, Rochdale, Lancashire. Provided rabbits for Francis Galton's transfusion experiments, and judged poultry shows. (*BMD* (*Birth index*, *Death index*); Census returns 1881 (RG11/4109/45); Galton 1871, p. 405; letter from Francis Galton, 8 April 1870.)

Royer, Clémence Auguste (1830–1902). French author and economist. Studied natural science and philosophy in Switzerland. In Lausanne in 1859, founded a course on logic for women. Translated *Origin* into French in 1862. (*Dictionnaire universel des contemporains*, Harvey 1997.)

Ruck, Arthur Ashley (1847–1939). Soldier and civic official. Entered the army in 1866; major, 1882; retired with the rank of lieutenant colonel, 1886. Chief Constable of Carnarvonshire after his retirement from the army until 1912. (*BMD* (*Birth index*), *Army list*, Records of the Carnarvonshire Constabulary (GB 0219 XJ, XS/1234), *The Times*, 15 July 1939, p. 1.)

Ruck, Lawrence (1819/20–96). Landowner. Married Mary Anne Matthews in 1841. Father of Amy Richenda Ruck, who married Francis Darwin in 1874. (1881 Wales Census (RG11/5478/64/18), *BMD* (*Death index*, *Marriage index*), *Montgomeryshire collections* 17: 52, *ODNB* s.v. Darwin, Bernard Richard Meirion and Darwin, Sir Francis.)

[after 29 April 1869?]

Ruck, Mary Anne (1821/2–1905). Née Matthews; married Lawrence Ruck in 1841. Lived at Pantlludw, near Machynlleth, Wales. Mother of Amy Richenda Ruck, who married Francis Darwin in 1874. (1861 England Census (RG9/15/12/28), *BMD* (*Death index*, *Marriage index*), *Darwin pedigree*, *ODNB* s.v. Ruck, Amy Roberta.)

Ruck, Richard Matthews (1851–1935). Army officer. Brother of Amy Ruck, the first wife of Francis Darwin. Attended the Royal Military Academy, Woolwich. Entered the Royal Engineers, 1871; captain, 1883, major, 1889; lieutenant-colonel, 1896; colonel, 1904; major-general, 1908. Knighted, 1920. (Ruck 1935, *WWW* 1929–40.)

Rucker, Sigismund (1809/10–76). East and West India broker with premises in Great Tower Street, City of London. Cultivated a major orchid collection at his residence at West Hill, Wandsworth, and frequently exhibited at Chiswick. Member of the council of the Royal Horticultural Society. (*Post Office London directory* 1861; *Post Office London suburban directory* 1860; *Gardeners' Chronicle*, 23 October 1875, pp. 532–3.)

Rudler, Frederick William (1840–1915). Geologist. Assistant curator, Museum of Practical Geology, 1860–76. Professor of natural science, Aberystwyth, 1876–9. Registrar, Royal School of Mines, 1879–80. Curator and librarian, Museum of Practical Geology, 1879–1902. FRS 1870. (*Mineralogical magazine* 17 (1916): 368–9, *Quarterly Journal of the Geological Society of London* 72 (1915): lviii–lix.)

Ruggles, Samuel Buckley (1800–81). American lawyer, land developer, and economist. During the 1830s and 1840s he developed large tracts of land in New York city. Trustee of Columbia College, 1836–81. American delegate to the Fifth International Statistical Congress in Berlin, 1863. (*ANB*.)

Russell, John, 1st Earl Russell (1792–1878). Statesman. Home secretary, 1835–41. Liberal prime minister, 1846–52 and 1865–6. Foreign secretary under Lord Palmerston, 1859–65. Created Earl Russell of Kingston Russell, 1861. In later life, occupied with literary work. FRS 1847. (*ODNB*, *Record of the Royal Society of London*.)

[10 July 1848]

Rütimeyer, Karl Ludwig (Ludwig) (1825–95). Swiss palaeozoologist and geographer. Professor of zoology and comparative anatomy, University of Basel, 1855; rector, 1865; professor in the medical and philosophical faculties, 1874–93. Made important contributions to the natural history and evolutionary palaeontology of ungulate mammals. (*DSB*.)

Rye, Edward Caldwell (1832–85). Entomologist. Contributed articles on Coleoptera to the *Weekly Intelligencer*, the *Zoologist*, and the *Entomologist's Annual*. Was offered a position on the staff of the *Entomologist's Monthly Magazine*, 1864. Contributed to *Zoological Record* from 1870; later became its editor. Elected fellow of the Zoological Society, 1874. Librarian of the Geographical Society, 1875–85. Married Isabella Sophia, second daughter of George Robert Waterhouse, in 1867. (*ODNB*.)

14 February 1870, 29 June 1870

Sabine, Edward (1788–1883). Physicist and army officer. Entered the army in 1803; general, 1870. Astronomer to expeditions in search of a north-west passage, 1818 and 1819–20. Appointed one of three scientific advisers to the Admiralty, 1828. General secretary of the British Association for the Advancement of Science, 1839–52 and 1853–9. Foreign secretary of the Royal Society of London, 1845; treasurer, 1850; president, 1861–71. Knighted, 1869. FRS 1818. (*DSB*, *ODNB*.)

Sachs, Julius (1832–97). German botanist and plant physiologist. PhD, Prague, 1856. Research assistant, forestry academy, Tharandt, 1859. Professor of botany, agricultural training institute, Poppelsdorf, 1861; full professor, Freiburg im Breisgau, 1867; Wurzburg 1868. Founded the institute of plant physiology, Wurzburg. Ennobled 1877. (*DBE*, *DSB*.)

St John, Charles George William (Charles) (1809–56). Sportsman and naturalist. Worked in the Treasury, 1828–33, then moved to Sutherland, where he studied animals and birds. Author of *Short sketches of the wild sports and natural history of the Highlands* (1846), which went through many reprints. (*ODNB*.)

Salimbeni, Leonardo (1830–89). Italian naturalist and political activist. Teacher of natural history and geography, Collegio San Carlo, Modena, 1861–7. Elected parliamentary deputy for the province of Modena, 1864. Translated the first Italian edition of *Origin* (1864) with Giovanni Canestrini. (Barbieri 1973; Commissione Araldica Modenese, busta 30, fascicolo 'Salimbeni', Archivio di Stato di Modena; 'Cronaca del Collegio San Carlo, redatta dal rettore dottor don Luigi Spallanzani dal 1° gennaio fino alla sua morte' and 'Registro II° degli atti dell'amministrazione del Collegio dal 1855 al 1881 inclusive', Archivio Storico del'Collegio San Carlo; Pancaldi 1991, pp. 79–81, 193–4.)

Salter, John William (1820–69). Geologist and palaeontologist. Apprenticed to James de Carle Sowerby in 1835. Assistant to Adam Sedgwick, 1842–6. Assistant to Edward Forbes on the Geological Survey of Great Britain, 1846; palaeontologist to the survey, 1854–63. (*ODNB*.)

28 February [1862]

Salvin, Osbert (1835–98). Ornithologist and entomologist. Undertook natural history explorations in North Africa, 1857, and Central America, 1857–60, 1861–3, 1873–4. Strickland Curator of ornithology, University of Cambridge, 1874–82. FRS 1873. (*DNB*.)

[1868?], 24 June 1870

Sandars, Edmund (1834–82). Tutor and poet. BA, Cambridge (Trinity Hall), 1858. Sometime in the civil service. Subsequently a private tutor at Bournemouth. Author of *By the sea* (poems). (*Alum. Cantab.*)

Schmarda, Ludwig Karl (1819–1908). Austrian naturalist and traveller. Studied medicine at Olmütz, Moravia; natural science at Vienna. DPhil 1841; MD 1843. Professor of natural history, and founder and director of the Zoological Museum, Graz, 1850. Professor of zoology, Prague, 1852. Made several scientific voyages from 1853 to 1857 and after his retirement. Professor of zoology, Vienna, 1862–83. Worked primarily on the geographical distribution of animals. (*OBL*.)

Schmick, Jacob Heinrich (1824–1905). Teacher. From 1848, taught at various levels in Germany, England, Ireland, and France. Dr. Phil., 1859, Jena. Published on spiritualism and topics in science (*BHGW* vol. 4, *NUC*.)

12 November 1870

Schmid, Theophilus (b. *c.* 1832). German missionary. Linguist for the South American Missionary Society (then the Patagonian Missionary Society), 1854. Arrived in South America in 1856, later taking up residence on Keppel Island. Travelled to mainland Patagonia with Allen Gardiner in 1858. Appointed

Schmid, Theophilus, cont.

catechist, 1859, and travelled to Punta Arenas, Chile, where he spent about six months researching the language, costume, and customs of the Patagonians. Made a second expedition to Patagonia with another catechist, Johannes Friedrich Hunziker, in 1861. Set up a mission at Weddell Bluff, on the Santa Cruz river, with Hunziker, in 1862. Later worked near Bahia Blanca (Argentina) and in Fray Bentos (Uruguay). Gave up his position and returned to Europe, 1874. (*Ecclesiastic and Theologian* 19 (1857): 524–6; Schmid 1964, pp. 14–18.)

Schroeder van der Kolk, Jacob Lodewijk Coenraad (Jacob Ludwig Conrad) (1797–1862). Dutch physician, anatomist, physiologist, and zoologist. Resident physician at the Buitengasthuis, Amsterdam, 1821. Professor of anatomy and physiology, University of Utrecht, 1827–62. Governor of the Utrechtsehe Dolhuis, 1827; inspector of lunatic asylums, 1842–62. Involved in reforms in the care of the mentally ill. (*DSB*, *NNBW*.)

Schultze, Max Johann Sigismund (1825–74). German anatomist. MD, Greifswald, 1849; privat-dozent, 1850–4. Assistant professor of anatomy, Halle, 1854–9. Professor of anatomy and director of the anatomical institute, Bonn, 1859. Founder and editor of *Archiv für mikroskopische Anatomie*, 1865–74. Noted for his work in microscopy, the reform of cell theory, and descriptive and taxonomic studies of rhizopods and sponges. (*DBE*, *DSB*.)

Schweizerbart, Christian Friedrich (1805–79). German publisher. Director of E. Schweizerbart'sche Verlagsbuchhandlung of Stuttgart, 1841–67. Publisher of the German translations of *Origin* (1860, 1863, 1867) and *Orchids* (1862). (*Jubiläums-Katalog*.)

Schweizerbart, Wilhelm Emanuel (1785–1870). German publisher. Founded E. Schweizerbart'sche Verlagsbuchhandlung in Stuttgart in 1830. Publisher of many German scientific works. Retired in 1841. (*Jubiläums-Katalog*.)

E. Schweizerbart'sche Verlagsbuchhandlung. German publishing company in Stuttgart; founded by Wilhelm Emanuel Schweizerbart in 1830; conducted by his nephew Christian Friedrich Schweizbart from 1841. (*Jubiläums-Katalog*.)

Sclater, Philip Lutley (1829–1913). Lawyer and ornithologist. One of the founders of *Ibis*, the journal of the British Ornithologists' Union, 1858; editor, 1858–65 and 1878–1912. Secretary of the Zoological Society of London, 1860–1903. FRS 1861. (*DSB*, Scherren 1905.)

30 August [1870], 4 November [1870], 9 November 1870, 11 November [1870], 21 November [1870], 5 December 1870, 26 December [1870], 29 December 1870

Scott, John (1836–80). Scottish botanist. Gardener at several different country estates, before becoming foreman of the propagating department at the Royal Botanic Garden, Edinburgh, in 1859. Through CD's patronage emigrated to India in 1864, and worked briefly on a *Cinchona* plantation before taking a position as curator of the Calcutta botanic garden in 1865. Seconded to the opium department, 1872–8. Carried out numerous botanical experiments and observations on CD's behalf. Fellow of the Linnean Society of London, 1873. (Lightman ed. 2004, *ODNB*.)

Scott, Walter, 1st baronet (1771–1832). Scottish poet and novelist. President of the Royal Society of Edinburgh, 1820. Created baronet, 1818. (*ODNB*.)

Scott, William Robson (1810/11–77). Teacher. Principal of the West of England Institution for the Education of the Deaf and Dumb, 1841–77. In his book, *The deaf and dumb*, he provided the earliest comprehensive linguistic analysis of British Sign Language. (*BMD* (*Death index*); Branson and Miller 2002, p. 158.)

28 June 1870 (to E. B. Tylor)

Scott-Siddons, Mary Frances (b. *c.* 1840, d. 1896). Actress. Born in India, educated in Germany. Her father was William Young Siddons, lieutenant in the Bengal Native Infantry and grandson of Sarah Siddons. Married T. Chanter, a naval officer, in 1862. Acted in London, Edinburgh, and New York; travelled in the US and Australia, 1872–9. (*Modern English biography*, *New York Times*, 20 November 1896, p. 5.)

Seabroke, George Mitchell (1838–1918). Astronomer. Became interested in astronomy while at Rugby. Entered a solicitor's office, but remained honorary curator of the Temple Observatory at Rugby. A founder of the British Astronomical Association. A member of the eclipse expedition to Sicily, 1870. (*Nature* 2 (1870): 409–10; *Notices of the Royal Astronomical Society* 79 (1919): 231–2.)

Sedgwick, Adam (1785–1873). Geologist and clergyman. Woodwardian Professor of geology, Cambridge University, 1818–73. Prebendary of Norwich Cathedral, 1834–73. President, Geological Society of London, 1829–31; British Association for the Advancement of Science, 1833. FRS 1821. (*DSB*, *ODNB*, *Record of the Royal Society of London*.)

30 May 1870, 1 June [1870]

Semper, Karl Gottfried (1832–93). German zoologist. Studied engineering, Hanover, 1851–4. Studied zoology, histology, and comparative anatomy, University of Würzburg; completed thesis, 1856. Traveled in the Philippines and Palau Islands, 1858–65, and acquired zoological and ethnographical collections. Appointed privat-dozent, University of Würzburg, 1866; professor and director of the Zoological Institute, 1869. Published on zoology (especially molluscs), geography, and ethnography. (*DSB*.)

Senior, Nassau William (1790–1864). Political economist. BA (Magdalen College), Oxford, 1811; probationary fellow, 1812. Called to the bar, 1819. Drummond Professor of political economy, Oxford, 1825–30 and 1847–52. Master of Chancery from 1836. Particularly interested in the relief of poverty. (*ODNB*.)

Settegast, Hermann Gustav (Hermann) (1819–1908). German agronomist. Studied agriculture for nine years in East Prussia, then natural science in Berlin from 1844 and rural economy at Hohenheim from 1845. Administrator and

Settegast, Hermann Gustav (Hermann), cont.
instructor, Proskau, 1847. Director of the agricultural academy, Waldau, 1858; Proskau, 1862. Professor of agriculture, Breslau, 1868. (*DBE*.)
29 September 1870

Sewa, Manga (d. 1884). Ruler in northern Sierra Leone. Son of Asan Yira (Sori Wuleng) former ruler of Solimana state. Ruler of Solimana state from around 1862. Maintained commercial links with other northern peoples in Sierra Leone and beyond; extended areas of influence to include smaller states in the country. Committed suicide rather than submit during the seige of his capital, Falaba, by Samori Touré. (*Dictionary of African biography* 2: 144.)

Sharpe, Richard Bowdler (1847–1909). Ornithologist and museum curator. Librarian of the Zoological Society of London, 1867–72. Wrote and edited scholarly works on birds, supplied the ornithological section of *Zoological Record*, and lectured. Published his *Monograph of the kingfishers* between 1868 and 1871. Senior assistant at the British Museum, 1872–95; assistant keeper of vertebrates from 1895. Worked on the *British Museum catalogue of birds*. Founded the British Ornithologists' Club in 1892. (*DNB*.)

Sharpey, William (1802–80). Scottish physiologist. MD, Edinburgh University, 1823. Joint lecturer on systematic anatomy, Edinburgh University, 1832. Professor of anatomy and physiology, University College, London, 1836–74. Examiner in anatomy, London University, from 1840. Secretary of the Royal Society of London, 1853–72. Member of the General Medical Council, 1861–76. FRS 1839. (*DNB*, *DSB*.)

Siddons, Sarah (1755–1831). Actress. The most acclaimed tragic actress of her time. (*ODNB*.)

Sidebotham, Joseph (1824–85). Calico printer in Manchester. Botanist and entomologist. One of the founders of the Manchester Field Naturalists' Society. Collected diatoms. (R. Desmond 1994, Gilbert 1977.)

Skirving, Robert Scot. Of Edinburgh. Author of several tracts on the labour question in Scotland. (*BLC*, *NUC*.)
16 November [1861–8]

Smith, Edward A. (b. 1842/3). Artist and civil servant. Son of the entomologist Frederick Smith. Provided insect illustrations for CD's *Descent* and other natural history books. Employed by the General Post Office, eventually attaining post of Assistant Controller by 1901. (Census records 1861 (Public Record Office RG9/38/33); Census records 1901 (Public Record Office RG13/435/65/11); *Correspondence* vol. 17, letter to E. A. Smith, 30 June [1869]; Peter Dance (personal communication).)
[11–31 March 1868]

Smith, Frederick (1805–79). Entomologist. Entomologist in the zoology department of the British Museum from 1849. Specialised in the Hymenoptera. President of the Entomological Society of London, 1862–3. (*Entomologist* 12 (1879): 89–92, Gilbert 1977.)

Smith, Henrietta (1831–1917). Née White. Bonnet sewer. Married George Worthington Smith in 1856. Had seven children; three survived early childhood. Resided in Dunstable. (*ODNB* s.v. Smith, George Worthington.)

Smith, James Edward (1759–1828). Botanist. Purchased the library and collections of Carl von Linné in 1784. Founded the Linnean Society of London in 1788 and served as president until his death. Knighted, 1814. FRS 1785. (*DSB*, *ODNB*.)

Smith, Julia. Aunt of Elinor Mary Bonham-Carter. Lived with her sister, Joanna Maria Bonham-Carter, at Keston House, two miles north-west of Down, Kent, from 1862. (Bonham-Carter 1960, *Post Office directory of the six home counties* 1862.)

Smith, Worthington George (1835–1917). Illustrator, botanist, and archaeologist. Apprentice to architects. Became a freelance architectural illustrator in 1861, then turned to botanical illustration. Studied and published in mycology. Beginning in 1878, made several significant discoveries of Palaeolithic sites, publishing two books. (*ODNB*.)

4 November 1870, 6 November 1870

Smith, Beck & Beck. Instrument-makers at 31 Cornhill, London; one of the leading manufacturers of microscopes in London. The partners, James Smith (died 1870), Richard Beck (1827–66), and Joseph Beck (1829–91), were members of the Microscopical Society of London. (*Post Office London directory* 1864, Turner 1989, p. 171.)

Snow, George (1820/1–85). Coal-dealer, Down, Kent. Operated a weekly carrier service between Down and London. (Census returns 1861 (Public Record Office RG9/462: 72); gravestone inscription, Down churchyard; *Post Office directory of the six home counties* 1862.)

Somerville, Mary (1780–1872). Writer on science and mathematics. Author of a condensed English version of Laplace's *Mécanique céleste* (1831), and of *On the connection of the physical sciences* (1834). Advocate of higher education for women and women's suffrage. (*BDWS*, *DSB*, *ODNB*.)

Sophocles, Evangelinus Apostolides (1805–83). Greek-born scholar of Greek in the United States. Educated in Cairo and Greece. Emigrated to the US in 1826. Taught Greek, Mount Pleasant Classical Institute, 1830–40; high school, Hartford, Conn.; Yale College, 1837–42. Tutor at Harvard University, 1842–9; assistant professor, 1849–60; professor of ancient, Byzantine, and modern Greek, 1860–83. (*ANB*.)

Sowerby, George Brettingham (1788–1854). Conchologist and artist. Described the *Beagle* fossil shells. (*DNB*, Sarjeant 1980–96.)

Sowerby, James de Carle (1787–1871). Naturalist and scientific illustrator. An expert on fossil conchology. Founding member of the Royal Botanic Gardens, Regent's Park, London, 1838; secretary, 1838–69. (R. Desmond 1994, *ODNB*.)

Spencer, Herbert (1820–1903). Philosopher. Apprenticed as a civil engineer on the railways, 1837–41. Became sub-editor of the *Pilot*, a newspaper devoted to the suffrage movement, in 1844. Sub-editor of the *Economist*, 1848–53. From 1852,

Spencer, Herbert, cont.
author of books and papers on transmutation theory, philosophy, and the social sciences. (*DSB*, *ODNB*.)

Sprengel, Christian Konrad (1750–1816). German botanist. Rector of the Great Lutheran Town School, Spandau, where he taught languages and natural science, 1780–94. Moved to Berlin, where he worked as a private tutor. Published his major work on insect-aided fertilisation of flowers in 1793. (*ADB*, *DSB*.)

Spring Rice, Thomas (1790–1866). Irish statesman. Secretary to the Treasury, 1830–4. Chancellor of the Exchequer, 1835–9. Created Baron Monteagle of Brandon in Kerry, 1839. FRS 1841. (*ODNB*, *Record of the Royal Society of London* s.v. Monteagle, Lord.)
[before 7 July 1838]

Squier, Ephraim George (1821–88). American journalist and archaeologist. With Edwin Hamilton Davis, excavated prehistoric Indian mounds in Ohio, 1845–7; made archaeological investigations in New York state, 1848. Chargé d'affaires to the Central American states, 1849–51; used his spare time to continue archaeological and ethnological research. Secretary of the Honduras Inter-Oceanic Railway Company, 1853 (the venture was never realised). Chief editor to Frank Leslie's New York publishing firm, 1857–73. US claims commissioner to Peru, 1863–5. Consul general of Honduras at New York City, 1868. (*ANB*.)

Stanhope, Leicester FitzGerald Charles (5th earl of Harrington) (1784–1862). Soldier. Cornet, First Life Guards, 1799; major, Forty-Seventh Foot, 1816–23; colonel in the army, 1837. (*Modern English biography*.)

Stanley, Edward (1779–1849). Clergyman. Bishop of Norwich, 1837–49. Regularly attended the House of Lords; a staunch supporter of Whig principles. (*ODNB*.)

Stanley, Edward George Geoffrey Smith, 14th earl of Derby (1799–1869). Scholar, orator, and Conservative prime minister. MP for North Lancashire, 1832–44. Chief secretary for Ireland, 1830–3. Colonial secretary, 1833–4 and 1841–5. Prime minister, 1852, 1858–9, 1866–8. Appointed chancellor of Oxford University, 1852. Supported gradual parliamentary reform. (*ODNB*.)

Stanley, Edward Henry, 15th earl of Derby (1826–93). Politician and diarist. BA (Trinity College), Cambridge, 1848. MP for King's Lynn, 1848–69. Visited the West Indies twice, 1848–50. First secretary of state for India from 1858. Foreign Secretary from 1874. Colonial secretary from 1882. (*ODNB*.)

Stark, James (1811–90). Scottish physician, statistician, and religious writer. MD, Edinburgh. Established a private practice in Edinburgh. First superintendent of statistics for the General Register Office, Edinburgh, 1855. Fellow of the Royal Society of Edinburgh. Fellow of the Royal College of Physicians of Edinburgh. (*CDEL*, *Scottish way of birth and death* (database), University of Glasgow, Department of Economic and Social History, http://www.gla.ac.uk/departments/scottishwayofbirthanddeath/leadingactors/superintendentsofstatistics/ (accessed May 2009).)

Steward, George William (1805–78). Clergyman. BA (Corpus Christi College), Cambridge, 1827; MA, 1830. Ordained deacon, 1828; priest, 1829. Rector of Caister-next-the-Sea, 1829–78. (*Alum. Cantab.*, *BMD* (*Death index*.)

Stewart, Alexander, 1st earl of Buchan (*c.* 1345–1405). Magnate. Known as the 'wolf of Badenoch'. Lord of Badenoch in the western Grampians, and the castle and forest of Lochindorb in Strathspey, from 1371. Attacked Forres and burned Elgin in 1390. (*ODNB*.)

Stilling, Benedict (1810–79). German anatomist and surgeon. Received his doctorate at Marburg, 1832. State surgeon, Cassel, 1833. Travelled extensively in France, Britain, and Italy, and collaborated with many well-known researchers. Developed new techniques for studying the spinal cord. (*BLA*.)

Stirling, Waite Hockin (1829–1923). Clergyman and missionary. Secretary, Patagonian Mission Society, 1857–62. Superintendent missionary of Tierra del Fuego, 1862–9. Bishop of the Falkland Islands, 1869–1900. Assistant bishop of Bath and Wells, 1901–11; canon of Wells, 1901–20; precentor, 1903–20. (*Alum. Oxon.*, *Crockford's clerical directory* 1923, Hazlewood 2000, F. C. Macdonald 1929.)

Stoddard, Charles Warren (1843–1909). Poet and travel writer. Travelled to Hawaii, 1864, Tahiti, 1870, Europe and the Middle East, 1873–7. Converted to Roman Catholism, 1867. Wrote poetry, and travel columns for newspapers. Taught English literature at Notre Dame, 1885–6; taught literature at the Catholic University of America, Washington DC, 1889–1902. (*ANB*.)

11 April 1870, 5 May [1870]

Stokes, Charles (1783–1853). Stockbroker in London; collector of coins and drawings, naturalist, and geologist. Collected fossil wood. FRS 1821. (R. Desmond 1994, *Modern English biography*, Sarjeant 1980–96.)

Stokes, George Gabriel, 1st baronet (1819–1903). Physicist. Lucasian Professor of mathematics, Cambridge University, 1849–1903. Secretary of the Royal Society of London, 1854–85; president, 1885–90. President of the British Association for the Advancement of Science, 1869. Conservative MP for Cambridge University, 1887–91. Master of Pembroke College, Cambridge, 1902–3. Created baronet, 1889. FRS 1851. (*DSB*, *ODNB*.)

Strahan, Alexander Stuart (1833–1918). Scottish publisher. Founded Strahan & Co. in partnership with William Isbister in 1858; moved to London in 1862. Founded the periodicals *Good Words*, the *Sunday Magazine*, *Argosy*, and the *Contemporary Review*. Forced to resign from Strahan & Co., 1872; in 1874 the firm's name was changed to Isbister & Co. (*ODNB*.)

29 November 1870

Strickland, Hugh Edwin (1811–53). Geologist and zoologist. An advocate of reform in zoological nomenclature. FRS 1852. (*ODNB*, Sarjeant 1980–96.)

Suess, Eduard (1831–1914). Austrian geologist, palaeontologist, educational reformer, and politician. Palaeontologist at the Kaiserlich-königliche Hofmineralien-Cabinett, Vienna, 1852. Professor extraordinarius of palaeontology, University of Vienna, 1857–62; professor extraordinarius of geology, 1862–7. Professor of

Suess, Eduard, cont.

geology, University of Vienna, 1867–1911. President of the Kaiserliche Akademie der Wissenschaften, Vienna, 1898–1911. Foreign member, Royal Society, 1894. (*Almanach der kaiserlichen Akademie der Wissenschaften* 63 (1914); Eisenberg 1893, *Mitteilungen der Geologischen Gesellschaft Wien* (1914): 1–32; *Neue Österreichische Biographie* I.)

Sulivan, Bartholomew James (1810–90). Naval officer and hydrographer. Lieutenant on HMS *Beagle*, 1831–6. Surveyed the Falkland Islands in HMS *Arrow*, 1838–9. Commander of HMS *Philomel*, 1842–6. Resided in the Falkland Islands, 1848–51. Commanded HMS *Lightning* in the Baltic, 1854–5. Naval officer in the marine department of the Board of Trade, 1856–65. Admiral, 1877. Knighted, 1869. (*ODNB*.)

27 June 1870, 30 June [1870], *1 July 1870*, 2 July [1870], *17 December 1870*, 20 December [1870]

Sulivan, Catherine Sabine. Daughter of Sophia and Bartholomew James Sulivan. Married William Wallace Trench in 1870. (*Burke's peerage* 1970, p. 123.)

Sulivan, Henry Norton (b. 1848/9). Son of Sophia and Bartholomew James Sulivan; edited his father's *Life and letters*. (Sulivan ed. 1896.)

Sulivan, James Young Falkland (1838/9–1901). Naval officer. Eldest son of Sophia and Bartholomew James Sulivan. Sub-lieutenant on HMS *Firefly* in the Mediterranean, 1864; lieutenant on HMS *Nassau*, surveying the Straits of Magellan, 1866–7; on HMS *Favorite* in North America and the West Indies, 1868–9; on HMS *Indus* at Devonport, 1870. Stationed in China, 1870–5; studied at the Royal Naval College, Greenwich, 1875–6; lieutenant and commander of HMS *Britomart* at Southampton, 1876–9; served in the Coast Guard, 1882–8. Commander, 1889. (*Navy list* 1864–1901, *ODNB* s.v. Sulivan, Bartholomew James.)

Sulivan, Sophia (1809/10–90). Daughter of vice-admiral James Young, of Barton End, near Stroud, Gloucestershire. Married Bartholomew James Sulivan in 1837. (*County families* 1871, s.v. Sulivan, Bartholomew James; *ODNB*, s.v. Sulivan, Bartholomew James; Sulivan ed. 1896, pp. xii, 395.)

Sulivan, Thomas Edward (d. 1873). Naval officer. Second son of Sophia and Bartholomew James Sulivan. Sub-lieutenant on HMS *Gannet*, 1865–6, and lieutenant on HMS *Duncan*, stationed in North America and the West Indies, 1866–7; Transport Service, 1867–8; surveyed the Mediterranean in HMS *Newport*, 1868–70; lieutenant on HMS *Pert*, stationed off the Cape of Good Hope, 1870, and the south-east coast of America, 1871–3. (*Navy list* 1865–73, Sulivan ed. 1896.)

Sutherland, Peter Cormack (1822–1900). Scottish physician, naturalist, and explorer. Undertook expeditions to the west coast of Africa and to Greenland. Accompanied an expedition in search of John Franklin, 1850. Government geologist and surveyor-general in Natal, South Africa, 1853–86. (R. Desmond 1994, *DSAB*.)

Sutton, Seth. A keeper at the zoological gardens of London. (Archives of the Zoological Society of London, 'Daily occurrences', September 1860; *Correspondence* vol. 15, letter from Seth Sutton, 8 August 1867; R. B. Freeman 1978.)

Swale, William (1816–75). Nurseryman. Emigrated from Norfolk to Canterbury, New Zealand, 1857. Wrote for the *Gardener's Chronicle.* Possibly employed by William Wilson, nurseryman, of Canterbury. Owner of the Kilmore Street nursery from 1859, and of the Avonside Botanical and Horticultural Garden from 1860. Specialised in trees and native plants. (Challenger 1979, pp. 55–8.)
16 February [1870?]

Swanzy, Andrew (b. 1817/18). Merchant. London-based merchant who traded with West Africa. Sponsored William Winwood Reade's explorations of the Niger region in 1868–70. Fellow of the Royal Geographical Society. (Census returns 1871 (Public Record Office RG10/922/53); Reade 1873, 2: 352–3.)

Swinhoe, Robert (1836–77). Diplomat and ornithologist. Attached to the British consulate in Hong Kong, 1854; in Amoy (Xiamen), China, 1855. British vice-consul, Formosa (Taiwan), 1860–5; consul, 1865–73. Acting consul, Amoy, 1865–71; Ning-po (Ningbo), 1871–3. Consul, Ning-po, 1873–5. Collected plants and animals in Eastern Asia. FRS 1876. (*Foreign Office list* 1877, P. B. Hall 1987, *ODNB*.)
2 February 1870, 10 February 1870

Sylvester, James Joseph (1814–97). Mathematician. Professor of natural philosophy, University College, London, 1837–41. Professor of mathematics, University of Virginia, 1841–5; Royal Military Academy, Woolwich, 1855–70; Johns Hopkins University, 1877–83. Savilian Professor of geometry, University of Oxford, 1883–94. FRS 1839. (*DNB*.)

Symonds, Hyacinth (1842–1921). Daughter of William Samuel Symonds. Married Sir William Jardine of Applegirth, seventh baronet, in 1871; widowed in 1874. Married Joseph Dalton Hooker in 1876. (Allan 1967 s.v. 'Hooker pedigree', *Burke's peerage* 1876.)

Symonds, William Samuel (1818–87). Geologist, author, and clergyman. Curate of Offenham, Worcestershire, 1843–5; rector of Pendock, Worcestershire, from 1845. Wrote papers on local archaeology and geology. Visited the Auvergne and the Ardèche, south-west France, 1874–6, to search for traces of ancient glaciers. (*ODNB*, Sarjeant 1980–96.)

Talandier, Pierre Theodore Alfred (1822–90). Political exile from France, 1848. Master of French at the Royal Military College, Sandhurst, 1859; professor, 1868. Returned to France in 1870 and edited a radical periodical. Translated several English works into French. (*Dictionnaire des parlementaires Français*, RMAS Archives.)

Tanner, John (*c.* 1780 – *c.* 1846). American interpreter and author. Son of John Tanner, a preacher who settled with his family in Kentucky in the 1780s. Captured at the age of 9 by two Ojibwa men, and adopted by Net-no-kwa, a female chief of the Ottawas. Lived with Ojibwa and other Indian peoples in the Great Lakes and Red River regions, 1790 – *circa* 1820. Interpreter at Makinac, 1825 or 1826, and later at Saute Ste Marie. Published an autobiography in 1830. Disappeared in 1846. (*ANB*.)

Taylor, John (1779–1863). Mining geologist and chemical manufacturer. Treasurer of the Geological Society of London, 1816–47; treasurer of the British Association for the Advancement of Science, 1831–61; treasurer of University College London, 1842–60. FRS 1825. (*Modern English biography*, *ODNB*.)

Teebay, Richard (1811/2–85). Farmer and poultry breeder. Resident at Fulwood, Lancashire. (*BMD* (*Death index*, Census returns 1851 (HO107/2268/595/13), *Post Office directory of Lancashire, Liverpool and Manchester*; *The Times*, 3 December 1856, p. 12, 13 November 1883, p. 3.)

Tegetmeier, William Bernhard (1816–1912). Editor, journalist, lecturer, and naturalist. Pigeon-fancier and expert on poultry. Pigeon and poultry editor of the *Field*, 1864–1907. Secretary of the Apiarian Society of London. (*Field*, 23 November 1912, p. 1070; *ODNB*; E. W. Richardson 1916.)

[28 March – 14 April 1861], 2 February [1864], *14 July 1870*, 15 July [1870]

Te Hapuku (Te Ika-nui-o-te-moana) (d. 1878). Maori leader. Leader of the Ngati Te Whatu-i-apiti. Kinship links made him influential throughout the region of Hawke's Bay. Signatory to the Treaty of Waitangi, 1840. Sold much land to the Europeans, which caused friction with other Maori leaders. Retired to Poukawa in 1858 after a siege at Whakawhiti. (*DNZB*.)

Tennent, James Emerson (1804–69). Traveller, politician, and author. Assumed the name Tennent on the death of his father-in-law, William Tennent, in 1832. MP for Belfast, 1832–37, 1838–41, 1842–45. Civil secretary to the colonial government of Ceylon (Sri Lanka), 1845–50. Appointed governor of St Helena in 1850, but never assumed office. FRS 1862. (*ODNB*.)

Theobald, James (b. 1795/6). Farmer. Born in Mountnessing, Essex. Retired by 1871, residing in Chelmsford. (Census returns 1871 (Public Record Office RG10/93/15.))

Thomson, Charles Wyville (1830–82). Scottish naturalist. Lectureship in botany, Aberdeen, 1850. Held series of professorships in natural history in Ireland from 1853. Professor of natural history, Edinburgh, 1870–82. Studied the marine environment, particularly deep sea invertebrates. Scientific director, *Challenger* expedition, 1872–6. Knighted 1876. FRS 1869. (*ODNB*.)

7 October 1870, [after 7 October 1870]

Thomson, William, Lord Kelvin (1824–1907). Scientist and inventor. Professor of natural philosophy, Glasgow, 1846–99. Formulated laws of equivalence and transformation in thermodynamics and a doctrine of available energy. Pioneered telegraphic systems and assisted in the laying of the first transatlantic cable. Wrote on the age and cooling of the earth. Proposed a hydroelectric scheme for Niagara. Created Baron Kelvin of Largs, 1892. FRS 1851. Awarded the Copley Medal, 1883. (*DSB*, *ODNB*.)

Thorell, Tord Tamerlan Teodor (1830–1901). Swedish zoologist. Specialist in spiders. Professor of zoology at the University of Upsala. (Tort 1996.)

20 April 1870

Thorpe, Thomas Edward (1845–1925). Chemist. Junior assistant, chemical department, Owens College, Manchester, 1863. Studied at the University of

Heidelberg, where he took a PhD, 1867–9. Professor of chemistry, Anderson's University, Glasgow, 1870; Yorkshire College of Science, Leeds, 1874; Normal School of Science (later the Royal College of Science), South Kensington, 1885. A member of the eclipse expedition to Sicily, 1870. Principal chemist of the Inland Revenue laboratory (later the Government Laboratory), 1894–1909. Professor of general chemistry, Imperial College, 1909–12. FRS 1876. Knighted, 1909. (*Philosophical Transactions of the Royal Society of London* 161: 467–76, *ODNB*.)

Tollet, Georgina (1808–72). Daughter of George Tollet. A close friend of the Wedgwoods and Darwins. Edited the manuscript of *Origin*. (M. E. Smith ed. 2008.)

1 October [1846?–71], 13 March [1870?] (from Charles Wicksted)

Torrey, John (1796–1873). American botanist and chemist. Collected and described plants from various government expeditions. Professor of chemistry at the College of Physicians and Surgeons, New York City, 1827–54. Assayer of the United States Mint in New York, 1853–73. Mentor of, and collaborator with, Asa Gray. (*DAB*, *DSB*.)

Traill, Wilhelmina Barbra (1785/6–1862). Niece of John Sinclair, 11th earl of Caithness (d. 1789). Mistress of Hayes Place, Hayes, Kent. (*BMD* (*Death index*); Census returns 1841 (Public Record Office HO107/483/6; Thompson 1935, p. 59.)

Trench, William Wallace (1840–83). Husband of Catherine Sabine Sulivan (daughter of Bartholomew James Sulivan), who became his second wife in 1870. (*Burke's peerage* 1970, pp. 122–3 (s.v. Ashtown); letter from B. J. Sulivan, 27 June 1870.)

Treviranus, Ludolph Christian (1779–1864). German botanist. Professor of botany, University of Breslau, 1816–30; University of Bonn, 1830–64. (*ADB*, *DSB*.)

Trimen, Roland (1840–1916). Zoologist and entomologist. Emigrated to South Africa in 1858. Arranged the Lepidoptera at the South African Museum. Held civil-service positions in the Commission of Land and Public Works, the governor's office, and the colonial secretary's office. Became part-time curator of the South African Museum in 1873; full-time curator, 1876. FRS 1883. (*DSAB*.)

Trousseau, Armand (1801–67). French physician. Studied with Pierre Bretonneau. Physician to the Hôpital St Antoine, 1839; Hôtel Dieu, 1850. A prolific author. (*Journal of Thrombosis and Haemostasis* 1 (2003): 2463–5), Pearce 2002.)

Tschermak, Gustav (1836–1927). German mineralogist. Studied at the Faculty of Philosophy, University of Vienna, from 1856. Received his doctorate from the University of Tübingen, 1860. Privat-dozent, chemistry and mineralogy, University of Vienna, 1862; associate professor of petrography, 1868; professor of mineralogy and petrography, 1873. Second assistant curator of the imperial mineralogical collection, 1862; director, 1868–77. Worked most notably on feldspar and meteorites. Retired and was ennobled, 1906. (*DSB*.)

Turner, William (1832–1916). Anatomist and administrator. Senior demonstrator to Sir John Goodsir, professor of anatomy, University of Edinburgh, 1854–67; professor of anatomy, 1867–1916; principal, 1903–16. Published papers on

Turner, William, cont.
anthropology and comparative anatomy from 1854. Knighted, 1886. FRS 1877. (*DSB*, *ODNB*.)

3 January 1870

Tylor, Edward Burnett (1832–1917). Anthropologist. Educated at the School of the Society of Friends, Grove House, Tottenham. Author of *Primitive culture* (1871). Keeper of Oxford University Museum, 1883; reader in anthropology, Oxford University, 1883; professor, 1896; professor emeritus, 1909. President of the Anthropological Society, 1879–80, 1891–2. Helped to establish anthropology as a legitimate field of scientific enquiry. Knighted, 1912. FRS 1871. (*Men and women of the time* 1899, *ODNB*.)

25 June 1870, 28 June 1870 (from W. R. Scott)

Tyndall, John (1820–93). Irish physicist, lecturer, and populariser of science. Studied in Marburg and Berlin, 1848–51. Professor of natural philosophy, Royal Institution of Great Britain, 1853–87; professor of natural philosophy, Royal School of Mines, 1859–68; superintendent of the Royal Institution, 1867–87. Scientific adviser to Trinity House and the Board of Trade, 1866–83. FRS 1852. (*DSB*, *ODNB*.)

7 September 1870, 8 September 1870

Usborne, Alexander Burns (1808/9–85). Naval officer. Master's assistant then master on HMS *Beagle*, 1831–5. Took command of a small schooner and surveyed the coast of Peru, 1835–6. Master on the *Beagle* in Australia, 1838–9. Surveyed the coasts of England and Ireland, 1847–65. Created staff commander, 1863; captain, 1867. Retired in 1868. (*BMD* (*Death index*), R. B. Freeman 1978, *Narrative* 2: 19–20, *Navy list* 1838–85.)

Ussher, Herbert Taylor (1836–80). Colonial administrator. Collector of customs, Gold Coast, 1866; administrator, 1867–72. Lieutenant governor of Tobago, 1872–5; consul-general of Borneo, 1875; governor of Labuan, 1875–9; governor of Gold Coast colony, 1879–80. (*Colonial Office list* 1880, *Modern English biography*.)

Veitch, James (1792–1863). Nurseryman. Proprietor of a nursery at Mount Radford, near Exeter; in partnership with his son, James Veitch (1815–69), proprietor of a nursery in Chelsea, London, from 1853. (R. Desmond 1994.)

Verney, Harry, second baronet (1801–94). Born Harry Calvert. Politician. Served with the thirty-first foot, 1819–20; with the seventh fusiliers, and the seventy-second and fifty-second foot regiments. Adjutant, Grenadier Guards. Returned to private studies in 1822. Succeeded to the baronetcy in 1826. Assumed the surname Verney in 1827 on inheriting the estates of a cousin. Travelled in South America until 1829. Studied at Downing College, Cambridge, 1831–3. MP for Buckingham from 1832. (*ODNB*.)

Vignoles, Charles Blacker (1793–1875). Civil engineer. Served in the army until 1817. City surveyor, St Augustine, Florida, 1821–3. In 1823, returned to England, where he worked as a surveyor, especially for the railways. Also worked in Ireland, France, Germany, Switzerland, Brazil, and Spain. Built the Kiev bridge in

Ukraine, 1847–53. A member of the eclipse expedition to Sicily, 1870. Fellow of the Royal Astronomical Society, 1829. FRS 1855. (*ODNB*.)

Vignoles, Hutton (1824–89). Civil engineer. Assistant to his father, Charles Blacker Vignoles. Accompanied him on the eclipse expedition to Sicily, 1870. (Vignoles 1982.)

Vivian, Henry Hussey, first Baron Swansea (1821–94). Industrialist and politician. Entered Trinity College, Cambridge, 1840. Became a partner in the family copper-smelting business, 1842. MP for Truro, 1852–7; Glamorgan, 1857–85; Swansea district, 1885–93. Fellow of the Geological Society. Created baronet 1882; became Baron Swansea, 1893. (*DNB*.)

[April or May 1870?], *10 May 1870*, 11 May [1870]

Vogel, Hermann Carl (1841–1907). German astrophysicist. Studied at Leipzig. Director of the observatory of F. G. von Bülow at Bothkamp, near Kiel, 1870. A member of the British eclipse expedition to Sicily, 1870. Worked on spectroscopic analysis of the stars. Observer at the astrophysical observatory at Potsdam, 1874; professor, 1879; director, 1882. (*DSB*, Eder *et al.* 1945, p. 463.)

Vogt, Carl (1817–95). German naturalist. After receiving a doctorate in Giessen, 1839, moved to Switzerland and worked with Louis Agassiz on a treatise on fossil and freshwater fish until 1846. Professor of zoology, Giessen, 1846. Forced to leave the German Federation for political reasons in 1849; settled in Geneva. Professor of geology, Geneva, 1852; director of the Institute of Zoology, 1872. (*ADB*, *DSB*, Judel 2004.)

Wagner, Moritz (1813–87). German zoologist and explorer. Studied at the natural history colleges in Erlangen and Munich, 1834–5. Made a scientific expedition to Algeria, 1836. Editor of the *Augsburger Allgemeine Zeitung*, 1838. Continued his studies in Göttingen, 1840. Made research trips to the Caucasus, Armenia, Persia, and South Asia, 1843–5; to North and Central America, 1852–5; to Panama and Ecuador, 1857–9. Professor at Munich University from 1862. Formulated the migration theory for fauna and flora in 1868. (*DBE*, Tort 1996.)

Wagner, Rudolph (1805–64). German anatomist, physiologist, and anthropologist. Professor of physiology, comparative anatomy, and natural history at Göttingen University, 1840. (*ADB*, *DSB*.)

Walker, Thomas. Scottish newspaper editor. Editor of the Fife Sentinel, 1843–5. Secretary of the Galashiels geological society. (North 1989; 'The Galashiels geological society', *The Scotsman*, 21 March 1840, p. 3.)

30 March 1840 (from Charles Maclaren), 5 May [1840]

Wallace, Alfred Russel (1823–1913). Naturalist. Collector in the Amazon, 1848–52; in the Malay Archipelago, 1854–62. Independently formulated a theory of evolution by natural selection in 1858. Lecturer and author of works on protective coloration, mimicry, and zoogeography. President of the Land Nationalisation Society, 1881–1913. Wrote on socialism, spiritualism, and vaccination. FRS 1893. (*DSB*, *ODNB*.)

22 January 1870, 26 January [1870], 7 March [1870?], 31 March [1870], 20 April

Wallace, Alfred Russel, cont.

[1870], 5 [July 1870], *6 July 1870*, 22 November [1870], *24 November 1870*

Wallace, Annie (1848–1914). Eldest daughter of William Mitten. Married Alfred Russel Wallace in 1866. (Marchant ed. 1916, 2: 252–3.)

Waterhouse, George Robert (1810–88). Naturalist. A founder of the Entomological Society of London, 1833. Curator of the Zoological Society of London, 1836–43. Assistant in the mineralogical branch of the natural history department of the British Museum, 1843–50; keeper, mineralogical and geological branch, 1851–6; keeper, geology department, 1857–80. Described CD's mammalian and entomological specimens from the *Beagle* voyage. (*DNB*.)

Watson, Hewett Cottrell (1804–81). Botanist, phytogeographer, and phrenologist. Edited the *Phrenological Journal*, 1837–40. Collected plants in the Azores in 1842. Wrote extensively on the geographical distribution of plants. (*DSB*, *ODNB*.)

Weale, James Philip Mansel (1838–1911). Naturalist, farmer, and writer. BA, Oxford, 1860. Resided in South Africa, approximately 1860–78. Supported the formation of a museum in Port Elizabeth. Farmed property (Brooklyn) near King William's Town. CD communicated his papers on orchid pollination to the Linnean Society of London; his papers on South African insects appeared in the *Transactions of the Entomological Society*. (Gunn and Codd 1981; *BMD* (*Death index*); *Royal Society catalogue of scientific papers*; *The Times*, 26 September 1899, p. 9; Weale 1877; Weale 1878; Weale 1891.)

[25 May 1870], 30 July 1870

Wedgwood, Alfred Allen (1842–92). Son of Hensleigh and Frances Emma Elizabeth Wedgwood. (R. B. Freeman 1978.)

Wedgwood, Caroline Sarah (1800–88). CD's sister. Married Josiah Wedgwood III, her cousin, in 1837. (*Darwin pedigree*.)

Wedgwood, Elizabeth (Bessy) (1764–1846). Eldest daughter of John Bartlett Allen. Married Josiah Wedgwood II in 1792. (*Emma Darwin* (1915).)

Wedgwood, Frances Julia (Snow) (1833–1913). Novelist, biographer, historian, and literary critic. Daughter of Hensleigh and Frances Emma Elizabeth Wedgwood. Published two novels in her mid-twenties, one under the pseudonym Florence Dawson. Wrote book reviews and an article on the theological significance of *Origin*. Conducted an intense friendship with Robert Browning between 1863 and 1870. Published a study of John Wesley (1870), and helped CD with translations of Linnaeus in the 1870s. Published *The moral ideal: a historical study* (1888). Active in the anti-vivisection movement. (B. Wedgwood and Wedgwood 1980, *ODNB*.)

11 July [1861]

Wedgwood, Hensleigh (1803–91). Philologist. Emma Darwin's brother. Qualified as a barrister in 1828, but never practised. Fellow, Christ's College, Cambridge, 1829–30. Police magistrate at Lambeth, 1831–7; registrar of metropolitan carriages, 1838–49. An original member of the Philological Society, 1842. Published

A dictionary of English etymology (1859–65). Married Frances Emma Elizabeth Mackintosh in 1832. (R. B. Freeman 1978, *ODNB*.)

[1868–70?], [March 1870]

Wedgwood, Josiah I (1730–95). Master-potter. Founded the Wedgwood pottery works at Etruria, Staffordshire. Grandfather of CD and Emma Darwin. Greatly interested in experimental chemistry. Contributed several papers on the measurement of high temperatures to the Royal Society of London's *Philosophical Transactions*. Actively associated with scientists and scientific societies. FRS 1783. (*DSB*, *ODNB*.)

Wedgwood, Josiah II (1769–1843). Master-potter. Partner in the Wedgwood pottery works at Etruria, Staffordshire, 1795–1841. Resident at Maer Hall, Staffordshire. Whig MP for Stoke-on-Trent, 1832–4. Emma Darwin's father. (*Burke's peerage* 1980, *Emma Darwin* (1915), R. B. Freeman 1978.)

Wedgwood, Josiah III (1795–1880). Master-potter. Partner in the Wedgwood pottery works at Etruria, Staffordshire, 1841–4; moved to Leith Hill Place, Surrey, in 1844. Emma Darwin's brother. Married CD's sister Caroline, his cousin, in 1837. (R. B. Freeman 1978.)

31 July 1844, 14 August [1847], [22 August? 1847]

Wedgwood, Katherine Elizabeth Sophy (Sophy) (1842–1911). Daughter of Caroline Sarah Wedgwood and Josiah Wedgwood III. CD's niece. (*Darwin pedigree*, R. B. Freeman 1978.)

Wedgwood, Lucy Caroline (1846–1919). Daughter of Caroline Wedgwood and Josiah Wedgwood III. CD's niece. Married Matthew James Harrison in 1874. (*Darwin pedigree*, R. B. Freeman 1978, B. Wedgwood and Wedgwood 1980.)

[25 April 1870], [5 May 1870]

Wedgwood, Margaret Susan. *See* Williams, Margaret Susan Vaughan.

Wedgwood, Sarah Elizabeth (Elizabeth) (1793–1880). Emma Darwin's sister. Resided at Maer Hall, Staffordshire, until 1847, then at The Ridge, Hartfield, Sussex, until 1862. Moved to London before settling in Down in 1868. (*Emma Darwin* (1915), R. B. Freeman 1978.)

Weierstrass, Karl Theodor Wilhelm (1815–97). German mathematician. Taught mathematics and other subjects, 1841–55. Professor of the Industry Institute, Berlin, 1856–64. Associate professor, University of Berlin, 1856–64; professor, 1864–97. (*DSB*.)

Weir, Harrison William (1824–1906). Painter and illustrator. Specialised in landscape and natural history subjects. Brother of John Jenner Weir. (*ODNB*.)

Weir, John Jenner (1822–94). Naturalist and accountant. Worked in HM customs as an accountant, 1839–85. Studied entomology, especially microlepidoptera; conducted experiments on the relations between insects and insectivorous birds and published papers in 1869 and 1870. Member of the Entomological Society of London from 1845, serving many times on the council. Fellow of the Linnean Society of London, 1865; Zoological Society, 1876. (*Science Gossip* n.s. 1 (1894): 49–50.)

Weir, John Jenner, cont.

17 March 1870, 17 March [1870], 14 June [1870], *27 June 1870*, 29 June [1870], *4 July 1870*

Wells, David Ames (1828–98). American economist. Appointed head of a national revenue commission, 1865. Special commissioner of the revenue, 1866–70. (*ANB*.)

Wells, William Charles (1757–1817). American-born physician. Born in South Carolina, of Scottish parents, and educated in Scotland, 1768–71. Elected assistant physician to St Thomas's Hospital, London, 1795; physician, 1800–17. Awarded the Rumford Medal of the Royal Society for his 'Essay on dew' (1814). Wrote *An account of a female of the white race of mankind, part of whose skin resembles that of a negro* (1818), which has been thought to anticipate CD's theory. FRS 1793. (*ODNB*.)

Welsh, Thomas. Scottish labourer. Galashiels carter. (*Correspondence* vol. 2, letter from William Kemp, 4 December 1843.)

Whewell, William (1794–1866). Mathematician, historian, and philosopher of science. Knightbridge Professor of moral philosophy, Cambridge University, 1838–55. Master of Trinity College, Cambridge, 1841–66. FRS 1820. (*DSB*, *ODNB*.)

White, Gilbert (1720–93). Naturalist and clergyman. Author of *The natural history and antiquities of Selborne* (1789). (*ODNB*.)

Wicksted, Charles (1796–1870). Landowner and breeder. Son of George Tollet and childhood friend of CD's. Took the name Wicksted in 1814. High-sheriff of Cheshire, 1822. Inherited Betley Hall, Staffordshire, in 1855; continued living at Shakenhurst, Worcestershire, the home inherited by his wife to which he had moved *circa* 1844. A noted fox-hunter and breeder of hounds. (Blagg 1902, *Burke's landed gentry* 1875, R. B. Freeman 1978, *Post Office directory of Birmingham* 1864.)

13 March [1870?] (to Georgina Tollet)

Wilberforce, Samuel (1805–73). Clergyman. Rector of Brighstone, Isle of Wight, 1830–40; of Alverstoke, Hampshire, 1840–3. Chaplain to Prince Albert, 1841. Dean of Westminster, 1845. Bishop of Oxford, 1845–69. Bishop of Winchester, 1869–73. FRS 1845. (*ODNB*, *Record of the Royal Society of London*.)

Wilkinson, John Gardner (1797–1875). Explorer and Egyptologist. FRS 1834. Knighted, 1839. (*DNB*.)

Williams, Arthur Charles Vaughan (1834–75). Clergyman. Student at Christ Church, Oxford, 1853–68; BA 1857. Vicar of Down Ampney, Wiltshire, 1868–75. (*Alum. Oxon.*, *ODNB* s.v. Williams, Ralph Vaughan.)

Williams, Edward Vaughan (1797–1875). Lawyer. Attended Trinity College, Cambridge; BA 1820. Entered Lincoln's Inn, 1818; called to the bar, 1823. Created serjeant-at-law and justice of the court of common pleas, 1846. Knighted, 1847. (*ODNB*.)

Williams, Hervey Wedgwood (1869–1944). Son of Margaret Susan Wedgwood and Arthur Vaughan Williams. (Birth certificate, General Register Office, Dorking; Milo Keynes 1998, p. 255.)

Williams, Margaret Susan Vaughan (1843–1937). Daughter of Caroline Wedgwood and Josiah Wedgwood III. CD's niece. Married Arthur Charles Vaughan Williams in 1869. Mother of Ralph Vaughan Williams (*ODNB*). (*Emma Darwin* (1915), R. B. Freeman 1978.)

Williams & Norgate. Booksellers and publishers specialising in foreign scientific literature, with premises at Covent Garden, London, and South Frederick Street, Edinburgh. A partnership between Edmund Sydney Williams (1817–91) and Frederick Norgate. Publishers of the *Natural History Review.* (*Modern English biography* s.v. Williams, Edmund Sidney, *Post Office Edinburgh directory.*)
 10 April [1860]

Wood, Thomas W. (*fl.* 1865–80). Illustrator. Son of animal artist Thomas W. Wood. Made some animal illustrations for *Expression.* (Engen 1985, Tort 1996.)
 14 June 1870, 16 June 1870, [after 26 June 1870]

Woodd, Basil George (1781–1872). Of Hillfield House, Hampstead. Father of Ellen Sophia, William Darwin Fox's second wife. (*Burke's landed gentry* 1952, *Post Office London suburban directory* 1868.)

Woolner, Thomas (1825–92). Sculptor and poet. Member of the Pre-Raphaelite brotherhood. Established his reputation in the 1850s with medallion portrait sculptures of Robert Browning, Thomas Carlyle, and William Wordsworth. Went on to make acclaimed busts of CD, Charles Dickens, Thomas Henry Huxley, Adam Sedgwick, and Alfred Tennyson, and life-size studies of Francis Bacon, John Stuart Mill, and William Whewell. (*DNB.*)
 10 March [1870]

Wright, Chauncey (1830–75). American mathematician and philosopher. Calculator for the newly established *American ephemeris and nautical almanac*, for which he devised new methods of calculation, 1852–72. Recording secretary, American Academy of Arts and Sciences, Boston, 1863–70. Published the first of a series of philosophical essays in the *North American Review* in 1864. (*ANB.*)

Wyman, Jeffries (1814–74). American comparative anatomist and ethnologist. Curator of the Lowell Institute, Boston, 1839–42. Travelled in Europe, 1841–2. Professor of anatomy and physiology, Hampden-Sydney College, Virginia, 1843–8. Hersey Professor of anatomy, Harvard College, 1847–74. Curator, Peabody Museum of Archaeology and Ethnology, Harvard, 1866–74. (*ANB, DSB.*)

Ykema (IJkema), Johan (*fl.* 1870s). Dutch bookseller, editor, and publisher. Publisher of Dutch translation of *Descent* (Hartogh Heijs van Zouteveen, trans. 1871–2). (Letter from Hermanus Hartogh Heijs van Zouteveen, 27 November 1870.)

Yorke, Henry Galgacus Redhead (d. 1848). Politician. Attended Christ's College, Cambridge, from 1825. MP for York, 1841–8. (*Alum. Cantab.*)

Youmans, Edward Livingston (1821–87). American writer, editor, and promoter of science education. Took a degree in medicine, *circa* 1852, but did not practise.

Youmans, Edward Livingston, cont.

Wrote and lectured on scientific subjects. Advised the Appleton company on scientific publications and translations; founded their International Scientific Series in 1871. Founded the *Popular Science Monthly*, 1872. (*ANB*.)

25 September 1870, 25 September 1870

Zanichelli, Nicola (1819–84). Italian bookseller and publisher based in Modena and Bologna. Published the first Italian edition of *Origin*. (*EI*; R. B. Freeman 1977, p. 105.)

Zollinger, Heinrich (1818–59). Swiss naturalist and traveller. Studied botany and geology at Geneva, 1837. Collected plants on several expeditions throughout the Dutch East Indies, 1841–6. Returned to Switzerland, 1847; director of a seminary at Küsnacht, 1849. Returned to Java, 1855. (*ADB*.)

BIBLIOGRAPHY

The following bibliography contains all the books and papers referred to in this volume by author–date reference or by short title. Short titles are used for some standard reference works (e.g., *ODNB*, *OED*), for CD's books and papers, and for editions of his letters and manuscripts (e.g., *Descent*, *LL*, *Notebooks*). Works referred to by short titles are listed in alphabetical order according to the author's surname. Notes on manuscript sources are given at the end of the bibliography.

Adam, William. 1865. Consanguinity in marriage. *Fortnightly Review* 2: 710–30, 3: 74–88.

ADB: *Allgemeine deutsche Biographie.* Under the auspices of the Historical Commission of the Royal Academy of Sciences. 56 vols. Leipzig: Duncker & Humblot. 1875–1912.

Agassiz, Alexander. 1864. *Embryology of the starfish.* From J. L. R. Agassiz's *Contributions to the Natural History of the United States* vol. 5, published 1877. Cambridge, Mass.: n.p.

Agassiz, Alexander, *et al.* 1869. Contributions to the fauna of the Gulf Stream at great depths. *Bulletin of the Museum of Comparative Zoology* 3d ser. (1869): 253–386.

Agassiz, George Russell, ed. 1913. *Letters and recollections of Alexander Agassiz, with a sketch of his life and work.* London: Constable & Co.

Agassiz, Jean Louis Rodolphe and Agassiz, Elizabeth. 1868. *A journey in Brazil.* Boston: Ticknor and Fields.

Allaby, Michael, ed. 1999. *A dictionary of zoology.* 2d edition. Oxford and New York: Oxford University Press.

Allan, Mea. 1967. *The Hookers of Kew, 1785–1911.* London: Michael Joseph.

Alter, Stephen. 1999. *Darwinism and the linguistic image: language, race, and natural theology in the nineteenth century.* Baltimore: The Johns Hopkins University Press.

Alum. Cantab.: *Alumni Cantabrigienses. A biographical list of all known students, graduates and holders of office at the University of Cambridge, from the earliest times to 1900.* Compiled by John Venn and J. A. Venn. 10 vols. Cambridge: Cambridge University Press. 1922–54.

Alum. Oxon.: *Alumni Oxonienses: the members of the University of Oxford, 1500–1886: … with a record of their degrees. Being the matriculation register of the university.* Alphabetically arranged, revised, and annotated by Joseph Foster. 8 vols. London and Oxford: Parker & Co. 1887–91.

ANB: *American national biography.* Edited by John A. Garraty and Mark C. Carnes. 24 vols. and supplement. New York and Oxford: Oxford University Press. 1999–2002.

Anderson, Nancy Fix. 1986. Cousin marriage in Victorian England. *Journal of Family History* 11: 285–301.

Andersson, Charles John. 1856. *Lake Ngami: or, explorations and discoveries during four years' wanderings in the wilds of South Western Africa.* London: Hurst and Blackett.

Annual register: *The annual register. A view of the history and politics of the year.* 1838–62. *The annual register. A review of public events at home and abroad.* N.s. 1863–1946. London: Longman & Co. [and others].

Appiah, Kwame Anthony and Gates, Henry Louis, Jr., eds. 2005. *Africana: the encyclopedia of the African and African American experience.* 2d edition. 5 vols. Oxford and New York: Oxford University Press.

Arkell, W. J. 1950. A classification of the Jurassic ammonites. *Journal of Paleontology* 24: 354–64.

Armstrong, Patrick. 1991. *Under the blue vault of heaven: a study of Charles Darwin's sojourn in the Cocos (Keeling) Islands.* Nedlands, Western Australia: Indian Ocean Centre for Peace Studies.

Ashton, Rosemary. 1991. *G. H. Lewes: a life.* Oxford: Clarendon Press.

Audubon, John James. 1831–[9]. *Ornithological biography, or an account of the habits of the birds of the United States of America accompanied by descriptions of the objects represented in the work entitled 'The birds of America' and interspersed with delineations of American scenery and manners.* 5 vols. Edinburgh: Adam Black; Adam & Charles Black.

——. 1840–4. *The birds of America; from drawings made in the United States and their territories.* 7 vols. New York: J.J. Audubon, and Philadelphia: J.B. Chevalier.

Aust. dict. biog.: *Australian dictionary of biography.* Edited by Douglas Pike *et al.* 14 vols. [Melbourne]: Melbourne University Press. London and New York: Cambridge University Press. 1966–96.

Autobiography: *The autobiography of Charles Darwin 1809–1882. With original omissions restored.* Edited with appendix and notes by Nora Barlow. London: Collins. 1958.

Axell, Severin. 1869. *Om anordningarna för de fanerogama växternas befruktning.* Stockholm: H Eggstrøm.

Azara, Félix d'. 1801. *Essais sur l'histoire naturelle des quadrupèdes de la province du Paraguay.* 2 vols. Paris: Madame Huzard.

——. 1802–5. *Apuntamientos para la historia natural de los páxaros del Paraguay y Rio de la Plata.* 3 vols. Madrid: Impr. de la viuda de Ibarra.

Babington, Charles Cardale. 1843. *Manual of British botany, containing the flowering plants and ferns arranged according to the natural orders.* London: John Van Voorst.

Bain, Alexander. 1868. *Mental and moral science: a compendium of psychology and ethics.* London: Longmans, Green, and Co.

Barbieri, Alberto. 1973. *Modenesi da ricordare.* Modena, Italy: Società Tipografica Editrice Modenese, Mucchi.

Barkly, Henry. 1870. Notes on the flora and fauna of Round Island. In *The Mauritius almanac and colonial register for 1870.* Edited by John B. Kyshe. Mauritius: [n.p.].

Barnhart, John Hendley, comp. 1965. *Biographical notes upon botanists … maintained in the New York Botanical Garden Library.* 3 vols. Boston, Mass.: G. K. Hall.

Barrande, Joachim. 1861–81. *Défense des colonies.* 5 vols. Prague and Paris: the author.
——. 1870. *Distribution des Céphalopodes dans les contrées siluriennes.* Prague and Paris: the author. [Extract from Barrande 1852–1911.]
Barrington, Daines. 1773. Experiments and observations on the singing of birds. [Read 22 April and 6 and 13 May 1773.] *Philosophical Transactions of the Royal Society* 63: 249–91.
Bassett, Thomas J. and Porter, Philip W. 1991. 'From the best authorities': the mountains of Kong in the cartography of West Africa. *Journal of African History* 32 (1991): 367–413.
Bastian, Henry Charlton. 1870. Facts and reasonings concerning the heterogenous evolution of living things. *Nature* 2: 170–7, 193–201, 219–28.
Bates, Henry Walter. 1861. Contributions to an insect fauna of the Amazon valley. *Lepidoptera*: *Heliconidæ.* [Read 21 November 1861.] *Transactions of the Linnean Society of London* 23 (1860–2): 495–566.
BDWS: *Biographical dictionary of women in science: pioneering lives from ancient times to the mid-20th century.* Edited by Marilyn Ogilvie and Joy Harvey. 2 vols. New York and London: Routledge. 2000.
Bean, William Jackson. 1970–88. *Trees and shrubs hardy in the British Isles.* 8th edition, fully revised by D. L. Clarke and George Taylor. 4 vols. and supplement. London: John Murray.
Bell, Charles. 1844. *The anatomy and philosophy of expression as connected with the fine arts.* Preface by George Bell, and an appendix on the nervous system by Alexander Shaw. 3d edition, enlarged. London: John Murray.
Beneden, Edouard van. 1868. Le genre dactycotyle, son organisation, et quelques remarques sur la formation de l'oeuf des trématodes. *Bulletins de l'Académie Royale de Belgique* 2d ser. 25: 22–37.
——. 1869a. Recherches sur l'embryogénie des crustacés: I. Observations sur le développement de l'Asellus aquaticus. *Bulletins de l'Académie Royale de Belgique* 2d ser 28: 54–87.
——. 1869b. Recherches sur l'embryogénie des crustacés: II. Développement de Mysis. *Bulletins de l'Académie Royale de Belgique* 2d ser 28: 232–49.
——. 1870a. Recherches sur l'embryogénie des crustacés: III. Développement de l'oeuf et de l'embryon des sacculines (*Sacculina carcini*, Thomps.). *Bulletins de l'Académie Royale de Belgique* 2d ser. 29: 99–112.
——. 1870b. Recherches sur l'embryogénie des crustacés: IV. Développement des genres Anchorella, Lerneopoda, Brachiella et Hessia. *Bulletins de l'Académie Royale de Belgique* 2d ser. 29: 223–54.
——. 1870c. On a new species of Gregarina to be called Gregarina gigantea. *Quarterly Journal of Microscopical Science* n.s. 10: 51–8.
——. 1870d. Etude zoologique et anatomique du genre *Macrostomum* comprenant la description de deux espèces nouvelles. *Bulletins de l'Académie Royale de Belgique* 2d ser. 30: 116–33.
Beneden, Edouard van and Bessels, E. 1868. Résumé d'un mémoire sur le mode

de formation du blastoderme dans quelques groupes de crustacés. *Bulletins de l'Académie Royale de Belgique* 2d ser. 25: 434–48.

Bennett, Alfred William. 1870. The theory of natural selection from a mathematical point of view. *Nature* 3: 30–3.

Bentham, George. 1858. *Handbook of the British flora; a description of the flowering plants and ferns indigenous to, or naturalized in, the British Isles.* London: Lovell Reeve.

——. 1870. Anniversary address. [Read 24 May 1870.] *Transactions of the Linnean Society of London* 27: lxxiv–xciv.

Bentham, George and Hooker, Joseph Dalton. 1862–83. *Genera plantarum. Ad exemplaria imprimis in herbariis Kewensibus servata definita.* 3 vols. in 7. London: A. Black [and others].

Berec, Ludek, *et al.* 2005. Sex determination in *Bonellia viridis* (Echiura: Bonelliidae): population dynamics and evolution. *Oikos: a Journal of Ecology* 108: 473–84.

Bergquist, Patricia R. 1978. *Sponges.* London: Hutchinson.

BHGW: *Biographisch-literarisches Handwörterbuch zur Geschichte der exacten Wissenschaften enthaltend Nachweisung über Lebensverhältnisse und Leistungen von Mathematikern, Astronomen, Physikern, Chemikern, Mineralogen, Geologen usw.* By Johann Christian Poggendorff. 5 vols. Leipzig: Johann Ambrosius Barth; Verlag Chemie. 1863–1926.

Birds: Pt 3 of *The zoology of the voyage of H.M.S. Beagle.* By John Gould. Edited and superintended by Charles Darwin. London: Smith, Elder and Co. 1838–41.

Bischoff, Theodor Ludwig Wilhelm. 1845. *Entwicklungsgeschichte des Hunde-Eies.* Brunswick: F. Vieweg und sohn.

BLA: *Biographisches Lexikon der hervorragenden Aerzte aller Zeiten und Völker.* Edited by A. Wernich *et al.* 6 vols. Vienna and Leipzig: Urban and Schwarzenberg. 1884–8.

Blackburn, Helen. 1902. *Women's suffrage: a record of the Women's Suffrage Movement in the British Isles with biographical sketches of Miss Becker.* London: Williams & Norgate.

Blagg, C. J. 1902. *A history of the North Staffordshire hounds and country 1825 to 1902.* London: Sampson Low, Marston & Company.

Blancpain, Jean-Pierre. 1990. *Les Araucans et la frontière dans l'histoire du Chili des origines au XIX°. siècle. Une épopée américaine.* Frankfurt am Main: Vervuert Verlag.

BLC: *The British Library general catalogues of printed books to 1975.* 360 vols. and supplement (6 vols.). London: Clive Bingley; K. G. Saur. 1979–88.

BMD: General Register Office, *England and Wales civil registration indexes. England & Wales birth index, 1837–1983. England and Wales marriage index, 1837–1983. England and Wales death index, 1837–1983.* Online database. Provo, Utah: The Generations Network. 2006. http://www.ancestry.co.uk

BNB: *Biographie nationale publiée par l'Académie Royale des Sciences, des Lettres et des Beaux-arts de Belgique.* 44 vols., including 16 supplements. Brussels: H. Thiry-Van Buggenhoudt [and others]. 1866–1986.

Bol'shaya entsiklopediya: *Bol'shaya entsiklopediya v shestidesyati dvukh tomakh.* Main editor: S.A. Kondratov. Scientific editorial advisor: G. A. Mesiyats. 62 vols. Moscow: Terra. 2006.

Bompas, George C. 1885. Life of Frank Buckland. London: Smith, Elder & Co.

Boner, Charles. 1860. *Chamois hunting in the mountains of Bavaria and the Tyrol.* New edition. London: Chapman and Hall.

——. 1861. *Forest creatures.* London: Longman, Green, Longman, and Roberts.

——. 1865. *Transylvania: its products and its people.* London: Longman, Green, Reader and Dyer.

Bonham-Carter, Victor. 1960. *In a liberal tradition: a social biography 1700–1950.* London: Constable.

Bosl, Karl, ed. 1983. *Bosl's Bayerische Biographie.* Regensburg: Friedrich Pustet.

Bostock, Edward Henry. 1927. *Menageries, circuses and theatres.* London: Chapman and Hall.

Bowlby, John. 1990. *Charles Darwin: a biography.* London: Hutchinson.

Branson, Jan and Miller, Don. 2002. *Damned for their difference: the cultural construction of deaf people as 'disabled'.* Washington DC: Gallaudet.

Brehm, Alfred Edmund, *et al.* 1864–9. *Illustrirtes Thierleben. Eine allgemeine Kunde des Thierreichs, etc.* 6 vols. Hildburghausen: Bibliographisches Institut.

Bridges, Esteban Lucas. 1948. *Uttermost part of the earth.* London: Hodder and Stoughton.

Bright, John Shenton. 1884. *A history of Dorking and the neighbouring parishes: with chapters on the literary associations, flora, fauna, geology, etc., of the district.* Dorking, London: R. J. Clark, Simpkin, Marshall.

Britten, Emma Hardinge. 1870. *Modern American spiritualism: a twenty years' record of the communion between earth and the world of spirits.* 2d edition. New York: the author.

Broca, Pierre Paul. 1870a. Sur le transformisme. *Revue des cours scientifiques de la France et de l'étranger* 7: 530–41, 550–8.

——. 1870b. Sur le transformisme. *Bulletins de la Société d'Anthropologie de Paris* 2d ser. 5: 168–239.

Bromfield, William Arnold. 1856. *Flora Vectensis: being a systematic description of the phænogamous or flowering plants and ferns indigenous to the Isle of Wight.* Edited by William Jackson Hooker and Thomas Bell Salter. London: William Pamplin.

Bronn, Heinrich Georg, trans. 1860. *Charles Darwin, über die Entstehung der Arten im Thier- und Pflanzen-Reich durch natürliche Züchtung, oder Erhaltung der vervollkommneten Rassen im Kampfe um's Daseyn.* Stuttgart: E. Schweizerbart.

——, trans. 1862. *Charles Darwin, über die Einrichtungen zur Befruchtung Britischer und ausländischer Orchideen durch Insekten und über die günstigen Erfolge der Wechselbefruchtung.* Stuttgart: E. Schweizerbart.

Bronn, Heinrich Georg and Carus, Julius Victor, trans. 1867. *Über die Entstehung der Arten durch natürliche Zuchtwahl oder die Erhaltung der begünstigten Rassen im Kampfe um's Dasein.* 3d edition. Translated by Heinrich Georg Bronn. Revised and corrected from the fourth English edition by Julius Victor Carus. Stuttgart: E. Schweizerbart'sche Verlagshandlung und Druckerei.

——, trans. 1870. *Charles Darwin über die Entstehung der Arten durch natürliche Zuchtwahl; oder, Die Erhaltung der begünstigten Rassen im Kampfe um's Dasein.* 4th edition. Stuttgart: E. Schweizerbart.

Brown, Robert. 1831. On the organs and mode of fecundation in Orchideæ and Asclepiadeæ. [Read 1 and 15 November 1831.] *Transactions of the Linnean Society of London* 16 (1833): 685–745.

Browne, Janet. 2002. *Charles Darwin. The power of place. Volume II of a biography.* London: Pimlico.

Büchner, Friedrich Karl Christian Ludwig. 1868. *Sechs Vorlesungen über die Darwin'sche Theorie der Verwandlung der Arten und die erste Entstehung der Organismenwelt.* Leipzig: Theodor Thomas.

Bucknill, John Charles. 1859. *The psychology of Shakespeare.* London: Longman, Brown, Green, Longmans & Roberts.

Bulmer, Michael. 2003. *Francis Galton: pioneer of heredity and biometry.* Baltimore: The Johns Hopkins University Press.

Burke's landed gentry: *A genealogical and heraldic history of the commoners of Great Britain and Ireland enjoying territorial possessions or high official rank but unvisited with heritable honours. Burke's genealogical and heraldic history of the landed gentry.* By John Burke, *et al.* 1st–18th edition. London: Henry Colburn [and others]. 1833–1969.

Burke's peerage: *A genealogical and heraldic dictionary of the peerage and baronetage of the United Kingdom. Burke's peerage and baronetage.* 1st– edition. London: Henry Colburn [& others]. 1826–.

Burton, Richard Francis. 1861. *The city of the saints, and across the Rocky Mountains to California.* London: Longman, Green, Longman, and Roberts.

BWN: *Biographisch woordenboek der Nederlanden, bevattende levensbeschrijvingen van zoodanige personen, die zich op eenigerlei wijze in ons vaderland hebben vermaard gemaakt.* Edited by K. J. R. van Harderwijk and G. D. J. Schotel. 21 vols. in 17. Haarlem: J. J. Van Brederode. 1852–78.

Calendar: *A calendar of the correspondence of Charles Darwin, 1821–1882. With supplement.* 2d edition. Edited by Frederick Burkhardt *et al.* Cambridge: Cambridge University Press. 1994.

Cambridge University calendar: *The Cambridge University calendar.* Cambridge: W. Page [& others]. 1796–1950.

Camões, Luiz Vaz de. 2001. *The Lusíads.* Translated by Landeg White. Oxford: Oxford University Press.

Candolle, Augustin Pyramus de and Candolle, Alphonse de. 1824–73. *Prodromus systematis naturalis regni vegetabilis, sive enumeratio contracta ordinum generum specierumque plantarum huc usque cognitarum, juxta methodi naturalis normas digesta.* 17 vols. Paris: Treuttel & Würtz [& others].

Canestrini, Giovanni. 1870. *Origine dell'uomo.* 2d edition. Milan: Gaetano Brigola.

——, trans. 1876. *Variazione degli animali e delle piante allo stato domestico.* By Charles Darwin. Translation of the English 2d edition. Turin: Unione Tipografico-Editrice.

Canestrini, Giovanni and Salimbeni, Leonardo, trans. 1864. *Carlo Darwin. Sull'origine delle specie per elezione naturale, ovvero conservazione delle razze perfezionate nella lotta per l'esistenza.* Modena, Italy: Nicola Zanichelli.

Caroe, Gwendolen Mary. 1985. *The Royal Institution: an informal history.* With a final chapter by Alban Caroe. London: John Murray.

Carpenter, William Benjamin. 1864. Additional note on the structure and affinities of Eozoön Canadense. [Read 23 November 1864.] *Quarterly Journal of the Geological Society of London* 21 (1865): 59–66.

——. 1866. Supplemental notes on the structure and affinities of Eozoon Canadense. [Read 10 January 1866.] *Quarterly Journal of the Geological Society of London* 22: 219–28.

Carpenter, William Benjamin and Thomson, Charles Wyville. 1868. Preliminary report of dredging operations in the seas to the north of the British islands, carried on in Her Majesty's steam-vessel 'Lightning'. [Read 17 December 1868.] *Proceedings of the Royal Society of London* 17 (1868–9): 168–200.

Carus, Julius Victor, trans. 1868. *Das Variiren der Thiere und Pflanzen im Zustande der Domestication.* By Charles Darwin. 2 vols. Stuttgart: E. Schweizerbart'sche Verlagshandlung (E. Koch).

——, trans. 1871. *Die Abstammung des Menschen und die geschlechtliche Zuchtwahl.* By Charles Darwin. 2 vols. Stuttgart: E. Schweizerbart'sche Verlagshandlung.

Catalogue of the library of the British Museum (Natural History): *Catalogue of the books, manuscripts, maps and drawings in the British Museum (Natural History).* 5 vols. London: Trustees of the British Museum. 1903–15.

Caton, John Dean. 1868. *American Cervus.* Read before the Ottawa Academy of Natural Sciences, 21 May 1868. Ottawa, Illinois: Osman and Hapeman.

——. 1877a. *The antelope and deer of America: a comprehensive scientific treatise upon the natural history, including the characteristics, habits, affinities, and capacity for domestication of the Antilocapra and Cervidae of North America.* New York: Hurd and Houghton. Boston: H. O. Houghton and Company.

——. 1877b. The wild turkey and its domestication. *American Naturalist* 11: 321–30.

——. 1880. *Miscellanies.* Boston: Houghton, Osgood.

CDEL: *A critical dictionary of English literature, and British and American authors, living and deceased, from the earliest accounts to the middle of the nineteenth century … with forty indexes of subjects.* By S. Austin Allibone. 3 vols. London: Trübner. Philadelphia: Childs & Peterson; J. B. Lippincott. 1859–71. *A supplement to Allibone's critical dictionary of English literature and British and American authors. Containing over thirty-seven thousand articles (authors), and enumerating over ninety-three thousand titles.* By John Foster Kirk. 2 vols. Philadelphia and London: J. B. Lippincott. 1891.

[Centlivre, Susanna.] 1718. *A bold stroke for a wife: a comedy … by the author of the Busie-body.* London: W. Mears, J. Browne and F. Clay.

Challenger, S. 1979. Pioneer nurserymen of Canterbury, New Zealand (1850–65). *Garden History* 7: 25–64.

Challinor, John. 1978. *A dictionary of geology.* 5th edition. Cardiff: University of Wales Press.

Chambers: *The Chambers dictionary.* Edinburgh: Chambers Harrap Publishers. 1998.

Child, Gilbert William. 1862. On marriages of consanguinity. *British and Foreign Medico-chirurgical Review* 29: 461–71.

——. 1868. *Essays on physiological subjects.* London: Longmans, Green, and Co.

——. 1869. *Essays on physiological subjects.* 2d edition. London: Longmans, Green, and Co.

Christie, Robert M. 2001. *Colour chemistry.* Cambridge: Royal Society of Cambridge.

Claparède, Edouard. 1869. Histologische Untersuchungen über den Regenwurm (Lumbricus terrestris *Linn.*). *Zeitschrift für wissenschaftliche Zoologie* 19: 563–624.

——. 1870. Remarques à propos de l'ouvrage de M. Alfred Russel Wallace sur la théorie de la sélection naturelle. *Archives des Sciences Physiques et Naturelles* n.s. 38: 160–89.

Clark, Ronald W. 1968. *The Huxleys.* London: Heinemann.

Clarke, Benjamin. 1866. *A new arrangement of phanerogamous plants: with especial references to relative position, including their relations with the cryptogamous.* London: n.p.

——. 1870. *On systematic botany and zoology, including a new arrangement of phanerogamous plants, with especial reference to relative position, and their relations with the cryptogamous; and a new arrangement of the classes of zoology.* London: n.p.

Claus, Carl Friedrich. 1872. Über den Bau und die systematische Stellung von Nebalia, nebst Bemerkungen über das seither unbekannte Männchen dieser Gattung. *Zeitschrift für wissenschaftliche Zoologie* 22: 323–30.

Cleave, Egbert. 1873. *Biographical cyclopaedia of homeopathic physicians and surgeons.* Philadelphia: Galaxy Publishing Co.

Clergy list: *The clergy list … containing an alphabetical list of the clergy.* London: C. Cox [and others]. 1841–89.

'Climbing plants': On the movements and habits of climbing plants. By Charles Darwin. [Read 2 February 1865.] *Journal of the Linnean Society (Botany)* 9 (1867): 1–118.

Climbing plants: *On the movements and habits of climbing plants.* By Charles Darwin. London: Longman, Green, Longman, Roberts & Green; Williams & Norgate. 1865.

[Cobbe, Frances Power.] 1855–7. *An essay on intuitive morals: being an attempt to popularise ethical science.* 2 vols. London: Longman.

——. 1870. Hereditary piety. [Review of Francis Galton, *Hereditary genius*, 1869, and Prosper Despine, *Pyschologie naturelle*, 1868.] *Theological Review* 7: 211–34.

——. 1904. *Life of Frances Power Cobbe as told by herself.* Posthumous edition. London: Swan Sonnenschein.

Collected papers: *The collected papers of Charles Darwin.* Edited by Paul H. Barrett. 2 vols. Chicago and London: University of Chicago Press. 1977.

Collins dictionary of geology. By Dorothy Farris Lapidus. London and Glasgow: HarperCollins. 1990.

Colonial Office list: *The Colonial Office list … or, general register of the colonial dependencies of Great Britain.* London: Edward Stanford; Harrison & Sons. 1862–99.

Colp, Ralph, Jr. 1977. *To be an invalid: the illness of Charles Darwin.* Chicago and London: University of Chicago Press.

——. 1998. *To be an invalid*, redux. *Journal of the History of Biology* 31: 211–40.

Columbia gazetteer of the world: *The Columbia gazetteer of the world.* Edited by Saul B. Cohen. 3 vols. New York: Columbia University Press. 1998.

Complete peerage: *The complete peerage of England, Scotland, Ireland, Great Britain, and the United Kingdom, extant, extinct, or dormant.* By George Edward Cokayne. Revised edition. Edited by Vicary Gibbs, *et al.* 12 vols. London: St Catherine Press. 1910–59.

Conry, Yvette. 1974. *L'introduction du Darwinisme en France au XIXe siècle.* Paris: J. Vrin.

Cooper, John W. [1869.] *Game fowls, their origin and history. With a description of the breeds, strains, and crosses.* West Chester, Pa.: the author.

Coral reefs: *The structure and distribution of coral reefs. Being the first part of the geology of the voyage of the* Beagle, *under the command of Capt. FitzRoy RN, during the years 1832 to 1836.* By Charles Darwin. London: Smith, Elder & Co. 1842.

Correns, C. 1916. Friedrich Hildebrand. *Berichte der deutschen botanischen Gesellschaft* 34 (pt 2): 28–49.

Correspondence: *The correspondence of Charles Darwin.* Edited by Frederick Burkhardt *et al.* 17 vols to date. Cambridge: Cambridge University Press. 1985–.

Corsi, Pietro and Weindling, Paul J. 1985. Darwinism in Germany, France and Italy. In *The Darwinian heritage*, edited by David Kohn. Princeton, N.J.: Princeton University Press in association with Nova Pacifica.

County families: *The county families of the United Kingdom; or, royal manual of the titled & untitled aristocracy of Great Britain & Ireland.* By Edward Walford. London: Robert Hardwicke; Chatto & Windus. 1860–93. *Walford's county families of the United Kingdom or royal manual of the titled and untitled aristocracy of England, Wales, Scotland, and Ireland.* London: Chatto & Windus; Spottiswoode & Co. 1894–1920.

Creuwels, Jeroen C. S. *et al.* 2007. Distribution and abundance of the southern fulmar *Fulmarus glacialoides. Polar Biology* 30: 1083–97.

Crockford's clerical directory: *The clerical directory, a biographical and statistical book of reference for facts relating to the clergy and the church. Crockford's clerical directory etc.* London: John Crockford [and others]. 1858–1900.

Cross and self fertilisation: *The effects of cross and self fertilisation in the vegetable kingdom.* By Charles Darwin. London: John Murray. 1876.

Cupples, George. 1894. *Scotch deer-hounds and their masters.* With a biographical sketch of the author by James Hutchison Stirling. Edinburgh and London: William Blackwood and Sons.

Curtis, Moses A. 1834. Enumeration of plants growing spontaneously around Wilmington, North Carolina, with remarks on some new and obscure species. *Boston Journal of Natural History* 1 (1834–7): 82–140.

Curtis, William. 1805. *Lectures on botany: as delivered in the botanic garden at Lambeth.* 3 vols. London: H. D. Symonds and Curtis.

Cuvier, Georges. 1849. *The animal kingdom, arranged after its organisation … by the late Baron Georges Cuvier.* Translated and adapted to the present state of science. Mammalia, birds, and reptiles by Edward Blyth. The fishes and radiata by Robert Mudie. The molluscous animals by George Johnston. The articulated animals by J. O. Westwood. A new edition with additions by W. B. Carpenter and J. O. Westwood. London: W. S. Orr and Co.

——. 1859. *The animal kingdom, arranged after its organisation … by the late Baron Georges Cuvier.* Translated and adapted to the present state of science. Mammalia, birds, and reptiles by Edward Blyth. The fishes and radiata by Robert Mudie. The molluscous animals by George Johnston. The articulated animals by J. O. Westwood. A new edition with additions by W. B. Carpenter and J. O. Westwood. London and Edinburgh: A. Fullerton & Co. [Reprint of Cuvier 1849, with added plates.]

DAB: *Dictionary of American biography.* Under the auspices of the American Council of Learned Societies. 20 vols., index, and 10 supplements. New York: Charles Scribner's Sons; Simon & Shuster Macmillan. London: Oxford University Press; Humphrey Milford. 1928–95.

Dallas, William Sweetland, trans. 1869. *Facts and arguments for Darwin.* By Fritz Müller. London: John Murray.

Dana, James Dwight. 1855. On American geological history: address before the American Association for the Advancement of Science. [Read August 1855.] *American Journal of Science and Arts* 2d ser. 22 (1856): 305–34.

——. 1856. On the plan of development in the geological history of North America. *American Journal of Science and Arts* 2d ser. 22: 335–49.

Dareste, Camille. 1867. *Rapport sur un veau monstreux.* (Extracted from the *Archives du Comice Agricole de l'Arrondisement Lille.*) Lille: Blocquel-Castiaux.

Darwin, Bernard. 1955. *The world that Fred made: an autobiography.* London: Chatto & Windus.

Darwin, Charles and Wallace, Alfred Russel. 1858. On the tendency of species to form varieties; and on the perpetuation of varieties and species by natural means of selection ... Communicated by Sir Charles Lyell ... and J. D. Hooker. [Read 1 July 1858.] *Journal of the Proceedings of the Linnean Society* (*Zoology*) 3 (1859): 45–62. [*Collected papers* 2: 3–19.]

Darwin, Erasmus. 1794–6. *Zoonomia; or, the laws of organic life.* 2 vols. London: J. Johnson.

Darwin, Francis. 1914. William Erasmus Darwin. *Christ's College Magazine* 29: 16–23.

Darwin, Robert Waring. 1786. New experiments on the ocular spectra of light and colours. [Read 23 March 1786.] *Philosophical Transactions of the Royal Society* 76: 313–48.

Darwin pedigree: *Pedigree of the family of Darwin.* Compiled by H. Farnham Burke. N.p.: privately printed. 1888. [Reprinted in facsimile in *Darwin pedigrees*, by Richard Broke Freeman. London: printed for the author. 1984.]

Dawkins, William Boyd. 1869. The Denbighshire caves. *Transactions of the Manchester Geological Society* 9 (1869–70): 31–7.

Dawkins, William Boyd and Busk, George. 1870. On the discovery of platycnemic men in Denbighshire; with notes on the human remains. [Abstract read 27 June 1870.] *Journal of the Ethnological Society of London* n.s. 2 (1869–70): 440–68.

Dawson, John William. 1864. On the structure of certain organic remains in the Laurentian limestones of Canada. [Read 23 November 1864.] *Quarterly Journal of the Geological Society of London* 21 (1865): 51–9.

——. 1869. Modern ideas of derivation. *Canadian Naturalist and Quarterly Journal of Science* n.s. 4: 121–38.

Dayman, Henry. 1869. *The beloved physician, a memoir of the late Dr. Joseph Bullar.* Southampton and London: Southampton Medical Society.

DBE: *Deutsche biographische Enzyklopädie.* Edited by Walter Killy *et al.* 12 vols. in 14. Munich: K. G. Saur. 1995–2000.

DBF: *Dictionnaire de biographie Française.* Under the direction of J. Balteau *et al.* 20 vols. (A–Lefèvre) to date. Paris: Librairie Letouzey & Ané. 1933–.

DBI: *Dizionario biografico degli Italiani.* Edited by Alberto M. Ghisalberti *et al.* 70 vols. (A–Marsilio) to date. Rome: Istituto della Enciclopedia Italiana. 1960–.

DCB: *Dictionary of Canadian biography.* Edited by George W. Brown *et al.* 13 vols. and index to first 12 vols. Toronto, Ontario: University of Toronto Press. 1966–94.

De Beer, Gavin, ed. 1958. *Charles Darwin and Alfred Russel Wallace. Evolution by natural selection.* Cambridge: Cambridge University Press. Reprint edition. New York and London: Johnson Reprint Corporation. 1971.

——, ed. 1959. Darwin's journal. *Bulletin of the British Museum (Natural History). Historical Series* 2 (1959–63): 3–21.

Delpino, Federico. 1867a. Sull'opera 'La distribuzione dei sessi nelle piante e la legge che osta alla perennità della fecondazione consanguinea' del prof. Hildebrand. Con note critiche. *Atti della Società Italiana delle Scienze Naturali in Milano* 10: 3–34.

——. 1867b. *Sugli apparecchi della fecondazione nelle piante antocarpe (fanerogame): sommario di osservazioni fatte negli anni 1865–1866.* Florence: Cellini.

——. 1867c. Pensieri sulla biologia vegetale, sulla tassonomia, sul valore tassonomico dei caratteri biologici, e proposta di un genere nuovo della famiglia delle labiate. *Il Nuovo Cimento. Giornale di Fisica, Chimica e Storia Naturali* 25: 284–304, 321–98.

——. 1868–75. *Ulteriori osservazioni sulla dicogamia nel regno vegetale.* 2 parts. Milan: Giuseppe Bernardoni.

——. 1869a. Alcuni appunti di geografia botanica a proposito delle tabelle fitografiche del Prof Hoffmann. *Bollettino della Società Geografica Italiana* (1869.)

——. 1869b. Breve cenno sulle relazioni biologiche e genealogiche delle Marantacee. *Nuovo Giornale Botanico Italiano* 1: 293–306.

——, trans. 1870a. *Applicazione della teoria Darwiniana ai fiori ed agli insetti visitatori dei fiori.* [Annotated translation of H. Müller 1869.] In *Bolletino della Società Entomologica Italiana.*

——. 1870b. Brief remarks on the biology and genealogy of the Marantaceæ. *Scientific Opinion* 3: 111–12, 135–7.

Dempster, W. J. 1996. *Natural selection and Patrick Matthew. Evolutionary concepts in the nineteenth century.* Durham: Pentland Press.

Descent: *The descent of man, and selection in relation to sex.* By Charles Darwin. 2 vols. London: John Murray. 1871.

Descent US ed.: *The descent of man, and selection in relation to sex.* By Charles Darwin. 2 vols. New York: D. Appleton. 1871.

Desmond, Adrian. 1994–7. *Huxley.* 2 vols. London: Michael Joseph.

Desmond, Ray. 1994. *Dictionary of British and Irish botanists and horticulturists including plant collectors, flower painters and garden designers.* New edition, revised with the assistance of Christine Ellwood. London: Taylor & Francis and the Natural History Museum. Bristol, Pa.: Taylor & Francis.

Desmond, Ray. 1995. *Kew: the history of the Royal Botanic Gardens.* London: Harvill Press with the Royal Botanic Gardens, Kew.

Despine, Prosper. 1868. *Psychologie naturelle: étude sur les facultés intellectuelles et morales dans leur état normal et dans leur manifestations anomales chez les aliénés et chez les criminels.* Paris: F. Savy.

Dictionary of African biography. Editor-in-chief, L. H. Ofusu-Appiah. 3 vols. to date. New York: Reference Publications. 1977–.

Dictionnaire historique & biographique de la Suisse. Under the direction of Marcel Godet *et al.* 7 vols. and supplement. Neuchâtel: Administration du Dictionnaire Historique & Biographique de la Suisse. 1921–34.

Dictionnaire des parlementaires français: *Dictionnaire des parlementaires français, comprenant tous les membres des assemblées françaises et tous les ministres français.* Edited by Adolphe Robert, Edgar Bourloton, and Gaston Cougny. 5 vols. Paris. 1889–91.

Dictionnaire universel des contemporains: *Dictionnaire universel des contemporains contenant toutes les personnes notables de la France et des pays étrangers … Ouvrage rédigé et continuellement tenu à jour avec le concours d'écrivains et des savants de tous les pays.* Edited by Louis Gustave Vapereau. Paris: Libraire Hachette. 1858. 3d edition, 1865. 4th edition, 1870. 5th edition, 1880. 6th edition, 1893.

'Dimorphic condition in *Primula*': On the two forms, or dimorphic condition, in the species of *Primula*, and on their remarkable sexual relations. By Charles Darwin. [Read 21 November 1861.] *Journal of the Proceedings of the Linnean Society* (*Botany*) 6 (1862): 77–96. [*Collected papers* 2: 45–63.]

Dixon, William Hepworth. 1865. *The Holy Land.* 2 vols. London: Chapman and Hall.

——. 1867. *New America.* 5th edition. 2 vols. London: Hurst and Blackett.

——. 1870. *Free Russia.* 2 vols. London: Hurst and Blackett.

DNB: *Dictionary of national biography.* Edited by Leslie Stephen and Sidney Lee. 63 vols. and 2 supplements (6 vols.). London: Smith, Elder & Co. 1885–1912. *Dictionary of national biography 1912–90.* Edited by H. W. C. Davis *et al.* 9 vols. London: Oxford University Press. 1927–96.

DNZB: *A dictionary of New Zealand biography.* Edited by G. H. Scholefield. 2 vols. Wellington, New Zealand: Department of Internal Affairs. 1940. *The dictionary of New Zealand biography.* Edited by W. H. Oliver *et al.* 5 vols. Auckland and Wellington, New Zealand: Department of Internal Affairs [and others]. 1990–2000.

Dobbs, David. 2005. *Reef madness: Charles Darwin, Alexander Agassiz, and the meaning of coral.* New York: Pantheon Books.

Dobson, Jessie. 1954. *William Clift.* London: William Heinemann medical books.

Dohrn, Anton. 1870. *Untersuchungen über Bau und Entwicklung der Arthropoden.* 2 parts. Leipzig: W. Engelmann.

——. 1875. *Der Ursprung der Wirbelthiere und das Princip des Functionswechsels. Genealogische Skizzen.* Leipzig: Wilhelm Engelmann. [Reprinted in *Theory in Biosciences* 125 (2007): 181–241.]

Donders, Frans Cornelis. 1868. Over de snelheid van psychische processen. *Onderzoekingen gedaan in het Physiologisch Laboratorium der Utrechtsche Hoogeschool* (1868–9): 92–120.

——. 1870a. De beweging van het oog toegelicht met het phaenophthalmotroop. *Nederlandsch Archief voor Genees- en Natuurkunde* 5: 222–42.

——. 1870b. On the action of the eye-lids in determination of blood from expiratory effort. Translated by William Daniel Moore. *Archives of Medicine* 5: 20–38.

——. 1969. On the speed of mental processes. *Acta Psychologica* 30: 412–31.

Douglas, George Sholto (Lord Morton). 1820. A communication of a singular fact in natural history. [Read 23 November 1820.] *Philosophical Transactions of the Royal Society of London* 111 (1821): 20–2.

Dressler, Robert L. 1993. *Phylogeny and classification of the orchid family.* Cambridge: Cambridge University Press.

Driver, Felix. 2001. *Geography militant: cultures of exploration and empire.* Oxford: Blackwell.

DSAB: *Dictionary of South African biography.* Edited by W. J. de Kock *et al.* 4 vols. Pretoria and Cape Town: Nasionale Boekhandel Beperk [and others]. 1968–81.

DSB: *Dictionary of scientific biography.* Edited by Charles Coulston Gillispie and Frederic L. Holmes. 18 vols. including index and supplements. New York: Charles Scribner's Sons. 1970–90.

Duchenne, Guillaume Benjamin Amand. 1862. *Mécanisme de la physionomie humaine, ou analyse électro-physiologique de l'expression des passions.* 1 vol. and 'Atlas' of plates. Paris: Ve Jules Renouard, Libraire.

Duncan, James Matthews. 1866. *Fecundity, fertility, sterility and allied topics.* Edinburgh: A. & C. Black.

——. 1871. *Fecundity, fertility, sterility and allied topics.* [2d edition.] Edinburgh: Adam & Charles Black.

Dunkin, Edwin. 1879. *Obituary notices of astronomers: Fellows and Associates of the Royal Astronomical Society.* London.

Duperrey, Louis Isidore. 1825–30. *Voyage autour du monde, exécuté par ordre du Roi, sur la corvette de Sa Majesté, La Coquille, pendant les années 1822, 1823, 1824 et 1825.* 7 vols. Paris: Arthus Bertrand.

Dupree, Anderson Hunter. 1959. *Asa Gray, 1810–1888.* Cambridge, Mass.: Belknap Press of Harvard University.

EB: *The Encyclopædia Britannica. A dictionary of arts, sciences, literature and general information.* 11th edition. 29 vols. Cambridge: Cambridge University Press. 1910–11.

EB (1970): *Encyclopaedia Britannica.* 24 vols. Chicago and London: Encyclopaedia Britannica. 1970.

Eder, Josef Maria, *et al.* 1945. *History of photography.* New York: Columbia University Press.

Ehrenberg, Christian Gottfried. 1868. Über die rothen Erden als Speise der Guinea-Neger. [Read 14 May 1868.] *Abhandlungen der königlichen Akademie der Wissenschaften zu Berlin* (*Physikalische Abhandlungen*) (1868): 1–55.

EI: *Enciclopedia Italiana di scienze, lettere ed arti.* 35 vols. Rome: Istituto della Enciclopedia Italiana. 1929–39.

Eisenberg, Ludwig. 1893. *Das geistige Wien. Künstler- und Schriftsteller-Lexikon.* 2 vols. Vienna: Daberkow.

Elie de Beaumont, Jean-Baptiste-Armand-Louis-Léonce. 1831. Researches on some of the revolutions which have taken place on the surface of the globe; presenting various examples of the coincidence between the elevation of beds in certain systems of mountains, and the sudden changes which have produced the lines of demarcation observable in certain stages of the sedimentary deposits. *Philosophical Magazine, or Annals of Chemistry [etc.]* n.s. 10: 241–64.

——. 1838. Recherches sur la structure et sur l'origine du Mont Etna. In *Mémoires pour servir à une description géologique de la France*, by Pierre Armand Dufrénoy and J.-B.-A.-L.-L. Elie de Beaumont, 4: 1–226. Paris: F.-G. Levrault.

Eliot, George. 1859. *Adam Bede.* 3 vols. Edinburgh: William Blackwood.

——. 2001. *Adam Bede.* Edited by Carol A. Martin. Oxford: Clarendon Press.

Ellegård, Alvar. 1990. *Darwin and the general reader: the reception of Darwin's theory of evolution in the British periodical press, 1859–1872.* Reprint edition. Chicago and London: University of Chicago Press.

Elliot, Clark A. 1979. *Biographical dictionary of American science: the seventeenth through the nineteenth centuries.* Westport, Conn., and London: Greenwood Press.

Emma Darwin (1904): *Emma Darwin, wife of Charles Darwin. A century of family letters.* Edited by Henrietta Litchfield. 2 vols. Cambridge: privately printed by Cambridge University Press. 1904.

Emma Darwin (1915): *Emma Darwin: a century of family letters, 1792–1896.* Edited by Henrietta Litchfield. 2 vols. London: John Murray. 1915.

Engen, Rodney K. 1985. *Dictionary of Victorian wood engravers.* Cambridge: Chadwyck-Healey.

English cyclopædia: *The English cyclopædia.: A new dictionary of universal knowledge.* Conducted by Charles Knight. 26 vols. London: Bradbury and Evans. 1854–73.

ENZ: *An encyclopedia of New Zealand.* Edited by A. H. McLintock. 3 vols. Wellington, New Zealand: R. E. Owen. 1966.

Eton school lists: *The Eton school lists, from 1791 to 1850.* With short notes by H. E. C. Stapylton. 2d edition. London: Williams. 1864.

Expression: *The expression of the emotions in man and animals.* By Charles Darwin. London: John Murray. 1872.

Falconer, Hugh. 1863. On the American fossil elephant of the regions bordering the Gulf of Mexico (*E. Columbi*, Falc.); with general observations on the living and extinct species. *Natural History Review* n.s. 3: 43–114.

Fawcett, Henry. 1860. A popular exposition of Mr Darwin on the origin of species. *Macmillan's Magazine* 3 (1861): 81–92.

'Fertilization of orchids': Notes on the fertilization of orchids. By Charles Darwin. *Annals and Magazine of Natural History* 4th ser. 4 (1869): 141–59. [*Collected papers* 2: 138–56.]

Flower, William Henry. 1870. Introductory lecture to the course of comparative anatomy . . . at the Royal College of Surgeons. [Read 14 February 1870.] *Medical Times and Gazette* 40: 195–200.

Forbes, David. 1860. On the geology of Bolivia and southern Peru. [Read 21 November 1860.] *Quarterly Journal of the Geological Society of London* 17 (1861): 7–62.

——. 1870a. On the Aymara Indians of Bolivia and Peru. [Read 21 June 1870.] *Journal of the Ethnological Society of London* n.s. 2 (1869–70): 193–298.

——. 1870b. On volcanoes. *Geological Magazine* 7: 314–28, 537–9.

Foreign Office list: *The Foreign Office list.* London: Harrison & Sons. 1852–1965.

Forms of flowers: *The different forms of flowers on plants of the same species.* By Charles Darwin. London: John Murray. 1877.

Fossil Cirripedia (1851): *A monograph on the fossil Lepadidæ, or, pedunculated cirripedes of Great Britain.* By Charles Darwin. London: Palaeontographical Society. 1851.

Fossil Cirripedia (1854): *A monograph of the fossil Balanidæ and Verrucidæ of Great Britain.* By Charles Darwin. London: Palaeontographical Society. 1854.

Foster, Mike. 1994. *Strange genius: the life of Ferdinand Vandeveer Hayden.* Niwot, Colo.: Roberts Rinehart Publishers.

Foundations: *The foundations of the Origin of Species. Two essays written in 1842 and 1844 by Charles Darwin.* Edited by Francis Darwin. Cambridge: Cambridge University Press. 1909. [Reprint edition. New York: Kraus Reprint Co. 1969. Also reprinted in De Beer ed. 1958.]

Freeman, Derek. 1981. *Some reflections on the nature of Iban society.* Canberra: Australian National University.

Freeman, Richard Broke. 1977. *The works of Charles Darwin: an annotated bibliographical handlist.* 2d edition. Folkestone, Kent: William Dawson & Sons. Hamden, Conn.: Archon Books, Shoe String Press.

——. 1978. *Charles Darwin: a companion.* Folkestone, Kent: William Dawson & Sons. Hamden, Conn.: Archon Books, Shoe String Press.

Freeman, Richard Broke and Gautrey, Peter Jack. 1972. Charles Darwin's *Queries about expression. Bulletin of the British Museum (Natural History) Historical Series* 4 (1970–5): 205–19.

——. 1975. Charles Darwin's *Queries about expression. Journal of the Society for the Bibliography of Natural History* 7 (1974–6): 259–63.

Frémont, John Charles. 1845. *Report of the exploring expedition to the Rocky Mountains in the year 1842, and to Oregon and North Carolina in the years 1843–'44.* Washington: United States Senate.

Galton, Francis. 1853. *The narrative of an explorer in tropical South Africa.* London: John Murray.

——. 1869. *Hereditary genius: an inquiry into its laws and consequences.* London: Macmillan.

——. 1871. Experiments in pangenesis, by breeding from rabbits of a pure variety, into whose circulation blood taken from other varieties had previously been largely transfused. [Read 30 March 1871.] *Proceedings of the Royal Society of London* 19 (1870–1): 393–410.

Gardner, James. 1999. *Sweet bells jangled out of tune: a history of the Sussex Lunatic Asylum (St Francis Hospital) Haywards Heath.* Brighton: James Gardner.

Garfield, Simon. 2000. *Mauve: how one man invented a colour that changed the world.* London: Faber and Faber.

Garretson, Martin S. 1938. *The American bison; the story of its extermination as a wild species and its restoration under federal protection.* New York: New York Zoological Society.

Gegenbaur, Carl. 1870. *Grundzüge der vergleichenden Anatomie.* 2d edition. Leipzig: Wilhelm Engelmann.

Geology of the 'Beagle': Geological observations on coral reefs, volcanic islands, and on South America. By Charles Darwin. London: Smith, Elder & Co. 1851.

Gilbert, Pamela. 1977. *A compendium of the biographical literature on deceased entomologists.* London: British Museum (Natural History).

Giles, Daniel. 1820. Particulars of a fact, nearly similar to that related by Lord Morton. [Read 23 November 1820.] *Philosophical Transactions of the Royal Society of London* 111 (1821): 23–4.

Goethe, Johann Wolfgang. 1989. *Zur Farbenlehre.* Edited by Peter Schmidt. Vol. 10 of *Johann Wolfgang Goethe. Sämtliche Werke nach Epochen seines Schaffens.* Munich: Carl Hanser Verlag.

Gohau, Gabriel. 2006. *A history of geology.* New edition, revised and translated by Albert V. Carozzi and Marguerite Carozzi. New Brunswick, N.J.: Rutgers University Press.

Gould, John. 1865. *Handbook to the birds of Australia.* 2 vols. London: the author.

Grande encyclopédie: La grande encyclopédie inventaire raisonné des sciences, des lettres et des arts. Edited by F. Camille Dreyfus *et al.* 31 vols. Paris: H. Lamirault; Société Anonyme de la Grande Encyclopédie. [1886–1902.]

Grant, Verne. 1998. Primary classification and phylogeny of the Polemoniaceae, with comments on molecular cladistics. *American Journal of Botany* 85: 741–52.

Gratiolet, Pierre. [1865.] *De la physionomie et des mouvements d'expression.* Paris: J. Hetzel.

Gray, Asa. 1856–7. Statistics of the flora of the northern United States. *American Journal of Science and Arts* 2d ser. 22: 204–32; 23: 62–84, 369–403.

——. 1861. *Natural selection not inconsistent with natural theology. A free examination of Darwin's treatise on the origin of species, and of its American reviewers.* London: Trübner. Boston, Mass.: Ticknor & Fields.

——. 1870. Revision of the North American Polemoniaceae. [Read 14 June 1870.] *Proceedings of the American Academy of Arts and Sciences* 8 (1873): 247–82.

——. 1878–84. *Synoptical flora of North America.* 2 vols. New York: Invison, Blakeman, Taylor & Co.

Gray, Jane Loring, ed. 1893. *Letters of Asa Gray.* 2 vols. London: Macmillan and Co.

[Greg, William Rathbone.] 1868. On the failure of 'natural selection' in the case of man. *Fraser's Magazine for Town and Country* 78: 353–62.

Groeben, Christiane. 1982. *Charles Darwin 1809–1882, Anton Dohrn 1840–1909: correspondence.* Naples: Macchiaroli.

Groom Napier, Charles Ottley. 1870. *The book of nature and the book of man, in which man is accepted as the type of creation—the microcosm,—the great pivot on which all lower forms of life turn.* With a preface by the late Lord Brougham. London: John Camden Hotten.

Gruber, Jacob W. 1960. *A conscience in conflict. The life of St. George Jackson Mivart.* New York: Columbia University Press for Temple University Publications.

GSE: *Great Soviet encyclopedia.* Edited by Jean Paradise *et al.* 31 vols. New York: Macmillan. London: Collier Macmillan. 1973–83. [Translation of the 3d edition of *Bol'shaia Sovetskaia entsiklopediia*, edited by A. M. Prokhorov.]

Gully, James Manby. 1846. *The water cure in chronic disease: an exposition of the causes, progress, & terminations of various chronic diseases of the digestive organs, lungs, nerves, limbs, & skin; and of their treatment by water, and other hygienic means.* London: J. Churchill.

Gunn, Mary and Codd, L. E. 1981. *Botanical exploration of Southern Africa.* Cape Town: A. A. Balkema.

Günther, Albert Charles Lewis Gotthilf. 1864–6. An account of the fishes of the states of Central America, based on collections made by Capt. J. M. Dow, F. Godman, Esq., and O. Salvin, Esq. [Read 22 March 1864 and 13 December 1866.] *Transactions of the Zoological Society of London* 6 (1863–7): 377–494.

Gunther, Albert E. 1972. The original drawings of George Henry Ford. *Journal of the Society for the Bibliography of Natural History* 6 (1971–4): 139–42.

——. 1975. *A century of zoology at the British Museum through the lives of two keepers, 1815–1914.* London: Dawsons of Pall Mall.

Haeckel, Ernst. 1866. *Generelle Morphologie der Organismen. Allgemeine Grundzüge der organischen Formen-Wissenschaft, mechanisch begründet durch die von Charles Darwin reformirte Descendenz-Theorie.* 2 vols. Berlin: Georg Reimer.

——. 1868. *Natürliche Schöpfungsgeschichte. Gemeinverständliche wissenschaftliche Vorträge über die Entwickelungslehre im Allgemeinen und diejenige von Darwin, Goethe und Lamarck im Besonderen, über die Anwendung derselben auf den Ursprung des Menschen und andere damit zusammenhängende Grundfragen der Naturwissenschaft.* Berlin: Georg Reimer.

——. 1870a. *Natürliche Schöpfungsgeschichte. Gemeinverständliche wissenschaftliche Vorträge über die Entwickelungslehre im allgemeinen und diejenige von Darwin, Goethe und Lamarck im Besonderen, über die Anwendung derselben auf den Ursprung des Menschen und andere damit zusammenhängende Grundfragen der Naturwissenschaft.* 2d edition. Berlin: Georg Reimer.

——. 1870b. Beiträge zur Plastidentheorie. *Jenaische Zeitschrift für Medizin und Naturwissenschaft* 5 (1869–70): 492–550.

——. 1870c. *Das Leben in den größten Meerestiefen.* Berlin: C. G. Lüderitzsche Verlagsbuchhandlung.

——. 1870–7. *Biologische Studien.* 2 vols. 1st vol.: Studien über Moneren und andere Protisten. 2nd vol.: Studien zur Gastraeatheorie. Leipzig: Engelmann. Jena: H. Dufft.

Haeckel, Ernst. 1872. *Die Kalkschwämme: eine Monographie.* 3 vols. Berlin: G. Reimer.

——. 1876. *The history of creation: or the development of the earth and its inhabitants by the action of natural causes. A popular exposition of the doctrine of evolution in general, and that of Darwin, Goethe, and Lamarck in particular.* Translation of Haeckel's *Natürliche Schöpfungsgeschichte.* Translation revised by E. Ray Lankester. 2 vols. London: Henry S. King & Co.

——. 1899. *Dei Welträthsel: gemeinverständliche Studien über monistische Philosophie.* Bonn: E. Strauss.

Hall, Philip B. 1987. Robert Swinhoe (1836–1877), FRS, FZS, FRGS: a Victorian naturalist in Treaty Port China. *Geographical Journal* 153: 37–47.

Hall, Robert. 1989. *History of Galashiels.* Galashiels: Alexander Walker & Son.

Halliday, Andrew. 1837. *The West Indies: the natural and physical history of the Windward and Leeward colonies.* London: John William Parker.

Hamerton, Philip Gilbert. 1869. *Wenderholme: a story of Lancashire and Yorkshire.* 3 vols. Edinburgh: W. Blackwood.

Hannavy, John, ed. 2008. *Encyclopedia of nineteenth-century photography.* 2 vols. New York and London: Routledge.

Harper, Francis. 1940. The nomenclature and type location of certain Old World mammals. *Journal of Mammology* 21 (1940): 191–203.

Hartogh Heijs van Zouteveen, Hermanus, trans. 1871–2. *De afstamming van den mensch, en de seksueele teeltkeus.* By Charles Darwin. 2 vols. (Dutch translation of *Descent.*) Delft: Van IJkema and Van Gijn.

Hauer, Franz von. 1860. Nachträge zur Kenntnis der Cephalopoden-Fauna der Hallstätter Schichten. *Sitzungberichte der Akademie der Wissenschaften in Wien* 41 (1860): 113–50.

Hayden, Ferdinand Vandeveer. 1869a. *Geological report of the exploration of the Yellowstone and Missouri rivers.* Washington, D.C.: Government Printing Office.

——. 1869b. *Preliminary field report of the United States Geological Survey of Colorado and New Mexico, conducted under the authority of Hon. J. D. Cox, secretary of the Interior.* Washington, D.C.: Government Printing Office.

——. 1871. *Preliminary report of the United States Geological Survey of Wyoming: and portions of contiguous territories.* Washington, D.C.: Government Printing Office.

Hazlewood, Nick. 2000. *Savage. The life and times of Jemmy Button.* London: Hodder and Stoughton.

Hennell, Sara Sophia. 1865–[87]. *Present religion as a faith owning fellowship with thought.* 3 vols. London: Trübner and Co.

Henslow, George. 1865. Note on the structure of *Medicago sativa*, as apparently affording facilities for the intercrossing of distinct flowers. [Read 16 November 1865.] *Journal of the Linnean Society (Botany)* 9 (1867): 327–9.

——. 1866. Note on the structure of *Indigofera*, as apparently offering facilities for the intercrossing of distinct flowers. [Read 19 April 1866.] *Journal of the Linnean Society (Botany)* 9 (1867): 355–8.

[Henslow, John Stevens, ed.] 1835. *Extracts from letters addressed to Professor Henslow by*

C. Darwin, esq, read at a meeting of the Cambridge Philosophical Society 16 November, 1835. Privately printed for the Cambridge Philosophical Society. [Reprinted 1960 and in *Collected papers* 1: 3–16.]

Herschel, John Frederick William, ed. 1849. *A manual of scientific enquiry; prepared for the use of Her Majesty's Navy: and adapted for travellers in general.* London: John Murray.

——, ed. 1851. *Admiralty manual of scientific enquiry.* 2d edition. London: John Murray.

Herzenstein, Ulrich. 1867. Zur Physiologie der Thränensecretion. *Archiv für Anatomie, Physiologie und wissenschaftliche Medicin* (1867): 651–67.

Heuss, Theodor. 1991. *Anton Dohrn: a life for science.* Translated from the German by Liselotte Dieckmann. Berlin: Springer Verlag.

Hildebrand, Friedrich. 1871. Experimente und Beobachtungen an einigen trimorphen Oxalis-Arten. *Botanische Zeitung* 29: 415–25, 431–42.

Historical register of the University of Cambridge: The historical register of the University of Cambridge, being a supplement to the Calendar *with a record of university offices, honours, and distinctions to the year 1910.* Edited by J. R. Tanner. Cambridge: Cambridge University Press. 1917.

Hodgson, Brian Houghton. 1833. Description of the wild dog of the Himalaya. *Asiatic Researches* 18, pt. 2: 221–37.

Hofacker, Johann Daniel. 1828. *Über die Eigenschaften welche sich bei Menschen und Thieren von den Eltern auf die Nachkommen vererben, mit besonderer Rücksicht auf die Pferdezucht, mit Beiträgen von F. Notter.* Tübingen: Osiander.

Holland, Henry. 1872. *Recollections of past life.* London: Longmans, Green, and Co.

Hooker, Joseph Dalton. 1860. Outlines of the distribution of Arctic plants. [Read 21 June 1860.] *Transactions of the Linnean Society of London* 23 (1862): 251–348.

——. 1870. *The student's flora of the British Islands.* London: Macmillan.

Hopkins, William. 1860. Physical theories of the phenomena of life. *Fraser's Magazine* 61: 739–52; 62: 74–90.

Horsburgh, E. L. S. 1980. *Bromley, Kent: from the earliest times to the present century, compiled from materials collected from original sources by various hands.* Reprint edition. Chislehurst, Kent: Lodgemark Press.

Howard, Michael. 1981. *The Franco-Prussian War; the German invasion of France, 1870–1871.* London and New York: Routledge.

Howsam, Leslie. 2000. An experiment with science for the nineteenth-century book trade: the International Scientific Series. *British Journal for the History of Science* 33: 187–207.

Hudson, William Henry. 1870. On the ornithology of Buenos Ayres. [Letters from W. H. Hudson to the Zoological Society of London read 10 and 24 February, 24 March, 26 May, 23 June, 1 and 15 November, and 6 December 1870] *Proceedings of the Zoological Society of London* (1870): 87–9, 112–14, 158–60, 332–4, 545–50, 671–3, 748–50, 798–802.

Hughes, John Scott. [1950.] *Kings of the Cocos; the story of the settlement on the atoll of the Keeling-Cocos in the Indian Ocean.* London: Methuen & Co.

Humboldt, Alexander von. 1814–29. *Personal narrative of travels to the equinoctial regions of the New Continent, during the years 1799–1804. By Alexander de Humboldt and Aimé Bonpland.* Translated into English by Helen Maria Williams. 7 vols. London: Longman, Hurst, Rees, Orme, & Brown; J. Murray; H. Colburn.

Hutchinson, Horace Gordon. 1914. *Life of Sir John Lubbock, Lord Avebury.* 2 vols. London: Macmillan.

Huxley, Leonard, ed. 1918. *Life and letters of Sir Joseph Dalton Hooker, OM, GCSI.* Based on materials collected and arranged by Lady Hooker. 2 vols. London: John Murray.

Huxley, Thomas Henry. 1856–7. Lectures on general natural history. *Medical Times & Gazette* n.s. 12: 429–32, 481–4, 507–11, 563–7, 618–23; 13: 27–30, 131–4, 157–60, 278–81, 383–6, 462–3, 537–8, 586–8, 635–9; 14: 133–5, 181–3, 255-7, 353–5, 505–8, 638–40; 15: 159–62, 186–9, 238–41, 467–71.

——. 1862. Notes upon human remains from the valley of the Trent, and from the Heathery Burn cave, Durham. *Geologist* 5: 201–4.

——. 1863. *Evidence as to man's place in nature.* London: Williams & Norgate.

——. 1866. *Lessons in elementary physiology.* London: Macmillan.

——. 1869a. On the physical basis of life. *Fortnightly Review* n.s. 5: 129–45.

——. 1869b. Anniversary address of the president. [Read 19 February 1869.] *Quarterly Journal of the Geological Society of London* 25: xxviii–liii.

——. 1870a. *Lay sermons, addresses, and reviews.* London: Macmillan.

——. 1870b. The anniversary address of the president. [Read 18 February 1870.] *Quarterly Journal of the Geological Society of London* 26: xxix–lxiv.

——. 1870c. Presidential address. *Report of the fortieth meeting of the British Association for the Advancement of Science; held at Liverpool in September 1870*, pp. lxxiii–lxxxix.

IGI: *International genealogical index.* Church of Jesus Christ of Latter-day Saints. 2008. http://www.familysearch.org.

India list: *The East-India register and directory.* 1803–44. *The East-India register and army list.* 1845–60. *The Indian Army and civil service list.* 1861–76. *The India list, civil and military.* 1877–95. *The India list and India Office list.* 1896–1917. London: Wm. H. Allen [and others].

Insectivorous plants. By Charles Darwin. London: John Murray. 1875.

ITIS: *Integrated Taxonomic Information System.* http://www.itis.usda.gov.

Jackson, Benjamin Daydon. 1906. *George Bentham.* London: J. M. Dent. New York: E. P. Dutton.

Jackson, Frank. 1990. *Crufts: the official history.* London: Pelham Books.

Jacobi, Carl Gustav Jakob. 1829. *Fundamenta nova theoriae functionum ellipticarum.* Königsberg: Borntraeger.

Jäger, Gustav. [1869.] *Die Darwin'sche Theorie und ihre Stellung zur Moral und Religion.* Stuttgart: Julius Hoffman.

Jerdon, Thomas Claverhill. 1862–4. *The birds of India; being a natural history of all the birds known to inhabit continental India, with descriptions of the species, genera, families, tribes, and orders, and a brief notice of such families as are not found in India, making it a manual of ornithology specially adapted for India.* 2 vols. in 3. Calcutta: the author.

Johns, R. J. 1995. Malesia—an introduction. *Curtis's Botanical Magazine* 12: 52–62.

Johnson, Robert. 1976. *Sir John Richardson: Arctic explorer, natural historian, naturalist.* London: Taylor & Francis.

Jones, Frederic Wood. 1910. *Coral and atolls. A history and description of the Keeling-Cocos Islands.* London: Lovell Reeve & Co.

Journal and remarks: *Journal and remarks. 1832–1836.* By Charles Darwin. Vol. 3 of *Narrative of the surveying voyages of His Majesty's ships Adventure and Beagle between the years 1826 and 1836, describing their examination of the southern shores of South America, and the Beagle's circumnavigation of the globe.* London: Henry Colburn. 1839. [Separately published as *Journal of researches.*]

Journal of researches: *Journal of researches into the geology and natural history of the various countries visited by HMS Beagle, under the command of Captain FitzRoy, RN, from 1832 to 1836.* By Charles Darwin. London: Henry Colburn. 1839.

Journal of researches (1840): *Journal of researches into the geology and natural history of the various countries visited by HMS Beagle, under the command of Captain FitzRoy, RN, from 1832 to 1836.* By Charles Darwin. London: Henry Colburn. 1840.

Journal of researches 2d ed.: *Journal of researches into the natural history and geology of the countries visited during the voyage of HMS Beagle round the world, under the command of Capt. FitzRoy RN.* 2d edition, corrected, with additions. By Charles Darwin. London: John Murray. 1845.

Journal of researches (1852): *Journal of researches into the natural history and geology of the countries visited during the voyage of HMS Beagle round the world, under the command of Capt. FitzRoy RN.* New edition [2d edition]. By Charles Darwin. London: John Murray. 1852.

Journal of researches (1860): *Journal of researches into the natural history and geology of the countries visited during the voyage of HMS Beagle around the world, under the command of Capt. FitzRoy RN.* By Charles Darwin. Reprint edition. London: John Murray. 1860.

Journal of researches (1870): *Journal of researches into the natural history and geology of the countries visited during the voyage of HMS Beagle round the world, under the command of Capt. FitzRoy RN.* By Charles Darwin. New edition. London: John Murray.

Journal of researches US ed.: *Journal of researches into the natural history and geology of the countries visited during the voyage of HMS Beagle round the world: under the command of Capt. FitzRoy, RN.* By Charles Darwin. 2 vols. New York: Harper & Brothers. 1846.

Jubiläums-Katalog: *Jubiläums-Katalog der E. Schweizerbart'schen Verlagsbuchhandlung (Erwin Nägele) G.m.b.H., Stuttgart, 1826–1926.* Stuttgart: E. Schweizerbart. 1926.

Judel, Claus Günther. 2004. Der Liebigschüler Carl Vogt als Wissenschaftlicher, Philosoph und Politiker. *Giessener Universitätsblätter* 37: 51–6.

Juniper, Tony and Parr, Mike. 1998. *Parrots. A guide to the parrots of the world.* Mountfield, Sussex: Pica Press.

Junker, Thomas. 1989. *Darwinismus und Botanik. Rezeption, Kritik und theoretische Alternativen im Deutschland des 19. Jahrhunderts.* Stuttgart: Deutscher Apotheker Verlag.

Kant, Immanuel. 1781. *Critik der reinen Vernunft.* Riga: J. F. Hartknoch.

——. 1836. *Metaphysics of ethics.* Translated by J. W. Semple. Edinburgh: T. Clark.

Kaur, Amarjit. 2001. *Historical dictionary of Malaysia.* 2d edition. Lanham, Md., and London: Scarecrow Press.

Keilhau, Baltazar Mathias. 1838–40. Theory of granite, and the other massive rocks; together with that of crystalline slate; proposed in lectures in geology, in the University of Christiania in Norway, in the year 1836. *Edinburgh New Philosophical Journal* 24 (1838): 387–403; 25 (1838): 80–101, 263–72; 28 (1840): 366–71.

——. 1844. On the mode of formation of crystalline limestone ... with preliminary observations on the present state of geology, and on the methods of investigation pursued in that science. *Edinburgh New Philosophical Journal* 36 (1844): 341–62; 37 (1844): 143–76.

Kemp, William. 1844. An account of some seeds buried in a sand-pit which germinated. By Mr William Kemp of Galashiels, in a letter to Charles Darwin, Esq. *Annals and Magazine of Natural History* 13: 89–91.

Keng, Hsuan. 1978. *Orders and families of Malayan seed plants.* Singapore: University of Singapore Press.

Kettle, Rosa Mackenzie ed. 1871. *Memoirs and letters of Charles Boner.* 2 vols. London: R. Bentley.

Keynes, Margaret. 1943. *Leonard Darwin, 1850–1943.* Cambridge: privately printed at Cambridge University Press.

Keynes, Milo. 1998. The Portland vase: Sir William Hamilton, Josiah Wedgwood and the Darwins. *Notes and Records of the Royal Society* 52: 237–59.

Keynes, Richard Darwin, ed. 2000. *Charles Darwin's zoology notes & specimen lists from H.M.S. Beagle.* Cambridge: Cambridge University Press.

King, William and Rowney, T. H. 1866. On the so-called 'Eozoonal Rock'. *Quarterly Journal of the Geological Society of London* 22: 185–218.

Kingsley, Charles. 1858. My winter garden. *Fraser's Magazine* 57: 408–25.

Kittredge, George F. 1871. *The origin of species and the antiquity and origin of man: with notes and references.* Buffalo: the author.

Koblitz, Ann Hibner. 1983. *A convergence of lives: Sofia Kovalevskaia, scientist, writer, revolutionary.* Boston: Birkhäuser.

Kohn, David. 1980. Theories to work by: rejected theories, reproduction, and Darwin's path to natural selection. *Studies in History of Biology* 4: 67–170.

Kölliker, Rudolf Albert von. 1847. Ueber die Struktur und die Verbreitung der glatten oder unwillkürlichen Muskeln. *Mitteilungen der Naturforschenden Gesellschaft in Zürich* 1: 18–28.

——. 1855. *Handbuch der Gewebelehre des Menschen. Für Aerzte und Studirende.* 2d edition. Leipzig: Wilhelm Engelmann.

——. 1869–72. Anatomisch-systematische Beschreibung der Alcyonarien: I. Die Pennatuliden. *Abhandlungen – Senckenbergische Naturforschende Gesellschaft* 7 (1869–70): 111–255, 487–602; 8 (1872): 85–275.

——. 1870. *Anatomisch-systematische Beschreibung der Alcyonarien: I. Die Pennatuliden.* Frankfurt am Main: Christian Winter.

Kovalevsky, Alexander Onufrievich. 1866. Entwickelungsgeschichte der einfachen

Ascidien. [Read 1 November 1866.] *Mémoires de L'Académie Impériale des Sciences de St.-Pétersbourg* 7th ser. 10: 1–19.

——. 1869. Embryologische Studien an Würmern und Arthropoden. [Read 18 November 1869.] *Mémoires de l'Académie Impériale des Sciences de St. Pétersbourg* 7th ser. 16 (1871): 1–70.

——. 1871. Weitere Studien über die Entwicklung der einfachen Ascidien. *Archiv für mikroskopische Anatomie* 7: 101–30.

Kovalevsky, Vladimir Onufrievich, trans. 1868–9. *Proiskhozhdenie vidov. Otdel 1. Izmeneniya zhivotnyk i rastenii vsledstvie priurucheniya. Priruchenniya zhivotniya i vozdelanniya rasteniya.* By Charles Darwin. (Russian translation of *Variation.*) Edited by Ivan Mikhailovich Sechenov, the botanical parts by Alexander Heard. 2 vols. St Petersburg: F. S. Sushchinskii.

Krause, Ernst. 1883. Prof. Dr. Hermann Müller von Lippstadt. Ein Gedenkblatt. *Kosmos* 13: 393–401.

Kricher, John C. 1998. *A field guide to eastern forests, North America.* 2d edition. Boston: Houghton Mifflin.

Kuper, Adam. 2002. Incest, cousin marriage, and the origin of the human sciences in nineteenth-century England. *Past and Present* 174: 158–83.

Kupffer, Carl. 1870. Die Stammverwandtschaft zwischen Ascidien and Wirbelthieren. Nach Untersuchungen über die Entwicklung der Ascidia canina. *Archiv für Mikroskopische Anatomie* 6: 115–72.

Lacaze-Duthiers, Félix Joseph Henri. 1858. Recherches sur la bonellie (*Bonellia viridis*). *Annales des Sciences Naturelles. Zoologie* 4th ser. 10: 49–110.

Lambert, James and Green, William. 1776. An account of a very extraordinary effect of lightning on a bullock, at Swanborow, in the parish of Iford near Lewes, in Sussex. [Read 1 May 1776.] *Philosophical Transactions of the Royal Society of London* 66: 463–503.

Landau, Sidney I., ed. 1986. *International dictionary of medicine and biology.* 3 vols. New York: John Wiley & Sons.

Langer, Karl von. 1852. Über den Bau und die Entwicklung der Milchdrüse bei beiden Geschechtern. *Denkschriften der kaiserlichen Akademie der Wissenschaften, Mathematisch-Naturwissenschaftliche Classe* 3 (pt 2): 25–38.

Lnkester, Edwin Ray. 1870. *On comparative longevity in man and the lower animals.* London: Macmillan.

Laplace, Pierre Simon marquis de. 1796. *Exposition du système du monde.* 2 vols. Paris: Imprimerie du Cercle-Social.

Lartet, Edouard. 1862. New researches concerning the co-existence of man with the great fossil mammals, regarded as characteristic of the latest geological period. *Natural History Review* n.s. 2: 53–71.

Laugel, Antoine Auguste. 1860. Nouvelle théorie d'histoire naturelle. L'origine des espèces. *Revue des Deux Mondes* 2d ser. 26: 644–71.

Law list: *Clarkes' new law list; being a list of the judges and officers of the different courts of justice … and a variety of other useful matter.* London: W. Clarke. 1816–40. *The law list; comprising the judges and officers of the different courts of justice: counsel, special pleaders, draftsmen, conveyancers, attorneys, notaries, &c., in England and Wales … and a variety of other useful matter.* London: V. & R. Stevens & G. S. Norton [and others]. 1841–1969.

Leighton, William Allport. 1841. *A flora of Shropshire.* London: John van Voorst. Shrewsbury: John Davies.

Lewes, George Henry. 1860. Studies in animal life. *Cornhill Magazine* 1: 61–74, 198–207, 283–95, 438–47, 598–607, 682–90.

Leydig, Franz von. 1864. *Vom Bau des thierischen Körpers: Handbuch der vergleichenden Anatomie.* Vol. 1. Tübingen: H. Laupp.

Lightman, Bernard, ed. 2004. *Dictionary of nineteenth-century British scientists.* 4 vols. Bristol: Thoemmes Press.

Lindley, John. 1840. *The theory of horticulture; or, an attempt to explain the principal operations of gardening upon physiological principles.* London: Longman, Orme, Brown, Green, and Longmans.

Linnaeus, Carolus (Carl von Linné). 1758–9. *Systema naturæ per regna tria naturæ, secundum classes, ordines, genera, species, cum characteribus, differentiis, synonymis, locis.* 10th edition. 2 vols. Stockholm: Laurentius Salvius.

List of the Linnean Society of London. London: [Linnean Society of London]. 1805–1939.

Living Cirripedia (1851): *A monograph of the sub-class Cirripedia, with figures of all the species. The Lepadidæ; or, pedunculated cirripedes.* By Charles Darwin. London: Ray Society. 1851.

Living Cirripedia (1854): *A monograph of the sub-class Cirripedia, with figures of all the species. The Balanidæ (or sessile cirripedes); the Verrucidæ, etc.* By Charles Darwin. London: Ray Society. 1854.

LL: *The life and letters of Charles Darwin, including an autobiographical chapter.* Edited by Francis Darwin. 3 vols. London: John Murray. 1887–8.

Lott, Dale F. 2002. *American bison: a natural history.* Berkeley, Calif. and London: University of California Press.

Lubbock, John. 1857. An account of the two methods of reproduction in *Daphnia*, and of the structure of the ephippium. *Philosophical Transactions of the Royal Society of London* 147: 79–100.

——. 1865. *Pre-historic times, as illustrated by ancient remains, and the manners and customs of modern savages.* London and Edinburgh: Williams & Norgate.

——. 1867. Notes on the *Thysanura.* Part III. [Read 6 June 1867.] *Transactions of the Linnean Society of London* 26 (1868): 295–304.

——. 1869. *Pre-historic times, as illustrated by ancient remains, and the manners and customs of modern savages.* 2d edition. London and Edinburgh: Williams & Norgate.

——. 1870a. *The origin of civilisation and the primitive condition of man.* London: Longmans, Green, and Co.

——. 1870b. Note on some stone implements from Africa and Syria. [Read 13 December 1870.] *Journal of the Anthropological Institute* 1 (1871–2): xcii–xcvii.

Lucas, Peter. 2007. Charles Darwin, 'little Dawkins' and the platycnemic Yale men:

introducing a bioarchaeological tale of the descent of man. *Archives of Natural History* 34: 318–45.

Lurie, Edward. 1960. *Louis Agassiz: a life in science.* Chicago: University of Chicago Press.

Lyell, Charles. 1830–3. *Principles of geology, being an attempt to explain the former changes of the earth's surface, by reference to causes now in operation.* 3 vols. London: John Murray.

——. 1871. *The student's elements of geology.* London: J. Murray.

M'Cann, James. 1869. *Anti-Darwinism. With Professor Huxley's reply.* Glasgow: Bryce & Co. Edinburgh: John Menzies and Andrew Elliot. London: Hamilton, Adams & Co.

MacDonald, David Whyte, ed. 2006. *The encyclopedia of mammals.* 2d edition. 3 vols. New York: Facts on File.

Macdonald, Frederick C. 1929. *Bishop Stirling of the Falklands. The adventurous life of a soldier of the cross whose humility hid the daring spirit of a hero & an inflexible will to face great risks.* London: Seely, Service & Co.

Mackintosh, James. 1846. *The miscellaneous works of Sir James Mackintosh.* Edited by R. J. Mackintosh. 3 vols. London: Longman, Brown, Green, and Longmans.

M'Lachlan, Robert. 1865. Observations on some remarkable varieties of *Sterrha sacraria* Linn., with general notes on variation in Lepidoptera. [Read 4 December 1865.] *Transactions of the Entomological Society of London* 3d ser. 2: 453–68.

Maclaren, Charles. 1839. *A sketch of the geology of Fife and the Lothians, including detailed descriptions of Arthur's Seat and Pentland Hills.* Edinburgh and London: Adam and Charles Black.

Macmillan dictionary of women's biography: *The Macmillan dictionary of women's biography.* 2d edition. Compiled and edited by Jennifer Uglow, assisted (1st edition) by Frances Hinton. London and Basingstoke, Hampshire: Macmillan Press. 1989.

Main, Robert, ed. 1859. *A manual of scientific enquiry; prepared for the use of officers in Her Majesty's Navy; and travellers in general.* 3d edition. Originally edited by Sir John Herschel. London: John Murray.

——, ed. 1871. *A manual of scientific enquiry: prepared for the use of officers in Her Majesty's navy, and travellers in general.* Fourth edition. Originally edited by Sir John Herschel. London: Her Majesty's Stationary Office.

Malthus, Thomas Robert. 1826. *An essay on the principle of population; or, a view of its past and present effects on human happiness; with an inquiry into our prospects respecting the future removal or mitigation of the evils which it occasions.* 6th edition. 2 vols. London: John Murray.

Marchant, James, ed. 1916. *Alfred Russel Wallace. Letters and reminiscences.* 2 vols. London: Cassell and Company.

Marcou, Jules. 1896. *Life, letters, and works of Louis Agassiz.* 2 vols. London and New York: Macmillan and Co.

Marginalia: *Charles Darwin's marginalia.* Edited by Mario A. Di Gregorio with the assistance of Nicholas W. Gill. Vol. 1. New York and London: Garland Publishing. 1990.

Martin, Stephen G. and Gavin, Thomas A. 1995. Bobolink (*Dolichonyx oryzivorus*). *The birds of North America online*, edited by A. Poole. Ithaca: Cornell Lab of Ornithology. http://bna.birds.cornell.edu/bna/species/176 doi:10.2173/bna.176.

Martin, William Charles Linnaeus. 1845. *The history of the horse.* London: Charles Knight & Co.

Matthew, Patrick. 1831. *On naval timber and arboriculture; with critical notes on authors who have recently treated the subject of planting.* London: Longman, Rees, Orme, Brown, & Green. Edinburgh: Adam Black.

Mayerhöfer, Josef. 1959–70. *Lexikon der Geschichte der Naturwissenschaften: Biographien, Sachwörter und Bibliographien.* Vol. 1. Vienna: Verlag Brüder Hollinek.

Medical directory: *The London medical directory … every physician, surgeon, and general practitioner resident in London.* London: C. Mitchell. 1845. *The London and provincial medical directory.* London: John Churchill. 1848–60. *The London & provincial medical directory, inclusive of the medical directory for Scotland, and the medical directory for Ireland, and general medical register.* London: John Churchill. 1861–9. *The medical directory … including the London and provincial medical directory, the medical directory for Scotland, the medical directory for Ireland.* London: J. & A. Churchill. 1870–1905.

Medical who's who: *The medical who's who.* London: London & Counties Press Association [and others]. 1913–27.

Medina, José Toribio. 1906. *Diccionario biográfico colonial de Chile.* Santiago: Elzeviriana.

Melvill, J. C. 1900. Lovell Reeve: a brief sketch of his life and career, with a fragment of an autobiography, extracts from his diary (1849), and correspondence. [Read 5 May 1900.] *Journal of Conchology* 9 (1898–1900): 344–57.

[Melville, Elizabeth Helen.] 1849. *A residence at Sierra Leone: described from a journal kept on the spot, and from letters written to friends at home by a lady.* Edited by Caroline Sheridan Norton. London: J. Murray.

Men and women of the time: *The men of the time in 1852 or sketches of living notables.* 2d edition, 1853. 3d edition, 1856. 4th edition, 1857. New edition, 1865. 7th edition, 1868. 8th edition, 1872. 9th edition, 1875. 10th edition, 1879. 11th edition, 1884. 12th edition, 1887. 13th edition, 1891. 14th edition, 1895. *Men and women of the time: a dictionary.* 15th edition. By Victor G. Plarr. 1899. London: David Bogue [and others]. 1852–99.

Men-at-the-bar: *Men-at-the-bar: a biographical hand-list of the members of the various inns of court, including Her Majesty's judges, etc.* By Joseph Foster. London: Reeves & Turner. 1885.

Mill, John Stuart. 1869. *The subjection of women.* London: Longmans, Green, Reader and Dyer.

Milne-Edwards, Henri. 1834–40. *Histoire naturelle des Crustacés, comprenant l'anatomie, la physiologie et la classification de ces animaux.* 4 vols. Paris: Librairie encyclopédique de Roret.

Miquel, Friedrich Anton Wilhelm. 1860. *Flora van Nederlandsch Indië. Erste bijvoegsel. Sumatra, zijne plantenwereld en hare voortbrengselen.* (*Flora Indiae Batavae. Supplementum primum. Prodromus florae Sumatranae.*) Amsterdam: C. G. van der Post. Utrecht: C. van der Post Jr.

Mitford, Mary Russell. 1870. *The life of Mary Russell Mitford, authoress of 'Our village', etc., related in a selection from her letters to her friends.* Edited by Alfred Guy Kingham L'estrange. 2d edition. 3 vols. London: Richard Bentley.

Mivart, St George Jackson. 1865. Contributions towards a more complete knowledge of the axial skeleton in the Primates. [Read 27 June 1865.] *Proceedings of the Zoological Society of London* (1865): 545–92.

——. 1866a. Contributions towards a more complete knowledge of the skeleton of the *Primates.* Part 1. The appendicular skeleton of *Simia.* [Read 13 December 1866.] *Transactions of the Zoological Society of London* 6 (1869): 175–225.

——. 1866b. On the Lemuroidea. [Read 27 March 1866.] *Transactions of the Zoological Society of London* 7 (1872): 1–113.

——. 1867. On the appendicular skeleton of the primates. [Read 10 January 1867.] *Philosophical Transactions of the Royal Society of London* 157: 299–429.

[——.] 1869. Difficulties of the theory of natural selection. *Month* 11: 35–53, 134–53, 274–89.

——. 1870. On the axial skeleton of the *Urodela.* Read 28 April 1870. *Proceedings of the Zoological Society of London* (1870): 260–78.

——. 1871. *On the genesis of species.* London: Macmillan and Co.

Modern English biography: *Modern English biography, containing many thousand concise memoirs of persons who have died since the year 1850.* By Frederick Boase. 3 vols. and supplement (3 vols.). Truro, Cornwall: printed for the author. 1892–1921.

Molina, Juan Ignacio. 1788–95. *Compendio de la historia geografica, natural y civil del reyno de Chile.* Translated by Domingo Joseph de Arquella da Mendoza and Nicolas de la Cruz. 2 vols. Madrid: A. de Sancha.

Möller, Alfred, ed. 1915–21. *Fritz Müller. Werke, Briefe und Leben.* 3 vols in 5. Jena: Gustav Fischer.

Montagu, George. 1831. *Ornithological dictionary of British birds by Colonel G. Montagu, F.L.S.* 2d edition by James Rennie. London: Hurst, Chance, and Co.

Moore, James Richard. 1985. Darwin of Down: the evolutionist as squarson-naturalist. In *The Darwinian heritage*, edited by David Kohn. Princeton, N.J.: Princeton University Press in association with Nova Pacifica.

Morgan, Lewis Henry. 1868. A conjectural solution of the origin of the classificatory system of relationship. [Read 11 February 1868.] *Proceedings of the American Academy of Arts and Sciences* 7: 436–77.

——. 1870. *Systems of consanguinity and affinity of the human family.* Washington: Smithsonian Institution.

Morris, John. 1843. *A catalogue of British fossils: comprising all the genera and species hitherto described with references to their geological distribution and to the localities in which they have been found.* London: John van Voorst.

Moulinié, Jean-Jacques, trans. 1868. *De la variation des animaux et des plantes sous l'action de la domestication.* By Charles Darwin. Preface by Carl Vogt. Paris: C. Reinwald.

Moulinié, Jean-Jacques, trans. 1872. *La descendance de l'homme et la sélection sexuelle.* By Charles Darwin. Preface by Carl Vogt. Paris: C. Reinwald & Cie.

——, trans. 1873. *L'origine des espèces au moyen de la sélection naturelle, ou, la lutte pour l'existence dans la nature.* By Charles Darwin. Paris: C. Reinwald.

Müller, Fritz. 1864. *Für Darwin.* Leipzig: Wilhelm Engelmann.

Müller, Hermann. 1868. Beobachtungen an westfälischen Orchideeen. *Verhandlungen des naturhistorischen Vereines der preussischen Rheinlande und Westphalens* (*Botanik*) 25: 1–62.

——. 1869. Die Anwendung der Darwin'schen Lehre auf Blumen und blumenbesuchende Insekten. [Read 18 May 1869.] *Verhandlungen des naturhistorischen Vereines der preussischen Rheinlande und Westphalens* (Botanik, Correspondenzblatt) 26: 43–66.

——. 1872. Anwendung der Darwin'schen Lehre auf Bienen. *Verhandlungen des naturhistorischen Vereins der preussischen Rheinlande und Westphalens* 29: 1–96.

Murch, Jerom. 1893. *Biographical sketches of Bath celebrities, ancient and modern, with some fragments of local history.* London: Isaac Pitman and Sons.

Murchison, Roderick Impey. 1870. Address to the geography section. *Report of the 40th Meeting of the British Association for the Advancement of Science; held at Liverpool in September 1870*, Transactions of the sections, pp. 158–66.

Murphy, Joseph John. 1869. *Habit and intelligence in their connexion with the laws of matter and force: a series of scientific essays.* 2 vols. London: Macmillan and Co.

Murray, Andrew. 1868. On the geographical relations of the chief coleopterous faunæ. [Read 17 December 1868.] *Journal of the Linnean Society* (*Zoology*) 11 (1873): 1–89.

Murray, John. 1908–9. Darwin and his publisher. *Science Progress in the Twentieth Century* 3: 537–42.

Musters, George Chaworth. 1871. *At home with the Patagonians: a year's wanderings over untrodden ground from the Straits of Magellan to the Rio Negro.* London: John Murray.

Narrative: *Narrative of the surveying voyages of His Majesty's ships Adventure and Beagle, between the years 1826 and 1836.* [Edited by Robert FitzRoy.] 3 vols. and appendix. London: Henry Colburn. 1839.

Natural selection: *Charles Darwin's Natural selection: being the second part of his big species book written from 1856 to 1858.* Edited by R. C. Stauffer. Cambridge: Cambridge University Press. 1975.

Naudin, Charles Victor. 1862. Cucurbitacées cultivées au Muséum d'Histoire Naturelle en 1862. Description d'espèces nouvelles et de quelques formes hybrides obtenues de plantes de cette famille. *Annales des Sciences Naturelles* (*Botanique*) 4th ser. 18: 159–208.

Navy list: *The navy list.* London: John Murray; Her Majesty's Stationery Office. 1815–1900.

NBL: *Norsk biografisk leksikon.* Edited by Edvard Bull *et al.* 19 vols. Oslo: H. Aschehoug & Co. 1923–83.

NBU: *Nouvelle biographie universelle depuis les temps les plus reculés jusqu'à nos jours, avec les renseignements bibliographiques et l'indication des sources a consulter.* Edited by Jean Chrétien Ferdinand Hoefer. 46 vols. in 23. Paris: Firmin Didot Frères. 1852–66.

NDB: *Neue deutsche Biographie.* Under the auspices of the Historical Commission of the Bavarian Academy of Sciences. 21 vols. (A–Rohlfs) to date. Berlin: Duncker & Humblot. 1953–.

Neue Österreichische Biographie: *Neue Österreichische Biographie 1815–1918.* 22 vols. and bibliography. Vienna: Amalthea-Verlag. 1923–87.

Newman, William. 1993. Darwin and cirripedology. *History of Carcinology. Crustacean Issues* 8: 349–434.

Nicholas, Thomas. 1872. *Annals and antiquities of the counties and county families of Wales.* 2 vols. London: Longmans, Green, Reader, & Co.

NNBW: *Nieuw Nederlandsch biografisch Woordenboek.* Edited by P. C. Molhuysen *et al.* 10 vols. Leiden: A. W. Sijthoff's Uitgevers-Maatschappij. 1911–37.

North, John S. 1989. *The Waterloo directory of Scottish newspapers and periodicals, 1800–1900.* 2 vols. Waterloo, Ontario: North Waterloo Academic Press.

Notebooks: *Charles Darwin's notebooks, 1836–1844. Geology, transmutation of species, metaphysical enquiries.* Transcribed and edited by Paul H. Barrett *et al.* Cambridge: Cambridge University Press for the British Museum (Natural History). 1987.

Nouveau dictionnaire des Belges: *Nouveau dictionnaire des Belges.* Edited by Thierry Denoel. Brussels: Le Cri. [1992.]

NUC: *The national union catalog. Pre-1956 imprints.* 685 vols. and supplement (69 vols.). London and Chicago: Mansell. 1968–81.

Nyhart, Lynn K. 1995. *Biology takes form. Animal morphology and the German universities, 1800–1900.* Chicago and London: University of Chicago Press.

OBL: *Österreichisches biographisches Lexikon 1815–1950.* Edited by Leo Santifaller *et al.* 12 vols. and 2 fascicles of vol. 13 (A–Stich Ignaz) to date. Vienna: Osterreichischen Akademie der Wissenschaften. 1957–.

O'Byrne, William R. 1849. *A naval biographical dictionary: comprising the life and services of every living officer in Her Majesty's Navy, from the rank of admiral of the fleet to that of lieutenant, inclusive.* London: John Murray.

ODNB: *Oxford dictionary of national biography: from the earliest times to the year 2000.* (Revised edition.) Edited by H. C. G. Matthew and Brian Harrison. 60 vols. and index. Oxford: Oxford University Press. 2004.

OED: *The Oxford English dictionary. Being a corrected re-issue with an introduction, supplement and bibliography of a new English dictionary.* Edited by James A. H. Murray, *et al.* 12 vols. and supplement. Oxford: Clarendon Press. 1970. *A supplement to the Oxford English dictionary.* 4 vols. Edited by R. W. Burchfield. Oxford: Clarendon Press. 1972–86. *The Oxford English dictionary.* 2d edition. 20 vols. Prepared by J. A. Simpson and E. S. C. Weiner. Oxford: Clarendon Press. 1989. *Oxford English dictionary additional series.* 3 vols. Edited by John Simpson *et al.* Oxford: Clarendon Press. 1993–7.

Ogle, William. 1869. The fertilisation of Salvia and of some other flowers. *Popular Science Review* 8: 261–73.

——. 1870a. Anosmia; or, cases illustrating the physiology and pathology of the sense of smell. *Medico-Chirurgical Transactions* 53: 263–90.

Ogle, William. 1870b. The fertilisation of certain plants (Didynamia). *Popular Science Review* 9: 45–56.

——. 1870c. The fertilisation of various flowers by insects. (Compositæ, Ericaceæ, &c.) *Popular Science Review* 10: 160–72.

O'Leary, Patrick. 1989. *Sir James Mackintosh: the Whig Cicero.* Aberdeen: Aberdeen University Press.

Oliver, John R. 1869. *The geology of St Helena.* St Helena: Benjamin Grant.

Orbigny, Alcide Charles Victor Dessalines d'. 1835–47. *Voyage dans l'Amérique Méridionale (le Brésil, la République orientale de l'Uruguay, la République Argentine, la Patagonie, la République du Chili, la République de Bolivia, la République du Pérou), exécuté pendant les années 1826 … 1833.* 6 vols. in 7 and 4 atlases. Paris and Strasbourg: Pitois-Levrault et Cie, P. Bertrand.

Orchids: *On the various contrivances by which British and foreign orchids are fertilised by insects, and on the good effects of intercrossing.* By Charles Darwin. London: John Murray. 1862.

Orchids 2d ed.: *The various contrivances by which British and foreign orchids are fertilised by insects, and the good effects of intercrossing.* By Charles Darwin. Revised. London: John Murray. 1877.

Origin: *On the origin of species by means of natural selection, or the preservation of favoured races in the struggle for life.* By Charles Darwin. London: John Murray. 1859.

Origin 2d ed.: *On the origin of species by means of natural selection, or the preservation of favoured races in the struggle for life.* By Charles Darwin. London: John Murray. 1860.

Origin 3d ed.: *On the origin of species by means of natural selection, or the preservation of favoured races in the struggle for life.* With additions and corrections. By Charles Darwin. London: John Murray. 1861.

Origin 4th ed.: *On the origin of species by means of natural selection, or the preservation of favoured races in the struggle for life.* With additions and corrections. By Charles Darwin. London: John Murray. 1866.

Origin 5th ed.: *On the origin of species by means of natural selection, or the preservation of favoured races in the struggle for life.* With additions and corrections. By Charles Darwin. London: John Murray. 1869.

Origin 6th ed.: *The origin of species by means of natural selection, or the preservation of favoured races in the struggle for life.* With additions and corrections. By Charles Darwin. London: John Murray. 1872.

Origin US ed.: *On the origin of species by means of natural selection, or the preservation of favoured races in the struggle for life.* A new edition, revised and augmented by the author. By Charles Darwin. New York: D. Appleton. 1860.

Origin 2d US ed.: *On the origin of species by means of natural selection, or the preservation of favoured races in the struggle for life.* By Charles Darwin. From the fifth London edition, with additions and corrections. New York: D. Appleton. 1870.

Orton, James. 1870. *The Andes and the Amazon; or, across the continent of South America.* New York: Harper.

[Owen, Richard.] 1853. *Descriptive catalogue of the osteological series contained in the museum of the Royal College of Surgeons of England.* 2 vols. London: [Royal College of Surgeons of England].

——. 1855. *Lectures on the comparative anatomy and physiology of the invertebrate animals, delivered at the Royal College of Surgeons.* 2d edition. London: Longman, Brown, Green, and Longmans.

——. 1866–8. *On the anatomy of vertebrates.* 3 vols. London: Longmans, Green & Co.

Oxford classical dictionary. 3d edition, revised. Edited by Simon Hornblower and Anthony Spawforth. Oxford: Oxford University Press. 2003.

Pancaldi, Giuliano. 1991. *Darwin in Italy. Science across cultural frontiers.* Translated by Ruey Brodine Morelli. Updated and expanded edition. Bloomington and Indianapolis, Ind.: Indiana University Press.

'Parallel roads of Glen Roy': Observations on the parallel roads of Glen Roy, and of other parts of Lochaber in Scotland, with an attempt to prove that they are of marine origin. By Charles Darwin. [Read 7 February 1839.] *Philosophical Transactions of the Royal Society of London* 129: 39–81. [*Collected papers* 1: 89–137.]

Park, Mungo. 1799. *Travels in the interior districts of Africa: performed under the direction and patronage of the African Association, in the years 1795, 1796, and 1797 … with an appendix, containing geographical illustrations of Africa by Major Rennell.* London: W. Bulmer and Co.

Paston, George. 1932. *At John Murray's: records of a literary circle, 1843–1892.* London: John Murray.

Pearce, J. M. 2002. Armand Trousseau—some of his contributions to neurology. *Journal of the History of the Neurosciences* 11: 125–35.

Pearson, Karl. 1914–30. *The life, letters and labours of Francis Galton.* 3 vols. in 4. Cambridge: Cambridge University Press.

Peckham, Morse, ed. 1959. *The Origin of Species by Charles Darwin: a variorum text.* Philadelphia: University of Pennsylvania Press.

Penzig, Ottone. 1905. Commemorazione di Federico Delpino. *Malpighia* 19: 294–310.

Peschel, Oscar Ferdinand. 1870. *Neue Probleme der vergleichenden Erdkunde als Versuch einer Morphologie der Erdoberfläche.* Leipzig: Duncker & Humblot.

Physicians: *The roll of the Royal College of Physicians of London.* By William Munk. 2d edition, revised and enlarged. 3 vols. London: Royal College of Physicians. 1878. *Lives of the fellows of the Royal College of Physicians of London.* Compiled by G. H. Brown *et al.* 5 vols. London: Royal College of Physicians. Oxford and Washington, D.C.: IRL Press. 1955–89.

Pictet de la Rive, François Jules. 1860. Sur l'origine de l'espèce par Charles Darwin. *Bibliothèque Universelle. Revue Suisse et étrangère* n.s. 7: 233–55.

Plarr, Victor Gustave. 1930. *Plarr's lives of the fellows of the Royal College of Surgeons of England.* Revised by Sir D'Arcy Power. 2 vols. London: Simpkin Marshall.

Post Office directory of Birmingham: *Post Office directory of Birmingham, Warwickshire, and part of Staffordshire. Kelly's directory of Birmingham, Staffordshire, Warwickshire, and Worcestershire.* London: W. Kelly; Kelly & Co. 1845–1928.

Post Office directory of Cambridge, Norfolk, and Suffolk: *Post Office directory of Cambridge, Norfolk, and Suffolk*. *Kelly's directory of Cambridgeshire, Norfolk and Suffolk*. London: W. Kelly; Kelly & Co. 1853–1900.

Post Office directory of the county of Durham: *The Post Office directory of the county of Durham, and the principal towns and adjacent places in Northumberland, Cumberland & Westmorland*. *The Post Office directory of Durham and Northumberland*. London: Kelly and Co. 1873–1929.

Post Office directory of Lancashire, Liverpool and Manchester: *Post Office directory of Lancashire*. *Post Office directory of Lancashire, Liverpool and Manchester*. London: Kelly & Co. 1858–92.

Post Office directory of the six home counties: *Post Office directory of the six home counties, viz., Essex, Herts, Kent, Middlesex, Surrey and Sussex*. London: W. Kelly & Co. 1845–78.

Post Office Edinburgh directory: *Post-Office annual directory and calendar*. *Post-Office Edinburgh and Leith directory*. Edinburgh: Ballantyne & Hughes [and others]. 1845–1908.

Post Office London directory: *Post-Office annual directory. . . . A list of the principal merchants, traders of eminence, &c. in the cities of London and Westminster, the borough of Southwark, and parts adjacent . . . general and special information relating to the Post Office*. *Post Office London directory*. London: His Majesty's Postmaster-General [and others]. 1802–1967.

Post Office London suburban directory: *The Post Office London suburban directory*. *Kelly's London suburban directory*. London: Kelly & Co. 1860–1903.

Prestwich, Joseph. 1862. Theoretical considerations on the conditions under which the (drift) deposits containing the remains of extinct mammalia and flint implements were accumulated, and on their geological age. On the Loess of the valleys of the south of England, and of the Somme and the Seine. [Read 27 March and 19 June 1862.] *Philosophical Transactions of the Royal Society of London* 154 (1864): 247–309.

Preyer, William. 1868. Ueber einige Eigenschaften des Hämoglobins und des Methämoglobins. *Archiv für die gesammte Physiologie des Menschen und der Thiere* 1: 395–454.

——. 1869. Vergiftungsversuche mit wasserfreier Blausäure und ihr Nachweis im vergifteten Blute. *Archiv für die gesammte Physiologie des Menschen und der Thiere* 2: 146–56.

——. 1870a. Die Verwandtschaft der Töne und Farben. *Jenaischen Zeitschrift für Medicin und Naturwissenschaft* 5: 376–88.

——. 1870b. Charles Darwin, eine biographische Skizze. *Das Ausland*, 2 April 1870, pp. 314–20.

Prichard, James Cowles. 1836–7. *Researches into the physical history of mankind*. 3d editon. 2 vols. London: Sherwood, Gilbert and Piper.

Quatrefages de Bréau, Jean Louis Armand de. 1867–9. Anthropologie. Cours de M. de Quatrefages. *Revue de cours scientifiques de la France et de l'étranger* 5 (1867–8): 366–9, 621–31, 655–64, 720–8, 730–44, 751–60; 6 (1868–9): 201–6, 219–24, 235–40.

——. 1868–9. Origines des espèces animales et végétales. *Revue des Deux Mondes* 78 (1868): 832–60; 79 (1869): 208–40; 80 (1869): 64–95, 397–432, 638–72.

——. 1870. *Charles Darwin et ses précurseurs français. Etude sur le transformisme*. Paris: Germer Baillière.

Quoy, Jean René Constant and Gaimard, Joseph Paul. 1830[–4]. *Zoologie.* 4 vols. In *Voyage de découvertes de l'Astrolabe. Exécuté par ordre du Roi, pendant les années 1826-1827-1828-1829, sous le commandement de J. Dumont d'Urville*, by Jules Sébastian César Dumont d'Urville. 15 vols. Paris: J. Tastu. 1830[–5].

Raby, Peter. 2001. *Alfred Russel Wallace: a life.* London: Chatto & Windus.

Reade, William Winwood. 1872. *The martyrdom of man.* London: Trübner & Co.

——. 1873. *The African sketch-book.* 2 vols. London: Smith, Elder, and Co.

Recht, Georg. 1869. *Die Entwicklung der Weltgesetze.* Munich: Schurich.

Record of the Royal Society of London: *The record of the Royal Society of London for the promotion of natural knowledge.* 4th edition. London: Royal Society. 1940.

Records of Lincoln's Inn: *The records of the honorable society of Lincoln's Inn: admissions.* Edited by William Paley Baildon. 2 vols. London: Lincoln's Inn. 1896.

Rérolle, Louis, trans. 1870. *De la fécondation des orchidées par les insectes et des bons résultats du croisement.* By Charles Darwin. Paris: C. Reinwald.

Richardson, Edmund William. 1916. *A veteran naturalist; being the life and work of W. B. Tegetmeier.* London: Witherby & Co.

Ritvo, Harriet. 1990. *The animal estate: the English and other creatures in the Victorian age.* London: Penguin Books.

[Robertson, John.] 1868. Review of *The variation of plants and animals under domestication. Athenæum*, 15 February 1868, pp. 243–4.

Robson, N. K. B. 1990. Proposal to reject the name *Hypericum quadrangulum* L. (Guttiferae). *Taxon* 39: 135–7.

Rohrbach, Paul. 1866. *Über den Blüthenbau und die Befruchtung von* Epipogium Gmelini. Göttingen: Universitäts-Buchdruckerei von E. A. Huth.

Rolleston, George. 1870a. Address to the biology section. *Report of the fortieth meeting of the British Association for the Advancement of Science; held at Liverpool in September 1870*, Transactions of the sections, pp. 91–09.

——. 1870b. *Forms of animal life: being outlines of zoological classification based upon anatomical investigation and illustrated by descriptions of specimens and of figures.* Oxford: Clarendon Press.

Roll of the graduates of the University of Aberdeen: *Roll of the graduates of the University of Aberdeen 1860–1900.* By William Johnston. Aberdeen: Aberdeen University Press. 1906.

Roll of the graduates of the University of Glasgow: *A roll of the graduates of the University of Glasgow. From 31st December, 1727. To 31st December, 1897.* Compiled by W. Innes Addison. Glasgow: James MacLehose & Sons. 1898.

Ross, Christel A. and Auge, Harald. 2008. Invasive *Mahonia* plants outgrow their native relatives. *Plant Ecology* 199: 21–31.

Roth, Henry Ling. 1896. *The natives of Sarawak and British North Borneo.* 2 vols. London: Truslove & Hanson.

Royal Society catalogue of scientific papers: *Catalogue of scientific papers (1800–1900). Compiled and published by the Royal Society of London.* 19 vols. and index (3 vols.). London: Royal Society of London. Cambridge: Cambridge University Press. 1867–1925.

Royer, Clémence Auguste, trans. 1862. *De l'origine des espèces ou des lois du progrès chez les êtres organisés.* By Charles Darwin. With preface and notes by the translator. [Translated from the 3d English edition.] Paris: Guillaumin & Cie; Victor Masson.

Royer, Clémence Auguste, trans. 1866. *De l'origine des espèces par sélection naturelle ou des lois de transformation des êtres organisés.* By Charles Darwin. With preface and notes by the translator. 2d edition. Augmented in accordance with notes by the author. Paris: Guillaumin et Cie; Victor Masson et fils.

——, trans. 1870. *De l'origine des espèces par sélection naturelle ou des lois de transformation des êtres organisés.* By Charles Darwin. With preface and notes by the translator. 3d edition. Paris: V. Masson et fils.

Ruck, Berta. 1935. *A story-teller tells the truth: reminiscences & notes.* London: Hutchinson & Co.

Russell-Gebbett, Jean. 1977. *Henslow of Hitcham: botanist, educationalist and clergyman.* Lavenham, Suffolk: Terence Dalton.

Rütimeyer, Ludwig. 1861. *Die Fauna der Pfahlbauten in der Schweiz. Untersuchungen über die Geschichte der wilden und der Haus-Säugethiere von Mittel-Europa.* Basel, Switzerland: Bahnmaier's Buchhandlung (C. Detloff).

——. 1862. Eocaene Säugethiere aus dem Gebiet des Schweizerischen Jura. *Neue Denkschriften der allgemeinen Schweizerischen Gesellschaft für die gesammten Naturwissenschaften* 19: 1–98.

——. 1866. Über Art und Raçe des zahmen europäischen Rindes. *Archiv für Anthropologie* 1: 217–50.

Sachs, Julius. 1870. *Lehrbuch der Botanik nach dem gegenwärtigen Stand der Wissenschaft.* 2d edition. Leipzig: Wilheln Engelmann.

St John, Charles. 1849. *Short sketches of the wild sports and natural history of the Highlands.* London: John Murray.

Salter, John William. 1862. On Peltocaris, a new genus of Silurian Crustacea. [Read 21 May 1862.] *Quarterly Journal of the Geological Society* 19 (1863): 87–92.

Salter, Samuel James Augustus. 1863. Note on the fertility, inter se, of hybrids of different species of the genus *Gallus. Natural History Review* n.s. 3: 276–9.

Sanderson, Michael, ed. 1975. *The universities in the nineteenth century.* London and Boston: Routledge and Kegan Paul.

Sarjeant, William A. S. 1980–96. *Geologists and the history of geology: an international bibliography.* 10 vols. including supplements. London: Macmillan. Malabar, Fla.: Robert E. Krieger Publishing.

SBL: *Svenskt biografiskt lexikon.* Edited by Bertil Boëthius *et al.* 32 vols. (A–Sparre) to date. Stockholm: Albert Bonnier and P. A. Norstedt. 1918–.

Schaub, Edward Leroy. 1925. *Immanuel Kant: papers read at Northwestern University on the bicentenary of Kant's birth.* Chicago; London: Open Court Publishing Co.

Scherren, Henry. 1905. *The Zoological Society of London: a sketch of its foundation and development and the story of its farm, museum, gardens, menagerie and library.* London: Cassell.

Schmarda, Ludwig Karl. 1852. Zur Naturgeschichte der Adria. I: Bonellia viridis.

Denkschriften der kaiserlichen Akademie der Wissenschaften. Mathematisch-naturwissenschaftliche Klasse (1852) part 2: 117–40.

Schmick, Jacob Heinrich. 1869. *Die Umsetzungen der Meere und die Eiszeiten der Halbkugeln der Erde, ihre Ursachen und Perioden.* Cologne: M. DuMont-Schauberg'schen Buchhandlung.

——. 1871. *Thatsachen und Beobachtungen zur weiteren Begründung seiner neunen theorie einer Umsetzung der Meere durch die Sonnenanziehung und eines gleichzeitigen Wechsels der Eiszeiten auf beiden Halbkugeln der Erde.* Görlitz: E. Remer's Buchhandlung.

Schmid, Theophilus. 1964. *Misionando por Patagonia austral 1858–1865. Usos y costumbres de los indios Patagones.* Buenos Aires: Academia Nacional de la Historia.

Schopf, J. William. 2000. Solution to Darwin's dilemma: discovery of the missing Precambrian record of life. *Proceedings of the National Academy of Sciences of the United States of America* 97: 6947–53.

Scott, John. 1864. Observations on the functions and structure of the reproductive organs in the *Primulaceæ.* [Read 4 February 1864.] *Journal of the Linnean Society* (*Botany*) 8 (1865): 78–126.

Scott, Walter. 1832. *Redgauntlet.* 2 vols. Edinburgh: Robert Cadell. London: Whittaker & Co.

Scott, William Robson. 1870. *The deaf and dumb: their education and social position.* 2d edition. London: Bell & Daldry.

[Sedgwick, Adam.] 1860. Objections to Mr Darwin's theory of the origin of species. *Spectator*, 24 March 1860, pp. 285–6. [Reprinted with revisions in *ibid.*, 7 April 1860, pp. 334–5.]

——. 1870. *Supplement to the memorial of the trustees of Cowgill Chapel, with an appendix &c. printed in 1868.* Cambridge: printed at the University Press (privately published).

Semper, Karl Gottfried, *et al.* 1868–1916. *Reisen im Archipel Philippinen.* Part 2: *Wissenschaftliche Resultate.* 10 vols. Leipzig: Wilhelm Engelmann. Wiesbaden: C. W. Kreidel.

Settegast, Hermann Gustav. 1868. *Die Thierzucht.* Breslau: Wilh. Gottl. Korn.

Sharpe, Richard Bowdler. 1868–71. *A monograph of the Alcedinidae: or, family of kingfishers.* London: the author.

Sheets-Pyenson, Susan. 1991. 'Pearls before swine': Sir William Dawson's Bakerian Lecture of 1870. *Notes and Records of the Royal Society of London* 45: 177–91.

Sidebotham, Joseph. 1847. Living heath and fossil tree. *Phytologist* 2 (pt 3): 778–9.

Slotten, Ross A. 2004. *The heretic in Darwin's court; the life of Alfred Russel Wallace.* New York: Columbia University Press.

Smith, Adam. 1759. *The theory of moral sentiments.* London: A. Miller, A. Kincaid, and J. Bell.

Smith, James Edward. 1800–4. *Flora Britannica.* 3 vols. London: J. White.

Smith, Kenneth G. V. 1987. Darwin's insects: Charles Darwin's entomological notes. *Bulletin of the British Museum (Natural History). Historical series* 14: 1–143.

Smith, Mavis E., ed. 2008. *Ellen Tollet of Betley Hall. Journals and letters from 1835.* Crewe: the author.

Somerville, Mary. 1870. *Physical geography.* 6th edition. Revised by Henry Walter Bates. London: John Murray.

South America: *Geological observations on South America. Being the third part of the geology of the voyage of the Beagle, under the command of Capt. FitzRoy RN, during the years 1832 to 1836.* By Charles Darwin. London: Smith, Elder & Co. 1846.

Southward, Alan J., ed. 1987. *Barnacle biology.* Rotterdam: Balkema.

Spencer, Herbert. 1855. *The principles of psychology.* London: Longman, Brown, Green, and Longmans.

Sprengel, Christian Konrad. 1793. *Das entdeckte Geheimniss der Natur im Bau und in der Befruchtung der Blumen.* Berlin: Friedrich Vieweg.

Stace, Clive Anthony. 1997. *New flora of the British Isles.* 2d edition. London and Cambridge: Cambridge University Press.

Stark, James. 1834. Historical account of experiments regarding the influence of colour on heat, the deposition of dew, and odours. *Edinburgh New Philosophical Journal* 17: 65–98.

Stecher, Robert M. and Klavins, Janis V. 1965. Charles Darwin and the Moscow Society of Naturalists. *Journal of the History of Medicine and Allied Sciences* 20: 157–61.

Stedman's medical eponyms: *Stedman's medical eponyms.* Edited by Pat Forbis and Susan L. Bartolucci. Baltimore and London: Williams & Wilkins. 1998.

Stenton, Michael. 1976. *Who's who of British members of Parliament, 1832–1885.* Brighton: Hanover Press.

Stephen, Leslie. 1885. *Life of Henry Fawcett.* London: Smith, Elder.

Stewart, John. 1996. *The British empire. An encyclopedia of the Crown's holdings, 1493 through 1995.* Jefferson, N.C., and London: McFarland & Company.

Stoddard, Charles Warren. [1885.] *The lepers of Molokai.* Notre Dame, Ind.: 'Ave Maria' Press.

Strick, James. 1999. Darwinism and the origin of life: the role of H. C. Bastian in the British spontaneous generation debates, 1868–1873. *Journal of the History of Biology* 32: 51–92.

Strickland, Hugh Edwin and Daubeny, Charles Giles Bridle. 1842. Second report of a committee consisting of Mr. H. E. Strickland, Prof. Daubeny, Prof. Henslow, and Prof. Lindley, appointed to make experiments on the growth and vitality of seeds. *Report of the 12th meeting of the British Association for the Advancement of Science; held at Manchester in June 1842*, p. 34.

Suess, Eduard. 1854. Ueber die Brachiopoden der Kössener Schichten. *Denkschriften der kaiserlichen Akademie der Wissenschaften* 7: 29–65.

——. 1855. Ueber die Brachiopoden der Hallstätter. *Denkschriften der kaiserlichen Akademie der Wissenschaften* 9: 23–32.

Sulivan, Henry Norton, ed. 1896. *Life and letters of the late Admiral Sir Bartholomew James Sulivan, KCB, 1810–1890.* London: John Murray.

Sutherland, Peter Cormack. 1852. *Journal of a voyage in Baffin's Bay and Barrow Straits, in the years 1850–1851, performed by HM Ships 'Lady Franklin' and 'Sophia', under the command of Mr William Penny, in search of the missing crews of HM Ships Erebus and Terror.* 2 vols. London.

Swinhoe, Robert. 1864. On a new rat from Formosa. [Read 10 May 1864.] *Proceedings of the Zoological Society of London* (1864): 185–7. [*Annals and Magazine of Natural History* 15 (1865): 71–3.]

Swinney, Geoffrey N. 1999. A natural history collection in transition; Wyville Thomson and the relationship between the University of Edinburgh and the Edinburgh Museum of Science and Art. *Journal of the History of Collections* 11: 51–70.

[Tait, Peter Guthrie.] 1869. Geological time. *North British Review* n.s. 11: 406–39.

Tanner, John. 1830. *A narrative of the captivity and adventures of John Tanner (U.S. interpreter at the Saut de Ste. Marie,) during thirty years residence among the Indians in the interior of North America.* Prepared for the press by Edwin James. London: Baldwin & Cradock.

Taxonomic literature: *Taxonomic literature. A selective guide to botanical publications and collections with dates, commentaries and types.* By Frans A. Stafleu and Richard S. Cowan. 2d edition. 7 vols. Utrecht, Netherlands: Bohn, Scheltema & Holkema. The Hague, Netherlands: W. Junk. 1976–88.

Taxonomic literature. Supplement: *Taxonomic literaure: a selective guide to botanical publications and collections with dates, commentaries and types.* Supplement. By Frans A. Stafleu and Erik A. Mennega. 7 vols. to date. Königstein: Koeltz Sceintific Books. 1992–.

Taylor, Richard. 1870. *Te ika a Maui.* London: W. Macintosh.

Tee, Garry J. 1976–7. Sof'ya Vasil'yevna Kovalevskaya. *Mathematical Chronicle* 5: 113–39.

Tegetmeier, William Bernhard. 1867. *The poultry book: comprising the breeding and management of profitable and ornamental poultry, their qualities and characteristics; to which is added 'The standard of excellence in exhibition birds', authorized by the Poultry Club.* London and New York: George Routledge & Sons.

——. 1873. *The poultry book: comprising the breeding and management of profitable and ornamental poultry; to which is added 'The standard of excellence in exhibition birds'.* New edition [2d]. London and New York: George Routledge & Sons.

Tennent, James Emerson. 1859. *Ceylon, an account of the island, physical, historical, and topographical, with notices of its natural history, antiquities and productions.* 3d edition. 2 vols. London: Longman, Green, Longman, and Roberts.

——. 1861. *Sketches of the natural history of Ceylon, with narratives and anecdotes illustrative of the habits and instincts of the mammalia, birds, reptiles, fishes, insects &c.* London: Longman, Green, Longman, and Roberts.

Thompson, Henry Percy. 1935. *A history of Hayes in the county of Kent.* London: Lovat Dickson & Thompson.

Thomson, Charles Wyville. 1861. On the embryology of *Asteracanthion violaceus* (L.). *Quarterly Journal of Microscopical Science* n.s. 1: 99–108.

——. 1862. On the development of *Synapta inhærens* (sp.). *Quarterly Journal of Microscopical Science* n.s. 2: 131–46.

Thomson, Charles Wyville. 1863. On the embryology of the echinodermata. *Natural History Review* n.s. 3: 395–415.

——. 1869. The depths of the sea. [Read 10 April 1869.] *Journal of the Royal Dublin Society* 5 (1870): 316–26.

——. 1873. *The depths of the sea. An account of the general results of the dredging cruises of the H.M.Ss. 'Porcupine' and 'Lightning' during the summers of 1868, 1869 and 1870.* London: Macmillan and Co.

Thomson, William. 1862. On the secular cooling of the earth. [Read 28 April 1862.] *Transactions of the Royal Society of Edinburgh* 23 (1864): 157–70.

Thorne, R. G. 1986. *The House of Commons 1790–1820.* 5 vols. London: Secker & Warburg for the History of Parliament Trust.

Thorell, Tord Tamerlan Teodor. 1869–70. *On European spiders. Part I, Review of the European genera of spiders, preceded by some observations on zoological nomenclature.* Upsala: Ed. Berling.

Torrey, John and Gray, Asa. 1838–43. *A flora of North America: containing abridged descriptions of all the known indigenous and naturalized plants growing north of Mexico; arranged according to the natural system.* 2 vols. New York and London: Wiley & Putnam.

Tort, Patrick. 1996. *Dictionnaire du Darwinisme et de l'evolution.* 3 vols. Paris: Presses Universitaires de France.

Trimen, Roland. 1864. On the structure of *Bonatea speciosa*, Linn. sp., with reference to its fertilisation. [Read 1 December 1864.] *Journal of the Linnean Society* (*Botany*) 9 (1867): 156–60.

Trousseau, Armand. 1868–72. *Lectures on clinical medicine: delivered at the Hôtel Dieu.* Translated by Pierre Victor Bazire and John Rose Cormack. 5 vols. London: New Sydenham Society.

Turner, Gerard L'E. 1989. *The great age of the microscope: the collection of the Royal Microscopical Society though 150 years.* Bristol and New York: Adam Hilger.

Tylor, Edward Burnett. 1865. *Researches into the early history of mankind and the development of civilization.* London: John Murray.

Tyndall, John. 1867. Miracles and special providences. *Fortnightly Review* n.s. 1: 645–60.

——. 1870. *On the scientific use of the imagination: a discourse delivered before the British Association at Liverpool, on Friday evening, 16th September 1870.* London: Longmans, Green & Co.

Uchida, Wakana, *et al.* 2003. Morphological development of anthers induced by the dimorphic smut fungus *Microbotryum violaceum* in female flowers of the dioecious plant *Silene latifolia*. *Planta* 218: 240–8.

Variation: *The variation of animals and plants under domestication.* By Charles Darwin. 2 vols. London: John Murray. 1868.

Variation 2d ed.: *The variation of animals and plants under domestication.* By Charles Darwin. 2 vols. London: John Murray. 1875.

Vignoles, K. H. 1982. *Charles Blacker Vignoles: romantic engineer.* Cambridge: Cambridge University Press.

Volcanic islands: *Geological observations on the volcanic islands, visited during the voyage of HMS Beagle, together with some brief notices on the geology of Australia and the Cape of Good Hope. Being the second part of the geology of the voyage of the Beagle, under the command of Capt. FitzRoy RN, during the years 1832 to 1836*. By Charles Darwin. London: Smith, Elder & Co. 1844.

Wagner, Moritz. 1870. *Naturwissenschaftliche Reisen im tropischen Amerika.* Stuttgart: J. G. Cotta'schen Buchhandlung.

Wagner, Rudolph. 1851–9. *Icones physiologicae. Erläuterungstafeln zur Physiologie und Entwickelungsgeschichte.* 2d edition, revised and edited by Alexander Ecker. Leipzig: Leopold Voss.

Wake, Charles Staniland. 1868. *Chapters on man. With the outlines of a science of comparative psychology.* London: Trübner & Co.

Wallace, Alfred Russel. 1864. The origin of human races and the antiquity of man deduced from the theory of 'natural selection'. [Read 1 March 1864.] *Anthropological Review* 2: clviii–clxx.

——. 1867a. Birds' nests and plumage, or the relation between sexual differences of colour and the mode of nidification in birds. *Gardeners' Chronicle and Agricultural Gazette*, 12 October 1867, pp. 1047–8.

[——.] 1867b. Mimicry and other protective resemblances among animals. *Westminster Review* n.s. 32: 1–43.

——. 1867c. The philosophy of birds' nests. *Intellectual Observer* 11: 413–20.

——. 1868. A theory of birds' nests: shewing the relation of certain sexual differences of colour in birds to their mode of nidification. *Journal of Travel and Natural History* 1 (1868–9): 73–89.

——. 1869a. *The Malay Archipelago: the land of the orang-utan, and the bird of paradise. A narrative of travel, with studies of man and nature.* 2 vols. London: Macmillan and Co.

[——.] 1869b. Geological climates and the origin of species. [Review of Charles Lyell's *Principles of geology*, 10th edition, and *Elements of geology*, 6th edition.] *Quarterly Review* 126: 359–94.

——. 1869c. *Der Malayische Archipel: die Heimath des Orang-Utan und des Paradiesvogels: Reiseerlebnisse und Studien über Land und Leute.* Translated by Adolf Bernhard Meyer. 2 vols. Brunswick: Westermann.

——. 1869d. The origin of species controversy. [Review of *Habit and intelligence, in their connection with the laws of matter and force*, by Joseph John Murphy.] *Nature* 1 (1869–70): 105–7, 132–3.

——. 1870a. *Contributions to the theory of natural selection. A series of essays.* London: Macmillan and Co.

——. 1870b. The measurement of geological time. *Nature* 1 (1869–70): 399–401, 452–55.

——. 1870c. [Review of Francis Galton, *Hereditary genius*, 1869.] *Nature* 1 (1869–70): 501–3.

——. 1870d. Natural selection—Mr. Wallace's reply to Mr. Bennett. *Nature* 3 (1870–1): 49–50.

Wallace, Alfred Russel. 1871a. Presidential address to the Entomological Society of London. [Read 23 January 1871.] *Transactions of the Entomological Society of London* (*Proceedings*) (1870): xliv–lxix.

——. 1871b. *Contributions to the theory of natural selection: a series of essays.* 2d edition, with corrections and revisions. London: Macmillan and Co.

Wallis, J. P. R. 1936. *Fortune my foe: the story of Charles John Andersson, African explorer (1827–1867).* London: Jonathan Cape.

Walters, Stuart Max and Stow, E. Anne. 2001. *Darwin's mentor. John Stevens Henslow 1796–1861.* Cambridge: Cambridge University Press.

Warwick, Andrew. 2003. *Masters of theory: Cambridge and the rise of mathematical physics.* Chicago: University of Chicago Press.

Watkins, A. H. 1977. William Baxter: Bromley antiquary. *Bromley Local History* 2: 30–5.

Watson, Hewett Cottrell. 1868–70. *Compendium of the Cybele Britannica; or, British plants in their geographical relations.* 3 vols. Thames Ditton: printed for private distribution.

——. 1870. *A compendium of the Cybele Britannica: or British plants in their geographical relations.* London: Longmans, Green, Reader & Dyer.

Wawro, Geoffrey. 2003. *The Franco-Prussian war: the German conquest of France in 1870–1871.* Cambridge: Cambridge University Press.

Weale, James Philip Mansel. 1870a. Notes on some species of *Habenaria* found in South Africa. [Read 3 November 1870.] *Journal of the Linnean Society* (*Botany*) 13 (1873): 47–8.

——. 1870b. Notes on a species of *Disperis* found on the Kagaberg, South Africa. [Read 3 November 1870.] *Journal of the Linnean Society* (*Botany*) 13 (1873): 42–5.

——. 1870c. Observations on the mode in which certain species of Asclepiadeæ are fertilized. [Read 3 November 1870.] *Journal of the Linnean Society* (*Botany*) 13 (1873): 48–58.

——. 1870d. Some observations on the fertilization of *Disa macrantha.* [Read 23 November 1870.] *Journal of the Linnean Society* (*Botany*) 13 (1873): 45–7.

——. 1877. On the variation of Rhopalocerous forms in South Africa. [Read 4 July 1877.] *Transactions of the Entomological Society of London* (1877): 265–75.

——. 1878. Notes on South African insects. [Read 3 April 1878.] *Transactions of the Entomological Society of London* (1878): 183–8.

——. 1891. *The truth about the Portuguese in Africa.* London: Swan Sonnenschein.

Wedgwood, Barbara and Wedgwood, Hensleigh. 1980. *The Wedgwood circle, 1730–1897: four generations of a family and their friends.* London: Studio Vista.

[Wedgwood, Frances Julia.] 1860–1. The boundaries of science, a dialogue. *Macmillan's Magazine* 2 (1860): 134–8; 4 (1861): 237–47.

Weel, Adriaan van der. 2003. W. H. Kirberger: early specialist importer of English books into the Netherlands. *Quaerendo* 32: 245–56.

Weinreich, Heinrich. 1993. *Duftstofftheorie: Gustav Jaeger (1832–1917); vom Biologen zum 'Seelenriecher'.* Stuttgart: Wissenschaftliche Verlagsgesellschaft.

Wellesley index: *The Wellesley index to Victorian periodicals 1824–1900.* Edited by Walter

E. Houghton *et al.* 5 vols. Toronto: University of Toronto Press. London: Routledge & Kegan Paul. 1966–89.

Wells, William Charles. 1818. *Two essays: one upon single vision with two eyes; the other on dew. A letter to the Right Hon. Lloyd, Lord Kenyon and an account of a female of the white race of mankind, part of whose skin resembles that of a negro; with some observations on the causes of the differences in colour and form between the white and negro races of men.* London: Archibald Constable and Co. [and others].

West, David A. 2003. *Fritz Müller. A naturalist in Brazil.* Blacksburg, Va.: Pocahontas Press.

White, Gilbert. 1825. *The natural history of Selborne.* New edition. 2 vols. London: C. and J. Rivington [and others].

[Wilberforce, Samuel.] 1860. [Review of *Origin.*] *Quarterly Review* 108: 225–64.

Williams, Herbert William. 1971. *A dictionary of the Maori language.* 7th edition. Wellington, New Zealand: A.R. Shearer.

Winsor, Mary Pickard. 1976. *Starfish, jellyfish and the order of life: issues in nineteenth-century science.* New Haven and London: Yale University Press.

Winstanley, Denys Arthur. 1947. *Later Victorian Cambridge.* Cambridge: Cambridge University Press.

Wood, Thomas W. 1870. The courtship of birds. *Student and Intellectual Observer* 5 (1870–1): 113–25.

Woodward, Horace B., ed. 1891. *Memorials of John Gunn, MA, FGS … Being some account of the Cromer Forest bed and its fossil mammalia … with a memoir of the author.* With the assistance of E. T. Newton. Norwich: W. A. Nudd.

——. 1907. *The history of the Geological Society of London.* London: Geological Society.

Wright, Chauncey. 1870. [Review of *Contributions to the theory of natural selection*, by Alfred Russel Wallace, 1870.] *North American Review* 111: 282–311.

WWW: *Who was who: a companion to Who's who, containing the biographies of those who died during the period* [*1897–1995*]. 9 vols. and cumulated index (1897–1990). London: Adam & Charles Black. 1920–96.

WWWS: *World who's who in science from antiquity to the present.* Edited by Allen G. Debus. Chicago: Marquis-Who's Who. 1968.

Wyman, Jeffries. 1867. Observations and experiments on living organisms in heated water. *American Journal of Science and Arts* 2d ser. 44: 152–69.

Zollinger, Heinrich. 1857. Over het begrip en den omvang eener flora Malesiana. *Natuurkundig Tijdschrift voor Nederlansch Indië* 13: 293–322.

Zoology: *The zoology of the voyage of HMS Beagle, under the command of Captain FitzRoy RN, during the years 1832 to 1836.* Edited and superintended by Charles Darwin. 5 pts. London: Smith, Elder & Co. 1838–43.

NOTES ON MANUSCRIPT SOURCES

The majority of the manuscript sources cited in the footnotes to the letters are either in the Darwin Archive, Cambridge University Library, or at Down House, Downe, Kent. Further details about the Darwin Archive are available in the *Handlist of Darwin papers at the University Library Cambridge* (Cambridge: Cambridge University Press, 1960) and the unpublished supplementary handlist available at the library; a new catalogue of the papers is currently being prepared. Further details about the manuscripts at Down House are available in Philip Titheradge, ed. *The Charles Darwin Memorial at Down House, Downe, Kent*, revised ed. ([Downe: Down House Museum], 1981) and from the curator (The Curator, Down House, Downe, Kent, BR6 7JT). In addition, there are a number of named sources that are commonly used in the footnotes: for each of these, the editors have provided brief descriptive notes. Many manuscript sources are now available online at *The complete work of Charles Darwin online* (http://darwin-online.org.uk).

CD's Account books (Down House MS). This series of seventeen account books begins on 12 February 1839, a fortnight after CD and Emma's marriage, and ends with CD's death. The books contain two sets of accounts. From the start, CD recorded his *cash account* according to a system of double-entry book-keeping. On each left-hand page he recorded credits (i.e., withdrawals from the bank, either in the form of cash paid to himself or cheques drawn for others), and on each right-hand page he recorded debits (i.e., cash or cheques paid to others). CD also recorded details of his *banking account* from the start, but only noted them down in a single column at the bottom of the left-hand page of his cash account. In August 1848, however, he began a system of detailing his banking account according to double-entry book-keeping, in a separate chronological section at the back of each account book. On the left, he recorded credits to the account in the form of income (i.e., investments, rent, book sales, etc.). On the right, he recorded debits to the account (i.e., cash or cheque withdrawals).

CD's Classed account books (Down House MS). This series of four account books, covering the years 1839–81, runs parallel to CD's Account books. For each year, September–August (after 1867, January–December), CD divided his expenditure into different classes; in addition, he made a tally for the year of his income, expenditure, cash in hand, and money in the bank. From 1843, CD also compiled at the back of each book a separate account of the total expenditure under the various headings in each year, and from 1844 he added a full account of his income in each year, and of capital invested and 'paid' up.

CD's Experimental notebook (DAR 157a). This notebook contains notes on some of the experiments carried out between 13 November 1855 (with some back references) and 20 May 1868; the majority of the notes date from before 1863. Often only the details of the experiment attempted are given, usually with cross-references to results recorded in CD's portfolios of notes. The notebook also contains a number of letters to CD.

CD's Investment book (Down House MS). This book records for each of CD's investments the income received during the period 1846–81.

CD's 'Journal'. *See* Appendix II.

CD's Library catalogue (DAR 240). This manuscript catalogue of CD's scientific library was compiled by Thomas W. Newton, assistant librarian of the Museum of Practical Geology, London, in August 1875. Additions to the catalogue were later made by Francis Darwin (who inherited most of his father's scientific library) and by H. W. Rutherford, who apparently used this catalogue as a basis for compiling his *Catalogue of the library of Charles Darwin now in the Botany School, Cambridge* (Cambridge: Cambridge University Press, 1908). However, there are items listed in this manuscript catalogue that do not appear in Rutherford's published catalogue, and which must have been dispersed after being listed.

Down Coal and Clothing Club account book (Down House MS). CD was for some years treasurer of this charitable organisation. The account book records subscriptions made by honorary subscribers between 1841 and 1876; between 1848 and 1869 the entries are in CD's handwriting. For the years 1841–8 and 1868–76, there is also a statement of expenditures, though not in CD's handwriting.

Emma Darwin's diary (DAR 242). This is a series of small pocket diaries, in which Emma recorded details of the health of family members, trips made by herself, CD, and their children, school holidays, and visits to Down by others. The collection at CUL comprises diaries for the years 1824, 1833–4, 1839–45, and 1848–96.

H. E. Litchfield's autobiography (DAR 246). This unfinished autobiography, written in 1926 on forty-two loose leaves, and chiefly concerning Henrietta Emma Darwin's childhood, has never been published.

List of pamphlets (DAR 252.4). This is a catalogue of CD's pamphlet collection prepared by CD and Francis Darwin in 1878 (see the letter from Emma Darwin to Henrietta Emma Litchfield, [June 1878] (DAR 219.9: 175)). From about 1878 CD began to arrange the articles, papers, and reprints he received into a numbered collection. CD maintained this reprint collection until his death, when it was taken over by Francis Darwin. Francis continued the collection, adding new items, the numbers running consecutively from those of his father. Evidently, until this catalogue was prepared, CD used a working index similar to that of his 'List of reviews'. The catalogue is in two sections, a list of the quarto collection and one of the general collection. Both sections are alphabetically arranged with the entries pasted on sheets in a loose-leaf folder.

List of reviews (DAR 262.8: 9–18 (English Heritage MS: 88206151–60)). This manuscript, headed 'List Reviews of Origin of Sp & of C. Darwins Books', was CD's working index to his collection of reviews of his own books. It corresponds approximately to the review collection in the Darwin Pamphlet Collection–CUL, but includes some items that were dispersed after being listed.

Reading notebooks. *See Correspondence* vol. 4, Appendix IV. These notebooks are divided into sections entitled 'Books Read' and 'Books to be Read'. CD's entries in 'Books Read' often include a brief opinion of the work.

Scrapbook of reviews (DAR 226.1 and 226.2). Many of the reviews contained in these two volumes bear CD's annotations and thus were evidently collected by CD. However, the scrapbook seems to have been assembled by Francis Darwin: the tables of contents are in the handwriting of H. W. Rutherford, an assistant at Cambridge University Library who acted as a copyist for Francis on several occasions (see *ML*, 1: x, and Francis Darwin, ed. *The foundations of the Origin of Species. Two essays written in 1842 and 1844 by Charles Darwin* (Cambridge: Cambridge University Press, 1909)). In addition, the scrapbook is identified as Francis's in a note (DAR 226.1:132a) made in 1935 by Arthur Keith, whose appeal led to the purchase of Down House as a Darwin memorial (see Arthur Keith, *An autobiography* (London: Watts & Co., 1950)). DAR 226.1 bears the inscription 'Reviews of C. Darwin's works' on the spine, and contains, among others, reviews of *Origin* and *Orchids*; DAR 226.2 is inscribed: 'Reviews. Descent. Expression. Insect. Pl. Eras. D.'

W. E. Darwin's botanical notebook (DAR 117). This notebook contains observational and experimental notes on plants made by William, often in consultation with CD. The first observation bears the date 13 July 1862, and, although the date of the last observation is 26 June 1870, most of the notes were made between 1862 and 1864. The notebook originally contained letters from CD, but these were later removed. William entered notes made from botanical textbooks in a separate notebook (DAR 234).

W. E. Darwin's botanical sketchbook (DAR 186: 43). This sketchbook, which contains entries dated 1862–72, was evidently begun in parallel to William's botanical notebook. It contains ink drawings of various parts of plants, and of sections, together with descriptions, which are sometimes very extensive.

INDEX

The dates of letters to and from Darwin's correspondents are listed in the biographical register and are not repeated here. Darwin's works are indexed under the short titles used throughout this volume and listed in the bibliography.

Table of

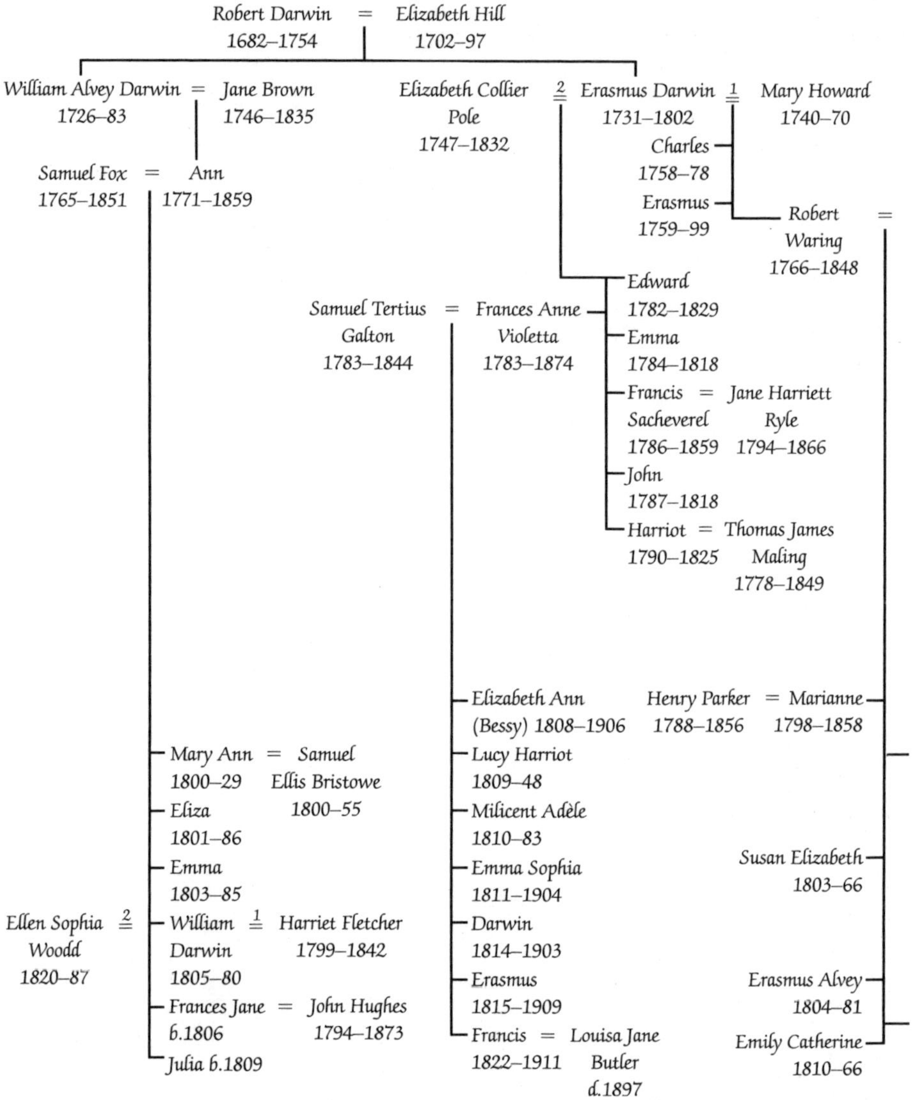

Robert Darwin = Elizabeth Hill
1682–1754 1702–97
William Alvey Darwin = Jane Brown
1726–83 1746–1835
Elizabeth Collier Pole 1747–1832
2 = Erasmus Darwin 1731–1802 1 = Mary Howard 1740–70
Charles 1758–78
Erasmus 1759–99
Robert Waring 1766–1848 =
Samuel Fox = Ann
1765–1851 1771–1859
Edward 1782–1829
Samuel Tertius Galton 1783–1844 = Frances Anne Violetta 1783–1874
Emma 1784–1818
Francis Sacheverel 1786–1859 = Jane Harriett Ryle 1794–1866
John 1787–1818
Harriot 1790–1825 = Thomas James Maling 1778–1849
Elizabeth Ann (Bessy) 1808–1906
Henry Parker 1788–1856 = Marianne 1798–1858
Mary Ann 1800–29 = Samuel Ellis Bristowe 1800–55
Lucy Harriot 1809–48
Eliza 1801–86
Milicent Adèle 1810–83
Emma 1803–85
Emma Sophia 1811–1904
Susan Elizabeth 1803–66
Ellen Sophia Woodd 1820–87 2 = William Darwin 1805–80 1 = Harriet Fletcher 1799–1842
Darwin 1814–1903
Erasmus 1815–1909
Erasmus Alvey 1804–81
Frances Jane b.1806 = John Hughes 1794–1873
Francis 1822–1911 = Louisa Jane Butler d.1897
Emily Catherine 1810–66
Julia b.1809